# Physics of the Sun

With an emphasis on numerical modeling, *Physics of the Sun: A First Course* presents a quantitative examination of the physical structure of the Sun and the conditions of its extended atmosphere. It gives step-by-step instructions for calculating the numerical values of various physical quantities in different regions of the Sun.

Fully updated throughout, with the latest results in solar physics, this second edition covers a wide range of topics on the Sun and stellar astrophysics, including the structure of the Sun, solar radiation, the solar atmosphere, and Sun–space interactions. It explores how the physical conditions in the visible surface of the Sun are determined by the opacity of the material in the atmosphere. It also presents the empirical properties of convection in the Sun, discusses the physical conditions that must be satisfied for nuclear reactions to occur in the core, and describes how radiation transports energy from the core outwards.

This text enables a practical appreciation of the physical models of solar processes. Numerical modeling problems and step-by-step instructions are featured throughout, to empower students to calculate, using their own codes, the interior structure of different parts of the Sun and the frequencies of $p$-modes and $g$-modes. They encourage a firm grasp of the numerical values of actual physical parameters as a function of radial location in the Sun.

It is an ideal introduction to solar physics for advanced undergraduate and graduate students in physics and astronomy, in addition to research professionals looking to incorporate modeling into their practices. Extensive bibliographies at the end of each chapter enable the reader to explore the latest research articles in the field.

**Features**

- Fully updated with the latest results from the spacecraft Hinode, STEREO, Solar Dynamics Observatory (SDO), Interface Region Imaging Spectrograph (IRIS), and Parker Solar Probe
- Presents step-by-step explanations for calculating numerical models of the photosphere, convection zone, and radiative interior with exercises and simulation problems to test learning
- Describes the structure of polytropic spheres and the acoustic power in the Sun and the process of thermal conduction in different physical conditions

# Physics of the Sun
## A First Course

Second Edition

Dermott J. Mullan

Second edition published 2023
by CRC Press
6000 Broken Sound Parkway NW, Suite 300, Boca Raton, FL 33487–2742

and by CRC Press
4 Park Square, Milton Park, Abingdon, Oxon, OX14 4RN

*CRC Press is an imprint of Taylor & Francis Group, LLC*

© 2023 Dermott J. Mullan

First edition published by CRC Press 2009

Reasonable efforts have been made to publish reliable data and information, but the author and publisher cannot assume responsibility for the validity of all materials or the consequences of their use. The authors and publishers have attempted to trace the copyright holders of all material reproduced in this publication and apologize to copyright holders if permission to publish in this form has not been obtained. If any copyright material has not been acknowledged please write and let us know so we may rectify in any future reprint.

Except as permitted under U.S. Copyright Law, no part of this book may be reprinted, reproduced, transmitted, or utilized in any form by any electronic, mechanical, or other means, now known or hereafter invented, including photocopying, microfilming, and recording, or in any information storage or retrieval system, without written permission from the publishers.

For permission to photocopy or use material electronically from this work, access www.copyright.com or contact the Copyright Clearance Center, Inc. (CCC), 222 Rosewood Drive, Danvers, MA 01923, 978–750–8400. For works that are not available on CCC please contact mpkbookspermissions@tandf.co.uk

*Trademark notice*: Product or corporate names may be trademarks or registered trademarks and are used only for identification and explanation without intent to infringe.

*Library of Congress Cataloging-in-Publication Data*
A Library of Congress catalog record has been requested for this book

ISBN: 978-0-367-71039-2 (hbk)
ISBN: 978-0-367-72032-2 (pbk)
ISBN: 978-1-003-15311-5 (ebk)

DOI: 10.1201/9781003153115

Typeset in Times
by Apex CoVantage, LLC

# Contents

Preface ..................................................................................................................................... xiii
Author ..................................................................................................................................... xix

**Chapter 1** The Global Parameters of the Sun ................................................................ 1

    1.1   Orbital Motion of the Earth ........................................................................ 1
    1.2   The Astronomical Unit (AU) ..................................................................... 3
    1.3   $GM_\odot$ and the Mass of the Sun ................................................................ 6
    1.4   Power Output of the Sun: The Solar Luminosity ..................................... 6
    1.5   The Radius of the Sun: $R_\odot$ ..................................................................... 8
    1.6   Acceleration due to Gravity at the Surface of the Sun ............................ 10
    1.7   The Mean Mass Density of the Sun ........................................................ 10
    1.8   Escape Speed from the Solar Surface ..................................................... 11
    1.9   Effective Temperature of the Sun ........................................................... 11
    1.10  The Oblateness of the Sun ...................................................................... 12
    1.11  The Observed Rotation of the Sun's Surface .......................................... 13
    1.12  A Characteristic Frequency for Solar Oscillations Due to Gravity ........ 16
    Exercises ............................................................................................................ 16
    References ......................................................................................................... 17

**Chapter 2** Radiation Flow through the Solar Atmosphere ........................................... 19

    2.1   Radiation Field in the Solar Atmosphere ................................................ 19
    2.2   Empirical Properties of the Radiant Energy from the Sun ..................... 24
    2.3   The Radiative Transfer Equation (RTE) ................................................. 27
    2.4   Optical Depth and the Concept of "the Photosphere" ............................. 29
    2.5   Special Solutions of the RTE .................................................................. 29
          2.5.1   $S$ = Constant at All Optical Depths ......................................... 30
          2.5.2   $S$ = Constant in a Slab of Finite Thickness ............................ 30
          2.5.3   Depth-Dependent $S$: Polynomial Form ................................... 31
          2.5.4   Depth-Dependent $S$: Exponential Form .................................. 32
    2.6   The "Eddington–Barbier" (or "Milne–Barbier–Unsöld") Relationship ....................................................................................... 32
    2.7   Is Limb Brightening Possible? ................................................................ 33
    2.8   Is $S(\tau) = a + b\tau$ Realistic? The Gray Atmosphere ................................ 33
    2.9   How Does Temperature Vary as a Function of $\tau$? .................................. 36
    2.10  Properties of the Eddington (Milne) Relation ........................................ 37
    Exercises ............................................................................................................ 37
    References ......................................................................................................... 38

**Chapter 3** Toward a Model of the Sun: Opacity .......................................................... 39

    3.1   Relationship between Optical Depth and Linear Absorption Coefficient ................................................................................................ 39
    3.2   Two Approaches to Opacity: Atomic and Astrophysical ....................... 41
          3.2.1   Energy Levels in Atomic Hydrogen ........................................ 42

| | 3.3 | Atomic Physics: (i) Opacity due to Hydrogen Atoms ........................... 44 |
|---|---|---|
| | | 3.3.1 Absorption from the Ground State: Dependence on Wavelength ...... 46 |
| | | 3.3.2 Absorption from Excited States: Dependence on Wavelength and $T$ ..... 48 |
| | 3.4 | Atomic Physics: (ii) Opacity due to Negative Hydrogen Ions ........................ 50 |
| | 3.5 | Atomic Physics: (iii) Opacity due to Helium Atoms and Ions ........................ 51 |
| | 3.6 | Astrophysics: The Rosseland Mean Opacity ........................... 51 |
| | | 3.6.1 Limit of Low Density and/or High $T$: Electron Scattering ................. 53 |
| | | 3.6.2 Low $T$ Limit ........................... 54 |
| | | 3.6.3 Higher Density: Free-Bound Absorptions ........................... 54 |
| | | 3.6.4 Magnitude of the Largest Opacity ........................... 55 |
| | 3.7 | Power-Law Approximations to the Rosseland Mean Opacity ........................ 55 |
| | 3.8 | Narrow Band Opacity: Absorption Lines in the Spectrum ........................... 56 |
| | | 3.8.1 Characterizing the Properties of Absorption Lines ........................ 57 |
| | | 3.8.2 Heights of Formation of Different Spectral Lines ........................ 60 |
| | | 3.8.3 Shape of an Absorption Line Profile: C-Shaped Bisectors ................. 63 |
| | | 3.8.4 Shape of an Absorption Line: Magnetic Fields ........................... 65 |
| | Exercises ........................... 66 | |
| | References ........................... 67 | |

**Chapter 4** Toward a Model of the Sun: Properties of Ionization ........................... 69

| | 4.1 | Statistical Weights of Free Electrons ........................... 69 |
|---|---|---|
| | 4.2 | Saha Equation ........................... 71 |
| | 4.3 | Application of the Saha Equation to Hydrogen in the Sun ........................... 73 |
| | 4.4 | Application of the Saha Equation to Helium in the Sun ........................... 75 |
| | 4.5 | Contours of Constant Ionization: The Two Limits ........................... 76 |
| | 4.6 | Application of the Saha Equation to the Negative Hydrogen Ion ................. 76 |
| | Exercises ........................... 77 | |
| | References ........................... 78 | |

**Chapter 5** Computing a Model of the Sun: The Photosphere ........................... 79

| | 5.1 | Hydrostatic Equilibrium: The Scale Height ........................... 79 |
|---|---|---|
| | 5.2 | Sharp Edge of the Sun's Disk ........................... 81 |
| | 5.3 | Preparing to Compute a Model of the Solar Photosphere ........................... 82 |
| | 5.4 | Computing a Model of the Photosphere: Step by Step ........................... 82 |
| | 5.5 | The Outcome of the Calculation ........................... 87 |
| | 5.6 | Overview of the Model of the Solar Photosphere ........................... 88 |
| | Exercises ........................... 90 | |
| | References ........................... 90 | |

**Chapter 6** Convection in the Sun: Empirical Properties ........................... 91

| | 6.1 | Nonuniform Brightness ........................... 91 |
|---|---|---|
| | 6.2 | Granule Shapes ........................... 93 |
| | 6.3 | Upflow and Downflow Velocities in Solar Convection ........................... 95 |
| | 6.4 | Linear Sizes of Granules ........................... 96 |
| | 6.5 | Circulation Time around a Granule ........................... 98 |
| | 6.6 | Temperature Differences between Bright and Dark Gas ........................... 99 |
| | 6.7 | Energy Flux Carried by Convection ........................... 100 |
| | | 6.7.1 Convective Energy Flux in the Photosphere ........................... 100 |
| | | 6.7.2 Convective Energy Flux above the Photosphere? ........................... 102 |

Contents   vii

|  |  | 6.7.3 | Convective Energy Flux in Gas That Lies *below* the Photosphere .................................................................. 102 |
|---|---|---|---|

    6.8    Onset of Convection: The Schwarzschild Criterion ....................... 104
    6.9    Onset of Convection: Beyond the Schwarzschild Criterion ........... 105
    6.10  Numerical Value of $g_{ad}$ .................................................................. 106
    6.11  Alternative Expression for $g_{ad}$ ..................................................... 107
    6.12  Supergranules .................................................................................. 108
    Exercises .................................................................................................... 111
    References .................................................................................................. 112

**Chapter 7**  Computing a Model of the Sun: The Convection Zone ............................ 115

    7.1    Quantifying the Physics of Convection: Vertical Acceleration ..... 115
    7.2    Vertical Velocities and Length-Scales ........................................... 116
    7.3    Mixing Length Theory (MLT) of Convection ............................... 117
    7.4    Temperature Excesses Associated with MLT Convection ............ 118
    7.5    MLT Convective Flux in the Photosphere ..................................... 119
    7.6    MLT Convective Flux *below* the Photosphere .............................. 119
    7.7    Adiabatic and Nonadiabatic Processes ........................................... 120
    7.8    Computing a Model of the Convection Zone: Step by Step .......... 122
    7.9    Overview of Our Model of the Convection Zone .......................... 123
    Exercises .................................................................................................... 125
    References .................................................................................................. 125

**Chapter 8**  Radiative Transfer in the Deep Interior of the Sun ..................................... 127

    8.1    Thermal Conductivity for *Photons* ................................................ 127
    8.2    Flux of Radiant Energy at Radius $r$ .............................................. 129
    8.3    Base of the Convection Zone ......................................................... 130
    8.4    Temperature Gradient in Terms of Luminosity .............................. 131
    8.5    Temperature Gradient in Terms of Pressure .................................. 131
    8.6    Integrating the Temperature Equation ............................................ 132
    Exercise ..................................................................................................... 133
    References .................................................................................................. 133

**Chapter 9**  Computing a Mechanical Model of the Sun: The Radiative Interior ............................................................................................................... 135

    9.1    Computational Procedure: Step by Step ........................................ 135
    9.2    Overview of Our Model of the Sun's Radiative Interior ............... 137
    9.3    Photons in the Sun: How Long before They Escape? ................... 139
    9.4    A Particular Global Property of the Solar Model .......................... 140
    9.5    Does the Material in the Sun Obey the Perfect Gas Law? ............ 141
    9.6    Summary of Our (Simplified) Solar Model ................................... 142
    Exercises .................................................................................................... 143
    References .................................................................................................. 143

**Chapter 10**  Polytropes .................................................................................................. 145

    10.1  Power-Law Behavior ...................................................................... 145
    10.2  Polytropic Gas Spheres .................................................................. 146

|  | 10.3 | Lane–Emden Equation: Dimensional Form | 147 |
|---|---|---|---|
|  | 10.4 | Lane–Emden Equation: Dimensionless Form | 148 |
|  | 10.5 | Boundary Conditions for the Lane–Emden Equation | 149 |
|  | 10.6 | Analytic Solutions of the Lane–Emden Equation | 149 |
|  |  | 10.6.1 Polytrope $n = 0$ | 150 |
|  |  | 10.6.2 Polytrope $n = 1$ | 150 |
|  |  | 10.6.3 Polytrope $n = 5$ | 150 |
|  | 10.7 | Are Polytropes in Any Way Relevant for "Real Stars"? | 151 |
|  | 10.8 | Calculating a Polytropic Model: Step by Step | 152 |
|  | 10.9 | Central Condensation of a Polytrope | 154 |
|  | Exercises | | 155 |
|  | References | | 155 |

**Chapter 11** Energy Generation in the Sun ........................................................... 157

|  | 11.1 | The *pp*-I Cycle of Nuclear Reactions | 158 |
|---|---|---|---|
|  | 11.2 | Reaction Rates in the Sun | 160 |
|  | 11.3 | Proton Collision Rates in the Sun | 160 |
|  | 11.4 | Conditions Required for Nuclear Reactions in the Sun | 162 |
|  |  | 11.4.1 Nuclear Forces: Short-Range | 162 |
|  |  | 11.4.2 Classical Physics: The "Coulomb Gap" | 164 |
|  |  | 11.4.3 Quantum Physics: Bridging the "Coulomb Gap" | 165 |
|  |  | 11.4.4 Center of the Sun: Thermal Protons Bridge the Coulomb Gap | 166 |
|  |  | 11.4.5 Other Stars: Bridging the Coulomb Gap | 167 |
|  |  | 11.4.6 Inside the Nuclear Radius | 167 |
|  | 11.5 | Rates of Thermonuclear Reactions: Two Contributing Factors | 167 |
|  |  | 11.5.1 Bridging the Coulomb Gap: "Quantum Tunneling" | 168 |
|  |  | 11.5.2 Post-Tunneling Processes | 169 |
|  |  | 11.5.3 Probability of *pp*-I Cycle in the Solar Core: Reactions (a) and (b) | 171 |
|  | 11.6 | Temperature Dependence of Thermonuclear Reaction Rates | 171 |
|  | 11.7 | Rate of Reaction (c) in the *pp*-I cycle | 173 |
|  | Exercises | | 173 |
|  | References | | 174 |

**Chapter 12** Neutrinos from the Sun ...................................................................... 175

|  | 12.1 | Generation and Propagation of Solar Neutrinos | 175 |
|---|---|---|---|
|  | 12.2 | Fluxes of *pp*-I Solar Neutrinos at the Earth's Orbit | 177 |
|  | 12.3 | Neutrinos from Reactions Other than *pp*-I | 177 |
|  |  | 12.3.1 *pp*-II and *pp*-III Chains | 178 |
|  |  | 12.3.2 Other Reactions That Occur in the Sun | 180 |
|  | 12.4 | Detecting Solar Neutrinos on Earth | 181 |
|  |  | 12.4.1 Chlorine Detector | 181 |
|  |  | 12.4.2 Cherenkov Emission | 182 |
|  |  | 12.4.3 Gallium Detectors | 183 |
|  |  | 12.4.4 Heavy Water Detector | 183 |
|  | 12.5 | Solution of the Solar Neutrino Problem | 185 |
|  | Exercises | | 186 |
|  | References | | 186 |

Contents

**Chapter 13** Oscillations in the Sun: The Observations .................................................. 187
    13.1    Variability in *Time* Only .................................................................. 188
    13.2    Variability in *Space* and *Time* ...................................................... 191
    13.3    Radial Order of a Mode ................................................................. 194
    13.4    Which *p*-Modes Have the Largest Amplitudes? .......................... 195
    13.5    Trapped and Untrapped Modes .................................................... 195
        13.5.1    Vertically Propagating Waves in a Stratified Atmosphere ............... 196
        13.5.2    Simplest Case: The Isothermal Atmosphere ..................... 197
        13.5.3    Critical Frequency and the "Cut-Off" Period ................... 199
        13.5.4    Physical Basis for a Cut-Off Period .................................. 199
        13.5.5    Numerical Value of the Cut-Off Period ............................ 199
    13.6    Waves Propagating in a *Non-Vertical* Direction ......................... 200
    13.7    Long-Period Oscillations in the Sun .............................................. 201
    13.8    *p*-mode Frequencies and the Sunspot Cycle ................................ 202
    Exercises ................................................................................................... 202
    References ............................................................................................... 203

**Chapter 14** Oscillations in the Sun: Theory ............................................................... 205
    14.1    Small Oscillations: Deriving the Equations ................................... 205
    14.2    Conversion to Dimensionless Variables ......................................... 208
    14.3    Overview of the Equations ............................................................. 209
    14.4    The Simplest Exercise: *p*-Mode Solutions for the Polytrope $n = 1$ ............ 210
        14.4.1    Procedure for Computation ............................................... 211
        14.4.2    Comments on the *p*-mode Results: Patterns in the Eigenfrequencies .......................................... 213
        14.4.3    Eigenfunctions ................................................................... 214
    14.5    What About *g*-Modes? ................................................................. 216
    14.6    Asymptotic Behavior of the Oscillation Equations ....................... 218
        14.6.1    *p*-modes ............................................................................. 218
        14.6.2    *g*-modes ............................................................................ 220
    14.7    Depth of Penetration of *p*-modes beneath the Surface of the Sun ............... 221
    14.8    Why Are Certain *p*-Modes Excited More than Others in the Sun? ............... 223
        14.8.1    Depths Where *p*-Modes Are Excited ............................... 223
        14.8.2    Properties of Convection at the Excitation Depth ........... 223
    14.9    Using *p*-Modes to Test a Solar Model ......................................... 225
        14.9.1    Global Sound Propagation ................................................ 225
        14.9.2    Radial Profile of the Sound Speed .................................... 225
        14.9.3    The Sun's Rotation ............................................................ 226
    14.10  *r*-Modes in the Sun ....................................................................... 229
    Exercises ................................................................................................... 231
    References ............................................................................................... 231

**Chapter 15** The Chromosphere .................................................................................... 233
    15.1    Definition of the Chromosphere ................................................... 234
    15.2    Linear Thickness of the Chromosphere ....................................... 236
    15.3    Observing the Chromosphere on the Solar Disk .......................... 236
    15.4    Supergranules Observed in the $H\alpha$ Line ..................................... 238
    15.5    The Two Principal Components of the Chromosphere ................ 240

| 15.6 | Temperature Increase from Photosphere to Chromosphere: Empirical Results | 240 |
| 15.7 | Temperature Increase into the Chromosphere: Mechanical Work | 242 |
| 15.8 | Modeling the Chromosphere: The Input Energy Flux | 243 |
| 15.9 | Modeling the Chromosphere: The Energy Deposition Rate | 245 |
| 15.10 | Modeling the Equilibrium Chromosphere: Radiating the Energy Away | 247 |
| | 15.10.1 Radiative Cooling Time-Scale | 247 |
| | 15.10.2 Magnitude of the Temperature Increase: The Low Chromosphere | 248 |
| | 15.10.3 Magnitude of the Temperature Increase: The Middle Chromosphere | 249 |
| | 15.10.4 Magnitude of the Temperature Increase: The Upper Chromosphere | 251 |
| 15.11 | The IRIS Satellite | 251 |
| 15.12 | A Variety of Wave Modes in the Chromosphere? | 253 |
| | 15.12.1 The "Plasma Beta" Parameter and Conversions between Wave Modes | 254 |
| Exercise | | 256 |
| References | | 256 |

## Chapter 16  Magnetic Fields in the Sun .......... 259

| 16.1 | Sunspots | 259 |
| | 16.1.1 Spot Temperatures | 260 |
| | 16.1.2 Why Are Sunspots Cooler than the Rest of the Photosphere? | 260 |
| | 16.1.3 Areas of Spots and Plages | 263 |
| | 16.1.4 Spot Numbers: The "11-Year" Cycle | 264 |
| | 16.1.5 Spot Lifetimes | 267 |
| | 16.1.6 Energy Deficits and Excesses | 267 |
| 16.2 | Chromospheric Emission | 269 |
| 16.3 | Magnetic Fields: The Source of Solar Activity | 270 |
| 16.4 | Measurements of Solar Magnetic Fields | 270 |
| | 16.4.1 Measurement of Magnetic Fields on the Sun: Optical Data | 271 |
| |    16.4.1.1 Zeeman Splitting | 271 |
| |    16.4.1.2 Zeeman Polarization: The Longitudinal Case | 273 |
| |    16.4.1.3 Zeeman Polarization: The Transverse Case | 275 |
| |    16.4.1.4 Babcock Magnetograph: Longitudinal Fields | 275 |
| |    16.4.1.5 Vector Magnetograph | 276 |
| | 16.4.2 Magnetic Field Strengths in Sunspot Umbrae | 277 |
| | 16.4.3 Orderly Properties of Sunspot Fields | 279 |
| | 16.4.4 Remote Sensing of Solar Magnetic Fields: Radio Observations | 279 |
| | 16.4.5 How Are Coronal Fields Related to Fields in the Photosphere? | 280 |
| | 16.4.6 Direct Magnetic Measurements in Space: The Global Field of the Sun | 280 |
| 16.5 | Empirical Properties of Global and Local Solar Magnetic Fields | 282 |
| 16.6 | Interactions between Magnetic Fields and Ionized Gas | 283 |
| | 16.6.1 Motion of a Single Particle | 283 |
| | 16.6.2 Motion of a Conducting Fluid | 286 |
| |    16.6.2.1 Magnetic Pressure and Tension | 286 |
| |    16.6.2.2 The Equations of Magnetohydrodynamics (MHD) | 287 |
| |    16.6.2.3 Time-Scales for Magnetic Diffusion in the Sun | 289 |

| | | | |
|---|---|---|---|
| | 16.7 | Understanding Magnetic Structures in the Sun | 290 |
| | | 16.7.1 Sunspot Umbrae: Inhibition of Convection | 290 |
| | | 16.7.2 Pores: The Smallest Sunspots | 291 |
| | | 16.7.3 Sunspots: The Wilson Depression | 291 |
| | | 16.7.4 Sunspots: What Determines Their Lifetimes? | 292 |
| | | 16.7.5 Prominences | 293 |
| | | 16.7.6 Faculae | 293 |
| | | 16.7.7 Excess Chromospheric Heating: Network and Plages | 293 |
| | | 16.7.8 Magnetic Field and Gas Motion: Which Is Dominant? | 295 |
| | 16.8 | Amplification of Strong Solar Magnetic Fields | 296 |
| | 16.9 | Why Does the Sun Have a Magnetic Cycle with $P \approx 10$ Years? | 298 |
| | 16.10 | Releases of Magnetic Energy | 300 |
| | 16.11 | Magnetic Helicity | 300 |
| | Exercises | | 302 |
| | References | | 302 |

**Chapter 17** The Corona .................................................................................................307

17.1 Electron Densities ........................................................................................ 309
17.2 Electron Temperatures ................................................................................. 311
    17.2.1 Optical Photons ............................................................................. 311
    17.2.2 X-ray Photons ............................................................................... 313
17.3 "The" Temperature of Line Formation ........................................................ 315
17.4 Emission Lines That Are Popular for Imaging the Corona ......................... 316
    17.4.1 SOHO/EIT ..................................................................................... 316
    17.4.2 SDO/AIA ....................................................................................... 317
    17.4.3 Hinode/EIS .................................................................................... 318
17.5 Quantitative Estimates of the "Emission Measure" of Coronal Gas ........... 319
17.6 The Solar Cycle in X-rays ............................................................................ 321
17.7 The Solar Cycle in Microwave Radio Emission .......................................... 322
17.8 Ion Temperatures .......................................................................................... 323
17.9 Densities and Temperatures: Quiet Sun versus Active Regions .................. 324
17.10 Gas Pressures in the Corona ........................................................................ 325
17.11 Spatial Structure in the X-ray Corona ......................................................... 325
17.12 Magnetic Structures: Loops in Active Regions ........................................... 326
17.13 Magnetic Structures: Coronal Holes ........................................................... 327
17.14 Magnetic Structures: The Quiet Sun ........................................................... 328
17.15 Why Are Quiet Coronal Temperatures of Order 1–2 MK? ......................... 328
    17.15.1 Thermal Conduction by Electrons ................................................ 329
    17.15.2 Radiative Losses ............................................................................ 330
    17.15.3 Combination of Radiative and Conductive Losses ....................... 331
17.16 Abrupt Transition from Chromosphere to Corona ...................................... 332
17.17 Rate of Mechanical Energy Deposition in the Corona ................................ 333
17.18 What Heats the Corona? .............................................................................. 334
    17.18.1 Wave Heating ................................................................................ 334
        17.18.1.1 Acoustic Waves? ........................................................... 335
        17.18.1.2 Alfven Waves? .............................................................. 335
    17.18.2 Non-Wave Heating: The Magnetic Carpet ................................... 336
17.19 Solar Flares .................................................................................................. 337
    17.19.1 General ........................................................................................... 337
    17.19.2 How Many Solar Flares Have Been Detected? ............................ 339

|  | 17.19.3 | Flare Temperatures and Densities | 341 |
|---|---|---|---|
|  | 17.19.4 | Spatial Location and Extent | 343 |
|  | 17.19.5 | Energy in Nonthermal Electrons | 343 |
|  | 17.19.6 | Where Are Flare Electrons Accelerated? | 344 |
|  | 17.19.7 | Other Channels of Flare Energy | 345 |
|  | 17.19.8 | Do Flares Perturb Solar Structure Significantly? | 346 |
|  | 17.19.9 | Energy Densities in Flares | 347 |
|  | 17.19.10 | Physics of Flares (Simplified): Magnetic Reconnection in 2-D | 347 |
|  | 17.19.11 | Physics of Flares (More Realistic): Magnetic Reconnection in 3-D | 349 |
|  | 17.19.12 | Consequences of Magnetic Reconnection | 351 |
|  | 17.19.13 | Can Flares Be Predicted? | 352 |
|  | References | | 352 |

**Chapter 18** The Solar Wind ..................................................................... 357

| | 18.1 | Global Breakdown of Hydrostatic Equilibrium in the Corona | 357 |
|---|---|---|---|
| | 18.2 | Localized Applicability of HSE | 359 |
| | 18.3 | Solar Wind Expansion: Parker's Model of a "Thermal Wind" | 360 |
| | 18.4 | Conservation of Energy | 362 |
| | 18.5 | Asymptotic Speed of the Solar Wind: The Magnetic Spiral | 363 |
| | 18.6 | Magnetic Field Effects: "High-Speed" Wind and "Slow" Wind | 364 |
| | 18.7 | Observations of Solar Wind Properties | 366 |
| | | 18.7.1 *In situ* Measurements: ≈ 1 AU and Beyond | 366 |
| | | 18.7.2 *In situ* Measurements in the Inner Wind: $r < 1$ AU | 368 |
| | | 18.7.3 Remote Sensing of the Solar Wind | 370 |
| | 18.8 | Rate of Mass Outflow from the Sun | 373 |
| | 18.9 | Coronal Mass Ejections (CMEs) | 375 |
| | | 18.9.1 Rates of CME Occurrence | 376 |
| | | 18.9.2 Masses of CMEs | 377 |
| | | 18.9.3 Speeds of CMEs | 377 |
| | | 18.9.4 Kinetic and Potential Energies of CMEs | 378 |
| | | 18.9.5 Comparison and Contrast between Flares and CMEs | 378 |
| | | 18.9.6 CME Contributions to Solar Mass Loss Rates | 380 |
| | | 18.9.7 CMEs and Magnetic Helicity | 380 |
| | 18.10 | How Far Does the Sun's Influence Extend in Space? | 381 |
| | | 18.10.1 Where Does the "True" Corona End and the "True" Wind Begin? | 382 |
| | | 18.10.2 The Outer Edge of the Heliosphere | 382 |
| | Exercises | | 385 |
| | References | | 385 |

**Appendix A:** symbols used in the text ..................................................... 389

**Appendix B:** instruments used to observe the Sun .................................. 393

**Index** ....................................................................................................... 399

# Preface

The goal of this course is to undertake a quantitative examination of the physical structure of the Sun. The text is aimed at upper-level physics undergraduates and at graduate students in their early years of graduate study. To achieve our goal, we explore how various laws of physics can be used to help us derive realistic information about different regions in the Sun. Although the material that makes up the Sun is gaseous throughout its volume, the physical conditions change dramatically from one part of the Sun to another: e.g., the gas pressure changes by some 12 orders of magnitude between center and "surface". As a result, different aspects of physics are found to be dominant in different regions. For someone who is interested in physics, the Sun provides a "real world" laboratory in which to test a broad range of topics, including fluid flow, heat transport, thermodynamics, radiative transfer, quantum mechanics, nuclear physics, plasma physics, and turbulence. Of special interest is the fact that the Sun presents us with striking examples of magnetohydrodynamic (MHD) processes, i.e., the interactions that occur between fluid flows and electromagnetic fields. In the later chapters of this book, we shall discuss how the Sun displays evidence for an array of MHD processes: specifically, in certain regions, the fields control the motions of the gas, whereas in other regions, the gas controls the motions of the field.

In the broadest context, the human race has one principal question that requires solar physicists to answer: how stable is the Sun's output of energy? Specifically, is the Sun's power output steady enough that we who live on Earth will *not* be subjected to chaotic fluctuations in the input heating rate that could lead to dangerous consequences for life?

In order to arrive at reliable answers to this question, we must begin with what we know about the Sun. And the place where reliable physical knowledge starts is with observations. Growing numbers of instrument types are currently in operation for observing the Sun (e.g., see Figure P.1). If we were limited to observations with the unaided eye, we could conclude that the Sun appears as a luminous yellowish-white circular object (a "disk") with an apparently sharp edge (the "limb"). The color suggests that material at the surface of the Sun (in the region called the "photosphere", i.e., the sphere from which light emerges) has a temperature of 5–6 thousand degrees K (or K for short). However, the eye is not especially useful in deriving further physical information about the Sun. Since the time of Galileo in the early 1600s, optical telescopes have been used to improve our knowledge of the Sun. These observations point to the presence of certain features in the photosphere. The best known among these features are dark regions called "sunspots".

Access to spectroscopy in the 1800s led to the discovery of a region of hotter gas above the photosphere that can be seen to shine briefly as a colored rim of light during a total solar eclipse. This region is called the "chromosphere" (literally: the "color" sphere), where the temperature of the gas rises to values of 10–20 thousand K. We shall be interested in quantifying the physical process(es) that cause this rise in temperature.

During a total solar eclipse, the human eye can see a faint outer extension of the Sun's atmosphere known as the "corona". The first indication that coronal material has temperatures as large as $10^6$ K was provided in 1941 by Bengt Edlen (Ark. Mat. Astron. Fys. 28B) in his interpretation of the strongest emission lines in the corona. These lines had previously been assigned to an unknown element ("coronium"), until Edlen applied the relatively new tools of quantum mechanics to calculate energy levels in atoms from which many electrons had been stripped away. Further discoveries about coronal physics emerged when new techniques in radio astronomy and X-ray astronomy were applied to the Sun. When the Sun is "viewed" at microwave radio frequencies, the Sun appears patchy, with certain "active regions" much brighter than others. When the Sun is "viewed" in X-rays, different images emerge depending on the X-ray energy. At low energies, diffuse emission is seen to extend over most of the disk, but there are also some dark areas that seem to be empty (coronal "holes"). When viewed in higher energy X-rays, the Sun is dominated by bright "active regions", and within each active region, one can often identify discrete structures ("loops"). The brightest of all features in the corona, and also in the chromosphere, are short-lived brightenings known as "flares".

Observations of the photosphere, the chromosphere, and the corona show beyond any doubt that the surface of the Sun and its outer atmosphere are subject to variability of different kinds. For example, the number of sunspots on the surface is observed to wax and wane every 11 years or so. At times, a large quantity of gas erupts from the Sun into interplanetary space, and there may also be flares that erupt unpredictably from time to time in certain active regions. Before the modern era of solar observations, these occasional "eruptive" events would have had minimal effects on Earth, apart from occasional more or less brilliant shows of "northern lights", when the sky would light up with "aurora borealis" ("northern dawn"). But in our day and age, eruptive solar events can have more serious effects, including damage to satellites and to the equipment that power companies rely on to distribute electricity across entire continents. These potential dangers have given rise to a field of study known as "space weather", the goal of which is to identify warning signs of impending solar events so that precautions may be put in place on Earth.

A good opportunity to examine graphical illustrations, examples, and videos of the broad variety of eruptive phenomena as they occur on the Sun can be found online at the site https://sdo.gsfc.nasa.gov (maintained by NASA, the National Aeronautics and Space Administration). This site presents images obtained by one particular spacecraft: the Solar Dynamics Observatory (SDO). This spacecraft, launched in 2010 into an orbit around Earth that ensures that SDO has an uninterrupted view of the Sun, provides continuous observations of the Sun with no interruptions due to day-night cycle or clouds. The SDO website allows anyone with web access to see what the Sun is currently doing at any time of the day or night.

Among the images that can be examined at NASA's SDO website, one is particularly important in understanding the origin of the most dynamic solar phenomena: this is an instrument known as the Helioseismic and Magnetic Imager (HMI). HMI displays *magnetic field* properties on the surface of the Sun. The data show that magnetic fields with a wide range of strengths are present in many locations on the Sun, and not only in sunspots (although sunspots have the strongest fields). On the same 11-year cycle that the numbers of sunspots exhibit, the numbers and sizes of magnetic areas on the solar surface also wax and wane. When we compare the magnetic fields that HMI measures on the Sun's surface, it becomes apparent that sunspots, active regions, the patchy radio corona, coronal loops, flares, and mass ejections are all related in different ways to regions where there are magnetic fields.

The question that concerns us Earthlings most is the following. Do the solar eruptions and outbursts that constitute such a spectacular component of the SDO images represent perturbations that should concern us in the context of life on Earth? Or do they constitute relatively minor disturbances against the backdrop of a much larger, and much steadier, output of energy that the Sun generates continuously as the days, years, and eons go by? In order to address these questions, we need to determine, on the one hand, the properties of the Sun as a whole, and on the other hand, the properties of the magnetically driven phenomena in the atmosphere. Specifically we must determine how hot and how dense the material is in the deep interior, in order to determine if the inertia of that material can offset the dynamic phenomena that attract our attention so spectacularly from time to time in the surface layers.

How are we to determine the physical conditions in the deep interior? We need to rely on the laws of physics. These laws indicate that the power output of the Sun depends on how the physical parameters temperature $T$, pressure $p$, and density $\rho$ vary as a function of radial location $r$ between the center of the Sun ($r = 0$) and a region that we will refer to as "the visible surface" ($r = R_\odot$). Here, and throughout this book, the subscript $\odot$ denotes a parameter of the Sun as a whole. An important goal in our study of the Sun is to use various laws of physics so that students can determine for themselves the radial "profiles" of $T(r)$, $p(r)$, and $\rho(r)$ in the range of radial locations between $r = 0$ and $r = R_\odot$.

The starting point for these profiles is provided by observations of certain parameters at the "visible surface" of the Sun. Photons from those visible layers reach us on Earth and carry information on local conditions at $r = R_\odot$. Once the global parameters of the Sun are determined, they serve as boundary conditions to help us get started on our first set of calculations, namely how do $T(r)$, $p(r)$, and $\rho(r)$ vary as a function of $r$? At first (in Chapter 5), we shall consider values of $r$ close to the surface, i.e., at $r \approx R_\odot$. Then we shall proceed in steps to move inward, going progressively farther below the visible surface (in Chapters 7 and 9). By applying a variety of physical laws, and also by a judicious choice of computational techniques, our goal is to calculate $T(r)$, $p(r)$, and $\rho(r)$ at all locations from the surface down to the very center of the Sun, i.e., down as far as $r = 0$. These calculations will provide us with physical details about the *internal* structure of the Sun.

How can we check that our calculations of the physical conditions deep inside the Sun are reliable? After all, we cannot possibly "see" directly down there with any telescope. Despite this handicap, we note that, fortunately, the gas in the Sun supports a number of "waves" (or "oscillations") of various types. The number of such waves that have already been identified is vast, in the millions. The area of solar research into the properties of waves inside the Sun is known as helioseismology: the analogy is with "seismology" on Earth that relies on the propagation of various types of waves generated by earthquakes to study the internal structure of Earth. Helioseismology has opened up remarkable vistas on the solar interior that were totally absent prior to the 1970s. Not only can the oscillations help us to check the structural calculations, but they can also help us determine how the Sun rotates at depths far below the surface. More recently, in 2018, a new class of oscillations (referred to as "*r*-modes") has been identified that owe their existence to the fact that the Sun is rotating.

In view of the fundamental advances in solar physics that helioseismology has enabled, it is important that, even in this "first course" in solar physics, attention should be paid to understanding how to calculate the basic properties of some of the solar oscillations.

Once we are satisfied with our study of the *interior* of the Sun, we can return to the conditions at the visible surface and recognize that they also serve as boundary conditions for another interesting physics problem: how do the physical parameters vary as a function of radial location as we explore higher and higher altitudes *above the surface*? A rich variety of physical phenomena occurs in these locations, which may be referred to as the "outer atmosphere of the Sun". Many of these phenomena, especially those that owe their existence to variable magnetic fields, are covered by the umbrella term "solar activity". This term includes sunspots, flares, and coronal mass ejections. How far up above the surface of the Sun can we carry out our study? Physical parameters of the gas that lies between $r = R_\odot$ and the Earth's orbit (at a distance $r$ that is defined to be 1 astronomical unit (AU) $\approx 215\ R_\odot$) can also be determined by applying the laws of physics to the increasingly rarefied environment that exists at greater and greater distances above the surface of the Sun. We shall see that, in the presence of a corona that is as hot as $10^6$ K, this rarefied environment can simply *not* remain static: it must expand away from Sun. This expanding medium was called the solar wind by Eugene Parker, who first proposed its existence in 1958 on theoretical grounds. In 1959, the Soviet mission Luna 1 proved Parker correct: such a wind does exist in "interplanetary space" surrounding Earth. (However, to reach the wind, one must be outside the region where the Earth's magnetic field shields the Earth from direct exposure to the wind). Subsequently, spacecraft in the vicinity of 1 AU, and also those that are traveling throughout the solar system (some out at distances of ≥150 AU from the Sun, while others are in at distances as close as ≤0.1 AU to the Sun), provide *in situ* values of the various physical parameters in the solar wind. These help to check our calculations.

In summary, our approach to studying the physics of the Sun consists of starting at the visible surface ($r = R_\odot$), where the local physical parameters can be reliably measured by means of visible light, and then proceeding in two distinct directions. First, we move inward, into the deep interior of the Sun. Second, we move outward into the tenuous gas that exists above the visible surface.

As regards the long-term stability of the Sun, two aspects of the physics are key. The first has to do with the pressure $p_c$ at the very center of the Sun: if $p_c$ can support the weight of the overlying material (i.e., the weight of the entire Sun), then a condition known as hydrostatic equilibrium is ensured. In this case, the Sun will be in a structural condition where, in a global sense, all the mechanical forces will be in balance. This is an important property that ensures that our Sun is able to remain for long periods of time in a state free of major disruptions. We shall have to check to see if this condition is indeed satisfied by our model calculations.

The second aspect is that energy must be generated at such a rate that the power output remains more or less steady on long time-scales (several billion years). The only known source of energy that will satisfy this is nuclear fusion. This requires that the central temperature $T_c$ be high enough that nuclear reactions can occur at a suitably rapid rate to warm the Earth (situated at a distance of 1 AU away). As we shall see, nuclear reactions in the Sun rely on quantum mechanics if they are to occur at all. To be sure, quantum mechanics is typically associated with events on atomic scales. Nevertheless, every time you feel the warming energy of the Sun, it is worthwhile to remember that that level of warmth would not be available if quantum mechanics were not at work deep in the core of the Sun.

Our goal in this book is to determine enough information about the physical conditions inside the Sun, and in its extended atmosphere, so that we may appreciate, from a global perspective, the amazing entity that enables life to survive on Earth for eons without serious disruptions.

In this book, a particular emphasis will be placed on numerical modeling. In five of the chapters (Chapters 5, 7, 9, 10, and 14), the reader is given step-by-step instructions for calculating, in a simplified manner, the numerical values of various physical quantities as a function of radial distance inside the Sun. In my over 30 years of teaching solar physics to students at the University of Delaware, I have found that there is a significant pedagogical advantage to this approach. A student can gain a lot of insight into the conditions in the Sun by watching, step by step, how the pressure, or the amplitude of an oscillation, varies as one moves from one radial position inside the Sun to another. The student, in a subsequent more advanced course, may eventually encounter the complete equations of stellar structure. These equations will include detailed expressions for the equation of state, the opacity, and the energy generation rate. However, the codes are so complicated that it is not easy to understand why the solutions behave the way they do: there are too many variables to keep track of. In the present book, my hope is that the student can obtain a firm grasp of the following questions. How does the pressure (in "physics" units, i.e., dynes cm$^{-2}$ in this book) vary as a function of radial location (also in "physics" units, i.e., cm in this book) from the center of the Sun, to the photosphere, to the chromosphere, to the corona, and eventually into the distant wind? Likewise, I would like the student to obtain a good grasp of the radial profiles of density and temperature. A feel for the actual physical length-scales and pressures can help a student to appreciate the true immensity of the Sun. And as the student will learn in Chapter 11, it is precisely this immensity that enables the nuclear reactions that make life on Earth possible.

Ultimately, theories need to be connected to the real world by means of observations. Thankfully, solar physicists are now living in a "golden age" of solar observatories, when instruments on the ground and in space are providing a flood of data that may be used to check the theories. In Figure P. 1, the acronyms for some of these observatories appear as labels on the various curves: we will discuss many of the acronyms in more detail in appropriate chapters in the book. (See also Appendix B for summaries.) The vertical axis in Figure P. 1 illustrates the volumes of data that have been (or will be) accumulated by the various observatories. But for now, I would like to highlight one particular example for a unique achievement: the project known as STEREO. This consists of two spacecraft (A and B) that were launched into orbits almost identical to Earth's orbit. However, STEREO A was placed in an orbit that would ensure that, as time went by, it would drift slowly *Ahead* of Earth in its orbit. At the same time, the spacecraft STEREO B was placed in an orbit that would ensure that, as time went by, it would drift slowly *Behind* Earth. The angle between A and B (as seen by an imaginary observer at the Sun) was designed to increase slowly, until, early in the

Preface                                                                                                    xvii

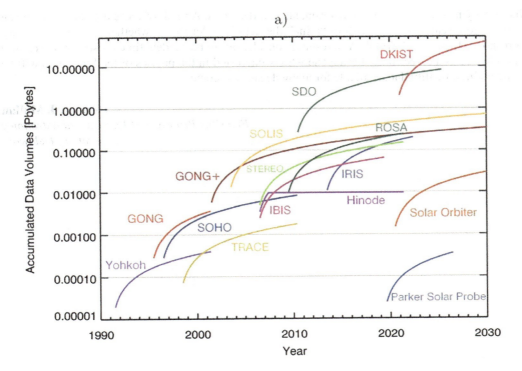

**FIGURE P. 1** Growth in the accumulated amount of data from solar observations (Lapenta et al. 2020, *Solar Phys.* 295, 103). The units on the vertical axis are petabytes: 1 Pbyte = $2^{50}$ bytes of data, i.e., about one million gigabytes. Each curve refers to a spacecraft or a ground-based instrument that will be mentioned in later chapters of the book and in Appendix B. (Used with permission of Springer.)

year 2011, about 5 years after launch, A was exactly 90 degrees ahead of Earth, while B was exactly 90 degrees behind Earth. As a result of these stereoscopic vantage points, and in combination with observations from Earth, it became possible *for the first time in human history* for *the entire surface of the Sun to be viewed simultaneously.* The capability to make such observations lasted until 2015, when unfortunately, STEREO B encountered difficulties in communicating with Earth: at the time of writing (2021), NASA has not been successful in resuscitating STEREO B. But for a span of 4 years, humans were in a position where there was no longer any need to speculate about what might be happening on the "hidden" side of the Sun. Instead, we are able to "see" these happenings with our very own "eyes". Amazing!

The time axis in Figure P. 1, stretching as it does across some 40 years, indicates that different instruments were available at different times. If one draws a vertical line upward from any particular date, one sees that there are occasions when multiple observatories (as many as six or more) may have been in simultaneous operation. This has led to the possibility that observations of solar phenomena may be conducted in multiple wavelengths simultaneously. This is not a simple process: coordination of spacecraft in different orbits requires detailed planning. For example, during the years 2012–2017, there were 6953 solar flares larger than a certain limit: of those flares, only 40 were observed by six or more instruments simultaneously, with each flare being observed by (on average) only 2.4 instruments (see Milligan, R. O., & Ireland, J., 2018, *Solar Phys.* 293, 18).

The data from some of the observatories in Figure P. 1 keep pouring into Earth in such great quantities that more and more solar physicists will be needed in the years ahead to keep up with the analysis of the data that are already available, not to speak of the data that will eventually be obtained by newer generations of instruments and detectors. And all of these data are being made available to researchers who wish to "mine" the data in order to test whether a particular theory corresponds to the "real world". As a result, I consider this to be an ideal time for students to choose solar physics as a career path: for anyone who is interested in the physics of the Sun, there will be abundant high-quality data available for many decades to come.

**Dermott J. Mullan**
*Emeritus Professor of Physics and Astronomy*
*University of Delaware*

# Author

**Dermott J. Mullan** has been a faculty member in the Bartol Research Center, Department of Physics and Astronomy, University of Delaware, since 1972. His research interests are in stellar structure, especially as the structure is altered by the presence of magnetic fields. His research, supported by several dozen grants from NASA, is based on observations with a number of satellites that detect photons from the Sun and stars over a broad range of wavelengths, from infrared to X-rays, and also satellites that detect particles and fields in interplanetary space. His research results have been published in almost 300 papers in the refereed literature. At the University of Delaware, he has taught graduate-level courses entitled Introduction to Astrophysics, Stellar Structure and Evolution, Plasma Astrophysics, and Solar Physics, as well as undergraduate courses entitled Concepts of the Universe, Black Holes and Cosmic Evolution, Introduction to Astronomy, Introduction to Physics: Electricity and Magnetism, and Astrophysics and the Origins of Life. During the years 2005–2015, he served as the Director of the Delaware NASA Space Grant Program, as well as Director of the NASA EPSCoR Program in the state of Delaware. In 2021, the Department of Physics and Astronomy voted to grant him the title of Emeritus Professor.

# 1 The Global Parameters of the Sun

In order to understand the physical processes that occur in the Sun, we need to know certain properties of the Sun, including mass, radius, and other quantities. In this chapter, we summarize the relevant information, with emphasis on describing *how* the information is obtained and *how precise* the current measurements actually are.

When it comes to astrophysical measurements, the quantity that can typically be measured with the greatest accuracy is *time*. As a result, we start our discussion of the determination of solar parameters by referring to measurements of certain intervals of time.

## 1.1 ORBITAL MOTION OF THE EARTH

The single most important property that determines the evolutionary behavior of any star is its mass; it is the mass that determines whether a star will eventually end its life quietly or explosively. The mass of the Sun can be determined by using a formula originally derived by Newton. This formula (see Equation 1.4 later) requires us to measure two quantities associated with the orbit of a planet: (i) the time it requires to complete one orbit and (ii) the (average) distance between the planet and the Sun. For present purposes, we will consider the Earth as the planet that will help us to estimate the Sun's mass $M_\odot$.

We begin by considering the Earth's orbit around the Sun: how long does it take for the Earth to complete one orbit? Determination of this time-scale $P$ (i.e., referred to as the period of the orbit) is achieved by observing the interval of time required for the Sun, starting at a given location relative to the "fixed" stars as seen by an observer on Earth, to return to that same location. In modern times, the "fixed stars" used for this determination are not true stars at all, but are instead a class of galaxies known as quasi-stellar radio sources ("quasars"). Why radio sources? Because an observing technique known as very long baseline interferometry (VLBI) uses radio emissions to provide what are currently the most precise positions of celestial objects. Moreover, quasars are so far away that no overall angular motion of a "defining group" of several hundred quasars has been detected even when the positions are measured with the highest available precision (better than 0.001 arcsec). Relative to this International Celestial Reference Frame (ICRF), the quasars are observed to be, as a group, stationary. When the Sun returns, after an elapsed time of $P$ to the same point in the ICRF, this elapsed time defines the orbital period of the Earth. The interval of time $P$ is defined to be one sidereal year. Measurements indicate the following value for $P$:

$$P = 1 \text{ sidereal year} = 365.25636 \text{ days} = 31{,}558{,}150 \text{ seconds}$$

This value, and other precise estimates of various parameters of interest to solar system dynamics, can be found on an informative website at https://ssd.jpl.nasa.gov/astro_par.html maintained by NASA's Jet Propulsion Laboratory.

In order to derive the relationship between $M_\odot$ and measureable quantities (including $P$), we turn now to the equation of motion of the Earth in its orbit. This equation can be written in terms of position vectors of Sun and Earth. Relative to a zero point that can be arbitrarily chosen, the position vector of the Sun at some instant is $\vec{r}(S)$ and the position vector of the Earth at the same instant is $\vec{r}(E)$. The position vector of the Earth relative to the Sun is then $\vec{r} = \vec{r}(E) - \vec{r}(S)$, where the magnitude of $\vec{r}$ has the value $r$. The unit vector, $\hat{r}$, associated with the relative position vector $\vec{r}$ is directed from the center of the Sun towards the center of the Earth. Why do we refer, in this definition, to the *center* of each body? Because Isaac Newton showed that the gravitational force between two objects behaves as if all of the mass of each object was located at the *center* of that object: this feature owes its existence to the fact that the gravitational force between two masses falls off as the inverse square of the distance between the masses.

The mutual forces that act on the Sun (with mass $M_\odot$) and on the Earth (with mass $m_\oplus$) are given by Newton's law of gravitation. The gravitational force causes the Earth to accelerate according to the equation

$$m_\oplus \frac{d^2 \vec{r}(E)}{dt^2} = -\frac{GM_\odot m_\oplus}{r^2} \hat{r} \qquad (1.1)$$

where $G$ is Newton's gravitational constant. The negative sign on the right-hand side of Equation 1.1 indicates that the force is directed *towards* the Sun, i.e., in the negative $\hat{r}$ direction. If the masses are expressed in units of gm, distances in units of cm, and time in units of seconds, then $G$ has the following numerical value: $6.67430(\pm 0.00015) \times 10^{-8}$ cm$^3$ gm$^{-1}$ sec$^{-2}$. This value is listed (in March 2021) as a physical constant in a table maintained by the National Institute of Standards and Technology at the following website: http://physics.nist.gov/cuu/Constants/

At the same time as the Earth is accelerating according to Equation 1.1, the mutual gravitational force also acts on the Sun, causing the Sun to accelerate according to the equation

$$M_\odot \frac{d^2 \vec{r}(S)}{dt^2} = +\frac{GM_\odot m_\oplus}{r^2} \hat{r} \qquad (1.2)$$

In Equation 1.2, the positive sign on the right-hand side now indicates that *this* force is directed *towards* the Earth, i.e., in the positive $\hat{r}$ direction.

In terms of the relative position vector $r$, Equations 1.1 and 1.2 can be combined to yield

$$\frac{d^2 \vec{r}}{dt^2} = -\frac{G(M_\odot + m_\oplus)}{r^2} \qquad (1.3)$$

Using several properties of vector algebra, it can be shown (e.g., Karttunen et al. 2017) that the solution of Equation 1.3 is an ellipse with the center of mass of the Sun and Earth at one focus. The Earth's orbital ellipse has a semi-major axis $D$, which is referred to as 1 astronomical unit (AU). In terms of $D$ (which is the mean distance between Earth and Sun, averaged over the orbit), the period $P$ of orbital motion that emerges from Equation 1.3 can be shown to be given by

$$P^2 = \frac{4\pi^2 D^3}{G(M_\odot + m_\oplus)} \qquad (1.4)$$

This allows us to write an expression for the mass of the Sun in terms of $D$ and $P$:

$$\frac{GM_\odot}{D^3} = \frac{4\pi^2}{P^2[1 + m_\oplus/M_\odot]} \qquad (1.5)$$

The ratio of $m_\oplus$ to $M_\odot$ is very small (we will evaluate it shortly). If we were to neglect the ratio $m_\oplus/M_\odot$ compared to unity, then we would get a fairly good first approximation to $GM_\odot/D^3$. According to this approximation, for each planet in the solar system (all of which are orbiting the same object [i.e., the Sun], of mass $M_\odot$), the square of the planet's period $P^2$ is proportional to the cube of the mean distance $D$ between the planet and the Sun. This property of the orbits of the planets in the solar system was first identified empirically by Johannes Kepler in the year 1618: it referred to historically as Kepler's third law of planetary motion. In order to arrive at this law, Kepler plotted, for the six planets that were known in his day, log($P$) against log($D$) using tables of logarithms that had been published for the first time by John Napier in 1614. The advantage of using a log-log plot of $P$ versus $D$ was immediately apparent to Kepler: the data were found to fall along a straight line with slope 3/2. (This led Kepler to refer to what we now call Kepler's third law as "the 3/2-power law".) Moreover, in this approximation, Kepler's third law can be written in an easily memorizable equality $P^2 = D^3$: this equality is true for *all* solar system objects (planets, dwarf planets, asteroids, comets) that are in orbit around the Sun *provided* that two important caveats are borne in mind: (i) $P$ must be expressed in units of sidereal years and (ii) $D$ must be expressed in units of AU.

However, even if we make the approximation of neglecting the Earth's mass in Equation 1.5, we still cannot determine the value of $M_\odot$ unless we first determine the numerical value of $D$ (= 1 AU).

## 1.2 THE ASTRONOMICAL UNIT (AU)

Once the orbital periods of the various planets are known from careful observations over a span of many years (such as those recorded with the then-available highest precision by Tycho Brahe between the 1570s and the 1590s), the application of Kepler's third law provides a scale model of the solar system. The scale model provides knowledge, at any given instant, of the distance between any two solar system objects *in units of AU*. As a result, at any given instant, we know how far away any solar system object is from Earth in terms of AU. To make the conversion from AU to linear units (e.g., cm), two methods have been used.

First, by observing the asteroid Eros during a close approach to Earth in 1930–1931, the parallax was determined by observing from two observatories in different countries on Earth: combining the parallax with the known distance between the two observatories, the linear distance between Earth and Eros was calculated. The result for the length of 1 AU was quoted in terms of a quantity called the "horizontal parallax (HP) of the Sun", i.e., the angle subtended by the Earth's radius at the Sun. The results of this analysis were reported by Spencer Jones (1941): HP = 8.790 ± 0.001 arc seconds. If there were no systematic error in the results, the formal uncertainty in HP would indicate that the AU was known with a precision of order one part in 10,000, i.e., with an error of order 15,000 km. As a result, if one wished to launch a spacecraft to Venus (for example), whose closest distance to Earth is some 40 million km, the uncertainty in the position of Venus would be $\sigma > 4000$ km. But in fact, it was later discovered that there was a systematic error: the value of 1 AU reported from the Eros observations in 1931 was actually in error by four times the stated uncertainty (Atkinson 1982). The corresponding error at the distance of Venus turned out to be 16,000 km, well in excess of the diameter of Venus: a satellite could well have missed the planet altogether.

Second, in favorable conditions, radar reflection can be used to determine the actual linear distance to the object at that instant. This has the advantage that a distance measurement (between, say, Earth and Venus) is performed in terms of a measurement of an interval of *time*, namely, the time of flight for round-trip travel from the transmitter on Earth to the object (e.g., Venus) and back to the receiver on Earth. Reliable radar reflection measurements off a solar system object were first made in an observing run between March 10 and May 10, 1961, using the planet Venus (Victor and Stevens 1961). If Venus had been at its closest possible distance to Earth (i.e., with Venus at aphelion, and Earth at perihelion), the round-trip time would have been 255 seconds. However, in 1961, Venus and Earth had their closest approach on April 10 (i.e., the midpoint of the observing run), and on that date, Earth was not at perihelion, nor was Venus at aphelion. Moreover, on other days in the observing interval from March 10 to May 10, Venus and Earth were by no means at their closest. As a result, the round-trip time for radar reflections during the observations was of order 300 seconds. The radar instruments that were available in 1961 allowed this round-trip time to be measured with a precision of about 0.001 second. Preliminary analysis of the data led Victor and Stevens (1961) to the conclusion that 1 AU = 149,599,000 km. They quoted an uncertainty of 1500 km, i.e., a relative uncertainty of about one part in 100,000. The technical achievement of the experiment was noteworthy: the reflected signal received from Venus had a power level at least 20 orders of magnitude weaker than the signal that was transmitted. The 10-fold improvement in the precision of our knowledge of the AU provided confidence for launching the first mission (Mariner II) to Venus in 1962.

During the 60 years or more that have elapsed since the first Venus radar measurements, repeated measurements of signals passing back and forth between Earth and transmitters on artificial satellites that passed close to (or went into orbit around) various planets, as well as improvements in detector sensitivity, have resulted in steady improvements in the value of the AU in linear units. In 2012, the General Assembly of the International Astronomical Union (IAU), in its Resolution B2, voted that, for the sake of uniformity, astronomers should adopt the following exact value for the AU in terms of centimeters:

$$1 AU \equiv D = 1.49597870691 \times 10^{13} \, cm$$

This is the *average* value of the distance from Earth to Sun: since the Earth actually follows an elliptical orbit with eccentricity 0.0167, the Earth is *closer* to the Sun than 1 AU by about 1.67% at a certain date each year (in early January: when Earth is at perihelion), and *farther* from the Sun than 1 AU by about 1.67% in early July of each year (aphelion). The effects of the Earth's orbital eccentricity can be detected by (for example) observers who chart the behavior of sunspots (see Chapter 16) over the course of several years. In one study, Pevtsov et al. (2019) report that the diameter of the Sun (in units of pixels on their detector) varies from a minimum of roughly 1235 pixels in early July to a maximum of roughly 1275 pixels in early January. With a mean diameter of 1255 pixels, the amplitude of the diameter variations (as viewed from Earth) is roughly ±20 pixels, i.e., roughly ±0.016 of the mean diameter, as expected from the orbital eccentricity.

On a historical note, the aforementioned value of 1 AU corresponds to a horizontal parallax of 8.794148 arcsec: this indicates that the value of 1 AU reported from the Eros observations in 1931 was in error by four times the stated uncertainty (Atkinson 1982). Nevertheless, the fact remains that 8.790 arcsec was the best estimate for HP when the space age began in 1957.

For future reference, we note that, given that the average distance between Earth and Sun is equal to $D$, we can say that if any feature on the Sun is observed to have an angular diameter of 1 arc second when viewed from Earth, then the linear diameter of that feature on the Sun is (on average) roughly 725.3 km. The largest solar telescope in the world, the Daniel K. Inouye Solar Telescope (DKIST), with its 4-meter primary mirror, is expected (with assistance of technology that cancels the "twinkling" caused by turbulence in the Earth's atmosphere) to resolve objects with angular sizes of ~0.03 arcsec. Thus, DKIST can in principle resolve solar features that have linear sizes as small as ~21 km, if such structures exist in the surface layers of the Sun. DKIST reported its first image of a region of the Sun's surface in January 2020: we will discuss this image later in Chapter 6.

Now that we know the value of $D$ (in units of cm) and also the value of the orbital period $P(s)$ (in units of seconds), we could (if we wished) use Equation 1.5 to obtain a first approximation to an important physical quantity that is relevant for all objects in orbit around the Sun: $GM_\odot = 4\pi^2 D(cm)^3/P(s)^2$. This leads to $M_\odot \approx 2 \times 10^{33}$ gm. But there is little reason to compute this first approximation.

Instead, we can go beyond the first approximation for $GM_\odot$ and obtain a more precise estimate of $M_\odot$, provided that we can first evaluate the ratio of the mass of the Earth to the mass of the Sun $m_\oplus/M_\odot$. To do this, we comparing the orbits of two objects, one in orbit around the Sun, the other in orbit around the Earth. For both objects, we need to determine two quantities: a period and a distance.

For the object (our own planet Earth) that is in orbit around the Sun, we write for clarity the period as $P(S)$ and the semi-major axis as $D(S)$. Using Equation 1.4 and omitting the constant coefficient $4\pi^2/G$, we arrive at the proportionality

$$P(S)^2 \sim \frac{D(S)^3}{M_\odot + m_\oplus} \tag{1.6}$$

Analogously, for an object (an artificial satellite) that is in orbit around the Earth, with period $P(E)$ (in units of seconds) and semi-major axis $D(E)$ (in units of cm), we also have that

$$P(E)^2 \sim \frac{D(E)^3}{m_\oplus} \tag{1.7}$$

In Equation 1.7, we have made the reasonable assumption that the mass of the artificial satellite is entirely negligible (by 20 orders of magnitude or more) compared to the mass of the Earth.

Combining the previous equations, we have that

$$\frac{M_\odot}{m_\oplus} + 1 = \frac{D(S)^3}{D(E)^3} \frac{P(E)^2}{P(S)^2} \tag{1.8}$$

We already have precise values for $D(S)$ and for $P(S)$. Can we also find precise values for $D(E)$ and $P(E)$? Indeed we can. There are a large number (almost 50,000 in March 2021) of choices that we can make for an artificial satellite in orbit around the Earth. Any one will suit our purpose. To obtain information about any particular artificial satellite, it is convenient to examine the Satellite Database at the website www.heavens-above.com/: there, orbital information is provided for all (tens of thousands) of the artificial satellites currently in orbit around the Earth. (Some of them were launched as long ago as 1958, and new launches continue to occur every year.) By way of example, I find that the database happens to list a satellite called IRIDIUM 175 (object #43928 in the online database, launched into orbit on January 11, 2019) with the following orbital details: the perigee (the point on the elliptical orbit where the satellite is closest to Earth) lies at an altitude of 652 km above the Earth's surface, and the apogee (the point on the orbit where the satellite is farthest from Earth) lies at an altitude of 656 km above the Earth's surface. This suggests that IRIDIUM 175 is in a nearly circular orbit with a mean altitude above the Earth's surface of $h = 654$ km. Note that only three significant digits are provided for these distances in the database: this will limit the precision of our evaluation of $m_\oplus/M_\odot$.

Since IRIDIUM 175 is in orbit around the Earth, Newton's law of gravitation informs us that the relevant semi-major axis $D(E)$ must be measured relative to the center of the Earth. With a mean altitude of $h = 654$ km, the semi-major axis of the orbit around the center of the Earth is $D(E) = R_\oplus + h$, where $R_\oplus$ is the radius of the Earth. The equatorial radius of the Earth has been accurately measured, by the International Union of Geodesy and Geophysics, to be $R_\oplus = 6378.137$ km. This leads to $D(E) = 7032.137$ km for IRIDIUM 175. With an orbit which is specified only to the nearest kilometer, we round off $D(E)$ to the value 7032 km. Comparing this with the value of $D(S) = 1$ AU, we see that $D(S)/D(E) = 21273.872$ for IRIDIUM 175. Thus, the first factor on the right-hand side of Equation 1.8 is $9.628079 \times 10^{12}$.

Turning now to the orbital period, information on the aforementioned website indicates that the IRIDIUM 175 satellite completes a total of 14.72248940 orbits around Earth per day (i.e., every 86,400 seconds) corresponding to a period $P(E) = 5868.57$ sec. Compared to the value of $P(S)$ (= 1 sidereal year, expressed in units of seconds), we find that the second factor on the right-hand side (r.h.s.) of Equation 1.8 has the value $3.45813 \times 10^{-8}$.

Combining the terms on the r.h.s. of Equation 1.8, we find that $m_\oplus/M_\odot = 1/332{,}950$. This is the mass ratio that we obtain when we use the orbital data for a single satellite (IRIDIUM 175), for which we know the altitude to only three significant digits. When multiple satellites are used, the currently accepted value of $m_\oplus/M_\odot$ is found to be more precisely $1/332{,}946$. Thus, our use of IRIDIUM 175 data alone leads to an error in the mass ratio of about one part in 80 thousand. Our calculations would have led to the currently accepted value of $m_\oplus/M_\odot$ if we were to use a value of $h$ which was only slightly different from the listed value (654 km) for the mean altitude of IRIDIUM 175.

## 1.3 $GM_\odot$ AND THE MASS OF THE SUN

We now have enough information to use Equation 1.5 to evaluate the product of the gravitational constant $G$ and the mass of the Sun to nine significant figures:

$$GM_\odot = 1.3271244 \times 10^{26} \, \text{cm}^3 \, \text{sec}^{-2} \tag{1.9}$$

The precision with which the product $GM_\odot$ is known has increased over the course of the space age, as more and more spacecraft have traveled throughout the solar system, always subject to a sunward acceleration that is proportional to the aforementioned product. Currently, the numerical value of $GM_\odot$ is actually known even more precisely than Equation 1.9 indicates: the best measured value currently (in 2021) extends to 12 significant digits (e.g., see Astrodynamic Parameters at nasa.gov). However, we do not need all those digits here, because the value of $G$ is not as precisely known.

To extract a value for the mass of the Sun, we need to divide the previous product by $G$. The numerical value of $G$ is among the most poorly measured of the fundamental constants of nature: $G$ is known with a precision of only one part in $4 \times 10^4$. In fact, one can find that "the value of $G$", as reported in the literature, changes from time to time in the sixth or fifth significant digit when new experiments become available (see Prsa et al. 2016). Using the value of $G = 6.67430(\pm 0.00015) \times 10^{-8}$ cm$^3$ gm$^{-1}$ sec$^{-2}$ cited in Section 1.1, we obtain the following estimate for the mass of the Sun, reliable to one part in $4 \times 10^4$:

$$M_\odot = 1.98841 \times 10^{33} \, \text{gm} \tag{1.10}$$

## 1.4 POWER OUTPUT OF THE SUN: THE SOLAR LUMINOSITY

Spacecraft equipped with radiometers are designed to measure the total flux of radiant energy coming from the Sun at all wavelengths in the electromagnetic spectrum, from gamma rays to long-wave radio emissions. This total flux, known as the total solar irradiance (TSI), has been measured by a number of spacecraft in recent decades. In Figure 1.1, we show results from the Total Irradiance Monitor (TIM) on the Solar Radiation and Climate Experiment (SORCE) satellite, which reported TSI values starting on February 25, 2003, and ending when the satellite was decommissioned on February 25, 2020.

The units customarily used to report TSI are watts per square meter (W m$^{-2}$). As seen in Figure 1.1, the TSI when averaged over days and weeks has values ranging from about 1360.5 (in 2009) to about 1362 W m$^{-2}$ (in 2014–2015). Prominent narrow dips in the data in Figure 1.1 are associated with the presence of transient dark sunspots traversing the visible disk: the numbers of such spots on the Sun's surface are observed to wax and wane in the course of a "sunspot cycle", which lasts for roughly 11 years (although this can vary between 9 and 12 years from one cycle to another). We will discuss the properties of sunspots in Chapter 16.

The deepest dip in the data in Figure 1.1 occurred on October 29, 2003, when the TSI dipped by almost 5 W m$^{-2}$. (Dips were also recorded during transits of Venus across the Sun in 2004 and 2012, but those had nothing to do with the sunspot cycle.) In 2015, the IAU, in its resolution B3, adopted a "nominal" value of 1361 W m$^{-2}$ for the solar TSI: Prsa et al. (2016) estimate that the uncertainty in this value is $\pm 0.5$ W m$^{-2}$, i.e., the fractional uncertainty is $3.7 \times 10^{-4}$. As can be seen in Figure 1.1, this nominal value is intermediate between the extreme highs and lows of the measured TSI, and so can be regarded as a sort of "average" value for the radiant energy that reaches Earth from the Sun over the course of a sunspot cycle. When the first edition of this book was being prepared in 2008, the data that were available from the Solar and Heliospheric Observatory (SOHO) satellite (and several other satellites) suggested that the TSI value was definitely larger than the above "nominal" value: the excesses were as much as 10 W/m$^2$. However, those data were subsequently found to require reductions because of scattered light inside the instruments. When corrections were made

# The Global Parameters of the Sun

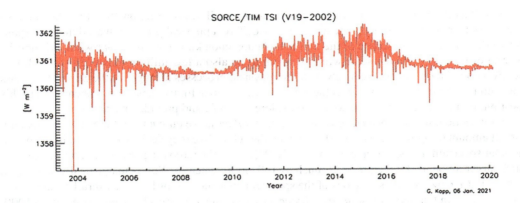

**FIGURE 1.1** The solar irradiance, normalized to a solar distance of 1 AU, measured over almost 20 years as measured by SORCE/TIM. Each data point represents an average over 6 hours. (The plot is taken from the website https://spot.colorado.edu/~koppg/TSI/#lower_TSI_value; used with permission of G. Kopp.)

for the scattered light, the data from the other satellites were found to come into satisfactory agreement with the results from SORCE/TIM shown in Figure 1.1.

As Figure 1.1 indicates, the magnitude of TSI is observed to vary in the course of a sunspot cycle by an amount of 1–1.5 W/m², i.e., a fractional change by about one part in 1000. The phase of the variation is noteworthy because it is unexpected: the flux of radiant energy from the Sun (the TSI) is observed to be *smallest* when the sunspot number is *smallest* (e.g., in 2009), while the TSI is observed to be *largest* when the sunspot number is *largest* (e.g., in 2014–2015). On the face of it, this seems counterintuitive: shouldn't the solar output of radiant energy be smaller when there are more sunspots to block the light from the Sun? We shall find an answer to this conundrum when we discuss (in Section 16.7.6) certain magnetic features on the Sun called faculae.

Given the distance from Earth to Sun ($D$), the nominal TSI transforms to an output power from the Sun of $L_\odot = 4\pi D^2 \times$ TSI. In terms of c.g.s. units, the nominal TSI can be written as $1.361 \times 10^6$ ergs cm$^{-2}$ sec$^{-1}$. Inserting the value of $D = 1$ AU in units of cm, we find

$$L_\odot = 3.828 \times 10^{33} \text{ ergs} \cdot \text{s}^{-1} \tag{1.11}$$

This power output from the Sun is also referred to by astronomers as the "solar luminosity". With a fractional uncertainty of $3.7 \times 10^{-4}$ in TSI, and a much smaller uncertainty in the value of $D$, the corresponding uncertainty in $L_\odot$ is $\pm 0.0014 \times 10^{33}$ ergs s$^{-1}$.

For future reference, comparing Equations 1.11 and 1.10, we note that the ratio of $L_\odot / M_\odot$ has a numerical value that is easy to remember in centimeter-gram-second (c.g.s.) units: about 2 ergs gm$^{-1}$ sec$^{-1}$. We will find it useful to use this value of the $L_\odot / M_\odot$ ratio when we calculate the internal structure of the radiative interior of the Sun (Section 8.5).

Also for future reference, we note that the power output from the Sun relies on the conversion of (nuclear) mass into energy in the deep inner core of the Sun. Using the conversion formula $E = mc^2$, we note that the value of the Sun's power output $L_\odot$ requires the conversion of mass to energy at a rate $(dM/dt)_{\text{nucl}}$ such that the product $c^2 \times (dM/dt)_{\text{nucl}}$ is equal to the value of $L_\odot$. Using Equation 1.11, and referring to https://ssd.jpl.nasa.gov/astro_par.html for the value of $c = 2.99792458 \times 10^{10}$ cm s$^{-1}$, we find $(dM/dt)_{\text{nucl}} = 4.259 \times 10^{12}$ gm sec$^{-1}$, i.e., roughly 4 million tons of the Sun's material is converted every second into energy. In the course of the Sun's lifetime, which is estimated to be about 4.6 Gy, the mass of the Sun has been reduced by nuclear processing by only a small amount, a few parts in $10^4$.

As well as losing mass due to nuclear reactions, we shall see (in Section 18.7) that the Sun also loses mass from its outer atmosphere due to a very different physical process: hydrodynamic expansion of the hot outer atmosphere of the Sun in a phenomenon known as the "solar wind". Curiously, the rate of mass loss via the solar wind turns out to be also a few million tons every second. Is it merely a coincidence that two entirely different physical processes arrange for the Sun to lose mass at essentially the same rate? As far as the author is aware, no definitive answer has been given to this question. Perhaps a reader of this book will explore the topic and provide an answer.

In this section, we have been discussing that the *total* radiative output from the Sun varies by only a small amount (roughly one part in 1000) during the solar 11-year cycle. However, certain regions of the solar spectrum (in the extreme ultraviolet [EUV] and in the radio ranges) can undergo changes by much larger amounts, by factors of as much as 100–1000, when the Sun is magnetically active (see Section 18.6). However, those regions of the spectrum contribute so little to the overall radiative output from the Sun that the *total solar irradiance* varies during a solar cycle by only one part in 1000.

## 1.5 THE RADIUS OF THE SUN: $R_\odot$

Now that the mean distance to the Sun is known, it would seem to be a simple matter to obtain the linear radius (or diameter) of the Sun: "simply" measure the angular radius (or angular diameter) in radians and multiply by $D$. But it is not a simple matter to measure the angular diameter of the Sun precisely. Relying on the human eye alone, we know from solar eclipses that the angular diameter of the Sun sometimes appears definitely larger than the angular size of the Moon (in an annular eclipse), but at other times the Sun appears smaller than the Moon (in a total eclipse). This tells us that the Sun has an angular diameter close to that of the Moon. The latter can be measured unambiguously because the Moon has a solid surface: on average, the Moon's angular diameter is about 31 arc minutes, i.e., the Moon's angular radius is about 930 arc seconds ("). But because the Moon's orbit has an eccentricity of 0.055, the Moon's angular radius can be as large as 977" or as small as 879". Therefore, eclipses tell us that the Sun's angular radius must lie in the range 879"–977". However, because the Sun does not have a solid surface, the definition of "the" radius of the Sun (or even the "limb" of the Sun) requires some care. Observers of the Sun's "limb-darkening curve" measure how the brightness of the Sun falls off with increasing radial distance from the center of the disk: the brightness is most intense at disk center, the brightness falls off as the line of sight approaches the limb, and when the line of sight goes "off the limb", the brightness falls steeply to essentially zero. Along the steeply falling part of the curve, there is an inflection point where the curvature changes from downward to upward. Observers of the "limb-darkening function" typically define the relevant angular radius of the Sun $R_a$ by the location of the inflection point. For example, Meftah et al. (2014) used this approach to report that 20,000 ground-based measurements between 2011 and 2013 yield $R_a = 959.78" \pm 0.19"$ (adjusted to a standard distance of 1 AU) at a wavelength of 5357 Å. Using the PICARD satellite, Meftah et al. (2018) reported values of $R_a$ in visible light ranging from 959.83" to 959.91". Using the HMI instrument on the Solar Dynamics Observatory (SDO), Hauchecorne et al. (2014) used a comparison between arcs on the limb of the Sun and on the limb of Venus (during the 2012 transit of Venus) to determine $R_a = 959.90" \pm 0.06"$ at a wavelength of 6173 Å. Since 1" corresponds (on average) to 725.3 km at the Sun, the above range of $R_a = 959.78"-959.91"$ corresponds to linear "limb radii" $R_L$ in the range from 696,128 km to 696,223 km. Clearly, all of these results lie well within the range of estimated angular radii 879"–977" mentioned earlier in the context of annular/total solar eclipses.

The problem with measuring the solar "limb" is that the line of sight used by an observer on Earth at the inflection point passes only tangentially through the Sun's atmosphere. Now, when we turn to consider (in Chapter 2) how radiation makes its way through the solar atmosphere, it will be useful to define a scale of "optical depth" $\tau$ which measures how difficult it is for light to escape from any particular layer in the solar atmosphere. The significance of $\tau$ is that, when we observe the *center* of the solar disk, values of $\tau$ close to the value $\tau=1$ correspond to the layers in the Sun where most of the light we see from the Sun in effect originates. And according to one model of radiative transfer (Section 2.10), the level where $\tau = 2/3$ has the advantage that the local temperature $T(2/3)$ has a special value called the effective temperature $T_{eff}$ such that in combination with the radius $R(2/3)$ at that level, the luminosity of the Sun can be written as $L_\odot = 4\pi R(2/3)\sigma_B T_{eff}^4$. Here, $\sigma_B = 5.67040 \times 10^{-5}$ ergs cm$^{-2}$ sec$^{-1}$ deg$^{-4}$ is known as the Stefan–Boltzmann constant. The difficulty with measurements of the limb is that the tangential line of sight does *not* penetrate all the way down to the layer of gas where $\tau=2/3$: instead, the line of sight passes no closer to the Sun than a finite distance $h(limb)$ *above* that layer. The question now is: how large is $h(limb)$? To answer that, detailed calculations of the propagation of radiation through the solar atmosphere are necessary: thus, Haberreiter et al. (2008) have shown that, at wavelengths of 5000 Å, the value of $h(limb) = 333$ km above $\tau=2/3$, and 347 km above $\tau=1$. Given that 1 arcsec at the Sun corresponds to a linear distance of 725 km, we see that angular measurements of the solar radius using the *limb* of the Sun are too large by 0.46" than the "true" radius at $\tau=2/3$. Applying these corrections to the "limb radii" listed in the previous paragraph (696,128 to 696,223 km), we find that the values of the "true" solar radius $R(2/3)$ are expected to lie in the range 695,795 km to 695,890 km.

An independent method of determining the (linear) radius of Sun can be obtained by using data on waves (oscillations) that exist in the Sun. In Chapters 13 and 14, we shall see that a class of oscillatory modes in the Sun known as *f*-modes are horizontally propagating waves with maximum kinetic energy densities in the surface layers of the Sun (where each *f*-mode has a peak in the plot of its eigenfunction versus radius). The *f*-modes have frequencies ω that are related to the wave number $k$ in an especially simple way: $\omega^2 = gk$. Here, $g = GM_\odot / r^2$ is the acceleration due to gravity in the radial location $r = r(f)$ where the *f*-modes have maximum kinetic energy density. Therefore, measurement of both ω and $k$ for any mode gives a value for $g$ at radius $r(f)$. And since we already know the value of $GM_\odot$ accurately (see Equation 1.9), we can use the value of $g$ to obtain a reliable value of $r(f)$. Ground-based observations of the frequencies of many solar modes of oscillation have been obtained by the GONG project: GONG, or Global Oscillations Network Group, is a group of observing stations with identical equipment distributed around the world so as to obtain continuous observing sequences of the Sun over time intervals that are as long as possible: the longer the interval, the more precisely can the frequencies be determined. Using 1 month of GONG data, Antia (1998) found that the observed values of ω at a given $k$ for the *f*-modes were systematically larger (by a factor of $1.000437 \pm 0.000005$) than predicted by a model in which the value of $R(2/3)$ had been assumed to equal 695,990 km. Using the scaling $\omega \sim r^{-1.5}$, Antia concluded that the solar radius in the model would need to be reduced by 203 km: thus a better choice for $R(2/3)$ would be 695,787 km. Independently of Antia, Schou et al. (1997) used 144 days of spacecraft data (from the Michelson Doppler Imager [MDI] on board SOHO) to conclude also that $r(f)$ was smaller than the "standard" radius of the Sun (695,990 km) by 310 km: this leads to

$$R_\odot = (6.9568 \pm 0.0003) \times 10^{10} \cdot \text{cm} \qquad (1.12)$$

This is about 100–200 km smaller than the values of $R(2/3)$ reported earlier from angular diameter data. The IAU 2015 Resolution B3 suggested that the nominal value of the solar radius should be taken to be 695,700 km, a value that lies between the seismic result and the "limb" results.

Given the striking change in the visual appearance of the Sun's surface between sunspot minimum (when the disk can be almost entirely free of spots for days on end) and sunspot maximum (when many spots can be found on the disk), it is natural to wonder: does the radius of the Sun undergo any changes in the course of the sunspot cycle? Antia et al. (2000) used *f*-mode frequencies to address this: using 3 years of GONG data, they found changes in the frequencies that suggest that the solar radius is *smaller at solar maximum* by 5 km. Lefebvre et al. (2009) used *f*-modes to show that the solar radius is *smaller at solar maximum* by $7 \pm 1$ km. Bush et al. (2010) used SOHO/MDI imaging data covering an entire solar cycle to derive an upper limit on the peak-to-peak change in solar radius during the cycle: 0.023", i.e., ≤17 km. Larson and Schou (2015) used SOHO/MDI data for the years 1997–2011 and found that the fractional change in seismic radius $\Delta R/R$ between solar minimum and solar maximum was about $2 \times 10^{-5}$, i.e., about 0.02", consistent with the amplitude obtained by Bush et al. Moreover, Larson and Schou were also able to show (see their Figure 3) that the seismic radius of the Sun is *smaller at solar maximum than at solar minimum*. Kosovichev and Rozelot (2018), using 21 years of SOHO and SDO data (i.e., extending over almost two solar cycles), reported that during solar maxima, the solar radius is *reduced* by 5–8 km at depths of $5\pm 2$ megameters (Mm). These papers are consistent in the following conclusion: during the 11-year sunspot cycle, the Sun's radius *decreases* (slightly) when sunspots are most abundant, and *increases* (slightly) when sunspots are least abundant. However, the changes in radius between solar minimum and maximum are small enough that they do not exceed the uncertainty (±30 km) listed in Equation 1.12. Therefore, the value of solar radius in Equation 1.12 is applicable throughout the solar cycle. Theoretical models of solar structure that incorporate variable magnetic fields with periods of order 11 years have also reported that the solar model becomes (slightly) smaller at solar maximum by amounts of 7–30 milliarcsec ≈ 7–30 km (Mullan et al. 2007) and by 30 km (Piau et al. 2014).

Interestingly, if the length of the solar cycle were not 11 years, but much longer, then the predicted change in solar radius due to magnetic effects would undergo a change in sign. In stars where the magnetic field is steady or varies on long enough time-scales (>3500 years), there is enough time for the interior of a star to adjust its entropy in such a way (Piau et al. 2014) that the radius *increases*. Models of low-mass stars that include *steady* magnetic fields are indeed found to exhibit larger radii than in nonmagnetic models (Mullan and MacDonald 2001).

For future reference, when we come to discuss the solar wind (Chapter 18), it will be helpful to know how far the Earth is from the Sun in units of the solar radius. Comparing $D$ with $R_\odot$, we can see that 1 AU is equivalent to 215.04 $R_\odot$.

## 1.6 ACCELERATION DUE TO GRAVITY AT THE SURFACE OF THE SUN

Now that we know the mass and radius of the Sun, we can calculate the acceleration due to gravity $g_s$ at the solar surface.

$$g_s = \frac{GM_\odot}{R_\odot^2} = \frac{1.3271244 \times 10^{26}}{(6.957 \times 10^{10})^2} = 27420 \text{ cm sec}^{-2} \tag{1.13}$$

For future reference, we note that a convenient way to remember this value is to recall the logarithmic value (in base 10): $\log g_s = 4.44$.

## 1.7 THE MEAN MASS DENSITY OF THE SUN

Knowing $M_\odot$ and $R_\odot$, we can calculate the mean mass density of the Sun:

$$\bar{\rho} = \frac{M_\odot}{(4/3)\pi R_\odot^3} \tag{1.14}$$

Inserting the values of $M_\odot$ and $R_\odot$ from Equations 1.10 and 1.12, we find $\bar{\rho}$ = 1.410 gm cm$^{-3}$, somewhat greater than the mean density of (liquid) water. Once we calculate a model for the interior of the Sun (Chapter 9), it will be a matter of interest to compare the density at the *center* of the Sun to the *mean* density. We shall find that the central density in the Sun is at least 100 times larger than the density of liquid water. Clearly, in order for the Sun to have a mean density of order 1 gm cm$^{-3}$, while its central density is so much larger, there must be a compensating decrease of density as we approach the surface of the Sun. We shall see in Chapter 5 that indeed the density of the gas in the

# The Global Parameters of the Sun

photosphere is of order $10^{-7}$ gm cm$^{-3}$: this is about four orders of magnitude smaller than the density of air on the surface of the Earth. We shall be interested in Section 5.6 to find out why the Sun has such a density in its photosphere: we shall find that the answer to this question has to do with the process that controls the escape of light from the photosphere through the overlying atmosphere.

Despite the large densities at the center of the Sun, it is important to note that the material of which the Sun is composed does *not* behave as a liquid or a solid: instead, we shall find that it obeys the laws that govern the behavior of a gas (Section 9.5).

## 1.8 ESCAPE SPEED FROM THE SOLAR SURFACE

The escape speed from the surface of the Sun is given by

$$V_{esc} = \sqrt{\frac{2GM_\odot}{R_\odot}} = 617.7 \text{ km sec}^{-1} \tag{1.15}$$

This escape speed is a measure of the depth of the gravitational potential well due to the mass of the material in the entire Sun. It is a measure of how strongly the Sun's weight crushes the gas in the core of the Sun. It is a law of physics that, if the Sun is to remain in pressure equilibrium, the crushing effects of the weight of the overlying material on the core have to be balanced by the effects of outward-directed pressure.

Now, the pressure that operates in a gas is determined by the momentum flux of the individual gas particles. As a result, the thermal pressure in the core is related to the mean square velocity of the thermal particles there (e.g., Sears 1959). Thermal particles, each with mass $m$ and in a medium with temperature $T$, have a root-mean-square (r.m.s.) velocity $V(rms) = \sqrt{(3kT/m)}$ where $k = 1.3806504 \times 10^{-16}$ ergs deg$^{-1}$ is Boltzmann's constant. We shall be especially interested in the value of $V(rms, center)$ at the center of the Sun. The existence of the two velocities, $V(rms, center)$ and $V_{esc}$, which are both characteristic of the Sun, suggests that if the Sun is to be in pressure equilibrium, then $V(rms, center)$ and $V_{esc}$ should have comparable magnitudes. We shall check on this expectation after we calculate a model of the Sun (Section 9.2).

For future reference, we note that for a gas consisting of hydrogen atoms, $1/m = 1/m_H$ (where $m_H$ is the mass of a hydrogen atom $= 1.6605389 \times 10^{-24}$ gm), and the value of $1/m$ equals Avogadro's number $N_a$, which is the number of particles in one mole. The combination $kN_a$ is referred to as the gas constant $R_g = 8.314472 \times 10^7$ ergs deg$^{-1}$ mole$^{-1}$. For a gas consisting of particles with atomic mass $\mu$ (in units of $m_H$), the r.m.s. velocity $V(rms) = \sqrt{(3R_g T/\mu)}$.

## 1.9 EFFECTIVE TEMPERATURE OF THE SUN

Now that we know the output power of the Sun as well as its radius, we are in a position to calculate the value of $T_{eff}$, i.e., the temperature of the equivalent blackbody that would radiate a flux equal to that emitted by the Sun. In Section 1.5 earlier, we mentioned that the solar luminosity can be written in terms of an effective temperature ($T_{eff}$) and the solar radius.

In Section 1.5, we labeled the radius of the layer where the local temperature $T$ equals $T_{eff}$ as $R(2/3)$. Now that we have a reliable value of $R_\odot$, we replace $R(2/3)$ with $R_\odot$ and write:

$$L_\odot = 4\pi R_\odot^2 \sigma_B T_{eff}^4 \tag{1.16}$$

The surface flux of radiant energy at the Sun, $F_\odot = L_\odot / 4\pi R_\odot^2$, can now be evaluated: $6.2939 \times 10^{10}$ ergs cm$^{-2}$ sec$^{-1}$. Using this, we find that the effective temperature of the Sun is

$$T_{eff} = 5772 K \tag{1.17}$$

In Chapter 2, we shall see that this value of $T_{eff}$ has the consequence that the light emitted by the Sun has a peak intensity *per unit wavelength* at a wavelength close to 5000 Å, i.e., in the green part of the visible spectrum where the sensitivity of the human eye is close to optimal.

## 1.10 THE OBLATENESS OF THE SUN

To the unaided eye, the disk of the Sun appears to be circular in shape. But careful measurements reveal a slight departure from circularity: the polar diameter is slightly smaller than the equatorial diameter. The fractional difference between the solar radius at the equator and the solar radius at the poles (i.e., the "ellipticity" or "oblateness") is $\varepsilon = (R_{eq} - R_{pole})/R_{eq}$. Oblateness can arise from rotation. From the earliest telescopic observations of sunspots in the early 1600s, Galileo observed that the spots would move systematically across the disk of the Sun as time went on: he attributed these motions to solar rotation, and he concluded that the Sun rotated about once per month.

Oblateness can be caused by rotational speed. If rotation were absent, the Sun's figure would settle into an equi-potential surface, for which the potential would be spherically symmetric: $\varphi = -GM_\odot/r$. With such a potential, the surface acceleration due to gravity $g = -d\varphi/dr$ would also be spherically symmetric. This would cause $\varepsilon$ to have a zero value. However, in the presence of rotation, the (inward) force due to gravity is counteracted to some extent by the (outward) centrifugal acceleration $r\Omega^2$, especially near the equator. If the Sun were to rotate with a uniform angular velocity $\Omega$ throughout its volume, then the net gravitational acceleration $g(rot)$ at colatitude $\theta$ would be given by

$$g(rot) = g - r\Omega^2 \sin^2\theta \tag{1.18}$$

corresponding to a gravitational potential

$$\varphi = -\frac{GM_\odot}{r} - 0.5 r^2 \Omega^2 \sin^2\theta \tag{1.19}$$

This leads to an equi-potential surface which, in the presence of an equatorial rotational velocity $V(eq) = r\Omega$, has an oblateness of $\varepsilon = 0.5\, V(eq)^2/gr$: in this case, the ellipticity is equal to one-half of the ratio of centrifugal acceleration at the equator to the acceleration due to gravitation.

To proceed, we need to know: how fast does the Sun rotate? This question can be answered as regards the surface of the Sun by means of direct observations of the wavelengths of light associated with the gas at the surface. To estimate the rotational velocity of this gas, we use a spectrometer to measure the wavelength $\lambda$ of a spectral line at the east limb of the solar disk (where the gas first rotates into our view from the "far side" of the Sun). Due to rotation, the wavelength will undergo a shift $\Delta\lambda_D$ away from the "true" wavelength that the line would have if the gas were at rest. According to the Doppler formula, a shift $\Delta\lambda_D$ in the wavelength $\lambda$ of a particular line corresponds to a speed $V$ such that $\Delta\lambda_D/\lambda = V/c$ where $c = 2.99792458 \times 10^{10}$ cm s$^{-1}$ is the speed of light. (This value of $c$ is listed on the JPL website cited earlier in Section 1.1.)

Observations on the solar equator show that this gas is moving towards us with velocity $V(eq) \approx 2$ km s$^{-1}$. And at the west limb (where the gas is rotating away from our view and moving onto the "far side"), the gas is rotating away from us, so the lines are Doppler shifted to redder wavelengths, by amounts that also correspond to gas speeds of about 2 km s$^{-1}$. Since $V(eq)$ can be written as $r\Omega(eq)$, we can insert the value of the solar radius (Equation 1.12), and determine that, at the equator, the Sun's angular velocity $\Omega(eq)$ is roughly $\approx 2.874 \times 10^{-6}$ sec$^{-1}$. The corresponding period of rotation of gas at the solar equator is $P(eq) = 2\pi/\Omega(eq) \approx 2.186 \times 10^6$ sec $\approx 25.3$ days. Galileo was not far off in his estimate of the Sun's rotational period.

Now, in order to convert to an initial estimate of the solar oblateness, let us suppose for simplicity that the Sun rotates *as if it were a solid body* with exactly the same value of $\Omega = \Omega(eq)$ in

# The Global Parameters of the Sun

all regions of the Sun: after all, solid body rotation is what happens on the astronomical object with which we are most familiar (the Earth). If this supposition were true, then the ellipticity of the solar surface

$$\varepsilon = 0.5 \frac{V(eq)^2}{gr} = 0.5 \frac{r\Omega(eq)^2}{g} \qquad (1.20)$$

could be evaluated by inserting the values of $g$ and $r$ from Equations 1.13 and 1.12: this would lead to a numerical value of $\varepsilon_S = 10.47 \times 10^{-6}$ where the subscript S indicates that this applies to an object in solid body rotation. If the value of oblateness $\varepsilon_S$ were in fact applicable to the Sun, then the value of $R_{eq}$ would differ from the value of $R_{pole}$ by an amount $\varepsilon_S R_{eq} \approx 7.3$ km. Given that at the average distance of the Sun, features of linear size 725 km have an angular size of 1 arc second, the aforementioned difference in equatorial and polar radii would cause the angular radius of the equator to exceed the angular radius at the poles by about 0.01 arcsec. This is so small that observations from the ground (where turbulence in the Earth's atmosphere typically caused the edge of the solar disk to fluctuate with an amplitude of 1 arcsec) are unreliable. Measurements from above the atmosphere are preferable.

A balloon-borne instrument aimed at measuring the solar diameter, the Solar Disk Sextant (SDS), reached altitudes of order 30 km above the ground, where the residual air pressure has fallen off to 0.003–0.005 times the pressure at ground level. By 2011, the SDS had made 12 flights (Sofia et al. 2013). Residual atmospheric refraction caused the vertical diameter to be subject to corrections of 0.005–0.025 arcsec. Results from four flights between 1992 and 1996 reported oblateness values of $\varepsilon = (4–10.3)\pm 2 \times 10^{-6}$ (Egidi et al. 2006). The upper limit of these observations is not far from $\varepsilon_S$.

Measurements of $\varepsilon$ from space were made by SOHO/MDI: the spacecraft was rolled through 360 degrees in small angular increments, each 0.7 degrees in extent, corresponding to 360/0.7 = 514 individual "pie slices" of data around the entire circumference. Each "pie slice" was fitted with a radial profile: taking a numerical radial derivative of each profile and squaring the derivative, the location of the peak of squared derivative was defined to be the location of the limb. With more than 500 samples in each spacecraft roll, and by repeating the roll multiple times in the space of several months, Kuhn et al. (1998) claimed that they could achieve a precision of 0.5 km in the angular radius at pole and equator. Using observations obtained in 1996–1997, Kuhn et al. reported a solar oblateness of

$$\varepsilon = (7.77 \pm 0.66) \times 10^{-6} \qquad (1.21)$$

This value of oblateness overlaps with the range reported by SDS, but with smaller error bars. The range of ellipticities reported in Equation 1.21 are sufficiently small that they *cannot* be considered consistent with $\varepsilon_S$: there is a discrepancy at the $4\sigma$ level. This implies that something is wrong with our assumption of solid body rotation for the solar surface. The entire surface of the Sun *cannot* be rotating with a period that is as short as 25.3 days: some regions must be rotating at a rate that is *slower* than the equatorial value of 25.3 days in order to drive the value of $\varepsilon$ down to a level that is clearly smaller than $\varepsilon_S$. We now turn to evidence supporting this conclusion.

## 1.11 THE OBSERVED ROTATION OF THE SUN'S SURFACE

An important observational finding is that, unlike what happens on a solid body such as the Earth, the rotational period of the Sun is *not* the same at all latitudes. Instead, the period is found to be *shortest* at the equator, and the period becomes *longer* as we observe at higher and higher latitudes, i.e., at points that lie closer to the poles. Thus, the gas on the solar surface rotates faster at the equator than at any other latitude. This behavior is called "latitudinal differential rotation" (LDR).

A functional fit to the solar rotation often used is the following expression for the angular velocity $\Omega$ of the Sun's surface as a function of latitude $\lambda$:

$$\Omega(\lambda) = \Omega(0)\left[1 - b\sin^2\lambda - c\sin^4\lambda\right] \tag{1.22}$$

In a study involving Doppler shift data for spectral lines formed in the solar gas at many points on the surface, obtained over the course of 14 years, Howard et al. (1983) reported average values for the parameters in this fit: $\Omega(0) = 2.867 \times 10^{-6}$ rad sec$^{-1}$, $b = 0.121$, and $c = 0.166$. (The value of $\Omega(0)$ is a measured quantity: previously, in Section 1.10, we calculated a value of $\Omega(eq)$ which was only approximate, based on an approximate value of $V(eq)$.) Converting to units of degrees per day, the value of $\Omega(0)$ corresponds to a gas rotation rate at the equator of $A(eq) = 14.19$ deg day$^{-1}$. This corresponds to an equatorial rotational period of $P(rot, eq) = 360/A(eq) = 25.37$ days. The equatorial rotational velocity $V(eq) = \Omega(0)R_\odot$ has a numerical value of 1.99 km sec$^{-1}$. At latitudes of 60°, the rotational period $P(rot, 60) = 31.3$ days. At the N and S poles, Equation 1.22 indicates that $\Omega(90) = 0.713\,\Omega(0) = 2.044 \times 10^{-6}$ rad sec$^{-1}$, corresponding to a polar rotational period $P(rot, poles) = 2\pi/\Omega(90) = 35.6$ days. Remarkably, the gas in the polar regions of the Sun rotates almost 30% *more slowly* than the gas near the equator. If we needed any reminder that the Sun is *not* a solid body (but is composed entirely of gas), the LDR would provide the "smoking gun" evidence.

The existence of LDR on the solar surface helps us to understand why the observed oblateness of the solar disk (Equation 1.21) is *smaller* than the oblateness predicted by assuming solid body rotation with the same $\Omega(eq)$ at all latitudes. In fact, the observations tell us that the actual value of $\Omega$ is a maximum ($\Omega(eq)$) at the equator, and at all other locations in the surface, $\Omega$ is smaller than $\Omega(eq)$. The observed value of solar oblateness (Equation 1.21) could be made consistent with the formula in Equation 1.19 if an "effective" average of the angular velocity over the solar surface were to have the value $\Omega(eff) = 2.474 \times 10^{-6}$ rad sec$^{-1}$. This value actually is intermediate between the values reported by Howard et al. (1983) for $\Omega(0)$ and $\Omega(90)$, i.e., it is an angular velocity that does in fact exist at some intermediate latitude on the solar surface. It appears therefore that most (or all) of the observed oblateness of the surface of the Sun can be ascribed without serious contradiction to rotational effects.

Another approach to studying the LDR in the Sun, not involving any spectroscopic data, is to observe the spatial locations of "tracers" such as sunspots or other features at a particular instant, and then, as time passes, watch the tracers rotating from east limb to west limb. Wohl et al. (2010) have reported on a study of small bright coronal structures (SBCS) observed by the Extreme-ultraviolet Imaging Telescope (EIT) instrument on board the SOHO spacecraft during the years 1998–2006: their sample includes 55,000 SBCSs. They find an equatorial rotation rate A(eq) ranging from 14.37 and 14.55 deg day$^{-1}$. Using the same units, Wohl et al. also summarize the results of using other tracers including sunspots (A(eq) = 14.393 to 14.551), H$\alpha$ filaments (14.45–14.48), magnetic fields (14.0–14.5), coronal bright points (14.19–17.6), and radio regions with high and low brightness in microwaves (13.92–14.91). Wohl et al. also cite Doppler studies, for which the corresponding values are found to be 13.76–14.05 deg day$^{-1}$. The ranges of values are bewildering, but when we compare magnetic features to Doppler measurements in (mainly) nonmagnetic gas, one particular aspect may be emerging: the Doppler data (such as those reported by Howard et al. [1983]) yield rotational rates A(eq) that are in general *smaller* than the A(eq) values for magnetic features. If this is a robust result, then magnetic structures on the solar surface are rotating systematically *faster* than the nonmagnetic gas that dominates in the Doppler data.

Why might that be so? One explanation is that magnetic fields in the Sun are thought to be generated by a process known as "dynamo action" in the solar interior: gas motions of certain kinds may cause charged particles to create currents that give rise to magnetic fields. If this is true, then the fields that we can observe on the surface of the Sun must originate in deeper layers and then rise (by some process) to the surface. In so doing, the field lines may retain (at least some) connection with the gas where they originated, i.e., with their "roots" somewhere below the surface. In this scenario, the faster motion observed for magnetic features on the solar surface could be understood if the Sun were to be rotating faster at the deep levels where the field lines are rooted. Observational evidence in favor of faster rotation beneath the surface will be discussed in Section 14.9.3.

We might ask the question: how "strong" is the differential rotation on the solar surface? That is, how much does the angular velocity of the Sun vary between different locations on the surface? In principle, the coefficients $b$ and $c$ in Equation 1.22 might be able to provide quantitative measures of the extent to which the solar surface departs from solid body rotation. In fact, many observers have sought to determine how (or if) the values of the coefficients change during the sunspot cycle by looking for changes in the values of $b$ and $c$ at different times. Some of these changes have been interpreted as evidence for systematic changes in LDR during the solar cycle, although the various results reported in the literature are somewhat contradictory. However, there may be a fundamental problem in the analyses: the choice of functional form in Equation 1.22 is mathematically incorrect (Bertello et al. 2020) because the basis functions used in the series expansion (1, $\sin^2 \lambda$, and $\sin^4 \lambda$) are not in fact orthogonal to one another. As a result, there can be numerical "cross-talk" between the coefficients that can lead to spurious variations when data are analyzed, especially if the data extend over only relatively short intervals. Bertello et al. suggest a different choice of basis functions (Gegenbauer polynomials, denoted by $T_j^i$) which are truly orthogonal: they then fit the observed $\Omega(\lambda)$ using "rectified" coefficients $\bar{A}, \bar{B}$, and $\bar{C}$ as follows:

$$\Omega(\lambda) = \bar{A} + \bar{B} T_2^1(\sin \lambda) + \bar{C} T_4^1(\sin^2 \lambda) \qquad (1.23)$$

Bertello et al. have used Equation 1.23 to fit the observed rotation rates in 70 years of images of chromospheric features ("plages", or active regions) obtained at Mt. Wilson Observatory in the K line ($\lambda = 3933$ Å) of ionized calcium. With some 14,000 images to work with, they conclude that over the course of 70 years, the coefficients $\bar{A}, \bar{B}$, and $\bar{C}$ do *not* exhibit any significant alteration with time. Moreover, some earlier analyses of various tracers had hinted at an asymmetry between northern and southern hemispheric behavior; however, Bertello et al. report that they did not detect any significant asymmetry between N and S hemispheres "during most of the twentieth century".

For future reference, we note that when we convert angular velocities to (temporal) frequencies, $\nu = \Omega/2\pi$, the equatorial rotation $\Omega(0)$ corresponds to $\nu(0) = \Omega(0)/2\pi = 456$ nHz, while at the poles, $\Omega(90)$ corresponds to $\nu(90) = 325$ nHz.

The question naturally arises: why is the Sun's equatorial region rotating faster than at other latitudes? Theoretical analysis (Canuto et al. 1994) as well as three-dimensional computational models (Miesch et al. 2006) have been used to examine the question; unfortunately, these explanations involve complicated fluid interactions that lie beyond the scope of a first course in solar physics.

In order to keep track of individual solar rotations over time spans of many decades (and centuries), solar astronomers refer to the "Carrington rotation number". This is assigned on the basis of the rotation period of the solar equator: although the "true" rotational period at the equator of the Sun is 25.37 days (see discussion after Equation 1.22), the equatorial region of the Sun *as viewed from Earth* requires an interval of 27.2753 days for the same physical location on the solar equator to return to the same position on the solar disk *as viewed from Earth*. The period of 27.2753 days was first suggested by the English astronomer R. C. Carrington (famous for his discovery of the first "white-light flare" in 1859) as the defined length of 1 Carrington rotation (CR): by convention, CR 1 started on November 9, 1853. For future reference, we note that CR 2092 spanned the time interval January 3–29 in the year 2010: that was the rotation during which, for the first time, a full set of measurements of the *vector* magnetic field over the entire surface of the Sun became available on a daily basis.

So far, the observations we have been discussing (Doppler shifts, "tracer" locations) allow us to quantify differential rotation only at the *surface* of the Sun. As it turns out, when we discuss the helioseismological data in Chapter 14, we shall discuss evidence revealing that it is not merely the surface layers of the Sun that depart from solid body rotation: it turns out that the gas at different *radial* locations regions inside the Sun also rotate with different periods. As a result, differential rotation is not merely a function of *latitude*: the Sun also exhibits *radial differential rotation* (RDR).

Furthermore, whereas the term "rotation of the Sun" typically has to do with gas motions occurring in the direction of increasing (or decreasing) *longitude*, there is also observational evidence that a different type of systematic motion occurs in the Sun in the direction of increasing (or decreasing) *latitude*. Motions of the latter kind are referred to as "meridional circulation" because they give rise to motions along the meridians that run between equator and pole (Komm et al. 2015): these motions occur within 10 Mm of the solar surface and have speeds of order 10 m sec$^{-1}$, i.e., some 100 times slower than the rotational speed. As regards the origin of meridional flows, Kitchatinov (2013) suggests that two competing forces (both caused by the Sun's rotation), one driving the flow poleward, the second driving it equatorward, almost exactly cancel each other. The observed flow is driven by slight excesses in the poleward (centrifugal) force. Despite the relative slowness of the meridional flows, this circulation nevertheless plays a role to determining the 11-year time-scale of sunspots (see Section 16.9).

## 1.12 A CHARACTERISTIC FREQUENCY FOR SOLAR OSCILLATIONS DUE TO GRAVITY

Now that we know the radius and mass of the Sun, a characteristic frequency can be constructed: this will be relevant when we come to discuss (in Chapters 13 and 14) the various modes of oscillation in the Sun. For a pendulum of length $d$ in the presence of gravitational acceleration $g$, there is a natural period $P_g = 2\pi\sqrt{(d/g)}$. By analogy, in the Sun (with radius $R_\odot$ and surface gravity $g_s$), there exists a natural gravitational period $P_g = 2\pi\sqrt{(R_\odot/g_s)}$. This can be written as

$$P_g = 2\pi\sqrt{\frac{R_\odot^3}{GM_\odot}} \tag{1.24}$$

Noting that the ratio of the Sun's mass to its radius cubed is proportional to the mean density of the Sun, we see that the time period $P_g$ can be written as $1/\sqrt{(G\rho)}$. Inserting solar values, we find $P_g$ is close to $10^4$ seconds, with an associated gravitational frequency $\nu_g \approx 100$ microHz.

Now that we have information on the relevant physical parameters on a global scale, we can turn to a study of the internal structure of the Sun.

## EXERCISES

1.1 Consult a table of the orbital period $P$ and mean distance (semi-major axis $a$) for each of the planets Mercury, Jupiter, and Neptune. Express $P$ in seconds and $a$ in cm in preparation for using Equation 1.5. Ignoring the ratio of planet mass to Sun's mass, use each planet to determine an independent value for the mass of the Sun in units of gm. How much fractional difference do you find between the three estimates of "the" solar mass?

1.2 We shall see (Chapter 18) that the Sun's influence over the surrounding space extends out as far as $D \approx 100$ AU. Determine the period (in years) that a planet would have if it were in an orbit with that value of $D$. Assuming the orbit is circular, determine the speed of such planet in its orbit in units of km s$^{-1}$.

1.3 The Sun is currently estimated to be some 4.6 Gy old. When the Sun was younger than 1 Gy, theory suggests that its luminosity was only 70% of what it is today. Assuming that the Sun's radius has not changed, calculate the effective temperature of the young Sun.

1.4 Using information provided in the text, determine the mean density of the Earth and the acceleration due to gravity at the Earth's surface.

1.5 Given that the Earth rotates with a period of 1 day, calculate the Earth's angular velocity $\Omega(E)$. Use that to calculate the expected oblateness of the Earth (assuming that the material of Earth is free to respond to centrifugal forces). How does your answer compare with the observed oblateness of Earth (1/297)?

1.6 Stars belonging to a feature called "the main sequence" have radii $R_*$ and masses $M_*$ that scale roughly as $R_* = R_\odot(M_*/M_\odot)^{0.7}$. For stars with masses 0.1, 0.3, 1, 3, and 10 $M_\odot$ on the main sequence, calculate $R_*$ and evaluate the surface gravity (Equation 1.13) and the escape speed $V_{esc}$ from the surface (Equation 1.15).

1.7 It has been observed that the masses and luminosities of stars of a certain class (the so-called main sequence stars) can be approximated by $L_* \sim M_*^{3.8}$. Using the formula for luminosity in Equation 1.16, and the $R_* - M_*$ formula in Exercise 6, show that $T_{eff}$ for main sequence stars scales as $M_*^{0.6}$. Using this scaling along with Equation 1.17, calculate $T_{eff}$ for main sequence stars with masses of 0.1, 0.3, 1, 3, and 10 $M_\odot$.

1.8 Assume that the Earth and Jupiter move in circular orbits, with radius = semi-major axis. Given the orbital period $P$ for each planet, calculate the orbital speed $v(orb)$ for each planet in units of cm s$^{-1}$.

1.9 Astronomers are actively searching for planets around other stars (hoping that one day they may find life out there). How can planets be detected around other stars? One way is to use Newton's law of gravitation: this law states that each planet follows an elliptical orbit relative to the center of mass of the sun–planet pair. This has the effect that it is not merely the planet that moves along an orbit: the Sun also moves along a (smaller) orbit relative to the same center of mass. (This small motion is called the Sun's "reflex" motion.) If we assume that there are no other planets present, then Newton says that the semi-major axis $a(Sun)$ of the Sun's orbit will be smaller than that for the planet's orbit $a(planet)$ by the factor $M(Sun)/M(planet)$. Now, the *period* required for the Sun to traverse its small orbit (also assumed to be circular) is exactly the same as the period required for the planet to orbit once around the Sun. Using the value of $M(sun)/M(Earth)$ given in the text, and the value of $v(orb)$ for Earth obtained earlier in Exercise 8, show that the Sun's reflex motion *due to Earth* is about 10 cm s$^{-1}$. (To confirm the presence of an Earth-like planet around a Sun-like star, astronomers will need to discover a reflex motion as small as 10 cm s$^{-1}$. This is very challenging for the technology that is available in the year 2021.)

1.10 Repeat the calculation of Exercise 9 for Jupiter. (Find the mass of Jupiter online.) Show that the Sun's reflex motion *due to Jupiter* is about 100 times larger than the reflex motion due to Earth. This helps to explain why the first detection of planets around other Sun-like stars (beginning with the Nobel Prize–winning work of Mayor and Queloz [1995]) were due to Jupiter-sized planets, not Earth-sized planets.

## REFERENCES

Antia, H. M., 1998. "Estimate of solar radius from *f*-mode frequencies", *Astron. & Astrophys.* 330, 336.

Antia, H. M., et al., 2000. "Solar cycle variation in solar *f*-mode frequencies and results", *Solar Phys.* 192, 459.

Atkinson, R. d'E., 1982. "The eros parallax 1930–31", *J. Histor. Astron.* 13, 77.

Bertello, L., Pevtsov, A. A., & Ulrich, R. K., 2020, "70 years of chromospheric activity and dynamics", *Astrophys. J.* 897, 181.

Bush, R. I., Emilio, M., & Kuhn, J. R., 2010. "On the constancy of the solar radius. III", *Astrophys. J.* 716, 1381.

Canuto, V. M., Minotti, F. O., & Schilling, O., 1994. "Differential rotation and turbulent convection: a new Reynolds stress model and comparison with solar data", *Astrophys. J.* 425, 303.

Egidi, A., Caccin, B., Sofia, S., et al., 2006. "High precision measurements of the solar diameter and oblateness by the SDS experiment", *Solar Phys.* 235, 407.

Haberreiter, M., Schmutz, W., & Kosovichev, A. G., 2008. "Solving the discrepancy between the seismic and photospheric radius", *Astrophys. J. Lett.* 675, L53.

Hauchecorne, A., Meftah, M., Irbah, A., et al., 2014. "Solar radius determination from SODISM/PICARD and HMI/SDO observations of the decrease of the spectral solar radiance during the 2012 June Venus transit", *Astrophys. J.* 783, 127.

Howard, R., Adkins, J. M., Boyden, J. E., et al., 1983. "Solar rotation results at Mount Wilson. IV. Results", *Solar Phys.* 83, 321.

Karttunen, H., et al., 2017. *Fundamental Astronomy*, 6th ed. Springer-Verlag, Berlin, p. 124.

Kitchatinov, L. L., 2013. "Theory of differential rotation and meridional circulation", *Proceedings of the International Astronomical Union Symposium # 294*, eds. A. G. Kosovichev et al., Cambridge University Press, Cambridge, UK, p. 399.

Komm, R., Hernandez, I. G., Howe, R., et al., 2015. "Solar-cycle variations of sub-surface meridional flow", *Solar Phys.* 290, 3113.

Kosovichev, A., & Rozelot, J.-P., 2018. "Cyclic changes of the Sun's seismic radius", *Astrophys. J.* 861, 90.

Kuhn, J. R., Bush, R. I., Scherrer, P., et al., 1998. "The Sun's shape and brightness", *Nature* 392, 155.

Larson, T. P., & Schou, J., 2015. "Improved helioseismic analysis of medium-$l$ data from MDI", *Solar Phys.* 290, 3221.

Lefebvre, S., Nghiem, P., & Turck-Chieze, S., 2009. "Impact of a radius and composition variation on stratification of the solar subsurface layers", *Astrophys. J.* 690, 1272.

Mayor, M., & Queloz, D., 1995. "A Jupiter-mass companion to a solar type star", *Nature*, 378, 355.

Meftah, M., Corbard, T., Hauchecome, A., et al., 2018. "Solar radius determined from PICARD/SODISM observations", *Astron. & Astrophys.* 616, A64.

Meftah, M., Corbard, T., Irbah, A. et al., 2014. "Ground-based measurements of the solar diameter during the rising phase of solar cycle 24", *Astron. & Astrophys.* 569, A60.

Miesch, M. S., Brun, A. S., & Toomre, J., 2006. "Solar differential rotation influenced by latitudinal entropy variations in the tachocline", *Astrophys. J.* 641, 618.

Mullan, D. J., & MacDonald, J., 2001. "Are magnetically active low-mass dwarfs completely convective", *Astrophys. J.* 559, 353.

Mullan, D. J., MacDonald, J., & Townsend, R. H. D., 2007. "Magnetic cycles in the Sun: Modeling the changes in radius, luminosity, and $p$-mode frequencies", *Astrophys. J.* 670, 1420.

Pevtsov, A. A., Tlatova, K. A., Pevtsov, A. A., et al., 2019. "Reconstructing solar magnetic fields from historical observations", *Astron. Astrophys.* 628, A103.

Piau, L., Collet, R., Stein, R. F., et al., 2014. "Models of solar surface dynamics: impact on eigenfrequencies and radius", *Mon. Not. Roy. Astron. Soc.* 437, 164.

Prsa, A., Harmanec, P., Torres, G., et al., 2016. "Nominal values for selected solar and planetary quantities: IAU 2015 resolution B3", *Astron. J.* 152, 41.

Schou, J., Kosovichev, A. G., Goode, P. R., et al., 1997. "Determination of the Sun's seismic radius from SOHO/MDI", *Astrophys. J. Lett.* 489, L197.

Sears, F. W., 1959. *An Introduction to Thermodynamics*, Addison-Wesley, Reading, MA, p. 236.

Sofia, S., Girard, T., Sofia, U., et al., 2013. "Variation of the diameter of the Sun as measured by SDS", *Monthly Not. Roy. Astron. Soc.* 436, 2151.

Spencer Jones, H., 1941. "The solar parallax from observations of eros at the opposition of 1931", *Mem. Roy. Astron. Soc.* lxvi, part 2.

Victor, W. K., & Stevens, R., 1961. "Exploration of Venus by radar", *Sci.* 134, 46.

Wohl, H., Brajsa, R., Hanslmeier, A., et al., 2010. "A precise measurement of the solar differential rotation by tracing small bright coronal structures in SOHO-EIT images", *Astron. Astrophys.* 520, A29.

# 2 Radiation Flow through the Solar Atmosphere

Now that we have knowledge of the global parameters of the Sun, we are in a position to turn to an interpretive study of the photons that are the principal means by which information comes to us from the Sun. If we can make certain measurements on the photons from the Sun, such as their distribution in wavelength, and the integrated flux of radiant energy, the goal is to extract quantitative information about the temperature and other physical quantities in the region from which the photons originated. The photons originate mainly in a region that can be considered roughly as "the (visible) surface of the Sun": a more precise definition of this region will emerge subsequently from a discussion of radiative transfer.

Our goal is to use the information carried by the solar photons to undertake a task of physical interpretation that will take us in two opposite directions away from the "surface": (i) into the deep inner regions of the Sun and (ii) outward toward the rarefied material lying above the visible surface.

We aim to use certain laws of physics to help us determine a "model of the Sun", i.e., to determine the radial profile of physical parameters such as temperature, density, and pressure.

## 2.1 RADIATION FIELD IN THE SOLAR ATMOSPHERE

The goal of radiative transfer in the solar atmosphere is to determine how radiation interacts with the medium as it passes through material with a particular set of physical properties. The interaction is mutual: on the one hand, the medium imprints certain properties on the radiation, and on the other hand, the material in the medium is affected (as far as its temperature and density are concerned) by the photons streaming outward from deep inside the star.

An important way to characterize the radiant energy is the intensity $I_\lambda$: this is the amount of radiant energy that flows through unit area per unit time per unit wavelength and per unit solid angle. The units of $I_\lambda$ that we will use to describe the visible spectrum of the Sun are the c.g.s. units: ergs cm$^{-2}$ sec$^{-1}$ cm$^{-1}$ steradian$^{-1}$.

An alternative approach to quantifying the radiant power is to specify the intensity *per unit frequency*: this is given the symbol $I_\nu$ in units of ergs cm$^{-2}$ sec$^{-1}$ Hz$^{-1}$ steradian$^{-1}$. Conservation of energy requires that $I_\lambda d\lambda = I_\nu d\nu$. Since $\lambda\nu = c$ (where $c = 2.99792458 \times 10^{10}$ cm sec$^{-1}$ is the speed of light), this means that $I_\lambda = I_\nu(c/\lambda^2)$.

The numerical value of $I_\lambda$ (or $I_\nu$) at any point inside a medium (whether it is inside the Sun or in a star, or inside an oven on Earth) depends on the local temperature: other things being equal, the higher the temperature, the larger the value of $I_\lambda$ (or $I_\nu$). The value of $I_\lambda$ (or $I_\nu$) also depends on the wavelength $\lambda$ at which observations are made.

The simplest example of $I_\lambda$ (or $I_\nu$) useful for astrophysical studies refers to the radiant energy field that is in thermal equilibrium inside a closed cavity (or oven). This leads to the so-called blackbody radiation. In thermal equilibrium, the radiation is in equilibrium with the walls, and equal numbers of radiant modes are being absorbed and emitted by the walls per unit time. Inside a 1 cubic cm cavity, the only radiant modes that are present with significant amplitudes have wavelengths such that an integral number of half-wavelengths fit into the cavity. Thus, in one dimension of the cavity, with length 1 cm, the longest wavelength that can be accommodated is $\lambda=2$ cm. At shorter wavelengths, the number of modes $n_1$ with wavelength $\lambda$ cm that can be fitted into 1 cm scales are $1/(0.5\lambda)$. In three dimensions, the number of modes of wavelength $\lambda$ cm that can be fitted in scales are $[1/\lambda]^3 \sim \nu^3/c^3$. The shorter the wavelength, the more modes can be fitted in. If we restrict our attention to a certain range of wavelengths (or a certain range of frequencies), it is possible to use classical physics to enumerate the numbers of radiant modes permitted to exist per unit volume inside the oven. The number of such modes *per unit frequency* can be shown to be equal to $8\pi v^2/c^3$ cm$^{-3}$ Hz$^{-1}$.

The close coupling of radiation and walls inside a closed oven suggests that, since the thermal energy of a single particle with temperature $T$ is $kT$, where $k = 1.381 \times 10^{-16}$ ergs (deg K)$^{-1}$ is Boltzmann's constant, it *might* also be appropriate to assign an energy of $kT$ to each radiant mode: after all, each mode is a standing wave in thermal contact with the walls of the oven. If we adopt such an assignment, the radiant energy density (i.e., energy per unit volume) *per unit frequency* would be $E_v = 8\pi v^2 kT/c^3$ ergs cm$^{-3}$ Hz$^{-1}$.

Evaluation of the *energy density* of the radiation field is the first step toward deriving an expression for the radiant intensity: the latter is associated specifically with the *flow of energy* across an element of surface area and into unit solid angle in a particular direction. (Note: "steradian" is a unit of solid angle, such that an entire sphere contains $4\pi$ steradians by definition.) To transform from energy *density* to *radiation intensity*, the energy density must be multiplied by the speed of propagation ($c$) and also by the factor 1/4: the latter includes a factor of 1/2 to allow for inward and outward propagation and a factor of 1/2 for geometric averaging over spherical angles. This leads to

$$I_v = 2\pi v^2 kT / c^2 \quad \text{ergs cm}^{-2} \text{ sec}^{-1} \text{ Hz}^{-1} \text{ steradian}^{-1} \tag{2.1}$$

Equation 2.1 describes the radiant intensity according to the Rayleigh–Jeans law: it in fact provides a good fit to the radiant flux inside an oven at *long* wavelengths, i.e., at $\lambda \gg 1.4/T$ cm. For radio-astronomical objects containing material hotter than (say) 100 K, Equation 2.1 is found to apply reliably at wavelengths in the centimeter or meter range (or longer). However, if we try to apply Equation 2.1 to progressively shorter (ultraviolet) wavelengths, i.e., as $v \to \infty$, the prior expression for $I_v$ diverges. This divergence of the radiation energy is known historically as the "ultraviolet catastrophe".

In order to avoid this catastrophe, Max Planck in 1900 suggested that, despite the arguments of classical physics, it is *no longer* correct to assign the same energy (namely, the mean thermal energy of a particle $kT$) to each and every mode of the radiation field. Instead, Planck offered the following radical postulate: for modes of a given frequency $v$, *only certain discrete energies are allowed*, namely the arithmetic sequence $E(i) = 0, hv, 2hv, 3hv, \ldots$ where $h$ (now known as Planck's constant) is a constant of nature. That is, photon energies are discrete, i.e., *quantized*. In the quantum world, modes of frequency $v$ with energies intermediate between the numbers in the $E(i)$ list *simply do not exist*, any more than there are places to "stand" in between the rungs of a ladder. In the scenario envisioned by Planck, the total energy available from the thermal energy of the cavity is distributed among a large number of modes. The question is: how many modes are expected to be present at any given frequency $v$? Planck suggested the following answer: in a medium where the temperature is $T$, the available thermal energy is $kT$. Therefore, modes with energies that are increasingly larger than $kT$ are less and less likely to be populated. Quantitatively, Ludwig Boltzmann had earlier shown in a discussion of the thermodynamics of a multiparticle system that, in a thermal distribution, particles with energy $E(i)$ would be present in numbers which are proportional to their Boltzmann factor $BF(i) = \exp(-E(i)/kT)$: this takes into account the fact that very few particles in the cavity can possibly have energies which are greatly in excess of $kT$. (Where would such particles ever get their energy from anyway?) Planck applied this line of reasoning to the radiation modes in a cavity with temperature $T$: the number of modes in the cavity with energy $E(i)$ should be proportional to $BF(i)$. The total energy of such modes in the cavity will be the sum of terms involving the product $E(i)*BF(i)$. Because of the exponentiation in $BF(i)$, the arithmetic sequence of $E(i)$ values now becomes, in the sum over all modes, predominantly a geometric series in $BF(i)$ as $i$ increases. To be sure, in the sum over all modes, a multiplicative factor $E(i)$ also increases with $i$: however, this factor increases only slowly with $v$, whereas the Boltzmann factor falls off exponentially as $v$ increases.

# Radiation Flow through Solar Atmosphere

Thus, each successive term in the geometric series is smaller than its predecessor by a factor which is essentially $f' = \exp(-h\nu/kT)$. Since $f'$ is always less than unity, the sum of this infinite geometric series is finite: $a(1)/(1-f')$ where $a(1)$ is the first nonzero term in the series, i.e., $a(1) \sim \exp(-h\nu/kT)$. Adding up the occupation numbers to determine the overall partition function, it is possible to calculate the mean energy per mode. This mean energy is found to be no longer $\langle E \rangle = kT$ (as had been assumed in the classical case): instead, Planck found the following

$$\langle E \rangle = \frac{h\nu}{e^{h\nu/kT} - 1} \tag{2.2}$$

In the limiting case of low-energy photons, $h\nu \ll kT$, Equation 2.2 has the desirable property that it reduces to the classical result: $\langle E \rangle = kT$. However, in the opposite limit, for photons with energies that are greatly in excess of $kT$, the mean energy per photon falls well below the classical value: when we consider the limit $h\nu \gg kT$, Equation 2.2 indicates that the mean photon energy $\langle E \rangle$ falls exponentially rapidly toward zero.

Using the revised estimate of mean energy in the quantum-limited radiation modes, Planck found that the classical energy density per unit frequency $E_\nu = 8\pi(\nu^2/c^3)kT$ should be replaced by $E_\nu = 8\pi(\nu^2/c^3)h\nu/[\exp(h\nu/kT) - 1]$. Multiplying this energy density by the factor $c/4$ to convert to intensity (as in the classical treatment), the radiant intensity $I_\nu$ which is emitted *per unit frequency* from a surface at temperature $T$ is found to be:

$$I_\nu = \frac{2\pi h\nu^3}{c^2} \frac{1}{e^{h\nu/kT} - 1} \text{ ergs cm}^{-2} \text{ sec}^{-1} \text{ Hz}^{-1} \text{ steradian}^{-1} \tag{2.3}$$

If we wish instead to express the radiation flow $I_\lambda$ in terms of intensity *per unit wavelength*, the corresponding result is

$$I_\lambda = \frac{2\pi hc^2}{\lambda^5} \frac{1}{e^{hc/\lambda kT} - 1} \text{ ergs cm}^{-2} \text{ sec}^{-1} \text{ cm}^{-1} \text{ steradian}^{-1} \tag{2.4}$$

These are the expressions (the "Planck functions": see Figure 2.1) for the intensity of radiation inside an absorbing/emitting cavity. Deep in the solar atmosphere, where local thermodynamic equilibrium holds, we shall find that the mean free path for photons is so short (typically a few km) that the photons within a "small" volume can be considered to zeroth order to be essentially contained in a "cavity" where the temperature changes only slightly from one part of the cavity to another. In such conditions, a unique temperature is not a bad fit to local conditions, and the Planck function provides a reasonable approximation to the properties of the local radiation field.

How well does a Planck curve actually fit the radiation that we receive on Earth from the Sun? After all, the Planck curve was derived for the case of a closed oven where photons are absorbed by the walls. But by definition, the radiation we see on Earth coming from the Sun has not been absorbed by any walls: it comes streaming to us at great intensity across more than 100 million km of space. Therefore, the radiation has not been confined by the Sun, but has emerged so that we can see it. This might lead us to expect that the radiation we receive from the Sun might have *nothing* to do with the predictions of the Planck function. But this expectation is, interestingly, not correct. In Figure 2.2, we show how the measured radiation from the Sun behaves as a function of wavelength: the units of wavelength in the figure are nanometers (nm) where 1 nm = 10 Å. It turns out that the Planck function (referred to in Figure 2.2 as "blackbody" spectrum) provides a fit to the observations which is "not too bad".

Two important characteristic properties of the Planck functions (see Figure 2.1) are relevant in the context of the Sun. First, the curve $I_\lambda$ peaks at a certain wavelength $\lambda_{\max}$ defined by the condition

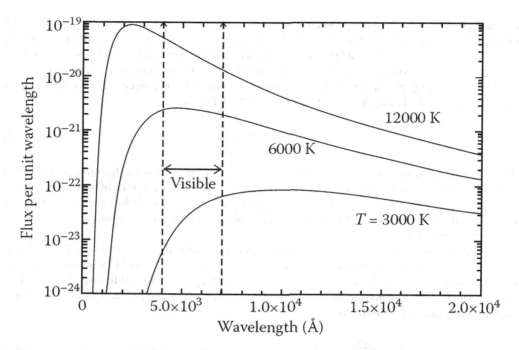

**FIGURE 2.1** Planck functions $I_\lambda$: radiant energy flux *per unit wavelength* calculated as a function of wavelength for cavities of different temperature. The wavelengths are expressed in units of angstrom (1 Å = $10^{-8}$ cm). The human eye is sensitive mainly to the "visible" region of the spectrum from about 4000 Å to about 7000 Å. The units of the ordinate are arbitrary, but the relative locations of the curves are quantitatively correct.

$dI_\lambda/d\lambda = 0$. Performing the differentiation of Equation 2.4, we find that $\lambda_{max}$ decreases as the temperature (in units of degrees K) increases according to the formula $\lambda_{max}(cm) = 0.288/T$. This formula was originally derived (using thermodynamics arguments) by Wilhelm Wien in 1893 long before the Planck function had been derived. In the case of the Sun, where the temperature in the vicinity of the "visible surface" is close to 6000 K, $\lambda_{max}$ occurs at a wavelength close to 5000 Å. Empirically, the solar spectrum, when plotted in the form of $I_\lambda$, is indeed found to exhibit a peak at wavelengths near 5000 Å (see Figure 2.2). Therefore, a Planck function provides a "not-too-bad" first-order fit to the radiation emerging from the Sun. Second, if the Planck function is integrated over all wavelengths and over all angles, the total energy density of the photons in a cavity with temperature $T$ is found to be $u(T) = a_R T^4$ ergs cm$^{-3}$. The radiation density constant $a_R$ has a numerical value of 7.5658 × $10^{-15}$ ergs cm$^{-3}$ deg$^{-4}$. Converting from total *energy* density to a total *flux* of radiation in a certain direction, the integral over all frequencies and over all solid angles leads to a total flux of $\sigma_B T^4$ ergs cm$^{-2}$ sec$^{-1}$. Here $\sigma_B = (c/4)a_R = 5.67040(\pm 0.00004) \times 10^{-5}$ ergs cm$^{-2}$ sec$^{-1}$ deg$^{-4}$ is referred to as the Stefan–Boltzmann constant.

For future reference, we note that the wavelength scale in Figure 2.2 is given in terms of nanometers, where 1 nm = 10 angstroms. Another unit also used for wavelength is microns, where 1 micron $\equiv$ 1μm = $10^{-4}$ cm = $10^4$ Å.

Finally, although this is somewhat removed from a study of solar physics, it is worth wondering: how did Planck in 1900 come up with the innovation mentioned earlier? In a book published shortly after his death, Planck (1948) admitted that he was guided by some of the reasoning that had been done earlier by Boltzmann. In the early developments of statistical thermodynamics, Boltzmann had considered a collection of $n$ particles and assigned a set of discrete energies (0, $\varepsilon$, $2\varepsilon$, ..., $p\varepsilon$) to the particles. The goal was to facilitate the probabilistic calculations of the permissible

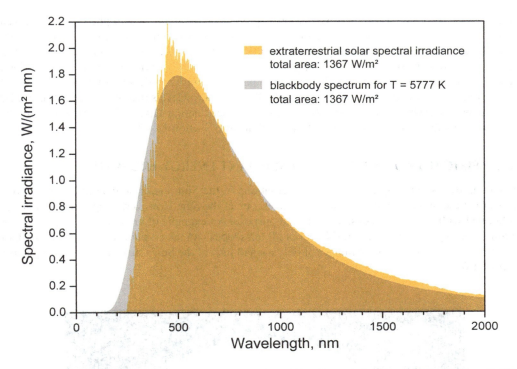

**FIGURE 2.2** Downloaded from the website: https://en.wikipedia.org/wiki/Planck%27s_law#/media/File:EffectiveTemperature_300dpi_e.png. Drawn by author Sch (identification unknown). The solar spectrum is the WRC spectrum provided by Iqbal (1983). The blackbody spectral irradiance has been computed from a blackbody spectrum for $T = 5777$ K and assuming a solid angle of 6.8e−5 steradian for the source (the solar disk). http://creativecommons.org/licenses/by-sa/3.0/ (uploaded May 5, 2006; last accessed February 17, 2022).

"complexions" (i.e., states) of an ensemble of classical particles (Cercignani 1998). Boltzmann wanted to obtain results that could then be extended to the continuous limit by considering a group of atoms where the number $n$ was so large that the ratio $p/n$ becomes infinitesimally small. In this limit, Boltzmann's approach had the effect that there would be in the final analysis no remaining discreteness of energy. Planck used Boltzmann's concept to start the process of counting photons, but Planck stopped short of Boltzmann's reasoning: Planck did *not* allow the discrete energies of the photons to merge together into a continuum. Instead, Planck's photons would retain discreteness, i.e., they would be quantized.

In this regard, the philosopher Karl Popper (1959) used strong language to reject what he called the "myth of the scientific method". Instead of "distilling" science out of "uninterpreted sense-experiences", Popper stated, "Bold ideas, unjustified anticipations, and speculative thought are our only means for interpreting nature". Planck's step beyond Boltzmann was a bold idea that helped to distinguish the quantum worldview from the classical worldview.

One final point should be made in connection with Planck's formula $E = h\nu$. When Albert Einstein developed his special relativity theory in 1905, he showed that a particle with rest mass $m$ and momentum $p$ has an energy given by $E^2 = (mc^2)^2 + (pc)^2$. Since photons have zero rest mass, we can write $E = pc$ for a photon. That is, we can associate a momentum with a photon as follows: $p = E/c = h\nu/c = h/\lambda$ where $\lambda = h/p$ is the electromagnetic wavelength associated with the photon. Two decades after Einstein's work, Louis de Broglie suggested that there exists in nature a wave-particle duality such that not only for photons could one write $\lambda = h/p$. De Broglie made the bold suggestion that a material particle with momentum $p$ also had a wave-like property associated with it: the wavelength of the associated wave $\lambda_D$ was predicted (by analogy with photons) to be given by the expression $\lambda_D = h/p$. The "de Broglie wave" associated with a material article is *not* an electromagnetic wave. Rather, it is interpreted to be a "probability wave" with amplitude $\psi$ at spatial location $x$ such that the probability that the particle is to be found at location $x$ is proportional to $|\psi|^2$. The existence of de Broglie waves will turn out to be important for understanding several physical issues in this book: why do electrons occupy only discrete energy levels in atoms (see Section 3.3.1)? What are the linear sizes of nuclei in different elements (see Section 11.4.1)? Why are protons in the core of the Sun able to participate in nuclear reactions (see Section 11.4.3)?

## 2.2 EMPIRICAL PROPERTIES OF THE RADIANT ENERGY FROM THE SUN

The human eye is not adapted for direct observations of the Sun: under no conditions should one *ever* stare at the Sun or point binoculars or a telescope at the Sun. However, images of the Sun can be obtained with instruments designed for that purpose. An example of an image of the *full disk* of the Sun is presented in Figure 2.3. (Note: not all telescopes can take a picture of the entire disk of the Sun. Some telescopes obtain images of only a small *part* of the Sun's surface, e.g., see Figure 6.1 in Chapter 6).

**FIGURE 2.3** Downloaded from the website: File:Sun white.jpg – Wikimedia Commons. Author: Geoff Elston. The image was taken through an 80 mm Vixen refractor mounted on a Super Polaris equatorial mounting. Sonnenfilter SF100 full aperture solar filter was used to reduce the light intensity. The camera was a Canon 550D DSLR at prime focus. Exposure was 1/400 sec at ISO100. (This file is licensed under the Creative Commons Attribution 4.0 International license. Used with permission of Geoff Elston. Uploaded October 27, 2013; last accessed February 18, 2022.)

# Radiation Flow through Solar Atmosphere

In the image in Figure 2.3, one's eyes are usually drawn at first to the localized dark spots ("sunspots"), but that is not the point in the present context. (We will deal with sunspots in detail in Chapter 16.) Here, we note that the solar disk is amenable to measurements of radiant intensity at all positions across the disk, from center to limb. Apart from sunspots, the intensity of the disk in visible light is azimuthally symmetric. However, inspection of Figure 2.3 shows that if we start at the center of the disk, and then move from disk center toward the limb, the intensity does *not* remain the same. One can see that the *limb* of the Sun is observed to be *fainter* than the center of the disk when observations are made in visible light: this phenomenon is known as "limb darkening".

In order to quantify this statement, we need to use a coordinate system best suited to the conditions that prevail because our observing platform (P) is situated on the Earth's surface. In this case, it is convenient (see Figure 2.4) to describe the location of a point S on the surface of the Sun in terms of the azimuthally symmetric angle $\psi$ between our line of sight and the local normal to the Sun's surface at point S.

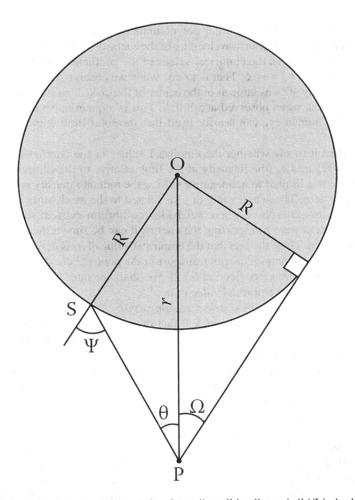

**FIGURE 2.4** Downloaded from the website https://en.wikipedia.org/wiki/Limb_darkening; author: unknown, source: unknown. Limb-darkening geometry: the star is centered at $O$ and has radius $R$. The observer is at point $P$ a distance $r$ from the center of the star, and is looking at point $S$ on the surface of the star. From the point of view of the observer, $S$ is at an angle $\theta$ from a line through the center of the star, and the edge or *limb* of the star is at angle $\Omega$. This file is licensed under the Creative Commons Attribution-Share Alike 3.0 Unported license. (Uploaded May 1, 2010; last accessed February 18, 2022.)

If it happens that the point S lies at the center of the disk (from the observer's vantage point), the line of sight from P to S enters the solar atmosphere along a line parallel to the local normal to the solar surface. As a result, the angle $\psi$ has a value of 0 at the center of the disk. On the other hand, if the point S lies close to the limb, the line of sight from P to S intersects the solar surface at S at a large angle to the local normal. As a result, as we attempt to make measurements of the intensity at positions lying progressively closer to the limb, our observing direction corresponds to the limit $\psi \to 90$ degrees.

In terms of the variable $\mu = \cos\psi$, observations at disk center correspond to $\mu = 1$, while observations close to the limb correspond to $\mu \to 0$.

Measurements of limb darkening have been made from many sites on Earth and at many different wavelengths. These measurements indicate that, for any particular wavelength, $I_\lambda$ varies across the solar disk in a way that can be described to first approximation as a *first-order polynomial in $\mu$*:

$$I_\lambda(\mu) \approx a_\lambda + b_\lambda \mu \qquad (2.5)$$

Some observers choose to fit higher order polynomials to the limb darkening, but the linear fit in Equation 2.5 often works well. For wavelengths in the vicinity of 5000 Å, if we normalize $I_\lambda$ to its value at disk center, $I_\lambda(\mu = 1)$, the empirical values of the coefficients in the linear fit are found to be roughly $a_\lambda = +0.4$ and $b_\lambda = +0.6$. That is to say, when we observe the Sun in visible light, the intensity at the limb is only 40% as large as at the center of the disk. That is, the limb is 60% fainter than the center of the disk when observed at 5000 Å. This is a quantitative restatement of the fact that in Figure 2.3, the human eye can see for itself that the solar limb is indeed fainter than the center of the disk.

*A priori*, it is difficult to say whether the empirical values of the coefficients in Equation 2.5 are "reasonable". As regards $a_\lambda$ (the intensity at the limb relative to the central intensity), we can confidently assert that it is limited to nonnegative values: the radiant intensity is a physical quantity that cannot fall below zero. The actual value of $a_\lambda$ is related to the mechanism of energy transport in the atmosphere: in an atmosphere where adiabatic equilibrium existed, such as would be the case if efficient convection were transporting the energy, it can be shown that the numerical value of $a_\lambda$ would approach zero. Thus, the fact that the empirical value of $a_\lambda$ is definitely nonzero already conveys useful physical information: energy transport in the photosphere does *not* occur primarily by means of efficient convection (see Section 6.7.2). We shall see later (Section 2.8) that radiative transport yields a good fit to the empirical value of $a_\lambda$.

As regards $b_\lambda$, there is no obvious *a priori* reason why $b_\lambda$ should necessarily be restricted to having a particular algebraic sign. The empirical fact that $b_\lambda$ is observed to be a positive number (for wavelengths around 5000 Å) indicates that the intensity at the limb, when observed at 5000 Å, is *less* than the intensity at disk center (hence the term: "limb darkening").

At near infrared wavelengths ($\lambda \approx 1$ μm), $b_\lambda$ is observed to be less than 0.6: thus, limb darkening is less severe in the infrared than at visible wavelengths. At wavelengths as long as 5 μm, limb darkening is no more than about 10%: the intensity at the limb is roughly 90% of the center intensity. At wavelengths in the near ultraviolet ($\lambda \approx 0.3$ μm), $b_\lambda$ is observed to be greater than 0.6, indicating that the limb darkening is more severe in the near ultraviolet than at visible wavelengths.

The fact that the Sun is limb darkened when observed at visible wavelengths does not exclude the possibility that at other wavelengths, the limb may be observed to be *brighter* than the disk center. In fact, at long radio wavelengths, limb brightening is observed. In such a case, $b_\lambda$ takes on a negative value: there is no mathematical difficulty with this as long as the sum of the two coefficients $a_\lambda + b_\lambda$ remains nonnegative.

The numerical values of the empirical coefficients $a_\lambda$ and $b_\lambda$, and the algebraic sign of $b_\lambda$, contain important information as to how the temperature $T(z)$ in the Sun's atmosphere varies as a function

# Radiation Flow through Solar Atmosphere

of the linear depth z. (Note that the depth z increases as we go *downward* into the interior of the Sun. We will also have occasion to use a linear height variable h that increases as we go *upward* in the solar atmosphere.) Obtaining this depth dependence of T(z) is the first step toward determining the radial profile of temperature inside the Sun. In order to extract T(z), it is first necessary to derive the radiative transfer equation (RTE), which describes how $I_\lambda$ varies as a function of a related coordinate known as the "optical depth" $\tau$.

## 2.3 THE RADIATIVE TRANSFER EQUATION (RTE)

In a radiant medium, such as the solar atmosphere, where the material has a finite temperature, each cubic centimeter of gas *emits* radiant energy at a certain rate $\varepsilon_\lambda$ ergs cm$^{-3}$ sec$^{-1}$ Å$^{-1}$ ster$^{-1}$. This quantity describes how much radiant energy is emitted into a certain region of the spectrum, across a width of spectrum equal to 1 Å per unit solid angle. The origin of this radiant emission can be traced ultimately to the pool of thermal energy residing in the particles of the gas (at temperature T): as particles in this thermal pool move past one another, the mutual accelerations of electric charges (electron, ions) or electric dipoles (atoms) give rise to electromagnetic emissions (i.e., photons) with energies related to the local thermal energy, of order $kT$ per particle. Because of this, the emissivity $\varepsilon_\lambda$ is a function of temperature.

The gas in the solar atmosphere also *absorbs* radiant energy at a rate that is described by a (*linear*) *absorption coefficient* $k_\lambda$ cm$^{-1}$. The subscript indicates that the absorption coefficient depends on the wavelength. The value of $k_\lambda$ may also depend on temperature. The wavelength-dependence of $k_\lambda$ is sometimes extremely rapid, e.g., in the vicinity of a strong spectral line, or near an "ionization edge". However, in other cases, $k_\lambda$ varies only slowly with wavelength: we refer to this as "continuum absorption". The optical spectrum that we receive from the Sun is dominated by continuum absorption, due mainly to a particular (and unusual) ion (see Section 3.4).

In the presence of absorption, when a beam of radiation with intensity $I_\lambda(0)$ enters a uniform slab of linear thickness x, the intensity that emerges is not as large as the intensity that entered. Instead, the emergent intensity is given by

$$I_\lambda(x) = I_\lambda(0)e^{-\tau} \qquad (2.6)$$

The quantity $\tau = k_\lambda x$ is a dimensionless number called the optical depth of the slab at wavelength $\lambda$. The quantity $1/k_\lambda$ is a linear distance such that a slab of thickness $x = 1/k_\lambda$ reduces the intensity of a beam by a factor of $1/e$: such a slab has an optical depth $\tau = 1$. If light passes through material with optical depth $\tau = 10$, the emergent intensity is attenuated below the initial value by a factor of order $2 \times 10^4$. This means that when we view a medium where radiation is coming from a variety of optical depths, it becomes progressively harder to detect a significant fraction of the radiation that originated in layers of gas with optical depths that are much larger than $\tau = 1$.

What numerical value is typical of the quantity $k_\lambda$? The answer depends on the medium. In the surface layers of the Sun, the value of $k_\lambda$ at $\lambda \approx 5000$ Å is found to be of order $10^{-6}$ cm$^{-1}$. This means that two points in those layers that are separated by a distance of 10 km (= $10^6$ cm) are separated by a medium with $\tau \approx 1$. In such a medium, radiation emitted by gas that is, say, 100 km deeper than at a reference point P′ arrives at P′ with an intensity of only 1/20,000 times its original value. If P′ lies on the visible surface of the Sun, this means that radiation emitted from 100 km below the Sun's surface is essentially all absorbed before we have a chance to see it.

If $k_\lambda$ is nonuniform in the slab, then the optical depth is given by $\tau = \int k_\lambda dx$. In the limit of a thin slab, with thickness $dx$, we see that

$$I_\lambda(x) = I_\lambda(0) - k_\lambda dx I_\lambda(0) \qquad (2.7)$$

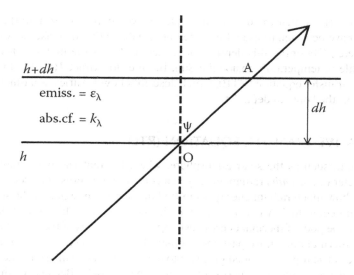

**FIGURE 2.5** Schematic of radiative transfer. The arrowed line represents the path of a ray of light propagating along a slanted path from below to above, passing through a slab of material with vertical thickness $dh$. Inside the slab, the material has absorption coefficient $k_\lambda$ and emissivity $\varepsilon_\lambda$. Dashed line indicates the local vertical direction. The ray enters the slab at point O, and exits the slab at point A. The ray path makes an angle $\psi$ relative to the local normal. The path length OA of the ray as it passes through the slab has a length $dh\sec\psi$.

Thus, the magnitude of the reduction in $I_\lambda$ associated with passing through a path length $dx$ is proportional to the path length and also to the intensity of the radiation.

Combining the concepts of emission and absorption, we can now derive the RTE using Figure 2.5.

Consider the slanted line OA along which radiation is propagating in a stellar atmosphere. The vertical dashed line denotes the local normal to the surface of the Sun. Let OA extend across a slab of gas located between heights $h$ and $h + dh$: in our notation, we assume that the numerical value of $h$ increases in the *upward* direction. Let the slanted line OA lie along a direction that makes an angle $\psi$ relative to the local normal. The intensity at height $h$ in the direction of the line is $I_\lambda(h, \psi)$: this is a measure of the energy flux that enters an element of unit area per unit time at height $h$. The element of area is perpendicular to the line OA. After traversing the atmosphere and arriving at the higher level, the intensity emerging from the element of unit area per unit time $I_\lambda(h + dh, \psi)$ differs from the value $I_\lambda(h, \psi)$ because of two processes: (i) *reduction* in intensity due to absorption in the gas that lies along the path OA, and (ii) *enhancement* in intensity due to emission from the gas along OA. Let us imagine that the elements of unit area at $h$ and at $h + dh$ are connected by a rectangular prism of unit area: the length $dl$ of such a prism is equal to the slant length $dl = dh\sec\psi$ along the axis of the prism. The *reduction* in energy along this path is equal to $-I_\lambda(h, \psi)k_\lambda dh\sec\psi$. The volume of the prism is $dh\sec\psi$. In such a volume, the increase in energy flux due to local emission is $\varepsilon_\lambda(h)dh\sec\psi$.

Combining the two terms (enhancement and reduction), we find that

$$I_\lambda(h+dh,\psi) - I_\lambda(h,\psi) = \varepsilon_\lambda(h)dh\sec\psi - I_\lambda(h,\psi)k_\lambda dh\sec\psi \qquad (2.8)$$

In the limit $dh \to 0$, and defining $\mu = \cos\psi$, this leads to

$$\mu\frac{dI_\lambda}{dh} = \varepsilon_\lambda - k_\lambda I_\lambda \qquad (2.9)$$

Note that the value of μ ranges from μ = +1 (i.e., ψ=0, in the direction of the *outward* normal) to μ= −1 (i.e., ψ= 180 deg, in the direction of the *inward* normal).

## 2.4 OPTICAL DEPTH AND THE CONCEPT OF "THE PHOTOSPHERE"

At this point, we introduce the concept of the *optical depth that characterizes a particular layer of the atmosphere* $\tau_\lambda$. The increment of this optical depth $d\tau_\lambda$ is defined by $d\tau_\lambda = -k_\lambda dh$, where the negative sign indicates that the numerical value of $\tau_\lambda$ *increases* as the height coordinate $h$ becomes more negative, i.e., as we move *deeper* into the star. At any height $h'$ in the atmosphere, the local optical depth is computed by integrating from a height of infinity down to the height $h'$:

$$\tau_\lambda(h') = \int_{h'}^{\infty} k_\lambda dh \tag{2.10}$$

The zero point of the optical depth scale lies at $h' \to \infty$ far above the visible surface of the Sun. For an observer P at a remote point (such as on Earth) (Figure 2.4), the optical depth along the line of sight between P and a point S' (which is closer to the Sun than P) is determined by the location of S'. If S' lies close to P, the point S' lies in gas of extremely low density. As a result, the optical depth of point S' (as viewed by observer P) remains very small. However, as point S' is moved progressively closer to the Sun, there is a monotonic increase in the optical depth of S' as viewed by P. Eventually, point S' becomes immersed in solar atmospheric gas where the density has a value that is large enough to make $k_\lambda$ appreciable: now the increments $d\tau_\lambda = k_\lambda dh$ start to build up appreciably in the integral of $\tau_\lambda(h')$. Eventually, the gas surrounding S' is dense enough, and deep enough in the solar atmosphere, that the integral $\tau_\lambda(h')$ approaches a value of order unity. We call the location where $\tau_\lambda(h') = 1$ the "photosphere" (from the Greek: photo = having to do with light). The numerical value of $h'$ at the photosphere depends on the wavelength: at certain wavelengths (e.g., $\lambda \approx 1.6$ μm), the absorption coefficient is smaller than at other wavelengths, and we can see deeper into the atmosphere.

Why is the photosphere significant as far as our study of solar radiation is concerned? Because as we go deeper into the Sun, below the photosphere, the optical depth of the deeper layers rapidly becomes so large that any radiation emitted from the deep layers is significantly reduced before it can reach our observational instruments. The photosphere can be regarded as more or less the deepest lying layer of gas from which we still have a good chance of seeing most of the radiation emitted by that layer.

Converting Equation 2.9 from an equation in which $h$ is the independent variable to an equation in which $\tau$ is the independent variable, we arrive at the following equation:

$$\mu \frac{dI_\lambda}{d\tau_\lambda} = I_\lambda - S_\lambda \tag{2.11}$$

where $S_\lambda = \varepsilon_\lambda / k_\lambda$ is referred to as the source function at wavelength $\lambda$. The units of $S_\lambda$ are the same as those of $I_\lambda$, namely, ergs cm$^{-2}$ sec$^{-1}$ Å$^{-1}$ ster$^{-1}$.

Equation 2.11 is referred to as the radiative transfer equation.

## 2.5 SPECIAL SOLUTIONS OF THE RTE

In the following illustrative solutions, we shall for simplicity omit the wavelength subscript on all variables, but it is implied.

Equation 2.11 is an ordinary differential equation that can be solved by multiplying both sides by the integrating factor $e-\tau/\mu\}$. Thus, we can rewrite the RTE as

$$\frac{d}{d\tau}(I(\tau)e^{-\tau/\mu}) = -\frac{S(\tau)e^{-\tau/\mu}}{\mu} \tag{2.12}$$

The solution of this equation yields the intensity an observer would "detect" if an instrument were located in a layer with optical depth $\tau$, and if the instrument were pointed in such a way as to be observing only the radiation propagating along a direction that makes an angle $\psi = \cos^{-1} \mu$ relative to the local normal.

The formal solution $I(\tau, \mu)$ of Equation 2.12 can be considered in the limit of two distinct regimes of the $\mu$ parameter, one for the radiant intensity that flows *into the upper hemisphere* (relative to the point where the optical depth is $\tau$), and the other for the radiation that flows *into the lower hemisphere* (also relative to the point where the optical depth is $\tau$).

First, for radiation flowing into the *upper* hemisphere, $\psi$ takes on values that range from 0 to 90 degrees (i.e., $\mu$ takes on positive values between 1 and 0). In this case, the local intensity $I(\tau,\mu)$ is due to radiation emerging from deeper layers (inside the Sun) and flowing *outward* toward free space. As a result, the local value of intensity at depth $\tau$ in the upper hemisphere (denoted by $\mu+$) involves an integration over all gas that lies *below* the level $\tau$, i.e., from $\tau \to \infty$ up to the level where the optical depth equals $\tau$:

$$I(\tau, \mu+) = -e^{\tau/\mu} \int_{\infty}^{\tau} \frac{S(t)}{\mu} e^{-t/\mu} dt \qquad (2.13)$$

Second, for radiation flowing into the *lower* hemisphere, $\psi$ takes on values from 90 to 180 degrees, (i.e., $\mu$ takes on negative values between $-1$ and 0). In this case, the local intensity $I(\tau, \mu)$ is due to radiation emerging from shallower layers (higher in the atmosphere) and flowing *inward* toward the interior of the Sun. As a result, the local value of intensity at depth $\tau$ in the lower hemisphere (denoted by $\mu-$) involves an integration over all gas that lies *above* the level $\tau$, from $\tau \to 0$ down to the level where the optical depth equals $\tau$:

$$I(\tau, \mu-) = -e^{\tau/\mu} \int_{0}^{\tau} \frac{S(t)}{\mu} e^{-t/\mu} dt \qquad (2.14)$$

Let us limit our considerations now to the outermost layers of the Sun. That is, let us move our radiation instrument to the upper atmosphere where $\tau \to 0$. This could include moving the instrument all the way to the Earth's orbit. In this way, we would be recording what is truly the "emergent intensity" from the Sun. In this case, since there is essentially zero source of radiation coming in from free space, the integral into the lower hemisphere $I(\tau, \mu-)$ vanishes. Only the intensity $I(\tau, \mu+)$ entering into the upper hemisphere retains a nonzero value. And this component, in the limit $\tau \to 0$, becomes

$$I(0, \mu+) = \int_{0}^{\infty} \frac{S(t)}{\mu} e^{-t/\mu} dt \qquad (2.15)$$

Let us consider some simple cases.

## 2.5.1 $S$ = Constant at All Optical Depths

If $S(\tau) = S$, independent of $\tau$, the integral in Equation 2.15 is straightforward: we find $I(0, \mu+) = S$. Thus, the emergent intensity of radiation is just equal to $S$ itself. Moreover, $I(0) = I(0, \mu+)$ is independent of $\mu$: in this case, there is neither limb darkening nor limb brightening.

## 2.5.2 $S$ = Constant in a Slab of Finite Thickness

In the case of a slab with finite optical depth $\tau'$, in which $S$ is constant, the emergent intensity is

# Radiation Flow through Solar Atmosphere

$$I(0,\mu+) = S\int_0^{\tau'} e^{-t/\mu} dt/\mu = S(1-e^{-\tau'/\mu}) \tag{2.16}$$

Thus, the emergent intensity is not as large as $S$, but is reduced by an optical depth term. In the special case where we observe perpendicular to the slab, we can set $\mu = 1$, and then find that

$$I(0,1+) = S(1-e^{-\tau'}) \tag{2.17}$$

In the limit of infinite thickness, $\tau' \to \infty$, we recover the solution in Section 2.5.1: $I(0,1+) \to S$. In the opposite limit, when the slab is optically thin, we find

$$I(0,1+) \to \tau'S \tag{2.18}$$

Thus, the emergent intensity from a very thin slab can take on values that are much smaller than the source function. The reduction factor is just the (small) optical depth of the slab.

In general, in the case of constant $S$, the emergent intensity cannot be greater than $S$, but it may be much smaller than $S$ if the optical depth is small. This is an important result in helping to interpret certain properties of the upper solar atmosphere.

## 2.5.3 Depth-Dependent $S$: Polynomial Form

We now revert to the case of an infinite atmosphere and consider a case where the source function depends on the optical depth. Specifically, we consider the polynomial form $S(\tau) = a + b\tau + c\tau^2$. (We shall see in Section 2.8 that there is some basis for such a choice in the solar atmosphere.) To obtain the emergent intensity from such an atmosphere, we insert this function into Equation 2.15.

The first term in $S(\tau)$ corresponds to the case in Section 2.5.1 (i.e., constant $S$): this term results in a contribution of $a$ to $I(0, \mu+)$. The term $b\tau$, when inserted in Equation 2.15, leads to an integral that can be integrated by parts: it contributes a term $b\mu$ to $I(0, \mu+)$. Finally, the term $c\tau^2$, when inserted in Equation 2.15, requires two integrations by parts: this leads to a term $2c\mu^2$ to $I(0,\mu+)$. Combining terms, we find that

$$I(0,\mu+) = a + b\mu + 2c\mu^2 \tag{2.19}$$

Clearly, this solution is of particular interest for the Sun's atmosphere since the empirical limb darkening of the Sun (Equation 2.5) is of precisely this form (in the special case $c = 0$, although empirical fits have been extended in some publications to include a term in $\mu^2$). It therefore appears that the source function at optical depth $\tau$ in the Sun (at visible wavelengths) can be described by the function $S(\tau) = a + b\tau$. In view of the fact that the empirical value of the coefficient $b$ is positive, this allows us to draw an important conclusion: the source function $S(\tau)$ in the solar atmosphere at visible wavelengths *increases with increasing* $\tau$. That is, if we were able to place a radiation detector at any depth we liked in the Sun's atmosphere, we would find that the source function increases as we move the detector inward to greater depths.

We shall see later that in certain situations, the source function increases as the local temperature increases. In view of this, the empirical observation that the Sun's disk undergoes limb *darkening* (i.e., $b > 0$) provides us with a nontrivial (in fact, significant) piece of information: in the visible layers of the solar atmosphere, the temperature *increases* as we penetrate *deeper* into the atmosphere. This property of the temperature sets the stage for the following conclusion: although temperatures of order $T_{\text{eff}}$ ($\approx 6000$ K) occur near the photosphere, larger temperatures are expected to exist in regions that lie deeper in the Sun.

If we switch to considering how the temperature varies as a function of *radial distance from the center of the Sun*, the temperature in the photosphere is *decreasing* as r increases. In other words, the radial gradient of temperature, $dT/dr$, is a negative quantity in the photosphere. When we come to consider the structure of the deep interior of the Sun (in Section 8.1), we will take advantage of the negative temperature gradient to express how much heat flux is passing out from the center of the Sun towards the surface.

### 2.5.4 Depth-Dependent S: Exponential Form

Suppose the source function has the form $S(\tau) = e^{\alpha\tau}$ where $\alpha < 1/\mu$. Inserting this into Equation 2.15, we find that

$$I(0, \mu+) = \frac{1}{1 - \alpha\mu} \qquad (2.20)$$

## 2.6 THE "EDDINGTON–BARBIER" (OR "MILNE–BARBIER–UNSÖLD") RELATIONSHIP

The fact that a linear source function $S(\tau) = a + b\tau$ yields a limb-darkening function $I(0, \mu+) = a + b\mu$, which is exactly the same linear function of $\mu$, leads to a result that is historically referred to as the "Eddington–Barbier" relationship (EBR): namely, the intensity that is observed at any value of $\mu$ equals the source function at the level where the local optical depth $\tau$ has the value $\tau = \mu$. In other words, at any particular location on the disk of the Sun, i.e., at a given value of $\mu$, the radiation that is observed comes effectively from gas situated at a height where the local optical depth is equal to $\mu$.

This means, in effect, that when one observes the Sun at disk center ($\mu = 1$), one's line of sight penetrates down essentially to the gas in the solar atmosphere where $\tau \approx 1$. On the other hand, when observing near the limb, say at $\mu = 0.1$, one's line of sight penetrates into the atmosphere only as far as the layer of gas where $\tau = 0.1$. In terms of a model of the solar atmosphere, we shall find (see Table 5.3 later) that the gas that lies in layers where $\tau = 0.1$ is situated about 130 kilometers higher up than the gas at $\tau = 1$. The deeper gas at $\tau = 1$ is hotter by about 1000 K than the gas at $\tau = 0.1$, and that fact contributes to making the Sun brighter at the center of the disk than near the limb.

Moreover, we have already noted (Section 1.5) that "limb observers", who try to measure the angular diameter of the Sun, do their work by using a line of sight that only "skims" the top of the solar atmosphere, penetrating into a rarefied layer of gas lying some 347 km above the level where $\tau = 1$. Again referring to Table 5.3, we see that this height corresponds to a level where $\tau \approx 0.002$. According to the EBR, the "limb observers" are in effect using a line of sight corresponding to $\mu = 0.002$, i.e., the angle between the line of sight and the local normal is not quite 90 degrees, but about 0.115 degrees smaller.

On a historical note, Paletou (2018) has recently pointed out, in an extensive study of the relevant literature, that the "EBR" should be more accurately referred to as the Milne–Barbier–Unsöld relationship. Paletou, citing what he calls "A lost contribution of Milne?", shows that in 1917, Milne derived the result that we have referred to in this subsection as the EBR, although Milne's paper did not appear in print until 1921. Eddington's first reference to what we have called the EBR appears in 1926 in his book on stellar structure (Eddington 1926). And Barbier, in a 1943 paper, explicitly cites the result that appears in Eddington's book. Later in 1948, Unsöld cited Barbier's work when he referred to what he called the "$\tau$-$\mu$ method" as it applies in both radiant intensity and flux. Moreover, Milne (1921) also derived a result that is historically referred to as the "Eddington approximation" (see first line after equation 2.31). Unfortunately, the weight of historical precedent has had the effect that Eddington's name is essentially universally used in textbooks on radiative transfer. It is

# Radiation Flow through Solar Atmosphere

to be hoped that the detective work of Paletou (2018) will in the future lead to authors giving more credit to Milne in this context. In the meantime, however, to add to the confusion, we shall have occasion to refer to a different result bearing the title of "Milne–Eddington" (ME) when we consider certain models of the solar atmosphere (see Section 3.8.3). The difference between the discussion of the EBR (or MBUR) (in this section) and the discussion of the ME models (in Chapter 3) is the following: here, we are interested solely in *continuum* opacity, whereas in the ME models, we will be interested in absorption lines each with its own particular (narrow-band) *line* opacity as seen against the background of a (broad-band) continuum opacity.

## 2.7 IS LIMB BRIGHTENING POSSIBLE?

Although limb darkening is certainly the feature which is most relevant to observations of the Sun in the visible part of the solar spectrum (see Figure 2.3), this does *not* exclude the possibility there might be limb *brightening* when the Sun is observed at certain other wavelengths.

The existence of limb brightening requires, by definition, that $I(0, \mu = 0)$ exceed $I(0, \mu = 1)$.

In the case of a polynomial source function, this possibility is formally excluded if all coefficients ($a, b, c$) are nonnegative. Limb brightening is possible only in cases where either $b$ or $c$ (or both) are sufficiently *negative* to ensure that $b + 2c$ is negative.

In the case of an exponential source function, the ratio $I(0, \mu = 0)/I(0, \mu = 1)$ is equal to $1 - \alpha$. To avoid nonphysical (negative) intensities, this requires that $\alpha$ have a value that is no greater than 1. This is stricter than the limit (already noted earlier) $\alpha < 1/\mu$. Thus, limb brightening is possible if $\alpha < 0$, i.e., if the source function *decreases* exponentially as the optical depth *increases*. Are we ever likely to encounter such a behavior in the Sun? Perhaps surprisingly, the answer to this question is a definite "Yes". As we shall see in Chapter 15 below, we shall see that there does indeed exist a region (called the "chromosphere") in the Sun's atmosphere where, as one observes down deeper into larger and larger values of $\tau$, the temperature is found to be *decreasing*. In this region, $dT/dr$ has a positive value, and the physical conditions are quite different from those in the deep interior of the Sun. If we could arrange to observe the Sun with a detector which is sensitive to radiation with $\tau \approx 1$ *in the chromosphere*, then we might expect to see limb brightening. An example of such a detector is the radio detector known as the Atacama Large Millimeter/Submillimeter Array (ALMA) located high up (16,000 feet altitude) in a desert in Chile: by observing at wavelengths between 3.6 mm and 0.32 mm, ALMA probes the chromosphere of the Sun in regions where the temperature is increasing with increasing height. Using observations at wavelengths of 3 mm and 1.2 mm, Sudar et al. (2019) have reported that indeed the limb is brighter than disk center by 10–15% at those wavelengths.

## 2.8 IS $S(T) = A + BT$ REALISTIC? THE GRAY ATMOSPHERE

We have seen that the observed limb darkening in the Sun, which can be described by $I(\mu) = a + b\mu$, agrees with the limb darkening that should be observed if the source function has a particular form: $S(\tau) = a + b\tau$. Now we ask, is there any physical reason why $S(\tau) = a + b\tau$ might be an acceptable description of conditions in the Sun's atmosphere?

The answer is "Yes", provided we consider a limiting case known as the *gray atmosphere*. In this case, the opacity is independent of wavelength, allowing immediate integration of Equation 2.11 (RTE) over wavelength.

$$\mu \frac{dI(\tau)}{d\tau} = I(\tau) - S(\tau) \qquad (2.21)$$

The unsubscripted $\tau$-dependent variables $I(\tau)$ and $S(\tau)$ in this section refer to quantities which, at any given optical depth, have been integrated over all wavelengths. Starting with Equation 2.21, and with the goal of deriving $S(\tau) = a + b\tau$, we now consider three steps to derive three distinct quantities which are functions of $\tau$: $F$, $J$, and $K$.

First, at optical depth $\tau$ in the atmosphere, the flux of radiation $F(\tau)$ (in units of ergs cm$^{-2}$ sec$^{-1}$) flowing along the outward normal can be obtained by considering the component of $I(\tau)$ along the normal: $I(\tau)\cos\psi = \mu I(\tau)$. We obtain $F(\tau)$ by integrating $\mu I(\tau)$ over all solid angles $d\omega$:

$$F(\tau) = \int \mu I(\tau) d\omega \qquad (2.22)$$

In conditions of radiative equilibrium, there are no new sources of energy within the atmosphere: the energy flux $F(\tau)$ is determined by processes which occur deep inside the star. As far as the atmosphere is concerned, the value of $F(\tau)$ is effectively a boundary condition: a certain quantity of energy flux "arrives" from the deep interior at the base of the atmosphere, and must be transported (somehow) through the atmosphere and released into the darkness of space. As a result, $F(\tau) = F_o$ is a constant at all optical depths in the atmosphere.

Second, also at optical depth $\tau$, we define the mean intensity of radiation $J(\tau)$ as

$$J(\tau) = \frac{1}{4\pi} \int I(\tau) d\omega \qquad (2.23)$$

Using these definitions of $F(\tau)$ and $J(\tau)$, we integrate both sides of Equation 2.11 over $d\omega$ and find that at any given location in the atmosphere, in a layer where the optical depth is $\tau$,

$$\frac{dF(\tau)}{d\tau} = 4\pi J(\tau) - 4\pi S(\tau) \qquad (2.24)$$

In performing the integrations over $d\omega$, we have assumed that the source function $S(\tau)$, which is determined by atomic processes in the immediate neighborhood of $\tau$, is spherically symmetric. In such a case, an integral over solid angle simply recovers the factor $4\pi$ as the number of steradians in a sphere. Inserting $F(\tau) = F_o$ in Equation 2.24, we find that

$$J(\tau) = S(\tau) \qquad (2.25)$$

Recall that in these expression, both $J(\tau)$ and $S(\tau)$ have been integrated over all wavelengths.

We now proceed to the third step in the derivation. Notice that, by definition, the quantities $J(\tau)$ and $F(\tau)$ represent zeroth and first moments of the radiation intensity at optical depth $\tau$. Now we introduce the second moment of the radiation intensity at depth $\tau$ (again integrated over all frequencies):

$$K(\tau) = \frac{1}{4\pi} \int \mu^2 I(\tau) d\omega \qquad (2.26)$$

With this definition, the quantity $K(\tau)$ is proportional to the radiation pressure $p_r$ at optical depth $\tau$. When the Planck functions are integrated over all wavelengths, the radiation pressure $p_r$ is related to the energy density $u(T)$ (see Section 2.1) by $p_r = u(T)/3 = a_R T^4/3$.

With these definitions of the three functions $F(\tau)$, $J(\tau)$, and $K(\tau)$, let us now return to Equation 2.11, multiply both sides by $(\mu/4\pi)$, and then integrate over $d\omega$. This operation leads to $dK(\tau)/d\tau$ on the left-hand side of RTE. On the right-hand side, the first term reduces to the constant $F_o/4\pi$. The second term, involving integration of $\mu S$ over all solid angles, reduces to zero due to the spherical symmetry of $S$. Thus we find

$$\frac{dK(\tau)}{d\tau} = \frac{F_o}{4\pi} \qquad (2.27)$$

# Radiation Flow through Solar Atmosphere

Since $F_o$ is a constant, this equation can be integrated to obtain:

$$K(\tau) = \frac{F_o}{4\pi}\tau + const. \tag{2.28}$$

We need to find a way to evaluate the constant of integration in Equation 2.28. To do this, we introduce the "two-stream approximation": the angular distribution of the radiant intensity is replaced by two streams, one with $\mu = +1$ going into the outer (upper) hemisphere $I_o(\tau)$, the other with $\mu = -1$ going into the inner (lower) hemisphere $I_i(\tau)$. In this approximation, and noting that the element of solid angle $d\omega$ can be written as $2\pi d\mu$, we find the following expression for the three moments of the radiation field:

$$J(\tau) = \frac{1}{2}(I_o(\tau) + I_i(\tau)) \tag{2.29}$$

$$F(\tau) \equiv F_o = \pi(I_o(\tau) - I_i(\tau)) \tag{2.30}$$

$$K(\tau) = \frac{1}{6}(I_o(\tau) + I_i(\tau)) \tag{2.31}$$

Comparing $J(\tau)$ and $K(\tau)$, we see that in this approximation, referred to as the Eddington approximation, $K(\tau) = J(\tau)/3$ at all optical depths in the atmosphere.

In particular, let us consider a special location in the atmosphere, namely, the top, where $\tau = 0$. At that location, the incoming flux of radiant energy $I_i(\tau = 0)$ is certainly zero (there is no radiation coming in from the darkness of space). As a result, $K(0) = I_o(0)/6$ and $F_o = \pi I_o(0)$. Reverting to Equation 2.28, we now have enough information to evaluate the constant of integration: it equals $F_o/6\pi$. Replacing $K(\tau)$ by $J(\tau)/3$, and multiplying both sides by three, we then find

$$J(\tau) = \frac{F_o}{2\pi} + \frac{3F_o}{4\pi}\tau \tag{2.32}$$

Since we know that $J(\tau) = S(\tau)$ (see Equation 2.25), we finally have

$$S(\tau) = (\frac{3F_o}{4\pi})(\frac{2}{3} + \tau) \tag{2.33}$$

This is referred to as the "Eddington solution" for the gray atmosphere.

Now we can answer the question: is there any physical basis for considering the function $S(\tau) = a + b\tau$ such as we suggested using in Section 2.5.3?

Indeed there is: the Eddington solution yields just such a solution, with a specific value of 2/3 for the ratio of $a/b$. The function in Equation 2.33 therefore leads to a limb darkening of the form $I(\mu) \sim (2/3) + \mu$. Thus, we see that the intensity at the limb (where $\mu = 0$), i.e., $I(0) \sim (2/3)$, is only 40% of the intensity at the center of the disk (where $\mu = 1$), i.e., $I(1) \sim (5/3)$. In this regard, we recall that the limb darkening of the Sun (Section 2.2) is in fact observed to be close to a linear function of $\mu$, with a limb intensity of about 40% of center intensity in visible wavelengths, just as the Eddington solution predicts. Apparently, the Eddington solution, with its assumption of a "gray atmosphere" (i.e., the opacity is essentially constant at all wavelengths), provide a valuable approach to replicating, in a quantitative manner, the observed limb darkening of the Sun at visible wavelengths. We shall return to why this assumption of constant opacity at all wavelengths might be "not too bad" when we discuss possible sources of opacity in the solar photosphere at visible wavelengths (Section 3.4).

A detailed solution of the RTE in a gray atmosphere (Chandrasekhar 1944) shows that, rather than the solution in Equation 2.33, where $S(\tau) \sim \tau + (2/3)$, a more exact solution yields $S(\tau) \sim \tau + q(\tau)$.

Here, $q(\tau)$ is a slowly-varying function of $\tau$, taking on values of 0.58 as $\tau \to 0$ and 0.71 as $\tau \to \infty$. The two-stream approximation, which replaces $q(\tau)$ with the constant 2/3 (intermediate between the limits of 0.58 and 0.71) is not far off from Chandrasekhar's more sophisticated solution.

## 2.9 HOW DOES TEMPERATURE VARY AS A FUNCTION OF $T$?

Now that we derived how the source function behaves as a function of $\tau$, our aim here is to derive how the temperature behaves as a function of $\tau$. To do this, we use the equality established above between $S(\tau)$ and $J(\tau)$. Recalling the frequency-dependent definition of $J_\nu(\tau) = (1/4\pi)\int I_\nu(\tau)d\omega$, we notice that $J_\nu(\tau)$ at any particular frequency is related to the energy density of the radiation $u_\nu$ at that frequency. In order to evaluate $u_\nu$ (with units of ergs cm$^{-3}$ Hz$^{-1}$) we must integrate $I_\nu(\tau)$ (with units of ergs cm$^{-2}$ sec$^{-1}$ ster$^{-1}$ Hz$^{-1}$) over solid angle, and divide by the speed of light:

$$u_\nu(\tau) = \frac{1}{c}\int I_\nu(\tau)d\omega \tag{2.34}$$

Comparing Equations 2.23 and 2.34, we see that

$$J_\nu(\tau) = \frac{c}{4\pi}u_\nu(\tau) \tag{2.35}$$

We have already noted (Section 2.1) that when the energy density $u_\nu$ of the Planck function is integrated over all frequencies for an object of temperature $T$, the resulting energy density of the radiation is $u(T) = a_R T^4$, where $a_R$ is the radiation density constant. Therefore, if we integrate $J_\nu(\tau)$ over all frequencies, we find that the integral (without any subscript $\nu$) is given by $J(\tau) = (c/4\pi)a_R[T(\tau)]^4$. Noting also that the combination of physical constants $(c/4)a_R$ is equal to another physical constant $\sigma_B$ (the Stefan–Boltzmann constant), we find that

$$J(\tau) = \sigma_B[T(\tau)]^4/\pi \tag{2.36}$$

Since $S(\tau) = J(\tau)$ (see Equation 2.25), Equation 2.36 can be written as

$$S(\tau) = \sigma_B[T(\tau)]^4/\pi \tag{2.37}$$

Inserting this into Equation 2.33 we find

$$\frac{\sigma_B[T(\tau)]^4}{\pi} = \left(\frac{3F_o}{4\pi}\right)\left(\frac{2}{3}+\tau\right) \tag{2.38}$$

The constant flux $F_o$ which propagates through the atmosphere can be expressed in terms of an effective temperature $T_{\text{eff}}$ by means of the definition

$$F_o \equiv \sigma_B T_{\text{eff}}^4 \tag{2.39}$$

Combining Equations 2.38 and 2.39, we finally arrive at an expression for how the temperature varies as a function of optical depth in an Eddington atmosphere:

$$[T(\tau)]^4 = \frac{T_{\text{eff}}^4}{4}(2+3\tau) \tag{2.40}$$

Radiation Flow through Solar Atmosphere

Equation 2.40 is usually referred to as the "Eddington relation", although Paletou (2018) has argued that it is more correctly referred to as the "Milne relation". (In what follows, in order to avoid confusion with references to the Milne–Eddington model (which deals with spectral line formation), we shall refer to Equation 2.40 as the Eddington (Milne) relation (which deals with continuum radiation). It is Equation 2.40 that will eventually (in Chapter 5) start us on the way to computing profiles of density and pressure at various heights in the photosphere of the Sun: in our computations in Chapter 5, we shall take $\tau$ to be the independent variable.

Note that when we come to describe other regions of the atmosphere of the Sun, including the convection zone (Chapter 6), radiative interior (Chapter 8), chromosphere (Chapter 15) and the corona (Chapter 17), optical depth will *not* be appropriate as an independent variable. We shall have to choose different parameters in those regions as independent variable.

Note that although the source function in Equation 2.37 at first sight appears similar to the expression for the flux $F_o$ (Equation 2.39), there are two important differences: (i) $S(\tau)$ varies with $\tau$, whereas $F_o$ is independent of $\tau$; (ii) $S(\tau)$ includes an extra factor of $\pi$ in the denominator.

## 2.10 PROPERTIES OF THE EDDINGTON (MILNE) RELATION

In the upper layers of the atmosphere, as $\tau \to 0$, the Eddington (Milne) relation predicts that the temperature does *not* by any means fall to zero. Instead, it approaches an asymptotic limit: this limit is referred to as the "boundary temperature":

$$T_{boundary} \equiv T(\tau = 0) = \frac{T_{eff}}{2^{1/4}} \qquad (2.41)$$

In the case of the Sun, with $T_{eff}$ = 5772 K, we find a boundary temperature of 4854 K. As a result, the Planck function in the outermost levels of the Eddington (Milne) atmosphere does *not* go to zero. Instead, the Planck function tends towards a finite nonzero value. When we subsequently (in Chapter 15) consider the level in the solar atmosphere where the chromosphere exists, we shall find that, in the gas that lies at a certain altitude above the photosphere, the temperature begins to increase. We shall find that the temperature of gas in the chromosphere rises to values well above the "boundary temperature" of 4854 K predicted by the Eddington (Milne) atmosphere. However, there is no necessary contradiction here: the physical process(es) that cause the chromosphere to become hot are (as we shall see) quite different from the process of radiative transfer that leads to the Eddington (Milne) atmosphere model.

According to Equation 2.40, there exists a particular optical depth at which the local temperature in the Eddington (Milne) atmosphere has the value $T_{eff}$: that optical depth is $\tau = 2/3$. This is consistent with the observation that the photons we see coming from the Sun, emerging (as they do) from layers where the optical depth cannot be much greater than $\tau \approx 1$, appear to emerge from a gas with a temperature that is only slightly greater than 5772 K. In the model that we shall calculate in Chapter 5, we shall find that at the level where $\tau \approx 1$ in the Sun, the gas has a temperature close to 6000 K.

## EXERCISES

The search for possible life on planets orbiting other stars is an exciting topic for modern astronomers. One of the major effects that the Sun has on life (apart from keeping the temperature warm on Earth) is to cause photosynthesis (PS): photons from the Sun add energy to $CO_2$ plus $H_2O$ and generate sugar as well as releasing free $O_2$ gas into the atmosphere. The question we pose here is:

how good are other stars at driving PS? The answer is: it all depends on the mix of photons that the star can provide. PS requires "good" photons with wavelengths between 4000 Å (the blue end) and 7000 Å (the red end). "Bad" photons are those with ultraviolet (UV) wavelengths that are so short (2000 Å) that they damage chemical bonds in the DNA molecule. In this exercise we compare the supply of photons from three different stars: the Sun, a hotter star, and a cooler star.

In view of the plot in Figure 2.2, assume that stars radiate light that follows the Planck function. Omitting constants, each star emits light with this intensity: $I_\lambda = 1/\{\lambda^5 [\exp(hc/\lambda kT) - 1]\}$. In this expression, when we use the c.g.s. system of units, the combination of constants of nature $hc/k$ has a well-defined constant value. Using $h = 6.626 \times 10^{-27}$ gm cm$^2$ sec$^{-1}$, $c = 2.998 \times 10^{10}$ cm sec$^{-1}$, and the value of $k$ listed in the text, we find $hc/k = 1.438$ cm deg K. When using this value to evaluate $I_\lambda$, it must be remembered that $\lambda$ must be expressed in units of cm, and $T$ must be expressed in units of deg K.

2.1 Consider a star with T = 5772 K. Calculate $I_\lambda$ at three wavelengths (i) $\lambda_1 = 7000$ Å $= 7 \times 10^{-5}$ cm, (ii) $\lambda_2 = 4000$ Å, and (iii) $\lambda_3 = 2000$ Å. Label your answers $R_1$(sun), $R_2$(sun), $R_3$(sun).

2.2 Now consider a hot star with T = $1.5 \times 10^4$ K. Using the same three wavelengths as in Exercise 1, calculate the fluxes $R_1$(hot), $R_2$(hot), $R_3$(hot).

2.3 Calculate the ratio of $R_3$(hot)/$R_3$(sun): this ratio is a measure of how much more dangerous it will be for life on a planet near the hot star.

2.4 Now consider a cool star with T = 3000 K. (Most of the stars in our galaxy belong to this group: there is widespread interest in looking for planets around these stars.) Using the same wavelengths as in Exercise 1, calculate the fluxes $R_1$(cool), $R_2$(cool), $R_3$(cool).

2.5 Calculate the ratio of $R_3$(cool)/$R_3$(sun): this ratio is a measure of how much safer it will be (as regards UV light) for life on a planet near the cool star.

2.6 Calculate the ratio of $R_2$(cool)/$R_2$(sun). This ratio is a measure of how ineffective PS (at its "blue end") will be on a planet near the cool star compared with a planet near the Sun.

2.7 Calculate the ratio of $R_1$(cool)/$R_1$(sun). This ratio is a measure of how ineffective PS (at its "red end") will be on a planet near the cool star compared with a planet near the Sun.

2.8 Conclusion: conditions for life (few UV photons, many PS photons) are probably most favorable around stars that have temperatures similar to those which occur on the Sun.

## REFERENCES

Barbier, D., 1943. "Sur la théorie du spectre continu des étoiles", *Annales d'Astrophys.* 6, 113.
Cercignani, C., 1998. *Ludwig Boltzmann: The Man who Trusted Atoms*, Oxford University Press, Oxford, UK, pp. 89, 121, 219.
Chandrasekhar, S., 1944. "On the radiative equilibrium of a stellar atmosphere. II", *Astrophys. J.* 100, 76.
Eddington, A. S., 1926. *The Internal Constitution of Stars*, Cambridge University Press, Cambridge, UK, p. 330.
Iqbal, M., 1983. *An Introduction to Solar Radiation*, Academic Press, Toronto, Table C1.
Milne, E. A., 1921. "Radiative equilibrium in the outer layers of a star", *Mon. Not. Roy. Astron. Soc.* 81, 361.
Paletou, F., 2018. "On Milne-Barbier-unsold relationships", *Open Astronomy*, 27, 76.
Planck, M., 1948. *Wissenshaftliche Selbstbiographie*, Johann Ambrosius Barth, Leipzig.
Popper, K., 1959. *The Logic of Scientific Discovery*, cited in Ferris, T., ed. 1991. *The World Treasury of Physics, Astronomy, and Mathematics*, Little Brown & Co., New York, p. 799.
Sudar, D., Brajsa, R., Skokic, I., et al., 2019. "Center-to-limb brightness variations from ALMA full-disk solar images", *Solar Phys.* 294, 163.

# 3 Toward a Model of the Sun
## *Opacity*

Now that we have information as to how the temperature in the vicinity of the solar photosphere behaves as a function of optical depth, we have taken the first step in achieving one of the principal goals of solar physics: to calculate how the physical quantities in the Sun behave as a function of radial location. We refer to such a radial profile as a "solar model".

When the only information that we have access to is limb darkening, the range of radial locations in the Sun that can be modeled reliably is quite restricted: we can extract information only for a range of heights in the vicinity of the photosphere. For present purposes, the "vicinity of the photosphere" refers to locations in the solar atmosphere that lie *above* the convection zone and *below* the chromosphere. For the sake of brevity, we refer to these limits as the "lower" photosphere and the "upper" photosphere, respectively. In what follows, given the physical conditions that exist in the solar atmosphere, we shall find that the "lower" and "upper" photosphere differ in height by $\Delta h$ = several hundred kilometers.

In subsequent chapters, when we wish to extend our modeling efforts deeper into the interior of the Sun or upward into the chromosphere, we shall need access to data over and above what limb darkening can provide. But as long as we are dealing with situations in which photons are trying to make their way through a more or less opaque medium, it is important that we address the following question quantitatively: how opaque is the medium to the photons that are of interest to us? The physical quantity that measures this opaqueness (i.e., lack of transparency) is referred to by astronomers as the opacity. The opacity is a physical property of the gas present in the Sun: the physics of opacity require us to understand in detail some of the properties of individual atoms and ions in the mixture of elements that make up the Sun. The study of these atomic properties is the subject of this chapter.

It is salutary to remember that in order to gain an understanding of the properties of a very large macroscopic object (i.e., the Sun, containing some $10^{57}$ atoms/ions), it is necessary to focus our attention at times on the details of individual atoms. The goal is to understand how the detailed properties of certain atoms play a role in determining the parameters of the large macroscopic object on which we depend for life. This interplay between the very small and the very large is one of the features that makes life interesting (and fun!) for a solar physicist.

## 3.1 RELATIONSHIP BETWEEN OPTICAL DEPTH AND LINEAR ABSORPTION COEFFICIENT

Up to this point, the absorption of radiation as it passes through a gas has been discussed in terms of $k_\lambda$ cm$^{-1}$, which is a *linear* absorption coefficient. More customary in astrophysics is the opacity, $\kappa_\lambda = k_\lambda/\rho$: this is a measure of how opaque a medium with mass density $\rho$ gm cm$^{-3}$ is for light of wavelength $\lambda$. Given the units of $k_\lambda$ and of $\rho$, we see that the units of $\kappa_\lambda$ are cm$^2$ gm$^{-1}$. These units suggest that the opacity is associated with a cross-sectional area that impedes the free passage of radiation as the radiation propagates through 1 gm of material.

Since opacity includes a cross-sectional area responsible for absorbing and/or scattering light out of an incoming beam, we start the discussion by considering a fundamental cross-section that is associated with what happens when a beam of photons, on its way through the Sun's atmosphere along what we will call the "forward direction", encounters a free electron. The electric field $E$ of a photon causes the electron (with electric charge $e$ and mass $m_e$) to undergo an acceleration $a = eE/m_e$. When an electric charge is accelerated, it emits electromagnetic radiation at a rate that is proportional to $a^2$: this emitted radiation emerges in multiple directions. The electron recoils slightly when it is "hit" by a passing photon, and this causes a reduction in the energy of the photon: as a result, the wavelength of the photon increases slightly by an amount $\Delta\lambda_C$ on the order of the Compton wavelength ($=h/m_ec = 0.024$ Å). For incoming photons with wavelengths in the physical range (4000–7000 Å), this shift is insignificant: the photon emerging from the photon-electron interaction has a wavelength that is essentially identical to that of the original photon.

As a result, an observer who is "looking for" the photons along their initial (forward) direction of travel will indeed see a reduction $\delta$ in the *amount* of photons traveling in that direction after the photons have propagated past the electron: but the energy of each photon in the system will *not* be significantly altered. The photons that have "gone missing" (in order to give rise to the reduction $\delta$ in flux in the forward direction) are to be found emerging in directions other than the initial direction of the beam. Since the photon flux *in the forward direction* has indeed been reduced by passing over the electron, an observer might be tempted to say, "some photons must have been absorbed", in order to explain why fewer photons are now being seen in the forward direction. However, the missing photons have not formally been "absorbed" by anything: rather, they have been "scattered" away from their original direction as a direct result of the interaction between a photon and an electron. Nevertheless, the scattering process does result in an effect (reduction in flux) that appears (to the observer in the forward direction) similar to what would be seen if the photons had passed through a medium with finite optical depth $\tau$: the reduction $\delta$ in intensity of the photon beam could be written as $e^{-\tau}$.

Reverting now to the case of the photon passing by an electron, we wish to derive a measure of how effective the scattering process is: to do this, we imagine that the electron presents an obstacle to the incoming photons. The larger is the cross-sectional area of the obstacle, the greater is the amount of reduction in photon flux. When a photon encounters a free electron, the photon in effect "sees" an obstacle with a finite "scattering" cross-section, which is given by the Thomson formula $\sigma_T = (8/3)\pi r_e^2$. Here, $r_e = e^2/m_ec^2$ is the classical radius of the electron. In discussions about the interactions between photons and matter, the numerical value of the Thomson cross-section is an important quantity:

$$\sigma_T = 6.6245873 \times 10^{-25}\,\text{cm}^2 \tag{3.1}$$

Note that $\sigma_T$ is independent of wavelength (at least for photons with energies less than $m_ec^2 = 0.5$ MeV).

In an ionized gas, there are always some free electrons that contribute to opacity with the aforementioned cross-section. In fact, in the limit of high temperatures, when almost no electrons are bound inside an atom/ion, essentially all electrons become free, and Equation 3.1 provides a good approximation to the absorption cross-section (see Section 3.6.1).

However, in the gas that exists at various locations in the Sun, not all of the electrons are free. There are also atoms and/or ions in which some electrons are in bound orbits around the nucleus. Each bound orbit has a well-defined value of the total energy $E_j$ (for a calculation of specific examples of such energies $E_i$, see Section 3.2.1). When a photon passes by an atom or ion where electrons are held in bound orbits, the photon may interact with the atom and undergo "scattering" of the same sort as described earlier: that is, a change in *direction* of the photon may occur but without any change in *wavelength*. How might this occur? If the incoming photon with energy $h\nu$ happens to encounter an atom/ion where the difference $\Delta E = E_u - E_l$ in energy between an upper bound level and a lower bound level is equal to $h\nu$, then the photon has a good chance of being absorbed by an electron in the lower bound level $E_l$. As a result, that electron undergoes a transition into $E_u$. Two outcomes are then possible. First, the electron may quickly return to $E_l$, thereby re-emitting a photon with exactly the same energy $h\nu = \Delta E$ as the incoming photon. In such a case, the emergent photon,

Toward a Model of the Sun

although it has the same *energy* as the incoming photon, would in general not emerge in the same *direction* as the incoming beam. The term "scattering" is used for this process just as we used it for the photon's interaction with a free electron.

However, in the case of an atom/ion with multiple energy levels, a second outcome is possible: the electron in $E_u$ may, with finite probability, make a transition to an energy level that is *not* $E_l$. In that case, the emergent photon will *differ* in wavelength from the incoming photon, i.e., the incoming photon "disappears", and a photon with different wavelength takes its place. Such a process is referred to as "absorption". Which of the two processes (scattering, absorption) is likely to occur in any given situation? The answer depends on the probability for an electron to make a quick transition from the upper to the lower energy level before "something else happens". Formal solutions of the RTE for conditions in which spectral lines are included as well as continuum require that both scattering and absorption be included in the theory. In this "first course", we will not deal with these complexities.

Instead, we shall confine our attention to opacity as a measure of how strongly photons interact with the atoms/ions/electrons of the medium through which the photons are passing. We shall find that quantitative aspects of the opacity, and especially its sensitivity to temperature, play a key role in modeling three regions of the Sun: the photosphere, the deep interior, and the chromosphere. Because of this, the more we understand the properties of opacity, its numerical values and its variations with temperature, the more physical insight we will have into the structure of the Sun.

*A note about "allowed" and "forbidden" transitions.* An electron in a certain energy level is *not* allowed to make a (quick) transition to any and all other energy levels in the atom. Only certain transitions are "allowed", based on quantum mechanical ("electric dipole") selection rules: "allowed" transitions can occur at rates of up to $10^8$ times per second. When the rules indicate that a certain transition is not allowed, a rarer type of transition may be allowed due to a different type of transition ("magnetic dipole", "electric quadrupole"). The photons emerging from these rarer transitions are referred to as "forbidden lines": such transitions occur at much slower rates, sometimes as slow as only 1–10 times per second. When we discuss the solar corona (Section 17.2.1), we shall see that "forbidden" lines played a major role in the earliest determinations of temperature and density in the corona.

## 3.2 TWO APPROACHES TO OPACITY: ATOMIC AND ASTROPHYSICAL

There are two different approaches to opacity, depending on one's interest: atomic physics and astrophysics. On the one hand, from the *atomic* point of view, the main goal is to understand the following: given a photon *with a specified wavelength* (but of unspecified origin), what is the cross-section for photon interaction (i.e., either absorption or scattering) by a particular atom? Quantum mechanics can be used to derive quantitatively the numerical value of the cross-section at any particular wavelength. We shall take this viewpoint in Sections 3.3, 3.4, and 3.5.

On the other hand, from the point of view of *astrophysics*, the main goal is to understand how photons distributed with a certain functional form over *a broad range of wavelengths* interact with the medium through which they are passing. In the solar atmosphere, opacity involves a broadband process of interaction between photons and atoms: it is not merely the atomic physics that is relevant, but also the "spectrum of radiation" in the atmosphere and how this spectrum overlaps with regions of large and small opacity. The key questions in astrophysics turn out to be: how many photons are present at wavelengths where the opacity is large, and how many photons are present at wavelengths where the opacity is small? It is all very well for the atomic physicist to report that hydrogen absorbs most strongly at wavelengths 912–1216 Å (see Figure 3.2), but if, on the one hand, researcher A is studying the flow of radiation through an atmosphere where there are essentially *no* photons at 912–1216 Å, then the peak in hydrogen absorption at 912–1216 Å (which undoubtedly exists) is of no great relevance to researcher A. In such a case, the radiation in the atmosphere, with most of its photons at (say) long wavelengths, may encounter very little effective opacity. On the other hand, if researcher B is considering an atmosphere where the photon spectrum happens to peak in the wavelength range 912–1216 Å, then the strongly absorbing behavior of hydrogen atoms in this wavelength range is highly relevant. In such a case, researcher B will find that the effective opacity is very large.

In astrophysics, the ease with which photons propagate through an atmosphere depends on an intricate process in which a wavelength-dependent photon spectrum (such as the Planck function with a peak at a wavelength that is determined solely by the temperature) interacts with a wavelength-dependent opacity (which may also contain peaks, narrow valleys, and sharp "edges"). Thus, we need to have a way of calculating a "mean opacity" of some sort, such that the two important features (atomic physics as well as spectral information) can be intertwined in a meaningful way. This will lead us to define a mean opacity named in honor of Svein Rosseland: we will introduce this "Rosseland mean opacity" in Section 3.6 after we have considered certain details of atomic physics. It is the Rosseland mean opacity that we will use in Chapter 5 to calculate a model of the solar photosphere.

### 3.2.1 ENERGY LEVELS IN ATOMIC HYDROGEN

In order to set the stage for a realistic physics discussion of what a photon "sees" when it passes by a hydrogen atom, it is important to quantify the energies of the various levels occupied by electrons inside such an atom. Neils Bohr (1913) was the first to estimate these energy levels for a hydrogen atom by suggesting two "bold ideas" (in Karl Popper's phrase): (i) the electron orbits a proton with speed $V$ in a circular orbit of radius $r$; (ii) the angular momentum of the electron in its orbit is allowed to have only discrete values, namely $nh/2\pi$, where $h$ is Planck's constant, and $n = 1, 2, 3 \ldots$ Thus, Bohr suggests that a simple two-step argument will lead to predictions of the stable energy levels in a hydrogen atom. The argument proceeds as follows. Using (i), we expect to find a stable orbit if the electrostatic force of attraction $e^2/r^2$ between proton and electron is balanced by the centrifugal force $mV^2/r$ of the electron (with mass $m$) in its orbit. (Note we use electrostatic units here, with the charge on the electron being given by $e = 4.8032 \times 10^{-10}$ electrostatic units (e.s.u.). Then using cm as the unit of $r$, the electrostatic force will be in units of dynes.) This force balance yields a first equation involving $r$ and $V$: $r = e^2/mV^2$ (where $m$ is in grams and $V$ is in cm sec$^{-1}$). Using (ii), we set the electron's orbital angular momentum ($=mVr$) equal to $nh/2\pi$ and obtain a second equation involving $r$ and $V$: $r = nh/2\pi mV$. With two equations to solve for two unknowns ($V$ and $r$), we readily find $V = 2\pi e^2/nh$ and $r = (nh/2\pi e)^2(1/m)$. Note that $V/c$ can be written as $\alpha/n$ where $\alpha = 2\pi e^2/hc$ is called the "fine structure constant", with a numerical value of about 1/137: thus in the ground state of hydrogen ($n=1$), the electron moves in its orbit with a speed of $c/137 \approx 2 \times 10^8$ cm sec$^{-1}$.

In order to calculate the total energy $E$ of the electron in its orbit, we need to consider two terms: kinetic energy $KE = mV^2/2$ and potential energy $PE = -e^2/r$. The $KE$ is a positive quantity, while the $PE$ is negative. Therefore the total energy $E$ will have a sign determined by which of the two terms, $KE$ or $PE$, has the larger magnitude. In order for an electron to be bound inside an orbit, the $PE$ must be larger in magnitude than the $KE$: that is, the total energy $E$ must be negative. Otherwise, the electron could not remain bound to the proton. Inserting the prior values of $r$ and $V$, we find that the total energy of the orbit having angular momentum $nh/2\pi$ is $E_n = -(1/2)mV^2 = -2\pi^2 me^4/h^2 n^2$. In terms of $\alpha$, it is sometimes convenient to write $E_n = -0.5\alpha^2 E_r/n^2$ where $E_r = mc^2$ is the rest-mass energy of the electron. Using c.g.s. units ($m = 9.109 \times 10^{-28}$ gm and $h = 6.626 \times 10^{-27}$ gm cm$^2$ sec$^{-1}$), we obtain the energy $E_n$ in units of ergs. A convenient unit for the energies of orbits in atoms turns out to be a quantity known as an "electron volt": 1 eV = $1.602 \times 10^{-12}$ ergs.

The lowest-lying energy level of the H atom (this level is referred to as the "ground state") has $n = 1$. Using the previous expression, we find that the energy $E_1$ of this level is $-13.61$ eV. Excited states in hydrogen with $n = 2, 3, 4 \ldots$ have energies of $E_n = -13.61/n^2$, i.e., $-3.40, -1.51, -0.851$ eV. Thus, the energy difference between $E_1$ and $E_2$ is 10.21 eV. If an electron makes a transition from $n = 2$ to $n = 1$, a photon will be emitted with energy $\Delta E_{21} = 10.21$ eV. According to Planck's assumption, this photon will have a frequency $\nu_{21}$ such that $h\nu_{21} = 10.21$ eV $= 1.636 \times 10^{-11}$ ergs. Therefore $\nu_{21} = 2.4691 \times 10^{15}$ Hz. Since photon frequency is related to wavelength by $\lambda = c/\nu$ (where $c$ is the speed of light), we find that the photon corresponding to $\Delta E_{21}$ has a well-defined wavelength $\lambda_{21} = 1.2156 \times 10^{-5}$ cm. Expressing this in units of Å, we find $\lambda_{21} = 1215.6$ Å. This corresponds to the wavelength of a very strong line in the Sun belonging to a series of lines called the Lyman series: the line in the Lyman series with the lowest frequency (i.e., with the longest wavelength, 1215.6 Å) is referred to as the Lyman-alpha line (or Ly-α) (see Figure 3.2).

Bohr's model can be applied in principle to elements other than hydrogen, provided that the following condition is met: we must consider an ion of the element from which *all except one* of the electrons have been stripped way. For example, if one considers the element neon (with atomic number 10, i.e., having $Z = 10$ protons in the nucleus), an atom of neon at low temperatures will contain 10 electrons that are bound to the nucleus: as a result, the neon atom appears to an outside observer as being electrically neutral, with a charge of 10+ in the nucleus, cancelled by a charge of 10− in the bound electrons. However, if by some means we increase the temperature of the gas where the neon atom is situated, the electrons are stripped away one by one. Eventually, at high enough temperatures, we reach the condition where nine of the electrons have been stripped away. If that happens, then the remaining ion (still with 10 protons in the nucleus) will have only one electron remaining in a bound orbit. An ion with only one bound electron is referred to as a "hydrogenic" ion with nuclear charge $+Ze$: in the presence of such a charge, a repeat of the prior derivation of energy levels indicates that $E_n$ is larger in magnitude by a factor of $Z^2$ than the value of $E_n$ derived earlier for hydrogen itself. Bohr's model can in principle be applied to such an ion, no matter where the element lies in the periodic table. Note that only one of the "hydrogenic ions" has an electrical charge of zero, i.e., it is a (neutral) atom: hydrogen itself. All of the other hydrogenic ions have a net electrical charge of $+(Z-1)e$.

*A word about notation.* In this and in subsequent chapters, we will have occasion to deal with ionized and neutral species of various elements. For each element X in the periodic table, the notation for the neutral atom is X I, where the "I" in the second position is the Roman numeral for the number "one". If an atom of element X loses one electron, the resulting ion is labeled X II. If more than one electron is lost, the X is followed by a Roman numeral that equals the number of lost electrons *plus one*. Thus, X VII is an ion of element X that has lost *six* electrons. Of course, if the atomic number of the element $Z(X)$ is smaller than six, then the neutral atom X I contains fewer than six electrons to start with: it would be impossible for such an element to lose as many as six electrons. Therefore, the "ion labeled X VII" would not exist. We shall be especially interested in "hydrogenic" ions that have lost *all but one* of the electrons present in the neutral atom. The hydrogenic ion of element He (where $Z = 2$) is He II. Other such ions in elements with (e.g.) $Z = 6, 8, 20$, and 26 are C VI, O VIII, Ca XX, and Fe XXVI respectively.

Historically speaking, Bohr's model of the hydrogen atom was subsequently supplanted by results from Erwin Schrödinger in 1926 using a wave equation and eventually by results from Paul Dirac in 1928 using relativistic quantum mechanics. These later developments showed that *some* of Bohr's results were in need of revision. For example, his vision of circular orbits with definite radii had to be replaced by smeared-out distributions of the probability that an electron would actually be present at radial location $r$. Moreover, the distribution of the electron in 3-D space were found to be complicated geometrical shapes describing the distribution of electron probability in space. As a result of the revisions emerging from Schrödinger's and Dirac's work, it can be tempting to believe that Bohr's model is of no use whatsoever as a predictor of physical quantities. However, despite getting the orbital size and shape wrong, it turns out (amazingly enough) that Bohr's model for a hydrogenic ion *does* yield reliable results for one important physical quantity: the *total energy of each electron energy level*. For example, Schrödinger's wave equation, when applied to the hydrogen atom, yielded energy levels with $E_n = -2\pi^2 me^4/h^2n^2$: this expression appears identical in form to Bohr's result. However, the integer $n$ (referred to by Schrödinger as the *total quantum* number) is no longer solely a measure of angular momentum as it was in Bohr's discussion: instead, Schrödinger's

$n$ is related to the sum of two integers, one ($l$) related to the orbital angular momentum $lh/2\pi$, and the other $n'$ related to the order of a polynomial which describes the probability wave as a function of radial distance from the nucleus. Specifically, in Schrödinger's formula for $E_n$, the subscript $n = n' + l + 1$ (Schiff 1955). However, there is still a close connection between $n$ and $l$: the integer $l$ is permitted to take on only the discrete integer values between 0 and $n-1$. The ground state of Schrödinger's hydrogen atom (with $n' = 0$ and $l = 0$) still is identified by the number $n = 1$, just as in Bohr's model. As regards Dirac's solution, the value of $E_n$ was found to be expressible as a series of terms. The leading term in the expression for $E_n$ (apart from the rest mass energy) was found to be $-0.5\alpha^2 E_r/n^2 = -13.61/n^2$ eV, identical to Bohr's and Schrödinger's result. In the Dirac model, $n$ is once again a combination of several factors. One of the factors is the order of the polynomial describing the radial portion of the probability distribution. Another factor is associated with the orbital angular momentum $l$ of the electron. However, because of relativistic effects, Dirac showed that an extra type of angular momentum now enters into the problem: this has to do with the spin $sh/2\pi$ of the electron (where $s = 1/2$). Relativistic effects also show up in the total energy as slight corrections to the Bohr energies. Dirac found that the next term in the expression for $E_1$ was smaller (in magnitude) than $E_1$ by a factor of $(3/4)\alpha^2 = 4 \times 10^{-5}$. Thus, even with all the complexities of the Dirac solution of the H atom, the energy level of the H atom ground state was found to differ from the Bohr energy by no more than one part in 10,000. As a result, Bohr's predicted wavelength for the photon that emerges from a transition between (say) $E_2$ and (say) $E_1$ in the H atom (i.e., the Lyman-$\alpha$ transition) turns out to be highly reliable.

To be specific, in the exercises at the end of this chapter, the reader is invited to consider how well Bohr's predictions of Lyman-$\alpha$ wavelengths have worked out as regards a sample of hydrogenic ions that have in recent decades been observed in various astrophysical objects. These tests of Bohr's predictions were impossible to perform until there were X-ray instruments in orbit high above the Earth (where Earth's atmosphere does not absorb the photons) that could measure X-ray wavelengths down to values as small as a few Å: such instruments were not available until 60 years (or more) after Bohr published his original paper. Bohr would be pleased to see that his predictions of the wavelengths of Lyman-$\alpha$ lines have been remarkably well validated. Doschek and Feldman (2010) provide a list of X-ray and UV lines with wavelengths ranging from 1.5 Å to 2000 Å that have been observed by an array of 16 different satellites between 1977 and 2007: included in their list are the Lyman-alpha lines of at least 10 hydrogenic ions from H I to Fe XXVI.

## 3.3 ATOMIC PHYSICS: (I) OPACITY DUE TO HYDROGEN ATOMS

To see how the linear absorption coefficient $k_\lambda$ is related to the properties of individual atoms, consider Figure 3.1. This shows what an observer sees when looking through the "endface" of a column of length $l$. The area of each "endface" of the column is 1 cm$^2$, and the observer's line of sight passes through a medium (inside the column) with density $\rho = n_a m_a$. Here, $n_a$ cm$^{-3}$ is the number density of absorbers, each of mean mass $m_a$ (gm).

Each absorber in the column has a cross-sectional area $\sigma$ for the absorption of light. The total number of absorbers in the column is the *number column density*: $N = n_a l$, with units of cm$^{-2}$. (An equivalent quantity is the *mass column density* $d_c = Nm_a = \rho l$ in units of gm cm$^{-2}$.) In the specific case shown in Figure 3.1, the total number of absorbers in the column is $N = 15$. In the limit $N\sigma \ll 1$, light, which enters the column at one "endface", travels along the column of length $l$, and emerges through the other "endface", encounters a total absorption area of $N\sigma$ cm$^2$. Comparing this to the 1 cm$^2$ area of the "endface", this means that the light emerging from the column is reduced by the fractional amount $N\sigma$. Recalling that, in the limit of small optical depth, $e^{-\tau} = 1 - \tau$, we see that the fractional reduction of light $N\sigma = n_a l\sigma$ can be set equal to $\tau$. We have already seen (in Section 2.3) that in the limit of small optical depth, a thin slab of gas of thickness $x$ has an optical depth $\tau = x k_\lambda$. In the present case, this means that in the limit of small $\tau$, we can set $\tau = lk_\lambda$. This indicates that $k_\lambda$ can be set equal to the product $n_a\sigma$. The latter product, combining the number of

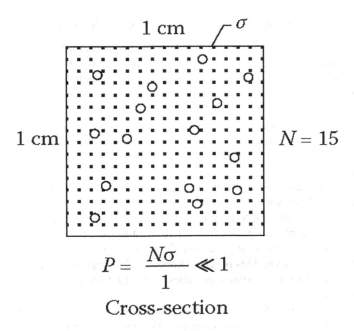

**FIGURE 3.1** Indicating how to estimate the blockage of light passing through a column of material. Each endface of the column is a square with dimensions 1 × 1 cm. In the figure, we are looking in at one endface, and the column extends a length $l$ behind the page. (Original figure was taken in 2009 from the website http://mysite.du.edu/~jcalvert/phys/scat.htm with permission of Dr J. Calvert. However, in the interim, some of Dr Calvert's websites have been taken down. The original figure is pedagogically valuable, and I retain it here. For an alternative figure which contains similar information, see Figure 16.2 in Karttunen et al. [2017].)

absorbers per cc with a cross-section in cm², has dimensions of cm⁻¹, as does the linear absorption coefficient $k_\lambda$. In the case of Equation 3.1, it happens that we were dealing with an especially simple case: the scattering cross-section $\sigma_T$ is independent of wavelength. But in other cases, we shall see (e.g., in Figure 3.2) that $\sigma$ may vary enormously (by 10 orders of magnitude or more) from one wavelength to another. If $\sigma$ depends on wavelength, we will make the wavelength-dependence explicit by using the symbol $\sigma_\lambda$. In such a case, the linear absorption coefficient has the form $k_\lambda = n_a \sigma_\lambda$.

Converting now from linear absorption coefficient to opacity, we find $\kappa_\lambda = k_\lambda/\rho = \sigma_\lambda/m_a$. This expression helps us to see an important aspect of opacity. Since $1/m_a$ is equal to the number of absorbers in 1 gm of the material, the opacity can be written as the product of two factors: (i) the *cross-section of an individual absorber* and (ii) the *number of absorbers per gram*. Thus, the units of opacity are cm² per gm. We shall find (see Table 5.3 later) that in the photosphere of the Sun, the mean opacity in these units has a numerical value of order 0.1–1. In Section 3.4, we shall identify the (unusual) atomic constituent that causes the opacity in the photosphere to have values of this order.

An alternative formula for optical depth can be obtained by combining $\kappa_\lambda = k_\lambda/\rho$ with the definition of mass column density $d_c = \rho l$. Given that the product $lk_\lambda$ is equal to $\tau$ (for $\tau \ll 1$), we now see that an alternative definition of $\tau$ is the following: $\kappa_\lambda d_c = lk_\lambda = \tau$.

For gas of a given composition, $m_a$ is a constant, independent of wavelength. So if we examine a plot of $\kappa_\lambda$ versus $\lambda$, then we will be able to trace how the cross-section $\sigma_\lambda$ of individual absorbers behaves at each wavelength. When the atoms (or ions) in the medium through which light is propagating contain electrons in bound energy levels, $\sigma_\lambda$ may vary by many orders of magnitude as a function of wavelength. To see why this is so, let us consider the simplest case: hydrogen atoms.

### 3.3.1 Absorption from the Ground State: Dependence on Wavelength

If hydrogen atoms are in a medium with low enough temperature, only the ground state (i.e., the energy level that has the lowest energy in the atom) has a significant population of electrons. Atomic physics demonstrates that the energy of the ground state lies at an energy of $E_1 = -13.6$ eV below the continuum. (Note: one unit of energy eV corresponds to $1.6 \times 10^{-12}$ ergs.) Now, the lowest excited state in hydrogen lies at an energy $E_2 = -3.4$ eV, i.e., an energy which lies 10.2 eV above the ground state. Because of this gap in energy between levels 1 and 2 in H, any photons which have energies less than 10.2 eV (i.e., with wavelength $\lambda > 1216$ Å) do not have sufficient energy to excite the electron from the ground state into any other energy level. As a result, interactions between photons and the ground state of an H atom become progressively more ineffective when the photons have $\lambda > 1216$ Å: therefore, if these were the only interactions that the photons could undergo, $\kappa_\lambda$ at all wavelengths longer than 1216 Å would be zero.

Other physical processes come into play for $\lambda > 1216$ Å. For example, weak scattering occurs when the electric field in a photon passing by an atom induces a transient dipole moment in the atom: the time-variable dipole moment generates electromagnetic radiation that has nothing to do with the energy levels of the atom. This process (known as Rayleigh scattering) leads to an opacity where the leading term in a series expansion varies as $\lambda^{-4}$ (see the dashed line in the lower right-hand side of Figure 3.2).

Notice the units of opacity in Figure 3.2. First, they are expressed *per hydrogen atom*. Also, to facilitate comparison with our prior discussion (see Equation 3.1) of free electron scattering, the results in Figure 3.2 are expressed in units of the Thomson cross-section $\sigma_T$. Along the vertical axis, we see plotted numbers that can be much larger than unity: the largest number in the plot is $10^{10}$. Thus, the plot gives a clear indication that an electron that is *bound inside a hydrogen atom* can be much more effective (by up to 9–10 orders of magnitude!) at interacting with a photon than a *free electron* can. This is a key concept for understanding the opacity arising in the solar interior, where

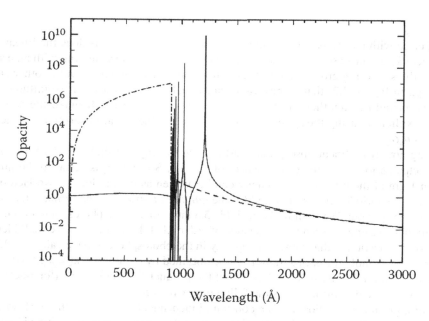

**FIGURE 3.2** Lyman scattering opacity *per hydrogen atom*, in units of the Thomson scattering cross-section per electron. At wavelengths longer than 912 Å, Lyman lines are plotted as solid "spiky" features. At wavelengths shorter than 912 Å, opacity due to the Lyman continuum is plotted as a dot-dashed line (Stenflo 2005; used with permission from ESO).

*many* electrons are ejected from certain atoms and become free when the temperature is measured in millions of degrees K. Nevertheless, even in these conditions, a small fraction of certain ions (e.g., iron) retain a few bound electrons even at such temperatures: because of the absorbing effectiveness of bound electrons, these ions, though few in number, can contribute to the overall opacity (see Sections 3.7 and 8.3).

By inspecting Figure 3.2, we see that, for a narrow range of wavelengths in the vicinity of the Lyman-$\alpha$ line ($\lambda$ = 1216 Å), $\sigma_\lambda$ increases to very large values, some $10^{10}$ times larger than $\sigma_T$. This very strong interaction is associated with the bound-bound transition between the two lowest-lying energy levels ($E_1$ [with total energy −13.6 eV] and $E_2$ [at −3.4 eV]) in the H atom. At wavelengths that are immediately shortward of Lyman-$\alpha$, over a range of about 200 Å in wavelength, there is another range of wavelengths in which $\sigma_\lambda$ falls to small values. In the vicinity of $\lambda$ = 1026 Å (Lyman-$\beta$), excitation from energy level $E_1$ to energy level $E_3$ (at −1.51 eV) gives rise once again to a locally significant value of $\sigma_\lambda$. A series of line absorptions (the Lyman series of lines), separated by regions of continuum where $\sigma_\lambda$ is small continues until the wavelength becomes as short as $\lambda_1$ = 912 Å. At that point, $\sigma_\lambda$ increases abruptly by several orders of magnitude: this is referred to as the "Lyman edge". At wavelengths lying shortward of this edge, $\sigma_\lambda$ remains large over a wavelength interval of hundreds of Å, although $\sigma_\lambda$ decreases systematically (∼$\lambda^3$) toward shorter wavelengths. Why does the Lyman edge exist? Because all photons with wavelengths shorter than $\lambda_1$ = 912 Å have energies larger than $h\nu_1 = hc/\lambda_1$ = 13.6 eV. Such photons have energies that are large enough to eject an electron from the ground state of the hydrogen atom. Such an ejection is called a "bound-free transition". All of the photons in the Lyman continuum (i.e., those with $\lambda < \lambda_1$) lead to ionization of the hydrogen atom into a free proton and a free electron.

How large does the cross-section for photon absorption become at the Lyman edge? Detailed quantum mechanical calculations indicate that the peak value is $\sigma_{912} = 6 \times 10^{-18}$ cm$^2$. Is this to be regarded as a "large" value or a "small" value for the cross-section? To answer that, we need to have some standard against which we can make a comparison. An important aspect of the value of $\sigma_{912}$ is the following: $\sigma_{912}$ exceeds the Thomson $\sigma_T$ by some *seven orders of magnitude*. Therefore, by occupying the ground state of hydrogen, an electron enhances its ability to interact with a photon of wavelength 912 Å by a factor of order 10 million. The conclusion is that an electron in the bound orbit with energy $E_1$ can be (if the wavelength is right) a powerful and effective absorber of photons, up to 10 million times more effective at absorbing a photon than a free electron.

Another way to look at the "largeness" of $\sigma_{912}$ is to consider the area of the ground state orbit. According to the Bohr model, the orbital radius of the ground state $r_1$ is $0.528 \times 10^{-8}$ cm. The corresponding orbital area is $0.88 \times 10^{-16}$ cm$^2$. The value of $\sigma_{912}$ amounts to almost 10% of this area. As far as a passing 912 Å photon is concerned, it is as if the ground state electron in the H atom has "spread itself out" over a significant fraction of the orbital area. This phenomenon is consistent with the Heisenberg uncertainty principle: if a particle is confined to a certain region of linear size $L$, the permitted energy levels are those in which the de Broglie wavelength $\lambda_D = h/p = h/mV$ (see Section 2.1) can be "fitted in" $n$ times into length $L$, where $n$ = 1, 2, 3 . . . In the ground state of H, the electron velocity is such that the length of $\lambda_D$ "fits in" exactly once into the circumference $2\pi r_1$. Given that the electron is in effect "smeared out" over a length of order $\lambda_D$, it is as if the electron in the H ground state orbit has become an "obstacle" to a passing photon with a cross-section that is a fraction of the area of the orbit.

The large absorbing efficiency of a bound electron is an important result for understanding why the numerical value of opacity varies strongly in different regions of the Sun. In any region where there is still an appreciable fraction of electrons bound to abundant nuclei, the opacity may be much larger than elsewhere in the star. This fact plays a key role in determining the internal structure of the Sun: in certain regions where hydrogen retains its electron, or where helium retains at least one of its electrons, the passage of photons may be rendered so difficult that radiation can no longer serve as an effective method of transporting energy. (We will explore the consequences of this in Chapter 6.)

Now that we know the cross-section at the Lyman edge, we can calculate (roughly) the opacity for all photons in the Lyman continuum. In a gas consisting only of hydrogen, the number of H atoms per gram is $1/m_H \approx 6 \times 10^{23}$. Multiplying this by $\sigma_{912}$, we find that the opacity at the Lyman edge is of order $\kappa_L \approx 3.6 \times 10^6$ cm$^2$ gm$^{-1}$: if we were to include all the photons in the Lyman continuum (where the intensity falls off as $\lambda^3$), we would increase the value of $\kappa_\lambda$, but probably not by much. We shall return to the value of $\kappa_L$ in Section 3.6.4.

Since we are dealing with the Lyman continuum, should we also include the opacity due to the Lyman-$\alpha$ line? After all, the peak of Lyman-$\alpha$ has an opacity that is about 1000 times larger than the opacity at the Lyman edge. However, recall that the overall opacity will involve the convolution of a spectral range and an atomic opacity: photons in the Lyman continuum contribute to the overall opacity over a range of hundreds of Å, whereas photons with large opacity near the peak of Ly-$\alpha$ occupy no more than a fraction of an Å. It seems likely that the Lyman continuum opacity will dominate.

We have referred to ways in which photons can lose/gain energy by interacting with atoms via bound-bound and bound-free transitions. There is a third class of interactions that also allow loss or gain of photon energy: these are "free-free" transitions. A free electron in a plasma can be thought of as being in an orbit (admittedly unbound) around a distant proton. If a passing photon can cause that electron to move farther away from, or nearer to, the distant proton, then the photon has caused the electron essentially to make a transition into a different (again unbound) orbit. If the total energy of the new orbit is greater than before, then the electron has gained energy from the photon: the photon experiences this interaction as "free-free" opacity.

### 3.3.2 Absorption from Excited States: Dependence on Wavelength and $T$

So far, we have considered hydrogen atoms at "low" temperatures. In such a case, there is essentially no interaction with photons at wavelengths longward of (roughly) 1500 Å. However, in a medium that is sufficiently hot, absorption at longer wavelengths becomes possible. Thermal excitation allows some electrons to populate excited levels at a more or less significant rate. As far as photon absorption is concerned, each excited level displays at a certain wavelength its own "edge" where a free-bound transition can occur. Each of these "edges" has similar characteristics to those of the Lyman "edge". That is, photons on the longward side of the "edge" pass essentially freely through the gas, but photons lying shortward of the "edge" can be effectively absorbed. The "edges" corresponding to the energy levels with principal quantum numbers $n$ lie at wavelengths $\lambda_n$(Å) $= 912n^2$. Two of these "edges" are of interest for the Sun because they lie in a part of the spectrum where the Sun emits much of its power: $\lambda_2 = 3648$ Å and $\lambda_3 = 8208$ Å. These are the Balmer and Paschen edges, respectively.

The peak opacity of hydrogen gas *at the Lyman edge* is essentially independent of temperature, at least as long as hydrogen is not significantly ionized. But this is not the case for the other "edges": in those cases, the magnitude of the peak opacity (in cm$^2$ *per gram*) depends on the fraction of the atoms that have an electron in the corresponding excited state. In a medium where the total number density of hydrogen is $n_H$, the number density $n_i$ of H atoms with electrons in the $n = i$ level is related to the number density in the ground state ($n_1$) by a Boltzmann factor (which we have already encountered in deriving Equation 2.2 earlier):

$$n_i = \frac{g_i}{g_1} n_1 \exp\left(\frac{-\Delta E_{i1}}{kT}\right) \quad (3.2)$$

The term $g_i$ is the statistical weight of energy level $i$: in a hydrogen atom, the bound levels have $g_i = 2i^2$. The quantity $\Delta E_{i1}$ is the energy difference between levels 1 and $i$. When the energy difference $\Delta E$ is expressed in units of eV, the exponential term is more conveniently written in the form $10^{-\theta \Delta E}$, where $\theta = 5040/T$.

# Toward a Model of the Sun

Consider, for example, the $n = 2$ level of hydrogen that can be ionized by photons below the Balmer edge (at $\lambda < 3648$ Å). For this level, $\Delta E_{21} = 10.2$ eV. If we consider by way of example a medium where the temperature $T = 10^4$ K, we find $n_2/n_1 = 2.89 \times 10^{-5}$. Thus, in a parcel of gas which contains 1 gm of hydrogen at $T = 10^4$ K, only one atom in roughly 35,000 is capable of absorbing photons at the Balmer edge. Now, in terms of quantum mechanics, the cross-section for a single H atom to undergo photoionization from $n = 2$ by means of a photon of wavelength $\lambda_2$ is not greatly different from that for photoionization from $n = 1$ by means of a photon of wavelength $\lambda_1$. As a result, when we convert to the absorption cross-section *per gram of material* (= opacity), the peak opacity $\kappa_B$ at the Balmer edge at $T = 10^4$ K is no more than $\zeta_{BL} = 2.89 \times 10^{-5}$ times the peak opacity $\kappa_L$ at the Lyman edge.

Because of the exponential factor in Equation 3.2, the magnitude of the reduction factor $\zeta_{BL}$ becomes rapidly *smaller* at *lower* temperatures. For example, in the lower photosphere, where $T = 6000$ K, $\zeta_{BL}$ is of order $10^{-8}$, while in the upper photosphere, where $T = 4900$ K, $\zeta_{BL}$ falls to $10^{-10}$. Conversely, if something causes the local temperature to *increase*, the exponential factor ensures that $\zeta_{BL}$ becomes rapidly larger: thus, if the local temperature can be raised (for example, by local heating by means of magnetic fields) from 6000 K to $10^4$ K, i.e., by less than a factor of two, then the factor $\zeta_{BL}$ will increase by a factor of more than 1000. This sensitivity of the population of the $n = 2$ level in hydrogen to temperature will be useful when we discuss the properties of small solar features called "spicules" in Section 15.4.

Since the opacity $\kappa_\lambda$ for a medium consisting of pure hydrogen has the value $\approx 3.6 \times 10^6$ cm$^2$ gm$^{-1}$ at the Lyman peak, the numerical value of opacity in the solar photosphere at the Balmer edge $\kappa_B$ does not exceed 0.036 cm$^2$ gm$^{-1}$. On the redward side of the edge, i.e., at wavelengths longer than 3648 Å, $\kappa_B$ is zero.

Another bound level of hydrogen relevant in a discussion of photons at visible wavelengths is the $n = 3$ level, which can be ionized by photons at the Paschen edge. For this level, $\Delta E = 12.1$ eV. As a result, in the lower photosphere, where $T = 6000$ K, $n_3/n_1 = 6 \times 10^{-10}$. As a result, the Paschen peak opacity is 10 to 20 times smaller than the Balmer peak opacity in the lower solar photosphere. Thus, $\kappa_P$ does not exceed 0.0036 cm$^2$ gm$^{-1}$ at wavelengths near 8200 Å. Applying the $\lambda^3$ law for a free-bound continuum, we see that at wavelengths close to 4000 Å, the Paschen continuum opacity in the photosphere does not exceed 0.0004 cm$^2$ gm$^{-1}$.

As a result, when we consider the "visible spectrum" of the Sun, which reaches its peak intensity between wavelengths of $\lambda \approx 5000$–6000 Å, the opacity due to atomic hydrogen $\kappa_v$ ranges from about 0.0004 to at most 0.002 cm$^2$ gm$^{-1}$. (At these wavelengths, the Lyman absorption is essentially zero.) Later, we shall see (Chapter 5, Section 5.1) that the mass column density in the photosphere (i.e., the mass of a column with horizontal cross-section 1 cm$^2$ extending upwards from the photosphere to "infinity") is roughly $d_c \approx 4$ gm cm$^{-2}$. Multiplying $\kappa_v$ and $d_c$, we find that absorption by atomic hydrogen contributes an optical depth $\tau_v = \kappa_v d_c$ of no more than 0.008 in the photosphere. Since by definition the photosphere is the region where the optical depth is of order unity, it appears that atomic hydrogen is *not* a significant contributor to the optical depth at visible wavelengths in the solar photosphere. We could repeat the exercise for atomic helium and arrive at a similar conclusion. And this is true despite the fact that hydrogen and helium contribute by far the dominant percentage (~99%) of atoms in the Sun's atmosphere. The photons emerging from the solar interior, with wavelengths predominantly in the range 5000–6000 Å, i.e., with energies predominantly in the range 2–2.5 eV, undergo only very slight interactions with H and He in the photosphere of the Sun. The reason for this behavior has to do with the details of atomic structure: the lowest energy levels of both H and He simply lie "in the wrong places" (as regards energy) to be able to interact effectively with solar photons that have dominant energies of 2–2.5 eV.

So what is it that provides most of the absorption of visible light in the Sun's photosphere? It turns out to be an unusual "atom/ion".

## 3.4 ATOMIC PHYSICS: (II) OPACITY DUE TO NEGATIVE HYDROGEN IONS

The principal absorber in the solar photosphere at visible wavelengths is the *negative hydrogen ion*, i.e., a hydrogen atom with an extra electron *attached*. The standard hydrogen atom consists of one electron and one proton bound (by means of an attractive central force) in a stable arrangement with an infinity of bound energy levels. But there also exists the possibility that, if free electrons are available in the surrounding medium, an extra electron can be added without the system necessarily being unstable. In essence, the two electrons in an $H^-$ ion arrange themselves (because of Coulomb repulsion) to remain on opposite sides of the proton, as far away from each other as possible. In this situation, the force acting on one of the electrons is no longer central and is no longer purely attractive. As a result, when a detailed quantum mechanical treatment is applied, it turns out that there is no longer an infinite set of bound levels. But a bound state *does* exist. Just one. The bound level lies at an energy of $E(H^-) = -0.754$ eV: this energy is more than an order of magnitude smaller (in absolute magnitude) than the energy (−13.61 eV) of the lowest bound level of the hydrogen atom.

Photons with the capacity to excite a free-bound transition in $H^-$ (removing one of the electrons and leaving a neutral hydrogen atom) have a wavelength $\lambda < \lambda(H^-) = 16{,}450$ Å. In contrast to the sharp edge that occurs in the case of free-bound transitions in the H atom, photoionization of the $H^-$ion shows a much more gradual wavelength dependence: the cross-section $\sigma_\lambda$ rises from zero at 16,450 Å to a maximum at wavelengths around $\lambda_{max} \approx 8500$ Å, i.e., at energies of about $2E(H^-)$. Experimental measurements of photons detaching the extra electron from $H^-$ show (Smith and Burch 1959) that the photo-detachment cross-section has a broad peak (extending over several thousand Å in wavelength) with a numerical value $\sigma_\lambda(max) = 4.5 \times 10^{-17}$ cm$^2$. This value is almost an order of magnitude larger than the maximum Lyman continuum cross-section for atomic hydrogen. As a result, the absorption due to $H^-$ turns out to be by no means a negligible process in the atmosphere of the Sun. On the redward side of $\lambda_{max}$, the value of $\sigma_\lambda$ falls to one-half of its maximum value at $\lambda \approx 1.3$ µm. On the blueward side, $\sigma_\lambda$ falls to $0.5\sigma_\lambda(max)$ at $\lambda \approx 0.4$ µm.

Another source of continuous opacity due to $H^-$ arises when free electrons pass by a "free" hydrogen atom. This "free-free" process (see Section 3.3.1) contributes opacity that is relatively small at visible wavelengths, but which increases monotonically toward longer wavelengths as $\lambda^2$.

The total opacity due to $H^-$ is the sum of the bound-free and free-free processes (Chandrasekhar and Breen 1946). The minimum opacity due to $H^-$ occurs at $\lambda \approx 1.6$ µm, where the free-bound process has its "edge". This minimum in $H^-$ opacity is a useful tool for observers who wish to probe deeper than the nominal photosphere where $\tau(5000$ Å$) = 1$. For example, in the photosphere, observations at 1.6 µm from the International Space Station (Meftah et al. 2020) reveal brightness temperatures close to 6400 K, clearly indicating penetration down to layers of gas lying deeper (and therefore hotter by some 400 K) than the gas at $\tau=0.9$ (see Table 5.3). Moreover, in sunspot umbrae, observations at $\lambda \approx 1.6$ µm reveal the conditions at depths that lie some 50 km deeper than the nominal photosphere (Andic et al. 2011). And in studies of simulated solar convective flows, speeds can be extracted reliably at 1.6 µm down to optical depths of $\tau(5000$ Å$) \approx 3$, but data from $\tau(5000$ Å$) \approx 10$ is too noisy to be reliable (see Section 6.3).

Across the visible portion of the solar spectrum, from about 4000 Å to about 7000 Å, where most of the solar energy flux emerges, it has been found that the free-bound opacity due to $H^-$ does not vary by more than a factor of two. There are no large discontinuities in opacity throughout the visible spectrum. As a result, for most of the photons passing through the solar photosphere (i.e., those in the visible spectrum, where the Sun emits most of its radiant energy), the opacity is "nearly" independent of wavelength. Because of this, the *gray* approximation (i.e., wavelength-independent opacity) turns out to be "not too bad" when we consider the solar photosphere. This could explain why the gray atmosphere solution ($S(\tau) = a + b\tau$) fits the solar limb darkening ($I(\mu) = a + b\mu$) quite well at visible wavelengths.

We shall find (Section 5.1) that the column density of atomic particles $N$ above the photosphere is about $2 \times 10^{24}$ cm$^{-2}$. Most of this column is composed of hydrogen *atoms*. But a (small) fraction of the atomic particles are negative H ions. Let the ratio of the abundance of $H^-$ ions to the abundance of H atoms in the photosphere be $\phi$. Then with a cross-section that does not depart significantly from $\sigma_\lambda(\max) = 4.5 \times 10^{-17}$ cm$^2$, the optical depth above the photosphere due to $H^-$ is $\tau \approx 9 \times 10^7 \phi$. This can attain values of order unity if the ratio $\phi$ is close to $10^{-8}$.

Is it plausible that the fractional number of H atoms that will capture a second electron in the solar atmosphere is of order $10^{-8}$? To answer that question quantitatively, we need to know how ionization equilibrium depends on temperature and pressure. For this, we need to consider the Saha equation, which will be the topic of Chapter 4. For now, we note that in the solar photosphere, we shall find (Chapter 4, Section 4.6) that it is entirely consistent with local temperature and pressure that $\phi$ should be of order $10^{-8}$. As a result, there are indeed enough $H^-$ ions in the solar atmosphere to cause the optical depth at visible wavelengths to be of order unity in the photosphere.

It is essentially the formation of $H^-$ ions that limits how deeply we can observe into the Sun at visible wavelengths. The favorable formation of $H^-$ in the solar photosphere requires the presence of both hydrogen atoms and free electrons. In the photosphere, free electrons are available mainly from the most abundant "metals" that have low ionization potentials (Mg, Si, S, and Fe). Together, these provide electrons with an abundance of a few times $10^{-4}$ relative to the abundance of H.

## 3.5 ATOMIC PHYSICS: (III) OPACITY DUE TO HELIUM ATOMS AND IONS

The spectrum of singly ionized helium (He II) is analogous to that of neutral atomic hydrogen (H I), except that all wavelengths are reduced by a factor of four. Thus, the He II-Lyman edge lies at $\lambda = 228$ Å.

For neutral helium (He I), with an ionization potential of 24.6 eV, the edge at which absorption is maximum lies at $\lambda = 504$ Å. The maximum absorption cross-section at the He I edge is $8 \times 10^{-18}$ cm$^2$, comparable to the maximum cross-section at the Lyman edge of H.

Is there any observational evidence for the existence of the various "edges" that we have mentioned in this section? Although our main emphasis in this section has been on *absorption* of radiation, there are also physical conditions in which the enhanced optical depth at the shortward side of a bound-free edge results in enhanced *emission* there. (Reasons for this can be found in Equations 2.17 and 2.18.) In the Sun, transient sources of hot dense plasma occur during events known as "flares" (see Section 17.19): during a flare, the spectrum of the Sun can show emission continua with clearly visible "bound-free edges" at 912 Å (due to H I) and at 504 Å (due to He I): see Figure 3.3. And during a large solar flare, the He II edge at 228 Å, as well as free-free continuum emission at wavelengths as short as 60 Å have been observed by the Extreme-ultraviolet Variability Experiment (EVE) instrument on SDO (Milligan and McElroy 2013). As well as the "edges" that are apparent in Figure 3.3, it is also worth noting that the flare Lyman continuum (which can be traced shortward of 912 Å down to perhaps 700 Å) can be fitted well in many flares by a Planck function with a temperature ranging from 9000 to 17,000 K in different flares (Machado et al. 2018): this Lyman continuum is emitted by a relatively thin layer (<100 km thick) where electron densities are large (>$10^{13}$ cm$^{-3}$).

## 3.6 ASTROPHYSICS: THE ROSSELAND MEAN OPACITY

Once we know how opacity varies as a function of wavelength ($\kappa_\lambda$) or as a function of frequency ($\kappa_\nu$), a mean opacity can in principle be obtained simply by taking an arithmetic average over all wavelengths or frequencies. But that would not be especially useful in our attempts to determine how photons make their way through an atmosphere. The derivation of a relevant "mean opacity" should also incorporate somehow the shape of the spectrum of radiation.

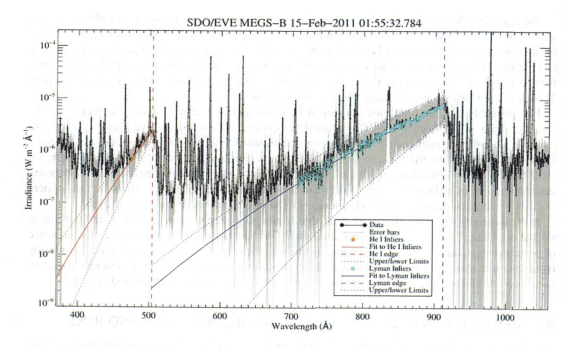

**FIGURE 3.3** Emission from a solar flare in the wavelength range 400–1000 Å. Data obtained by EVE on SDO. Note the Lyman edge at λ = 912 Å (marked by a vertical dashed blue line) and the He I edge at 504 Å (marked by a vertical dashed orange line) (Milligan 2015; used with permission from Springer.)

The most common method for calculating a mean opacity relevant to the passage of radiation in the deep interior of a star is called the Rosseland mean $\kappa_R$, defined by:

$$\frac{1}{\kappa_R}\int \frac{dB_\nu}{dT} d\nu = \int \frac{1}{\kappa_\nu}\frac{dB_\nu}{dT} d\nu \qquad (3.3)$$

In deriving Equation 3.3, it is assumed that the shape of the spectrum is related to the local Planck function $B_\nu$. The appearance in Equation 3.3 of (a) $1/\kappa_\nu$ and (b) the first derivative of $B_\nu$ with respect to temperature can be traced ultimately to the fact that the RTE provides an expression for $1/\kappa_\nu$ times the first spatial derivative of the intensity $(dI_\nu/dx)$: in a given atmosphere, the spatial gradient can be converted to a derivative with respect to temperature.

The Rosseland mean as defined in Equation 3.3 is a "transparency mean": as far as opacity is concerned, the right-hand side of Equation 3.3 gives maximal weight to regions in the spectrum where the opacity is *smallest*. In the atmosphere of the Sun, photons will tend to "leak out" through such regions. Also, Equation 3.3 weighs more heavily those parts of the spectrum where $dB_\nu/dT$ is larger, i.e., at wavelengths that are shorter than the Wien maximum (see Section 2.1).

The units of $\kappa_R$ are the same as those of $\kappa_\nu$, i.e., cm$^2$ gm$^{-1}$.

For a medium containing a certain mixture of elements, the Rosseland mean is calculated by first determining the frequency-dependence of the opacity due to each atomic species in the mixture. At each frequency, the total opacity is obtained by summing the contributions before performing the integral in Equation 3.3. For absorption from the respective ground states of the various types of atoms in the mixture, there is no dependence on temperature. But absorption due to excited states introduces significant temperature dependence at longer wavelengths. Further temperature

# Toward a Model of the Sun

dependence enters as a result of the Planck function that enters in Equation 3.3. As a result, it is not surprising that $\kappa_R$ varies significantly with temperature.

To illustrate how $\kappa_R$ depends on temperature, we show in Figure 3.4 the results for a gas consisting of H and He (with mass fractions $X = 0.7$ for H and $Y = 0.28$ for He) plus all heavier atoms (referred to by astronomers as "metals": these have mass fraction $Z = 0.02$). Using logarithms to base 10, we see that the range of temperatures along the abscissa extends from $T = 1000$ K ($\log T = 3$) (a lower temperature than is observed anywhere in the Sun) to $\log T = 8$ (a temperature that is higher than anywhere in the Sun). For the sake of compactness and to demonstrate that $\kappa_R$ also depends on the density of the gas, each $\kappa_R$ curve in Figure 3.4 is labeled with a value of $\log(R)$ where the parameter $R = \rho/T_6^3$ is a combination of density $\rho$ (in units of gm cm$^{-3}$) and $T_6$, the temperature in units of $10^6$ K.

How can we understand the behavior of $\kappa_R$ appearing in Figure 3.4? Let us examine certain limiting behaviors. Two aspects of the figure stand out: (i) at the highest temperatures, all curves converge to a single finite value (especially at low densities); (ii) at the lowest temperatures, all curves plunge steeply toward zero opacity.

### 3.6.1 Limit of Low Density and/or High $T$: Electron Scattering

In the limit of very low density, e.g., when the value of $\log(R)$ has its smallest value ($= -6$ in Figure 3.4), and also in the limit of the highest temperatures, H and He are essentially completely ionized. There are essentially no bound states available for electrons to occupy: all electrons are free particles, and electron scattering is the sole source of photon interaction. Using the Thomson cross-section $\sigma_e = 6.6 \times 10^{-25}$ cm$^2$ per electron and a mean molecular weight of

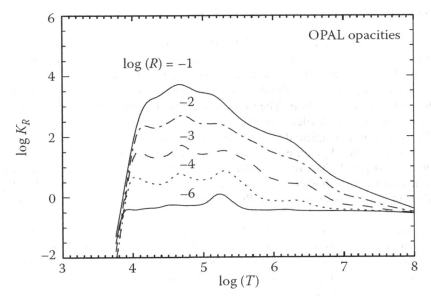

**FIGURE 3.4** Rosseland mean opacity ($\kappa_R$, in units of cm$^2$ gm$^{-1}$) as a function of the logarithm (to base 10) of the temperature $T$ (deg K) for a number of densities. The parameter on each curve is the quantity $R = \rho/T_6^3$, where $\rho$ is density in units of gm cm$^{-3}$, and $T_6$ is temperature in millions of deg K. The data plotted here are taken from a table of OPAL opacities for near-solar composition (hydrogen mass fraction $X = 0.7$, helium mass fraction $Y = 0.28$, metals mass fraction $Z = 0.02$). (More extensive tables of OPAL opacities are publicly available at the Lawrence Livermore website: https://opalopacity.llnl.gov/existing.html)

1.3 (corresponding to $n_{eg} \approx 5 \times 10^{23}$ electrons per gram), we expect that in conditions of complete ionization $\kappa_R = n_{eg} \sigma_e \approx 0.3$ cm$^2$ gm$^{-1}$. Thus, $\log(\kappa_R) = -0.5$ when ionization is complete. Indeed, the curve labeled −6 in Figure 3.4 lies close to this value at all temperatures except for the very coolest (less than 6000 K).

### 3.6.2 Low $T$ Limit

As $T$ falls below $10^4$ K, all curves in Figure 3.4 fall steeply to very small values. The reason for this behavior is readily understood: as $T \to 0$, the gas becomes electrically neutral (no free electrons), and all electrons are bound in atoms of H I or He I, where they fall down to the lowest energies (i.e., the ground states). The only photons that can be absorbed effectively require wavelengths shorter than 912 Å and 228 Å, respectively. But at temperatures below $10^4$ K, the Wien maximum lies at $\lambda_{max} > 2880$ Å, and the flux of photons at $\lambda \leq 912$ Å is exponentially small. The lower the temperature, the more drastic is the exponential reduction in photon flux at $\lambda \leq 912$ Å. For example, in conditions that apply to the upper photosphere, i.e., $T = 4900$ K, the Rosseland mean opacity falls to values of order 0.001 cm$^2$ gm$^{-1}$: such a low opacity lies below the lower boundary of the plot in Figure 3.4.

### 3.6.3 Higher Density: Free-Bound Absorptions

At a fixed temperature, increasing density in Figure 3.4 corresponds to an increase in $\log(R)$. At the highest temperatures in Figure 3.4 ($\log(T) > 7.5$), density effects are minimal: at such high temperatures ionization is almost complete at all densities relevant to the Sun. As a result, electron scattering dominates the opacity, as shown by the curves in Figure 3.4 converging towards $\kappa_R = 0.3$ cm$^2$ gm$^{-1}$. However, at intermediate temperatures, increasing the density leads to increased recombination rates in the gas. In such conditions, the number of atoms with populated bound states grows, and this causes significant increases in opacity, especially at the various bound-free "edges".

For the lowest density curve in Figure 3.4 (labeled −6), the most significant departure from electron scattering (i.e., departure from $\log(\kappa_R) = -0.5$) is the "bump" at temperatures $\log(T)$ in the range 5.0–5.5. In this range, Wien's law (see Section 2.1) indicates that the peak of the Planck function lies at wavelengths of 100–300 Å. It is noteworthy that this range of wavelengths overlaps with the He II-Lyman edge at 228 Å. The overlap of the peak in the spectrum with the large peak in atomic opacity at the Lyman edge is conducive to creating enhanced opacity. Of course, the ionization of He II is almost complete due to the low densities, so there are relatively few He II ions available to contribute their Lyman-edge absorptions: this explains why the "bump" reaches maximum opacities ($\log(\kappa_R) = 0$) that are larger than electron scattering opacity, although not by much. However, when we examine higher densities, e.g., on the curves labeled −4 and −3 in Figure 3.4, larger numbers of He II ions survive at $\log(T) = 5.0$–5.5, and the "bumps" in opacity grow to larger amplitudes.

There is a second "bump" on the curve labeled −6 in the range $\log(T) = 4.5$–4.75: at such temperatures, Wien's law predicts a peak in the Planck function around $\lambda \approx 500$–900 Å. Such photons have energies approaching those required to cause bound-free absorptions from He I. This peak also becomes more prominent at higher densities, where the number of He I atoms per gram (at a given temperature) increases significantly.

A third "bump" in $\log(\kappa_R)$, most prominent on the curves in Figure 3.4 labeled −4 and −3, occurs at $\log(T) \approx 4.0$. At such temperatures, excited states in H I are rapidly being populated: the populations grow exponentially with increasing $T$. Each of these states contributes absorption due to its free-bound "edge". In particular the Balmer "edge" overlaps with the peak of the blackbody function at $\log(T) \approx 4.0$. The exponential growth in bound populations causes $\kappa_R$ to increase rapidly with

# Toward a Model of the Sun

increasing $T$. However, as $\log(T) \to 4.0$, hydrogen is also ionizing rapidly. The competition between excitation and ionization leads to a rather narrow peak in the opacity curves in Figure 3.4.

### 3.6.4 MAGNITUDE OF THE LARGEST OPACITY

As density increases, the ranges of temperatures at which significant ionizations of H I, He I, and He II occur begin to overlap more and more. The various relatively narrow individual "bumps" in Figure 3.4 that are apparent at relatively low densities tend to merge into a single broad "bump", although "shoulders" are still apparent on both sides of the broad "bump". The broad "bump" lies at temperatures of $\log(T) = 4.5$–5.

For the density range entering into the data shown in Figure 3.4, the largest values of opacity have numerical values in the range $\log(\kappa_R) = 3$–4.

What is the maximum density for the results in Figure 3.4? At a temperature of $\log(T) = 4.75$ (i.e., $T_6 = 0.06$), the maximum density, corresponding to the curve $\log(R) = -1$, is of order $2 \times 10^{-5}$ gm cm$^{-3}$.

Other investigations (e.g., Ezer and Cameron 1963) have extended the calculation of Rosseland mean opacity to higher densities. With $\rho = 10^{-3}$ gm cm$^{-3}$, the peak in opacity at $\log(T_{max}) = 4.5$–4.75 has a value $\log(\kappa_R) \approx 5$. If we attempt a rough extrapolation of the results of Ezer and Cameron, it appears that the peak numerical value of $\kappa_R$ might be as large as $10^6$ cm$^2$ gm$^{-1}$. Recalling that the (wavelength-dependent) opacity at the Lyman edge has the numerical value $\kappa_\lambda(\text{max}) \approx 3.6 \times 10^6$ cm$^2$ gm$^{-1}$ (see Section 3.3.1), it is difficult to imagine how the maximum permissible value of $\kappa_R$ in a mixture of elements where H and He are dominant could be much larger than $\kappa_\lambda$ (max). If a blackbody curve were to be matched optimally in wavelength such that its peak overlapped with the Lyman edge, then we might expect to see $\kappa_R$ values approaching $\kappa_\lambda$ (max): the temperature required for such matching would be such that $\lambda_{max} = 0.288(\text{cm})/T_{max} = 912$ Å. This leads to $\log(T_{max}) = 4.5$. This is indeed consistent with the peak in Figure 3.4.

The results in Figure 3.4 refer to a gas where H and He are the principal constituents. There are a few "metals" included in the calculations, and these alter the H/He opacity curves slightly, giving rise to some "bumps" appearing at $\log(T) = 5$–6. The reason that the changes are only slight has to do with the relatively small abundances of the "metals" compared to H and He: even the most abundant "metals" have fractional abundances of no more than 0.001 times H (by number).

## 3.7 POWER-LAW APPROXIMATIONS TO THE ROSSELAND MEAN OPACITY

For future reference, we note that it is at times convenient to fit the Rosseland mean opacity to power laws of temperature and density. This allows analytic solutions to be extracted for certain problems. Different power laws apply in different parameter regimes.

In the earliest attempts to model the deep interior of the Sun, at temperatures in excess of roughly $10^6$ K, certain functional forms were examined in order to describe how the opacity depends on temperature and density. In this temperature range, the dominant constituents of the Sun are almost completely ionized. The fractional abundances of incompletely ionized atoms are becoming rapidly smaller as the temperature increases. As a result, the strongest contributors to opacity (bound electrons) are becoming progressively scarcer in the gas. This leads to a rapid *decrease* in $\kappa_R$ as the temperature *increases*. Valid for both bound-free and free-free transitions, certain approximations lead to the functional form $\kappa_R = \kappa_o \rho/T^{3.5}$. We shall find this form (referred to as the Kramers' opacity law in honor of Hendrik Kramers who first derived the formula) useful later (in Section 8.3) when we model the radiative interior of the Sun.

In a different limit, at temperatures below $10^4$ K, the temperature sensitivity is very different. Starting at the lowest temperatures, at say a few thousand degrees, essentially all H and He atoms are in their ground states, while most photons in the local Planck function are at long wavelengths. If $T =$ (say) 3000 K, the Wien peak lies at $\lambda$(max) $\approx$ 1μm, where photons have mean energies of about 1 eV. Photons with such energies are much too small to raise electrons out of the ground state of either H I or He I. As a result, most of the photons (i.e., those near the Wien peak) simply stream through the gas essentially unimpeded, and the opacity falls off to very low levels. However, increasing the temperature has two effects: (i) it populates excited states of H and He, creating opportunities for longer wavelength photons to be absorbed; (ii) the peak in the blackbody spectrum moves toward shorter wavelengths. Both effects combine to cause opacity to *increase* rapidly as the temperature *increases*. The opacity also increases as the density increases, although the sensitivity to density is much less pronounced than the temperature-sensitivity. A power law fit to opacities in this temperature regime suggests that $\kappa = \kappa_1 \rho^a T^b$ could provide a reasonable zeroth order fit. In order to determine what the power-law indices $a$ and $b$ are, we refer to a specific table of opacities calculated by Robert Kurucz (1992). Some of the opacities from that table will be used later (in Chapter 5) to calculate a model of the photosphere. For present purposes, we note that Kurucz lists log ($\kappa_R$) for a series of log(temp) and log(press). (A portion of this table appears in Chapter 5 in Table 5.1.). In Kurucz's complete table, at each value of $\log(T)$ and $\log(p)$, the local density is also listed. Using the values of $\log(\kappa)$ tabulated by Kurucz at all values of $\log(T) \leq 4.0$, we have obtained least squares fits for the coefficients in the relationship $\log(\kappa) = c + a \log(\rho) + b \log(T)$. We find $a = 0.343$, $b = 9.0583$, and $c = -31.97$. The steeply rising dependence on increasing temperature is noteworthy. We shall find this formulation of the opacity useful in Chapter 15 when we model the rise in temperature between photosphere and chromosphere.

## 3.8 NARROW BAND OPACITY: ABSORPTION LINES IN THE SPECTRUM

So far, the opacity we have discussed occurs mainly over broad ("continuum") regions of the spectrum. For example, the negative hydrogen ion contributes significant opacity at wavelengths from as short as 4000 Å to as long as 1–2 μm. And bound-free absorption from the $n = 2$ level of hydrogen contributes opacity at all wavelengths shortward of 3648 Å. These are truly "broadband" sources of opacity, and they help to determine, in conjunction with the atmospheric temperature profile, the overall "rainbow" of the solar spectrum, extending from peak intensity in the yellow-green region toward the red and toward the violet, where the intensity gradually fades from human eyesight.

However, when one is presented with a spectrum of the Sun (see Figure 3.5), the first thing that attracts one's attention is the presence of a multitude of more-or-less narrow dark "lines" distributed across the entire range of visible wavelengths. These are the features that Joseph von Fraunhofer first discovered in 1814 when he fed sunlight into the entrance slit of a spectroscope. Fraunhofer drew up a list of the strongest lines that he could discern in the spectrum of the Sun and labeled them with a series of *upper* case letters from A to K (in order of increasing frequency). A list of weaker lines was subsequently labeled with *lower* case letters, starting with a in the red and using subsequent letters as the wavelength became shorter. Although Fraunhofer did not identify the origin of his lines, many of his labels persist in common use to this day: e.g., Fraunhofer's D line near 5900 Å is now known to be a close doublet (prominently visible in the yellow-orange region [Figure 3.5], about 1/3 the way down from the top of the figure): the lines are referred to today as the $D_1$ and $D_2$ lines and are now known to be due to neutral sodium. And Fraunhofer's K line at 3934 Å, the strongest line in the solar spectrum, is now known to arise from calcium atoms that have lost one of their electrons. It is now known that the features labeled A, B, and a by Fraunhofer (at wavelengths near 7590, 6870, and 6280 Å, respectively) have nothing to do with the Sun: instead, they are caused by diatomic molecules of oxygen in the Earth's atmosphere. Fraunhofer's C line is a strong line in the red in Figure 3.5, about 1/6 the way from the top, and about 1/3 the way from the right-hand side: this line is now known to be due

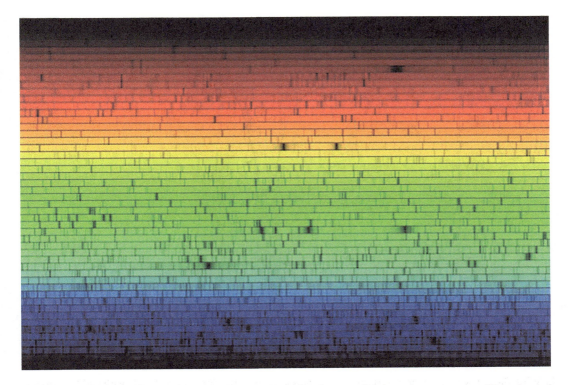

**FIGURE 3.5** Fraunhofer lines: dark narrow absorption features in the solar spectrum. (Downloaded from https://calgary.rasc.ca/redshift.htm. Credit: National Optical Astronomy Observatory/Association of Universities for Research in Astronomy/National Science Foundation. With permission.)

to hydrogen, and the line is called *Hα* (where α indicates that this is the first in a series of lines called the Balmer series).

The occurrence of the solar "Fraunhofer lines", so striking in their darkness against the backdrop of the rainbow, indicates the presence in the solar atmosphere of atoms and ions of particular elements in particular stages of excitation. Some astronomers (including, famously, Cecilia Payne, see Section 4.2) have put a lot of work into using the strength of the absorption lines to derive the relative abundances of the elements in the chemical mix that makes up the Sun. We choose to leave a description of those inquiries to other authors in more advanced texts (e.g., Aller 1953). In the present book, instead, we will look to the lines to tell us about other physical properties of the Sun: the properties of most interest to us here are the *motions* of the gas and the *magnetic fields* that permeate the gas in certain locations in the Sun.

### 3.8.1 CHARACTERIZING THE PROPERTIES OF ABSORPTION LINES

It is helpful to consider quantitative measures of absorption lines. To do this, we examine the "line profile", i.e., how does the radiant intensity (or flux) vary as a function of wavelength?

In each line, when one plots the radiation flux $F_\lambda$ as a function of wavelength (see Figure 3.6), one starts far from line center on (say) the blueward side, with an intensity essentially equal to the continuum $F_c$. For all lines, if one chooses a wavelength that is far enough from line center, the

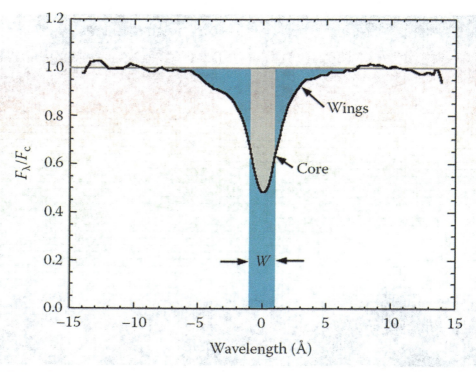

**FIGURE 3.6** The shape of an absorption line in the spectrum of the Sun. The rectangle with width $W$ has an area equal to the total area absorbed by the line. $W$ is referred to as the equivalent width of the absorption line. (From http://web.njit.edu/~gary/321/equiv_width.gif. With permission.)

ratio $F_\lambda/F_c$ approaches unity (apart from some "noise" associated with other spectral lines). As wavelength increases, and one enters into the line, the intensity decreases more or less rapidly: this decrease gives rise to what are called the "wings" of the line (Figure 3.6). At a certain wavelength, the radiation flux reaches a minimum value. This location, where the depth of the line is maximum, is defined to be line center. Each line arises when an electron makes a transition from one energy level to another. Quantum mechanics allows physicists to calculate what the energy levels are, and therefore, each line can be assigned a precise "rest wavelength" $\lambda_o$. For example, a neutral carbon atom can emit a line at $\lambda_o = 5380.3308$ Å, and a neutral iron atom can emit a line at $\lambda_o = 5250.2084$ Å. If the carbon and iron atoms are emitting these lines in a stationary laboratory experiment, the wavelengths of the line centers will be found to lie at the aforementioned wavelengths.

The precision with which the wavelength $\lambda_o$ is known for many lines provides us with a valuable tool in studying the atmosphere of the Sun: it allows us to measure the *velocity* of the atoms producing that line. To do this, we use an effect discovered by Christian Doppler in 1842: if the emitting atoms are moving systematically towards or away from the observer at a speed of $V$, then the center of the line will be shifted in wavelength by an amount $\Delta\lambda_o$ which is given by

$$\frac{\Delta\lambda_o}{\lambda_o} = \frac{V}{c} \tag{3.4}$$

If the atoms are moving *away* from the observer, $V$ is assigned a positive value and $\Delta\lambda_o$ is positive: the line is shifted towards a longer wavelength, i.e., the line experiences a redshift. If the atoms are moving *towards* the observer, the line is shifted towards the blue. In a gas with finite temperature $T$, thermal motions of atoms with mass $m_a = \mu_a m_H$ (where $m_H = 1.66 \times 10^{-24}$ gm is the mass of a hydrogen atom) have a mean speed $V_t \approx \sqrt{(3kT/m_a)}$. These motions cause one-half of the atoms to move towards the observer at any time, while the other half are moving away: the accompanying "thermal Doppler" shifts, $\Delta\lambda_D = \pm \lambda_o V_t/c$, have the effect that a spectral line is said to be "thermally broadened". For example, in the photosphere of the Sun, where $T \approx 5800$ K, atoms of carbon ($\mu_a = 12$) have $V_t \approx 3$ km s$^{-1}$: therefore, a carbon line emitted with a wavelength of 5380 Å has a thermal width $\Delta\lambda_D \approx \pm 0.05$ Å. In similar conditions, an iron atom ($\mu_a = 56$) has $V_t \approx 1.4$ km s$^{-1}$. If thermal motions were the only motions being experienced by the gas in the photosphere of the Sun, then each line would be broadened solely by its thermal width. In such cases, the heavier atoms would be observed to have definitely smaller widths than would light atoms.

However, as we shall see (Chapter 6), the gas in the photosphere of the Sun does not undergo merely the random motions that are associated with heat: the gas is also subject to systematic organized flows in three dimensions due to turbulent convection, involving upflows and downflows at speeds that can be as large as $\pm 3$ km s$^{-1}$. Moreover, turbulent convective flows inevitably act as a source of sound waves in the atmosphere. Convective flows plus sound waves contributes to line broadening. A generic way to allow for these line broadenings when we approximate the transfer of radiation in a stellar atmosphere as a one-dimensional process is to introduce the concept of "microturbulence": this is assigned a mean speed of $\xi$ km s$^{-1}$ such that the total velocity broadening of a line due to *all* motions is $V = \sqrt{(V_t^2 + \xi^2)}$. For the Sun, analysis of a large sample of lines due to Fe I and Fe II (Pavlenko et al. 2012) indicates that the best fit to Fe I lines requires $\xi = 0.75$ km s$^{-1}$, while the best fit to Fe II lines requires $\xi = 1.5$ km s$^{-1}$. There is not necessarily a contradiction between these values of $\xi$: the Fe I lines may sample (on average) a higher region of the atmosphere where convection flows and/or sound waves are weaker, while the Fe II lines (which on average require higher temperatures) may be sampling deeper layers where convective flows and/or sound waves are better developed (see Section 3.8.3) The point is, absorption lines in the solar atmosphere are observed to be subject to finite broadening (with amplitudes of order 1 km s$^{-1}$) over and above thermal broadening.

Moreover, the concept of microturbulence is not useful merely in the photosphere: we shall also find that when we consider spectral lines formed in the corona (Section 17.6), those lines also exhibit broadening that is definitely larger than thermal motions alone can explain. In the coronal case, the excess broadening may be due to waves of a magnetic nature.

The depth at the center of a spectral line varies from one line to another: some lines are so weak that they dip to no more than a percent or so below the continuum: such weak lines, with central intensities of 0.99 times the local continuum, can be difficult to identify against the brightness of the continuum. At the other extreme, the strongest lines have depths in excess of 90%. In the center of such deep lines, the radiant intensity may amount to only a few percent of the continuum: e.g., in the $D_1$ and $D_2$ lines, the centerline intensities in the Sun are only 5.0% and 4.4% of the continuum (Waddell 1962). What is it that determines the depth of a line? It depends on the ratio $\eta = k_{\lambda o}/k_\lambda$ between the absorption coefficient $k_{\lambda o}$ at the line center and the absorption coefficient $k_\lambda$ in the local continuum. (A complicating factor is that in some lines, scattering of photons is more important than absorption. However, inclusion of scattering effects would take us too far afield in this "first course", so we ignore it here.) The value of $k_{\lambda o}$ is determined by the number of atoms along the line of sight, as well as by an atomic quantity called the "oscillator strength". The latter is a quantum mechanical parameter that depends on the probability that an electron will actually be able to make a transition between the lower and the upper energy levels of the line: the oscillator strength increases with the strength of the line. In the strongest lines in the visible part of the solar spectrum, the ratio $\eta = k_{\lambda o}/k_\lambda$ becomes especially large close to line center: e.g., the Ca II $K$ line has $\eta = 316$ at a wavelength interval $\Delta\lambda = 0.5$Å away from line center (e.g., Aller 1953), and $\eta \approx 2000$ at $\Delta\lambda = 0.2$Å. As we move even closer to line center, e.g., to values of $\Delta\lambda \leq 0.1$ Å, the value of $\eta$ increases by further factors of 10 or more (Aller 1953): at the center of the Ca II $K$ line, the line opacity may exceed

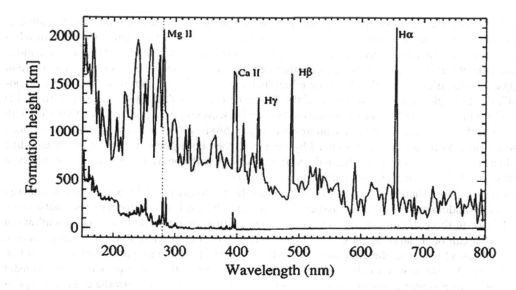

**FIGURE 3.7** Upper curve: heights of formation of spectral lines relative to the photosphere in the solar atmosphere. Lower curve: heights of formation in the continuum. (Thuillier et al. 2012; used with permission of Springer.)

the continuum opacity by a factor of order $10^4$. This result will be important when we discuss how best to observe the Sun in a region known as the chromosphere (see Chapter 15).

At line center, the optical depth $\tau_o$ in the line has its maximum value: as one moves way from line center, the line optical depth decreases, at first rapidly (out to a distance of $\Delta\lambda_D$ from line center), and then more slowly as one moves farther from line center (see Aller 1953, p. 256). At a certain distance from line center, the line no longer has any appreciable optical depth. At that point, the line profile merges smoothly back into the local continuum: this occurs at the outermost parts of the "wings" of the line in Figure 3.6.

### 3.8.2 Heights of Formation of Different Spectral Lines

Because of the variation in optical depth across the line profile, when we observe at line center, our line of sight into the solar atmosphere does *not* reach down all the way to the photosphere (where the continuum optical depth $\tau \approx 1$). Instead, our line of sight at the center of a line reaches in only down to a level where the optical depth *in the center of the line* $\tau_o$ is of order unity. Since $\tau_o$ can be larger than the local continuum optical depth $\tau$ by large factors, the gas we are "viewing" when we observe at line center does *not* lie at the same layer of the atmosphere where the continuum originates. When we observe at line center, we are seeing roughly a layer where the *continuum optical depth* $\tau$ is $\approx 1/\eta$. Depending on the value of $\eta$, the line center material may be situated hundreds of kilometers higher up in the atmosphere than the continuum level $\tau \approx 1$.

An example of the heights of formation of almost one thousand lines in the solar spectrum has been obtained by Thuillier et al. (2012): see Figure 3.7. These results were obtained by using a one-dimensional radiative transfer code (COSI) in spherical geometry to obtain non-LTE populations of the atomic levels that give rise to almost one thousand of the strongest lines in the solar spectrum. The formation heights of these lines range from <100 km in the red portion of the solar spectrum to about 2000 km in the resonance lines of Mg II at 2796 and 2805 Å and in the longest-wavelength line of the Balmer series, $H\alpha$, at 6563 Å. Among the lines that (in their core) probe gas at heights of

# Toward a Model of the Sun

1500 km or more, two are labeled in Figure 3.7: the Ca II K line at 3933 Å, and the second line in the Balmer series, $H\beta$, at 4861 Å. The third line in the Balmer series, $H\gamma$, at 4341 Å, probes the gas up to heights of about 1300 km in its core. As we shall see in Chapter 15, gas that lies at altitudes of 500 km or more above the solar photosphere has unusual properties, which sets it apart from the photosphere: gas at these altitudes belong to the "chromosphere". Thus, many of the lines in Figure 3.7 are formed (at least in their cores) in the chromosphere.

A more complicated study of the heights of formation at the center of several strong solar lines leads to the results shown in Figure 3.8 (Leenaarts et al. 2013). To obtain the results in the figure, the authors first computed a 3-D model of the solar atmosphere extending 24 Mm × 24 Mm in $x$ and $y$, and 16.8 Mm in $z$ (1 Mm = 1000 km). The 3-D chromosphere is highly dynamic: in some locations, the gas temperature reaches 20 kK at a height as low as 2 Mm, while in other locations, $T$ does not reach a value as high as 20 kK until the height is ≥3 Mm. Once the 3-D model is available, a 2-D slice from the computational results (in the $y$-$z$ plane) can be selected to solve the RTE and calculate the intensity in the core of several strong lines. This process provides a more careful RTE solution in a spectral line than the approach used by Thuillier et al. (2012) in Figure 3.7. The gray-black background in Figure 3.8 shows the temperature of the gas in the solar atmosphere up to a maximum value of 20 kK: the corrugations correspond to material elements that rise to different heights $z$ at different horizontal locations $y$. The features that rise to the greatest heights are related to solar structures known as "spicules" (see Section 15.6). The heights where the cores of four of the strongest spectral lines in the Sun have $\tau_o = 1$ are plotted with different colors. We see that the core of the Mg II k line (at $\lambda = 2796$ Å) is formed at heights that can be as large as 3–4 Mm above the continuum level ($z=0$): these heights are larger than the ~2 Mm suggested by a 1-D radiation code

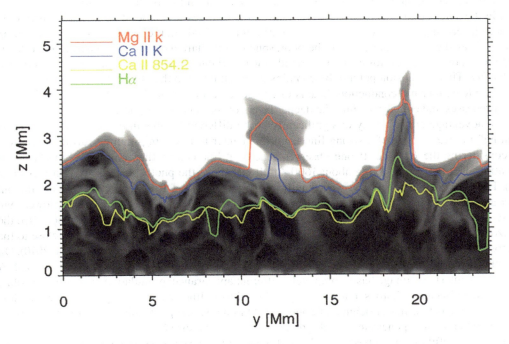

**FIGURE 3.8** A 2-D illustration of the height of formation in the *core of some of the strongest spectral lines* in the Sun. To obtain these results, fully 3-D computations of gas motions in the solar atmosphere were first obtained, and then radiative transfer was performed in various lines. The vertical scale has its zero point $z=0$ at the level where the average optical depth *in the continuum* is $\tau=1$. The corrugations are due to the turbulent conditions existing in the solar photosphere, corresponding roughly to wave crests of different heights on the ocean surface during a storm (Leenaarts et al. 2013; used with permission of J. Leenaarts).

(Figure 3.7). The core of the Ca II $K$ line (at $\lambda$ = 3934 Å) is formed at levels that are a few hundred km lower than the Mg II k line: as a result, the density of the gas where Ca II $K$ is formed is larger by a factor of 10 (or so) than the density of the gas where Mg II k is formed. The core of the $H\alpha$ line (at $\lambda$ = 6563 Å) is formed at heights of 1–2.5 Mm above $z$ = 0: this is not too different from what the 1-D RTE solution predicts (Figure 3.7). In contrast to these strong lines, we shall discuss a weak line in Section 3.8.3 where the core is formed at a height of only 0.04 Mm above $z$=0.

It is an amazing aspect of solar physics that if an observer decides to "step" the observing instrument in wavelength from the center of any particular spectral line out to the far wings, the data obtained at each "step" in wavelength are in effect "probing" the conditions that exist at increasingly low altitudes in the solar atmosphere. For example, in a study of a particular "event" (a small flare) in the Sun, Kleint (2012) used the 8542 Å line of Ca II (one of the lines plotted in Figure 3.8) and obtained images of the event at 41 different steps across the line profile: the amplitude of each step in wavelength was ≥0.07Å. The core of the 8542Å line is formed in the "chromosphere" (see Chapter 15), while the wings are formed in gas that lies closer to the photosphere. Simultaneously, Kleint (2012) also observed the event in the 6302.5 Å line of Fe I, and used 26 steps to go from core to wing: each step in wavelength was ≥0.02 Å. The properties of the Fe I line are such that the line is formed mainly in the photosphere. Interestingly, Kleint's clever choice of lines had the following beneficial effect: when the Fe I line was sampled close to its *core* (at 6302.49 Å: at the *maximum* altitude that can be probed in this line), the features in the image were found to be similar to the features observed in the *wing* of the Ca II line (at 8540.95 Å: at the *minimum* altitude that can be probed in this line). Remarkably, different parts of these two lines are actually formed in the *same layers* of the solar atmosphere. Using this ability to "step through" the Sun's atmosphere at different altitudes, Kleint demonstrated that a certain feature in the flare had observable effects in the *chromosphere* but not in the *photosphere*. Apparently, conditions in this particular flare (that originated high in the solar atmosphere) gave rise to measurable changes in temperature and/or density only in those layers of gas located *above* a certain minimum altitude in the Sun's chromosphere. But in layers of gas that lay deeper (e.g., in the photosphere), this flare gave rise to *no* detectable effects. Apparently, the effects of this event penetrated *downward* into the solar atmosphere *only so far, and no farther*. This conclusion potentially contains information as to the physical processes occurring in a flare (e.g., thermal conduction, beams of energetic particles) and how far down the effects of such processes make their way into the deeper layers of the solar atmosphere.

The advantage of the ability to significantly probe different heights in the solar atmosphere by "tuning" the wavelength of a strong line from line center to line wings can be further illustrated by considering the $H\alpha$ line. If one observes exactly at line center, then one is probing layers of the atmosphere at heights lying about 1600–2000 km above the photosphere (see Figures 3.7 and 4.2). However, if one tunes to a wavelength in $H\alpha$ that is offset from line center by an amount $\Delta\lambda$ = ±0.84Å from line center, one is then probing gas lying much lower down, certainly lower than the temperature minimum, and perhaps as low as only 200–300 km above the photosphere (see the horizontal line labeled $H\alpha$ in Figure 4.2). In fact, at $\Delta\lambda$ = ±0.84 Å, one is observing so close to the photosphere that granulation can actually be identified in the images (Kontogiannis et al. 2010). On the other hand, if one tunes to a wavelength lying close to the core of the line (e.g., $\Delta\lambda$ = ±0.35 Å from line center), the image has a completely different appearance: granulation is no longer visible. Why is that? Because photons at a wavelength shifted from line center by only $\Delta\lambda$ = ±0.35Å allow one to probe gas that lies at heights of about 1000 km above the photosphere: convective motions are not fast enough to penetrate to such great heights (see Exercise 6.1).

It is very useful in our studies of the Sun's photosphere to have access to the previously described approach, namely, probing different heights in the solar atmosphere by "tuning" the wavelength all the way from the core of a line out to its wings. This approach depends on the fact that, in an optically thick absorption line, the *opacity* at line center is much larger than in the wings. Later in this book (Chapter 17), in order to probe different layers of the Sun's corona, we shall not be able to take advantage of changes in opacity across a line. Instead, we shall examine *emission* lines that

Toward a Model of the Sun

originate from various ions: these lines are formed in coronal gas where densities are nine (or more) orders of magnitude smaller than in the photosphere. In such conditions, the lines are optically thin and variations of opacity are not particularly useful as diagnostic tools. Instead, we will use the fact that lines emitted by different ions are formed preferentially at different *temperatures* (see Section 17.4): by a judicious choice of lines, we can probe various regions of coronal plasma that differ not in opacity but in temperature.

### 3.8.3 SHAPE OF AN ABSORPTION LINE PROFILE: C-SHAPED BISECTORS

Random thermal motions are *not* the only contributor to the widths of lines in the solar spectrum. We shall see (in Chapter 6) that organized flows of gas (i.e., convective flows) exist in certain regions of the Sun in order to transport heat. In such regions, gas that is hotter than average rises, while gas that is cooler than average sinks. The regions of rising and falling gas organize themselves into "cells" or "granules" with bright material rising at the center and dark material sinking around the periphery. Because the rising gas is brighter than the falling gas, a careful inspection of the shape of a spectral line will preferentially detect the upward motions of the brighter gas.

To undertake such an inspection, we use the fact that by choosing different depths in a line, we are probing different layers of the atmosphere. Suppose we start near the continuum and pick the part of the line where the depth is only (say) 10% below the continuum. There will be two points on the line profile that have that depth, one on the blue side of line center, one on the red side of

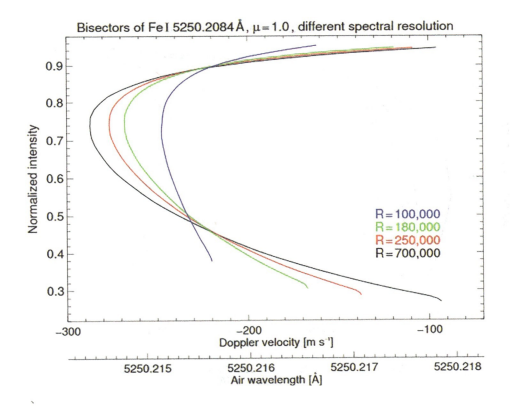

**FIGURE 3.9** Doppler velocities obtained from bisectors of the neutral iron line at $\lambda_o = 5250.2084$ Å at the center of the solar disk (Löhner-Böttcher et al. 2019, *Astron. Astrophys.* 624, A57; used with permission of ESO).

line center. We draw a line between those two points and bisect it: at the bisector, we measure the wavelength as precisely as our instrumentation allows. Typically, we find that the bisector does *not* have the same wavelength as line center ($\lambda_o$): instead, the bisector is found to have a wavelength $\lambda_B$ that is shifted from $\lambda_o$ by $\Delta\lambda_B = \lambda_B - \lambda_o$. Interpreting this $\Delta\lambda_B$ as a Doppler shift will give us a velocity associated with the 10% depth point. Then we move deeper into the line, say to the part of the line that lies 20% below the continuum. We measure the Doppler shift of the bisector at that level, and obtain another value of velocity. We then proceed downward until we reach line center. At each depth, we will have (in general) found different values of the velocity. What do we see when we plot the velocity versus line depth? Results are shown in Figure 3.9.

The line used to obtain the data in Figure 3.9 is a deep iron line with a central depth as low as 0.2–0.3. Thus, this line is not the strongest line in the spectrum, but it is strong enough to give rise to a deep absorption line. The characteristic that emerges in bisector plots of the Sun is that (i) all velocities are negative, i.e., gas is preferentially moving towards the observer, and (ii) the curves are shaped like the letter C, i.e., the curves are concave towards the right. Why are the shifts systematically negative? Because hotter (brighter) gas is rising upward in the Sun, and these photons dominate over the fainter photons moving downward into the Sun. Near the continuum (i.e., high up in the profile), the Doppler shift is relatively small. Also, the Doppler shift is relatively small near the bottom of the line, at the lowest part of the profile. At intermediate depths (0.8–0.7), the magnitude of the Doppler shift reaches a maximum. The maximum velocity observed in this iron line is

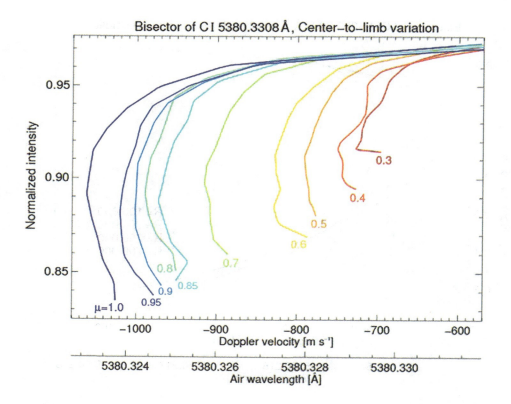

**FIGURE 3.10** Doppler velocities obtained from bisectors of the neutral carbon line at $\lambda_o = 5380.3308$ Å at the center of the solar disk (Löhner-Böttcher et al. 2019, *Astron. Astrophys.* 624, A57; used with permission of ESO).

Toward a Model of the Sun

found to be almost 0.3 km s⁻¹. On an observational point, we note that the amplitude of the velocity depends on how large the wavelength resolving power ($R = \lambda/\Delta\lambda$) of the instrument is: values of $R$ up to 700,000 are shown in Figure 3.9. (The enormous resolving power $R = 700,000$ is achieved by using a "laser comb" technique to measure wavelengths with exceptional precision.) Velocities of 0.3 km s⁻¹ are some sort of average velocity of the convective motions that exist in relatively high layers of the Sun corresponding to the altitudes where the gas giving rise to depths of 0.8–0.7 in the 5250 Å line profile is situated.

Should we conclude from Figure 3.9 that 0.3 km s⁻¹ is a reliable measure of the speeds of convective flows in the Sun? To answer that, we consider another line and examine the C-shaped bisectors in that line as well (Figure 3.10). The line in Figure 3.10 is due to neutral carbon at $\lambda_o = 5380.3308$ Å. Results plotted in Figure 3.10 refer to measurements not only at the center of the solar disk (labeled $\mu \equiv \cos\psi = 1$, where the angle $\psi$ is defined in Figure 2.3), but also at other positions on the disk approaching the limb: $\mu = 0.3$ corresponds to an angle of $\psi = 73$ heliographic degrees from disk center. Here, we confine attention to the disk center measurement ($\mu=1$, shown as a dark purple curve): we see that the bisector velocity is blueshifted there by more than 1000 m s⁻¹. We also note that this line is a weak line: the intensity at line center "goes down" only to levels of 0.83–0.84 times the continuum. Such a weak line has an η value that is relatively small, and therefore the line is formed quite close to the photosphere: quantitative modeling in fact indicates that the center of the 5380 Å carbon line is formed only about 40 km higher than the level $\tau=1$ in the Sun. The results in Figure 3.10 indicate that convective flow speeds close to the photosphere have magnitudes that are no less than 1 km s⁻¹. We need to recall that the C-shaped bisectors give Doppler shifts that are "averaged" in some way over a mixture of bright rising gas and dark falling gas: therefore, we will not be surprised to find out below (Chapter 6) that other observations (with greater *spatial* resolution) indicate convective flows that are in certain locations in excess of 1 km s⁻¹.

### 3.8.4 Shape of an Absorption Line: Magnetic Fields

In the presence of magnetic fields, the energy levels within an atom or ion become split into several distinct sublevels. Only certain transitions between the sublevels are permitted. As a result, an absorption line, which in nonmagnetic gas is a single narrow feature in the spectrum, can break up into a group of distinct, but closely spaced, components when the gas is placed in a magnetic field. The amount of spacing increases as the field becomes *stronger* (see Section 16.4.1). The polarization of the various components contains information as to the *direction* of the field lines relative to the line of sight. In order to handle the radiative transfer of line radiation in a magnetic field, it is necessary to solve four separate equations of radiative transfer. There is one equation for each of the four Stokes parameters $I$, $Q$, $U$, $V$: respectively the intensity of light propagating along the $z$ direction, the linear polarization along the $x$ direction, the linear polarization along the $y$ direction, and the circular polarization vector pointing along the $z$ direction. Wasaburo Unno (1956) derived the four RTE equations and showed how to obtain solutions in one special case. The special case considered by Unno involves what is referred to as the Milne–Eddington (ME) approximation for solving the RTE for a spectral line in nonmagnetic gas: the central assumption of the ME approximation is that $\eta = k_{\lambda o}/k_\lambda$ (i.e., the ratio of line to continuum opacities) retains a *constant numerical value* at all levels in the atmosphere. The advantage of this is that an analytic solution of the RTE can be obtained (e.g., Aller 1953). In modern observations of polarization in the Sun, the ME approximation continues to be used in order to calculate the forward RTE problem: i.e., given a specific choice of numerical values for several key physical parameters at a number of specified altitudes ("nodes") in a stellar atmosphere, use the Unno solution to calculate a large array of line profiles of $I$, $Q$, $U$, and $V$. The eight physical quantities that are to be chosen in order to proceed with this forward solution include the source function ($= S_o + S_1\tau$); three field components $B_x$, $B_y$, and $B_z$; η; $\Delta\lambda_D$; and a velocity along the line of sight. Clearly, with so many parameters to specify, even if we select no more than (say) three or four nodes, the array size will be as large as $10^{4-5}$. Once the array of line profiles is complete, any observational data set can in principle be "inverted" by searching through the array for the best fit to the four Stokes profiles: various codes have been developed to do the inversion, including codes with acronyms SIR (since 1992), SPINOR (since 2000), and NICOLE (since 2015). A great improvement in testing the inversion codes has become available with access to large realistic 3-D MHD simulations of the solar atmosphere. These simulations yield spatially variable values of all eight physical quantities at each instant: therefore, with a snapshot of the simulation at hand, one has all the information needed to do the forward calculation of any spectral line profile. Subjecting that profile to an inversion code will then lead to "solutions" that can be compared directly with the actual values of all eight parameters in the snapshot (e.g., Danilovic et al. 2016). Only when such a "solution" is found to be satisfactory should one proceed to apply the inversion to solar data.

In summary, absorption lines in the spectrum of the Sun can be used to extract information about velocities and magnetic fields at various altitudes in the solar atmosphere.

## EXERCISES

3.1 Lines due to hydrogen, sodium, calcium, and iron occur in the solar spectrum at wavelengths of 6563 Å, 5890 Å, 3933 Å, and 5250 Å, respectively. Assuming that each line arises in a gas with $T = 6000$ K, use the various atomic weights (find them online) to calculate the thermal width in Å for each line. What resolving power is required to resolve the thermally broadened line profile in each case?

3.2 Consider a neutral hydrogen atom. Calculate the *speed* of the electron (in units of cm s$^{-1}$) in each of the Bohr orbits with $n = 1, 2,$ and 3. Using those speeds, calculate the de Broglie wavelength of an electron in each of the three orbits. Express your answers in units of Å.

3.3 Consider a neutral hydrogen atom. Calculate the *radius* (in units of Å) of the lowest three Bohr orbits, i.e., those with $n = 1, 2,$ and 3. Using those radii and the results of Exercise 3.2, calculate how many de Broglie wavelengths of an electron can be "fitted in" along the circumference of each of the three orbits.

3.4 Bohr's hydrogenic ion is one in which a single electron is in orbit around a nucleus with charge $+Ze$. Repeat the two-step argument in Section 3.2.1. Derive a formula for the electron speed in orbit $n$. Derive the predicted wavelengths $\lambda_{21}(Z)$ (in units of Å) of the Lyman-$\alpha$ line in the hydrogenic ion of elements with $Z = 11, 12, 14, 16, 26,$ and 28. Compare your predicted results with the observed wavelengths $\lambda_{21}$(obs) reported by Torrejon et al. (2012)*.

3.5 Starting with the $\lambda_{21}$(obs) values reported by Torrejon et al. (2012) for six ions with various $Z$ values, multiply each $\lambda_{21}$(obs) by the value of $Z^2$ for that element. If Bohr's model is exact, then your values of $Z^2\lambda_{21}$(obs) should all equal the value of $\lambda_{21}$(obs) = 1215.6 Å for the hydrogen atom. What trend do you see in your results? By what percentage does $Z^2\lambda_{21}$(obs) for $Z = 28$ differ from the $\lambda_{21}$(obs) = 1215.6 Å for the hydrogen atom (with $Z=1$)?

3.6 Relativistic effects (approximate correction): according to Einstein's special theory of relativity, the mass of an electron moving at speed $V$ should be larger than the rest mass of the electron by a factor $\gamma = 1/\sqrt{(1-V^2/c^2)}$. (Here, $c$ is the speed of light.) As a result, the magnitude of $E_n$ should be increased by a roughly a factor of $\gamma$ above the results in Section 3.2.1. Using the results from Exercise 3.2 for levels n = 1 and 2 in the H atom, calculate the values of $\gamma$ for the n=1 orbit, and also for the n=2 orbit. Use these two values of $\gamma$ to calculate an "improved" value of $\lambda_{21}$(rel) for the H atom.

3.7 Relativistic effects in hydrogenic ions (approximate correction). Using the electron speeds from Exercise 3.4 for levels n = 1 and 2 in a hydrogenic ion with charge Z = 28, calculate the values of $\gamma$ for the n=1 orbit, and also for the n=2 orbit in this ion. Use these values of $\gamma$, calculate an "improved" value of $\lambda_{21}$(rel) for this ion. By what percentage does $Z^2\lambda_{21}$(rel) for Z = 28 differ from the $\lambda_{21}$(obs) = 1215.6 Å for the hydrogen atom?

*For further astrophysical measurements of Lyman-α line wavelengths in hydrogenic ions, see also Hanke et al. (2009), Lopes de Oliveira et al. (2010), and Phillips et al. (2015). The reader may also find it an interesting exercise to search the astrophysical literature for further examples, especially for the rarer elements that have *odd* values of Z.

## REFERENCES

Aller, L. H., 1953. *The Atmospheres of the Sun and Stars*, Ronald Press Company, New York, pp. 256, 263, 269.

Andic, A., Cao, W., & Goode, P. R., 2011. "Umbral dynamics in the near infrared continuum", *Astrophys. J.* 736, 79.

Bohr, N., 1913. "On the constitutions of atoms and molecules. Part I", *Philosophical Magazine* (Series 6), 26, 1.

Chandrasekhar, S., & Breen, F. H., 1946 "On the continuous absorption coefficient of the negative hydrogen ion. III", *Astrophys. J.* 104, 430.

Danilovic, S., van Noort, M., & Rempel, M., 2016. "Internetwork magnetic field as revealed by 2-D inversions", *Astron. Astrophys.* 593, 93.

Doschek, G., & Feldman, U., 2010. "The solar UV-X-ray spectrum from 1.5 to 2000 Å", *J Phys. B. At. Mol. Op. Phys.* 43, 232001.

Ezer, D., & Cameron, A. G. W., 1963. "The early evolution of the Sun", *Icarus*, 1, 422.

Hanke, M., Wilms, J., Nowak, M. A., et al., 2009. "*Chandra* X-ray spectroscopy of the focused wind in the Cyg X-1 system", *Astrophys. J.* 690, 330.

Karttunen, H., Kroger, P., Oja, H. et al., 2017. *Fundamental Astronomy*, 6th ed., Springer Verlag, New York, p. 329.

Kleint, L., 2012. "Spectropolarimetry of C-class flare footpoints", *Astrophys. J.* 748, 138.

Kontogiannis, I., Tsiropoula, G., & Tziotziou, K., 2010. "Power halo and magnetic shadow in a solar quiet region observed in the *H*α line", *Astron. Astrophys.* 510, A41.

Kurucz, R. L., 1992. "Atomic and molecular data for opacity calculations", *Rev. Mexicana Astron. Astrof.*, 23, 45. The Rosseland mean *Opacity tables* in the present book (Tables 5.1 and 5.2) were made available to the author upon request.

Leenaarts, J., Pereira, T., Carlsson, M., et al., 2013. "The formation of IRIS diagnostics. II. The formation of the MgII h&k lines in the solar atmosphere", *Astrophys. J.* 772, 90.

Löhner-Böttcher, J., Schmidt, W., Schlichenmaier, R., et al., 2019. "Convective blueshifts in the solar atmosphere. III. High accuracy observations of spectral lines in the visible", *Astron. Astrophys.* 624, A57.

Lopes de Oliveira, R., Smith, M. A., & Motch, C., 2010. "Gamma Cass: An X-ray Be star with personality", *Astron. Astrophy.* 512, A22.

Machado, M. E., Milligan, R. O., & Simoes, P. J. A., 2018. "Lyman continuum observations of solar flares using SDO/EVE", *Astrophys. J.* 869, 63.

Meftah, M., Dame, L., Bolsee, D., et al., 2020. "A new version of the SOLAR-ISS spectrum covering the 165–3000 nm spectral region", *Solar Phys.* 295, 14.

Milligan, R. O., 2015. "EUV spectroscopy of the lower solar atmosphere during solar flares", *Solar Phys.* 290, 3399.

Milligan, R. O., & McElroy, S. A., 2013. "Continuum contributions to the SDO/AIA passbands during an X-class solar flare", *Astrophys. J.* 777, 12.

Pavlenko, Ya. V., Jenkins, J. S., Jones, H. R. A., et al., 2012. "Effective temperatures, rotational velocities, microturbulence velocities and abundances in the atmospheres of the Sun, HD 1835, and HD 10700", *Mon. Not. Roy. Astron. Soc.* 422, 542.

Phillips, K. J. H., Sylwester, B., & Sylwester, J., 2015. "The X-ray line feature at 3.5 keV in galaxy cluster spectra", *Astrophys. J.* 809, 50.

Schiff, L. I., 1955. *Quantum Mechanics*, McGraw-Hill, New York, p. 83.

Smith, S. J., & Burch, D. S., 1959. "Relative measurement of the photo-detachment cross-section for H⁻", *Phys. Rev.* 116, 1125.
Stenflo, J. O., 2005. "Polarization of the Sun's continuous spectrum", *Astron. Astrophys.* 429, 713.
Thuillier, G., DeLand, M., Shapiro, A., et al., 2012. "The solar spectral irradiance as a function of the MgII index for atmosphere and climate modelling", *Solar Phys.* 277, 245.
Torrejon, J. M., Schulz, N. S., & Nowak, M. A., 2012. "*Chandra* and *Suzaku* observations of the Be/X-ray star HD 110432", *Astrophys. J.* 750, 73.
Unno, W., 1956. "Line formation of a normal Zeeman triplet", *Publ. Astron. Soc. Japan*, 8, 108.
Waddell, J., 1962. "Center-to-limb observations of the sodium D lines", *Astrophys. J.* 136, 223.

# 4 Toward a Model of the Sun
## *Properties of Ionization*

The properties of opacity at high temperature (and low pressure) are controlled by the physical process of ionization. Other physical properties of the gas, including thermodynamic quantities such as the specific heats (which are important for transport of energy), are also significantly affected by the ionization process. In order to have a clear understanding of the physics of certain regions in the Sun, it is important to have a quantitative model of ionization. This leads us to consider an equation originally derived by Meghnad Saha (1921) for ionization equilibrium.

We have already mentioned (see Equation 3.2) how the numbers of *bound* electrons inside an atom are distributed among energy levels: the number density $n_i$ of atoms with electrons in the $i^{\text{th}}$ energy level is related to the number density of atoms in the ground state $n_1$ according to a Boltzmann distribution:

$$\frac{n_i}{n_1} = \left(\frac{g_i}{g_1}\right) \exp\left(\frac{-\Delta E_{i1}}{kT}\right) \tag{4.1}$$

where $\Delta E_{i1}$ is the difference in energy between level $i$ and the ground state. The coefficient $g_i$ refers to the "statistical weight" of the level $i$, i.e., the number of distinct sublevels that are available to an electron with principal quantum number $i$. The value of $g_i$ in the H atom is derived by noting that for principal quantum number $i$, the angular momentum sublevels take on integer quantum numbers $L$ from zero up to $i - 1$: each of these sublevels has orbital multiplicity $2L + 1$, as well as multiplicity 2 for electron spin. Summing the combined multiplicity $4L + 2$ per sublevel over $L$ sublevels from $L = 0$ to $L = i - 1$ yields a total of $g_i = 2i^2$.

## 4.1 STATISTICAL WEIGHTS OF FREE ELECTRONS

Now, when we wish to move on to the case of ionization, we need to write down an expression for the population of *unbound ions and electrons*. Analogously to the Boltzmann formula, which we used in Equation 3.2 for bound states, the Boltzmann distribution for ions and electrons can be written as:

$$\frac{n_i}{n_a} = \frac{g_{i+e}}{g_a} \exp\left(\frac{-I_p}{kT}\right) = \frac{g_i g_e}{g_a} \exp\left(\frac{-I_p}{kT}\right) \tag{4.2}$$

Here, $n_i$ and $n_a$ are number densities of ions and atoms respectively, and $I_p$ is the "ionization potential", i.e., the energy required to ionize an atom from the ground state. The statistical weight of the ion+(free)electron "system" is labeled $g_{i+e}$. The statistical weight of the atom $g_a$ is essentially that of the ground state, i.e., $g_a = 2$ for a hydrogenic atom.

When we consider ionization, in contrast to considering bound electrons, the principal difference is that the statistical weight of the ionized system, consisting of ion plus electron, must not only include the statistical weight of the ion $g_i$ (also mainly in its ground state), but must also include the statistical weight $g_e$ of the free electron. Because the electron is now free of all attachments to the ion, i.e., the ion and electron now operate independently of each other, the overall statistical weight for ion and electron $g_{i+e}$ can be written as the product of two terms: $g_{i+e} = g_i g_e$. In contrast to an electron that occupies a *bound* energy level, where the number of available sublevels is small (leading therefore to small statistical weights), a free electron has (as we shall see later) access to an enormous number of states. This has the striking effect that, in Equation 4.2, the right-hand side of Equation 4.2 may grow to values of order unity even when the exponential term is small, i.e., even when $kT$ is much smaller than $I$.

For example, consider the case of hydrogen, where $I_p = 13.6$ eV. What do we have to do in order to achieve significant ionization? By the term "significant", we choose the following line of reasoning. We start at low temperatures with a number of atoms in a certain volume, and essentially no ions in that volume. As we increase the temperature, gradually, more and more of the original atoms will lose an electron and become an ion. Thus, the number of ions in the volume increases as time goes on, while the number of atoms decreases. At some point, we will find that $n_i \approx n_a$, i.e., about one-half of all the original atoms have ionized, and each one of the ionized atoms is included in the number $n_i$. We describe the gas at that point as "significantly" ionized. (Clearly, we could have chosen a different criterion: as $T$ continues to increase, more and more of the atoms become ionized and $n_i$ becomes much larger than $n_a$. But the "50% point" is satisfactory for us here.)

So, returning to our question "What do we have to do in order to achieve significant ionization?" the answer can be seen by inspection of Equation 4.2: we need to make the right-hand side of Equation 4.2 approach unity. At first, let us suppose that the statistical weight factors are all of order unity, i.e., suppose that the only relevant term on the right-hand side of Equation 4.2 is the exponential factor alone. In such a case, if we are to have any chance of approaching the limit $n_i \approx n_a$, the temperature would have to satisfy $kT \geq 13.6$ eV. (Actually, in this case, the r.h.s. could never formally be as large as unity: but it could "get close" to unity at high enough $T$.) The condition $kT \geq 13.6$ eV corresponds to $T \geq 158{,}000$ K. On the other hand, when statistical weights are included (as of course they must be in order to be realistic), we shall find that it is possible to achieve $n_i \approx n_a$ even when $T$ is less than 10,000 K (at low pressure). How can it be possible for a gas with a temperature as low as 10,000 K (i.e., $kT = 1.38 \times 10^{-12}$ ergs = 0.86 eV) to be able to ionize an H atom that requires 13.6 eV to become ionized? Shouldn't the Boltzmann factor of $\exp(-13.6/0.86) = 1.35 \times 10^{-7}$ render this outcome highly unlikely? The solution to this conundrum lies in the fact that $g_e$, the statistical weight of a free electron, can take on (very) large values.

In order to demonstrate this fact, we need to evaluate $g_e$. That is, we need to know the number of states that are available to a free electron. Such an electron moves in a six-dimensional (6-D) $p - r$ phase space, where $p$ represents the momentum vector (in 3-D), and $r$ represents the position vector in (3-D) coordinate space. The uncertainty principle, as originally stated by Heisenberg (1927), in one dimension (1-D) restricts the uncertainties in 1-D momentum and 1-D position such that the product of those uncertainties $dp_x dr_x$ cannot be less than a quantity of order $h$, the Planck constant. When the 1-D result is extended to 3-D, this principle leads to the concept that the 6-D phase space *cannot* be regarded as infinitesimally finely "grained": instead, there must exist minimally occupiable "cells" in 6-D phase space, each with volume $d^3p d^3x \approx h^3$. Allowing for electron spin, the Pauli (1925) exclusion principle indicates that each "cell" in phase space cannot be occupied by more than two electrons. As a result, in a gas occupying a volume $V(r)$ of coordinate space and a volume $V(p)$ in momentum space, the total number of states available to a free electron is $g_e = 2V(p)V(r)/h^3$.

What are we to use for $V(r)$ and $V(p)$? In a medium where the number density of electrons has a known value ($n_e$), the mean volume of coordinate space $V(r)$ occupied by a single electron is readily determined: $V(r) = 1/n_e$. Turning our attention now to the 3-D momentum space, each electron has access to a large volume in this space. In momentum space, few electrons have momenta faster than the momentum $p_{th} = mV_{th}$ corresponding to the mean thermal speed: $V_{th} = \sqrt{(8kT/\pi m)}$. As a result, momentum space is (roughly) filled up inside a sphere with a radius of

# Toward a Model of the Sun

order $r(p) = p_{th} = mV_{th}$. The associated volume $V(p)$ filled up in momentum space is (roughly) $(4/3)\pi r(p)^3$. Thus, $V(p) = (4\pi/3)p_{th}^3$.

Now we combine $V(r)$ and $V(p)$ to find

$$g_e = \frac{8\pi}{3}\left(\frac{8km}{\pi}\right)^{3/2}\frac{1}{h^3}\frac{T^{3/2}}{n_e} \quad (4.3)$$

The coefficient of $T^{3/2}/n_e$ in Equation 4.3 can be evaluated by inserting appropriate values for the natural constants $k$, $m$, and $h$. The result is $5.22 \times 10^{15}$. The largeness of this number is noteworthy in the context of the possible largeness of $g_e$. Of course, we do not yet know the value of $g_e$ because we still need to evaluate the quantity $T^{3/2}/n_e$. How large might the latter quantity become in the solar atmosphere? In the upper part of the atmosphere, in the corona, $T$ reaches values of order a few times $10^6$ K, while $n_e$ is no larger than $10^9$ cm$^{-3}$ (see Chapter 17). In such conditions, $g_e$ is of order $10^{16}$ or more. Such large values make it easy to significantly overcome the effects of the Boltzmann factor in Equation 4.2. In the chromosphere, where $T \approx 10^4$ K, $n_e$ may be of order $10^{11}$ cm$^{-3}$ (Vernazza et al. 1973). In such conditions, $g_e$ is of order $10^{11}$, still large enough to more than compensate for the Boltzmann factor in Equation 4.2.

For practical purposes, it is usual to rewrite Equation 4.3 in terms of electron *pressure* $p_e$ rather than in terms of electron *number density* $n_e$. Replacing $n_e$ with $p_e/kT$ in Equation 4.3, we find that Equation 4.2 becomes

$$\frac{n_i p_e}{n_a} = \frac{g_i}{g_a}C_i T^{2.5}\exp\left(\frac{-I_p}{kT}\right) \quad (4.4)$$

where $C_i$ is a combination of numerical and physical constants. The physical constants in $C_i$ occur in the combination $m_e^{1.5}k^{2.5}/h^3$: in c.g.s. units, the numerical value of this combination is 0.021. Including the remaining numerical constants in Equation 4.3, the value of $C_i$ turns out to be (coincidentally, but as it turns out, conveniently) close to unity (0.72).

## 4.2 SAHA EQUATION

In logarithmic form, we can now write Equation 4.4 as

$$\log\left(\frac{n_i}{n_a}\right) = -\log(p_e) + 2.5\log T - \theta I_p + \log\left(\frac{g_i}{g_a}\right) - 0.14 \quad (4.5)$$

In Equation 4.5, $I_p$ is the ionization potential expressed in units of eV, the quantity $\theta$ is related to the temperature according to $\theta = 5040/T$, and the logarithms are to base 10. As an aid to memory, we note that we will be interested in this chapter in cases where $n_i \approx n_a$: in such cases, the left-hand side (l.h.s.) of Equation 4.5 is close to zero. The question is: which of the five terms on the r.h.s. of Equation 4.5 are likely to be dominant? We shall be interested in locations where $T$ is within a factor of a few of $T = 10^4$ K: therefore, the term $2.5\log T$ is of order 10. Also, the quantity $\theta$ will be within a factor of a few of the value 0.5. Many elements have $I_p \approx 10$–20. Therefore, the magnitude of the term $\theta I_p$ will be of order 10. For many atoms/ions, it is found that $g_i \approx g_a$: therefore, the fourth term on the r.h.s. is in many cases close to zero. The fifth term on the r.h.s. is some two orders of magnitude smaller than the second and third terms on the r.h.s. In view of these relative magnitudes of the various terms, no significant error is made if we make the approximation of retaining only the first three terms on the right-hand side of Equation 4.5. Thus, in simplified form, we refer to "the Saha equation" as the following expression:

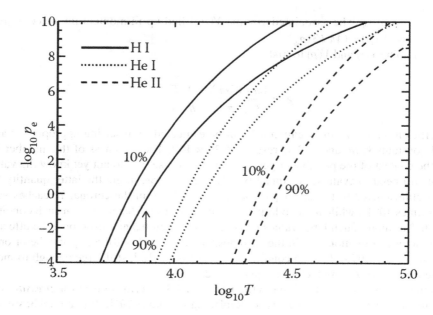

**FIGURE 4.1** Solid, dotted, and dashed lines: ionization "strips" of neutral hydrogen H I, neutral helium He I, and singly ionized helium He II in the (log $p_e$ versus log $T$) plane. Units of $T$ are K, and units of $p_e$ are dyn cm$^{-2}$. The solid curves labeled 10% and 90% indicate the loci along which hydrogen is 10% ionized and 90% ionized, respectively. The dashed curves labeled 10% and 90% indicate the loci along which He II is 10% ionized and 90% ionized, respectively. The dotted lines refer to He I, but the 10% and 90% labels are omitted from the upper and lower lines (respectively) for the sake of clarity.

$$\log\left(\frac{n_i}{n_a}\right) = -\log(p_e) + 2.5\log T - \theta I_p \tag{4.5'}$$

where the label (4.5') denotes a simplified version. In what follows, we adopt this simplifying approximation.

Equation 4.5 is the Saha equation: it allows us to evaluate the degree of ionization for each element separately in a medium of given $T$ and $p_e$. If a particular medium is considered (e.g., the photosphere in the Sun), then it will contain some elements that are mainly in the neutral atom state, while some other elements are ionized. In general, the elements that remain neutral in the solar photosphere are those with *large* values of $I_p$ (e.g., He and Ne, with $I_p$ = 24 eV and 21 eV respectively). And the elements that tend to be ionized in the solar atmosphere are those with *small* values of $I_p$ (e.g., Na and K, both with $I_p \approx$ 5 eV). Of course when we move to a different location in the Sun, e.g., into the corona (or even more so, into a flare), the temperature is so much larger than in the photosphere that all of the elements are observed to be ionized, some to the extent that only a single electron may be left remaining attached to the nucleus. And in each case, if you know the values of $T$ and $p_e$, the Saha equation allows you to determine the degree of ionization of each element separately.

It is worth reiterating why considerations of ionization are important as far as opacity is concerned. If the atoms in a gas are highly ionized, the lack of bound states for electrons leads to low values of opacity. Because a bound electron may be up to seven orders of magnitude more efficient than a free electron as regards interacting with photons, the presence of even a few bound electrons in the gas can cause the opacity to be enhanced significantly. It is the Saha equation that allows us to determine quantitatively if there are a lot of bound electrons in the gas of interest to us, or only a few.

On a historical note, it was precisely with the help of the Saha equation that Cecilia Payne (1925) first established that hydrogen and helium are by far the most abundant elements in most stellar atmospheres. Interpretation of the observed strength of an H or He spectral line in the visible spectrum of a star requires careful inclusion of the (large) Boltzmann factors that are relevant for the population of the lower levels of such transitions. Analyses of spectra performed before Payne's work did not properly incorporate the Boltzmann factors: the conclusion of the earlier work had been that the stars had roughly the same composition as Earth! Payne's painstaking application of the Saha equation to the analysis of the lines in the spectra of stars of many different types in the Harvard plate collection led her to a result that was eventually recognized as a stunning breakthrough in stellar astrophysics. Although Payne could not have known it in 1925, her results paved the way for future developments in cosmology, where it became clear that her results concerning the dominant abundances of hydrogen and helium can ultimately be traced back to the fact that H and He are essentially the only elements that emerged from the Big Bang: all of the other 90 elements found in nature were generated subsequently when various generations of stars returned the ashes of their nuclear "factories" to interstellar space.

## 4.3 APPLICATION OF THE SAHA EQUATION TO HYDROGEN IN THE SUN

There are two distinct locations in the Sun where hydrogen makes a transition from mostly neutral to significant ionization. Since the photosphere is the location in the Sun where the temperature is close to its minimum value, the abundance of neutral hydrogen is largest in the vicinity of the photosphere. In view of this, it is not surprising to find that the two distinct locations where H is ionizing lie on either side of the photosphere. One lies above the photosphere, in the low-density gas of the upper chromosphere. The second lies well below the surface, in the denser gas of the convection zone. Let us use the Saha equation to determine the temperatures of these locations.

In order to discuss the Saha equation, it is convenient to plot the Saha equation in the $(p_e - T)$ plane, and to introduce the concept of an "ionization strip", as shown in Figure 4.1. Along the curves labeled X% in Figure 4.1, H and He are X% ionized. To avoid crowding in the figure, we plot only two curves for each element: $X = 10\%$ (essentially the onset of ionization), and $X = 90\%$ (essentially the near completion of ionization). The area between the 10% and 90% ionization levels can be considered an "ionization strip", where either H or He is in the process of transitioning from mostly neutral to mostly ionized. Within each ionization strip, where heat input tends to increase the degree of ionization rather than increase the temperature, the specific heat of the element in question becomes much larger (by an order of magnitude or more) than the standard value from kinetic theory (see Chapters 6 and 7).

To be specific, let us quantify the onset of "significant" ionization as the location where $n_i \approx n_a$, corresponding to 50% ionization. According to the simplified version of Equation 4.5', this occurs for hydrogen when the temperature satisfies the equation

$$\theta I_p - 2.5 \log_{10} T \approx -\log_{10} p_e \qquad (4.6)$$

where we must set $I_p = 13.6$ for hydrogen.

Let us consider the upper chromosphere, where the pressure is relatively low: $\log p_e \approx 0$ (see Chapter 15). In this case, Equation 4.6 reduces to $\theta = (2.5/13.6) \log T$, i.e., $T \log T = 27,400$. To solve this, we can use a handheld calculator that has a feature to solve transcendental equations. Otherwise, we can guess at the slowly varying logarithm term and then iterate. For example, if we initially guess $\log T = 4$ (i.e., $T = 10^4$ K), then the first iteration at the solution would be $T = 27,400/4 = 6850$, too low to be consistent with the initial guess $\log T = 4$. If we make an initial guess $\log T = 3.7$ (i.e., $T = 5000$ K), then the solution would be $T = 7400$ K, too high to be consistent with the initial guess. An iterative solution to this equation yields $T = 7100$–$7200$ K. Thus, when the

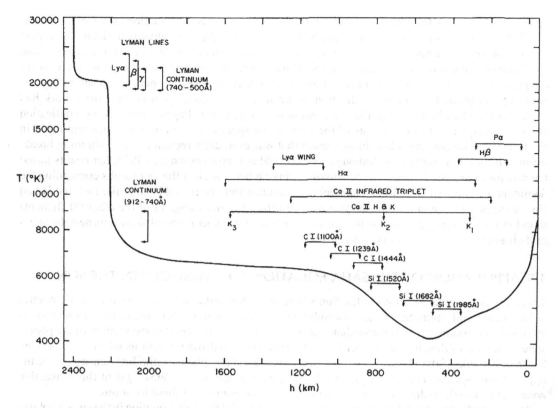

**FIGURE 4.2** Illustrative example of the temperature $T$ versus height $h$ in a particular one-dimensional model of the solar photosphere/chromosphere due to Vernazza et al. (1973). The height scale is chosen to have the value $h = 0$ in the photosphere (near the right-hand side of the figure). The temperature minimum in this model (with $T \approx 4000$ K) occurs at $h \approx 500$ km. As we move towards the left-hand side of the figure, the chromosphere (a region where temperatures become hotter than the photosphere) extends up to $h \approx 2000$–2200 km (see Chapter 15). At heights above $h \approx 2200$ km, the temperature rises steeply into the corona (Chapter 17). Lines in the solar spectrum are formed over a finite range of heights in the solar atmosphere: horizontal lines with vertical arrows at their ends are used to illustrate the height ranges where various lines are formed. Some of these ranges are quite narrow: e.g., the Si I line at 1985 Å is formed between $h = 350$ and 500 km. But some of the ranges are broad: e.g., the $H\alpha$ line and also the Ca II $K$ line are formed between $h = 300$ and $h = 1600$ km. (Used with permission of E. Avrett.)

electron pressure is as low as it is in the upper chromosphere, hydrogen begins to ionize significantly when the temperature rises above 7100–7200 K.

It is remarkable that such a simple approach to the Saha equation helps us to understand an important feature of the solar chromosphere (see Chapter 15): briefly, the chromosphere is found in the range of heights where the local gas temperature rises from a low of about 4000 K (in the upper photosphere) to temperatures hotter than that by several thousand degrees K. (We will discuss in Chapter 15 what kind of energy/work input might be causing this rise in temperature.) Observational data pertaining to the chromospheric gas suggest, upon analyzing the data in terms of a one-dimensional model (where height $h$ above the photosphere is the independent variable), that there exists a plateau in temperature close to 7000 K over an extended range of heights (more than 1000 km) in the solar chromosphere (see Figure 4.2). Whatever is inputting energy into the chromosphere is able to raise the temperature from its coolest value (close to 4000 K) up to about 6500 K over a range of heights from about 500 km to about 1000 km above the photosphere. But then, over

# Toward a Model of the Sun

the height range from about 1000 km to about 2000 km, there is essentially *no further increase* in the temperature of the chromospheric gas. Why would that be so? Has the input of energy stopped? No: energy is still being supplied, but instead of going into increasing the *temperature* of the gas, the inputted energy is diverted to perform a very different process, namely, the *ionization of hydrogen*. Because ionization of H consumes so much energy (13.6 eV for each H atom, much larger than the thermal energy $kT \approx 0.5$ eV), there is little or no energy left over to do the work of raising the local temperature. This situation continues until H becomes essentially completely ionized, at temperatures just above 7000 K. Above that level, the gas once again is supplied with input energy that is no longer diverted to do the work of ionizing: instead, the work can go back to its original task of raising the temperature.

Now let us apply the Saha equation to the gas lying *below* the photosphere. Photospheric models (see Chapter 5) indicate that the gas pressure in the photosphere is $\log p_g \approx 5$, and it rises as we move deeper below the photosphere. In the photosphere, electron number densities are less than atom number densities by about 2000. (Reasons for this number will be discussed later.) As a result, $\log p_e \approx 1.7$ in the photosphere. But with increasing depth and increasing temperature, increasing ionization of hydrogen causes the value of $p_e$ to approach closer to the value of $p_g$. At a depth where $\log p_g \approx 7$, the electron pressure is within an order of magnitude of $p_g$. In this case, the 50% ionization point occurs when

$$\theta \approx 0.18 \log_{10} T - 0.53 \qquad (4.7)$$

Iterative solution of this equation yields $T = 20{,}000$–$21{,}000$ K.

The contrast between the ease of ionization in the chromosphere and in the subphotosphere is important to note. In the *low-pressure* conditions of the chromosphere, it is relatively easy to ionize hydrogen: a temperature of just above 7000 K will suffice. But in the *high-pressure* gas below the surface, the ionization of hydrogen is "postponed" to higher temperatures: 50% ionization of hydrogen does not occur until the temperature has risen above 20,000 K.

As a final application of the Saha equation to hydrogen in the Sun, let us check the fractional ionization of hydrogen in the photosphere. In the lower photosphere, inserting $\log p_e = 1.7$, $T = 6000$ K (i.e., $\theta = 0.84$), we find $\log(n_i/n_a) = -3.7$. In the upper photosphere, close to the chromosphere, inserting $\log p_e = 0$, $T = 4900$ K (i.e., $\theta = 1.03$), we find $\log(n_i/n_a) = -4.8$. Thus, the average degree of hydrogen ionization in the solar photosphere is about $10^{-4.25}$: only one H in some 20,000 is ionized. In contrast, all elements in the solar gas with ionization potentials of about 9 eV and less exist more or less completely in the singly ionized state. This includes (in order of decreasing abundances) the elements Si, Mg, Fe, Al, Ca, and Na. Using the standard abundances by number of each of these elements from a table of cosmic abundances and summing over them, we find that their ionization provides about $10^{-4}$ electrons for every hydrogen atom. This somewhat exceeds the average degree of ionization of hydrogen in the upper photosphere: therefore, the "metals" are the primary source of free electrons in the upper photosphere. In the lower photosphere, the supply of free electrons comes from metals and hydrogen in roughly comparable amounts.

## 4.4 APPLICATION OF THE SAHA EQUATION TO HELIUM IN THE SUN

The second ionization of helium requires $I_p = 54$ eV. This means that the equation for 50% ionization is

$$54\theta - 2.5 \log_{10} T \approx -\log_{10} p_e \qquad (4.8)$$

Solutions to this equation occur at higher temperatures than those for hydrogen. In the chromosphere ($\log p_e \approx 0$), the solution satisfies $T \log T = 1.1 \times 10^5$: this corresponds to $T \approx 25{,}000$ K.

In the case of the subphotosphere, we must go to deeper layers than those where hydrogen ionization reaches the 50% level. In the deeper layers, $\log p_e$ may be as high as 10–12. This leads to

$$\theta \approx 0.046 \log_{10} T - 0.2 \tag{4.9}$$

The solution of this is $T \approx 140{,}000$ K.

Again, the contrast between the ease of ionization of He II in the chromosphere and the difficulty of ionization of He II in the subphotosphere is apparent. In the low-pressure conditions of the chromosphere, it is relatively easy to ionize helium: a temperature of just above 25,000 K will suffice. But in the high-pressure gas below the surface, the ionization of helium is "postponed" to higher temperatures: 50% ionization of helium does not occur until the temperature has risen above 140,000 K. This behavior is reminiscent of the differences that we found in the ease of ionization of hydrogen (see prior discussion of Equations 4.5 and 4.6): H is easy to ionize in the chromosphere but more difficult to ionize in the high pressure of the subphotospheric gas. It is as if higher density forces electrons and ions closer together, thereby improving the chances of electrons becoming bound once again into a bound energy state: and the more bound states there are, the higher the opacity will be.

These are quantitative illustrations of a point we made already in discussing the numerical values of opacity at various temperatures and densities plotted in Figure 3.4: at a given temperature, increasing pressure leads to *lower* degrees of ionization, and therefore more bound states that cause higher opacity. The Kramers' opacity law in the deep interior of a star attempts to find a functional form of the Rosseland mean opacity as a function of density and temperature that captures the different effects of density and temperature. On the one hand, the higher the temperature, the fewer the number of bound states, and therefore the lower the opacity should be. Therefore, the Kramers' opacity requires that the opacity must *decrease* as the temperature increases, i.e., the Kramers' opacity needs to have a *negative* exponent for the temperature variation. On the other hand, the higher the pressure, the *more* bound states are present in the gas, and the larger the opacity should be. Therefore, the Kramers' opacity needs to have a *positive* exponent for the density variation. The functional form $\kappa_R = \kappa_o \rho/T^{3.5}$ (see Section 3.7) satisfies both of these requirements.

## 4.5 CONTOURS OF CONSTANT IONIZATION: THE TWO LIMITS

Another way to look at Equation 4.5 is in terms of contours in the $(p_e - T)$ plane along which the degree of ionization is constant. Such contours are shown in Figure 4.1 for 10% and 90% degrees of ionization. There are two principal segments of each contour, with different dependences on temperature. At low $T$, where $\theta = 5040/T$ has a *larger value*, the term $\theta I$ dominates on the left-hand side of Equation 4.6. As a result, each ionization contour is described essentially by $\theta I = -\log p_e$, i.e., $p_e \sim \exp(-1/T)$. This is a curve that falls off steeply in the $(p_e - T)$ plane at low $T$. In the opposite limit, where temperatures are high, $\theta$ has a smaller value, and in the limit of high temperature, the term $\theta I \to 0$. In this limit, each contour in Figure 4.1 is described essentially by $p_e \sim T^{2.5}$. This is a line that slopes gently up and to the right. The transition from one segment to the other occurs at the location where the terms $\theta I$ and $2.5 \log T$ are comparable in magnitude.

## 4.6 APPLICATION OF THE SAHA EQUATION TO THE NEGATIVE HYDROGEN ION

What about $H^-$? Can we apply the Saha equation to this ion? Yes, except that in this case, we start with a charged particle and end up with one neutral particle plus one electron. But the principle

Toward a Model of the Sun

remains the same, as long as we replace $n_a$ in Equation 4.2 by $n(H^-)$, and replace $n_i$ in Equation 4.2 by $n(H)$. (Here we use the symbol $n(H)$ to emphasis that the resulting "ionized particle" is actually a neutral hydrogen atom.) Inserting the known ionization potential of the negative H ion ($I_p = 0.754$ eV: see Section 3.4) into Equation 4.5', we find

$$\log_{10}\left(\frac{n(H)}{n(H^-)}\right) = -\log_{10} p_e + 2.5\log_{10} T - 0.754\theta \qquad (4.10)$$

Rather than concentrating on the (large) ratio of numbers of H atoms to numbers of $H^-$ ions, it is more common to invert the ratio and focus on the (small) ratio $\phi = n(H^-)/n(H)$. So we rewrite Equation 4.10 as

$$\log_{10} \phi = \log_{10} p_e - 2.5\log_{10} T + 0.754\theta \qquad (4.11)$$

In the solar photosphere, where $T \approx 6000$ K (i.e., $\theta \equiv 5040/T \approx 0.84$), we have already seen (Section 4.3) that the electron pressure is estimated to be given by $\log p_e \approx 1.7$. Inserting numerical values in Equation 4.11, we find $\log \phi \approx -7.1$. In the upper photosphere, where $T \approx 4900$ K ($\theta \approx 1.03$), and $\log p_e \approx 0$ (Section 4.3), $\log \phi \approx -8.45$. In view of these estimates for upper and lower photosphere, we conclude that the average value of $\log \phi$ in the solar photosphere is not far from $-8$. It was a value of precisely this order, $\phi \approx 10^{-8}$, which we found in Chapter 3 (Section 3.4) to be necessary to have $H^-$ contribute the dominant opacity in the solar photosphere. This is a satisfying closure of the physics that suggest why the Sun's atmosphere behaves the way it does as regards interactions between photons and the material atoms/ions that make up the gas.

It is striking that the gross properties of the atmosphere of a macroscopic object such as the Sun are controlled by the detailed properties of an ion that is not only rarely alluded to ($H^-$) but that is also present in only tiny amounts (one part per 100 million) relative to the number of hydrogen atoms in the solar photosphere.

## EXERCISES

4.1 The ionization strips in Figure 4.1 are defined by the somewhat arbitrary percentages of 10% and 90% ionization. Determine where the ionization strips lie in the $\log p_e - \log T$ plane for the cases where the ionization percentages are 0.1% and 99.9%. Do this for H I, He I, and He II.

4.2 In Section 4.5, there is a definition of a transition point between the two segments of the ionization contours. Use the appropriate ionization potentials to evaluate the temperature of the transition point for H I, He I, and He II.

4.3 Using the simplified equation (4.5'), show that the fraction of neon ($I_p = 21.56$ eV) that is singly ionized in a gas with $T = 5700$ K and $\log p_e = 1.5$ is less than $10^{-11}$.

4.4 The strongest absorption lines in the visible solar spectrum are two lines labeled H and K by Frauhofer. They are due to transitions from the ground state of singly ionized calcium (Ca II) ($I_p = 6.113$ eV). In the photosphere of the Sun ($T = 5772$ K, $\log p_e = 1.7$), use Equation 4.5' to show that the number density of Ca II ions exceeds the number density of neutral Ca I atoms by a factor of about 250.

4.5 Comparing the results of Exercises 4.3 and 4.4, it can be seen that as we go from $I_p = 6.113$ eV to 21.56 eV in conditions that are close to those that prevail in the photosphere of the Sun, the degree of ionization decreases by more than 13 orders of magnitude. In those conditions (say, $T = 5750$ K and $\log p_e = -1.6$), show that a value of $I_p \approx 9.0$ eV would be required for an element to be 50% ionized, i.e., $n_i = n_a$. Consult a table of ionization potentials to determine which two elements in the periodic table comes closest (within ±0.01 eV) to satisfying $I_p = 9.0$ eV. (Both of these elements are found to be extremely rare in the Sun, with the numbers of atoms in any given volume being less than $10^{-10}$ times the number of hydrogen atoms. Therefore, even though the atoms are 50% ionized, neither of the two elements contributes in any significant way to the supply of free electrons that exist in the solar photosphere.)

## REFERENCES

Heisenberg, W., 1927. "Über der anschaulichen Inhalt der quantentheoretischen Kinematik und Mechanik", *Zeits. f. Phys.* 43, 172.

Pauli, W., 1925. "Über den Zusammenhang der Abschlusses der Elektronengruppen im Atom mit der Komplexstructure der Spectren", *Zeits. f. Phys.* 31, 765.

Payne, C. H., 1925. "Stellar atmospheres: A contribution to the observational study of high temperature in the reversing layers of stars", *Harvard Obs. Monograph Number 1*, published by Harvard Obs., Cambridge, MA.

Saha, M. N., 1921. "On a physical theory of stellar spectra", *Proc. Roy. Soc London A*, 99, 135.

Vernazza, J. E., Avrett., E. H., & Loeser, R., 1973. "Structure of the solar chromosphere. Basic computations and summary of the results", *Astrophys. J.* 184, 605.

# 5 Computing a Model of the Sun
## *The Photosphere*

Now that we have information about opacity, we are almost ready to undertake the calculation of a model of the first segment of the Sun accessible to modeling: the photosphere. In calculating this model, we do not inquire into the *origin of the energy* that flows through the solar atmosphere. Instead, we simply take the luminosity (or flux of radiant energy passing through 1 square centimeter of solar surface every second) as a given, and calculate how the physical parameters of the medium arrange themselves so as to "handle" the energy passing through.

But first we require the answer to two questions pertaining to the physical properties of the gas in the photosphere: (i) how is the pressure related to the temperature? (ii) How does the pressure vary with height? As regards (i), the gas in the solar photosphere is of sufficiently low density $\rho$ and of sufficiently high temperature $T$ that the gas can be taken to behave as a perfect gas, with pressure given by the formula $p = R_g T \rho / \mu$. Here $\mu$ is the mean molecular weight, and $R_g = 8.31448 \times 10^7$ ergs deg$^{-1}$ mole$^{-1}$ is the gas constant. The chemical composition of the solar photosphere, consisting of some 90% hydrogen (by number), about 9% of helium, and about 1% of heavier elements ("metals"), leads to $\mu \approx 1.3$ in the photosphere. Moreover, between the upper and lower photosphere (as defined earlier), the temperature ranges from the boundary value $T_o \approx 4900$ K (in the upper photosphere) to a temperature of order 6000 K (at the with photosphere): the latter will be considered to be the deepest layer of gas that we are studying in this chapter. Over such a range of temperature, variations in the degree of ionization of hydrogen and helium are very small. This allows us to assume, without significant error, that $\mu_a$ remains essentially constant throughout the photosphere.

## 5.1 HYDROSTATIC EQUILIBRIUM: THE SCALE HEIGHT

As for question (ii), the variation of gas pressure with height in any medium may be determined readily if the medium satisfies the condition of hydrostatic equilibrium (HSE). The HSE condition is applicable if the pressure $p(h)$ of the atmospheric material at any location (at height $h$) supports the weight of all of the atmospheric material located at heights above $h$. The equation that describes HSE is

$$\frac{dp}{dh} = -g\rho \qquad (5.1)$$

where $g$ is the local acceleration due to gravity, acting downward, in the direction of decreasing $h$. In the photosphere of the Sun, we have already mentioned (see Equation 1.13) that $g = 27{,}420$ cm sec$^{-2}$. The value of $g$ at any particular height $h$ decreases as $h$ increases in proportion to $1/(R_\odot + h)^2$. When we consider a photospheric model in which the height $h$ varies over a range of (say) $\Delta h = 500$ km, the relative change in gravity $\Delta g/g$ from lower to upper photosphere is given by

$$\frac{\Delta g}{g} \approx \frac{2\Delta h}{R_\odot} \approx 10^{-3} \qquad (5.2)$$

The ratio $\Delta g/g$ is so small that, in a model where we are considering only the gas that is in the photosphere, $g$ may safely be taken to be a constant without significant error.

DOI: 10.1201/9781003153115-5

In a medium where $g$ is constant, a particular solution of HSE provides a useful length-scale characteristic of the distance one must travel vertically in order to make the pressure change by a factor of $e = 2.71828\ldots$ To see this, consider a medium that is an isothermal perfect gas, with $p = R_g T\rho/\mu$. Then the solution of Equation 5.1 for the pressure $p(h)$ at height $h$ is as follows:

$$p(h) = p(0)\exp\left(\frac{-h}{H_p}\right) \quad (5.3)$$

And for the density at height $h$, the solution is analogous:

$$\rho(h) = \rho(0)\exp\left(\frac{-h}{H_p}\right) \quad (5.4)$$

That is, the pressure and the density both decrease exponentially as the height $h$ increases in the upward direction. At some arbitrary height, which is chosen as the zero point of $h$, the local pressure and density are $p(0)$ and $\rho(0)$, respectively. The characteristic length scale $H_p$ is referred to as the "scale height", "pressure scale height", or "density scale height" of the isothermal atmosphere. The formula for $H_p$ is $R_g T/g\mu$. Inserting numerical values of $T = 4900$–$6000$ K, $g = 27420$ cm sec$^{-2}$, and $\mu = 1.3$, we find that in the photosphere, $H_p$ varies over the range 114–140 km, i.e.,

$$H_p = (1.14 - 1.4)\times 10^7\, cm \quad (5.5)$$

Therefore, *if* the gas in the solar photosphere is in fact in HSE, then we should expect to find, observationally, that the gas density/pressure fall off exponentially as a function of height with a scale height of 114–140 km. Is there any observational evidence that the gas in the solar photosphere is actually obeying this behavior? To address this, Saint-Hilaire et al. (2010) analyzed X-rays from a sample of almost 1000 flares observed by the RHESSI spacecraft with X-ray energies of at least 25 keV. Assuming that a beam of electrons was involved in each flare, with the beam emitting X-rays by penetrating downward into denser and denser gas (in a so-called thick-target model), Saint-Hilaire et al. were able to derive how the local gas density varied as a function of height. They found that an exponential variation with height fits their data well. Averaging over all flares in their sample, they found that in the "average flaring atmosphere" in the Sun, the density scale height at low altitudes was found to have a value of $131 \pm 16$ km. This range overlaps well with the theoretical prediction for the photospheric gas in Equation 5.5. The RHESSI X-ray data suggest that HSE is indeed a good approximation for the gas in the Sun's photosphere.

The atmosphere lying above the level where $h = 0$ presses down on the gas at $h = 0$ due to the gravitational pull of all the matter in the Sun that lies between the level $h=0$ and the center of the Sun. The weight of the overlying material exerts a pressure on the gas at $h = 0$. In order to evaluate the pressure, let us consider a 1 cm$^2$ horizontal element of area at $h = 0$, and let us imagine a column with cross-sectional area 1 cm$^2$ extending upward to infinity from that element. The total amount of mass in that column can be obtained by integrating Equation 5.4 from $h = 0$ to $h = \infty$. The result is a mass column density $d_c$ equal to $\rho(0)H_p$ gm cm$^{-2}$. An alternative way to state this information is to note that the column density, i.e., the number $N_c$ of atoms in a square centimeter column above $h = 0$ equals $n(0)H_p$ cm$^{-2}$, where $n(0)$ is the number density of atoms (per cubic centimeter) at $h = 0$.

We shall find that in the solar photosphere, $\rho(0) \approx (2-3) \times 10^{-7}$ gm cm$^{-3}$, i.e., $n(0) \approx (1-2) \times 10^{17}$ cm$^{-3}$. Combining $\rho(0)$ and $n(0)$ with a mean $H_p \approx 130$ km, we find $d_c \approx (3-4)$ gm cm$^{-2}$, and $N_c \approx (1-3) \times 10^{24}$ cm$^{-2}$. In HSE, the pressure that occurs due to the weight of this column is $p(0) = d_c g \approx 10^5$ dyn cm$^{-2}$. (We shall want to check, when we compute a model of the photosphere, that our model yields pressures of this order: see Section 5.6.)

For comparison, we note that the atmospheric pressure on the surface of the Earth is about 10 times larger than the photospheric $p(0)$. Of course, the processes that determine the atmospheric density and pressure at the surface of the Earth are very different from those that determine $\rho(0)$ in the Sun: the latter is determined by the requirement that the optical depth $\tau(0)$ be of order unity (see Sections 2.4 and 3.4). There is no such requirement for the Earth: the fact that an observer standing on the surface of the Earth can see the Sun and stars clearly indicates that the optical depth $\tau_E$ of cloud-free atmosphere (in visible light) is actually considerably less than unity.

## 5.2 SHARP EDGE OF THE SUN'S DISK

Before moving on to the photospheric model, we make a short diversion here to address a problem that we now have enough information to solve. Combining the scale height in the photosphere with the results of Chapter 2 (Section 2.5.2) helps us to understand why the Sun, although a gaseous body, has nevertheless an edge that appears sharp when we observe the Sun from our Earth-based vantage point.

It is a fact of life on Earth that the atmosphere we breathe is in turbulent motion: variations in temperature and pressure from one location on Earth to another have the effect that the atmospheric gas is forced into motion (i.e., winds are generated). Typical wind speeds in large-scale "weather patterns" (hurricane, jet stream) can reach values of more than 100 km hr$^{-1}$. Even in "mild conditions", on length-scales $L$ of order hundreds of meters, wind speeds $u$ of order meters sec$^{-1}$ are not uncommon. The Reynolds number for such air flow, $Re = uL/\nu$, depends on the kinematic viscosity $\nu$ of air. With typical values of $\nu \approx 0.001–0.01$ m$^2$ sec$^{-1}$, even mild winds have $Re$ values in excess of the critical value for the onset of turbulence ($Re \approx 10^3$) (cf. "Laminar-turbulent transition" in Wikipedia). When we observe a distant object through the atmosphere, the turbulence causes smearing of the object. This is referred to as "the effects of seeing". As a result of "seeing", it is typically true that an observer on the Earth cannot distinguish two objects that are closer together than about 1 arcsec.

Suppose an observer wishes to make two measurements of solar intensity, $I_1$ and $I_2$, near the limb of the Sun. The first measurement $I_1$ is along a line of sight that is as nearly as possible "on the limb". This line of sight, at its closest approach to the Sun, passes through gas at a certain height $h_1$ in the upper photosphere. This is the measurement which, in visible light, yields an intensity $I_1 = a_\lambda = 0.4$ (relative to disk center) (see Equation 2.5). According to the results of Chapter 2 (Section 2.5.2), the value of $I_1$ is determined by the product of the local source function $S_\lambda$ times the optical depth $\tau_1$ along the line of sight: $I_1 = \tau_1 S_\lambda$. To make the second measurement of intensity, $I_2$, the observer chooses a line of sight that is displaced off the limb by the smallest possible amount permitted by "seeing". This second line of sight will be shifted by 1 arcsec relative to the first (because of "seeing"). Therefore, the second line of sight, at its closest approach to the Sun, will pass through gas lying at a height $h_2 = h_1 + 730$ km (see Chapter 1, Section 1.2). The gas lying at height $h_2$ has a density that is reduced below that at height $h_1$ by a factor $\exp[(h_2 - h_1)/H_p] \approx e^{5.6} = 10^{2.44} \approx 270$. The reduction in density has the effect that the optical depth through such gas $\tau_2$ is less than $\tau_1$ by a factor of 270. Therefore the intensity $I_2 = \tau_1 S_\lambda/270$. We have seen (Chapter 2, Section 2.10) that the temperature in the upper photosphere approaches a constant value as height increases. Thus, $T$ does not change significantly between $h_2$ and $h_1$. To the extent that the source function can be identified with the Planck function, this means that $S_\lambda$ is essentially the same along both lines of sight. Therefore, $I_2 = I_1/270 = 0.0015$.

That is, by shifting my line of sight by a mere 1 arcsec away from the limb, I measure that the observed intensity falls off by a factor of almost 300. This is in contrast to what happens on the solar disk: as the line of sight is moved from disk center to the limb, i.e., as the line of sight traverses some 960 arcsec, the intensity decreases gradually from 1.0 to 0.4. But with a further shift of only 1 arcsec in the line of sight, the intensity falls by a further factor of almost 300. It is this fact that gives the Sun its sharp edge when viewed from a distance of 1 AU.

## 5.3 PREPARING TO COMPUTE A MODEL OF THE SOLAR PHOTOSPHERE

The aim of this exercise is to combine HSE and the temperature structure of the gray atmosphere to calculate a table of values of various physical parameters as a function of the vertical height coordinate (increasing upward). The model begins by tabulating temperature as a function of optical depth $\tau$ (increasing downward). Transformations between $\tau$ and $h$ require knowledge of the opacity as a function of relevant physical parameters.

The HSE equation (Equation 5.1) can be converted to an optical depth scale by noting the definition $d\tau = -\kappa\rho dh$ where $\kappa$ is the (gray) opacity: we shall use the Rosseland mean opacity, and we shall set $\kappa = \kappa_R$ in our calculations. This leads to the central equation for the present chapter:

$$\frac{dp}{d\tau} = \frac{g}{\kappa} \quad (5.6)$$

In order to solve this equation, we need to have access to values of Rosseland mean opacities, $\kappa_R$ as a function of temperature and pressure. A table of such values (see Kurucz, R. L. 1992) was kindly made available by Dr. R. L. Kurucz of the Harvard-Smithsonian Center for Astrophysics. For the convenience of the reader, these are presented in Tables 5.1 and 5.2. (The reader may also be able to find results obtained by other researchers on the internet, or the reader could in principle extract Rosseland mean opacities at different densities and temperatures by examining Figure 3.4 in detail.) In the following tables, the (log) opacities are tabulated as functions of temperature and gas pressure for a mixture of elemental abundances that is a "standard" solar mixture. In order to calculate the opacity, bound-bound, bound-free, and free-free transitions are included for many stages of ionization of all elements in the mixture. Negative hydrogen ions and hydrogen molecules are also included. For bound-bound transitions, the lines are assumed to be broadened with a microturbulent velocity of 2 km sec$^{-1}$. Each *row* of Tables 5.1 and 5.2 is labeled with LT, which is equal to the logarithm (to base 10) of the temperature (in degrees K): the temperatures range from close to 2000 K to almost 100,000 K. Each *column* of Tables 5.1 and 5.2 is labeled with LP, which is equal to the logarithm (to base 10) of the gas pressure (in dyn cm$^{-2}$): the pressures range over five orders of magnitude. In order to fit the results into a standard page width, the opacities are presented in the form of two tables (Tables 5.1 and 5.2), corresponding to a low subrange of pressure and a high subrange of pressure, respectively. The tabulated values of opacity exhibit the overall behavior described earlier in Chapter 3 (Figure 3.3) (where the results were presented in a different format): (i) in the limit of high temperature (and low pressure), $\log(\kappa) \to -0.5$; (ii) in the limit of low temperature, $\log(\kappa)$ tends to very small values; (iii) numerical values of opacity reach maximum values at $\log(T) = 4.0-4.5$; (iv) maximum values of opacity in the tables are $10^4-10^5$ cm$^2$ gm$^{-1}$.

The goal of the present chapter is to calculate a tabulated model of the solar photosphere. This means that we wish to obtain a table of values where each row of the table refers to a particular *optical depth* in the atmosphere. On that row, our goal is to provide numerical values for the temperature, pressure, density, and height in the solar atmosphere.

## 5.4 COMPUTING A MODEL OF THE PHOTOSPHERE: STEP BY STEP

The calculation proceeds by way of the following steps.

1. Choose a value of $\tau$ for the first row in the tabulated model: e.g., $\tau(1) = 10^{-4}$.
2. For row 1, choose the vertical depth coordinate to be $z(1) = 0$. (This is an arbitrary choice and is done merely for convenience. Afterward, you may change the zero point of height if you choose.) We will use $h$ (increasing upward) and $-z$ (where $z$ increases downward) interchangeably in the calculation.
3. Calculate the temperature in row 1, $T(1)$, from the Eddington solution (Chapter 2, Equation 2.40), using $T_{\text{eff}} = 5772$ K (see Chapter 1, Equation 1.17).

**TABLE 5.1**

$\log_{10}(\kappa)$ in Units of cm² gm⁻¹ for Pressures (dyn cm²) in the Lower Subrange $LP = \log_{10}(p) = 3.00$–$5.20$ and for temperatures in the range $LT = \log_{10}(T) = 3.32$–$4.90$

| LP→ | 3.00 | 3.20 | 3.40 | 3.60 | 3.80 | 4.00 | 4.20 | 4.40 | 4.60 | 4.80 | 5.00 | 5.20 |
|---|---|---|---|---|---|---|---|---|---|---|---|---|
| LT |  |  |  |  |  |  |  |  |  |  |  |  |
| 3.32 | −5.16 | −5.13 | −5.08 | −5.04 | −4.99 | −4.93 | −4.87 | −4.80 | −4.74 | −4.66 | −4.58 | −4.49 |
| 3.34 | −4.88 | −4.85 | −4.82 | −4.78 | −4.73 | −4.68 | −4.63 | −4.58 | −4.52 | −4.47 | −4.40 | −4.34 |
| 3.36 | 4.56 | −4.53 | −4.49 | −4.46 | −4.42 | −4.37 | −4.33 | −4.28 | −4.23 | −4.18 | −4.12 | −4.06 |
| 3.38 | −4.23 | −4.19 | −4.15 | −4.11 | −4.07 | −4.03 | −3.99 | −3.95 | −3.91 | −3.86 | −3.81 | −3.76 |
| 3.40 | −3.94 | −3.88 | −3.82 | −3.77 | −3.72 | −3.68 | −3.64 | −3.61 | −3.57 | −3.53 | −3.49 | −3.44 |
| 3.42 | −3.70 | −3.61 | −3.54 | −3.46 | −3.40 | −3.35 | −3.30 | −3.26 | −3.22 | −3.18 | −3.14 | −3.11 |
| 3.44 | −3.50 | −3.40 | −3.30 | −3.21 | −3.12 | −3.05 | −2.99 | −2.93 | −2.88 | −2.84 | −2.80 | −2.76 |
| 3.46 | −3.35 | −3.22 | −3.10 | −2.99 | −2.89 | −2.80 | −2.72 | −2.64 | −2.58 | −2.53 | −2.48 | −2.43 |
| 3.48 | −3.21 | −3.08 | −2.95 | −2.82 | −2.70 | −2.59 | −2.48 | −2.39 | −2.32 | −2.24 | −2.18 | −2.13 |
| 3.50 | −3.07 | −2.94 | −2.81 | −2.67 | −2.54 | −2.41 | −2.30 | −2.19 | −2.09 | −2.00 | −1.92 | −1.85 |
| 3.52 | −2.91 | −2.79 | −2.66 | −2.53 | −2.40 | −2.27 | −2.14 | −2.02 | −1.90 | −1.80 | −1.70 | −1.62 |
| 3.54 | −2.72 | −2.61 | −2.49 | −2.37 | −2.24 | −2.12 | −1.99 | −1.87 | −1.75 | −1.63 | −1.52 | −1.42 |
| 3.56 | −2.54 | −2.43 | −2.31 | −2.19 | −2.07 | −1.96 | −1.84 | −1.71 | −1.59 | −1.48 | −1.36 | −1.25 |
| 3.58 | −2.39 | −2.27 | −2.14 | −2.02 | −1.90 | −1.79 | −1.67 | −1.55 | −1.43 | −1.32 | −1.20 | −1.09 |
| 3.60 | −2.28 | −2.14 | −2.01 | −1.88 | −1.75 | −1.63 | −1.51 | −1.39 | −1.27 | −1.15 | −1.04 | −0.92 |
| 3.62 | −2.22 | −2.07 | −1.92 | −1.77 | −1.64 | −1.50 | −1.37 | −1.25 | −1.12 | −1.00 | −0.88 | −0.77 |
| 3.64 | −2.20 | −2.03 | −1.87 | −1.71 | −1.56 | −1.41 | −1.27 | −1.13 | −1.00 | −0.87 | −0.75 | −0.62 |
| 3.66 | −2.21 | −2.03 | −1.86 | −1.69 | −1.52 | −1.36 | −1.21 | −1.05 | −0.91 | −0.77 | −0.63 | −0.50 |
| 3.68 | −2.19 | −2.03 | −1.86 | −1.69 | −1.52 | −1.35 | −1.18 | −1.01 | −0.86 | −0.70 | −0.56 | −0.41 |
| 3.70 | −2.11 | −1.97 | −1.83 | −1.67 | −1.51 | −1.34 | −1.17 | −1.00 | −0.83 | −0.67 | −0.51 | −0.35 |
| 3.73 | −1.86 | −1.75 | −1.63 | −1.51 | −1.39 | −1.25 | −1.11 | −0.96 | −0.80 | −0.64 | −0.48 | −0.31 |
| 3.76 | −1.53 | −1.44 | −1.34 | −1.24 | −1.13 | −1.02 | −0.91 | −0.79 | −0.66 | −0.53 | −0.39 | −0.25 |
| 3.79 | −1.18 | −1.10 | −1.01 | −0.92 | −0.82 | −0.73 | −0.63 | −0.53 | −0.42 | −0.31 | −0.20 | −0.08 |
| 3.82 | −0.81 | −0.74 | −0.66 | −0.58 | −0.50 | −0.41 | −0.32 | −0.23 | −0.14 | −0.04 | 0.06 | 0.16 |

*(Continued)*

**TABLE 5.1 (Continued)**

| LP→ | 3.00 | 3.20 | 3.40 | 3.60 | 3.80 | 4.00 | 4.20 | 4.40 | 4.60 | 4.80 | 5.00 | 5.20 |
|---|---|---|---|---|---|---|---|---|---|---|---|---|
| 3.85 | −0.43 | −0.37 | −0.30 | −0.23 | −0.16 | −0.08 | 0.00 | 0.08 | 0.16 | 0.25 | 0.34 | 0.43 |
| 3.88 | −0.03 | 0.02 | 0.07 | 0.13 | 0.19 | 0.26 | 0.33 | 0.40 | 0.47 | 0.55 | 0.63 | 0.71 |
| 3.91 | 0.37 | 0.41 | 0.46 | 0.51 | 0.56 | 0.61 | 0.67 | 0.73 | 0.79 | 0.86 | 0.93 | 1.00 |
| 3.94 | 0.76 | 0.80 | 0.84 | 0.88 | 0.93 | 0.97 | 1.02 | 1.07 | 1.12 | 1.18 | 1.23 | 1.30 |
| 3.97 | 1.08 | 1.13 | 1.18 | 1.23 | 1.27 | 1.32 | 1.36 | 1.40 | 1.45 | 1.49 | 1.54 | 1.59 |
| 4.00 | 1.28 | 1.37 | 1.45 | 1.51 | 1.57 | 1.62 | 1.66 | 1.71 | 1.75 | 1.79 | 1.83 | 1.88 |
| 4.05 | 1.30 | 1.46 | 1.60 | 1.73 | 1.85 | 1.95 | 2.03 | 2.10 | 2.16 | 2.21 | 2.26 | 2.30 |
| 4.10 | 1.07 | 1.26 | 1.44 | 1.62 | 1.79 | 1.95 | 2.10 | 2.23 | 2.35 | 2.45 | 2.53 | 2.60 |
| 4.15 | 0.78 | 0.96 | 1.16 | 1.35 | 1.54 | 1.73 | 1.92 | 2.10 | 2.27 | 2.43 | 2.58 | 2.71 |
| 4.20 | 0.54 | 0.71 | 0.88 | 1.06 | 1.25 | 1.45 | 1.64 | 1.84 | 2.04 | 2.23 | 2.42 | 2.60 |
| 4.25 | 0.36 | 0.51 | 0.67 | 0.83 | 1.01 | 1.20 | 1.38 | 1.58 | 1.77 | 1.97 | 2.17 | 2.38 |
| 4.30 | 0.21 | 0.34 | 0.48 | 0.63 | 0.80 | 0.98 | 1.16 | 1.35 | 1.55 | 1.74 | 1.94 | 2.14 |
| 4.35 | 0.09 | 0.20 | 0.32 | 0.47 | 0.62 | 0.78 | 0.96 | 1.15 | 1.34 | 1.54 | 1.74 | 1.94 |
| 4.40 | 0.00 | 0.09 | 0.19 | 0.32 | 0.46 | 0.61 | 0.78 | 0.96 | 1.15 | 1.35 | 1.54 | 1.75 |
| 4.45 | −0.04 | 0.04 | 0.12 | 0.23 | 0.35 | 0.48 | 0.64 | 0.81 | 0.99 | 1.18 | 1.37 | 1.58 |
| 4.50 | −0.04 | 0.02 | 0.10 | 0.19 | 0.30 | 0.41 | 0.55 | 0.70 | 0.87 | 1.04 | 1.23 | 1.43 |
| 4.55 | −0.08 | 0.00 | 0.08 | 0.17 | 0.27 | 0.38 | 0.51 | 0.64 | 0.80 | 0.96 | 1.13 | 1.32 |
| 4.60 | −0.15 | −0.08 | 0.00 | 0.10 | 0.21 | 0.33 | 0.46 | 0.60 | 0.75 | 0.90 | 1.07 | 1.24 |
| 4.65 | −0.22 | −0.17 | −0.10 | −0.02 | 0.08 | 0.20 | 0.33 | 0.47 | 0.63 | 0.80 | 0.97 | 1.15 |
| 4.70 | −0.29 | −0.25 | −0.20 | −0.13 | −0.05 | 0.05 | 0.16 | 0.29 | 0.44 | 0.60 | 0.77 | 0.95 |
| 4.75 | −0.33 | −0.30 | −0.27 | −0.22 | −0.16 | −0.09 | 0.00 | 0.11 | 0.24 | 0.38 | 0.54 | 0.71 |
| 4.80 | −0.35 | −0.33 | −0.31 | −0.28 | −0.24 | −0.19 | −0.13 | −0.05 | 0.05 | 0.17 | 0.31 | 0.46 |
| 4.85 | −0.35 | −0.34 | −0.32 | −0.31 | −0.28 | −0.25 | −0.21 | −0.15 | −0.08 | 0.01 | 0.12 | 0.24 |
| 4.90 | −0.36 | −0.35 | −0.34 | −0.32 | −0.30 | −0.27 | −0.24 | −0.20 | −0.15 | −0.09 | −0.01 | 0.08 |

*Source:* Used with the permission of R. L. Kurucz (Harvard-Smithsonian Center for Astrophysics).

## TABLE 5.2
$\log_{10}(\kappa_R)$ in Units of cm$^2$ gm$^{-1}$ for Pressures (dyn cm$^{-2}$) in the Upper Subrange $LP = \log_{10}(p) = 5.40$–$8.00$ and for temperatures in the range $LT = \log_{10}(T) = 3.32$–$4.90$

| LP→<br>LT | 5.40 | 5.60 | 5.80 | 6.00 | 6.25 | 6.50 | 6.75 | 7.00 | 7.25 | 7.50 | 7.75 | 8.00 |
|---|---|---|---|---|---|---|---|---|---|---|---|---|
| 3.32 | −4.39 | −4.27 | −4.14 | −3.99 | −3.78 | −3.51 | −3.19 | −2.98 | −2.88 | −2.77 | −2.67 | −2.57 |
| 3.34 | −4.28 | −4.21 | −4.14 | −4.07 | −3.99 | −3.90 | −3.81 | −3.72 | −3.63 | −3.54 | −3.44 | −3.36 |
| 3.36 | −4.00 | −3.93 | −3.87 | −3.80 | −3.71 | −3.62 | −3.53 | −3.44 | −3.35 | −3.26 | −3.17 | −3.08 |
| 3.38 | −3.70 | −3.64 | −3.58 | −3.51 | −3.43 | −3.34 | −3.26 | −3.17 | −3.08 | −2.99 | −2.90 | −2.81 |
| 3.40 | −3.39 | −3.34 | −3.28 | −3.22 | −3.14 | −3.06 | −2.98 | −2.89 | −2.80 | −2.72 | −2.63 | −2.55 |
| 3.42 | −3.06 | −3.02 | −2.97 | −2.92 | −2.85 | −2.77 | −2.69 | −2.61 | −2.53 | −2.45 | −2.36 | −2.28 |
| 3.44 | −2.73 | −2.69 | −2.64 | −2.60 | −2.54 | −2.47 | −2.40 | −2.33 | −2.25 | −2.17 | −2.10 | −2.02 |
| 3.46 | −2.39 | −2.35 | −2.32 | −2.28 | −2.22 | −2.17 | −2.11 | −2.04 | −1.97 | −1.90 | −1.83 | −1.75 |
| 3.48 | −2.08 | 2.03 | −1.99 | −1.96 | −1.91 | −1.86 | −1.81 | −1.75 | −1.69 | −1.63 | −1.56 | −1.49 |
| 3.50 | −1.79 | −1.74 | −1.69 | −1.65 | −1.60 | −1.55 | −1.50 | −1.45 | −1.40 | −1.35 | −1.29 | −1.23 |
| 3.52 | −1.54 | −1.48 | −1.42 | −1.37 | −1.31 | −1.26 | −1.21 | −1.16 | −1.12 | −1.07 | −1.02 | −0.96 |
| 3.54 | −1.33 | −1.25 | −1.18 | −1.11 | −1.04 | −0.98 | −0.93 | −0.88 | −0.84 | −0.79 | −0.75 | −0.70 |
| 3.56 | −1.15 | −1.05 | −0.97 | −0.89 | −0.81 | −0.74 | −0.67 | −0.62 | −0.57 | −0.53 | −0.49 | −0.44 |
| 3.58 | −0.98 | −0.88 | −0.78 | −0.69 | −0.60 | −0.51 | −0.44 | −0.38 | −0.32 | −0.28 | −0.23 | −0.19 |
| 3.60 | −0.81 | −0.71 | −0.61 | −0.51 | −0.41 | −0.31 | −0.23 | −0.16 | −0.09 | −0.04 | 0.00 | 0.05 |
| 3.62 | −0.65 | −0.55 | −0.44 | −0.34 | −0.23 | −0.13 | −0.04 | 0.04 | 0.11 | 0.17 | 0.22 | 0.27 |
| 3.64 | −0.51 | −0.39 | −0.28 | −0.18 | −0.06 | 0.05 | 0.14 | 0.23 | 0.30 | 0.37 | 0.43 | 0.48 |
| 3.66 | −0.38 | −0.26 | −0.14 | −0.04 | 0.09 | 0.21 | 0.31 | 0.40 | 0.48 | 0.56 | 0.62 | 0.67 |
| 3.68 | −0.28 | −0.15 | −0.03 | 0.09 | 0.23 | 0.35 | 0.47 | 0.57 | 0.65 | 0.73 | 0.79 | 0.85 |
| 3.70 | −0.21 | −0.07 | 0.07 | 0.19 | 0.34 | 0.48 | 0.60 | 0.71 | 0.81 | 0.89 | 0.96 | 1.02 |
| 3.73 | −0.15 | 0.00 | 0.15 | 0.30 | 0.46 | 0.62 | 0.76 | 0.89 | 1.01 | 1.11 | 1.19 | 1.25 |
| 3.76 | −0.10 | 0.06 | 0.21 | 0.36 | 0.54 | 0.71 | 0.87 | 1.02 | 1.16 | 1.28 | 1.38 | 1.46 |
| 3.79 | 0.05 | 0.18 | 0.31 | 0.45 | 0.62 | 0.79 | 0.96 | 1.12 | 1.27 | 1.41 | 1.53 | 1.63 |

*(Continued)*

**TABLE 5.2 (Continued)**

| LP→ | 5.40 | 5.60 | 5.80 | 6.00 | 6.25 | 6.50 | 6.75 | 7.00 | 7.25 | 7.50 | 7.75 | 8.00 |
|---|---|---|---|---|---|---|---|---|---|---|---|---|
| 3.82 | 0.27 | 0.38 | 0.50 | 0.62 | 0.76 | 0.91 | 1.06 | 1.22 | 1.37 | 1.52 | 1.65 | 1.77 |
| 3.85 | 0.53 | 0.63 | 0.73 | 0.83 | 0.96 | 1.09 | 1.23 | 1.37 | 1.50 | 1.64 | 1.78 | 1.90 |
| 3.88 | 0.80 | 0.89 | 0.98 | 1.07 | 1.19 | 1.31 | 1.43 | 1.55 | 1.68 | 1.81 | 1.93 | 2.06 |
| 3.91 | 1.08 | 1.15 | 1.24 | 1.32 | 1.43 | 1.54 | 1.65 | 1.76 | 1.88 | 1.99 | 2.11 | 2.23 |
| 3.94 | 1.36 | 1.43 | 1.50 | 1.57 | 1.67 | 1.77 | 1.87 | 1.97 | 2.08 | 2.19 | 2.30 | 2.41 |
| 3.97 | 1.65 | 1.71 | 1.76 | 1.83 | 1.91 | 2.00 | 2.09 | 2.18 | 2.28 | 2.38 | 2.49 | 2.59 |
| 4.00 | 1.92 | 1.97 | 2.02 | 2.08 | 2.15 | 2.23 | 2.31 | 2.39 | 2.48 | 2.57 | 2.67 | 2.77 |
| 4.05 | 2.35 | 2.39 | 2.43 | 2.48 | 2.54 | 2.60 | 2.66 | 2.73 | 2.81 | 2.89 | 2.97 | 3.07 |
| 4.10 | 2.67 | 2.72 | 2.78 | 2.82 | 2.88 | 2.94 | 3.00 | 3.06 | 3.13 | 3.20 | 3.28 | 3.36 |
| 4.15 | 2.82 | 2.92 | 3.00 | 3.08 | 3.16 | 3.23 | 3.30 | 3.36 | 3.43 | 3.50 | 3.57 | 3.65 |
| 4.20 | 2.77 | 2.92 | 3.06 | 3.19 | 3.32 | 3.43 | 3.52 | 3.61 | 3.69 | 3.76 | 3.84 | 3.92 |
| 4.25 | 2.57 | 2.76 | 2.95 | 3.13 | 3.32 | 3.50 | 3.64 | 3.77 | 3.88 | 3.98 | 4.07 | 4.16 |
| 4.30 | 2.35 | 2.55 | 2.76 | 2.95 | 3.20 | 3.43 | 3.63 | 3.82 | 3.98 | 4.12 | 4.24 | 4.35 |
| 4.35 | 2.14 | 2.35 | 2.56 | 2.76 | 3.03 | 3.28 | 3.53 | 3.76 | 3.97 | 4.16 | 4.33 | 4.48 |
| 4.40 | 1.96 | 2.16 | 2.38 | 2.59 | 2.85 | 3.12 | 3.38 | 3.64 | 3.88 | 4.12 | 4.33 | 4.52 |
| 4.45 | 1.78 | 1.99 | 2.20 | 2.41 | 2.68 | 2.95 | 3.23 | 3.50 | 3.77 | 4.03 | 4.26 | 4.49 |
| 4.50 | 1.63 | 1.83 | 2.05 | 2.26 | 2.53 | 2.79 | 3.07 | 3.35 | 3.63 | 3.90 | 4.17 | 4.41 |
| 4.55 | 1.51 | 1.70 | 1.91 | 2.11 | 2.38 | 2.64 | 2.92 | 3.19 | 3.47 | 3.75 | 4.03 | 4.29 |
| 4.60 | 1.41 | 1.60 | 1.79 | 1.99 | 2.24 | 2.50 | 2.76 | 3.03 | 3.30 | 3.58 | 3.85 | 4.12 |
| 4.65 | 1.33 | 1.50 | 1.69 | 1.87 | 2.11 | 2.36 | 2.61 | 2.86 | 3.12 | 3.38 | 3.65 | 3.91 |
| 4.70 | 1.14 | 1.33 | 1.53 | 1.72 | 1.96 | 2.20 | 2.44 | 2.67 | 2.91 | 3.16 | 3.42 | 3.67 |
| 4.75 | 0.89 | 1.08 | 1.27 | 1.47 | 1.72 | 1.96 | 2.21 | 2.45 | 2.69 | 2.93 | 3.17 | 3.41 |
| 4.80 | 0.63 | 0.81 | 0.99 | 1.19 | 1.43 | 1.67 | 1.92 | 2.17 | 2.41 | 2.66 | 2.90 | 3.14 |
| 4.85 | 0.39 | 0.55 | 0.72 | 0.90 | 1.14 | 1.38 | 1.62 | 1.87 | 2.11 | 2.36 | 2.60 | 2.84 |
| 4.90 | 0.20 | 0.33 | 0.48 | 0.64 | 0.87 | 1.10 | 1.34 | 1.58 | 1.82 | 2.07 | 2.31 | 2.55 |

*Source*: Used with the permission of R. L. Kurucz (Harvard-Smithsonian Center for Astrophysics).

# Computing a Model of the Sun

4. To obtain the pressure $p(1)$ in row 1, one could guess any finite starting value and then iterate. To avoid complications, I suggest that you simply choose the following guess: $\log p(1) = 3.0$. Note that in all cases, the log function refers to logarithms to base 10, and the physical quantities are in c.g.s. units.
5. Now that you know $T(1)$ and $p(1)$, calculate the density in row 1, $\rho(1)$, from the perfect gas expression $\rho(1) = p(1)\mu/(R_g T(1))$. Here, $R_g$ is the gas constant (see Chapter 1, Section 1.7), and $\mu$, the mean molecular weight, can be set equal to a constant value 1.3 for the material of the solar photosphere. (The value 1.3 arises because solar material is roughly 90% H, 10% He by numbers, plus less than 1% heavier elements.)
6. Now that you know $\log T(1)$ and $\log p(1)$, interpolate in Tables 5.1 and/or 5.2 to find a local value for the opacity $\kappa(1)$.
7. Step forward to the second row of the table, i.e., to the next value of $\tau$. In order to reduce numerical errors, I suggest that you keep the step size small. For example, consider using $\tau(2) = 2 \times 10^{-4}$. This means that the interval in optical depth between rows 1 and 2 is $\Delta\tau = 10^{-4}$.
8. With the new value of $\tau$, calculate the new value of $T$ from the Eddington solution. Call this $T(2)$.
9. Calculate the increase in pressure between rows 1 and 2 using an approximation to Equation 5.6: $\Delta p = g\Delta\tau/\kappa(1)$, where $g = 2.7 \times 10^4$ cm sec$^{-2}$. This then gives $p(2) = p(1) + \Delta p$.
10. Knowing $T(2)$ and $p(2)$, interpolate in the opacity table for $\kappa(2)$.
11. Calculate the density $\rho(2)$ from $T(2)$ and $p(2)$.
12. Convert the step in optical depth to a step in linear depth: $\Delta z = +\Delta\tau/(\rho(2)\kappa(2))$. If you want to be more precise, replace the denominator by the mean value of $\rho\kappa$ between row 1 and row 2: $\rho\kappa \approx 0.5(\rho(1)\kappa(1) + \rho(2)\kappa(2))$. Once $\Delta z$ is calculated, you can calculate the depth $z(2) = z(1) + \Delta z$ which is appropriate for row 2 of the tabulated model.
13. Calculate the local temperature gradient $dT/dz = (T(2) - T(1))/\Delta z$. (This gradient will be used later when we wish to calculate a model for the convection zone.)
   At this point, there should be seven entries in row 2: $\tau, T, p, \rho, z, \kappa$, and $(dT/dz)$.
   Use those values to step forward to row 3. For generality, we refer to the quantities in the row we have just calculated as row $i$.
14. Step forward to row $i + 1$. To start this step, choose a new value of $\tau(i + 1) = \tau(i) + \Delta\tau$. What step size should be used? Plausible choices might be $\Delta\tau = 10^{-4}$ until the optical depth $\tau$ reaches a value of $10^{-3}$. Then use a step size of $\Delta\tau = 10^{-3}$ until $\tau = 10^{-2}$. Then use a step size of $\Delta\tau = 10^{-2}$ until $\tau = 10^{-1}$. Then use a step size of $\Delta\tau = 10^{-1}$ until $\tau = 1$. Finally use a step size $\Delta\tau = 1$ until $\tau = 10$.
15. Repeat steps 8–14 multiple times until $\tau$ reaches a value of about 10. In each iteration, replace $T(1)$ in the previous instructions by $T(i)$ and $T(2)$ by $T(i + 1)$. Do the same replacements for the other variables.

## 5.5 THE OUTCOME OF THE CALCULATION

The outcome of the exercise is a table of physical quantities as a function of height: this is called a "model of the photosphere". An example of such an exercise, using the previous tables of opacity, is given in Table 5.3. At each tabulated value of optical depth (column headed "$\tau$"), we list the temperature (in units of K), pressure (units=dyn cm$^{-2}$), density (in gm cm$^{-3}$), depth (in cm, relative to an initial $h = 0$ at the top), the logarithm of the Rosseland mean opacity (in units of cm$^2$ gm$^{-1}$). In the seventh column, we give the temperature gradient, $dT/dz$ (in units of deg K cm$^{-1}$). The reason for including this quantity will be explained when we discuss the physical process known as Convection in Chapter 6.

## TABLE 5.3
## A Model of the Solar Atmosphere

| τ | Temperature | Pressure | Density | z | log(κ) | grad T |
|---|---|---|---|---|---|---|
| 2.00E-04 | 4.86E+03 | 1.77E+03 | 5.69E-09 | 1.91E+06 | $-2.04E+00$ | 9.57E-08 |
| 3.00E-04 | 4.86E+03 | 2.02E+03 | 6.52E-09 | 3.35E+06 | $-1.97E+00$ | 1.26E-07 |
| 4.00E-04 | 4.86E+03 | 2.26E+03 | 7.26E-09 | 4.51E+06 | $-1.93E+00$ | 1.57E-07 |
| 5.00E-04 | 4.86E+03 | 2.47E+03 | 7.94E-09 | 5.48E+06 | $-1.89E+00$ | 1.87E-07 |
| 6.00E-04 | 4.86E+03 | 2.66E+03 | 8.57E-09 | 6.32E+06 | $-1.86E+00$ | 2.17E-07 |
| 7.00E-04 | 4.86E+03 | 2.85E+03 | 9.17E-09 | 7.06E+06 | $-1.83E+00$ | 2.48E-07 |
| 8.00E-04 | 4.86E+03 | 3.02E+03 | 9.73E-09 | 7.71E+06 | $-1.80E+00$ | 2.78E-07 |
| 9.00E-04 | 4.86E+03 | 3.19E+03 | 1.03E-08 | 8.30E+06 | $-1.78E+00$ | 3.07E-07 |
| 1.00E-03 | 4.86E+03 | 3.35E+03 | 1.08E-08 | 8.84E+06 | $-1.76E+00$ | 3.38E-07 |
| 1.10E-03 | 4.86E+03 | 3.50E+03 | 1.13E-08 | 9.33E+06 | $-1.75E+00$ | 3.68E-07 |
| 2.10E-03 | 4.86E+03 | 4.97E+03 | 1.60E-08 | 1.27E+07 | $-1.73E+00$ | 5.42E-07 |
| 3.10E-03 | 4.86E+03 | 6.07E+03 | 1.95E-08 | 1.47E+07 | $-1.60E+00$ | 8.85E-07 |
| 4.10E-03 | 4.87E+03 | 7.00E+03 | 2.25E-08 | 1.62E+07 | $-1.53E+00$ | 1.20E-06 |
| 5.10E-03 | 4.87E+03 | 7.83E+03 | 2.51E-08 | 1.74E+07 | $-1.48E+00$ | 1.52E-06 |
| 6.10E-03 | 4.87E+03 | 8.58E+03 | 2.75E-08 | 1.84E+07 | $-1.44E+00$ | 1.82E-06 |
| 7.10E-03 | 4.87E+03 | 9.27E+03 | 2.98E-08 | 1.93E+07 | $-1.40E+00$ | 2.13E-06 |
| 8.10E-03 | 4.87E+03 | 9.92E+03 | 3.18E-08 | 2.00E+07 | $-1.37E+00$ | 2.43E-06 |
| 9.10E-03 | 4.87E+03 | 1.05E+04 | 3.38E-08 | 2.07E+07 | $-1.35E+00$ | 2.73E-06 |
| 1.01E-02 | 4.88E+03 | 1.11E+04 | 3.56E-08 | 2.13E+07 | $-1.33E+00$ | 3.02E-06 |
| 2.01E-02 | 4.89E+03 | 1.67E+04 | 5.32E-08 | 2.51E+07 | $-1.31E+00$ | 4.71E-06 |
| 3.01E-02 | 4.91E+03 | 2.06E+04 | 6.55E-08 | 2.73E+07 | $-1.16E+00$ | 8.11E-06 |
| 4.01E-02 | 4.93E+03 | 2.39E+04 | 7.57E-08 | 2.89E+07 | $-1.08E+00$ | 1.11E-05 |
| 5.01E-02 | 4.95E+03 | 2.67E+04 | 8.45E-08 | 3.01E+07 | $-1.02E+00$ | 1.39E-05 |
| 6.01E-02 | 4.96E+03 | 2.94E+04 | 9.25E-08 | 3.11E+07 | $-9.79E-01$ | 1.66E-05 |
| 7.01E-02 | 4.98E+03 | 3.18E+04 | 9.97E-08 | 3.20E+07 | $-9.45E-01$ | 1.92E-05 |
| 8.01E-02 | 5.00E+03 | 3.40E+04 | 1.06E-07 | 3.28E+07 | $-9.14E-01$ | 2.18E-05 |
| 9.01E-02 | 5.01E+03 | 3.61E+04 | 1.13E-07 | 3.35E+07 | $-8.88E-01$ | 2.43E-05 |
| 1.00E-01 | 5.03E+03 | 3.82E+04 | 1.19E-07 | 3.41E+07 | $-8.64E-01$ | 2.67E-05 |
| 2.00E-01 | 5.19E+03 | 5.67E+04 | 1.71E-07 | 3.80E+07 | $-8.30E-01$ | 3.95E-05 |
| 3.00E-01 | 5.33E+03 | 6.98E+04 | 2.05E-07 | 4.04E+07 | $-6.80E-01$ | 6.14E-05 |
| 4.00E-01 | 5.46E+03 | 8.02E+04 | 2.30E-07 | 4.20E+07 | $-5.78E-01$ | 8.05E-05 |
| 5.00E-01 | 5.59E+03 | 8.89E+04 | 2.49E-07 | 4.33E+07 | $-4.99E-01$ | 9.75E-05 |
| 6.00E-01 | 5.70E+03 | 9.64E+04 | 2.64E-07 | 4.43E+07 | $-4.38E-01$ | 1.12E-04 |
| 7.00E-01 | 5.81E+03 | 1.03E+05 | 2.77E-07 | 4.52E+07 | $-3.73E-01$ | 1.28E-04 |
| 8.00E-01 | 5.92E+03 | 1.08E+05 | 2.87E-07 | 4.59E+07 | $-3.06E-01$ | 1.47E-04 |
| 9.00E-01 | 6.01E+03 | 1.13E+05 | 2.94E-07 | 4.65E+07 | $-2.46E-01$ | 1.64E-04 |

## 5.6 OVERVIEW OF THE MODEL OF THE SOLAR PHOTOSPHERE

Now we have a table that lists certain physical properties of the gas over a range of heights in the solar photosphere. We refer to such a table as a simplified "model" of the solar photosphere. It is worthwhile to take a look at the properties of this "model" that we have obtained.

First, the gas temperature in the photosphere ($\tau = 0.667$) is 5772 K. This is of course a natural consequence of the Eddington solution (Chapter 2, Equation 2.40). But it is useful to remember that the effective temperature of the *radiation* (i.e., the photons) that comes to us from the Sun (Chapter 1, Equation 1.17) has a direct connection with the local thermodynamic temperature of the atoms *in the photosphere*. The photons we see on Earth have energies that are roughly characteristic of the thermal energies of the atoms with which the photons last interacted back in the photosphere before they started off on their free-streaming journey from the Sun to the Earth through (almost) empty space. The reason for this is that the radiation and the gas in the photosphere have strong enough interactions that radiation and gas are close to local thermodynamic *equilibrium*. (This is very different from the condition on the surface of the Earth, where the dominant photons [sunlight] have energies of a few electron volts [eV], while the gases in Earth's atmosphere have thermal energies of only 0.03 eV: this is far from thermodynamic equilibrium.)

Second, the density in the photosphere $\rho(\mathrm{ph})$ is $(2-3) \times 10^{-7}$ gm cm$^{-3}$. If the gas were purely hydrogen, the corresponding number density of atoms would be $n(\mathrm{ph}) = (1.2-1.8) \times 10^{17}$ cm$^{-3}$. (Given the presence of He and other heavier elements, the true value of $n(\mathrm{ph})$ is somewhat smaller than this.) The number column density above the photosphere $N(\mathrm{ph})$ is $n(\mathrm{ph})H_p$. Inserting $H_p = (1.14-1.4) \times 10^7$ cm (Equation 5.5), we find $N(\mathrm{ph}) = (1.4-2.5) \times 10^{24}$ cm$^{-2}$. This is the number of atoms that lie above each square centimeter of the solar photosphere. Of these, roughly 1 in $10^8$ is an $H^-$ ion. Thus, there are some $N(H^-) = (1.4-2.5) \times 10^{16}$ $H^-$ ions lying above each square centimeter of the photosphere.

Third, the pressure in the photosphere $p(\mathrm{ph})$ is of order $10^5$ dyn cm$^{-2}$. This is the pressure necessary to support the weight of the overlying gas. Recall that the mass column density $d_c$ above a 1 sq cm patch of the solar photosphere is equal to $\rho(0)H_p$ gm cm$^{-2}$. Inserting $\rho(0) = (2-3) \times 10^{-7}$ gm cm$^{-3}$ and $H_p = (1.14-1.4) \times 10^7$, we see that the exponents of $\rho(0)$ and $H_p$ cancel each other: as a result, the mass of gas $d_c$ that lies above each square centimeter of the photosphere is about 4 gm. In the presence of gravity with $g = 27,420$ cm sec$^{-2}$, the corresponding force, $d_c g$, pressing down on each square centimeter is therefore about 109,680 dyn cm$^{-2}$: this is close to the value of $10^5$ dyn cm$^{-2}$ mentioned in the first sentence of this paragraph. The fact that the value of the pressure $p(\mathrm{ph})$ is equal to $d_c g$ is not an accident; it indicates that the vertical forces (weight directed *downward*, pressure gradient directed *upward*) are actually in balance in the photosphere. This is another way of stating that in deriving the model atmosphere, we have assumed HSE.

Fourth, the range of depths $\Delta z$ between the photosphere and the "top" of the model photosphere (which we have chosen to be at $\tau = 10^{-4}$) is 400–500 km. The actual height range depends on the choice of opacity. If we had used a different opacity table from Tables 5.1 and 5.2, then the height range could have been somewhat different. Other parameters would also have changed somewhat. But the values just cited give a reliable zeroth order overview of the physical parameters in the solar photosphere.

Fifth, the opacity around $\sigma = 1$ cm$^2$ gm$^{-1}$. We recall that the principal contributor to opacity in the photosphere is the negative hydrogen ion. The numerical values of $p(\mathrm{ph})$ and $\rho(\mathrm{ph})$ cited earlier take on the numerical values they do mainly because of the particular cross-section ($\sigma = 4.5 \times 10^{-17}$ cm$^2$) that is presented by an $H^-$ ion to the photons that are most abundant in the solar spectrum. Given that our model has $N(H^-) = (1.4-2.5) \times 10^{16}$ $H^-$ ions lying above each square centimeter of the photosphere, the column density and the cross-section combine to yield an optical depth $N(H^-)\sigma$ of order unity in the solar photosphere. Thus, it is the presence of $H^-$ ions that dominate in determining how deeply we can see into the solar atmosphere.

An important question arises concerning the last line of Table 5.3. What reason can we have for stopping the calculation in Table 5.3 at a depth where $\tau = 0.9$? This seems like a rather random choice for the location where we stop the computation of a model photosphere. Shouldn't we keep going deeper? The answer is No, and the reason for this answer has to do with the numerical value of the temperature gradient $dT/dz$. We shall see in Chapter 6 that when the magnitude of $dT/dz$

increases above a certain critical gradient $g_{ad}$, convection sets in. As a result, radiative transfer is no longer the dominant mode of energy transport in the atmosphere. Now, the computation that led to the results in Table 5.3 is based on a particular solution (Equation 2.40) of the equation of radiative transfer, i.e., radiation carries the entire energy flux through the atmosphere. There is little meaning in applying such a computation to gas where convection is occurring. In Chapter 6, we shall show that the critical value $g_{ad}$ is about $1.7 \times 10^{-4}$ deg cm$^{-1}$. Inspection of Table 5.3 above shows that $|dT/dz|$ increases with increasing depth and is approaching this critical value as we approach the bottom of Table 5.3. In fact, if we were to continue the calculation of Table 5.3 to greater depths, we would find that at optical depth $\tau = 1.0$, the local value of $|dT/dz|$ would exceed $g_{ad}$. Thus, our results suggest that convection sets in at optical depths between 0.9 and 1.0. This corresponds to a depth of only a few tens of kilometers below the formal definition of the location of the photosphere ($\tau = 2/3$).

## EXERCISES

5.1 Evaluate the pressure scale height in regions of the solar atmosphere where the gas has temperature of $10^4$, $10^6$, and $10^7$ K. (Such temperatures exist in the chromosphere, in the corona, and in flares: see Chapters 15 and 17.) Use molecular weight $\mu \approx 0.5$.

5.2 Perform the step-by-step calculation described in Section 5.4, using the opacities given in Tables 5.1 and 5.2. Compare your results with those in Table 5.3.

5.3 Repeat the calculations for different choices of various parameters. For example, use a starting pressure $\log p(1) = 2$ or 4. What differences do you find compared to the results in Table 5.3? Are some parameters more sensitive than others are to the alteration in starting pressure? As a further example, use $\Delta\tau$ values that are twice as large as those suggested in Section 5.4. Then repeat the calculations using $\Delta\tau$ values that are one-half of the values suggested in Section 5.4. What differences do you find in the various cases?

5.4 The opacity tables given in Tables 5.1 and 5.2 were computed (by Dr. R. Kurucz) using a number of choices of parameters (chemical mixture, microturbulence, etc.). Other tables of Rosseland mean opacities, using different choices of some parameters, exist in the literature (e.g., Iglesias and Rogers 1996 and references therein). Use one of those tables to repeat the calculations in Section 5.4. Which parameters are altered most compared to the results in Table 5.3?

Note: in Tables 5.1 and 5.2, opacities are listed as functions of $T$ and of *gas pressure* $p$. Opacity tables in the literature *may* list the opacity as functions of the $R$ parameter (Chapter 3, Section 3.6), or as a function of $T$ and *electron pressure* $p_e$. In order to use such tables in the procedure described in Section 5.4, you will need to convert from density to pressure (assuming a perfect gas), or you will need to find auxiliary tables that first convert from $p_e$ to gas pressure $p$.

## REFERENCES

Iglesias, C. A., & Rogers, F. J., 1996. "Updated OPAL opacities", *Astrophys. J.* 464, 943.
Kurucz, R. L., 1992. "Atomic and molecular data for opacity calculations", *Rev. Mexicana Astron. Astrof.*, 23, 45. The Rosseland mean *Opacity tables* in the present book (Tables 5.1 and 5.2) were made available to the author upon request.
Saint-Hilaire, P., Krucker, S., & Lin, R. P., 2010. "Statistically derived flaring chromospheric-coronal density from non-thermal X-ray observations of the Sun", *Astrophys. J.* 721, 1933.

# 6 Convection in the Sun
## *Empirical Properties*

So far, we have been restricting attention to the upper parts of the *photosphere* in the Sun, where energy is transported through the gas almost entirely by means of radiation. The gas in the photosphere (at least in all parts except the deepest regions at optical depths in excess of $\tau \approx 1$) does not move: the reason for this lack of motion is the fact that the gas is in hydro*static* equilibrium. This static gas simply "processes" the photons, absorbing, emitting, and scattering them in such a way that the radiation (which includes both outward and inward streams of photons: see Section 2.8) produces a net transport of energy in the *outward* (radial) direction. Because of the existence of radiative equilibrium, the radiative transfer equation (RTE) allows us to extract reliable physical properties of the gas in the photosphere, where there are no systematic gas motions.

Now we turn our attention to a certain region of the Sun where radiative equilibrium becomes progressively less important. In this new region, called the convection zone (which we shall see extends from the photosphere down to a depth of order $10^5$ km below the photosphere), photons play only a minor role in transporting energy. In the deeper layers that now draw our attention, energy eventually is transported essentially completely by means of convection. Convection occurs when the material itself experiences bulk flows in the vertical direction, both upward and downward. However, just as we found in the photosphere that inward and outward streams of photons lead to a net outward flow of energy, in the convection zone the flows of gas upward and downward also results in a net outward flow of heat.

In order to obtain a quantitative physical model of the Sun, we shall eventually have to develop a theory that will allow us to model convective heat transport in solar conditions. The details of one such theory will be the subject of Chapter 7. Before embarking on the task of developing such a theory, however, we will describe what can be learned about the empirical properties of convection as they present themselves to us in the visible layers of the solar atmosphere: these empirical properties will guide us in developing a convective model.

## 6.1 NONUNIFORM BRIGHTNESS

Evidence for the presence of convection in the Sun can be seen in an image of the solar surface if the image has sufficiently high spatial resolution. The Daniel K. Inouye Solar Telescope (DKIST), a 4-meter-diameter telescope on the summit of Mt. Haleakala in Hawai'i is one such telescope. This telescope has the largest mirror that has ever been used to study the Sun directly. The influx of direct sunlight leads directly to serious difficulties in keeping the telescope and its optics from heat damage. We already know (see Section 1.4) that solar energy enters the top of Earth's atmosphere at a rate of 1.361 kW m$^{-2}$. Some of this amount is absorbed and scattered by Earth's atmosphere before it reaches the summit of Haleakala. With a mirror having a radius of 2 meters, i.e., an area of about 12 m$^2$, the DKIST telescope collects about 13 kilowatts of solar power and focuses some of it into various instruments. Keeping the telescope and optics cool in the presence of this powerful stream of heat requires more than 10 kilometers of piping that circulates coolant. In order to obtain the sharpest possible images, the telescope is equipped with an "adaptive optics" (AO) system that compensates in real time (using 1600 "actuators", each of which causes a small localized distortion of the mirror) for the "seeing" (i.e., blurring) caused by the turbulence in the Earth's atmosphere. As a result of AO, the telescope can reach the diffraction limit of the mirror: according to a formula known as the "Dawes criterion", a mirror with a diameter of $D$ inches can resolve two points of equal brightness that are separated by an angle of $\theta = 4.56/D$ arcsec. For a 4-meter mirror (i.e., $D = 156$ inches), this leads to $\theta = 0.029$ arcsec. At the distance of the Sun (where 1 arcsec corresponds to a linear distance of 725.3 km [see Section 1.2]), this corresponds to a linear diameter of ~20 km. If features of such a size do in fact exist in the solar photosphere, DKIST expects to be able to identify them.

Figure 6.1 shows immediately that a snapshot of the surface of the Sun is *not* uniformly bright. However, the departure from uniformity is *not* the same as we discussed in the context of Figure 2.3: in the latter, where we examined an image of the *entire* solar disk, a *large*-scale darkening could be seen as we approach the limb of the Sun. But in Figure 6.1, the image covers only a small section of the Sun's surface measuring 36.5 × 36.5 Mm: i.e., the width (or the height) of the section of Sun in Figure 6.1 extends (in linear measure) over less than 3% of the diameter of the Sun. As a result,

**FIGURE 6.1** A close-up view of a portion of the solar surface obtained by DKIST. The field of view is 36.5 × 36.5 Mm (where 1 Mm = 1000 km). The essential feature in the present context is the "granular" structure, consisting of brighter patches surrounded by darker boundaries. (Credit: NSO/AURA/NSF; used with permission.)

Convection in the Sun

what we see in Figure 6.1 occupies only about 0.1% of the Sun's surface area. We are truly dealing with *small*-scale departures from uniform brightness in Figure 6.1.

Morphologically, we see in Figure 6.1 that the solar surface is covered with two distinct types of features. First, we see small bright patches (called granules). And second, we also see small darker areas (called intergranular lanes). The bright granules contain hotter gas than do the dark intergranular lanes. The difference in temperature between the bright and the dark gas will be discussed quantitatively later. The granules/intergranules are found with essentially the same properties at all regions on the solar surface, except in sunspots. Since spots never occupy more than ~1% of the solar area, we can draw an important conclusion: any physical phenomenon that relies on granules will be present in essentially equal amounts at all parts of the Sun's surface. This will be relevant when we consider the corona (Chapter 17).

How large in physical terms are the granules on the Sun? We will discuss this topic in Section 6.4. But for now, we note that bright granules have linear diameters $D_g$ that are preferentially of order 1 Mm. Since the Sun has a diameter $D = 2R_\odot$, which is of order 1400 Mm (see Equation 1.12), we see that the number of granules that can be "fitted into" the visible disk of the Sun at any one time is of order $N_g \approx (D/D_g)^2$, i.e., of order $10^6$. In Figure 6.1, the image covers about 0.1% of the Sun's surface: therefore, the number of granules that can "fit into" the image is of order $10^3$.

In principle, an image of the entire Sun (such as that in Figure 2.3) *might* be able to detect about a million granules. But this is unlikely: the telescope used for Figure 2.3 has an objective lens with a diameter of only 88 mm, i.e., 3.5 inches. For such a lens, the Dawes limit is 1.3 arcsec, which is just barely able to resolve a granule with linear diameter of 1 Mm ($\approx$1.4 arcsec). Moreover, atmospheric "seeing" at sea level even on a "good night" is typically about 2 arcsec, and during daylight hours (which are necessary in order to observe the Sun!), typical seeing is often (due to solar heating!) even worse than 2 arcsec. This "seeing" therefore also contributes to blurring out granules with sizes of order 1 Mm.

Visual inspection of Figure 6.1 allows us to note the differing topology between brighter and darker features. It is possible to start in one location in an intergranular lane and move to other dark lanes without traversing a bright granule. But bright granules are for the most part isolated from one another: one cannot go from one bright granule to another without traversing an intergranular lane. As one stares at the image, one might almost imagine that one were looking down on an area of "bocage" country on Earth where "fields" (akin to bright granules) are surrounded by "hedgerows" (akin to dark intergranular lanes). However, a big difference is that although fields and hedgerows on Earth retain their shapes as time passes, this is not true of the Sun: any particular granule or intergranule lane survives for only a time, and then is replaced with a different feature.

As regards DKIST, we have already noted (see Preface, Figure P. 1) that this instrument may be destined to occupy a place of distinction in the world of solar data acquisition. If all goes well in the course of a 10-year planned lifetime, DKIST is expected to accumulate some 10 million gigabytes of solar data. The image in Figure 6.1 was among the first images released by DKIST in January 2020. How many pixels are included in Figure 6.1? If the AO system was operating at maximum effectiveness when the image was taken, the linear resolution was nominally 20 km. Since each side of the image spans 36,500 km, there are more than 3 million pixels in the image.

## 6.2 GRANULE SHAPES

Even a casual inspection of Figure 6.1 shows that individual granules in the snapshot image in Figure 6.1 have irregular shapes. But on closer inspection, one notices that the shapes of individual granules are not completely random. Instead, some of the boundaries of many granules consist of lines that are nearly straight. Rather than looking at a situation that is completely chaotic, one seems to be looking at a collection of shapes hinting at the geometrical structures known as polygons. This is not to say that we are looking at regular polygons: no one would claim that all of the straight edges are of the same length, nor is there the same number of straight edges in all granules. Nevertheless, the impression that the eye gets is that the granules have shapes that are closer to polygonal shapes than to completely irregular shapes. Is there any physical reason why polygons might be of interest when we speak of convection? Indeed there is.

From an empirical standpoint, polygons have been found to be the preferred spatial pattern of cells that occur in laboratory convection under certain controlled conditions. When a layer of liquid is heated from below and the temperature gradient between bottom and top is not too large, heat can be carried up through the liquid by a process called thermal conduction. In such conditions, each molecule is pursuing its own purposes, and each molecule carries heat upward as an individual. In this condition, the liquid itself is not in (bulk) motion, although each molecule in the liquid is certainly moving randomly at quite a high speed (hundreds of meters per second at room temperature).

But at some point, the lower boundary of liquid becomes so hot that the temperature gradient $|dT/dz|$ becomes larger than a critical value. When that happens, the laws of physics show that it becomes energetically more favorable for the liquid to alter its mode of heat transport away from conduction: instead, the liquid begins to move in a macroscopic way so as to transport heat upwards by convection. In conditions that were studied by Bénard (1900), the motions of the liquid were observed to organize themselves into a geometrical pattern of "cells" consisting of polygons. In a theoretical analysis of the onset of (laminar) thermal convection (driven by buoyancy), Rayleigh (1916) demonstrated that the preferred pattern would have a hexagonal pattern in any given horizontal plane. Inside each "cell", trillions of atoms or molecules move in a highly organized pattern. No longer do the molecules behave as individuals in order to carry heat: instead, a lower energy condition can be reached if molecules cooperate with many of their nearest neighbors in a *macroscopic* pattern of motion, with hot liquid rising in some locations, while cool liquid sinks in other locations. This method of transporting heat by means of organized fluid flow is called convection. In carefully controlled conditions, the individual Bénard cells are found to be long-lived polygonal structures. Thus, Bénard cells in a laboratory setting provide an example of upward heat transfer by means of steady-state convection.

In the Sun, the hint of polygonal structures among some granules, although not a rigorous conclusion, is nevertheless an intriguing reminder of the cells which Bénard found in his experiments. This leads to the conclusion that convection is occurring in the Sun to transport heat upwards. However, the principal difference between Bénard's polygons and the features which we have called "polygons" in the Sun is the following: whereas Bénard's polygons existed in steady state for long periods of time, the "polygons" in the Sun are by no means in steady state. Quite the contrary: individual granules are observed to live for only a finite time. When one performs correlation studies on images of granules over a large area of the Sun, one finds that on average, individual granules can no longer be clearly identified after a time-scale of 5–10 minutes (Title et al. 1989). This time-scale can be regarded as a sort of average "lifetime" of a granule. We shall return to the significance of this time-scale after we have discussed spatial scales and velocities.

The contrast between Bénard cells and solar granules contains important information about fluid dynamics. In Bénard cells, the material moves in an organized steady pattern known as "laminar flow": this is appropriate in conditions where a fluid of relatively high viscosity flows at relatively low speeds. In such conditions, the relevant quantity to determine whether the flow will be laminar or turbulent is a quantity known as the Reynolds number $Re = uL/\nu$ (where $u$ is the flow speed in m s$^{-1}$, $\nu$ is the kinematic viscosity in m$^2$ s$^{-1}$, and $L$ is a typical length scale of the flow in meters). In laboratory conditions, where $u \sim 0.01$, $L \sim 0.01$, and $\nu \sim 10^{-5}$, the value of $Re$ (~10) remains smaller than a critical value $Re_c$ ($\approx 10^3$), with the result that flow remains laminar and cells survive for long times. However, in solar granules, flow speeds are of order $10^3$ m s$^{-1}$ (see Section 6.3), and $L$ values are of order $10^6$ m (see Section 6.1). With $\nu$ values again of order $10^{-5}$ for the gas in the solar atmosphere, the solar value of $Re$ is found to be ~$10^{10}$. In such conditions, the flows cannot be laminar:

Convection in the Sun

instead, they are highly turbulent. The nonsteady turbulent flows in solar convection are reminiscent of eddies in fast-flowing water: each eddy lives for only a finite time.

## 6.3 UPFLOW AND DOWNFLOW VELOCITIES IN SOLAR CONVECTION

We have already discussed (Section 3.8.2) how spectral lines contain information about flow speeds of systematic flows at different levels of the atmosphere. In view of what we have discussed in the present chapter, we can now state with confidence that the flow speeds mentioned in Section 3.8.3 are associated with the systematic motions of convection. Already, we have seen that velocities of up to at least 1 km s$^{-1}$ are detectable in the C-shaped bisectors of various spectral lines. However, these "speeds" involve some kind of averaging over a line profile and therefore cannot be regarded as indicating how large the convective speeds actually can become. Our goal in this section is to describe the observations that indicate how fast the convective flows actually are in the solar atmosphere.

For now, we need to note an important correlation that exists in the bright and dark gas. When spectra of individual granules and intergranular lanes are obtained, it is found that the *brighter* regions are systematically associated with *upflows*, while the *darker lanes* are associated with *downflows*. As far as physics is concerned, we may ask: do the upflows and downflows cancel each other out in some way? The answer to this question depends on what physical quantity we are considering. If we consider the *mass of material* in the flows, there is no observed buildup of more and more mass in the solar atmosphere as time goes on: therefore, the amount of *mass* flowing up in the bright granules must be equal to the amount of *mass* flowing down in the intergranular lanes. But if we consider the *heat energy* in the flows, the answer is quite different. The fact that *upflows* are correlated with *excess temperature* is an obvious indication that heat is being transported *upward* in the solar atmosphere by the upwardly moving gas. Perhaps less obviously, the correlation of *downflows* with *reduced temperature* also has the effect of contributing to the net flow of heat *in the upward direction*. The mutually reinforcing combination of hot upflows and cool downflows is the essential aspect of convective heat transport. There is a significant difference between "up" and "down" in the solar atmosphere when it comes to heat flow: heat does indeed flow in the *upward* direction even though there is no net upflow of material.

As far as the quantitative flux of energy is concerned, an important physical parameter is the algebraic difference in *vertical velocity* between upflows and downflows. Spectroscopic data are required to evaluate these velocities. In some case, individual estimates of the amplitude as large as 6 km sec$^{-1}$ have been reported using lines which are maximally sensitive to gas near and above the photosphere (Beckers 1968). The average amplitude of the vertical velocity difference was reported to be about 2 km sec$^{-1}$ (Bray et al. 1976). Relevant data with both high spatial and spectral resolution (120 km and 85 mÅ, respectively) have been obtained in a balloon experiment called SUNRISE, which in a typical flight spends 5 days observing the Sun (using a 1-meter mirror) at stratospheric altitudes as the balloon drifts from Sweden to Canada. An example of a dopplergram (a map of vertical velocities) constructed from 20–30 minutes of data is shown in Figure 6.2. One clearly sees the spatial pattern corresponding to the granule/intergranule network, although of course the spatial resolution is not nearly as good as in DKIST data (Figure 6.1). For present purposes, our main interest is to note that the velocity scale extends up to ≥ 3 km s$^{-1}$ in the downward direction, while the upflows have magnitudes extending up to ≥ 2 km s$^{-1}$. Thus, the difference between up and down speeds can be as large as 5–6 km s$^{-1}$, consistent with the results of Beckers (1968).

Can convective velocities be measured reliably in gas that lies *below* the photosphere? To do this, lines that are formed at infrared wavelengths near the opacity minimum of $H^-$ (at 16,000 Å) are best suited: Milic et al. (2019) used Fe lines at 15,700 Å to show that, in a 3-D simulation of solar convection, flow speeds can be extracted reliably down to depths of τ(5000 Å) ≈ 3. But attempts to extract speeds at depths as deep as τ(5000 Å) ≈ 10 are too noisy to be reliable.

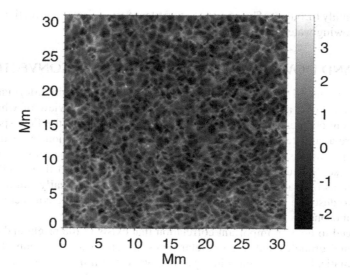

**FIGURE 6.2** Dopplergram covering the center of the solar disk obtained with data from SUNRISE flight on June 9, 2009, when solar activity was at an extremely low level. The velocity scale of the convective flows is shown (in units of km s$^{-1}$) by the vertical scale along the right-hand side: positive velocities (lighter shades) denote downflows, while negative velocities (darker shades) denote upflows. The granule/intergranule network is clearly visible with downflows in the intergranule lanes and upflows in the granules (Yelles Chaouche et al. 2014; used with permission of ESO).

In exceptional circumstances, the SUNRISE balloon experiment has seen Doppler "flashes" where the speed of gas motions in the photosphere departs from the mean by 4σ (McClure et al. 2019): these are attributed to regions where *p*-mode oscillations (see Chapter 13) happen to achieve local coherence. In combination with local granules, vertical speeds can then increase briefly to as much as 6–8 km sec$^{-1}$. In general, however, the observational evidence indicates that gas in the solar granulation moves upwards and downwards with speeds *V* of typically a few km sec$^{-1}$.

Knowing the magnitude of these convective speeds in the quiet Sun is important for two reasons. First, it provides a key piece of information that will allow us to estimate the amount of energy that is being transported upward by convection in the Sun. Second, it will serve as an important reference point when we compare these speeds with convective speeds inside certain regions on the Sun where convection is greatly *reduced* in amplitude (especially in sunspots).

Although not directly associated with upward energy transport, it is of interest to mention that *horizontal* velocities associated with granules can be measured by tracking algorithms which imagine that "corks" have been dumped in the flows, and the computer tracker follows each "cork" as it is pushed along. The root-mean-square (r.m.s.) horizontal velocities are found to be as large as 1.5 km sec$^{-1}$ (Title et al. 1989).

## 6.4 LINEAR SIZES OF GRANULES

Until recent decades, it has not been a trivial matter to observe granules on the surface of the Sun: they are small features that were not at all apparent to the early telescopic observers of the Sun. The first clear images of granules were not reported for almost three centuries after Galileo turned his small telescope to the Sun. The granules do not become detectable until (a) the resolution of the telescope becomes large enough, and (b) the disturbing effects of Earth's atmosphere are reduced to a minimum (or circumvented by observing from space).

Convection in the Sun

How high does the angular resolution have to be in order to distinguish clearly the bright and dark areas? The empirical answer is: the observing instrument must be able to resolve angles of 1 arcsec or better. Observations from favorable locations on the Earth's surface may occasionally satisfy this criterion. But observations from space routinely satisfy the criterion.

What are the horizontal spatial scales (i.e., linear diameters) $D_g$ associated with granules? Extensive information in this regard has been obtained by instruments that have observed the Sun with high spatial resolution from space. One such instrument was flown on Spacelab 2 on board a Shuttle mission lasting 8–9 days in 1985 using the Solar Optical Universal Polarimeter (SOUP) instrument on a 30-cm telescope. Pointing at a solar feature was maintained stable with remarkably high precision (0.003 arcsec). Properties of granules were found to depend on their surroundings, e.g., in magnetic regions or in nonmagnetic regions. Also, the $p$-mode oscillations present in the solar atmosphere (see Chapter 13) have to be removed in order to determine the properties of the granules more reliably. The largest granules have angular diameters as large as 2–3 arcsec, while others have diameters as small as the limiting resolution of the telescope. Title et al. (1989) conclude, "it is fair to say that there is a characteristic granule (angular) size in the vicinity of $d_g$ = 1.2–1.4 arcseconds". The corresponding characteristic linear dimensions are $D_g$ = 900–1000 km. Using the Hinode spacecraft, Yu et al. (2011) studied the properties of 71,538 granules in quiet regions of the Sun: they found two populations of granules, small (with angular diameters 0.31" ≤ $d_g$ ≤ 1.44") and large ($d_g$ = 1.44"–3.75"). For the small granules, the size distribution was found to be flat. But in the large granules, the distribution of sizes falls off towards the large granules: there are 10 times fewer granules with angular sizes of 3" than those with angular sizes of 1.5". Using high-quality ground-based data, with a diffraction limit as small as 77 km on the Sun, Abramenko et al. (2012) also reported on the existence of two populations of granules (See Section 5 in their paper, second paragraph.). One of these ("regular granules") has a Gaussian distribution of sizes with a mode of 1050 km and a standard deviation of 480 km: this Gaussian distribution has a characteristic linear size of $D_g$ = 1080–1300 km, i.e., a characteristic angular size of $d_g$ = 1.49–1.79 arc seconds. The second population consists of "mini-granules" with a power-law distribution at length-scales of 600 km and smaller. Abramenko et al. state, "The mini granules are mainly confined to broad intergranular lanes" and they may be "fragments of regular granules, which are subject to highly turbulent plasma flows in the intergranular lanes, where the intensity of turbulence is enhanced".

The fact that the size distribution of "normal granules" extends up to angular values as large as 3"–4" indicates that the largest granules can be up to 2–3 Mm in linear diameter. It is noteworthy that granules are not observed on the Sun with diameters larger than 2–3 Mm.

What determines that granules have horizontal sizes that are predominantly of order 1–2 Mm? Three factors are involved: convection, radiation, and density stratification (Hanasoge and Sreenivasan 2014). As regards radiation: when hotter gas rises due to convection in a granule, an effective way for the gas to lose its heat (so that it can become cooler gas and sink back down, thereby completing the convective cycle) is for the gas to radiate its excess heat when it rises up to a certain height. In order to do that, the photons need to have a good chance of escaping from the hotter gas, i.e., they have to be in a medium that does *not* have large optical depth. Thus, the loss of heat occurs most effectively in regions close to the layer in the photosphere where τ is of order unity. Moreover, in order to complete the up-down cycle of convection, the down-flowing cooler gas must not have to "fight its way through" the hotter gas: instead, the cooler gas takes a different path as it moves downward. To arrange for this, it is important that rising gas finds itself entering into a medium where the local density is becoming progressively smaller than at the level where the hotter gas started its upflow. This stratification in density in the near-surface layers is associated with hydrostatic equilibrium, where density falls off by a factor of $e$ when the height increases by a linear distance $H$ (the scale height). As a hot granule rises upwards into stratified gas (i.e., where density and pressure are decreasing more and more), the granule (with its higher internal pressure due to the presence of hotter gas) will expand. Expansion involves horizontal flows, and these carry gas "to the side", where it can find a path for downflow that will not interfere too much with the upflow. Moreover, as downflows enter progressively denser gas in the course of their descent, the downflows contract in the horizontal direction, leading to "fingers" of dense gas ("downdrafts") penetrating into the deeper layers of the convection zone. Mathematical models of 3-D convection that include the effects of radiation in a stratified medium have successfully generated images of the solar surface bearing a striking resemblance to the patterns of granules/intergranular lanes observed on the surface of the Sun (e.g., Stein and Nordlund 1998).

As regards the lifetimes of granules, data from the SOUP images on Spacelab 2 were used to perform autocorrelation function (ACF) analysis on granules in order to estimate lifetimes. In quiet regions of the Sun, the ACF was found to decrease by a factor of $1/e$ in a time of about 300 seconds. However, when the $p$-mode oscillations were removed, the $1/e$ lifetime increased to more than 400 sec. In the smallest granules, the ACF fell to essentially zero on time-scales of order 600 seconds. Among the largest granules (with angular diameters 1.3–1.8 arcsec), lifetimes were found to be as long as 1000 seconds. It seems that lifetimes of most granules lie in the range 300–600 sec.

In this section, we have described the sizes of granules that occur in *quiet* Sun regions (i.e., regions where there are only weak fields). However, in regions where magnetic fields are stronger (i.e., plages, active regions), granules are observed to be smaller in size. Narayan and Scharmer (2010) reported that in a plage where the mean field was 600 G, the granules were observed by the Swedish Solar Telescope to have horizontal diameters $D_g$, which were some four times *smaller* than in nonmagnetic regions. This reduction in horizontal scale is a clear indication of an important property of gas flow in the presence of a magnetic field: although an ionized gas can move freely *along* a field line, it is restricted in its ability to move *transverse* to the field line (see Section 16.6.1).

## 6.5 CIRCULATION TIME AROUND A GRANULE

Now that we know (1) the horizontal diameter $D_g$ associated with the top of a granule, and (2) a velocity of the gas flow in the cell, it is of some interest to estimate how much time it takes the gas to circulate around a cell.

To make the estimate, we need to know also the vertical depth $H$ of the granule. Let us estimate the distance an element of gas travels as it starts at the bottom of the cell, rises to the top (*distance H*), spreads out horizontally to the edge (*distance $D_g/2$*), sinks to the bottom (*distance H*), and then returns to the center of the cell (*distance $D_g/2$*). The total distance traversed by the element of gas is $C = D_g + 2H$. The time required for the gas to complete one circulation is $t(\text{circ}) = C/V$. With $D_g \approx 1000$ km (an empirical value in the Sun) and $V \approx 2$ km sec$^{-1}$ (also determined empirically in the Sun), we find $t(\text{circ}) \approx 500 + H$ sec, if $H$ is expressed in km.

What vertical depth should we consider for a convection cell in the Sun? We have already identified a natural length-scale that exists in the stratified solar atmosphere: the scale height $H_p$. We have already seen (Chapter 5, Section 5.1) that $H_p \approx 114$–$140$ km in photosphere. Because stratification of the ambient medium plays an essential role in convective flow (see Section 6.4), it would not be surprising if $H$ were to be related to $H_p$. In astrophysics, it is often assumed that $H = \alpha H_p$ where $\alpha$ is a number of order unity: $\alpha$ is referred to as the "mixing length parameter" in the context of a model for convection, which we will describe in detail in Chapter 7. With such a choice, we find $t(\text{circ}) \approx 500 + (114$–$140)\alpha$ sec.

We have already seen that granules live for 300–600 sec. This range of lifetimes overlaps with $t(\text{circ})$ as long as $\alpha$ does not exceed unity by a significant amount. Detailed modeling of the Sun suggests $\alpha \approx 1.5$ might be appropriate (e.g., Mullan et al. 2007). This leads to $t(\text{circ}) \approx 670$–$710$ sec.

Thus, it appears that solar granules may manage to live only long enough to almost (but perhaps not quite) complete *a single complete circulation of the cell.*

Convection in the Sun

This is in significant contrast with Bénard cells: under the carefully controlled (laminar flow) conditions of a laboratory experiment, convection cells can survive for many circulation times. On the other hand, in the highly turbulent solar granule flow, an individual cell lives only for a short time (5–10 minutes).

## 6.6 TEMPERATURE DIFFERENCES BETWEEN BRIGHT AND DARK GAS

The amount of heat transported by convection depends on the temperature difference between rising and falling gas. This temperature difference gives rise to an intensity contrast $\Delta I$ between hot gas (where the intensity is $I_h$) and cold gas (where the intensity is $I_c$). The earliest measurements from space (Title et al. 1989) suggest that the r.m.s. values of $\Delta I/I$ are up to ±16% in quiet Sun. However, when the effects of acoustic oscillations are allowed for, the r.m.s. values of $\Delta I/I$ are found to be ±10% at wavelengths around 6000 Å (Title et al. 1989). Using the highest quality ground-based observations prior to DKIST, namely, those obtained by the Swedish Solar Telescope (SST) with an angular resolution of 0.16 arcsec, Scharmer et al. (2019) reported on measurements of $\Delta I/I$ as follows: 14%, 13%, 12%, and 8% at $\lambda$(Å) = 5250, 5580, 6300, and 8535 respectively. The values of contrasts reported by the Solar Optical Telescope (SOT) on the Hinode spacecraft were found to be smaller than the above percentages: e.g., $\Delta I/I$ (SOT) = 8% at $\lambda$(Å) = 5550. Although space-based observations (Hinode) are in general expected to be more reliable than those from the ground (SST), the smaller contrasts in Hinode data are caused in part by the poorer angular resolution (0.32 arcsec) (Afram et al. 2011). 3-D simulations suggest that contrasts could be as much as +20% in the granules and −20% in the intergranular lanes, but when smoothed for finite resolution, the r.m.s. contrasts are about ±10% (Stein and Nordlund 1998).

Taken as a whole, the data and simulations suggest that the bright granules are observed to have an r.m.s. intensity that is some 10% in excess of the average, while the dark intergranular lanes have an r.m.s. intensity that is some 10% smaller than the average.

For purposes of calculating how much energy is transported by convection, we need to convert the intensity difference to a temperature difference between the temperatures in the hot and cold gases, $T_h$ and $T_c$. When we observe bright granules, we are seeing down into the solar atmosphere essentially to an optical depth of $\tau_h \approx 1$ in the hot rising gas. When we observe dark intergranular lanes, we are seeing into the solar atmosphere to an optical depth of $\tau_c \approx 1$ in the cold sinking gas. Although the *optical* depths are the same in hot and cold gas, we are not observing gas at the same *vertical height*: we see more deeply into the cold gas (where the opacity is lower).

Since we see to equal optical depths in both bright and dark gas, the emergent intensities $I_h$ and $I_c$ are proportional to the respective source functions $S_h$ and $S_c$ at optical depth unity. Thus, $\Delta I/I = \Delta S/S$ to a rough approximation. To the extent that the continuum source function at $\tau \approx 1$ can be equated with the Planck function, we expect that

$$\frac{I_h}{I_c} \approx \frac{\exp(c_2/\lambda T_c)-1}{\exp(c_2/\lambda T_h)-1} \qquad (6.1)$$

where $c_2 = hc/k$ = 1.44 cm deg is referred to as the second radiation constant. We know from Figure 2.2 earlier that although the Planck function is a fairly good fit to solar radiation, it is not perfect. Therefore, our assumption that the source function is identical to the Planck function is only an approximation. Whatever answer we extract will need to be compared to a more precise calculation.

In view of ±10% fluctuations in r.m.s. intensity, we have that $I_h \approx 1.1$ while $I_c \approx 0.9$. For purposes of this calculation, we assume that the dark intergranular gas has a temperature at $\tau \approx 1$ equal to the effective temperature, i.e., $T_c$ = 5772 K. Then we find that, according to Equation 6.1 (assuming an observing wavelength $\lambda$ = 5500 Å), the "r.m.s." temperature for the hot gases is $T_h \sim$ 6050 K. Thus, in the Planck approximation, the r.m.s. temperature difference $\Delta T$ between bright granules and dark intergranular material at $\tau \approx 1$ in both materials is roughly 300 K.

More careful treatment of radiative transfer, using a fully 3-D radiative-hydrodynamic code (Stein and Nordlund 1998), indicates that the temperatures at $\tau \approx 1$ in the coldest and hottest elements of gas can reach extreme values that range from 5800 to 7000 K. Thus, the extreme temperature differences at $\tau \approx 1$ according to the Stein-Nordlund model is 1200 K. The r.m.s. temperature differences would certainly be smaller than this extreme value: from the results of Stein and Nordlund we estimate $\Delta T_{rms} \approx 500$–600 K. Thus, our rough estimates of $\Delta T \approx 300$ K at $\tau \approx 1$ using the Planck approximation to interpret the observed intensity fluctuations are probably too small by a factor of about two.

As we move upward and downward from the level $\tau \approx 1$, the results of 3-D modeling (Stein and Nordlund 1998) have suggested that the range of temperature differences $\Delta T$ increases. In the upper atmosphere, the extreme range may be as large as 1500 K at $\tau \approx 0.001$, while in the deeper layers, at $\tau \approx 1000$, the extreme ranges of $\Delta T$ may be as large as 2500 K.

If we were to compare the rising and sinking gas at equal geometric depths (rather than at equal optical depths), the values of $\Delta T$ would be larger than the aforementioned estimates. In fact, the hydrodynamical modeling of Stein and Nordlund (1998) suggests that at equal depths close to the photosphere, the extremes of $\Delta T$ may rise to values as large as 4000 K. However, it has been suggested (Kalkofen 2012), based on a comparison with steady-state models of the chromosphere, that these extremes of $\Delta T$ in the hydro models may be unrealistically large because the treatment of radiation losses in certain heights in the model may have been incomplete.

As a large-scale spatial average of temperature differences between rising and falling gas, we will assume that, for our approximate estimates, the values of $\Delta T_{rms}$ are roughly 500–600 K.

## 6.7 ENERGY FLUX CARRIED BY CONVECTION

Now that we know how fast the gas is moving and how much temperature difference exists between the hot rising material and the cold sinking material, we can turn to a quantitative consideration of the key question relevant for solar physics: how much heat energy is being carried upward by the convective motions of the gas in the solar atmosphere?

In a parcel of gas, a temperature difference of $\Delta T$ corresponds to a difference in the heat content of $C_p$ times $\Delta T$ ergs gm$^{-1}$. Here, $C_p$ is the specific heat at constant pressure, in ergs gm$^{-1}$ K$^{-1}$. This is the excess amount of internal heat energy that the hot rising gas contains compared to the cool sinking gas.

Now, to calculate the upward flux of energy, in units of ergs cm$^{-2}$ sec$^{-1}$, we need to multiply the heat content (in ergs gm$^{-1}$) by the mass flux, $F_m = \rho \times V$ (in gm cm$^{-2}$ sec$^{-1}$). This leads us to the following estimate for the upward heat flux due to convection:

$$F(conv) = \rho\, V\, C_p\, \Delta T \qquad (6.2)$$

This formula is applicable in any region of the Sun where bulk gas motions are present.

### 6.7.1 CONVECTIVE ENERGY FLUX IN THE PHOTOSPHERE

To start off a discussion of convective energy transport in the Sun, we consider the particular case of a region in the Sun where we already know the magnitudes of all the parameters that enter into the prior formula: this particular region is the photosphere. Subsequently, we shall consider other layers as well.

Let us see what the magnitude of the convective energy flux is in the photosphere. The model of the solar atmosphere that we derived in Chapter 5 indicates that in the photosphere, $\rho \approx (2\text{–}3) \times 10^{-7}$ gm cm$^{-3}$. The discussion given in the earlier sections of the present chapter suggests that velocity differences

# Convection in the Sun

between hot and cold gas are of order a few km sec$^{-1}$, say $V \approx 3 \times 10^5$ cm sec$^{-1}$, and that $\Delta T \approx 500\text{–}600$ K. The final quantity that we need to evaluate in order to estimate $F(conv)$ is $C_p$.

To evaluate the specific heat, we note that in a perfect gas, where the particles are monatomic and are not undergoing ionization, the internal energy of an atom consists of a single term, due to thermal motion, i.e., $(3/2)kT$ per atom, where $k$ is Boltzmann's constant. In a gas composed of hydrogen atoms only, the internal energy per gram is $U = (3/2)kT/m_H = (3/2)R_g T$, where $R_g = k/m_H = 8.3145 \times 10^7$ ergs gm$^{-1}$ K$^{-1}$ is the gas constant. For a gas mixture with mean molecular weight $\mu$, we find $U = (3/2)R_g T/\mu$ per gram. This leads to a specific heat per gram at constant volume $C_v$ as follows:

$$C_v = \frac{dU}{dT} = \frac{3R_g}{2\mu} \qquad (6.3)$$

The specific heat at constant pressure, $C_p$, contains two terms, one related to the internal energy, $U$, and the other related to the work done on compressing the gas: $C_p = dU/dT + p(dV/dT)_p$ (where subscript $p$ denotes constant pressure). In a perfect gas that is nonionizing gas and monatomic, this leads to

$$C_p = C_v + \frac{R_g}{\mu} = \frac{5R_g}{2\mu} \qquad (6.4)$$

For the gas in the solar photosphere, consisting of a nonionizing mixture of H (90%) and He (10%), we have $\mu \approx 1.3$. This leads to $C_p \approx 1.6 \times 10^8$ ergs gm$^{-1}$ deg$^{-1}$. In arriving at Equation 6.2, we note that the rising and falling gas can adjust their internal pressure to be equal to the external pressure by means of sound (pressure) waves: the convective speeds (~1 km sec$^{-1}$) are smaller than the sound speed (~10 km sec$^{-1}$ in photospheric gas), and as a result, the adjustment of pressures inside and outside the convective elements can be considered as essentially instantaneous. Therefore, $C_p$ is the appropriate specific heat to use in calculating the convective heat flux in the photosphere.

In a monatomic, nonionizing gas, the ratio of specific heats $\gamma = C_p/C_v$ has the numerical value of 5/3. In terms of $\gamma$, the value of $C_p$ can be written as

$$C_p = \frac{\gamma}{\gamma - 1} \frac{R_g}{\mu} \qquad (6.5)$$

Combining Equation 6.2 with the various parameters, we find that in the photosphere, the numerical value of the flux of energy being transported by convection is roughly given by the following estimate:

$$F(\text{conv,ph}) \approx 7 \times 10^9 \text{ ergs cm}^{-2}\text{sec}^{-1} \qquad (6.6)$$

Is this flux of convective energy a "large" quantity or a "small" one? Well, large and small are relative terms. We need to compare the convective energy flux to some other flux of energy passing through the solar atmosphere in order to decide whether the convective flux is "large" or "small". The relevant flux that passes through the Sun's atmosphere is the flux of energy $F_\odot$ that eventually leaves the Sun and travels out into space. In radiative terms, we have already specified (Chapter 1, Section 1.9) what this is: $F_\odot = \sigma T_{eff}^4 = 6.2939 \times 10^{10}$ ergs cm$^{-2}$sec$^{-1}$.

Comparing $F(\text{conv, ph})$ with $F_\odot$, we can see that in the photosphere, convection is carrying about 10% of the total energy flux that passes upward through the solar atmosphere. This shows us that although radiation dominates in the process of transporting the energy flux up through the photosphere (and therefore we are justified in using the radiative solution $T^4 \sim \tau + 2/3$ in calculating the photospheric model in Chapter 5), radiation is not the *only* process that contributes to transporting energy though the photosphere. Gas motions associated with convection are *also* of material assistance in the photosphere. This explains why granules and intergranular lanes can be seen in the Sun when a picture is taken of the visible Sun (e.g., Figures 6.1 and 6.2). If it were true (as it is in hot stars) that convection is present but is transporting only a small fraction (say, <1%) of the heat flux, then the patterns of upflows and downflows would not stand out clearly if we were able to obtain an image of the photosphere.

### 6.7.2 Convective Energy Flux *above* the Photosphere?

Once we move away from the particular case of the photosphere and consider gas that lies either shallower or deeper in the Sun, we anticipate that $F(\text{conv})$ will not necessarily retain the value we have estimated in the photosphere.

In what sense might we expect the numerical values of $F(\text{conv}) = \rho V C_p \Delta T$ to change as we move upward into shallower layers of the atmosphere? Well, let us examine the four contributing factors to $F(conv)$ in order to answer this question. First, we have already seen (Chapter 5) that the density $\rho$ of the gas falls off exponentially with increasing height, that is, $\rho$ will be *decreased* in the upper photosphere. Second, as the gas density falls off, it becomes increasingly difficult for the temperature difference between rising and sinking gas to be maintained: leakage of photons in the increasingly rarefied gas has the effect that the rising and sinking gases more readily exchange photons so as to tend toward the same temperatures. As a result, $\Delta T$ is expected to *decrease* as we move up into the upper photosphere. On the other hand, $C_p$, which depends on the quantities $R_g$ and $\mu$, is expected to retain its photospheric value $\approx 1.6 \times 10^8$ ergs gm$^{-1}$ deg$^{-1}$. Finally, as regards the speed $V$, in the photospheric gas that lies above the convection region, the temperature gradient (see Table 5.3) is expected to be stable against convection (we will take this up in Section 7.4). As a result, gas motions above the convection region are no longer subject to the upward buoyancy forces that prevail inside the convection region. Instead, in the upper photosphere, convective speeds $V$ are expected to *decrease* relative to conditions in the photosphere. Thus, in view of the four factors in $F(conv) = \rho V C_p \Delta T$, three of which decrease in the upper photosphere while the fourth remains constant, we see that $F(\text{conv})$ *decreases* as we examine gas that lies *above* the photosphere. At a height of order 100 km above the photosphere, detailed models of convection indicate that $F(conv)$ has fallen to negligible values (e.g., Nordlund et al. 2009).

### 6.7.3 Convective Energy Flux in Gas That Lies *below* the Photosphere

What about the deeper gas? How large is the convective flux down there? In these layers, densities increase exponentially rapidly as the depth increases. This favors more effective convection. Moreover, there is a further factor, associated with ionization, that helps the gas to transport convective flux more easily. As the temperature rises in the deeper gas, the atoms begin to experience an increasing amount of ionization; the gas enters into one of the "ionization strips" in Figure 4.1. When ionization is in process, the internal energy of an atom of the gas $U$ is no longer due solely to thermal motions: instead, there is an extra term associated with the ionization potential energy $I$. For hydrogen, $I = 13.6$ eV. This is larger by more than an order of magnitude than the thermal energy: at temperatures corresponding to gas just below the photosphere (where $T = 6000–10,000$ K), the thermal energy per atom is only of order 0.6–1 eV. The occurrence of ionization energy of 13.6 eV represents the addition of such a large temperature-sensitive contribution to internal energy $U$ that the numerical value of specific heat $C_v$ ($= dU/dT$) increases significantly compared to the values cited earlier for a monatomic (and nonionizing) gas. The specific heat reaches a maximum when the gas is roughly 50% ionized: as we noted in Chapter 4, this occurs for hydrogen at depths where the temperature is about 20,000 K. In the vicinity of 50% hydrogen ionization, the value of $C_v$ is found to be enhanced by ≈ 36 times its "normal" value [$(3/2)R_g/\mu$], while the value of $C_p$ is found to be enhanced to ≈ 27 times its "normal" value [$(5/2)R_g/\mu$] cited earlier for a monatomic nonionizing gas (Clayton 1968). These enhancements are a quantitative indication of how important it is to allow for ionization effects in order to arrive at a realistic view of solar structure. Atomic effects have a measurable effect on the structure of the Sun itself.

When the gas is undergoing ionization and $C_p$ and $C_v$ both increase in value by significant amounts, the ratio of specific heats $\gamma = C_p/C_v$ is no longer as large as 5/3 (the value for a nonionizing monatomic gas). Moreover, in adiabatic conditions, the pressure and density are no longer related by the simple relationship $p \sim \rho^\gamma$. Instead, the pressure and density are related by $p \sim \rho^\Gamma$, where the numerical value of the generalized exponent $\Gamma$ is no longer strictly equal to $C_p/C_v$. When the degree of ionization is 50%, $\Gamma$ for a gas composed of hydrogen alone falls to minimum values of ≈ 1.135 (Clayton 1968). In the Sun, where hydrogen is not the sole constituent but helium contributes almost 10% by number, the minimum value of $\Gamma$ is not so small: when hydrogen is 50% ionized, helium is still essentially neutral, with $\Gamma(\text{He}) = 5/3$. The combination of (roughly) 90% of the atoms with $\Gamma(\text{H}) = 1.135$ and (roughly) 10% of the atoms with $\Gamma(\text{He}) = 5/3$ lead to an overall minimum value of $\Gamma(\text{H} + \text{He})$ of 1.19. We will return to this minimum value of $\Gamma$ when we consider the calculation of a model of the convection zone in Chapter 7.

In even deeper layers, where hydrogen ionization is approaching completion, the energy that is being diverted into ionization energy becomes less important as a contributing term in the internal energy. Thermal energy once again dominates. As a result, the specific heat reverts (almost) to the value cited earlier. But now there is a difference: for every hydrogen atom in the photosphere, there are now two particles at great depth (a proton and an electron), each with its own equal share ($kT$) of thermal energy. (In the real Sun, helium atoms are also present at about 10% abundance by number: it is as if there are 1.1 "atoms" by number at the solar surface. When hydrogen ionizes at a certain temperature, helium does not ionize until the temperature becomes significantly larger: therefore, at the base of the convection zone, each atom of hydrogen has converted to a proton and an electron, whereas the He remains as a single atom. Therefore, the number of particles has increased from 1.1 [at the top of the convection zone] to 2.1 [at the base of the convection zone].) As a result, the internal energy per gram, and therefore the value of $C_v$, is about *twice as large* as in the photosphere. It is also roughly true that $C_p$ is about twice as large at the base of the convection zone as at the surface. (In a detailed solar model [Baker and Temesvary 1966], $C_p$ at the base if the convection zone turns out to have a numerical value larger than the surface value by 2.1. This is consistent with our statement of "about twice as large".)

Combining the greatly increased factors of $\rho$ and $C_p$, we expect that $F(\text{conv})$ probably increases rapidly (compared to the photospheric value) as we go below the surface. The only possibilities for offsetting this rapid increase in convective flux would be either the temperature differential $\Delta T$ or the velocity $V$ undergoes a dramatic reduction as depth increases. However, this does not seem likely: the models of Stein and Nordlund (1998) suggest that $\Delta T$ may increase below the surface. Also, one-dimensional convective models (e.g., Vitense 1953) suggest that $V$ may remain as large as 1 km sec$^{-1}$ even at depths as great as 4–5 scale heights (i.e., up to ~1000 km) below the photosphere. In view of this, it is difficult to avoid the conclusion that $F(\text{conv})$ increases significantly above its photospheric value (Equation 6.6) as we examine the gas which lies deeper than the photosphere. In fact, the model of Vitense (1953) suggests that convection is already able to carry more than 90% of the total energy flux at a depth where the gas pressure has a value of $3 \times 10^5$ dyn cm$^{-2}$, i.e., at a depth of only a couple of hundred kilometers below the photosphere.

We conclude that convection provides a remarkably efficient method for the Sun to transport energy in the layers of gas that lie not far beneath the photosphere.

Now that we have seen how effective convection is for energy transport below the photosphere, this raises the question: does convection dominate the transport of energy within the *entire* interior of the Sun? We shall arrive at the important conclusion that the answer to this question is a firm

"No". But in order to arrive at this answer, we need to understand in more detail the causes driving convection. We now turn to a consideration of those causes.

## 6.8 ONSET OF CONVECTION: THE SCHWARZSCHILD CRITERION

Let us perform the following thought experiment. Consider an atmosphere in which $T$ is increasing as the depth increases. In the context of optical depth, we have already found such a case when we discussed radiative equilibrium in a gray atmosphere: $T^4 \sim (\tau + 2/3)$. Once we know the opacity, we can convert this into the functional form that indicates how $T$ varies with the linear depth $z$. A specific case of this procedure led us to the tabulated model of the solar photosphere that we obtained in Chapter 5.

In order to set the stage for a discussion of convection, it is important to remember that knowing how physical quantities vary *as a function of optical depth* $\tau$ is relevant in the part of an atmosphere where *radiation* is the dominant mode of energy transport. But when we come to the case of convection, where radiative transfer plays only a minor role, optical depth is not a useful parameter. Instead, it is more helpful to ask the question: how does $T$ vary as a function of *linear* depth $z$? If we have that information, then at each depth, the slope of $T$ versus $z$ has a certain numerical value. We refer to that slope as the local temperature gradient $g_T = dT/dz$ in units of degrees per cm. Because we are considering an atmosphere in which $T$ increases as $z$ increases, the sign of $g_T$ is positive. (This is in contrast to the case we considered in Equation 5.1, where we used height $h$ [increasing upward] rather than depth $z$ [increasing downward].) The quantity $g_T$ plays an essential role in determining the conditions in which convection can occur. To see why this is so, we argue as follows.

Consider a parcel of gas that lies initially at a depth $z$. The gas has a well-defined temperature $T$. Suppose this parcel is displaced vertically by some means (e.g., buoyancy forces). Now we ask: will the displaced parcel of gas be stable or unstable? That is, will the parcel return to its starting point, or will it keep on moving vertically? To address this, we consider the *change* in energy that occurs as a result of the displacement. This change in energy can be either positive or negative, and the algebraic *sign* of this energy change plays a key role in what follows.

Suppose we displace the parcel *up*ward along a vertical path of length $dz$. The local ambient temperature at the new depth $z - dz$ is $T - dT$ where $dT = |g_T dz|$. Let the upward displacement of the parcel to the final depth $z - dz$ be performed in a time that is so short that the parcel has no time to lose any of its internal energy by leakage to the ambient gas. When the parcel arrives at $z - dz$, it will still have its initial temperature $T$: therefore, the parcel finds itself hotter than its surroundings by $dT$. That is, the parcel will contain internal thermal energy that is in excess of that in the ambient gas. Now let enough time elapse that the parcel releases all of its excess thermal energy into the local gas under conditions where pressure is maintained constant: the amount of thermal energy it will release per gram of material is given by $C_p dT$ ergs.

Is this energy release significant as far as the displacement of the parcel is concerned? To answer that, we must compare the amount of thermal energy that has been released with another energy term that arises as a result of the vertical displacement: gravitational potential energy. In order to displace the parcel upward by an amount $dh (= -dz)$, something has to supply the energy needed to perform work against gravity. The amount of that work is $gdh (= -gdz)$ per gram of material.

Now we ask: what has happened to the total energy of the parcel in the course of its displacement to its final position? On the one hand, the gas has released $C_p dT$ ergs gm$^{-1}$. On the other hand, energy had to be found (from somewhere) to increase the potential energy by $-gdz$ ergs gm$^{-1}$. The total amount of energy $\Delta W$ associated with the displacement of 1 gm of material is therefore given by

$$\Delta W = C_p\, dT - g dz \qquad (6.7)$$

Note that ΔW is not constrained as to its algebraic sign: ΔW may have numerical values that can be either positive or negative. And whether ΔW turns out, in any given situation, to be positive or negative makes all the difference as far as convection is concerned.

To see why, suppose that the magnitude of $gdz$ exceeds the magnitude of $C_p dT$. In this case, ΔW is negative: in order to displace the parcel upward to its final position, we would have to supply *more* work than is released by the thermal excess. That is, there is simply not enough thermal energy released to compensate for the work that would have to be done in order to lift the parcel through the interval $-dz$. As a result of a lack of available energy, the parcel of gas cannot in fact reach the new position at $-dz$. Instead, the parcel will sink back down to its initial position. There is no incentive for the parcel to move upward. The parcel stays in the place where it started. This situation is referred to as "convective stability".

On the other hand, suppose that conditions are such that the magnitude of $C_p dT$ exceeds the magnitude of $gdz$. In this case, release of thermal excess at the end of the displacement is *more than enough* to compensate for the work of lifting the parcel. All of that lifting work can be provided for by releasing the excess internal energy of the parcel. In fact, after the work of lifting has been performed, there is even some internal energy left over to make sure that the parcel still remains hotter than the ambient gas. In other words, the gas itself contains more than enough internal energy to do the work of lifting the parcel against gravity. Therefore, the parcel, once displaced up by $-dz$, keeps on moving upward. We refer to this as "convective instability".

The boundary between stability and instability as far as convection is concerned occurs when $\Delta W$ has a value that is neither positive nor negative, i.e., when the total energy exchange between the parcel and its surroundings is zero. In such a situation, the parcel undergoes a change that is referred to as "adiabatic". This particular case occurs when $gdz = C_p dT$, i.e., when the temperature gradient $g_T = dT/dz$ takes on the particular value known as the "adiabatic gradient":

$$g_T = g_{ad} \equiv g/C_p \qquad (6.8)$$

Therefore, gas in which the local temperature gradient happens to have a particular value $g_T$ is said to be convectively *stable* if $g_T < g_{ad}$. In gas where the opposite holds, i.e., where $g_T > g_{ad}$, the gas is said to be convectively *unstable*.

This reminds us that the *algebraic sign* of $g_T$ is important: on the right-hand side of Equation 6.8, the quantities g and $C_p$ are both positive definite. If $g_T$ is a negative quantity, i.e., if the temperature *decreases* as the linear depth *increases*, it is *impossible* to satisfy the condition for convective *instability*: $g_T > g_{ad}$. As a result, such gas is always convectively *stable*. We shall find that in a certain region of the solar atmosphere (the chromosphere: see Chapter 15), $g_T$ is in fact negative: there is no convection in that part of the Sun.

The conclusion of the present section is the following. Will convection set in (or not), in a particular region of the Sun? It all depends on the answer to the question: is the local temperature gradient $dT/dz$ larger (steeper) or smaller (shallower) than the value of the adiabatic temperature gradient $g_{ad}$?

The criterion $g_T > g_{ad}$ that determines the onset of convection in a gas is referred to as the Schwarzschild criterion, named after a German scientist Karl Schwarzschild who published an article on the stability of the gas in the solar atmosphere in 1906. (Schwarzschild is best known for his work on general relativity that he did in 1915 while on active military service in World War I. Unfortunately, as a result of illness, he did not survive the war.)

## 6.9 ONSET OF CONVECTION: BEYOND THE SCHWARZSCHILD CRITERION

The Schwarzschild criterion suggests that if the temperature gradient exceeds the adiabatic gradient *however slightly*, then convection should set in. However, Schwarzschild was referring to convection in a medium where there is no mechanism at work that would prevent (or interfere with) the material in the medium from leaving a stationary condition and starting to move. In the real world, there is always *some* process that can slow down, or stop, the motion. In view of this, Rayleigh (1916) derived a more complicated criterion for the onset of convection in a liquid medium, where viscosity and thermal conduction can hinder the onset of movement of material. Rayleigh's criterion for the onset of convection in these conditions involves a combination of parameters in a quantity that is now known as the "Rayleigh number" $Ra$. Rayleigh showed that convection will occur only if $Ra$ (which is inversely proportional to viscosity and thermal conductivity) exceeds a critical value $Ra_c$. If a highly viscous fluid tries to move, the energy that must be expended to overcome viscosity may be so large that convection cannot occur unless the viscosity can somehow be reduced. (In the Earth's mantle, for example, ocean water that seeps into cracks in the upper mantle may reduce the viscosity enough to permit mantle convection to occur [cf. Schaefer and Sasselov 2015]. See also Exercise 6.2(b) at the end of this chapter.) When Rayleigh's result is applied to a medium in which convection can occur, it implies that convection should *not* automatically set in when $g_T$ exceeds $g_{ad}$ *by only an infinitesimal amount*. Instead, the Schwarzschild criterion must be modified to a new criterion, the Rayleigh criterion, which can be written approximately as follows: convection will set in when $g_T > g_{ad} + \Xi$. That is, the temperature gradient must exceed the adiabatic gradient by a finite amount $\Xi$ before convection sets in. The larger the viscosity, and the larger the thermal conductivity, the larger $\Xi$ becomes, and the harder it is for convection to set in. This is a general result for any viscous and thermally conductive fluid. To be sure, this result is not of particular importance in the Sun in general: viscosity of the gas in the Sun does not present a significant obstacle to convective onset (see Exercise 6.2(a) at the end of this chapter).

In view of that, one might well wonder: what is the point of raising the issue of the Rayleigh criterion in a book on solar physics? To be honest, the answer is: there would be no real need to raise the issue of viscosity at all if "normal" kinematic viscosity was the only "viscous" process at work in the Sun. However, a peculiar case of "effective viscosity" *can* occur in certain regions, specifically in regions where magnetic fields are present. To see the importance of this, we note that Chandrasekhar (1952) used an extension of Rayleigh's treatment to show that a vertical magnetic field inhibits the onset of convection in a medium where the *electrical conductivity* is large. In such a medium, gas is not permitted to move easily in a direction that is *perpendicular* to the field lines (we will discuss the physical reason for this in Sections 16.6.1 and 16.6.2.2). As a result, if a vertical magnetic field threads through a convection cell such as that described earlier in Section 6.5, the gas can flow up and down along the vertical field lines without any difficulty. But the circulation of the convective flow in a "cell" may be more or less (depending on how strong the field is) seriously impeded at the top and bottom surfaces of the cell. Gough and Tayler (1966) have shown that in such conditions, the Schwarzschild criterion for onset of convection should be replaced as follows: $g_T > g_{ad} + \Delta$. (This is reminiscent of the case of the viscous fluid studied by Rayleigh: in essence, a magnetic field makes an ionized gas extremely "viscous" in directions perpendicular to the field lines.) Thus, in the presence of a vertical magnetic field, $g_T$ must exceed $g_{ad}$ by *a finite amount* before convection can set in. The stronger the magnetic field, the larger the excess $\Delta$ becomes, and the harder it is for convection to set in. The results of Gough and Tayler (1966) are highly pertinent in regions of the Sun where magnetic fields are strong. We will discuss (in Chapter 16) the observational evidence showing that convection is indeed seriously impeded by vertical magnetic fields in certain solar features. In particular, we shall find (in Section 16.1) that the results of Gough and Tayler (1966) may help us understand why there exists a well-defined boundary between the umbra and the penumbra in a sunspot.

## 6.10 NUMERICAL VALUE OF $g_{ad}$

In the solar photosphere, we have seen (Chapter 1, Section 1.13) that $g = 27,420$ cm sec$^{-2}$. And in Section 6.7.1, we have seen that $C_p \approx 1.6 \times 10^8$ ergs gm$^{-1}$ in the photosphere. Combining these values, we find an important result regarding the adiabatic temperature gradient in the photosphere:

$$g_{ad} \approx 1.7 \times 10^{-4} \text{ deg cm}^{-1} \qquad (6.9)$$

Convection in the Sun

This is the numerical value of the critical temperature gradient that must be exceeded if convection is to occur in the solar photosphere.

Now we can see why, when we were calculating a model solar atmosphere in Chapter 5, it was important for us to tabulate the numerical value of the local (vertical) temperature gradient: see the column labeled "grad T" in Table 5.3. Interestingly, as the model calculation proceeds from top to bottom, we eventually did reach a layer of gas (with optical depth $\tau = 0.9$–$1.0$, i.e., near the photosphere) where the local temperature gradient $dT/dz$ increased to a numerical value in excess of the above critical gradient $g_{ad}$. This means that, in the model presented in Table 5.3 in Chapter 5, the gas is indeed convectively stable for all levels of the photosphere listed in the table. Therefore, radiation does indeed dominate the energy transport in the regions of the atmosphere listed in Table 5.3. According to the model in Table 5.3, the Schwarzschild criterion indicates that convection in the Sun does not set in until optical depth $\tau$ reaches values of $\geq 0.9$–$1.0$, i.e., just slightly deeper (by a vertical distance of ~20 km) than the formal definition of the photospheric level ($\tau_{ph} = 2/3$).

Why is this a noteworthy result? It means that we Earth-based observers are lucky enough to see down into the Sun deep enough to catch a glimpse of at least the uppermost layers of convection (see Figure 6.1). There is nothing to say *a priori* that this *must* happen: it is certainly possible that the onset of convection might have occurred so deep below the photosphere that Earth-based observers would be able to see nothing whatsoever of the convective motions. (For example, if we lived near a hot star of spectral class O or B, we would probably see little or no evidence for convection: there *is* a convection zone near the surface associated with helium ionization, but it carries such a small fraction of the stellar energy that the contrast between granules and intergranule lanes could be so weak that we might not detect any nonuniformity.)

As it is, we Earthlings *are* able to see the Sun's convection, with its up-and-down gas motions of hot and cold gas. Without this privilege, we might have to work a lot harder to learn about convection in the Sun.

The value of $g_{ad}$ in Equation 6.9 applies to the gas that lies in the photosphere of the Sun. What happens to the numerical value of $g_{ad}$ as we examine the gas that lies deeper inside the Sun? Below the photosphere, where hydrogen begins to undergo appreciable ionization, there is a rapid increase of $C_p$ (by factors of up to roughly 30: [Clayton 1968]). In the layers where this is happening, we are still quite close to the surface of the Sun. As a result, the local value of gravity is essentially the same as at the surface. As a result, the numerical value of $g_{ad}$ takes on values that are numerically as much as 30 times *smaller* than in the photosphere. Because of this, it is much easier for the local temperature gradient to exceed the local value of $g_{ad}$ when the ambient gas is undergoing ionization. As a result, it is much easier to satisfy the convective instability condition $g_T > g_{ad}$. Therefore, in a region of a star where the gas contains a majority constituent element that is undergoing ionization, we are likely to find convection. The convection we see in Figure 6.1 is associated with the ionization of hydrogen. Deeper down, there are in principle (see Figure 4.1) more convection zones due to the ionization of He I to He II, and ionization of He II to He III. In some hot stars, there may also be a convection zone associated with ionization of iron-group elements. However, in the Sun, detailed models of the internal structure of the Sun indicate that the convection zones overlap. In the Sun, there is in essence only a single convection zone.

## 6.11 ALTERNATIVE EXPRESSION FOR $g_{ad}$

As mentioned earlier, an alternative expression for $C_p$ is $[\gamma/(\gamma - 1)]R_g/\mu$. Using this, we can rewrite $g_{ad}$ as

$$g_{ad} \equiv (dT/dz)_{ad} = g\mu(\gamma - 1)/\gamma R_g \tag{6.10}$$

In the uppermost parts of the convection zone, where the convective speeds are significant fractions of the sound speed, local conditions are obviously not in hydrostatic equilibrium (HSE).

However, in the deeper layers of the convection zone, where convective speeds have fallen to values that are much smaller than the sound speed, the assumption of HSE is not too bad. Now, in HSE, we know that the pressure gradient is given by Equation 5.1. In the present context, where we are using the independent variable $z$ (i.e., the depth), rather than (as in Equation 5.1) the height $h$, the HSE equation is written as

$$dp/dz = +g\rho \qquad (6.11)$$

Dividing Equation 6.10 by Equation 6.11, we find that

$$\left(\frac{dT}{dp}\right)_{ad} = \frac{1}{\rho C_p} = \frac{(\gamma-1)\mu}{\gamma \rho R_g} \qquad (6.12)$$

Among the terms on the right-hand side, we note that for a perfect gas, $\mu/\rho R_g$ equals $T/p$. Carrying $T/p$ over to the left-hand side of the equation, we find

$$\left(\frac{d\log T}{d\log p}\right)_{ad} = \frac{\gamma-1}{\gamma} \qquad (6.13)$$

We note that if the local conditions in a gas in any region of the Sun are in fact adiabatic, then the local temperature and pressure will vary in such a way that the local gradient of temperature relative to pressure, $d\log T/d\log p$, will take on the value $(d\log T/d\log p)_{ad}$ as given by Equation 6.13. This has the effect that the pressure $p$ in that region of the Sun will vary as a power law of $T$. In the presence of ionization, we need to replace $\gamma$ in Equation 6.13 with the more generalized exponent $\Gamma$ (see Section 6.7.3):

$$p \sim T^{\Gamma/(\Gamma-1)} \qquad (6.14)$$

In a monatomic nonionizing gas, where $\Gamma = \gamma = 5/3$, the right-hand side of Equation 6.13 has the numerical value 0.4. In such a gas, adiabatic processes lead to a pressure-temperature relationship of the form $p \sim T^{2.5}$.

But if ionization is at work, the power-law relationship becomes steeper. For example, in a gas composed of pure hydrogen where the degree of ionization is 50%, $\gamma \approx 1.135$. In such a case, Equation 6.13 indicates that $p \sim T^{8.4}$. In such conditions, small increases in temperature would be associated with much larger increases in pressure than in the nonionizing limit $p \sim T^{2.5}$. These results will be applied to the solar convection zone in the next chapter when we try to calculate how pressure and temperature vary as a function of depth in the solar convection zone.

## 6.12 SUPERGRANULES

The most prominent sign of convection in the Sun is the presence of granules such as those in Figure 6.1. As we have seen, granules are convection cells where hotter gas rises in the center while cooler gas sinks at the periphery. Even a casual inspection of Figure 6.1 suggests that the multitude of granules (about one thousand or so can be identified in Figure 6.1) can reasonably be assigned a rather well-defined mean diameter of 900–1000 km (see Section 6.4). This is the preferred length-scale of convective structures at the solar surface.

The question we now raise is: does the solar convection exhibit any other features that have a different preferred length scale? The answer is a definite "Yes". When the Sun is viewed by an instrument that is sensitive to the Doppler shifts of organized flows on the surface, a characteristic feature emerges (see Figure 6.3): over most of the solar surface, there are multiple pairs of light and dark features indicating paired flows towards and away from the observer. But in the center of the solar

Convection in the Sun 109

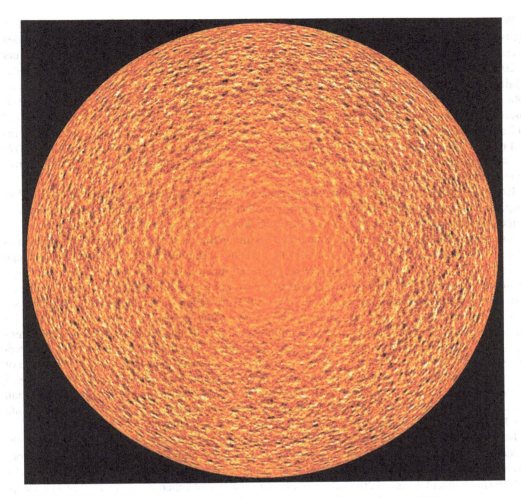

**FIGURE 6.3** Supergranulation pattern on the Sun: image of Doppler shifts made by SOHO/MDI. White and dark patches are associated with gas moving towards and away from the observer. (Courtesy of SOHO/MDI consortium. SOHO is a project of international cooperation between ESA and NASA.)

disk, it is obvious that these light/dark pairs disappear: this indicates that the motions that are causing the Doppler shifts in Figure 6.3 are mainly *horizontal* on the Sun. Each light/dark pair of horizontal flows defines a "supergranule" in which gas rises up at the center, flows horizontally outwards to a certain distance over most of the surface of the supergranule, and then sinks back into the Sun.

The Hinode and SDO/HMI instruments have permitted extensive studies of supergranules with access to uninterrupted series of observations spread out over many days. Thus, Svanda et al. (2014), in a study of the relationship between supergranules and sunspots, identified 222,796 individual supergranules in HMI images obtained at various times over a 38-month interval. The outflow region over an average supergranule was found to be very symmetric about the center of the cell. The average distance between the centers of neighboring supergranule cells was found to be 38 Mm. In a shorter data sample from HMI, spanning 7 days without interruption, Roudier et al. (2014) observed 14,321 supergranules and found an average lifetime of 1.5 days, with an average diameter of 25 Mm. In a review of supergranule properties, Rincon and Rieutord (2018) state that the mean horizontal flow speeds are 0.3–0.4 km s$^{-1}$, while the vertical flow speeds (of order 0.03–0.04 km s$^{-1}$) are 10 times smaller than the horizontal speeds.

Williams and Pesnell (2014) used SOHO/MDI to compare properties of supergranules during two different solar activity minima (1996, 2008). Observing more than 6000 supergranules in each minimum, the $1/e$ lifetime was found to be $17.8 \pm 0.5$ hours in 1996 and $17.7 \pm 0.4$ hours in 2008. Thus, there is *no significant change in supergranule lifetime* between these two solar minima. After the launch of SDO in 2010, Williams and Pesnell (2014) obtained simultaneous data over a 5-day interval with the two independent instruments SOHO/MDI and SDO/HMI. These data showed that the mean diameter of supergranules is $36.8 \pm 0.3$ Mm in MDI and $33.2 \pm 0.3$ Mm in HMI: the difference in these diameters is related to the higher spatial resolution of HMI. In another study using HMI over a 6-year period 2010–2016, Roudier et al. (2017) reported no significant changes in horizontal or vertical flows as the solar cycle rose from minimum to maximum.

How many granules can "fit into" one supergranule? With linear diameters of granules of order 1 Mm, a supergranule with diameter 30 Mm has room for of order 1000 granules. With about 1 million granules on the solar surface, this means that the surface of the Sun has room for about 1000 supergranules.

Knowing velocities $V$ and diameters $D$, we can estimate the circulation time of a supergranule as of order $D/V \approx 30$ Mm/300 m sec$^{-1}$ $\approx 10^5$ sec $\approx 1$ day. Thus, with lifetimes reported to be in the range 17 hours to 1.5 days, i.e., of order 1 day, we see that each supergranule lives long enough to undergo about one circulation time. This is reminiscent of the case of granules (Section 6.5). Could this be an indication that supergranules are convective features that are simply larger analogs of granules? If so, then the center of the supergranule should be hotter than the periphery. Measurements of intensity contrast in a sample of $10^4$ supergranules (Langfellner et al. 2016) do indeed suggest that the intensity at the center exceeds that at the periphery by a fractional amount of $(7.8 \pm 0.6) \times 10^{-4}$. This corresponds to a temperature difference of $\Delta T = 1.1 \pm 0.1$ K. Compared to granules, where $\Delta T$ can be of order hundreds of degrees or more (Section 6.6), and where vertical velocities are only 30–40 m sec$^{-1}$ in supergranules, compared to 1 km sec$^{-1}$ in granules, supergranules are much less effective than granules at transporting heat vertically through the solar atmosphere. In fact, the supergranule rate is at least three orders of magnitude *smaller* than the rate in granules.

However, computational models of convection, which reproduce very well the observed length-scales ($\approx 1$ Mm) of granules (Stein and Nordlund 1998), have found it more difficult to show convincingly that convection in the Sun should also exhibit a *second* preferred length-scale of order 30 Mm (Rincon and Rieutord 2018). As a result, the existence of supergranules has been a challenging puzzle for theorists in solar physics for many decades.

Recently, an intriguing suggestion by Featherstone and Hindman (2016) is that supergranulation might arise as a natural consequence of *rotationally constrained* convection. As is well known on Earth, Coriolis forces interfere with the flows of fluids provided that the length-scale of the flows is large enough. (On smaller length-scales, other forces, e.g., pressure gradients or nonlinear terms, dominate.) By comparing convection properties in models of a nonrotating and a rotating Sun, Featherstone and Hindman (2016) show clearly that larger scale convection cells that exist in the nonrotating model are simply *not present* in the rotating model. Coriolis forces in effect "wipe out" the larger scales of convection, giving rise in effect to a peak in the convective power spectrum of the rotating Sun: in one particular rotational model, Featherstone and Hindman find that the peak occurs at a length-scale corresponding to spherical harmonic degree $l = 71$. According to Equation 13.1 (which we shall derive later in Chapter 13), this corresponds to a "wavelength" of about 60 Mm. This is somewhat larger than supergranule diameters observed in the Sun, but at least the rotating model contains a physical basis for the existence of a second "preferred" length-scale on the order of tens of Mm.

In a recent observational development, we shall see later (Section 14.10) that rotational modes ($r$-modes) of oscillation in the Sun have been discovered to have vorticity that is in fact comparable in magnitude to the vorticity of large-scale convection cells. In such a situation, $r$-modes may serve to "drain" some vorticity off the largest convective cells. As a result, $r$-modes may serve as an essential component of solar dynamics.

# Convection in the Sun

In view of the suggestion of Featherstone and Hindman (2016) that rotational effects may contribute to the existence of certain length-scales in the solar convection flows, we may ask: is there any evidence that rotational effects are at work in other convection flows? In this regard, we note that in the Earth's atmosphere, convection contributes to the formation of certain cloud systems. The linear (horizontal) sizes of clouds have been determined over a range of nearly five orders of magnitude (from 0.1 km to 8000 km) using a variety of different observational platforms (Wood and Field 2011): the size distribution is found to be a single power law extending from scales of order 0.1 km up to a scale of about 1000 km. However, on scales that are larger than 1500 km (or so), the data indicate that there is a statistically significant *decrease* in the number of clouds relative to the number expected from an extrapolation of the power law. That is, there seems to be a "cut-off" in cloud sizes above a certain critical length scale. Wood and Field suggest some possible explanations for the existence of a "cut-off" in cloud numbers for length-scales larger than ~1500 km. They suggest that Rossby waves might play a role in setting an upper limit on the permissible sizes of clouds on Earth. Since Rossby waves are governed in part by Coriolis forces, it is tempting to speculate that the "cut-off" on cloud sizes on Earth might be related (*mutatis mutandis*) to the kind of physics that (according to Featherstone and Hindman) contributes to the existence of supergranule length-scales in the Sun. It will be interesting to see whether future studies help to determine the validity of this speculation.

## EXERCISES

6.1 Consider gas which flows with vertical speeds $v$ of 1, 3, 6, and 10 km sec$^{-1}$ occurring at the surface of the Sun ($h = 0$). Given the gravity $g$ at the surface of the Sun (Equation 1.13), calculate the maximum heights $s_{max}$ ($= v^2/2g$) to which these flows can rise above the surface. Compare your answers to the plot in Figure 4.2 to see that the maximum heights do not reach as high as the temperature minimum, even for the fastest of the aforementioned vertical speeds (which is already considerably larger than the maximum observed convective velocity in the Sun).

6.2 In order to define the Rayleigh number $Ra$, consider the problem that Rayleigh was analyzing: a layer of fluid of thickness $d$ is heated from below, with temperatures $T_1$ at the top, and $T_2 > T_1$ at the bottom, with $\Delta T = T_2 - T_1$. In these conditions, $Ra$ is defined as follows: $Ra = g\, d^3 \alpha (\Delta T)/(\nu \kappa)$. Here, $g$ is the acceleration due to gravity; $\alpha$ is the thermal expansion coefficient, i.e., the fractional amount that a length of material expands when the temperature is raised by 1 deg K; $\nu$ is the kinematic viscosity, and $\kappa$ is the thermal diffusivity. Rayleigh showed that if there were free boundaries at top and bottom, then the onset of convection requires that $Ra$ exceed the critical value $Ra(crit) = (27/4)\pi^4 = 658$.

(a) Consider the case of convection in the gas that is present near the photosphere in the Sun. There, $g = 27420$ cm s$^{-2}$. The depth of a convection cell is estimated to be at most a few scale heights: i.e., $d =$ (say) $3 \times 10^7$ cm. In a perfect gas, $\alpha \approx 1/T$: in the solar photosphere, $\alpha$ is of order $10^{-4}$ per degree. In the Sun, the temperature difference between bottom and top of the layer is of order 1000 K. Using the value of $\nu$ cited in Section 6.2 in units of m$^2$ s$^{-1}$, we convert to cm$^2$ s$^{-1}$ for the present calculation: $\nu \approx 0.1$ cm$^2$ s$^{-1}$. Finally, the thermal diffusivity $\kappa$ for a variety of gases is found online to be of order 1 cm$^2$ s$^{-1}$. (This value applies to room temperature, but we use it here for purposes of estimation.) Using these numerical values, calculate the Rayleigh number $Ra(sun)$ for solar convection. Show that $Ra(sun)$ exceed $Ra(crit)$ given earlier by at least 20 orders of magnitude. This shows that convection in the Sun certainly sets in.

(b) Consider the case of the Earth's mantle. Here, $g \approx 10^3$ cm s$^{-2}$. The depth of the part of the mantle where convection is most efficient is of order $10^8$ cm. The expansion coefficient $\alpha$ is of order $10^{-5}$ per degree, and $\Delta T$ may be of order 1000 K (Schaefer and Sasselov 2015). The online values of $\kappa$ in various solid materials are listed as 0.01–0.1 cm$^2$ s$^{-1}$. But now we come to the big difference between Earth and Sun: the value of ν in rocks in the Earth's mantle can be as high as $10^{23-24}$ in units of Pascal seconds. Converting Pa to c.g.s. units (1 Pa sec = 10 ergs cm$^{-3}$), ν may be as large as $10^{24-25}$ cm$^2$ s$^{-1}$. Using the aforementioned ranges of values, calculate $Ra$ for the Earth's mantle. Show that, depending on which combination of values you use, $Ra$ may fall short of $Ra(crit)$ (and convection will not occur), or $Ra$ may exceed $Ra(crit) = 658$, although not by orders of magnitude (and very slow convection can occur).

# REFERENCES

Abramenko, V. I., Yurchyshyn, V. B., Goode, P. R., et al., 2012. "Detection of small-scale granular structures in the quiet sun with the new solar telescope", *Astrophys. J. Lett.* 756, L27.

Afram, N., Unruh, Y. C., Solanki, S. K., et al., 2011. "Intensity contrasts between MHD simulations and Hinode observations", *Astron. Astrophys.* 526, A120.

Baker, N. H., & Temesvary, S., 1966. "A solar model", *Table of Convective Stellar Envelope Models*, 2nd ed. NASA Institute for Space Studies, New York, pp. 18–28.

Beckers, J. M., 1968. "High-resolution measurements of photospheric and sunspot velocity and magnetic fields using a narrow-band birefringent filter", *Solar Phys.* 3, 258.

Benard, H., 1900. "Les tourbillons cellulaires dans une nappe liquide", *Rev. Gen. des Sciences Pures Appl.* 11, 1261 and 1309.

Bray, R. J., Loughhead, R. E., & Tappere, E. J., 1976. "Convective velocities derived from granule contrast profiles in Fe I at 6569.2 Å", *Solar Phys.* 49, 3.

Chandrasekhar, S., 1952. "On the inhibition of convection by a magnetic field", *Phil. Mag. & J. of Sci.* 43, 501.

Clayton, D. D., 1968. "Thermodynamic state of the stellar interior", *Principles of Stellar Evolution and Nucleosynthesis*. McGraw-Hill, New York, pp. 77–165.

Featherstone, N. A., & Hindman, B. W., 2016. "The emergence of solar supergranulation as a natural consequence of rotationally constrained interior convection", *Astrophys. J. Lett.* 830, L15.

Gough, D. O., & Tayler, R. J., 1966. "The influence of a magnetic field on Schwarzschild's criterion for convective instability in an ideally conducting fluid", *Mon. Not. Roy. Astron. Soc.* 133, 85.

Hanasoge, S. M., and Sreenivasan, K. R., 2014. "The quest to understand supergranulation and large-scale convection in the Sun", *Solar Phys.* 289, 403.

Kalkofen, W., 2012. "The validity of dynamical models of the solar atmosphere", *Solar Phys.* 276, 75.

Langfellner, J., Birch, A. C., & Gizon, L., 2016. "Intensity contrast of the average supergranule", *Astron. Astrophys.* 596, A66.

McClure, R. L., Rast, M. P., & Martinez Pillet, V., 2019. "Doppler effects in the solar photosphere: Coincident superposition of fast granular flows and p-mode coherence patches", *Solar Phys.* 294, 18.

Milic, I., Smitha, H. N., & Lagg, A., 2019. "Using the infrared iron lines to probe solar sub-surface convection", *Astron. Astrophys.* 630, A133.

Mullan, D. J., MacDonald, J., & Townsend, R. D. H., 2007. "Magnetic cycles in the Sun: Modeling the changes in radius, luminosity, and p-mode frequencies", *Astrophys. J.* 670, 1420.

Narayan, G., & Scharmer, G. B., 2010. "Small-scale convection signatures associated with a strong plage solar magnetic field", *Astron. Astrophys.* 524, A3.

Nordlund, A., Stein, R. F., & Asplund, M., 2009. "Solar surface convection", *Living Rev. in Solar Phys.* 6, 2.

Rayleigh, Lord, 1916. "On convection currents in a horizontal layer of fluid when the higher temperature is on the under side", *Philos. Mag. Ser. 6.* 32(192), 529–546.

Rincon, F., & Rieutord, M., 2018. "The Sun's supergranulation", *Living Reviews in Solar Phys.* 15, 6.

Roudier, Th., Malherbe, J. M., & Mirouh, G. M., 2017. "Dynamics of the photosphere along the solar cycle from SDO/HMI", *Astron Astrophys.* 598, A99.

Schaefer, L., & Sasselov, D., 2015. "The persistence of oceans on Earth-like planets: Insights from the deep-water cycle", *Astrophys. J.* 801, 40.

Scharmer, G. B., Löfdahl, M. G., Sliepen, G., et al., 2019. "Is the sky the limit? Performance of the re-vamped SST 1-m", *Astron. Astrophys.* 626, A55.

Stein R. F., & Nordlund, A., 1998. "Simulations of solar granulation. I. General properties", *Astrophys. J.* 499, 914.

Svanda, M., Sobotka, M., & Bárta, T., 2014. "Moat flow system around sunspots in shallow sub-surface layers", *Astrophys. J.* 790, 135.

Title, A. M., Tarbell, T. D., Topka, K. P., et al., 1989. "Statistical properties of solar granulation derived from the SOUP instrument on Spacelab 2", *Astrophys. J.* 336, 475.

Vitense, E., 1953. "Die Wasserstoffkonvectionszone der Sonne", *Zeits. f. Astrophysik* 32, 135.
Williams, P. E., & Pesnell, W. D., 2014. "Time-series analysis of supergranule characteristics at solar minimum", *Solar Phys.* 289, 1101.
Wood, R., & Field, P. R., 2011. "The distribution of cloud horizontal sizes", *J. of Climate* 24, 4800.
Yelles Chaouche, L. Moreno-Insertis, F., & Bonet, J. A., 2014. "The power spectrum of solar convection flows from high resolution observations and 3D simulations", *Astron. Astrophys.* 563, A93.
Yu, D., Xie, Z., Hu, Q., et al., 2011. "Physical properties of large and small granules in solar quiet regions", *Astrophys. J.* 743, 58.

# 7 Computing a Model of the Sun
## *The Convection Zone*

In this chapter, we wish to calculate the structure of the region in the Sun where convection dominates the transport of energy. As in Chapter 5, we will not yet discuss the *origin of the energy that is flowing through the convection zone*. (We postpone that discussion to Chapter 11.) In this chapter, we again accept the total luminosity (or flux) of the Sun as a boundary condition and seek to determine how the material arranges itself so as to "handle" the energy that is passing through. We will examine the forces that act on the medium and determine how the medium responds. In this sense, the model we will derive is better referred to as a mechanical model rather than a complete model.

Based on empirical evidence, the gas in the photosphere of the Sun is moving (up and down) with speeds of a few km sec$^{-1}$. In order to determine the equations that will allow us to describe solar convection in plausible physical terms, we need first to understand why the convective motions in the surface layers of the Sun have speeds of this order of magnitude. Why are the motions not of order a few cm sec$^{-1}$? Or hundreds of km sec$^{-1}$? What is the determining factor that sets the scale of the speeds?

## 7.1 QUANTIFYING THE PHYSICS OF CONVECTION: VERTICAL ACCELERATION

We have seen that certain parcels of gas in the Sun are observed to be rising, while others are sinking. The rising parcels are hotter than the sinking ones, and the r.m.s. temperature differences are of order $\Delta T \approx 500–600$ K in the photosphere.

From a physics perspective, it is important to note that the speeds of convective motion are less than the local (adiabatic) speed of sound, $c_s = \sqrt{(\gamma R_g T/\mu)}$, where $\gamma$ is the ratio of specific heats and $\mu$ is the mean molecular weight. (In the solar photosphere, $c_s \approx 9$ km sec$^{-1}$.) This has the effect that sound waves can propagate quickly between hot and cold gas and equalize the pressures. Thus, the differences in pressure between hot and cold gas at any height are not significant. Now, for material that obeys the equation of state of a perfect gas, $p = R_g \rho T/\mu$, the pressure difference $\Delta p$ is related to the temperature difference by $\Delta p/p = \Delta T/T + \Delta \rho/\rho - \Delta \mu/\mu$. In the photosphere, there is no significant difference in the degree of ionization between hot and cold gas: therefore we will make no significant error if we assume there is no difference in the molecular weights in hot and cold gas, i.e., $\Delta \mu/\mu = 0$.

Using this, and setting $\Delta p/p = 0$, we see that the observed temperature difference $\Delta T$ between rising and sinking gas in the photosphere corresponds to a density difference $\Delta \rho/\rho = -\Delta T/T$. The negative sign indicates that the hotter (rising) gas has lower density than the cooler (sinking) gas. With an empirical fractional temperature difference in the photosphere observed to be $\Delta T/T \approx (500–600)/5772 \approx 0.1$, we expect that hotter gas has a density which is about 10% smaller than the density of the cooler gas. For purposes of calculation in the present section, we shall use the value $\Delta T/T \approx 0.1$ as representative.

Now the photosphere of the Sun is for the most part in hydrostatic equilibrium: this means that there are no net forces acting on the gas in the photosphere. This is not to say that there are no forces whatsoever acting on the gas: it means only that whatever forces *are* at work, they are in general balanced in the photosphere. On the one hand, there is a vertically upward force (per unit volume) due to the vertical gradient $dp/dz$ of the ambient pressure. On the other hand, there is a vertically downward force (per unit volume) due to the weight of the gas, $\rho g$. When these forces are in balance, there is no net acceleration in the vertical direction, and the gas remains at rest. This is the situation throughout the model of the photosphere that was presented in Chapter 5. That is, given a photospheric model where, at depth $z$, the density is $\rho_o$ and the pressure is $p_o$, then $dp_o/dz$ has a numerical value that is precisely equal to $\rho_o g$ at all heights in the photospheric model: i.e., $dp_o/dz = g\rho_o$.

This is the equation of HSE (cf. Equation 5.1), rewritten in terms of the depth $z$ (which increases in the *downward* direction) rather than the height parameter $h$ (which increases *upward*). Note that the force of gravity is also directed in the downward direction, i.e., parallel to the depth $z$.

But in a gas where convection is possible, the vertical forces are no longer balanced. And in the presence of such unbalanced forces, the material will begin to move in the vertical direction. Let us consider the unbalanced forces and the vertical accelerations that they cause.

Suppose a certain parcel of gas is hotter than the ambient medium. The density $\rho'$ in the parcel will be *lower* than the ambient density $\rho_o$. As a result, the vertically downward force on unit volume of gas in the parcel due to its weight $\rho'g$ is now *less* than the local upward force due to ambient pressure $dp/dz$. That is, the upward vertical force exceeds the downward vertical force. The unbalanced upward force $dp/dz - \rho'g$ acting on a parcel of gas with unit volume leads to a vertical acceleration of that parcel in the upward direction. Since unit volume of the gas has a mass of $\rho'$, Newton's second law of motion (force = mass times acceleration) tells us that the unit volume will move with vertical speed $V$ such that the upward acceleration $dV/dt$ satisfies

$$\rho' dV/dt = dp/dz - g\rho' \tag{7.1}$$

This equation expresses the law of conservation of momentum. Notice that in the absence of vertical flows ($V = 0$), the conservation of momentum in Equation 7.1 reduces to HSE.

As we have seen, the motions observed in the solar photosphere are such that pressure remains equalized between hot and cold gas. That is, the pressure of the gas remains relatively unchanged in hot or cold gas compared to the ambient medium. This means that we can, without serious error, replace $dp/dz$ with $dp_o/dz$. But we already know that $dp_o/dz = \rho_o g$. Therefore, the vertically upward acceleration $dV/dt$ experienced by the low-density gas parcel is given by $(dV/dt)_u = g(\rho_o - \rho')/\rho'$. The fact that $\rho'$ is *less* than $\rho_o$ has the effect that the sign of the right-hand side is positive. Therefore, the vertical acceleration is in the *upward* direction. Buoyancy forces create this upward acceleration.

If a parcel of gas is locally cooler (and denser) than ambient, then the local density $\rho''$ will be *greater* than $\rho_o$. Then the acceleration, with a magnitude $(dV/dt)_d = g(\rho_o - \rho'')/\rho''$ will be a negative quantity. Therefore, in this case the acceleration will be in the *downward* direction. Once again, this downward acceleration is due to buoyancy forces.

Since the differences in density between the ambient medium and the hot (upgoing) and cold (downgoing) gas are not large, we can write the relative acceleration $a_{hc}$ between hot and cold gas as

$$a_{hc} \equiv (dV/dt)_u - (dV/dt)_d = g(\rho'' - \rho')/\rho_o \approx g\Delta\rho/\rho = -g\Delta T/T \tag{7.2}$$

Inserting the empirical result $\Delta T/T \approx 0.1$, we find that the magnitude of the relative acceleration $a_{hc}$ between hot and cold gas in the solar photosphere is expected to be given by $a_{hc} \approx 0.1g \approx 2.7 \times 10^3$ cm sec$^{-2}$.

## 7.2 VERTICAL VELOCITIES AND LENGTH-SCALES

Now that we have an estimate for the relative vertical acceleration between hot and cold gas, we can ask: over what vertical length-scale $s_v$ must the acceleration $a_{hc}$ be allowed to operate in order to build up a vertical velocity difference $V$ that is comparable to the observed values, i.e., a few km sec$^{-1}$?

The relevant formula is $V^2 = 2a_{hc} s_v \approx 5.4 \times 10^3 s_v$. Setting $V = (2-4) \times 10^5$ cm sec$^{-1}$, we find $s_v = 74-296$ km. Thus, if the buoyancy forces due to the density differences between hot and cold gas in the solar photosphere are allowed to operate over vertical distances of 74–296 km, the vertical velocities that can be produced will be comparable to the observed values.

Is there any physical significance to length-scales that lie in the range 74–296 km? Well, we have seen (Chapter 5, Section 5.1) that the pressure (and density) scale height $H_p$ in the photosphere is 115–140 km. We note that this range of $H_p$ values overlaps the range of values we have obtained for the vertical distance $s_v$. This suggests that the dynamics of convection in the solar photosphere are constrained in such a way that the vertical acceleration due to buoyancy is allowed to operate over vertical length-scales that are comparable to $H_p$. Specifically, with the aforementioned numbers, it appears that $s_v$ should range from somewhat less than $1H_p$ to about $2H_p$.

This is an empirical conclusion. It is based on the *observed* temperature differences between hot and cold gas, and on the *observed* relative differences in velocity between rising and sinking gas. If we were not able to resolve the granulation in the Sun, thereby measuring differences in temperature and velocity between rising and sinking gas, we would have to rely on indirect arguments in order to decide what might be the best choice for $s_v$.

## 7.3 MIXING LENGTH THEORY (MLT) OF CONVECTION

It was mentioned earlier (Chapter 6, Section 6.2) that granules in the Sun have properties that are similar to eddies in a fast-flowing river: such eddies survive for a finite time and then dissolve into the ambient water. During their lifetime, they travel a finite distance before they "mix" their contents back into the river. This finite distance is called the "mixing length".

In solar convection, by analogy, it is imagined that a hot parcel of upward convective flow can preserve its identity for a finite time only. During that time, the material (which is moving at a finite speed $V$) can travel a finite distance $L$ (the "mixing length") in the vertical direction (buoyancy forces determine that the motion is preferentially vertical), and then the parcel mixes its material and its excess heat energy in with the ambient gas. When the mixing occurs, the original parcel of convective flow loses its identity. Based on the discussion in the previous section, it seems plausible to equate $L$ with the vertical distance $s_v$, which is (as we have seen) of order $H_p$ times a number that is close to unity. In solar convection, a "mixing length parameter" $\alpha$ is defined as the ratio between the mixing length and the local scale height: $\alpha = L/H_p$. Based on the discussion in Section 7.2, an appropriate choice for $\alpha$ in the solar photosphere is expected to be in the range 1–2. Again, this is an empirical conclusion, based on measured velocities and temperature differences. Note that since MLT discussions are based on vertical motions, we can regard the MLT as essentially a one-dimensional model of convection.

Is there any theoretical reason why the mixing length might be expected to be of order $H_p$? Well, when a parcel of gas starts its upward "lifetime" at depth $z$, it has a density that is only slightly smaller than that of the ambient gas. Once the parcel has risen to a new depth $z - L$, it finds itself in an ambient medium where the gas has a lower density than the ambient medium had at the starting depth $z$: specifically, if $L = H_p$, the ambient density is smaller by a factor of $e = 2.718\ldots$ than the density was at depth $z$. In order to reduce buoyancy forces to zero, the parcel must adjust its density to the ambient value at depth $z - L$. This requires that the parcel must expand in volume. If the vertical distance $L$ were as large as, say, $2H_p$, the parcel would find itself at the top of its path (at depth $z - L$) in a medium with an ambient density which is $e^2 \approx 7-8$ times smaller than the initial density. This would lead to a seven- or eight-fold increase in the parcel's volume, along with a roughly four-fold increase in surface area. As a result, if at depth $z$ at any instant, a snapshot of the gas at that depth showed the aggregate of all rising parcels occupying, say, 10%–25% of the available surface area, then at depth $z - L$, the parcels would have expanded to occupy 40%–100% of the available surface area. There would be no more room for further expansion. To be sure, the argument here is very approximate, but the existence of expansion of a parcel during upward motion is qualitatively reliable.

This leads us to suspect that the existence of the empirical limit $\alpha = 1$–$2$ may be related to a self-regulating process: there is simply not enough room for parcels that would expand, in the course of their lifetime, to 10 or more times their initial volume.

However, since convection in the Sun is a highly turbulent and time-dependent process, it has to be admitted that a physically realistic description of convection in the Sun cannot be confined to 1-D models such as the MLT. Instead, a correct model of the solar convection zone requires in principle a method that permits numerical modeling of hydrodynamics in three dimensions. Moreover, since radiation is the mechanism by which an upward parcel loses its energy into the ambient medium, the 3-D hydrodynamical model must include the effects of 3-D radiative transfer. Also to be included are the equation of state (EOS) of an ionizing medium, opacities with line blanketing, and a realistic specification of boundary conditions at the bottom and at the top of the computational zone. Finally, for numerical reasons, artificial dissipation terms (e.g., "hyperviscosity") may need to be included so as to avoid the development of instabilities in the code. Inclusion of such a multitude of physical effects would take us far beyond the limits of a first course in solar physics. Interested readers can find more details about the work that has already been done in modeling 3-D solar convection in, e.g., Trampedach et al. (2014).

For our present purposes, it is important to note that, as far as we are concerned here, a valuable result has emerged from the work of Trampedach et al. (2014). Comparing 1-D models with full 3-D models, Trampedach et al. (2014) find that for stars with $T_{eff}$ values within (roughly) ±1000 K of the solar value and with log $g$ values within (roughly) ±1.0 of the solar value, the properties of the 1-D model envelopes match best with those of the 3-D models if the MLT parameter is assigned the value $\alpha \approx 1.76$. Remarkably, this value lies in the range $\alpha = 1$–$2$ that we estimated (Section 7.2) based entirely on empirical arguments. Therefore, despite all its simplicity, it is important that the MLT *can* replicate important aspects of convection in the Sun. In the next few sections, we apply MLT to convection in the Sun.

## 7.4 TEMPERATURE EXCESSES ASSOCIATED WITH MLT CONVECTION

As a check on the plausibility of choosing $\alpha \approx 1$, let us estimate how large the temperature excess is expected to become between rising gas and the ambient medium. In other words, what is the temperature excess relative to ambient after a parcel of gas has traveled a length $L$?

To answer this question, consider a parcel of gas that rises from an initial depth $z$ to a new (upper) depth $z - L$, and rises so fast that it preserves its initial temperature along the way. At the upper depth, if the gas in the parcel were not called upon to perform any work, the parcel would have a temperature in excess of the ambient temperature by an amount $\Delta T_o = L g_o$. (Here, $g_o = dT/dz$ is the local temperature gradient in the ambient medium.)

Now, our discussion of the adiabatic gradient in Chapter 6 shows that, in a convective region, some of the internal energy of the parcel of gas is used to do the work of raising the parcel a distance $L$ against gravity. Specifically, the work against gravity, i.e., $gL$ per gram, can be performed by extracting the amount $C_{p\Delta}T_{ad} = Lg$ from the internal energy per gram of the gas. (Note, $g_{ad} = g/C_p$ is the adiabatic temperature gradient.)

As a result, when the parcel reaches its upper position, $z - L$, it finds itself with a temperature that exceeds the ambient by an amount $\Delta T$ that is not as large as the $\Delta T_o$ mentioned earlier. Instead, the temperature excess $\Delta T$ is given by the reduced quantity $\Delta T = \Delta T_o - \Delta T_{ad}$. Expressing $\Delta T_o$ and $\Delta T_{ad}$ in terms of the temperature gradients, we can write

$$\Delta T = L(g_o - g_{ad}) \equiv L(\Delta g_T) \qquad (7.3)$$

Here we define the quantity $\Delta g_T$ to be the "superadiabatic gradient", i.e., the amount by which the ambient temperature gradient $g_o$ exceeds the adiabatic gradient $g_{ad}$.

Computing a Model of the Sun

Let us estimate the numerical value of the temperature excess $\Delta T$. We have already seen that in the solar photosphere, $g_{ad}$ has a numerical value of about $1.7 \times 10^{-4}$ deg cm$^{-1}$ (see Equation 6.9). In regions of vigorous convection below the photosphere, the local temperature gradient $g_o$ may exceed $g_{ad}$ by an amount that is not necessarily small. In such conditions, there is no reason to exclude the possibility that that $g_o$ might exceed $g_{ad}$ by an amount that is comparable to $g_{ad}$ itself. This suggests that $\Delta g_T$ could have a value of order $10^{-4}$ deg cm$^{-1}$. In such a case, and setting $L \approx H_p \approx 10^7$ cm, we find that the mixing length theory would predict $\Delta T \approx 10^3$ K in the photosphere.

How does this compare with the temperature differences that exist in the solar granulation? We have seen, from rough analysis of the empirical brightness fluctuations in the granulation, that the r.m.s. temperature differences are estimated to be in the range 500–600 K. These are consistent, within factors of two, with the prior estimate of $\Delta T$.

It seems that estimates of temperature excesses based on MLT are not inconsistent with empirical data by significant amounts.

The fact that the velocities of solar granulation, as well as the temperature differences between hot and cold gas, can be replicated, at least roughly, in the context of MLT suggests that the theory can be of service when we attempt to model the complexities of turbulent solar convection.

## 7.5 MLT CONVECTIVE FLUX IN THE PHOTOSPHERE

The convective heat flux can be expressed as $F(conv) \approx \rho V C_p \Delta T$ (see Equation 6.2). In the context of MLT, let us see what this expression leads to. We replace $V = \sqrt{(2a_{hc}s_v)}$ with the expression $\sqrt{(2Lg\Delta T/T)}$, where we have used Equation 7.2 for $a_{hc}$. Now, using Equation 7.3, we replace $\Delta T$ by the expression $L\Delta g_T$. This leads to

$$F(conv) \approx \rho C_p \sqrt{\frac{2g}{T}} L^2 \left(\Delta g_T\right)^{3/2} \qquad (7.4)$$

Near the photosphere, substitution of appropriate quantities ($\rho \approx (2-3) \times 10^{-7}$ gm cm$^{-3}$, $T \approx 6000$ K, $L \approx 10^7$ cm, $\Delta g_T \approx 10^{-4}$ deg cm$^{-1}$) leads to $F(conv) \approx (1-2) \times 10^{10}$ ergs cm$^{-2}$ sec$^{-1}$. This result is consistent, within a factor of two, with the estimate given in Chapter 6 (Section 6.7.1).

## 7.6 MLT CONVECTIVE FLUX *BELOW* THE PHOTOSPHERE

An advantage of MLT is that it provides us with a key piece of information about how the temperature varies as a function of depth in the convection zone. This key piece of information will allow us to obtain a model of the convection zone that, in this first course, will be good enough to lead us fairly reliably into the deep interior of the Sun. In order to demonstrate how we arrive at this key piece of information, we need to examine what happens to $F(conv)$ as we examine material that lies deeper inside the Sun.

Specifically, as we go deeper beneath the surface, temperatures increase greatly, and the mean molecular weight decreases (by a factor of about two). As a result, the mixing length $L = \alpha H_p \sim T/\mu$ increases to values that are much greater than those near the surface. For example, we shall find that in the deepest layers where convection is at work, the gas has temperatures $T \approx 10^6$ K. In this gas, where $T$ exceeds the surface values by a factor of more than 100, the value of $H_p$ exceeds the value of $H_p$ in the photosphere (where $H_p \approx 100$–200 km) by factors of several hundred. As a result, in the deep convection zone, $H_p$ approaches values as large as $10^{10}$ cm. In such gas, the density is also much larger than the photospheric value: detailed models of the Sun find that $\rho$ approaches values of order 1 gm cm$^{-3}$. Moreover, $C_p$ increases above the surface value by a factor of at least two (due to the conversion of single particles (atoms) in the upper convection zone to two particles (ion plus electron) in the deep convection zone). Let us see how these values of the various parameters affect the expression for $F(conv)$.

Some of the parameters cause the value of F(conv) to increase, while other parameters cause F(conv) to decrease. Now, there is a definite upper limit on how large F(conv) can become: it must not exceed the overall flux of energy that emerges from the deep interior of the Sun. At the surface, we can measure what this emergent flux actually is: using the symbol $F_\odot = L_\odot/4\pi R_\odot^2$ for the emergent flux at the surface, we know that $F_\odot = 6.2939 \times 10^{10}$ ergs cm$^{-2}$ sec$^{-1}$ (Chapter 1, Section 1.9). As we go inward deeper into the convection zone to a location where the local radius has a value $r<R_\odot$, the local surface area $A(r) = 4\pi r^2$ decreases, but the total power from the Sun (the "luminosity" $L_\odot$) remains constant (as long as we do not approach too close to the center of the Sun). Therefore, the local energy flux that crosses unit area $F(r) = L_\odot/A(r)$ increases as *we go deeper into the convection zone*. By the time we reach a radial location of about $0.7R_\odot$ (where we shall find that the convection zone reaches its deepest extent), the value of $F(r)$ increases to about $10^{11}$ ergs cm$^{-2}$ sec$^{-1}$, i.e., about twice as large as the flux at the solar surface. And as long as we are inside the convection zone, this flux is transported essentially entirely by convection, i.e., F(conv) rises to a value of order $10^{11}$ ergs cm$^{-2}$ sec$^{-1}$ in the deepest part of the convection zone.

Now that we know F(conv) deep in the convection zone, we can return to Equation 7.4 and find an answer to an important question: how large must the super-adiabaticity $\Delta g_T$ be in order to transport a flux of $10^{11}$ ergs cm$^{-2}$ sec$^{-1}$ deep inside the Sun? Substituting the prior numerical values for the gas deep in the convection zone, we can evaluate the $\Delta g_T$ needed: we find $\Delta g_T \approx 10^{-11}$ deg cm$^{-1}$.

What is the significance of this result? The answer depends on what we compare $\Delta g_T$ to. Since $\Delta g_T$ has the dimensions of a temperature gradient, it is natural to ask: is there another temperature gradient that is relevant to convection in the Sun? Indeed there is (see Chapter 6, Section 6.8): it is the adiabatic gradient $g_{ad}$. Deep in the Sun, $g_{ad} = g/C_p$ still has a numerical value of order $10^{-4}$ deg cm$^{-1}$: the subsurface increase in $C_p$ is offset by the subsurface increase in $g$. Compared to $g_{ad}$, we see that *the superadiabaticity $\Delta g_T$ is seven orders of magnitude* smaller than $g_{ad}$.

This is the "key piece of information" that was mentioned in the opening paragraph of the present section. It means that, for all practical purposes, the numerical value of the superadiabaticity is zero. That is, the temperature gradient in the deeper layers of the convection zone is essentially *equal to the adiabatic gradient $g_{ad}$*. Since we already have a simple expression for $g_{ad}$, this provides an enormous simplification in our task of obtaining a model of the convection zone. We do not have to be too concerned with how the opacity, density, or pressure behaves as a function of depth: instead, we simply accept that (to a high degree of precision) $dT/dz$ is effectively identical to the ratio $g/C_p$. In regions where $g$ and $C_p$ are constant, this allows us to perform an immediate integral:

$$T(z) = T(z_o) + (z - z_o)\left(\frac{g}{C_p}\right) \quad (7.5)$$

The fact that the temperature gradient in the deep convection zone equals the adiabatic gradient means that the processes that occur in the solar convection zone are essentially adiabatic in nature. This valuable conclusion will help us determine how pressure and density vary with depth as we go down deeper and deeper into the convection zone.

## 7.7 ADIABATIC AND NONADIABATIC PROCESSES

Once it has been determined that the temperature profile in the Sun's deep convection zone is essentially the adiabatic profile, we can in principle apply the laws of adiabatic processes to the variations of density and pressure. Thus, if the density varies as a function of depth according to $\rho(z)$, then the pressure at depth $z$ is related to $\rho(z)$ according to $p(z) \sim [\rho(z)]^\Gamma$ (see Chapter 6, Section 6.7.3.) The index $\Gamma \equiv d(\log p)/d(\log \rho)_{ad}$ is the adiabatic exponent for pressure-density variations. For monatomic gases, under conditions where ionization is not occurring (or is essentially complete), the numerical value of $\Gamma$ is well defined: $\Gamma = C_p/C_v = 5/3$.

For a perfect gas, $p \sim T\rho$, and so the density $\rho(z)$ in an adiabatic region is related to $T(z)$ by $\rho(z) \sim [T(z)]^{1/(\Gamma-1)}$. Also, as we have seen already (Equation 6.14), the pressure $p(z)$ in the adiabatic region is related to $T(z)$ by $p(z) \sim [T(z)]^{\Gamma/(\Gamma-1)}$.

Thus, if we happened to be considering an adiabatic medium where $\Gamma = 5/3$ at all depths, then, given a temperature, density, and pressure at a reference depth $z_o$, the quantities at depth z would be given by Equation 7.5 plus the following two equations:

$$\rho(z) = \rho(z_o)\left[\frac{T(z)}{T(z_o)}\right]^{1.5} \equiv K_d [T(z)]^{1.5} \qquad (7.6)$$

$$p(z) = p(z_o)\left[\frac{T(z)}{T(z_o)}\right]^{2.5} \equiv K_p [T(z)]^{2.5} \qquad (7.7)$$

In Equations 7.6 and 7.7, we have introduced proportionality constants $K_d = \rho(z_o)/T(z_o)^{1.5}$ and $K_p = p(z_o)/T(z_o)^{2.5}$ for density and pressure, respectively. The constants $K_d$ and $K_p$ are related to the specific entropy of the gas at the top of the solar convection zone. Since the top of the convection is also the base of the photosphere, we could use the values of density, temperature, and pressure that occur in the last line of Table 5.3 to calculate numerical values of $K_d$ and $K_p$.

If Equations 7.5 through 7.7 were all that we needed to describe solar convection, then the computation of a model of the convection zone would be relatively simple. We would start with our model of the photosphere (Chapter 5), evaluate the constants $K_d$ and $K_p$ using the conditions at the base of the photosphere (where convection sets in), and then proceed to deeper layers by increasing the depth z.

Unfortunately, things are not quite so simple in the Sun.

Two effects are particularly important in seriously modifying the properties of the convection zone in a relatively narrow layer near its upper boundary. First, radiative losses near the solar surface from convective elements (granules and intergranular regions) are severe. As a result, processes in the granulation are highly *non*adiabatic within the uppermost 1–2 megameters (Mm) of the convection zone. Nonadiabaticity has the effect that the local temperature gradient in the uppermost 1–2 Mm rises to values that are well in excess of the adiabatic gradient. (We have already used this information in Section 7.4, when we estimated temperature differences between rising and falling material.) We simply cannot assume that, as soon as convection sets in, the processes instantaneously become adiabatic.

The second important effect is that, as a result of the onset of significant ionization in H and He, the value of the exponent $\Gamma$ departs significantly from the monatomic value of 5/3. To be sure, $\Gamma$ *is* close to 5/3 in the photosphere, and $\Gamma$ again reverts to values within a few percent of 5/3 at depths in excess of 20–30 Mm below the photosphere. In such regions, the exponents that appear in Equations 7.6 and 7.7 are entirely appropriate. However, at depths of a few megameters, where the degree of ionization of hydrogen is greater than (say) 10% and less than (say) 90% (i.e., the gas lies in the "ionization strip", which was discussed in Chapter 4, Figure 4.1), the numerical values of $\Gamma$ fall well below 5/3. As was mentioned earlier (Chapter 6, Section 6.7.3), $\Gamma$ may fall as low as ~1.19 in certain regions in the Sun. Now, in a medium where $\Gamma = 1.19$, the exponents in Equations 7.6 and 7.7 would take on values of 5.3 and 6.3, respectively. In such a medium, given an increase in temperature from depth $z_o$ to depth z, the accompanying increases in density and pressure under adiabatic conditions would be significantly *larger* than Equations 7.6 and 7.7 would predict. The reason for this behavior has to do with the increase in entropy associated with ionization. Because the ionization energy is large compared to the thermal energy, a large input of energy $dQ$ is required to cause ionization in unit mass of material, without any significant increase in temperature. This leads to a significant increase in the specific entropy $dS = dQ/T$.

Fully consistent modeling of solar convection requires inclusion of 3-D radiative transfer as well as a detailed treatment of the ionization of hydrogen (e.g., Stein and Nordlund 1998). The results of such calculations indicate that if we use the conditions at the top of the convection zone to calculate $K_d \equiv p(z_o)/T(z_o)^{1.5}$ and $K_p \equiv p(z_o)/T(z_o)^{2.5}$, we will make large numerical errors. The errors are in the following sense: if we were to use the prior numerical values of $K_d$ and $K_p$ in Equations 7.6 and 7.7, the pressures and densities we would calculate in the deep convection zone would turn out to be *too small* by two to three orders of magnitude.

In a complete model of the solar convection zone, we should include the full effects of radiative losses and include ionization effects at all depths. Such a model would demonstrate a behavior where $p \sim T^{2.5}$ at the shallowest depths near the surface ($z < 1$ Mm), then a narrow region of intermediate depths (a few Mm) where the exponent would be significantly larger than 2.5, followed by a deeper region where the exponent would decrease to approach 2.5 once more. At depths $z \geq 20$–30 Mm, conditions would revert to $p \sim T^{2.5}$. We shall see that the convection zone has a depth of order 200 Mm. Thus, the functional form $p \sim T^{2.5}$, as in Equation 7.7, applies throughout some 90% of the depth of the convection zone, although $K_p$ takes on different values in the upper and lower portions of the convection zone.

## 7.8 COMPUTING A MODEL OF THE CONVECTION ZONE: STEP BY STEP

How can we make allowance for the aforementioned properties of the solar material? In this first course in solar physics, rather than following in detail the complicated calculations of radiative transfer and of the ionization of hydrogen at each depth, we make the following simplification: we use a single "effective" value for the exponents in Equations 7.6 and 7.7 throughout the convection zone. To select the effective values, we use an effective value of $\Gamma$ which is given by the arithmetic mean of the minimum (1.19) and maximum (1.67) values cited earlier, i.e., $\Gamma(\text{eff}) = 0.5(1.19 + 1.67) = 1.43$. With this choice, the exponents in Equations 7.6 and 7.7 become 2.3 and 3.3. Therefore, rather than relying on Equations 7.6 and 7.7, we use instead the following depth dependences of density and pressure:

$$\rho(z) = \rho(z_o)\left[\frac{T(z)}{T(z_o)}\right]^{2.3} \tag{7.8}$$

$$p(z) = p(z_o)\left[\frac{T(z)}{T(z_o)}\right]^{3.3} \tag{7.9}$$

These equations, together with Equation 7.5, are the equations that we use to compute a model of the solar convection zone. In this part of the Sun, the independent variable that we will use for computations of structure is the linear depth $z$ below the photosphere. We proceed as follows.

Start at the deepest layer in the photospheric model (Chapter 5), which also corresponds to the top of the convection zone. There, the temperature, depth, pressure, and density are already known: since these are the first (topmost) values in our model of the convection zone, we refer to these as $T(1) = 6010$ K, $z(1) = 465$ km, $p(1) = 1.13 \times 10^5$ dyn cm$^{-2}$, and $\rho(1) = 2.94 \times 10^{-7}$ gm cm$^{-3}$.

Step (1): Step down below the photosphere by taking a step of say $\Delta z = 1000$ km = 1 Mm. Assuming adiabatic conditions, the increase in temperature across the step $\Delta z$ is

$$\Delta T = \Delta z \times \left(\frac{g}{C_p}\right) \tag{7.10}$$

Computing a Model of the Sun

What $g$ should we use in estimating $\Delta T$? We shall find that the convection zone occupies a spherical shell that extends inward to significant depths inside the Sun, as deep as 20%–30% of the solar radius. Within the convective shell, the total amount of mass is small compared to the total mass of the Sun. As a result, most of the mass of the Sun lies interior to the convection zone. Because of this, the value of $g$ varies as $1/r^2$. Thus, at depth $z$, the local acceleration due to gravity can be calculated from

$$g(z) = 27,420 \times \left(\frac{R_\odot}{R_\odot - z}\right)^2 \text{ cm sec}^{-2} \quad (7.11)$$

where we have inserted the acceleration due to gravity that occurs at the surface of the Sun (see Equation 1.13). Using Equation 7.11, we find that, at the base of the convection zone, where $z \approx 0.3 R_\odot$, $g(z)$ has a value that is about twice as large as the surface gravity.

What value of $C_p$ should be used? Equation 6.5 provides a starting point. In the photosphere of the Sun, where $\gamma = 5/3$ and the mean molecular weight $\mu \approx 1.3$, we find $C_p \approx 1.6 \times 10^8$ ergs gm$^{-1}$ K$^{-1}$. Both quantities $\gamma$ and $\mu$ vary with depth. Let us consider $\mu$ first. Deep inside the Sun, where H and He are completely ionized, there are two particles for each H nucleus, and three particles for each He nucleus. As a result, the mean molecular weight per particle is 1/2 for H and 4/3 for He. In a mixture of 90% H and 10% He, $\mu \approx (0.5*0.9) + (1.33*0.1) \approx 0.58$. This is the value of $\mu$ that we shall use in the deep interior of the Sun (Chapter 9). In the convection zone, where ionization is underway, causing $\mu$ to vary from 1.3 (at the top) to 0.58 (at the bottom), we shall approximate the value of $\mu$ by the average of these limits, i.e., $\mu_{conv} = 0.94$. Let us now consider $\gamma$. As in Section 6.7.3, we replace $\gamma$ in the convection zone with the generalized $\Gamma$. Specifically, Equations 7.8 and 7.9 are based on the effective value $\Gamma(\text{eff}) = 1.43$. In order to preserve consistency, in Equation 6.5 we replace $\gamma/(\gamma - 1)$ with $\Gamma/(\Gamma-1) = 3.3$. For simplicity, we assign a constant value to $C_p$ throughout the convection zone, namely $C_p = 3.3 R_g/\mu_{conv}$. Of course, this does not take into account the largest values that $C_p$ takes on at certain depths in the solar convection zone. We shall therefore not be surprised if our simplified model of the solar convection zone will be defective in certain ways: in particular, we shall find that our estimate of the depth of the convection zone will be too shallow.

Step (2): Now that $g$ and $C_p$ can be evaluated at depth $z(1)$, we can calculate $\Delta T$ using Equation 7.10. Therefore, at the new depth $z(2) = z(1) + \Delta z$, the temperature $T(2)$ has the value $T(1) + \Delta T$.

Step (3): Knowing the temperature $T(2)$ at $z(2)$, we calculate the local pressure and density using $p(2) = p(1)[T(2)/T(1)]^{3.3}$ and $\rho(2) = \rho(1)[T(2)/T(1)]^{2.3}$.

Step (4): Repeating the calculation at a greater depth, $z(3) = z(2) + \Delta z$, we step inwards into the Sun, evaluating temperature, pressure, and density at each step according to Equation 7.5 and Equations 7.8 and 7.9.

We continue increasing the depth until the temperature rises to a certain value, $T_b \approx 2 \times 10^6$ K. At that point, we stop the calculation. Why? Because we shall find in Chapter 8 that the base of the convection zone lies at a well-defined temperature $T_b$ that is close to 2 million K.

This step-by-step procedure leads to a table of values of $z$, $T$, $p$, and $\rho$ down to the base of the convection zone. Examples of values of parameters selected from such a table, which we have calculated according to the aforementioned steps, are shown in Table 7.1.

## 7.9 OVERVIEW OF OUR MODEL OF THE CONVECTION ZONE

Examining Table 7.1, we see that at the base of the convection zone, i.e., at the location where the temperature $T_b$ equals $2 \times 10^6$ K, our simplified model yields a pressure $p_b$ of order $3 \times 10^{13}$ dyn cm$^{-2}$ and a density $\rho_b$ of order 0.2 gm cm$^{-3}$. These values compare favorably with results from a sample of 10,000 models of the Sun in which all of the relevant parameters are allowed to take on values within the ranges of known error bars (Bahcall et al. 2006): $T_b = 2.01 \times 10^6$ K, $p_b = 4.3 \times 10^{13}$ dyn cm$^{-2}$, and $\rho_b = 0.16$ gm cm$^{-3}$.

**TABLE 7.1**

**A Simplified Model of the Solar Convection Zone**

| Depth $z$ (cm) | $T$ (K) | $p$ (dyn cm$^{-2}$) | $\rho$ (gm cm$^{-3}$) |
|---|---|---|---|
| 9.6492E+07 | 1.0819E+04 | 7.8964E+05 | 1.1412E-06 |
| 1.9649E+08 | 2.0260E+04 | 6.3889E+06 | 4.9305E-06 |
| 3.9649E+08 | 3.9224E+04 | 5.7757E+07 | 2.3023E-05 |
| 6.9649E+08 | 6.7877E+04 | 3.5920E+08 | 8.2740E-05 |
| 9.9649E+08 | 9.6780E+04 | 1.1716E+09 | 1.8928E-04 |
| 1.4965E+09 | 1.4552E+05 | 4.5614E+09 | 4.9010E-04 |
| 2.0465E+09 | 1.9997E+05 | 1.3156E+10 | 1.0287E-03 |
| 3.0465E+09 | 3.0127E+05 | 5.1560E+10 | 2.6759E-03 |
| 4.0465E+09 | 4.0566E+05 | 1.3897E+11 | 5.3565E-03 |
| 5.0465E+09 | 5.1329E+05 | 3.0446E+11 | 9.2742E-03 |
| 6.0465E+09 | 6.2432E+05 | 5.8471E+11 | 1.4643E-02 |
| 7.0465E+09 | 7.3889E+05 | 1.0252E+12 | 2.1694E-02 |
| 8.4965E+09 | 9.1169E+05 | 2.0652E+12 | 3.5418E-02 |
| 1.0046E+10 | 1.1057E+06 | 3.9286E+12 | 5.5552E-02 |
| 1.1546E+10 | 1.3034E+06 | 6.7961E+12 | 8.1527E-02 |
| 1.3046E+10 | 1.5115E+06 | 1.1135E+13 | 1.1518E-01 |
| 1.4546E+10 | 1.7310E+06 | 1.7496E+13 | 1.5804E-01 |
| 1.6046E+10 | 1.9628E+06 | 2.6598E+13 | 2.1187E-01 |
| 1.6296E+10 | 2.0027E+06 | 2.8443E+13 | 2.2205E-01 |

As far as the depth $z_b$ of the convection zone is concerned, our model indicates a depth of 163 Mm. In terms of the solar radius, this is a depth of 23%–24% of $R_\odot$. That is, the convection zone occupies about one-quarter of the distance from the solar surface to the center. This indicates clearly that convection in the Sun is by no means confined to a thin shell. Instead, we can properly refer to a thick "convective envelope" that penetrates inward by some 25% of the solar radius in the outermost layers of the Sun.

Actually, according to inversions of helioseismic data (Chapter 13), the Sun's convective envelope is somewhat thicker than 25% of the radius. The base of the convection zone is found to lie at a depth $z_b$ = 197–202 Mm, corresponding to $28.7 \pm 0.3\%$ of $R_\odot$ (Christensen-Dalsgaard et al. 1991). The simplified approach that we have used in calculating Table 7.1 yields a shallower convection zone than the helioseismic result by some 5% of $R_\odot$, i.e., by $\approx 35$ Mm. How can we understand such a discrepancy? It is due in large part to our neglect of the large increases in $C_p$ that occur in regions where hydrogen is undergoing ionization. In such ionization regions, the true values of $C_p$ may rise to become almost 10 larger than the value of 3.3 $R_g/\mu_{conv}$ that we have adopted here throughout the convection zone. As a result, for a given step in depth $\Delta z$, our computed $\Delta T = (g/C_p)\Delta z$ in the ionization zone may be almost 10 times too *large* compared to the true value. Conversely, for a given *temperature* interval (and we are, after all, aiming for a region where we are specifying what value the *temperature* must have there, namely 2 million K), our estimated value of the corresponding $\Delta z$ is too *small* in the ionization zone by a factor of order 10. Thus, in an ionization region that spans a depth range of 1–3 Mm in the "real" Sun, our method has the effect that an interval of depth of order 10–30 Mm is in effect "missing" by the time the integrated value of temperature reaches the limit $T_b$.

When solar models are computed with state-of-the-art computing techniques (e.g., Bahcall et al. 2006), the models yield estimates of the convection zone thickness that depend on the chemical

# Computing a Model of the Sun

composition that one assumes for the model. With two different choices of the solar composition, Bahcall et al. compute that the convective envelope has a thickness of 28.7% and 27.2% of $R_\odot$.

## EXERCISES

7.1 Perform the step-by-step calculation of the convection zone described in Section 7.8, using values of $T(1)$, $z(1)$, $p(1)$, and $\rho(1)$ that you obtained in one of your models of the photosphere (Chapter 5). What differences do you find from the results in Table 7.1?

7.2 Use your computer code to repeat the calculation of Section 7.8 using a different value of the step size, $\Delta z$, e.g., 500 km and 2000 km. How much do the various parameters differ from those in Table 7.1?

7.3 Repeat the calculations of Section 7.8 using different values of $\Gamma(\text{eff})$. Instead of using $\Gamma(\text{eff}) = 1.43$, consider $\Gamma(\text{eff}) = 1.3$ and 1.6. Each of these will lead to changes in the exponents in Equations 7.8 and 7.9. Proceed in each case to the depth $z_b$ where $T = T_b = 2$ MK. In each case, how do your values of $z_b$ compare with the value obtained from helioseismology (200 Mm)?

7.4 (More complicated) The model in Table 7.1 is based on an assumption that the specific heat $C_p$ retains a constant value at all depths in the convection zone, and that $\Gamma$ also retains a constant value (1.43) at all depths. But in the real Sun, $C_p$ varies with depth, as does $\Gamma$.

Let us try to incorporate in a simple way the depth dependences of $C_p$, $\Gamma$, and $\mu$. At the top of the convection zone, we can set $C_p = 1.6 \times 10^8$ ergs gm$^{-1}$ K$^{-1}$, $\Gamma = 1.67$ and $\mu = 1.3$. One example of how $C_p$ varies with depth is provided by Spruit (1974), in his Table II. In that table, $C_p$ rises to a maximum value of $1.486 \times 10^9$ ergs gm$^{-1}$ K$^{-1}$ at a depth of 1.3 Mm, where $\Gamma$ has a value of 1.19. At the base of the convection zone, Spruit finds $C_p = 3.4 \times 10^8$ ergs gm$^{-1}$ K$^{-1}$. (Note that, as expected, $C_p$ at the base of the convection zone is close to twice the value at the top.)

Suggested approach: At each depth between $z = 0$ and $z = 1.3$ Mm, use Spruit's values to linearly interpolate, at each value of $z$, a local value of $C_p$ and a local value of $\Gamma$. Then at depths between $z = 1.3$ Mm and (say) 163 Mm, use another linear interpolation to calculate a local value at each $z$ for $C_p$ and $\Gamma$. Start at the base of the photosphere (where $T = T(1)$), and use Equations 7.10 and 7.11 to take the first step downwards in $z$ to calculate $\Delta T$ using the local values of $C_p$ and $g$. Once you know $\Delta T$, calculate the local $T(2) = T(1) + \Delta T$. Now, knowing the updated temperature $T(2)$, enter Equations 7.8 and 7.9 to calculate the local $\rho(2)$ and $p(2)$: but in order to calculate these, it is important to use the proper exponents, which will vary as the local $\Gamma$ varies. In Equation 7.8, the exponent, rather than being fixed at 2.3, should be given the local value $1/(\Gamma-1)$. In Equation 7.9, the exponent (rather than being fixed at 3.3), should be given the local value $\Gamma/(\Gamma-1)$.

Proceed downward as described in the step-by-step process in Section 7.8 until you reach a temperature of $2 \times 10^6$ K. This process will in general lead to a deeper convection zone that the one we obtained in Table 7.1.

## REFERENCES

Bahcall, J. N., Serenelli, A. M., & Basu, S., 2006. "10,000 standard solar models: A Monte Carlo simulation", *Astrophys. J. Suppl.* 165, 400.

Christensen-Dalsgaard, J., Gough, D. O., & Thompson, M. J., 1991. "The depth of the solar convection zone", *Astrophys. J.* 378, 413.

Spruit, H. C., 1974. "A model of the solar convection zone", *Solar Phys.* 34, 277.

Stein, R. F., & Nordlund, A., 1998. "Simulations of solar granulation. I. General properties", *Astrophys. J.* 499, 914.

Trampedach, R., Stein, R. F., Christensen-Dalsgaard, et al., 2014. "Improvement to stellar structure models based on a grid of 3D convection simulations. II. Calibrating the mixing-length formalism", *Mon. Not. Roy. Astron. Soc.* 445, 4366.

# 8 Radiative Transfer in the Deep Interior of the Sun

Continuing inward to the deep interior of the Sun, we note that, below the Sun's convection zone, energy is transported once again by means of radiation. We refer to this region as the radiative interior of the Sun. The aim of this chapter is to derive the equations that determine the radial profiles of temperature, pressure, and density in the radiative interior.

In this region of the Sun, hydrogen and helium are essentially completely ionized. As a result, there are no longer many bound electrons available. At temperatures in excess of 2 million K, the only remaining bound electrons belong to some of the metals, and their relative abundances are small. Therefore, because of the lack of strong bound-bound and bound-free absorbers in the material, photons are not as strongly absorbed in the radiative interior as they are in the cooler gas in the convection zone. As a result, the opacity decreases rapidly in the radiative interior.

With reduced opacity, radiation can more readily carry the energy flux outward through the Sun without requiring the temperature gradient to become large. That is, radiation once again takes over as the preferred means of energy transport. Thus, we can consider energy transport through the Sun in an overall sense in terms of a "sandwich": there are two regions in which radiation is the principal agent for outward transport of energy (the photosphere and the radiative interior), separated by a region where convection is the principal agent for outward transport of energy. Another feature of the sandwich that involves the operation of different physical laws is this: hydrostatic equilibrium (HSE) applies in the innermost and outermost regions (radiative interior, photosphere) but HSE does *not* apply in the middle region (convection). In the latter region, we recall (from Equation 7.1) that bulk motions of gas arise there precisely because HSE is *not* applicable.

In the present chapter, we shall once again make use of HSE as we go about the process of determining how pressure, temperature, and density vary as a function of radius $r$ below the base of the convection zone (where $T(r) \approx 2 \times 10^6$ K). Our goal in this chapter is to calculate $p(r)$, $T(r)$, and $\rho(r)$ at radial locations $r$ that extend from the base of the convection zone all the way down to the center of the Sun (i.e., down to $r = 0$).

## 8.1 THERMAL CONDUCTIVITY FOR *PHOTONS*

When we turn our attention to considering how radiation travels deep inside the Sun, we find that it is easier to describe the flow of radiative energy there than was the case in the surface layers. As was described in Chapter 2 (especially Equations 2.29 through 2.31), when we considered radiative transfer *in the surface layers*, we had to give careful consideration to the large relative difference between the intensity of the outgoing stream of radiant energy $I_o$ and the intensity of the incoming stream of radiant energy $I_i$. In fact, in the extreme case of $\tau \to 0$, as we approach the uppermost layers of the photosphere, the $I_i$ can be set equal to zero, while $I_o$ carries the full outward-directed radiant energy generated by the Sun.

On the other hand, when we consider conditions in the gas that lies deep in the interior of the Sun, the situation is quantitatively different: $I_o$ and $I_i$ at any given point have numerical values that are enormous compared to their numerical values in the photosphere. However, in the deep interior, the difference between outgoing and incoming intensities is *very small* compared to the magnitude of either one of these intensities. In this situation, there is no real advantage to the Eddington "two-stream approximation" (see Section 2.8) in which $I_o$ and $I_i$ have distinctly different numerical values. Instead, in the deep interior, where $|I_o - I_i| \ll I_o$ or $I_l$, it makes more sense to adopt a different approach. Specifically, in the deep interior, local conditions are such that the photons flow down the temperature gradient in a manner that can be well described in *diffusive* terms.

This means that the flux $F$ of radiant energy can be described in the form of a generalized Fick's law: the flux $F$ flows down the temperature gradient from regions of hot gas (at small radial locations) to regions of cooler gas (at larger radial locations). And the magnitude of the flux $F$ is linearly proportional to (minus) the local radial gradient of temperature. That is

$$F(r) = -k_{th} \frac{dT}{dr} \qquad (8.1)$$

where $k_{th}$ is a physical parameter called thermal conductivity (with units of ergs cm$^{-1}$ sec$^{-1}$ deg$^{-1}$).

Referring to the kinetic theory of gases, we find (e.g., Roberts and Miller 1960) that in a medium where *particles* are responsible for the transport of heat, a general formula for thermal conductivity can be written in the form

$$k_{th} = \frac{1}{3} \lambda V_t \rho C_v \qquad (8.2)$$

Here, $V_t$ is the mean thermal speed of the particles that are transporting the heat, $\lambda$ is the mean free path of the particle (i.e., the mean distance a particle travels between collisions with another particle), $\rho$ is the mass density of the medium, and $C_v$ is the specific heat per gram at constant volume of the medium transporting the heat. Inserting the appropriate units, it is readily seen that the units of $k_{th}$ are ergs cm$^{-1}$ sec$^{-1}$ deg$^{-1}$, as required by Equation 8.1.

Now we come to an interesting and unusual aspect of the material that exists in the deep radiative interior of the Sun. Up to this point in the book, in the photosphere and in the convection zone, when we discuss quantities such as the specific heat and the mean speed of the particles, we have been dealing with material dominated by the atoms (or ions) in the local gas. But in the deep interior layers of the Sun, we encounter a different regime. In these layers, we are dealing with a medium consisting of two very different components: photons and material particles. The two components are closely coupled by means of emission and absorption of radiation. As regards the local pressure, the photons do not contribute much: the ratio of radiation pressure to gas pressure is of order <0.001 (see Section 9.2).

But despite this small contribution to pressure, the photons outstrip the particles in their contribution to the thermal conductivity. Let us therefore apply Equation 8.2 to this regime, where photons are considered to be the "particles" that transport energy. We consider separately the four physical parameters that enter into the r.h.s. of Equation 8.2.

For photons, the mean speed of the "particles" is the speed of light: therefore, we replace $V_t$ in Equation 8.2 with $c = 3 \times 10^{10}$ cm sec$^{-1}$.

The mean free path for a photon is determined by the length-scale $\lambda$ corresponding, at any particular radial location, to optical depth of order unity in the material situated at that location. Using the definitions in Chapter 3, Section 3.1, we see that this length-scale is given by the condition $\lambda \kappa \rho = 1$, where $\kappa$ is the local opacity. Thus, in Equation 8.2, we replace $\lambda$ by the quantity $1/\kappa\rho$. We shall see (Section 9.3) that near the center of the Sun, the magnitude of $\lambda$ is no more than 0.001 cm. This is a measure of how far the photons can travel before they interact with the local material. Compared with the size of the Sun (of order $10^{11}$ cm: see Equation 1.12), this is such a small distance that we can safely say the following: the photons and the material are closely coupled in the deep interior of the Sun, so closely coupled, in fact, that we can regard the photons+gas (in an approximate sense) as a single "fluid".

What shall we use for the density $\rho$? This *cannot* be contributed by photons, which have zero rest mass. Instead, the density is contributed by the material constituents (atoms/ions) of the local

# Radiative Transfer in the Deep Interior of the Sun

"fluid". Thus, for $\rho$, we use the local gas density, which can reach values in excess of 100 gm cm$^{-3}$ near the center of the Sun (see Table 9.1).

How do we estimate the term $C_v$ for the case of a photon-material fluid in which the photons are the carriers of energy? We start by recalling (Chapter 2, Section 2.1) that the energy contained in radiation *per unit volume* is $u = a_R T^4$ ergs cm$^{-3}$ where $a_R = 7.5658 \times 10^{-15}$ erg cm$^{-3}$ deg$^{-4}$ is the radiation density constant. In terms of units, we note that $C_v = (dU/dT)_v$ refers to an energy content $U$ *per gram* of the medium. To convert from energy per unit volume to energy per gram, we divide $u$ by the local mass density. This leads to $U = a_R T^4/\rho$ ergs gm$^{-1}$. Inserting this in $C_v = (dU/dT)_v$, we find $C_v = 4a_R T^3/\rho$ ergs gm$^{-1}$ deg$^{-1}$.

Combining the four parameters on the r.h.s. of Equation 8.2, we find the following:

$$k_{th} = \frac{4a_R c T^3}{3\kappa\rho} \tag{8.3}$$

This is the thermal conductivity that is appropriate for the fluid composed of photons and atoms/ions in the deep interior of the Sun.

It is important to understand in detail why photons (rather than particles) are the dominant contributor to the value of $k_{th}$. To see why this is so, consider the four parameters on the r.h.s. of Equation 8.2. The density $\rho$ is the same for both particles and photons, so neither particles nor photons have any advantage in that regard. With the specific heat $C_v = 3R_g/2\mu$ for the particles and $4a_R T^3/\rho$ for the photons and setting $\mu \approx 0.5$, we find $C_v(part) \approx 2.4 \times 10^8$ ergs gm$^{-1}$ deg$^{-1}$. Setting $T \approx 5 \times 10^6$ K and $\rho \approx 10$ gm cm$^{-3}$ as typical values in the radiative interior of the Sun, we find $C_v(phot) \approx 4 \times 10^5$ ergs gm$^{-1}$ deg$^{-1}$. Therefore, the particles have an advantage of about 1000 over photons as regards $C_v$. What about $V_t$? The mean thermal speed of the particles in the hottest part of the Sun is of order $V_t(part) = 6 \times 10^7$ cm sec$^{-1}$ (see Section 9.2), while photons have speed $c = 3 \times 10^{10}$ cm sec$^{-1}$: therefore, the photons have an advantage by a factor of almost 1000 over the particles as regards $V_t$. If $k_{th}$ depended only on $V_t$, $\rho$, and $C_v$, then the photons and particles would contribute comparably to the value of $k_{th}$. However, when we consider the fourth parameter $\lambda$, the key to the dominance of photons in $k_{th}$ emerges. We find that near the center of the Sun, the opacity is such that photons can travel a mean free path of $\lambda(phot) \approx 10^{-3}$ cm (see Section 9.3). On the other hand, proton-proton collisions occur at a rate of order $10^{15}$ sec$^{-1}$ (see Section 11.3): that is, the mean free time $\Delta t$ between particle collisions is of order $10^{-15}$ sec. In this short time, the protons can travel a distance of $\lambda(part) = \Delta t \times V_t(part) \approx 6 \times 10^{-8}$ cm. Therefore, the mean free path of photons in the deep interior of the Sun *exceeds* the mean free path of particles *by more than four orders of magnitude*: of the four parameters on the r.h.s. of Equation 8.2, $\lambda$ is the main reason why photons are the dominant contributor to $k_{th}$ in the deep interior of the Sun.

As regards Equation 8.2, we shall subsequently (Section 17.15.1) use it in a very different environment (the solar corona), where *electrons* are the dominant contributor to $k_{th}$.

## 8.2 FLUX OF RADIANT ENERGY AT RADIUS $R$

We have already noted (Section 2.1) that the Stefan–Boltzmann constant $\sigma_B$ is related to the radiation constant $a_R$ by the formula $\sigma_B = a_R c/4$. Using this in Equation 8.3, we find

$$k_{th} = \frac{16\sigma_B T^3}{3\kappa\rho} \tag{8.4}$$

This is the thermal conductivity of a medium of density $\rho$ in which photons are transporting energy. As we can see, the smaller the opacity (i.e., the more transparent the medium is), the greater is the thermal conductivity. Photons are better at transporting heat if the medium allows the photons to pass through with minimum obstruction.

Now that we know the thermal conductivity, we can write down the local flux of radiant energy $F(r)$ at radial location $r$ in terms of the local temperature gradient and in terms of the local values of $T$, opacity, and density:

$$F(r) = -\frac{16\sigma_B T(r)^3}{3\kappa(r)\rho(r)} \frac{dT(r)}{dr} \qquad (8.5)$$

We can use this equation in two ways, depending on what information is already available to us. On the one hand, if $F(r)$ is specified (somehow), then we can determine the local value which $dT/dr$ must have in order to transport that flux through the local medium. On the other hand, if $dT/dr$ is specified (somehow), then we can determine the local value that $F(r)$ must have.

## 8.3 BASE OF THE CONVECTION ZONE

At this point, we can determine a quantity to which we have already referred in Chapter 7: the temperature $T_b$ at the base of the convection zone. The value of $T_b$ is determined by the location where the temperature gradient due to radiation (given by Equation 8.5) becomes as large as the local adiabatic gradient $g/C_p$. (Recall that in the deepest parts of the convection zone, the local temperature gradient is essentially equal to the adiabatic gradient: see Section 7.6.) As was mentioned in Chapter 7 (see Equation 7.11), the numerical value of $g$ at the base of the convection zone (at a radial location of about $0.7R_\odot$) is larger than the surface $g$ by a factor of about two. However, at the base of the convection zone, the value of the specific heat is also larger by a factor of about two compared to its value at the solar surface (see Section 6.7.3). Thus, coincidentally, the surface value of $g/C_p$ ($\approx 1.7 \times 10^{-4}$ deg cm$^{-1}$: see Equation 6.9) can be inserted for $dT/dr$ in Equation 8.5 at the base of the convection zone.

Also at the base of the convection zone, where $r \approx 0.7R_\odot$, $F(r)$ is larger than the surface flux $F_\odot$ ($= 6.2939 \times 10^{10}$ ergs cm$^{-2}$ sec$^{-1}$: Chapter 1, Section 1.9) by a factor of $(R_\odot/r)^2 \approx 2$. Thus, we can set $F(r_b) \approx 1.3 \times 10^{11}$ ergs cm$^{-2}$ sec$^{-1}$.

As regards the opacity, we have already noted (Chapter 3, Section 3.7) that at temperatures in excess of about $10^6$ K, a reasonable fit to the opacities can be obtained by the Kramers' "law": $\kappa = \kappa_o \rho/T^{3.5}$ cm$^2$ gm$^{-1}$. By fitting to tabulated values of opacity in conditions that are relevant to the solar interior (e.g., Harwit 1973), we have determined that a plausible numerical value of $\kappa_o$ is roughly $10^{24}$ when $\rho$ is in units of gm cm$^{-3}$ and $T$ is in units of K. (This choice leads to a value of $\kappa \approx 10^3$ cm$^2$ gm$^{-1}$ in gas where $\rho \approx 1$ gm cm$^{-3}$ and $T \approx 10^6$ K. (Is this consistent with the graph of opacities which was shown in Figure 3.4? To answer that, given $\rho \approx 1$ and $T_6 \approx 1$, note that the parameter $R = \rho/T_6^3$ in Figure 3.4 has the value $\log R = 0$. Therefore, we need to extrapolate the curves in Figure 3.4 upwards by one unit in $\log R$: when we do that at $\log T = 6$, we see that $\log \kappa \approx 3$.)

Inserting the values of $F(r_b)$ and $dT/dr = 1.7 \times 10^{-4}$ into Equation 8.5, we find that the temperature $T_b$ and density $\rho_b$ at the base of the convection zone are related by

$$\frac{T_b^{6.5}}{\rho^2} \approx 3 \times 10^{42} \qquad (8.6)$$

The uncertainties in estimating the numerical values of the various parameters entering into Equation 8.5 are such that we retain only one significant digit in Equation 8.6.

To proceed further, we need to know the relationship between $T_b$ and $\rho_b$. Such a relationship is already available: in the deep convection zone, we have seen (Chapter 7) that the density and temperature are related by an adiabatic function. This function, Equation 7.8, when applied to the base of the convection zone, indicates that $\rho_b = K_d T_b^{2.3}$, where $K_d$ is related to the parameters at the top of the convection zone by $K_d = \rho(z_o)/T(z_o)^{2.3}$. Inserting values of $\rho(z_o) = 2.9 \times 10^{-7}$ gm cm$^{-3}$

and $T(z_o) = 6010$ K (from Chapter 5, Table 5.3), we find that $K_d \approx 6 \times 10^{-16}$, where we again retain only one significant digit in view of the simplification that enters into the choice of the exponent 2.3 (see Chapter 7).

Inserting these values into Equation 8.5, we find $T_b^{1.9} = 1 \times 10^{12}$. This leads finally to $T_b \approx 2.1 \times 10^6$ K. This is the origin of our approximate choice of temperature ($T_b \approx 2 \times 10^6$ K) at the base of the convection zone when we computed a model of the convection zone in Chapter 7.

As shown by our retention of only one significant figure in Equation 8.6, we admit that the approximations we have made in order to calculate a model of the interior of the Sun are merely that: approximations. However, we can test how good our approximations are by comparing with results obtained by researchers who did not make those approximations. For example, Bahcall et al. (2006) calculated 10,000 separate models of the Sun using the full details of energy generation, equation of state, metal abundances in the surface layers, opacities (including effects of individual abundances of each element), and the mixing length model of convection: 21 input parameters were randomly drawn for each model from separate probability distributions for every parameter. The paper by Bahcall et al. (2006) gives no information as to how much time was required to perform all of the requisite calculations, but it must have involved hundreds of hours of computing time. The preferred solutions of Bahcall et al. were found to have $T_b = 2.006$–$2.184 \times 10^6$ K. Our choice of $T_b \approx 2 \times 10^6$ K is certainly consistent with this range. And as regards density at the base of the convection zone, Bahcall et al. found $\rho_b = 0.1555$–$0.1862$ gm cm$^{-3}$: our calculation (Table 7.1) leads to $\rho_b \approx 0.22$ gm cm$^{-3}$. Although not quite as good as the fit to $T_b$, nevertheless, in the limit of one significant figure, our calculated value of density is consistent.

## 8.4 TEMPERATURE GRADIENT IN TERMS OF LUMINOSITY

It is useful to convert from units of flux to units of power (i.e., luminosity). At any radial location inside the Sun, the luminosity $L(r)$ (in units of ergs sec$^{-1}$) has a value that is determined by the summation of energy sources lying interior to radial location $r$. The value of $L(r)$ is zero near the center of the Sun, and it increases rapidly in magnitude as one moves out through the energy-generating core.

Detailed models indicate that $L(r)$ rises to > 90% of its surface value at a radial location of about 0.2 $R_\odot$. The local flux of radiant energy $F(r)$ (in units of ergs cm$^{-2}$ sec$^{-1}$) is related to $L(r)$ by $F(r) = L(r)/4\pi r^2$.

Combining this with Equation 8.5, we see that we can write

$$\frac{T(r)^3}{\kappa(r)} \frac{dT}{dr} = -\frac{3L(r)}{64\pi\sigma_B} \frac{\rho(r)}{r^2} \qquad (8.7)$$

## 8.5 TEMPERATURE GRADIENT IN TERMS OF PRESSURE

The usefulness of Equation 8.7 can be seen by comparing it with the equation of HSE:

$$\frac{dp}{dr} = -g(r)\rho(r) = -GM(r)\frac{\rho(r)}{r^2} \qquad (8.8)$$

where $M(r)$ is the mass interior to radial location $r$. Is it permissible to apply HSE to the radiative interior of the Sun? Yes: there are no bulk flows of gas in that part of the Sun (apart from rotation), indicating that forces are all in equilibrium.

Notice that on the right-hand sides, both Equations 8.7 and 8.8 contain the factor $-\rho(r)/r^2$. Therefore, if we take the ratio of Equations 8.8 and 8.7, the terms $\rho(r)$ and $r^2$ disappear, as well as the minus sign on the r.h.s. This leads to an equation that relates the temperature $T$ and the pressure $p$ at any radial location in the star:

$$\frac{T(r)^3}{\kappa(r)}\frac{dT}{dp} = \frac{3}{64\pi\sigma_B G}\frac{L(r)}{M(r)} \qquad (8.9)$$

As already mentioned, detailed solar models indicate that $L(r)$ builds up rapidly to its asymptotic value as $r$ increases from $r = 0$, reaching 90% of $L_\odot$ at $r \approx 0.2R_\odot$. For the mass function, $M(r)$ also rises from zero at $r = 0$ and tends to the asymptotic value $M_\odot$ as $r$ increases. The rate of rise in $M(r)$ is not as rapid as for $L(r)$: $M(r)$ reaches 90% of its asymptotic value $M_\odot$ around $r \approx 0.5R_\odot$. In the outer parts of the radiative interior, where both $M(r)$ and $L(r)$ are within 10% of their asymptotic values, the ratio $L(r)/M(r)$ can be well approximated with the asymptotic value $(L/M)_a = L_\odot/M_\odot \approx$ 2 ergs sec$^{-1}$ gm$^{-1}$ (see Chapter 1, Section 1.4). Closer to the center of the Sun, where $L(r)$ remains large while $M(r)$ decreases, the ratio $L(r)/M(r)$ becomes larger than $(L/M)_a$. Examination of detailed models suggests that $L(r)/M(r)$ exceeds $(L/M)_a$ by factors of 2, 4, and 6 at $r \approx 0.25R_\odot$, $r \approx 0.15R_\odot$, and $r < 0.1R_\odot$. Thus, throughout >98% of the volume of the Sun, the right-hand side of Equation 8.9 retains a constant value, within a factor of two.

For purposes of the simplified solar model we are considering here, we shall set the right-hand side of Equation 8.9 equal to a constant, $C_1 = 8 \times 10^9$ c.g.s. This is the appropriate value (to one significant digit) for regions of the Sun where $L(r)/M(r) = L_\odot/M_\odot$.

## 8.6 INTEGRATING THE TEMPERATURE EQUATION

Once we have assigned a constant numerical value to the r.h.s. of Equation 8.9, we can now proceed with the integration of Equation 8.9. To do this analytically, we use the Kramers' opacity law, as described earlier: $\kappa = 10^{24}\,\rho/T^{3.5}$ cm$^2$ gm$^{-1}$. Substituting this in Equation 8.9, we find

$$T^{6.5}dT = 10^{24}C_1\rho dp \qquad (8.10)$$

Using the perfect gas law, $\rho = p\mu/R_g T$, Equation 8.10 can be written as

$$T^{7.5}dT = C_2 p\,dp \qquad (8.11)$$

where $C_2 = 10^{24}\,C_1\,\mu/R_g \approx 10^{26}\mu$. In the radiative interior of the Sun, $\mu \approx 0.5$. Integration yields an expression for pressure in terms of the temperature:

$$p^2 = C_3 T^{8.5} + const. \qquad (8.12)$$

where $C_3 = 1/(4.25\,C_2)$. In order to avoid the use of large numbers, it is convenient to express the temperature in units of millions of degrees K: $T_6 = T/10^6$ K. In these units, we find

$$p^2 = 5\times 10^{24} T_6^{8.5} + const. \qquad (8.13)$$

To evaluate the constant, we can make use of the conditions that have been computed for the base of the convection zone: according to Table 7.1 (Chapter 7), we see that at that location, $T_6 = 2.0027$ and $p = 2.84 \times 10^{13}$ dyn cm$^{-2}$. Inserting these in Equation 8.13, we find that the constant has the value $-1.02 \times 10^{27}$ c.g.s. We shall use this in the next chapter.

## EXERCISE

8.1 The Kramers' "law" is not a perfect fit to the opacities in the solar interior. Other possible fits to the opacities include the cases $\kappa = \kappa'\rho/T^3$ cm$^2$ gm$^{-1}$ and $\kappa = \kappa''\rho/T^4$ cm$^2$gm$^{-1}$. For both these cases, evaluate $\kappa'$ and $\kappa''$ by fitting (in both cases) $\kappa = 10^3$ cm$^2$gm$^{-1}$ at $\rho = 1$ and $T = 10^6$ K. Starting at Equation 8.10, and keeping $C_1 = 8 \times 10^9$ c.g.s., obtain revised versions of Equation 8.13, including revised values for the constant of integration.

## REFERENCES

Bahcall, J. N., Serenelli, A. M., & Basu, S., 2006. "10,000 standard solar models: A Monte-Carlo simulation", *Astrophys. J. Suppl.* 165, 400.

Harwit, M., 1973. "Stars", *Astrophysical Concepts*, J. Wiley and Sons, New York, pp. 303–373.

Roberts, J. K., & Miller, A. R., 1960. "The transfer of heat by conduction and convection", *Heat and Thermodynamics*, Blackie and Son, London, UK, pp. 280–315.

# 9 Computing a Mechanical Model of the Sun
## *The Radiative Interior*

Following the spirit of Chapters 5 and 7, we now proceed deeper into the Sun and calculate a radial profile for the physical variables. As in the earlier chapters, we still refrain from considering the *origin of the energy* that is passing through. Our aim is purely mechanical: given a total luminosity, how does the medium arrange itself so as to "handle" the energy passing through? A complete model of the Sun would of course include a description of the processes whereby the energy is generated: we will discuss that in Chapter 11. But in this chapter, we do not attempt to calculate a complete model. Our goal is as follows: given the solar luminosity as a boundary condition, what can we deduce about the structure of the Sun?

In Chapter 8 (Section 8.6), we derived the following relationship between pressure and temperature:

$$p^2 = 5 \times 10^{24} T_6^{8.5} - 1.02 \times 10^{27} \tag{9.1}$$

This equation applies (within our simplification of constant $L/M$ ratio) to the radiative interior. We use Equation 9.1 to continue our computation of a solar model. The model will consist of a table in which each line refers to a particular depth (i.e., radial location), at which we calculate the local temperature, pressure, and density.

## 9.1 COMPUTATIONAL PROCEDURE: STEP BY STEP

We start at the base of the convection zone, where we already (see Table 7.1) have numerical values for the quantities $z_b$, $T_b (\approx 2 \times 10^6$ K), $p_b$, and $\rho_b$. In terms of the independent variable to be used in this region of the Sun, it is convenient to convert now from depth $z$ to radial distance from the center of the Sun: $r = R_\odot - z$. Thus, the starting values for the four parameters in the table we wish to compute for the radiative interior are $r(1) = R_\odot - z_b$, $T_6(1) = 2.0027$, $p(1) = p_b$, and $\rho(1) = \rho_b$. These are the parameters we enter into the first line of our table of the solar radiative interior. We shall use the temperature as the independent variable in this chapter.

The computation proceeds by means of the following steps:

1. Choose an increase in temperature of (say) $\Delta T_6 = 0.01$. Thus, $T_6(2)$, the temperature (in units of $10^6$ K) of the second row in the table, is given by $T_6(2) = T_6(1) + \Delta T_6$.
2. Using $T_6(2)$ in Equation 9.1, the pressure $p(2)$ on the second line of the table can be calculated.
3. The pressure increment between lines 1 and 2 is $\Delta p = p(2) - p(1)$.
4. The density on the second line of the table is calculated from the perfect gas law: $\rho(2) = p(2)\mu/(R_g T(2))$. Here, $\mu$ can be set equal to 0.58 (see Chapter 7, Section 7.8). The average density between lines 1 and 2 is $\rho(a) = 0.5(\rho(1) + \rho(2))$.
5. The linear distance between lines 1 and 2 can be derived from the equation of hydrostatic equilibrium:

$$\Delta r = -\frac{\Delta p}{g(r)\rho(a)} \tag{9.2}$$

Notice that there is a negative sign on the r.h.s. of Equation 9.2: this denotes that as the calculation proceeds, the radial coordinate will be found to *decrease* steadily as we move closer to the center of the Sun. However, before we can apply Equation 9.2, we need to discuss what value we should use for the acceleration $g(r)$.

6. In order to calculate $g(r)$, the physics of the solar interior tells us that there are two different zones in the radial coordinate that we need to distinguish. In the first zone, situated in the outermost parts of the Sun, at radial location $r$, the local gas density is small enough that the mass $M(r)$ enclosed within radius $r$ is essentially constant. As a result, $g(r) = GM(r)/r^2$ in the outermost parts of the Sun can be written without serious error as $g(r) = GM_\odot/r^2$. As a result, in this outer zone of the radial coordinate, $g(r)$ *increases* as the radial location *decreases*, according to the inverse square law, $1/r^2$. (We have already encountered the effects of this behavior when we evaluated the value of $g$ in increasingly deep layers of the convection zone (see Equation 7.11): at the base of the convection zone (where $r \approx 0.7 R_\odot$), the value of $g$ has already increased to about twice its surface value.) In the second zone, near the center of the Sun, the density does not change rapidly: within the inner 10% of the solar radius, detailed models indicate that the density changes by a factor of only about two. In the limit of constant density $\rho_c$ near the center, the local acceleration due to gravity tends towards the following functional form:

$$g(r) \equiv \frac{GM(r)}{r^2} \approx \frac{4\pi G \rho_c}{3} r \qquad (9.3)$$

That is, near the center of the Sun, the acceleration $g(r)$ due to gravity at radius $r$ *increases* linearly with *increasing* $r$.

The existence of the two distinct zones of $g(r)$ as a function of $r$ means that the radial profile of $g(r)$ inside the Sun is *not* monotonic. Instead, there exists, inside the Sun, a radial location $r_m$ where $g(r)$ takes on a *maximum value*.

In order to include this feature, and in the spirit of simplicity that informs our approach to modeling the interior of the Sun, we assume that the behavior of $g(r)$ inside the Sun can be captured adequately by a composite of two functions, depending on the radial location.

At radial locations that lie *outside* a certain critical radius $r_m$, we use the inverse square law (just as we did in Chapter 7 when we dealt with the convection zone):

$$g(r) = g_s \left(\frac{R_\odot}{r}\right)^2 \qquad (9.4)$$

where $g_s = 27{,}420$ cm sec$^{-2}$ is the acceleration at the surface of the Sun (see Equation 1.13). Note that Equation 9.4 is applicable only for radial locations $r$ that satisfy the condition $r \geq r_m$.

At radial locations that lie *inside* the critical radius $r_m$, we replace Equation 9.4 with the following linear law:

$$g(r) = g(r_m)\left(\frac{r}{r_m}\right) \qquad (9.5)$$

where $g(r_m) = g_s (R_\odot/r_m)^2$ is chosen so that $g(r)$ merges continuously (although the slope is discontinuous) onto Equation 9.4 at $r = r_m$. Note that Equation 9.5 is applicable only for radial locations $r$ that satisfy the condition $r \leq r_m$.

Now the question is: what is an appropriate estimate for $r_m$? Various values can be chosen in order to determine what effect the choice would have on the solar model. One possibility that we have used for the tabulated model to be reported later is to identify $r_m$ with the radial location where the mass interior to $r = r_m$ is 50% of the solar mass, i.e.,

$M(r_m) \approx 0.5 M_\odot$. Using information from detailed solar models, it turns out that this condition corresponds to $r_m \approx 0.25 R_\odot$. With this choice, as we move inward into the solar interior, Equation 9.4 indicates that the value of $g$ at first *increases* from its surface value $g_s = 27{,}420$ cm sec$^{-2}$ to a peak of $g(r_m) = 16 g_s$. Then in the inner zone, between $r_m$ and the center of the Sun, $g$ *decreases* along a linear ramp toward a value of zero at the center. This sharply peaked functional form of $g(r)$ is of course only an approximation that we adopt in the interests of simplicity. In detailed solar models (see Exercise 9.5 at the end of this chapter), the radial profile of $g(r)$ is *not* sharply peaked, but has a curved shape (with no discontinuities in slope) and has a less extreme peak value, namely, 8–9 $g_s$.

7. Now that $g(r)$ can be evaluated at any radial location using Equations 9.4 and 9.5, we can evaluate $g(r)$ at $r = r(1)$ and then calculate $\Delta r$ using Equation 9.2. Once we know the value of the (negative) quantity $\Delta r$, we can calculate $r(2) = r(1) + \Delta r$. In this way, as we assign an ever *increasing* value to the independent variable $T$ at step $n$, the radial location $r(n)$ at step $n$ will take on increasingly *smaller* numerical values.

8. Repeat steps 1–7 multiple times until the computed radial location $r(n+1) = r(n) - \Delta r$ reaches the value zero (or a negative value). At this point, the model will have reached the center of the Sun, and the tabulated parameters at the last line $n$ where the radial coordinate $r$ is nonnegative will refer essentially to conditions at the center of the Sun.

An example of an abbreviated table computed according to the prior prescription is given in Table 9.1. The results in the table were obtained by assuming that the critical radius $r_m \approx 0.25 R_\odot$.

## 9.2 OVERVIEW OF OUR MODEL OF THE SUN'S RADIATIVE INTERIOR

What do the results of the model in Table 9.1 tell us about conditions in the deep interior of the Sun? Note that the last entry in the table lies at a formal radial location of $r = 0.003 R_\odot$, but this can be considered as being equivalent (within the accuracy of our calculation) to the center of the Sun.

According to this model, the gas at the center of the Sun has a temperature of roughly $T_c = 16.5$ million K, a density of $\rho_c = 141$ gm cm$^{-3}$, and a pressure of $p_c = 3.34 \times 10^{17}$ dyn cm$^{-2}$. Protons at a temperature of $T_c$ have an r.m.s. speed $V_{rms} = \sqrt{(3 R_g T_c)} \approx 640$ km sec$^{-1}$.

It is important to compare this proton speed at the center of the Sun with another characteristic speed associated with the Sun as a whole: the escape speed $V_{esc}$ from the surface. In Chapter 1, Section 1.7, we saw that $V_{esc} = 617.7$ km sec$^{-1}$. The latter is a measure of the strength of the inward pull of gravity, which holds the hot gas at the center of the Sun together by means of the crushing weight of the overlying gas. We see that in our simplified mechanical model of the Sun, we have found that the conditions at the center are such that $V_{rms}$ agrees with $V_{esc}$ within about 4%. This agreement indicates that in our model, the inward pull of the gravitational forces and the outward force of pressure are close to achieving a balance. The error of 4% could be improved if we were to use a more realistic radial profile for the acceleration due to gravity. This is an important confirmation that our model of the Sun as a whole is in hydrostatic equilibrium.

How does our mechanical model in Table 9.1 compare with models that have been computed by including many more details of the physics (including energy generation by nuclear fusion)? To answer that, we may refer again to the series of 10,000 "standard solar models" reported by Bahcall et al. (2006): in these models, the best estimates of temperature, density and pressure at the center are found to be 15.48 million K, 150.4 gm cm$^{-3}$, and $2.34 \times 10^{17}$ dyn cm$^{-2}$. Compared with these, our mechanical model yields a central temperature that is too high by about 7%, a density that is too low by about 6%, and a pressure that is too high by about 40%. The main reason for the error in pressure has to do with the mean molecular weight: we have assumed $\mu = 0.58$ throughout the radiative interior, whereas in the "real Sun", nuclear reactions build up more and more helium in the core as time goes on. As a result, in the Bahcall et al. model, the central value of $\mu$ equals 0.83, i.e., some 40% larger

**TABLE 9.1**

**A Mechanical Model for the Radiative Interior of the Sun:** $r_m = 0.25 R_\odot$

| r(cm) | T(K) | p(dyn cm$^{-2}$) | ρ(gm cm$^{-3}$) | g(cm sec$^{-2}$) |
|---|---|---|---|---|
| 5.3093E+10 | 2.0127E+06 | 2.9803E+13 | 1.0329E−01 | 4.6766E+04 |
| 5.1046E+10 | 2.1027E+06 | 4.1817E+13 | 1.3873E−01 | 5.0548E+04 |
| 4.9373E+10 | 2.2027E+06 | 5.5592E+13 | 1.7605E−01 | 5.4111E+04 |
| 4.7997E+10 | 2.3027E+06 | 7.0541E+13 | 2.1369E−01 | 5.7301E+04 |
| 4.5697E+10 | 2.5027E+06 | 1.0561E+14 | 2.9436E−01 | 6.3262E+04 |
| 4.3737E+10 | 2.7027E+06 | 1.4960E+14 | 3.8613E−01 | 6.9084E+04 |
| 4.1179E+10 | 3.0027E+06 | 2.3715E+14 | 5.5092E−01 | 7.7965E+04 |
| 3.8249E+10 | 3.4027E+06 | 4.0589E+14 | 8.3209E−01 | 9.0393E+04 |
| 3.5726E+10 | 3.8027E+06 | 6.5218E+14 | 1.1964E+00 | 1.0364E+05 |
| 3.3520E+10 | 4.2027E+06 | 9.9832E+14 | 1.6570E+00 | 1.1775E+05 |
| 3.1572E+10 | 4.6027E+06 | 1.4696E+15 | 2.2273E+00 | 1.3275E+05 |
| 2.9839E+10 | 5.0027E+06 | 2.0944E+15 | 2.9204E+00 | 1.4865E+05 |
| 2.7924E+10 | 5.5027E+06 | 3.1399E+15 | 3.9805E+00 | 1.6977E+05 |
| 2.6240E+10 | 6.0027E+06 | 4.5442E+15 | 5.2808E+00 | 1.9229E+05 |
| 2.4747E+10 | 6.5027E+06 | 6.3846E+15 | 6.8491E+00 | 2.1621E+05 |
| 2.3416E+10 | 7.0027E+06 | 8.7472E+15 | 8.7135E+00 | 2.4153E+05 |
| 2.2220E+10 | 7.5027E+06 | 1.1726E+16 | 1.0903E+01 | 2.6825E+05 |
| 2.1141E+10 | 8.0027E+06 | 1.5426E+16 | 1.3446E+01 | 2.9636E+05 |
| 2.0162E+10 | 8.5027E+06 | 1.9958E+16 | 1.6373E+01 | 3.2588E+05 |
| 1.9269E+10 | 9.0027E+06 | 2.5443E+16 | 1.9715E+01 | 3.5680E+05 |
| 1.8452E+10 | 9.5027E+06 | 3.2014E+16 | 2.3501E+01 | 3.8912E+05 |
| 1.7702E+10 | 1.0003E+07 | 3.9810E+16 | 2.7763E+01 | 4.2284E+05 |
| 1.6998E+10 | 1.0503E+07 | 4.8980E+16 | 3.2532E+01 | 4.2916E+05 |
| 1.6272E+10 | 1.1003E+07 | 5.9684E+16 | 3.7840E+01 | 4.1087E+05 |
| 1.5513E+10 | 1.1503E+07 | 7.2092E+16 | 4.3720E+01 | 3.9173E+05 |
| 1.4714E+10 | 1.2003E+07 | 8.6382E+16 | 5.0204E+01 | 3.7161E+05 |
| 1.3870E+10 | 1.2503E+07 | 1.0274E+17 | 5.7325E+01 | 3.5033E+05 |
| 1.2970E+10 | 1.3003E+07 | 1.2138E+17 | 6.5116E+01 | 3.2768E+05 |
| 1.2004E+10 | 1.3503E+07 | 1.4249E+17 | 7.3612E+01 | 3.0334E+05 |
| 1.0953E+10 | 1.4003E+07 | 1.6630E+17 | 8.2845E+01 | 2.7687E+05 |
| 9.7901E+09 | 1.4503E+07 | 1.9304E+17 | 9.2851E+01 | 2.4760E+05 |
| 8.4694E+09 | 1.5003E+07 | 2.2295E+17 | 1.0366E+02 | 2.1438E+05 |
| 6.9014E+09 | 1.5503E+07 | 2.5628E+17 | 1.1532E+02 | 1.7498E+05 |
| 4.8543E+09 | 1.6003E+07 | 2.9330E+17 | 1.2785E+02 | 1.2370E+05 |
| 2.0805E+08 | 1.6503E+07 | 3.3427E+17 | 1.4130E+02 | 1.5297E+04 |

than our assumed value. In the best estimate models of Bahcall et al., protons at the center of the Sun have $V_{rms} = 621$ km sec$^{-1}$, agreeing with $V_{esc}$ to better than 1%. With a central density of 150 gm cm$^{-3}$ and a mean nuclear weight of about 2 (He is building up in abundance in the core because of the reactions), the corresponding number density of nuclei is of order $0.5 \times 10^{26}$ cm$^{-3}$.

The central density in the Sun (150 gm cm$^{-3}$) exceeds the mean density of the Sun (1.41 gm cm$^{-3}$: see Equation 1.14) by a factor of slightly more than 100. This is a measure of the "central condensation" of the Sun to which we shall return in Section 10.9.

Another aspect of our model that deserves attention concerns the ratio of radiation pressure $p_r$ to gas pressure. The value of $p_r$ at the center of the Sun can be determined from the result (see Section 2.8) $p_r = a_R T^4/3$: our model yields $p_r \approx 2 \times 10^{14}$ dyn cm$^{-2}$. The gas pressure at the center of the Sun exceeds $p_r$ by a factor of more than 1000. Elsewhere in the Sun, the gas pressure exceeds the radiation pressure by even greater factors. For example, in the photosphere, the gas pressure ($\approx 10^5$ dyn cm$^{-2}$) exceeds the radiation pressure by a factor of more than $3 \times 10^4$. These numerical values indicate that we are justified in neglecting radiation pressure compared to the gas pressure when we calculate a first model of the Sun: when we wrote down the equation of hydrostatic equilibrium (Equation 5.1), the quantity $p$ in Equation 5.1 includes only the gas pressure. In certain stars other than the Sun, especially in stars with hotter surface temperatures than those of the Sun, this neglect of radiation pressure may not be an acceptable approximation: but in the case of the Sun, radiation pressure does not contribute significantly to supporting the Sun against gravity.

However, it should not be too surprising that our model (with its various approximations and simplifications) is not perfect. Because we use different modeling techniques and different $\mu$ values in different regions of the Sun, our model includes an artificial "step" in density between the bottom of the convection zone (see Table 7.1, bottom line) and the top of the radiative interior (see Table 9.1, top line).

## 9.3 PHOTONS IN THE SUN: HOW LONG BEFORE THEY ESCAPE?

Now that we have obtained a model of the interior of the Sun where photons transport the energy, it is worthwhile to ask: how long does it take for a photon to propagate from the center of the Sun to the surface? Subsequently we shall compare the photon time-scale with the time-scale for the escape of a very different type of elementary particle (the neutrino: Chapter 12) that is also generated in the core of the Sun.

To estimate the photon time-scale, we note that photons generated at the center of the Sun make their way outward by diffusing through the material of the solar interior. In this process, the photons make their way outward in radius by means of a random walk: if the length of each step in the random walk is on average $l_p$, then after $N$ steps, the photon will have moved outward in the radial direction by a net distance of $r_N \approx l_p \sqrt{N}$.

As a first step toward estimating the diffusion time-scale, we consider the mean free path $l_p$ that a photon travels between interactions with the solar material. In view of the definition of opacity, it is clear that $1/\kappa\rho$ is an appropriate length-scale (see Chapter 8, Section 8.1). Therefore, we may take $l_p \approx 1/\kappa\rho$.

What is a typical numerical value for this length scale? Near the center of the Sun, the model in Table 9.1 indicates $\rho \approx 140$ gm cm$^{-3}$. Moreover, with opacity in the solar interior given by the Kramers' law, $\kappa = \kappa_o \rho T^{-3.5}$, and using the value $\kappa_o = 10^{24}$ (see Chapter 8), we find that near the center of the Sun, where $T \approx 1.6 \times 10^7$ K, the numerical value of the opacity $\kappa$ is of order 10 cm$^2$ gm$^{-1}$. This leads to $l_p \approx 0.7 \times 10^{-3}$ cm. In the center of the Sun, photons are restricted to mean free paths with lengths of only a few microns.

How is $l_p$ expected to vary with increasing radial distance from the center of the Sun? In the radiative interior, we have seen (see Equation 9.1) that the pressure (at least near the center, where the constant of integration does not contribute significantly) satisfies $p^2$ varies as $T^{8.5}$, i.e., $p \sim T^{4.25}$. In such conditions, the perfect gas law indicates that $\rho \sim p/T \sim T^{3.25}$. Moreover, assuming Kramers' opacity, $\kappa \sim \rho/T^{3.5}$, we find that the mean free path $l_p \approx 1/\kappa\rho \sim T^{3.5}/\rho^2 \sim 1/T^3$. As a result, the value of $l_p$ *increases* as we move away from the center of the Sun. However, even when we reach the base of the convection zone, at a radial location of $r_b \approx 0.7 R_\odot$, where the temperature has fallen to 2 million K (i.e., $T$ has decreased by a factor of about eight below its value at the center), $l_p$ has increased by a factor of about 512, i.e., $l_p \approx 0.4$ cm. Throughout the radiative interior the mean free time $t_p = l_p/c$ between photon-atom collisions is no more than $l_p/c \approx 10^{-11}$ sec.

In terms of the random walk argument given earlier, the number of "steps" $N_b$ that the photon must take in order to move outward from the center of the Sun to the base of the convection zone is given by $N_b \approx (r_b/l_p)^2$. The time required for this number of steps is $t_b = N_b t_p = r_b^2/l_p c$. Most of the time required by a photon to random walk to the base of the convection zone is spent in the core of the Sun, where $l_p$ is smallest. Inserting the values $r_b \approx 5 \times 10^{10}$ cm, $l_p \approx 0.001$ cm we find $t_b \approx 8 \times 10^{13}$ sec, i.e., $2-3 \times 10^6$ years. Thus, photons require on average a few million years to propagate from the core of the Sun out to the base of the convection zone. From there, their energies are transported to the surface by fluid flow on much shorter time-scales: near the top of the convection zone, flows of order 1 km sec$^{-1}$ carry material up through one cell depth (of order the local pressure scale height, 140 km) in a time-scale of minutes. And deep in the convection zone, where convective flow speeds may become much smaller than the surface speeds, say $10^2$ cm sec$^{-1}$, the local cell heights may again be of order $H_p \sim T/\mu g$. Between the surface and the bottom of the convection zone, $\mu$ *decreases* by a factor of about two while $g$ *increases* by a similar mount, i.e., $\mu g$ remains constant. Therefore, with $T$ increasing from roughly 5000 K at the surface to $2 \times 10^6$ K at the base of the convection zone, $H_p$ is larger at the base if the convection zone than at the surface by a factor of about 400, i.e., $H_p \approx 5 \times 10^9$ cm. The time for flows to carry material from the bottom of such a cell to the top is of order 1–2 years. Compared to the time required for photons to diffuse through the radiative region in the solar interior, the time for energy to get from the base of the convection zone to the surface is negligible.

This indicates that when we observe the Sun today, the energy entering our eyes was actually generated several million years ago. It will be a matter of interest in a subsequent chapter to compare this photon time-scale with the corresponding value for neutrinos.

## 9.4 A PARTICULAR GLOBAL PROPERTY OF THE SOLAR MODEL

We can obtain a complete model of the Sun, based on our simplified approach, by combining Tables 5.3, 7.1, and 9.1.

In view of the large ranges of physical parameters between surface and center, is there some way that we can check our calculations in some global sense? There is one test that we can apply.

In Chapters 13 and 14, we shall be interested in the topic of helioseismology, i.e., the study of eigenmodes of oscillation within the Sun. One class of eigenmodes, relying on pressure as the restoring force, is referred to as *p*-modes. Each eigenmode has an eigenfunction which, when plotted as a function of radius from the center of the Sun to the surface, exhibits a definite number of "nodes" (where the eigenfunction passes through 0). (See Figure 14.3 for the plot of a small section of certain eigenfunctions close to the solar surface.) The number of such nodes in the radial direction, $n_r$, helps to define each mode.

We shall find that at high frequencies, the *p*-modes display a well-defined "asymptotic behavior": for modes of a given angular degree (*l*, related to the number of nodes on the surface between north pole and south pole), the frequencies of modes that differ by unity in the value of $n_r$ differ from each other by a characteristic frequency spacing $\Delta\nu$. Theory indicates that the asymptotic frequency spacing $\Delta\nu$ is related to the time $t_s$ required for sound to propagate from the center of the star to a reflection point at radial location $R(r)$ near the photosphere:

$$t_s = \int_0^{R(r)} \frac{dr}{c_s(r)} \qquad (9.6)$$

where $c_s(r)$ is the sound speed at radial location *r*. Specifically, the frequency spacing $\Delta\nu$ can be shown to be equal to $1/(2t_s)$.

Now that we have obtained a model of the Sun, albeit only a simplified model, it is of interest to inquire: what is the sound travel time from center to photosphere according to our model? The integration in Equation 9.6 can be performed using the combined information in Tables 5.3, 7.1, and 9.1. When we do this, we find that the sound crossing time from the center of the Sun to the photosphere of our combined model is $t_s = 3804$ sec, i.e., a few minutes longer than one hour. This leads to $\Delta\nu = 131.5$ µHz.

Empirically, $p$-modes in the Sun with low values of $l(= 0, 1, 2)$, are found to have asymptotic spacings of $\Delta\nu = 134.8$–$135.1$ µHz (Appourchaux et al. 1998). Thus, our mechanical model of the Sun replicates the solar asymptotic spacings within 2%–3%. It is a gratifying feature of our model that the run of temperature between center and surface replicates so well the empirical asymptotic frequency spacing between $p$-modes.

## 9.5 DOES THE MATERIAL IN THE SUN OBEY THE PERFECT GAS LAW?

In computing the model of the three regions of the Sun (Chapters 5, 7, and 9), we have used the equation of state for a perfect gas. Now that we have calculated the conditions in the interior of the Sun, we need to perform a consistency check and ask: does the gas in the Sun really obey the perfect gas law? After all, we have found that the density at the center of the Sun exceeds 140 gm cm$^{-3}$: this is denser than solid gold or solid lead, and the latter materials certainly do not obey the perfect gas law.

What criterion can we use in order to test whether the perfect gas law, which follows from the classical kinetic theory of gases, is actually obeyed inside the Sun? The answer is: classical theories of matter are acceptable as long as quantum mechanical effects are negligible.

According to quantum mechanics, particles in certain circumstances behave with wave-like properties. The wavelength $\lambda_p$ associated with a particle of mass $m$, moving with speed $V$, is given by de Broglie's formula: $\lambda_p = h/(mV)$, where $h = 6.62606896 \times 10^{-27}$ gm cm$^2$ sec$^{-1}$ is Planck's constant.

Classical physics provides a reliable description of the behavior of matter as long as the de Broglie waves of individual particles do not overlap one another significantly. But the laws of classical physics break down if the de Broglie wave of one particle is so large that it overlaps significantly with the de Broglie waves of a number of the neighboring particles. Since the de Broglie wavelength is inversely proportional to mass, the particles with the smallest masses (electrons) will be the ones most likely to have a chance to have their wavelength overlap with their neighbors. How large should the number of overlaps be? It must be at least two, because the existence of electron spin allows two electrons to occupy the same element in phase space without contradicting Pauli's exclusion principle. Once the wave of one electron overlaps the waves of (say) 10 or more neighboring electrons, then the electrons with overlapping de Broglie waves begin to "feel the pressure" of the Pauli exclusion principle. In a very real sense, the electrons are subject to a physical pressure that "drives them away" from their neighbors in phase space. This pressure is quite different from ordinary gas pressure that is governed by the perfect gas law $p \sim \rho T$. In cases where the perfect gas law breaks down, quantum effects must be taken into account, and the electrons are said to be "degenerate", and the pressure that they exert is called "electron degeneracy pressure".

In conditions where quantum effects are important, the pressure associated with electron degeneracy can be strong enough to support the overlying weight of an object with a mass of order the Sun's mass. In such an object, the pressure of the *thermal gas* is no longer the physical agent that supports the star against its own weight. As a result, even if HSE were to hold, there would no longer be any reason why the r.m.s. *thermal speed* of the particles at the center of the star should be equal to the escape speed from the surface.

Let us see what happens to de Broglie waves at the center of the Sun. In order to make the quantum effects as large as possible, we consider electrons. For electrons at the center of the Sun, where the r.m.s. velocity is $\sqrt{(3kT/m_e)}$, the de Broglie wavelength has a mean value of $\lambda_e = h/\sqrt{(3kTm_e)}$. Inserting the temperature at the center of the Sun, $T = 1.6 \times 10^7$ K, we find $\lambda_e = 2.7 \times 10^{-9}$ cm.

Now that we know the value of the de Broglie wavelength of an electron at the center of the Sun, we can determine under what conditions electron degeneracy could become an important contributor to the pressure. In order that the de Broglie wave of one particular electron would extend throughout a volume in which there are 10 other electrons, the mean distance between the electrons at the center of the Sun must be less than $\lambda_e$ by a factor of about $10^{1/3}$. Thus if electron degeneracy is to be important in the center of the Sun, the mean distance between electrons in the center of the Sun must be no larger than $d_e < 1.3 \times 10^{-9}$ cm.

In a medium where the electron density is $n$ cm$^{-3}$, the mean distance $d_e$ between an electron and its neighbors is roughly $1/n^{1/3}$ cm. Thus, for electron degeneracy to be important at the center of the Sun, the number density of electrons would have to be at least as large as $5 \times 10^{26}$ cm$^{-3}$. At the center of the Sun, where hydrogen burning has been going on for several billion years, helium has increased to such an extent that the helium abundance is comparable to that of hydrogen. Corresponding to each electron from He, there are two nucleons to contribute mass, each with a mass of $1.67 \times 10^{-24}$ gm. Therefore, the mass density in a helium-dominated region with $n_e > 5 \times 10^{26}$ cm$^{-3}$ would exceed 1700 gm cm$^{-3}$. This is more than 10 times larger than the best estimates for the density at the center of the Sun (150.4 gm cm$^{-3}$ in the model of Bahcall et al. 2006).

Thus, despite the high gas density at the center of the Sun, the central temperature is so large that the mean thermal speed $V_t$ is of order $3 \times 10^9$ cm sec$^{-1}$. It is this large speed that makes the de Broglie wavelength of each electron relatively short. As a result, there is no significant overlap of the electron de Broglie waves at the center of the Sun.

This means that the laws of classical physics are adequate to describe the gas at the center of the Sun. In particular, we are justified in assuming that the material in the Sun, even at the very center where the material is denser than gold or lead, still obeys the equation of state of a perfect gas. The high temperature in the central material is what allows this to happen.

## 9.6 SUMMARY OF OUR (SIMPLIFIED) SOLAR MODEL

We have found that, even without considering the generation of energy in detail, it is nevertheless possible, using appropriate laws of physics, to calculate the radial profiles of various physical parameters from the center of the Sun all the way to the visible surface. These are contained in our Tables 5.3, 7.1, and 9.1.

Our results allow us to appreciate the great range that is spanned by the various physical parameters in the Sun. Of the three principal parameters ($T$, $p$, and $\rho$), all are found to fall off monotonically from the center of the Sun to the surface. We note that the smallest range in numerical value of a parameter in going from center to surface is exhibited by the temperature: the central temperature exceeds the photospheric temperature by only about 3.5 orders of magnitude. Pressure exhibits the widest range: the central pressure exceeds the photospheric pressure by at least 12 orders of magnitude. Density presents an intermediate case: the central density (~150 gm cm$^{-3}$) exceeds the photospheric density (~$3 \times 10^{-7}$ gm cm$^{-3}$) by eight to nine orders of magnitude. Thus, the Sun is far removed from being like a billiard ball where the density remains constant throughout the volume. The mean density of the Sun as a whole has already been noted: 1.410 gm cm$^{-3}$ (see Section 1.7). This mean density is significantly closer to the central density than to the density of the gas that we can see in the visible photosphere of the Sun.

It is encouraging to find that, despite these great ranges in the three main physical parameters, we have arrived at a model of the Sun that is consistent with a variety of observational data, despite the fact that the model is purely mechanical, i.e., our model has included the equation of momentum conservation, but we have not paid much attention to the energy equation (except to assume that the ratio $L(r)/M(r)$ is independent of $r$).

Of course, it is precisely the generation of energy that sets the Sun (and stars) apart from other celestial bodies. Now that we have derived estimates of certain physical parameters inside the Sun, we need to examine the processes by which energy (which we have treated so far as a boundary condition) is actually generated.

However, before we enter into the details of nuclear reactions, we make a digression into a topic that at first sight seems to be rather idealized and far afield from a study of the Sun. The topic has to do with mathematical entities called polytropes. However idealized these may seem, we shall find that we have already been working with models (in the convection zone and in the radiative interior) that are actually not far removed from polytropes. Moreover, the discussion of polytropes will stand us in good stead in a subsequent chapter when we consider oscillations in the Sun.

## EXERCISES

9.1 Perform the computation described in Section 9.1, using whatever computational technique you prefer for the numerical work. How do your numbers at the center of the Sun compare with those in Table 9.1?

9.2 In step 6 of the procedure in Section 9.1, there is a recommendation for an approximation to the nonmonotonic radial profile of gravity inside the Sun. In this exercise, we ask the reader to experiment with different choices for the parameters of the approximation, for example, choose a smaller value of $r_m$, e.g., $0.20R_\odot$ or $0.15R_\odot$, and recalculate the model according to Section 9.1. How do the parameters at the center change? Then choose a larger value of $r_m$, e.g., $0.30R_\odot$ or $0.35R_\odot$, and see how the central parameters change.

9.3 The formula in Equation 9.1 is based on the Kramers' opacity law. Use the two revised versions of Equation 9.1 that you obtained in Exercise 8.1 of Chapter 8. For each case of the revised opacity "law", repeat the calculation of Section 9.1. How do your revised results for the parameters at the center of the Sun agree with the results of Bahcall et al. (2006)?

9.4 For each of your models, calculate the sound crossing time (Equation 9.6) and the associated frequency interval $\Delta \nu$ (express the frequency in µHz). How well does your value of $\Delta \nu$ agree with the observed solar value of (about) 135 µHz?

9.5 An alternative approach to modeling the radial profile of the gravity is to examine the profile of the mass parameter $M(r)$ in a detailed solar model, such as that on the website of J. Christensen-Dalsgaard. The tabulated values of the model are contained in the file www.phys.au.dk/~jcd/solar_models/fgong.l5bi.d.15c. A description of the different columns and rows can be found at www.phys.au.dk/~jcd/solar_models/file-format.pdf. Extract $M(r)$ at a number of points in the tabulated model, and plot $g(r) = GM(r)/r^2$ as a function of $\log(r)$. A peak ($\approx 8.4$ times the surface gravity) occurs at a certain value of $\log(r/R_\odot)(\approx -0.8)$. Is there a simple functional form (such as a parabola) that you can find to fit the peak? Use that functional form to obtain a better estimate of the local value of $g$ in step 6 of the procedure in Section 9.1. How do the central parameters of the Sun change as a result?

## REFERENCES

Appourchaux, T., Rabello-Soares, M.-C., & Gizon, L., 1998. "The art of fitting p-mode spectra: II. Leakage and noise covariance matrices", *Astron. Astrophys. Suppl.* 132, 121.

Bahcall, J. N., Serenelli, A. M., & Basu, S., 2006. "10,000 standard solar models: A Monte Carlo simulation", *Astrophys. J. Suppl.* 165, 400.

# 10 Polytropes

Now that we have computed a model of the solar interior, albeit a simplified one, it is worthwhile to pay a certain amount of attention to a particular aspect of our solutions. This digression will be valuable in a later chapter when we come to consider how to compute the properties of oscillations in a solar model.

## 10.1 POWER-LAW BEHAVIOR

Inspection of Equation 9.1 indicates that, when we consider regions in the radiative interior of the Sun where the temperature is sufficiently large, the constant term in Equation 9.1 can be neglected. In this limit, we see that the pressure $p$ varies as a power law of $T$: $p \sim T^\beta$. In this particular case, the exponent $\beta$ in the power law between $p$ and $T$ has the numerical value of 4.25. The origin of this particular numerical value for the exponent can be traced to two simplifications: (i) $L(r)/M(r)$ remains constant as a function of radial location and (ii) the opacity depends on density and temperature according to "Kramers' law": $\kappa \sim \rho T^{-3.5}$. These are simplifications that deal with matters of (i) energy generation and (ii) energy transport. The fact that the radial profiles of pressure and temperature are related to each other by a power law therefore depends on certain assumptions we have made concerning the *energy equation*.

In a very different context, a power-law relation $p \sim T^\beta$ also emerged when we were modeling the deep convection zone: see Equations 6.14 and 7.9. Detailed models of the solar convection zone indicate that throughout some 90% of the depth of the convection zone, the power law in the $p \sim T$ relationship is close to 2.5. In the case of the convection zone, the reason for the power-law behavior can be traced to the physics of adiabatic processes, i.e., on processes that have do to do with the gain or less of energy as the fluid moves. Once again, certain assumptions that we make concerning the *energy equation* lead to a power-law behavior between $p$ and $T$.

It is noteworthy that we have encountered a power-law relationship between pressure and temperature in two quite different regions of the Sun: one region lies in the deep interior (where radiation dominates energy transport), while the second region lies in the convection zone (where radiation contributes almost nothing to the energy transport). In both regions of the Sun, the emergence of the power law has to do with *certain assumptions about the energy equation*, although the assumptions are quite different in the two different regions.

In this chapter, we discuss the properties of equilibrium gas spheres ("polytropes") in which no explicit attention whatsoever is given to the energy equation. We will pay explicit attention to the law of conservation of mass and also to the law of conservation of momentum. But we will make no attempt to include explicitly the law of conservation of energy. Instead of attending to the details of the energy equation, a certain functional form will be *assumed to exist* for the relationship between pressure and density. Specifically, pressure and density are assumed to be related by means of a *power law*. For a perfect gas, this means that pressure and temperature are also related by a *power law*. A power-law relationship between pressure and density is referred to as a "polytropic" equation of state.

In our search for deriving radial profiles of physical parameters inside the Sun (or for that matter inside any star), the only reason that polytropes have any claim on our attention is that in "real stars", the generation and transport of energy occur in fact in such a way that, in certain cases, the pressure and density *do* turn out to obey a polytropic functional form, at least over certain ranges of the radial coordinate.

## 10.2 POLYTROPIC GAS SPHERES

A polytrope is defined to be a medium in which the pressure and density are related by the following relationship:

$$p = K\rho^{(n+1)/n} \tag{10.1}$$

where $K$ and $n$ are constants. Equation 10.1 is referred to as the "polytropic equation of state". The constant $n$ is referred to as the "polytropic index". (In previous chapters, lower case $n$ has been used on occasion to denote number densities of atoms/electrons in a gas. In the present chapter, and also in Chapter 14, there are historical reasons for using $n$ as the polytropic index; but in the case of a polytrope, the constant $n$ is a dimensionless number and has nothing to do with density.) The properties of polytropic spheres of gas have been discussed by a number of authors, including Lane, Ritter, Kelvin, Emden, and Fowler. A detailed study can be found in a book by Chandrasekhar (1958). Among the results found in that book, we mention two in particular: (i) for each value of $n$, there is a well-defined numerical value (called the "central condensation") for the ratio of the central density to the mean density; (ii) in the particular polytrope with $n = 3$, the polytrope has a unique mass.

Why are polytropes relevant to our study of the Sun? Because, in a nonionizing medium that obeys the perfect gas equation of state, $p \sim \rho T$, the polytropic relationship can be written in the form of a power-law relationship between pressure and temperature: $p \sim T^\beta$, where $\beta$ takes on a specific value: $\beta(\text{polytrope}) = n + 1$.

This leads us to consider an application of polytropic concepts to the two portions of the Sun in which we have already identified a power-law relationship between pressure and temperature.

First, in the adiabatic portions of the convection zone, the fact that $p$ varies as $T^{2.5}$ suggests that the radial profile of the physical properties of those portions of the Sun is related to the radial profile of a polytrope with index $n = 1.5$. We may say that the structure of the Sun in the adiabatic portions of the convection zone corresponds to an "effective" polytropic index of 1.5.

Second, in the Sun's radiative interior, Equation 9.1 indicates that when $p$ and $T$ are large, $p$ varies as $T^{4.25}$, suggesting that a polytrope with index $n = 3.25$ can provide useful information on the radial profile of physical parameters. In this case, we may say that the structure of the Sun in the radiative interior corresponds to an "effective" polytropic index of 3.25.

To make this more quantitative, it is instructive to calculate models of polytropes. As was mentioned earlier, in this chapter we continue the practice of not referring to the energy equation explicitly: all details of energy generation and transport are implicit in Equation 10.1. Then, a polytrope model is obtained by solving the equation for conservation of mass and the (static version of the) equation for conservation of momentum. In this way, we are using information about the *mechanical* properties of the star without considering the *thermal* properties explicitly. Despite this limitation, since the speed of sound at any radial location $r$ depends on the ratio of $p(r)$ to $\rho(r)$ (both of which we shall calculate), our polytrope models (simple though they are) will still provide sufficient information to allow us to study quantitatively the propagation of acoustic waves through the (polytropic model of the) star.

The equation of mass conservation is:

$$\frac{dM(r)}{dr} = 4\pi r^2 \rho(r) \tag{10.2}$$

As we have already seen (Equation 7.1), the equation of momentum conservation (in the limit of zero velocity) is simply the equation of HSE:

$$\frac{dp(r)}{dr} = -\frac{GM(r)\rho(r)}{r^2} \tag{10.3}$$

# Polytropes

Equations 10.2 and 10.3 can be combined into a single second-order equation:

$$\frac{1}{r^2}\frac{d}{dr}\left(\frac{r^2}{\rho(r)}\frac{dp(r)}{dr}\right) = -4\pi G \rho(r) \quad (10.4)$$

This is Poisson's equation for a self-gravitating sphere. It is a second-order equation that includes two unknown functions of the radial coordinate: $p(r)$ and $\rho(r)$. At this point, the polytropic assumption, i.e., that $p$ is related to $\rho$ at each and every value of $r$ by the relation $p(r) = K\rho(r)^{(n+1)/n}$, is introduced, and this will allow Equation 10.4 to be reduced to an equation for a single function of $r$.

Let the density and pressure at the center of the polytrope by $\rho_c$ and $p_c$. Once those two parameters are specified, the constant $K$ in the polytropic equation of state can be expressed in terms of the quantities at the center. Then at all values of radial location $r$, the local pressure $p(r)$ and the local density $\rho(r)$ satisfy the relation

$$\frac{p(r)}{\rho(r)^{(n+1)/n}} = \frac{p_c}{\rho_c^{(n+1)/n}} \quad (10.5)$$

## 10.3 LANE–EMDEN EQUATION: DIMENSIONAL FORM

Equation 10.4 includes the physical parameters $p$ and $\rho$. It is now convenient to introduce a dimensionless function $y$ of the radial coordinate according to the following definition:

$$y^n = \frac{\rho(r)}{\rho_c} \quad (10.6)$$

The function $y$ is referred to as the Lane–Emden function. The goal of the polytropic exercise is to derive analytically, or compute, the function $y$ as a function of the radial location from the center of the polytrope (i.e., at $r = 0$) out to the surface (where $\rho$ falls to a value of zero for the first time). In view of the definition of $y$, it is clear that the boundary condition on $y$ at the center of the polytrope is $y(r = 0) = 1$.

Inserting Equation 10.6 into Equation 10.5, we find that

$$\frac{p(r)}{p_c} = y^{n+1} \quad (10.7)$$

From Equations 10.6 and 10.7, we see that at any radial location, $y \sim p(r)/\rho(r)$. From the equations of thermodynamics, we know that the ratio of pressure to density in a medium is related to the (square of the) sound speed. Thus, the numerical value of $\sqrt{y}$ at any radial location is proportional to the local sound speed. Because of this, in our subsequent study of helioseismology (Chapter 14), we shall be able to use polytropes to determine realistic global properties of oscillation modes that involve the propagation of acoustic waves in a sphere.

If the material of which the polytrope is composed happens to obey the perfect gas equation of state, then at any given radial location, $y \sim p(r)/\rho(r)$ is also proportional to $T(r)$, the local temperature. In fact, $y(r) = T(r)/T_c$, where $T_c$ is the central temperature. In this case, Equations 10.6 and 10.7 indicate that $\rho(r)$ scales as $T(r)^n$, while $p(r)$ scales as one higher power $T(r)^{n+1}$. We have seen scalings of this kind earlier: see Equations 7.8 and 7.9.

Using Equations 10.6 and 10.7 to replace $p(r)$ and $\rho(r)$ in Equation 10.4, and collecting all the constants on the left-hand side of the equation, we find

$$\frac{(n+1)p_c}{4\pi G \rho_c^2}\left(\frac{1}{r^2}\right)\frac{d}{dr}\left[r^2\frac{dy}{dr}\right] = -y^n \quad (10.8)$$

For any given value of the polytropic index $n$, the radial profile of the function $y$ can be obtained by solving Equation 10.8.

## 10.4 LANE–EMDEN EQUATION: DIMENSIONLESS FORM

In order to convert the Lane–Emden equation to dimensionless form, we introduce a new unit $r_o$, the Emden unit of length, which is defined by the combination of constants that appear on the left-hand side of Equation 10.8:

$$r_o^2 = \frac{(n+1)p_c}{4\pi G \rho_c^2} \tag{10.9}$$

How can we be sure that $r_o$ has the dimensions of length? To answer this, we note that on the right-hand side of Equation 10.9, we can first isolate the ratio of $p_c$ to the first power of $\rho_c$: dimensionally, this is the ratio of ergs cm$^{-3}$ to gm cm$^{-3}$, i.e., ergs/gm, i.e., the square of a speed. Depending on a numerical coefficient, the speed in this case turns out to be the speed of sound. Thus, the ratio $p_c/\rho_c$ has dimensions of [length]$^2$/[time]$^2$. The remaining dimensional units are those belonging to $1/G\rho_c$. In Equation 1.24, we mentioned a characteristic period $P_g$ associated with the gravitational field of the Sun. The value of $P_g$ is proportional to $\sqrt{(R_\odot^3/GM_\odot)}$. The dimensions of $M/R^3$ are those of density, indicating that the combination $1/\sqrt{(G\rho)}$ has the dimensions of [time]. Therefore, the factor $1/G\rho_c$ in Equation 10.9 has the dimensions of [time]$^2$. Combining the dimensions, we see that the dimensions of the right-hand side of Equation 10.9 are indeed [length]$^2$. Therefore, $r_o$ has the dimensions of length.

Is the Emden unit of length related to a length-scale that might be relevant in the context of the structure of "real stars"? In particular, would it be useful to consider the dimensions of a star such as the Sun in terms of $r_o$? Or does the radius of a "real star" differ from $r_o$ by many orders of magnitude? To answer these questions, we evaluate $r_o$ using the values of central density and pressure that we have already obtained in our simplified solar model. Substituting $p_c = 3.34 \times 10^{17}$ dyn cm$^{-2}$ and $\rho_c = 141$ gm cm$^{-3}$ from Table 9.1, we find that $r_o = 4.5 \times 10^9 \sqrt{(n+1)}$ cm. We shall see later that the radius $R_p$ of a polytropic star is larger than $r_o$ by a factor $x_1$ where the numerical value of $x_1$ depends on the $n$ value. For example, with $n = 3.25$, the numerical value of $x_1$ is about 8 (see Table 10.1). Moreover, for $n = 3.25$, we see that the Emden unit of length $r_o$ has a numerical value of about $9.2 \times 10^9$ cm. Multiplying $r_o$ by $x_1$, we find that $R_p$ is about $7.4 \times 10^{10}$ cm. Remarkably, this value of the polytropic star radius is within a few percent of the actual solar radius. So it appears that the linear dimensions of polytropes in which central pressures and densities overlap with those of the "real Sun" provide a realistic and useful unit of length for characterizing a star such as the Sun. Applicability of polytropes to other stars will be discussed later.

In terms of the Emden unit of length $r_o$, we convert the radial (dimensional) coordinate into a dimensionless variable $x$ for the radial coordinate as follows: $x = r/r_o$. This allows us to rewrite Equation 10.8 in dimensionless form as follows:

$$\frac{1}{x^2}\frac{d}{dx}\left[x^2 \frac{dy}{dx}\right] = -y^n \tag{10.10}$$

This is the dimensionless form of the *Lane–Emden equation*. It is an ordinary differential equation of second order containing one unknown, $y(x)$.

In certain cases, it is convenient to rewrite Equation 10.10 in terms of an auxiliary function $z$ defined by $z \equiv xy$. Inserting this into Equation 10.10, we find that the Lane–Emden equation can also be written in the form

$$\frac{d^2z}{dx^2} = -\frac{z^n}{x^{n-1}} \tag{10.11}$$

# 10.5 BOUNDARY CONDITIONS FOR THE LANE–EMDEN EQUATION

When we set out to calculate a polytropic model of a spherical "star", the aim of the exercise is to determine how physical parameters vary between the center of the sphere ($x = 0$) and the surface ($x = x_1$). To do this, we must solve Equation 10.10 for $y$ as a function of the radial coordinate $x$. Once we have such a solution, a plot of $y$ as a function of $x$ will show, for a perfect gas, a curve that is proportional to the radial profile of *temperature* from center to surface. According to Equation 10.6, the radial profile of the *density* will be obtained by raising the local value of $y$ at each value of $x$ to the power $n$ (the polytropic index). According to Equation 10.7, the radial profile of the pressure will be obtained if the local value of $y$ at each value of $x$ is raised to the power $n + 1$.

Since Equation 10.10 is second order, we need two boundary conditions (BCs) in order to obtain a unique solution for any given value of $n$. One BC is readily available from the definition in Equation 10.6: $y = 1$ at $x = 0$.

To obtain a second BC, it is helpful to consider how the acceleration due to gravity $g(r)$ is related to the Lane–Emden function $y(r)$. To derive such a relation, we recall that the value of the acceleration due to gravity $g(r)$ at any radial position $r$ is related to the local gravitational potential $\Phi$ by the formula $g = -d\Phi/dr$. This allows us to rewrite the equation of hydrostatic equilibrium (HSE) in the form $dp/dr = \rho d\Phi/dr$, leading to $dp = \rho d\Phi$. In view of the definition of a polytrope (Equation 10.1), we can also write $dp = [(n + 1)/n] K \rho^{1/n} d\rho$. This leads to the following differential equation relating $p$ and $\Phi$ in HSE: $d\Phi \sim \rho^\delta d\rho$ where the exponent $\delta = -1 + (1/n)$. Integrating the equation, we find $\Phi \sim \rho^{1/n} + constant$. Typically, the gravitational potential is set to zero at infinity, where $\rho \to 0$. This choice leads to $\Phi \sim \rho^{1/n}$. Recalling the definition of $y$ in Equation 10.6, we see that $\Phi \sim y$.

Now at the center of the Sun, where density approaches a constant value, we have already seen (Section 9.1, step 6, Equation 9.5) that $g \to 0$ as $r \to 0$. In other words, $d\Phi/dr \to 0$ as $r \to 0$. Converting to the dimensionless length parameter $x$, this is equivalent to $dy/dx \to 0$ as $x \to 0$. This provides us with the second BC, which we need in order to obtain a unique solution for Equation 10.10 for any specified value of the polytropic index $n$.

In order to satisfy the two BCs, a series expansion is helpful near the origin. To satisfy the BC $y = 1$ at $x=0$, the leading term in this series must be 1. And to satisfy the BC $dy/dx = 0$ at $x = 0$, the series must not contain a term which is first order in $x$. For a polytrope with index $n$, the result is found to be (e.g., Chandrasekhar 1958, p. 95)

$$y = 1 - \frac{x^2}{6} + \left(\frac{n}{120}\right) x^4 - \ldots \tag{10.12}$$

# 10.6 ANALYTIC SOLUTIONS OF THE LANE–EMDEN EQUATION

Since the boundary conditions both apply at $x = 0$, we obtain a solution for $y(x)$ (for any given value of $n$) by starting at the center of the polytrope and integrating outward.

We note that since $dy/dx \sim d\Phi/dr$, and $d\Phi/dr = -g$ (a negative number), the slope $dy/dx$ is negative. Therefore, although $y$ starts with the value $y = 1$ at $x = 0$, the value of $y$ *decreases* as $x$ increases, for all values of $n$. Since $y$ decreases as we move outward from the center, there exists a certain radial location, $x = x_1$ (which is different for different values of $n$), at which the value of $y$ passes through zero *for the first time*. At that radial location, pressure and density are both equal to zero. The ratio of $p/\rho$ (i.e., the temperature, if the medium obeys the perfect gas law) is also zero at $x = x_1$. Compared to the values of unity at the *center* of the polytrope, it is natural to consider that the first zero point of $y$ corresponds to the "*surface*" of the polytrope.

Analytic solutions are known for the Lane–Emden equation for three particular values of $n$.

### 10.6.1 POLYTROPE $n = 0$

In this case, Equation 10.10 becomes

$$\frac{d}{dx}\left(x^2 \frac{dy}{dx}\right) = -x^2 \tag{10.13}$$

Integrating once, we find

$$x^2 \frac{dy}{dx} = -\frac{x^3}{3} + const. \tag{10.14}$$

In order to satisfy the boundary condition $dy/dx = 0$ at $x = 0$, the constant must be zero. This leads to

$$\frac{dy}{dx} = -\frac{x}{3} \tag{10.15}$$

Integrating again and applying the condition $y = 1$ at $x = 0$, we find

$$y(n = 0) = 1 - \frac{x^2}{6} \tag{10.16}$$

The first zero of $y(n = 0)$ occurs at $x_1 = \sqrt{6}$. Thus, the (dimensional) radial coordinate associated with the "surface" of the $n=0$ polytrope is $r_o \sqrt{6}$, where $r_o$ in this case is obtained from Equation 10.9 by setting $n = 0$.

### 10.6.2 POLYTROPE $n = 1$

In this case, it is convenient to use Equation 10.11, which reduces, in the case $n = 1$, to the simple form

$$\frac{d^2 z}{dx^2} = -z \tag{10.17}$$

The solution of this equation, consistent with both boundary conditions at $x = 0$ is $z = \sin(x)$. Reverting to the solution for $y$, we have

$$y(n = 1) = \frac{\sin(x)}{x} \tag{10.18}$$

The first zero of $y(n = 1)$ occurs at $x_1 = \pi$.

### 10.6.3 POLYTROPE $n = 5$

Derivation of the solution in this case is more complicated than the two prior cases. (See Chandrasekhar's 1958 book, pp. 93–94, for a derivation.) Here we simply state the result:

$$y(n = 5) = \frac{1}{\sqrt{(1 + (x^2 / 3)}} \tag{10.19}$$

The first zero of $y(n = 5)$ occurs at $x_1 \to \infty$. Thus, for the case $n = 5$, the equilibrium configuration of the polytrope is infinitely extended.

Polytropes

The process of obtaining numerical solutions of the Lane–Emden equation for arbitrary values of $n$ will be discussed in Section 10.8.

## 10.7 ARE POLYTROPES IN ANY WAY RELEVANT FOR "REAL STARS"?

Note that in all three polytropes for which analytic solutions exist, inspection of the solutions indicates that $y$ is a monotonically decreasing function of $x$ for all values of $x$ between 0 and $x_1$. This property also emerges from numerical solutions of the other polytropes, where only nonanalytic solutions exist. Recalling that $y$ is proportional to temperature (in a perfect gas) (see Section 10.3), the fact that temperature decreases monotonically from center to surface indicates that if energy is generated at the center (by an unspecified mechanism), then that energy will find itself in a medium that has a negative temperature gradient: this facilitates the transport of energy outward toward the surface (i.e., in the direction of increasing $r$) in accordance with Fick's law (Equation 8.1). Here again, we come across a feature that makes it attractive to consider polytropes as structures which, although highly idealized, nevertheless have properties that are physically relevant in the context of modeling "real stars".

The most successful application of polytropes to stellar structure is found when one is modeling a star in which the equation of state in fact obeys the polytropic relation (Equation 10.1). Do such stars exist? The answer is a definite "Yes". We can summarize four examples.

First, the polytrope $n = 1.5$ is relevant to low-mass stars. We recall (Chapter 7) that the Sun has a convection zone that occupies a spherical shell with a finite thickness: the shell extends inwards (below the surface) until it comes to an end at a well-defined radial location ($r \approx 0.7 R_\odot$). As a result, only an outer envelope of the Sun is convective. It turns out that when models are computed for stars with masses that are progressively smaller than the Sun's mass, the convective envelope becomes progressively deeper, reaching ever farther into the star as we consider stars with lower and lower masses. Eventually, a mass $M_c$ is reached (in the range 0.33–0.24$M_\odot$: see e.g., Mullan et al. 2015) where the convective "envelope" extends all the way to the center of the star. For stars with masses less than $M_c$, the entire star is convective, and the adiabatic limit of convection applies throughout essentially the entire star. Such stars can be represented quite well by the $n = 1.5$ polytrope. Such stars have a much smaller value of central condensation that the Sun does: whereas the latter has a central density that exceeds the mean density by a factor of more than 100 (see Section 9.2), the central density in an $n = 1.5$ polytrope is only about six times larger than the mean density (see Table 10.2).

Second, for quite different reasons, the polytrope $n = 1.5$ also turns out to be relevant to old stars called "white dwarfs". These are objects where so much time has elapsed that the supply of nuclear fuel is exhausted: in these stars, electron degeneracy pressure (see Chapter 9, Section 9.5) supports the star against gravity. In such cases, if the electrons are nonrelativistic, Equation 10.1 applies with $n = 1.5$ and a value of $K$ that depends only on certain physical constants (Planck's constant and the mass of the electron). In this case, it can be shown that white dwarfs should obey a mass-radius relationship $R \sim M^{-1/3}$. This is a very different relationship from that which applies to solar-like stars: for the latter, the radius $R$ increases as the mass increases, roughly as $R \sim M$. But white dwarfs are predicted to have radii that *decrease* as we consider objects with increasingly large masses. There is observational evidence to support this prediction (e.g., Provencal et al. 1998).

Third, if the degenerate electrons supporting a white dwarf star are relativistic, it turns out that the equation of state is again given by Equation 10.1, but now with the value $n = 3.0$. Now, $K$ has a different value, again determined by a (different) combination of physical constants. For the particular value $n = 3.0$, an interesting outcome emerges: a polytrope with $n=3$ has a unique mass $M_3$, determined by physical constants (Chandrasekhar 1958). For typical stellar compositions, $M_3$ is found to be close to $1.4M$(sun): a star with mass $M_3$ is the most massive object that can exist in hydrostatic equilibrium with support from degenerate electrons. It is remarkable that a polytrope corresponding to an object being supported against gravity by the pressure of relativistic electrons has a unique mass of the same order as a "real star" such as the Sun. Nevertheless, this conclusion has emerged as of fundamental importance in observational attempts to probe the evolution of the stellar universe in its earliest stages. The stars that can be observed farthest away in space (and therefore farthest back in time) are exploding stars called supernovae. One class of supernova occurs when a white dwarf accumulates so much mass that it exceeds $M_3$: when that happens, the star cannot exist in equilibrium but collapses and releases gravitational energy in an explosion so large that it can be seen all the way across the universe. The fact that each member of this class of supernova relies on the same physical principles allows cosmologists to assume that each member of the class is (more or less) a "standard candle", with a unique output power. This allows a distance to be assigned to each such event.

Fourth, we have already seen (Chapter 9) that the Sun consists of distinct regions in which a polytropic equation is "not too bad": the convective envelope has $n = 1.5$ and the radiative core has $n = 3.25$. The Sun can therefore not be regarded as a "true polytrope" in the strict sense of the word. But how about considering the possibility of approximating the Sun as having a single "effective polytropic index" from surface to center? Might this help us to understand some of the global properties of the Sun? Let us see. From the results that have emerged from our model of the Sun (see Chapter 9, Section 9.6), we have seen that from surface to center, the temperatures, densities, and pressures increase by (roughly) 3, 9, and 12 orders of magnitude respectively. Now, if a single "effective polytropic index" $n_e$ could be considered as applying to the Sun as a whole, let us recall that in a polytrope, $\rho$ scales as $T$ to the power of $n_e$, while $p$ scales as $T$ to the power of $n_e + 1$ (see discussion between Equations 10.7 and 10.8). Therefore, an increase in $T$ by $10^3$ would be accompanied by increases in $\rho$ and in $p$ by $10^9$ and $10^{12}$ respectively if $n_e \approx 3$. This value of $n_e$ has a value that is intermediate between the values of 1.5 and 3.25, which are applicable to the Sun's envelope and core, respectively. As a result, even in the case of a composite object such as the Sun, the concept of a polytrope helps us (roughly) to understand why some of the global physical properties of the Sun behave in the way that they do as we go from the surface of the star inwards to the center.

In summary, the study of polytropes is not at all irrelevant as far as "real stars" are concerned. To be sure, the treatment is not complete: it tells us nothing about the sources of opacity or the sources of energy. Nevertheless, there is useful information to be gained in this "first course" by considering the mechanical properties that polytropes allow us to describe.

## 10.8 CALCULATING A POLYTROPIC MODEL: STEP BY STEP

For arbitrary values of the polytropic index $n$, numerical solutions can be obtained for the Lane–Emden equation. These numerical solutions (e.g., Chandrasekhar 1958) indicate that the first zeroes of polytropes with $n = 1.5$, 3.0, 3.25, and 4.0 occur at $x_1 \approx 3.65$, 6.90, 8.02, and 15.0, respectively. In dimensional units, the radius of the corresponding polytrope is $R(n) = x_1 r_o$ where $r_o$ is the Emden unit of length corresponding to the particular polytropic index.

By way of illustration, and because we shall use this particular case in discussing certain oscillations in the Sun (Chapter 14, Section 14.5), let us consider the polytrope $n = 3.25$. In this case, we have already pointed out (Section 10.4) that the value of $r_o$ is $9.3 \times 10^9$ cm. Combining this with the appropriate value of $x_1$, we have seen that the radius of a complete $n = 3.25$ polytrope with a central pressure and density equal to that of our simplified solar model would be $R(3.25) = 7.4 \times 10^{10}$ cm. Of course, the Sun is not a complete polytrope, with a constant $n$ value all the way from center to surface. Nevertheless, the dimensional radius that we determine for such a polytrope is within 6% of the radius of the "real Sun". It is amazing that a structure as simple as a polytrope (and in which the energy equation is replaced by a gross simplification) can have macroscopic properties that are not far removed from those of an actual star.

# Polytropes

To calculate the structure of a polytrope for arbitrary $n$, the aim is to compute the value of $y$ at each of a tabulated list of values of $x$. Also, at each value of $x$, we wish to calculate the *slope* of $y$ as a function of $x$, i.e., the quantity $y' = dy/dx$. For numerical purposes, it is convenient to start with the version of the Lane–Emden equation given in Equation 10.11, where the function $z$ is defined by $z = xy$. Then we can rewrite Equation 10.11 in the form of two coupled first-order differential equations for the functions $f_1 = z$ and $f_2 = dz/dx$. In terms of these functions, the Lane–Emden equation can be replaced by two equations for two unknowns:

$$f_2 = \frac{df_1}{dx} \tag{10.20}$$

$$\frac{df_2}{dx} = -\frac{f_1^n}{x^{n-1}} \tag{10.21}$$

We start to integrate these equations at $x = 0$ using the BCs $f_1 = 0$ and $f_2 = 1$. In order to start off the numerical integration correctly, we use the series expansion for the Lane–Emden equation near the origin (Equation 10.12). Then we find

$$f_1 = x - \frac{x^3}{6} + \left(\frac{n}{120}\right)x^5 \tag{10.22}$$

and

$$f_2 = 1 - \frac{x^2}{2} + \left(\frac{n}{24}\right)x^4 \tag{10.23}$$

The step-by-step procedure for calculating a polytrope, especially one that will be useful when we come to determining the oscillation properties (see Chapter 14), proceeds as follows. The goal is to obtain a table of reliable values of three quantities ($x$, $y$, and $y'$), extending from $x = 0$ (the center of the "star") to $x = x_1$ (the surface of the "star"). The process is as follows.

1. Choose a value for the polytropic index $n$.
2. The first entries in the table refer to the center of the star. They are $x(1) = 0$, $y(1) = 1$, and $y'(1) = 0$.
3. Choose a step size $\Delta x$, which may be as small as you like. A value $\Delta x = 0.01$ will eventually lead to a table of values which, for $n = 3.25$, contains about 800 rows.
4. Advance the $x$ value to its value for the second row in the table: $x(2) = \Delta x$. Use $x(2)$ in Equations 10.22 and 10.23 to calculate the corresponding values of $f_1(2)$ and $f_2(2)$. Then the value of $y(2)$ is given by $f_1(2)/x(2)$. And the value of $y'(2)$ is given by $y'(2) = (f_2(2) - y(2))/x(2)$.
5. For the third row of the table, we advance to $x(3) = x(2) + \Delta x$. Now we have enough information to start to use an integrator (such as a Runge-Kutta routine) to step forward the solution of Equations 10.20 and 10.21. This leads to values of $f_1(3)$ and $f_2(3)$, which we then convert to $y(3)$ and $y'(3)$ using the expressions in step 4.
6. For each new row of the table, increase the $x$ value by $\Delta x$, and compute the updated values of $y$ and $y'$.
7. It is easy to see when the integration must be stopped: $y$ (which depends on the density to a certain power) cannot take on negative values. (There is no such thing as negative density.) Therefore, the last row of the table should contain a $y$ value that is close to zero, say, $y < 0.001$. The last row in the table will therefore contain an $x$ value that is close

to $x_1$ for the polytrope you have chosen, e.g., for $n = 3.25$, the value of $x_1$ is known to be 8.01894 (Chandrasekhar 1958). All values of $y'$ in the table will be negative numbers.

An example of an abbreviated table for the polytrope $n = 3.25$ is given in Table 10.1. We will have occasion to use the results in (an expanded version of) Table 10.1 in Chapter 14, when we calculate the periods of a certain class of oscillations known as $g$-modes in a polytrope. It will be instructive to compare the periods to the values observed for certain oscillations in the Sun. We shall find that once again, the use of a polytrope, however idealized, to describe the structure of a star (e.g., the Sun) provides information that may be useful in interpreting data from the "real Sun".

## 10.9 CENTRAL CONDENSATION OF A POLYTROPE

A polytrope has the property that, when one evaluates the gradient $y'$ at the surface of the "star", one can then calculate (Chandrasekhar 1958) the ratio $C_c$ of the central density $\rho_c$ to the mean density $\rho_m = M/(4/3)\pi R^3$. The quantity $C_c$ is referred to as the "central condensation". Values of $C_c$ for some polytropes (taken from Chandrasekhar 1958, his Table 4) are given in Table 10.2.

**TABLE 10.1**
**Solution of Lane–Emden Equation for the Polytrope $n = 3.25$ (Notation: $a.bDsyy = a.b$ times $10^{syy}$ where $s$ is the algebraic sign, and dot in a.b is the decimal point.)**

| x | y | y' |
|---|---|---|
| 0.00 | 1.0 | 0.0 |
| 0.02 | 0.99993D+00 | −0.66662D-02 |
| 0.10 | 0.99833D+00 | −0.31573D-01 |
| 0.20 | 0.99337D+00 | −0.64203D-01 |
| 0.30 | 0.98521D+00 | −0.95610D-01 |
| 0.40 | 0.97400D+00 | −0.12524D+00 |
| 0.50 | 0.95995D+00 | −0.15262D+00 |
| 0.70 | 0.92434D+00 | −0.19925D+00 |
| 1.00 | 0.85655D+00 | −0.24651D+00 |
| 1.50 | 0.72480D+00 | −0.27000D+00 |
| 2.00 | 0.59385D+00 | −0.24945D+00 |
| 2.50 | 0.47832D+00 | −0.21231D+00 |
| 3.00 | 0.38202D+00 | −0.17418D+00 |
| 3.50 | 0.30362D+00 | −0.14111D+00 |
| 4.00 | 0.24015D+00 | −0.11434D+00 |
| 4.50 | 0.18856D+00 | −0.93307D-01 |
| 5.00 | 0.14626D+00 | −0.76917D-01 |
| 5.50 | 0.11119D+00 | −0.64141D-01 |
| 6.00 | 0.81774D-01 | −0.54120D-01 |
| 6.50 | 0.56812D-01 | −0.46189D-01 |
| 7.00 | 0.35392D-01 | −0.39844D-01 |
| 7.50 | 0.16823D-01 | −0.34709D-01 |
| 8.00 | 0.63652D-03 | −0.30492D-01 |
| 8.02 | 0.28430D-04 | −0.30332D-01 |

# Polytropes

**TABLE 10.2**
**Central condensation in various polytropes**

| $n = 1.0$ | 1.5 | 2.0 | 3.0 | 3.25 | 3.5 |
|---|---|---|---|---|---|
| $C_c = 3.29$ | 5.99 | 11.40 | 54.18 | 88.15 | 152.9 |

We have already noted (Chapter 9, Section 9.2) that the "real Sun" has $C_c \approx 100$. Therefore, in terms of central condensation, the Sun behaves as if it were a polytrope with an index $n$ slightly larger than 3.25. Recall (Chapter 10, Section 10.2) that for the radiative interior of the Sun, there are physical reasons (related to Kramers' opacity) that the polytrope $n = 3.25$ is relevant to the relationship between pressure and temperature.

## EXERCISES

10.1 Use the step-by-step procedure in Section 10.8 to calculate a table of values $x_i$ ($i = 1, 2, 3, \ldots$) of $y_i = y(x_i)$ and $y' = dy/dx$ from center to surface for the polytropes $n = 1.0$, 1.5, and 3.25.

10.2 For the case $n = 1$, also evaluate the analytic solution $y_a(x) = \sin(x)/x$ for each $x_i$. For each entry in the table, $x_i$, calculate the fractional difference $\delta y/y$ between your numerical $y(x_i)$ and the analytic solution $y_a(x_i)$. Repeat the calculation with a smaller and a larger choice of step size $\Delta x$. How do the fractional differences $\delta y/y$ change?

10.3 At the "surface" of each polytrope in Exercise 1, your table will give you the local values of $x$ and $y'$. For each polytrope, use those surface values to evaluate the quantity $-x/3y'$ at the surface. Compare the results with the central condensations $C_c$ listed in Table 10.2.

## REFERENCES

Chandrasekhar, S., 1958. "Polytropic and isothermal gas spheres", *An Introduction to the Study of Stellar Structure*, Dover Publications, New York, pp. 84–182.

Mullan, D. J., Houdebine, E. R., & MacDonald, J., 2015. "A model for interface dynamos in late K and early M dwarfs", *Astrophys. J. Lett.* 810, L18.

Provencal, J. L., Shipman, H. L., Hog, E., et al., 1998. "Testing the white dwarf mass-radius relation with HIPPARCOS", *Astrophys. J.* 494, 759.

# 11 Energy Generation in the Sun

Historically, the source of energy generation on the Sun has been attributed to a number of causes, including gravitational collapse and radioactive decay. The possibility that nuclear fusion *might* be the source of solar energy could not be evaluated quantitatively until certain key pieces of physical information were available. In particular, the masses of the relevant isotopes had to be measured to at least three or four significant digits before it became evident that atomic masses, although close to integer values, actually deviated from integers by small, but measurable, amounts.

Measurements of atomic weights for multiple isotopes were obtained by Francis Aston using three increasingly precise mass spectrometers that he constructed in the years 1919–1937. The construction of the spectrometers, including the combined effects of strong magnetic and electric fields, was not an easy task. In an obituary for Aston, it was stated, "Aston was a superb experimenter: his first mass spectrograph was a triumph: few but he could have got it to work at all" (Thomson 1946). In measuring isotopic masses, Aston achieved precisions of 1.5 parts in $10^4$ in 1927 and 2.5 parts in $10^5$ in 1937 (see Squires 1998). The deviations in mass from integer values were found to be at most only a few parts per thousand. But, as it turns out, those small deviations are at the very heart of nuclear energy generation in the Sun: if those deviations were absent, Earth would *not* be a hospitable place for life as we know it. In recognition of his work, Aston was awarded the Nobel Prize in Chemistry in 1922.

The characteristic that sets the Sun (and stars in general) apart from other structures in the universe is precisely the fact that *the Sun is able to generate its own supply of energy by means of nuclear fusion reactions in the deep interior*. On Earth, nuclear reactions can be made to happen by accelerating particles to energies of millions of electron volts (MeV), and then "slamming" the fast particles into a target nucleus. But there are no MeV accelerators in the Sun. Instead, the only available particles are those belonging to a *thermal population* in which the mean energies are much smaller than 1 MeV: in the Sun's core (where, as we have seen earlier [see Chapter 9], $T = 15$–$16$ MK), the mean thermal energy of the ions is of order $kT \approx 1$ keV only. Despite mean energies of mere keV, i.e., of order 1000 times less than 1 MeV, the fact remains that the solar particles *can* (and do) participate in nuclear fusion reactions. The fact that the reacting particles are *thermal* gives rise to the term "thermonuclear reactions" to describe the process whereby light nuclei in the Sun find a way to undergo fusion so as to form heavier nuclei. The (slight) loss of mass that occurs in the fusion reactions that build up helium nuclei from hydrogen nuclei emerges in the form of kinetic energy of particles and radiant energy of energetic photons. It is this emergent energy that makes the Sun a power generator.

It will not be sufficient to demonstrate that nuclear reactions *can* occur in the Sun. In addition, in order to determine the luminosity of the Sun (in units of ergs *per second*), it is important to ask: *at what rate will the reactions occur?* The answer to that question depends in part on how difficult it is for two nuclei to approach each other in the presence of Coulomb repulsion. But the rate also depends on how protons and neutrons interact once they are both *inside* the nucleus: the latter depends on the operation of two "forces", the strong and the weak. The strong force holds the nucleus together, while the weak force causes the nucleus to decay into lighter particles. Decays that involve emission of electrons ("beta decay") can be described by a theory that was first proposed by Enrico Fermi (1934) and subsequently developed by George Gamow and Edward Teller (1936).

Once the precise masses of isotopes of H, He, C, N, and O became available in the 1920s–1930s, and once the theory of nuclear beta decay had reached a state where the beta decay rate could be calculated reliably (in the mid-1930s), two distinct cycles of reactions were identified as being energetically permitted processes of nuclear energy generation in the Sun. Both cycles were identified

### TABLE 11.1
### Isotope Nuclear Masses in Atomic Mass Units (a.m.u.)

| | |
|---|---|
| Proton ($H^1$) | 1.007276467 |
| Neutron (n) | 1.008664916 |
| Deuteron (D = nucleus of $H^2$) | 2.013553213 |
| Helium-3 ($He^3$) | 3.0149321 |
| Helium-4 ($He^4$ = "alpha particle") | 4.001506179 |

by Hans Bethe. The cycles are referred to as the *pp*-cycle (Bethe and Critchfield 1938, using the Gamow-Teller [G-T] theory of β-decay) and the CNO cycle (Bethe 1939). Recent measurements of certain neutrinos (see Chapter 12) which originate in the CNO cycle (but not in the *pp*-cycle) indicate that "the relative contribution of CNO fusion in the Sun [is] on the order of 1%" (The Borexino Collaboration 2020). Because it has now been established experimentally that the *pp*-cycle is by far the dominant energy-producing process in the Sun, we focus, in this chapter, on the *pp*-cycle.

The important questions in the context of solar energy generation are: (i) which reactions occur? (ii) How much energy is liberated in each reaction? (iii) How many reactions occur per second? Now that we know certain physical parameters in the Sun, we can address these questions in turn.

In the following discussion, masses of the relevant nuclei (see Table 11.1) will be cited in terms of atomic mass units (a.m.u.). Four of the entries in Table 11.1 were obtained from the 2018 NIST Reference list https://physics.nist.gov/cuu/Constants/index.html: these are the entries with 10 significant figures in Table 11.1. In the case of the nucleus helium-3, we started with *atomic* masses (Audi and Wapstra 1993) and then subtracted two electron masses ($1m_e$ = 0.00054858 a.m.u.) to obtain the *nuclear* mass. In c.g.s. units, the 2018 NIST Reference list states that one a.m.u. corresponds to a mass of $1.660539067 \times 10^{-24}$ gm. The rest-mass energy equivalent of 1 a.m.u. is $E(1) = 1.492418086 \times 10^{-3}$ ergs. Expressed in units of electron volts (1 eV = $1.602176634 \times 10^{-12}$ ergs), we find $E(1)$ = 931.494102 MeV. For the electron, the rest mass energy is 0.510999 MeV.

## 11.1 THE *PP*-I CYCLE OF NUCLEAR REACTIONS

In the Sun, the most common set of reactions that occur are referred to as the *pp*-I cycle. There are also less common cycles referred to as *pp*-II and *pp*-III, but all have the same overall end result, namely, four protons are fused into one helium nucleus. We shall return to the *pp*-II and *pp*-III cycles in Chapter 12 when we discuss neutrinos. In the present Chapter, where energy generation is the principal focus of our discussion, we confine our attention to the *pp*-I cycle.

There are three nuclear reactions to be taken into account in the *pp*-I cycle. We label these as (a), (b), and (c) in what follows.

$$p + p \rightarrow D + e^+ + \nu (E = 1.442 MeV) \quad \text{(a)}$$

Here, $p + p$ denotes the reaction of two protons, both of which belong to the thermal distribution that exists at any given radial location with local temperature $T(r)$. The reaction products include a deuteron (a nucleus $D$ consisting of one proton and one neutron), a positron ($e^+$), and a very low-mass particle known as a neutrino ($\nu$: see Chapter 12).

In order to determine the amount of energy that is released in reaction (a), we use the masses of the various particles in Table 11.1. Using these, we find that the total mass on the left-hand side of reaction (a) is 2.014552934 a.m.u. This exceeds the deuteron mass (on the r.h.s. of reaction (a)) by

# Energy Generation in the Sun

$\Delta m = 0.000999721$ a.m.u. The fractional excess in mass is small, less than 0.1% of an a.m.u., but the existence of an excess (however small) ensures that the reaction is exothermic. In energy units, the corresponding energy is $c^2 \Delta m = 0.931234$ MeV. The positron is an antiparticle that requires an equivalent rest-mass energy equal to that of the electron, i.e., 0.510999 MeV. The net energy available for the neutrino and D from reaction (a) is the remaining energy $0.931234 - 0.510999 = 0.420235$ MeV: this is the maximum ("endpoint") energy that the neutrino can carry away. Actually, because the reaction energy is shared by three particles, the neutrino energy has a continuous spectrum between zero and 0.420235 MeV: the average energy carried off by the neutrino is about one-half of the endpoint energy, i.e., about 0.2 MeV. The positron quickly annihilates on an ambient electron (of which there are roughly $10^{26}$ in each cm$^3$ in the Sun's core, one electron for each proton: see Section 9.2), releasing an energy of 1.022 MeV. Adding this to the remaining energy estimated earlier (0.420 MeV), we see that the total amount of energy released into the core of the Sun by reaction (a) is 1.442 MeV. However, as we shall see in Chapter 12, the neutrino does not contribute significantly to energy deposition in the Sun: each neutrino escapes so easily from the Sun that its KE escapes from the Sun essentially instantaneously. Allowing for the average energy carried off by the neutrino, the amount of energy that is available (on average) to be deposited in the core of the Sun, thereby contributing to the thermal energy pool, is about 1.2 MeV.

The second reaction in the *pp*-I cycle is

$$p + D \rightarrow He^3 + \gamma (E = 5.494 MeV) \tag{b}$$

In this second step of the *pp*-I cycle, a third proton from the thermal population reacts with the deuteron that was produced in reaction (a). Reaction (b) results in a nucleus of He$^3$ plus an energetic photon: the photon is designated by the letter γ, because the photon's energy is a few MeV, i.e., in the gamma-ray range. Referring to Table 11.1, we see that the combined masses of *p* and *D* on the left-hand side of reaction (b) (= 3.02082968 a.m.u.) exceeds the mass of the He$^3$ nucleus on the right-hand side by $\Delta m = 0.0058976$ a.m.u. Once again, the fractional mass excess is relatively small, but it is finite. Therefore, the reaction is exothermic, with an energy release $c^2 \Delta m$ of 5.494 MeV. This energy is carried away from the reaction site by the fast-moving He$^3$ nucleus and the photon.

The final reaction in the *pp*-I cycle is

$$He^3 + He^3 \rightarrow He^4 + 2p(E = 12.859 MeV) \tag{c}$$

In this third step of the *pp*-I cycle, after reactions (a) and (b) have each occurred twice (thereby involving an "intake" of six protons), the two He$^3$ nuclei fuse to create one nucleus of He$^4$, releasing two protons in the "exhaust". The net effect of the *pp*-I cycle is to have four protons fuse into a single nucleus of He$^4$. The sum of the rest masses of two He$^3$ on the left-hand side (6.0298642 a.m.u.) exceeds the sum of the rest masses of the three particles on the right-hand side (6.016059113 a.m.u.) by $\Delta m = 0.0138051$ a.m.u. The corresponding energy release $c^2 \Delta m$ is 12.859 MeV.

For the Sun to produce energy by hydrogen fusion, it is *essential* that in each reaction of the aforementioned cycle, the combined mass of the products is *less* than the combined mass of the reactants. In the early days of measuring atomic weights, when the masses of the isotopes were known with a precision of only two significant digits, the mass of the reactants would be found to be equal to the mass of the products: in such a case, no energy generation would be possible. It was only when the atomic weight measurements reached a precision of at least three (or preferably four) significant digits that the mass difference $2m(p) - m(D)$ was found to have a positive value. And in order to derive the energy release in the reaction with a precision of $N$ significant digits, the portions of the various isotopic masses to the right of the decimal point (i.e., the deviations of isotopic masses from whole numbers) have to be measured with precisions of $N + 2$ significant digits. Since the mass of $He^3$ is listed in Table 11.1 with seven digits to the right of the decimal point, we can trust the energies listed in (b) and (c) to no more than five significant digits. Even less confidence can be assigned to the energy figure listed in (a): there, the main barrier to high precision in the estimate of energy is determining how much energy the neutrino carries off.

## 11.2 REACTION RATES IN THE SUN

Altogether, in a complete *pp*-I cycle, consisting of two reactions each of (a) and (b) plus one reaction (c), the total energy released is $\Delta E(pp\text{-I}) = 2(1.442 + 5.494) + 12.859 = 26.731$ MeV. However, some of the energy released in (a) (perhaps 0.2 MeV) is carried off by neutrinos. The amount of energy which is deposited into the thermal pool of the Sun's core, and which can therefore contribute to the radiant output power of the Sun, is roughly $\Delta E(pp\text{-I}) = 2(1.2 + 5.494) + 12.859 = 26.25$ MeV.

This is (roughly) the amount of thermal energy that is released into the thermal pool in the Sun's core when four protons fuse into one helium nucleus. Converting to c.g.s. units, each *pp*-I cycle generates $\Delta E(pp\text{-I}) = 4.206 \times 10^{-5}$ ergs. The main source of uncertainty in this result comes from the estimate of the mean energy of the neutrinos that escape from the Sun: this estimate cannot be more uncertain than (roughly) ±0.2 MeV (correspond to neutrinos carrying away zero energy, or carrying away the maximum possible energy of the neutrinos). All other quantities entering into $\Delta E(pp\text{-I})$ are known precisely. Therefore the value of $\Delta E(pp\text{-I})$ can be written as $26.25 \pm 0.4$ MeV, i.e., an error of 1%–2%.

Now, we already know the total output power of the Sun (Chapter 1, Section 1.4): $L_\odot = 3.828 \pm 0.0014 \times 10^{33}$ ergs sec$^{-1}$. Therefore, since *pp*-I cycles are by far the largest source of energy generation in the Sun, the number of these reactions that occur in the Sun every second (i.e., the frequency of the reactions) must equal

$$F_r = \frac{L_\odot}{\Delta E(pp-\text{I})} = 0.91 \times 10^{38} \text{ reactions sec}^{-1} \tag{11.1}$$

The uncertainty in the value of $F_r$ is dominated by the 1%–2% uncertainty in $\Delta E(pp\text{-I})$. As a result, we can write $F_r = (0.91 \pm 0.02) \times 10^{38}$ reactions sec$^{-1}$.

The Sun also relies, in a small percentage of cases, on *pp*-II and *pp*-III cycles (see Section 12.3). However, both of those cycles also begin with reactions (a) and (b) listed earlier, and their rates are controlled primarily by (the slowness of) reaction (a). Moreover, some ($\leq 1\%$) of the solar energy output comes from the CNO (see Section 12.3.2): in this cycle, carbon acts as a catalyst to bring about the same overall effect as in the *pp*-cycle, in effect fusing four protons into one He$^4$ nucleus.

In summary, we will not make a significant error if we take $F_r \approx 10^{38}$ per second as the number of *pp*-I chains that must occur in the Sun every second to account for the observed luminosity. The value of $F_r$ is based on the ratio of two quantities that are reliably known: the Sun's luminosity and the energy that is released by a single *pp*-I chain of reactions. Can the value of $F_r$ be checked observationally? Yes: each *pp*-I cycle emits two neutrinos due to reaction (a), which occurs twice in each *pp*-I cycle. Therefore, the Sun is predicted to emit (from the *pp*-I cycle alone) some $2 \times 10^{38}$ neutrinos every second. Detectors have been set up at various locations on Earth since 1967 to try to detect these neutrinos: reliable detection is a significant challenge because neutrinos can pass through vast numbers of atoms without interacting. But the searches eventually were successful (see Chapter 12).

## 11.3 PROTON COLLISION RATES IN THE SUN

In order to set the reaction rate $F_r$ in context, let us compare $F_r$ to the overall rate $F_c$ at which collisions between protons occur in the nuclear-generating core of the Sun. By the word "collision", we mean an event in which the momenta of the individual particles are altered significantly, in a manner analogous to the collision of two billiard balls. Two protons that at first happen to be approaching each other feel an increasingly strong Coulomb repulsion, which eventually causes the two to move apart, changing directions and speeds in such a way as to conserve energy and momentum.

# Energy Generation in the Sun

With a mean velocity of $V$ and a number density of $n_p$ protons cm$^{-3}$, the rate at which a single proton in the Sun experiences momentum-altering collisions with other protons is $f_c = n_p V \sigma$ per second, where $\sigma$ is the momentum collision cross-section. (Lower case $f$ denotes the collision rate for a *single* proton. Upper case $F$ denotes the total number of collisions experienced every second by all of the protons in the Sun.) Between two protons, the value of $\sigma$ is determined by the Coulomb force. To calculate the Coulomb cross-section $\sigma_c$, we note that in a gas with temperature $T$, the mean kinetic energy of thermal motion, of order $kT$, allows two protons to approach one another within a minimum distance $r_m$ such that $e^2/r_m \approx kT$. Such close collisions result in large deflections of the protons from their original motions. The cross-sectional area associated with $r_m$ (i.e., $\pi r_m^2$) would be a reasonable estimate for $\sigma_c$ if large deflections were the only contributors to deflecting protons in their motion. But because the Coulomb force is a long-range force, protons are also subject to a multitude of small deflections as a result of *distant* collisions. The net effect of these is to yield a cross-section that is larger than the estimate based on $r_m$ by a multiplying factor called the Coulomb logarithm. It is conventional to write $\sigma = \pi r_m^2 \Lambda$, where $\Lambda$ is a logarithmic term that includes the effects of distant collisions (e.g., Spitzer 1962). Thus, $\sigma_c \approx \pi e^4 \Lambda / (kT)^2$. (Note that, for thermal particles, with $V_{th} \sim \sqrt{T}$, the Coulomb cross-section $\sigma_c$ scales as $1/V_{th}^4$, i.e., fast particles have significantly *fewer* collisions in any particular time interval than slow particles.) In the core of the Sun, where $n_p \approx 10^{26}$ cm$^{-3}$ (see Section 9.2) and $T = (1.5-1.6) \times 10^7$ K, Table 5.1 in Spitzer (1962) indicates that the value of $\Lambda \approx 3-4$. This leads to $\sigma_c \approx (1-2) \times 10^{-19}$ cm$^2$. Since the mean thermal velocity of a proton in the core of the Sun is $V_{th} \approx 6 \times 10^7$ cm sec$^{-1}$ (see Section 9.2), we see that *each proton* undergoes $f_c = n_p V \sigma \approx 10^{15}$ momentum-changing collisions per second. Our estimate of $f_c$ is subject to uncertainties in the various factors that occur in $\sigma$, $V$, and $n_p$: as a result, we would not be surprised if the true value of $f_c$ might differ by a factor of a few above or below the value $10^{15}$ sec$^{-1}$.

The overall rate $F_c$ of Coulomb collisions in the core of the Sun is given by $f_c$ times the total number of protons $N_p(c)$ in the core. The core of the Sun, in which nuclear reactions occur, is confined, according to detailed models, within the innermost 20% (or so) of $R_\odot$. Although the volume of this core is a small fraction of the total solar volume, the high densities in the core have the effect that the mass of the core may be as large as $\approx 0.1 M_\odot$, i.e., about $2 \times 10^{32}$ gm. Dividing this by the mass of a proton, we find, $N_p(c) \approx 10^{56}$. This leads to $F_c = f_c N_p(c) \approx 10^{71}$ momentum-changing collisions occurring every second in the nuclear-generating core of the Sun. Due to uncertainties in the various factors entering into the calculation, the numerical value of $F_c$, which we have derived using various simplifications, is subject to uncertainties. Although we have not conducted a full error analysis, we estimate that the uncertainties might be as large as $\pm 1$ *in the exponent*.

The number $F_c$ of collisions per second in the Sun's core is by any reckoning a large number. But in order to judge the true significance of $F_c$, we need to compare it with another frequency. The natural frequency to compare it to is the frequency $F_r$ that we estimated earlier in Section 11.2, i.e., the number of nuclear reactions that must occur every second in the Sun in order to provide the Sun with energy. Comparing the rate at which the *pp*-I chain of reactions occur in the Sun, i.e., $F_r \approx 10^{38}$ sec$^{-1}$, with the momentum-changing collision rate $F_c$ in the core, we arrive at a noteworthy result: a proton undergoes (on average) a huge number of momentum-changing collisions $N_c(react)$ before that proton ever participates in a *pp*-I chain of reactions in the Sun's core. Specifically,

$$N_c(react) \approx \frac{F_c}{F_r} \approx 10^{33} \text{collisions} \qquad (11.2)$$

The main uncertainty in $N_c(react)$ is due to uncertainties in $F_c$: as a result, $N_c(react)$ might be as small as $10^{32}$ or as large as $10^{34}$.

Since an individual proton experiences on average $f_c \approx 10^{15}$ momentum-changing collisions per second (uncertain by a factor of $\pm$ a few), each proton in the Sun will participate in a *pp*-chain only after an average time span $t_{pp}$ of about $N_c(react)/f_c$. The smallest value of $t_{pp} \approx 10^{32}/(\text{a few times } 10^{15})$ is $\approx 3 \times 10^{16}$ sec, i.e., about one gigayear (Gy). The mean value of $t_{pp}$ is $\approx 3 \times 10^{17}$ sec, i.e., roughly

10 Gy. These values suggest that the lifetime of the Sun in its *pp*-I phase will *not* be conveniently measured in units of millions of years or in trillions of years: instead, the H-burning phase will last for a time for which the appropriate unit is expected to be gigayears. Observational support for this conclusion is provided by the fact that the Sun has already existed for 4.567 Gy. How do we know the age of the Sun? From studies of *p*-mode oscillations (Bonanno and Frohlich 2015) and from studies of isotopic ratios in special inclusions extracted from certain meteorites dating back to the early solar system (Connelly et al. 2012): these entirely independent approaches yield ages of $4.569 \pm 0.006$ and $4.56730 \pm 0.00016$ Gy respectively. The consistency in these independent estimates of ages of the current Sun is excellent. Detailed models of the Sun (e.g., Bahcall et al. 2006) suggest that in the core of the Sun, nuclear fusion has already converted roughly 50% of the H into He: this suggests that a rough estimate of the lifetime of the Sun in its *pp*-I nuclear phase can be obtained by doubling the current age of the Sun, i.e., 9–10 Gy.

The numerical value of the ratio in Equation 11.2 is strikingly large. The occurrence of nuclear reactions in the *pp*-cycle is a very rare event indeed in the conditions of the Sun's core: *only one collision in* (roughly) $10^{33}$ *p-p collisions results in a nuclear reaction*. It is worthwhile examining why, based on the laws of physics, the reaction rate is so small compared to the collision rate: part of the answer will lead us to understand why the rates of thermonuclear reactions in the Sun (as well as in stars, and in thermonuclear weapons) are very sensitive to temperature.

## 11.4 CONDITIONS REQUIRED FOR NUCLEAR REACTIONS IN THE SUN

Nuclear reactions provide the only physically realistic source for solar energy generation that has been occurring already for a time interval that is measured in gigayears. In this section we examine the following issue: what physical requirements must be satisfied before a (thermo)nuclear reaction can occur at all in the Sun?

In order to have a nuclear reaction occur, whether in the Sun or in the laboratory, certain conditions have to occur. First, two nuclei must undergo a "collision" with each other. The collision must be of a particular kind. We are not interested merely in momentum-changing collisions where the particles stay far apart and experience only a "glancing" blow off each other. Such "distant" collisions are certainly important in a plasma when we wish to evaluate certain transport coefficients in the plasma: because the Coulomb force is long range, the overall effect of many distant collisions can dominate over the rare large-angle collisions. (This is in fact the origin of the factor $\Lambda(>1)$ in the Coulomb cross-section mentioned earlier in Section 11.3.) However, distant collisions of this kind contribute nothing to nuclear reactions.

### 11.4.1 Nuclear Forces: Short-Range

Instead, in order for a nuclear reaction to have any chance of occurring, it is essential that two nuclei must approach one another so closely that the strong force, which binds nucleons (protons, neutrons) together inside a nucleus, can come into play. How close do such collisions have to be?

The answer depends on the range of distances over which the strong force can actually be "felt". Since nuclei are held together by the strong force, we know that the strong force must have a range that is at least as large as the size of a nucleus. But the interesting quantitative question is this: what is the size of a nucleus? When Ernest Rutherford did his experiments of "shooting" fast alpha particles (i.e., helium-4 nuclei, essentially "point-size bullets") into a thin foil of gold atoms, he found that essentially the entire mass of a gold atom must be concentrated in a massive point at the center of the atom. He estimated that the nucleus had a size about 10,000 times smaller than the atom, i.e., with a size of order $10^{-12}$ cm. Rutherford predicted that most of the incoming alpha particles would pass straight through the gold atom, but some would be deflected by an angle $\theta$: the probability $P(\theta)$ of being deflected by $\theta$ decreases with increasing $\theta$. For a point size nucleus, $P(\theta)$ is predicted to vary as $1/\sin^4(\theta/2)$.

How might one go about the task of measuring the size of an object as small as a nucleus? By relying on a concept that was first introduced in Section 2.1 According to de Broglie, the "bullet" particles in the Rutherford experiment are not exactly "point-size". Instead, they are "smeared out" over the finite extent of $\lambda_D = h/mV = h/p$. Therefore, if one could perform a Rutherford experiment in such a way that $\lambda_D$ for the "bullets" was comparable to the size of the nucleus, then the process of scattering would no longer involve one point-like ("bullet") object interacting with another point-like ("target") object. Instead, at least one of the objects involved in scattering would deviate significantly from "point-like" characteristics. In such circumstances, one would expect to see deviations from the predicted law of $1/\sin^4(\theta/2)$.

One such experiment was performed by Robert Hofstadter et al. (1953) when he obtained access to "bullets" in the form of electrons with energies $E$ of up to 100–200 MeV generated by the Stanford linear accelerator. At $E = 200$ MeV ($\gg m_e c^2$), an electron is highly relativistic, with the result that the momentum $p$ is essentially given by $p = E/c$. Therefore $\lambda_D = h/p = hc/E$. Hofstadter et al. (1953) "shot" these "bullets" at nuclei of various elements ranging from beryllium to gold and lead. They observed that the experimental data for gold did indeed depart noticeably from the predicted law of $1/\sin^4(\theta/2)$. For example, when $\theta$ increased from 35 deg to 90 deg, the predicted law says that the number of counts should fall off by a factor of $\approx 30$, but the experimental fall off turned out to be much larger, by a factor of >1000. Thus, the point-mass-on-point-mass concept did *not* work well in the Hofstadter et al. experiment: instead, the "bullets" are smeared out over finite sizes $\lambda_D$ that must have been comparable to the r.m.s. radius of a gold nucleus. Now, knowing the energy of the "bullets", the value of $\lambda_D$ can be calculated: it is $7 \times 10^{-13}$ cm. Therefore, the gold nucleus must have a radius of this order. Beryllium nuclei were found to have sizes of order $2 \times 10^{-13}$ cm. The radius of a nucleus with atomic weight $A$ could be fitted roughly as $1.45 \times 10^{-13} A^{1/3}$ cm.

As a result of these experiments and others, it is now known that nuclei have radii that range from about 1 to a few times $10^{-13}$ cm. This indicates that the strong force operates only within a finite length-scale, of order $r_N \approx 1$ to a few times $10^{-13}$ cm. (Hofstadter [1956] suggested that the unit of length $10^{-13}$ cm should be referred to as 1 "fermi" [fm] in honor of Fermi, a pioneer in nuclear physics who had died in 1954. The abbreviation fm is also used for the length 1 femtometer = $10^{-15}$ meter = $10^{-13}$ cm.) When two nucleons approach each other at distances closer than (or of order) 1 fm, the strong force has a chance to operate, giving rise to a force between the two nucleons which is strongly attractive. That is, at distances of order 1 fm and less, a proton is strongly *attracted* to a neutron, a neutron is strongly *attracted* to another neutron, and a proton is strongly *attracted* to another proton.

The last part of the previous sentence is especially noteworthy: in the macroscopic world, classical electrostatics teaches clearly that a proton situated at a distance $d$ ($\gg 1$ fm) from another proton experiences a *repulsive* (Coulomb) force $F_{es} = e^2/d^2$. Something fundamentally different from classical electrostatics comes into play when two protons approach each other closer than a distance of order 1 fm: in such circumstances, the *strong force* comes into play, overwhelming the repulsive Coulomb force. The (attractive) strong force is what holds a nucleus together. The necessity of such a force can be seen by considering a nucleus of uranium, which contains 92 protons packed into a sphere with a radius of no more than several fm: the repulsive force between all those protons would, under classical conditions, tear the nucleus apart.

What causes the strong force to be so strongly attractive as to overcome the Coulomb repulsion? The Japanese physicist Hideki Yukawa in 1934 introduced the novel idea that the strong force could be understood if two nucleons "exchange" a special (short-lived) particle called a "meson" back and forth between them at speeds approaching the speed of light. Yukawa estimated that in order to explain the shortness of the range of the strong force (i.e., of order 1 fm), this meson should have a mass that is intermediate between electron and proton masses: Yukawa's mass estimate for the exchange particle was of order 200 $m_e$. Later experiments (in 1947) indicated that 270 $m_e$ is a better estimate of the meson mass. In order to "explain" the concept of "exchange force", according to one analogy, two tennis players can be considered to be "held together" on a court by the tennis ball that they exchange back and forth from one side of the net to the other: the ball in essence creates a sort of "exchange force" between the players.

Whether or not this tennis analogy is physically realistic, the fact remains that nuclear reactions *cannot* occur if two nucleons remain too far apart. Specifically, a nuclear reaction between two nuclei becomes possible only if the particles are brought (somehow) as close together as a distance of order $r_N \approx 1$ fm. *This result is fundamental to understanding how thermonuclear reactions are even possible in the Sun.*

The strong force is not the only force that operates in the *pp*-chain in the Sun: there is also the weak force which enters into reaction (a) in Section 11.1. In order for reaction (a) to occur when two protons collide, the weak force must cause one of the protons to "decay" into a neutron. Under ordinary circumstances, if a proton is free in the laboratory, such a decay is impossible: the neutron mass exceeds the proton mass by a finite amount $\Delta m_{np}$ (see Table 11.1). If a proton is ever to "decay" into a neutron, the proton must have access to an energy that is at least as large as $c^2 \Delta m_{np} \approx 1.3$ MeV. Where could a proton have access to such an energy? The answer is: only if the proton is inside the deep potential well created by the strong nuclear force, which can be as deep as 20–30 MeV (e.g., Schiff 1955). In effect, both the strong force and the weak force in the Sun can operate only when particles are within a distance of order $r_N$.

As a result, if two nuclei have a "collision" in which the nuclei approach each other no closer than, say, $10^{-11}$ cm = 100 fm (or more), neither the nuclear force nor the weak force has a chance to come into play in that particular collision. The two nuclei would simply have a momentum-changing collision, bouncing off each other and continuing on their way, but completely unchanged as far as their nuclear properties are concerned.

How strong is the attractive force that holds two nucleons together? Well, it certainly has to be strong enough to overcome the Coulomb repulsion. The Coulomb repulsion between two protons separated by only $r = 1$ fm has a potential energy which can readily be calculated: $e^2/r \approx 1.5$ MeV. So the strong force must be larger than that. Moreover, the nucleons inside a nucleus must not allow their de Broglie waves to "leak out of the nucleus": this requires that each nucleon inside a nucleus must be moving with a speed $V_n$, which is so fast that the de Broglie wavelength of the nucleon $h/mV_n$ is no larger than a few fermi. To achieve this goal, a proton needs to have a kinetic energy (KE) of about 10 MeV. The effects of KE are such that they tend to disrupt the nucleus: therefore, in order to overcome this disruptive tendency, the strong force has to have an attractive energy of at least 10 MeV per nucleon. Detailed calculations suggest that the strong force in fact has an attractive energy of order 20–30 MeV (e.g., Schiff 1955).

## 11.4.2 Classical Physics: The "Coulomb Gap"

Consider two protons moving in such a way that their paths will cross (or at least "come close") at some point in space. According to classical physics, when two positive point charges $+Z_1 e$ and $+Z_2 e$ are separated from each other by a distance $r$, they experience a Coulomb repulsive force. The potential energy (PE) of the repulsion is $Z_1 Z_2 e^2/r$. The closer the two particles approach each other, the stronger the repulsion becomes.

The question on which we concentrate here is the following: how closely can such particles be made to approach each other? In classical terms, the answer is straightforward: the two can come no closer than a distance $r_c$, where their relative kinetic energy is equal to the repulsive PE. Let the masses be $A_1 m_p$ and $A_2 m_p$ where $m_p$ is the proton mass. In terms of the reduced mass $A m_p$ of the two nuclei ($A = A_1 A_2/(A_1 + A_2)$), the average KE is given by $0.5 A m_p V^2$.

This leads to the following expression for the distance of closest approach:

$$r_c = \frac{2 Z_1 Z_2 e^2}{A m_p V^2} \tag{11.3}$$

# Energy Generation in the Sun

For the collision of two protons (i.e., $Z_1 = Z_2 = 1$), this reduces to

$$r_c = \frac{4e^2}{m_p V^2} \tag{11.4}$$

In order to appreciate how nuclear reactions can occur in the Sun, and in order to appreciate that something beyond classical physics is at work, we need to ask a specific *quantitative* question: what is the magnitude of $r_c$ for two protons near the center of the Sun? Setting $V = 6.21 \times 10^7$ cm sec$^{-1}$ (see Section 9.2), and inserting the values of $e = 4.8032 \times 10^{-10}$ e.s.u. and $m_p = 1.673 \text{x} \times 10^{-24}$ gm, we readily find that $r_c \approx 1.43 \times 10^{-10}$ cm $\approx 1430$ fm.

The critical point of this result is that $r_c$ *greatly exceeds* the range of the nuclear force $r_N$ ($\approx 1$ fm). Specifically, with the values we use earlier, the classical distance of closest approach of two protons in the center of the Sun is more than 1000 times larger than the nuclear force range.

To be sure, not all of the protons have velocities equal to the r.m.s. speed. There are some faster ones. For example, in a thermal distribution, one proton in $e^{10}$ (i.e., one proton in 20,000) has a speed that exceeds the mean by a factor of 3.2. If two such protons collide, then their distance of closest approach, based on classical physics, would be reduced to $\approx 143$ fm. Even so, this is still more than 100 times larger than $r_N$, much too far apart for the strong force to operate.

Because of the Coulomb repulsive force, classical physics indicates that protons moving at the average thermal speed in the core of the Sun simply *cannot* approach each other closely enough to allow the nuclear force to come into play. In classical terms, two such protons will always remain separated by a distance that is at least as large as $r_c$. In what follows, we refer to $r_c$ (see Equation 11.4), i.e., the classical distance of closest approach of two protons in the Sun, as the "Coulomb gap".

If the Sun were governed by classical physics alone, the Coulomb gap would be an insuperable barrier that would prevent two protons from ever getting close enough together for any nuclear reactions to occur in the Sun in its present condition. In order to understand why the Sun shines at all by nuclear processes, we are forced to the following important conclusion: we need to go beyond classical physics. We must admit that the Sun is an object in which quantum physics plays an essential role.

### 11.4.3 Quantum Physics: Bridging the "Coulomb Gap"

So, let us enter the world of quantum mechanics. In this world, particles in certain circumstances no longer behave as points (see Section 2.1): instead, according to de Broglie (1924), a particle of mass $m$ moving with speed $V$ has a finite probability of occupying an extended region of space. This nonpoint-like behavior is modeled by saying that the particle can be represented by an associated "probability wave". The wavelength $\lambda_p$ is given by a formula first derived by de Broglie (1924):

$$\lambda_p = \frac{h}{m_p V} \tag{11.5}$$

where $h$ is Planck's constant. According to Equation 11.5, the proton can be considered as being "spread out" over a finite distance of order $\lambda_p$.

Now we come to the heart of the matter of nuclear fusion in the Sun: the fact that any individual particle is actually "spread out" *over a finite length-scale* is precisely the property that gives rise to the possibility of "bridging the Coulomb gap". When classical physics has reached its limit, and two particles can come no closer than the Coulomb gap, we appear to be faced with two "point particles" separated by $r_c$. But now quantum mechanics steps in and replaces each particle by a structure that is no longer point-like: instead, each "particle" has a finite size, of order $\lambda_p$. When the two protons approach each other to a critical separation of $2\lambda_p$, the wave of one proton extends far enough to "touch" the wave of the other proton. Since the reduced mass of two protons is $Am_p = 0.5m_p$, the critical separation equals the de Broglie wavelength $\lambda_p(Am_p)$ for a single particle with a mass equal to the reduced mass, $Am_p$.

We now have two key length-scales in the problem: $\lambda_p(Am_p)$ and $r_c(Am_p)$. The two scales depend on different physical constants, and (in particular) on different powers of the particle speed. As regards numerical values of these two lengths, there is no *a priori* reason why, in any particular environment, they might not differ from each other by orders of magnitude: the ratio $r_c/\lambda_p$ in general might be much greater than unity or much less than unity.

But let us consider a particular location where physical parameters have the values necessary to make $\lambda_p$ comparable to $r_c$. What happens then? Each particle "spreads out" and, in effect, the particles "reach across" the Coulomb gap, bridging the gap and "touching each other", i.e., they in effect come so close together that the distance between them is essentially zero. In particular, the two particles effectively approach each other within a distance of $r_N$, the range of the nuclear force. This is the essential physical process that, in the quantum world, sets the stage for nuclear reactions to occur.

The conclusion is that quantum effects allow the "Coulomb gap" to be "bridged" if $\lambda_p$ becomes large enough to be comparable to $r_c$. Since $\lambda_p$ and $r_c$ both depend on the particle speed $V$ (although to different powers), the "bridging" condition reduces to a condition on $V$. For collisions between two protons, the critical speed $V_c$ is the speed for which the Coulomb gap $2e^2/Am_pV^2$ is equal to the de Broglie wavelength $h/Am_pV$. This leads to the following expression for the critical thermal mean speed required for thermonuclear reactions to become possible:

$$V_c = \frac{2e^2}{h} \tag{11.6}$$

It is noteworthy that the critical speed which allows for "bridging the Coulomb gap" between two nuclei is determined by two of the fundamental constants of nature.

Even more interesting is the numerical value of the critical speed. Inserting constants into Equation 11.6, we find $V_c \approx 696$ km sec$^{-1}$. This is a significant speed in the context of the inner regions of the Sun.

### 11.4.4 CENTER OF THE SUN: THERMAL PROTONS BRIDGE THE COULOMB GAP

We note that the critical speed $V_c$ is not too different from $V_{th}$, the r.m.s. speed of *protons* at the center of the Sun ($\approx 621$ km sec$^{-1}$). Specifically, the ratio $r_c/\lambda_p = V_c/V_{th}$ in the core of the Sun has a numerical value of about 1.1, i.e., within 10% of unity.

In any gaseous object in hydrostatic equilibrium, gravitational effects ensure that the central temperature is such that the r.m.s. speed of the dominant constituent in the core is comparable to the escape speed from the surface of the object. In order for the object to further qualify for the special title of "star", this r.m.s. speed in the core must be large enough *to allow the Coulomb gap to be bridged by quantum effects* (Mullan 2006). Once this condition is satisfied, at least within roughly 10%, nuclear reactions *between thermal protons* can occur in the core. The Sun satisfies this condition. Therefore, the Sun can have access to proton nuclear reactions and the energy that emerges from such reactions. *It is this that makes the Sun a star.*

In a thermal population, the particle speeds are distributed over a finite range of values. Thus, not all protons in the core of the Sun have the same speed. However, the possibility that thermonuclear reactions will set in is quite sensitive to the proton speed. On the one hand, if the proton speed is a factor of (say) two *less* than $V_c$, then the Coulomb gap $r_c \sim 1/V^2$ opens up to a value that is four times *wider* than estimated earlier. At the same time, the wavelength $\lambda_p \sim 1/V$ increases by a factor of only two. Thus, the Coulomb gap is now *too wide to be bridged by the de Broglie wave*. On the other hand, if the proton speed is two times *larger* than $V_c$, then the wavelength $\lambda_p$ decreases by a factor of two, but the Coulomb gap is now four times *smaller*. Therefore, the gap can still be

# Energy Generation in the Sun

bridged. This indicates that once the temperature reaches a value that is high enough to ensure that the r.m.s. speed is of order $V_c$ (within 10% or so), nuclear reactions will occur. But if the temperature is too small to allow the r.m.s. speed to have a value that is large enough to equal $V_c$, then nuclear reactions will *not* occur.

## 11.4.5 OTHER STARS: BRIDGING THE COULOMB GAP

In a global sense, the Sun's mass $M$ and radius $R$ have values which have the effect that the crushing effects of gravity [as measured by $V_{esc} \sim \sqrt{(2GM/R)}$] provide enough "thermo" at the center of the Sun to create a certain temperature. At that temperature, thermal protons have mean speeds $V_{th}$ of order $V_{esc}$. When conditions are such that $V_{th}$ is comparable to $V_c$, then quantum mechanics bridges the Coulomb gap between two protons, and *pp*-nuclear reactions can set in.

Since $V_c$ is determined by physical constants only, *any star* which has the same $M/R$ ratio as the Sun will satisfy $V_{th} \approx V_c$, and will therefore also have *pp*-reactions in its core. Now, astronomers discovered in the 1920s that if the stars we see in the night sky are plotted in a diagram of luminosity versus effective temperature, 90% of the stars lie close to a band known as the "main sequence". After decades of study, astronomers also determined masses $M$ and radii $R$ for many of the stars. A striking result emerged from these data: although the masses and radii vary by factors of 100–1000 along the main sequence, the *ratio $M/R$ is almost constant* from one end of the main sequence to the other. This means that the main sequence is occupied by objects (stars) in which the mean thermal velocity in the core $V_{th} \approx V_{esc} \sim \sqrt{M/R}$ remains almost unchanged and equal to $V_{th}$ in the Sun. But the latter is, as we have seen, close to $V_c$: therefore, along the main sequence, all stars have $r_c \approx \lambda_p$. In such objects, *pp*-reactions can occur in the core. Therefore, *the main sequence is the locus of stars that have just the right conditions to allow hydrogen nuclei to undergo fusion in their core.*

Strictly speaking, the value of $M/R$ does not remain exactly constant all the way along the main sequence. On the one hand, at the high mass end, $M/R$ exceeds the solar value by a factor of a few, and $r_c/\lambda_p$ falls to values that are smaller than unity, thereby enhancing the probability of nuclear reactions. Such stars emit radiation with a power that exceeds the solar value by factors of several orders of magnitude. On the other hand, in low-mass stars, $M/R$ becomes smaller than the solar value. In fact, at a certain low mass (about $0.1 M_\odot$), $M/R$ becomes so small that $V_{th}$ falls to a value that is significantly smaller than the value of $V_c$: in such a case, the *pp*-I chain has an increasingly small probability of being able to occur. Without access to the energy-generating power of *pp* fusion reactions, such an object is no longer called a "star": instead, it is referred to as a "brown dwarf".

## 11.4.6 INSIDE THE NUCLEAR RADIUS

Once two protons approach each other closer than $r_N$, nuclear reactions become possible in principle. That is, the strong force between nucleons can now operate, and the weak force also can operate (if necessary). As a result of these forces, nuclear reactions occur on a certain time-scale. For example, in the Sun, once a deuterium nucleus is formed by reaction (a) in the *pp*-cycle (see Section 11.1), an ambient proton will interact with the deuteron *via* reaction (b): the latter reaction involves the strong force, and it occurs within time-scales of a few seconds. But reaction (a) requires the weak force to operate: as a result, this reaction is slower (see Section 11.5.2.).

## 11.5 RATES OF THERMONUCLEAR REACTIONS: TWO CONTRIBUTING FACTORS

The overall rate of any particular thermonuclear reaction in thermal plasma depends on two factors. One has to do with bridging the Coulomb gap: this factor is sensitive to the temperature. The second has to do with the operation of forces within the nuclear radius: this factor is independent of temperature.

### 11.5.1 Bridging the Coulomb Gap: "Quantum Tunneling"

We have described the process of bridging the Coulomb gap in terms of the comparative equality of the two lengths $r_c$ and $\lambda_p$. More formally, quantum mechanics treats the process in terms of "tunneling" through a potential barrier.

To quantify the tunneling, we first note that in quantum mechanics, the dynamics of particles (based on the ideas of de Broglie) can be described by a wave equation obtained by Erwin Schrödinger (1926) (based on the wave-particle ideas of de Broglie). According to this equation, a particle that is traveling in free space has a propagating wave-like character $\psi$ which is described, in 1-D motion, by a sinusoidal relation in space and time, i.e., an exponential with an imaginary argument:

$$\psi(x,t) \sim \exp[2\pi i \left(\frac{x}{\lambda_p} - ft\right)] \quad (11.7)$$

In Equation 11.7, the spatial (de Broglie) wavelength is $\lambda_p$, and $f = E/h$ is the frequency associated with a particle with energy $E$. When such a wave encounters a vertical wall (or "mountain") that is too high for a particle of energy $E$ to surmount, the sinusoidal solution of the Schrödinger equation is replaced by a damped (non-propagating) exponential:

$$\psi(x) \sim \exp[-2\pi \left(\frac{x}{\lambda_p}\right)] \quad (11.8)$$

In the Sun, *the very heart of energy generation depends on applying Equation 11.8 to the "mountain" that is caused by the Coulomb gap*, i.e., to the (huge) obstacle that prevents two thermal protons from approaching each other any closer than $r_c$.

According to quantum mechanics, the probability $P(V)$ that a particle with speed $V$ (and associated de Broglie wavelength $\lambda_p$) can penetrate a 1-D barrier with spatial width $x = r_c$ is proportional to

$$P(V) \approx |\psi(r_c)|^2 \approx \exp[-4\pi \left(\frac{r_c}{\lambda_p}\right)] \quad (11.9)$$

The fact that $|\psi(r_c)|^2$ is nonzero as long as $r_c/\lambda_p$ is finite, means that, in the quantum world, there is a *finite chance* that a particle can penetrate through a wall (or a Coulomb "mountain") that would be completely insurmountable in the classical world. This process is known as "quantum tunneling".

When the tunneling calculation is done rigorously, in 3-D and in the presence of a "mountain" that has the particular shape of the Coulomb barrier, it is found that the numerical coefficient $4\pi$ (= 12.6) in the exponent in Equation 11.9 must be replaced by the somewhat larger number $2\pi^2$ (= 19.7). That is, $P(V) \approx \exp(-2\pi^2 r_c/\lambda_p)$. Inserting the expressions given earlier for $r_c$ and $\lambda_p$, we find that the probability $P(V)$ for Coulomb barrier penetration is given by

$$P_G(V) = \exp\left[\frac{-4\pi^2 Z_1 Z_2 e^2}{Vh}\right] \quad (11.10)$$

This expression for the probability is known as the Gamow factor, in honor of the physicist who first performed the tunneling integral (Gamow 1928). In recognition of Gamow's role, we use subscript $G$ in Equation 11.10.

# Energy Generation in the Sun

What is the numerical value of the tunneling probability in the core of the Sun? We have seen (Section 11.4.4) that in the core, $r_c/\lambda_p \approx 1.1$. In that case, Equation 11.10 tells us that $P_G(V) \approx \exp(-2.2\pi^2) \approx 4 \times 10^{-10}$.

It is important to note that $P_G(V)$ is quite sensitive to the particle speed $V$. For example, suppose that, instead of considering particles moving with speed $V_{th}$, we were to consider the collisions of two particles, each of which moves with speed $2V_{th}$. In such a case, the tunneling probability $P_G(2V_{th}, \text{Sun})$ would be $\approx \exp(-1.1\pi^2) = 2 \times 10^{-5}$. Thus, by doubling the speed, we have increased the $pp$-tunneling probability by a large amount ($5 \times 10^4$). At first sight, this sensitivity to speed suggests that we might have made an error of many orders of magnitude by evaluating the tunneling probability at the particular speed $V_{th}$. But upon further consideration, we can see that the error is much less serious.

To see why this is so, we note that in a thermal velocity distribution, where $f(V) \sim V^2 \exp(-V^2/V_{th}^2)$, there are *fewer* particles moving at faster speeds. For example, for every particle that moves with speed $V_{th}$, there are only $4e^{-4} \approx 0.07$ particles in a Maxwellian distribution moving with $2V_{th}$. For this reason alone, the number of possible interactions that might occur every second between particles, each of which moves with speed $2V_{th}$, is smaller by $0.07^2 \approx 1/200$ than the collision rate between two particles moving with speed $V_{th}$. Furthermore, the cross-section for Coulomb collisions is smaller for faster particles: $\sigma_c \sim 1/V^4$ (see Section 11.3). This further reduces the collision rate by a factor of 16 when we compare particles with speed $2V_{th}$ to particles with speed $V_{th}$. Combining the Coulomb and Maxwellian factors, we see that the increase in $pp$-tunneling probability by $5 \times 10^4$ is offset by $200 \times 16 \approx 3 \times 10^3$. Therefore, as far as the actual rate of tunneling, particles with speed $2V_{th}$ are indeed more effective than particles with speed $V_{th}$, but not by many orders of magnitude. The increase in effectiveness is a factor of $\sim 17$.

If we were to repeat this exercise for particles moving even faster, say $4V_{th}$, we would find that the increase in tunneling probability (by a factor of $\approx 10^7$) is more than offset by the combined Maxwellian and Coulomb factors. The relative number of Maxwellian particles is $16e^{-16} \approx 2 \times 10^{-6}$, and Coulomb collisions occur 256 times less frequently. Thus, despite the increased Gamow factor, particles with speed $4V_{th}$ are about 10 times *less* effective than particles with speed $V_{th}$. Overall, the peak in $pp$-tunneling probability in a thermal distribution of protons occurs for particles with speeds of $2$–$3V_{th}$, and closer to $2V_{th}$ than to $3V_{th}$.

This suggests that our estimate of tunneling probability obtained earlier for particles moving with speed $V_{th} (\approx 4 \times 10^{-10})$ is a lower limit on the actual probability in the Sun. The lower limit should be increased by a factor of perhaps 20 in order to obtain a more realistic $pp$-tunneling probability for a Maxwellian distribution in the Sun: $P_G(\text{Sun}) \approx 8 \times 10^{-9}$.

We can now see an important conclusion of this discussion. Even in the "favorable" conditions that exist in the core of the Sun, only one collision in (roughly) 125 million results in one proton tunneling close enough to another to "feel" the nuclear force. On the other hand, as we have seen (Section 11.3), each proton in the core undergoes some $2 \times 10^{15}$ collisions *every second*. Therefore, each proton in the Sun's core experiences roughly $10^7$ tunneling events every second. When combined with the relevant post-tunneling processes (see Section 11.5.2), this rate of tunneling suffices to provide the Sun with its mighty output power.

We shall return later to examine how the functional form of $P_G(V)$ has the effect that the rates of thermonuclear reactions increase rapidly with increasing temperature. But for now, we turn to what happens inside the nucleus once the tunneling has occurred.

## 11.5.2 Post-Tunneling Processes

Once tunneling has occurred, the two particles are close enough together that they can be regarded as being together inside a nucleus. The processes that then occur in such conditions depend on which forces come into play.

We have already mentioned (Section 11.4.6) that the strong force is at work in reaction (b) of the *pp*-cycle. The strong force is also at work in reaction (c) of the *pp*-cycle. However, even though reaction (b) occurs on a time-scale of a few seconds in the Sun, reaction (c) requires on average several million years to occur. The principal reason that reaction (c) is so much slower than reaction (b) in the Sun has to do with the tunneling factor: referring to Equation 11.10, we see that the product $Z_1 Z_2$ is four times larger for reaction (c) than for (b). (We will return to this in Section 11.7.)

But reaction (a) in the *pp*-cycle is different. The strong force is *not* the predominant factor controlling this reaction. When two protons interact via the strong force, they might be expected at first to attempt to form a nucleus consisting of two protons and nothing else. Such a nucleus could be referred to as a "di-proton". However, calculations of nuclear structure indicate that such a nucleus is not stable: the combination of kinetic energy, Coulomb repulsion, and exchange forces overwhelms the attractive nuclear energy. As a result, the di-proton is unbound in the "real world". The strong force is simply not strong enough to bind the two protons in reaction (a) together in a stable nucleus.

So how does reaction (a) proceed? We note that the product of the reaction (i.e., the deuteron), *is* a stable (bound) nucleus consisting of one proton and one neutron. To form such a nucleus, one of the protons that enters into reaction (a) must become a neutron. During the course of a collision of two protons, during the (very) brief interval of "collision time" when the two protons are within a distance of $r_N$ of each other, one of the protons must become transformed into a neutron.

How long does the "collision time" last? The duration of a collision is $t_c \approx r_N/V$ where $V \approx 6 \times 10^7$ cm sec$^{-1}$ is the mean thermal speed of protons (and therefore neutrons) in the core of the Sun. Setting $r_N \approx 10^{-13}$ cm, we find $t_c \approx 2 \times 10^{-21}$ sec.

What is the chance that a proton-to-neutron transformation will happen during an interval of duration $t_c$? If we were considering a free proton in the Sun, the answer would be straightforward: the chance would be zero. It is impossible for a free proton in the Sun to decay into a neutron because the proton would have to *gain* a mass of 0.0014 a.m.u. (see Table 11.1). This is equivalent to an energy gain of 1.3 MeV, about 1000 times larger than the thermal energies in the Sun. However, inside a nucleus, in the presence of the strong force, with an attractive energy of 20–30 MeV, the transformation of a proton into a neutron becomes possible: in such an environment, in a potential well some 20–30 MeV deep, the possibility of "picking up" 1.3 MeV is no longer out of the question. As a result, the "decay" of a proton into a neutron *inside the nucleus* is no longer excluded: the weak force *can* do its work.

This requires that the weak force must work its transforming effects precisely during the "collision time". Now, a first estimate of the strength of the weak force is provided by the empirical result that free neutrons decay with a half-life $t_{1/2}$ of about 650 sec.

What is the probability $P_d(p)$ that a proton will decay into a neutron during the "collision time"? The correct answer to this question requires a theory of beta decay: the Gamow-Teller (G-T) version of Fermi's theory was used by Bethe and Critchfield (1938) in their calculation of the rate of the *pp*-cycle in the Sun. In the G-T theory, the conversion of a proton into a neutron is more effective than in the Fermi theory because the G-T theory allows the spin vector of the proton to flip, whereas the Fermi theory does not allow such a flip to occur.

Without going into the details of beta-decay theory, we can estimate an upper limit to the probability by considering a hypothetical analog to proton-proton collisions. Suppose two free *neutrons* were available in the thermal population in the Sun's core, and suppose they were to undergo a collision in which the distance of closest approach happened to be $r_N$. A free neutron always has the option of decaying into a proton. So, what is the probability $P_c(n)$ that one of the neutrons would decay into a proton *during the collision time $t_c$*? The answer is: $P_c(n)$ can be estimated roughly by the ratio of $t_c$ to the neutron half-life, $t_{1/2}$. This leads to $t_c/t_{1/2}$ of order $3 \times 10^{-24}$. Thus, the probability that a (free) neutron in the Sun's core would decay into a proton during the collision with another neutron is $P_c(n) \approx 3 \times 10^{-24}$. Even with the advantage of free neutron decay, this is still a very small probability.

# Energy Generation in the Sun

Returning now to the case of proton-proton collisions, we recall that the proton and the neutron are both nucleons with similar properties. (In the technical language of nuclear physics, protons and neutrons are members of the same "isospin doublet".) As a result, they are expected to behave to a certain extent in similar ways when they are within a distance of $r_N$ of each other. However, there is a difference in the energy $\Delta E$ that is released in the reaction: whereas reaction (a) earlier releases an energy of 1.44 MeV, the excess mass energy of 1.3 MeV of the neutron relative to the proton would have the effect that the energy released in the reaction $n + n \to D$ would equal 1.44 + 1.3 = 2.74 MeV. Now according to a general rule in particle decays (known as the "Sargent rule"), the rate of beta decay scales as $(\Delta E)^\alpha$, where $\alpha = 5$ in the limit that the decay products are relativistic. As a result, the reaction $p + p \to D$ is predicted to be less probable than $n + n \to D$ by a factor of order $(2.74/1.44)^5 \approx 25$ in the relativistic limit. Even in the nonrelativistic limit, the probability $P_c(p)$ that a proton will decay into a neutron during the collision is expected to be smaller than $P_c(n)$. Defining the ratio of $P_c(n)/P_c(p)$ as $\xi(>1)$, we write $P_c(p) \approx (3/\xi) \times 10^{-24}$.

### 11.5.3 Probability of pp-I Cycle in the Solar Core: Reactions (a) and (b)

Combining the probability factors for quantum tunneling and for the post-tunneling process of proton transformation, we see that in the center of the Sun, the overall probability $P(pp)$ of a $pp$-nuclear reaction (i.e., reaction (a) in the $pp$-I cycle) in a collision in the solar core is given by the product of the Gamow factor $P_G(\text{Sun})$ (= $8 \times 10^{-9}$: see Section 11.5.1.) and $P_c(p)$. Using the estimates given earlier, we find $P(pp) \approx (24/\xi) \times 10^{-33}$.

We recall that the observed properties of the Sun indicate that a $pp$-cycle occurs on average only once in every $N_c(react) \approx 10^{33}$ collisions in the Sun's nuclear-burning core (Equation 11.2). That is, the empirical probability of a nuclear reaction is of order $10^{-33}$ per collision. Compared with our estimates of $P(pp)$, we see that we can replicate the empirical probability of nuclear reaction in the Sun as long as $P_c(n)$ does not exceed $P_c(p)$ by a factor of more than $\approx 25$. This is almost exactly the factor that is available based on the Sargent rule.

Thus, of the 33 orders of magnitude that occur in the empirical reaction probability $1/N_c$ (react), the process of tunneling through the Coulomb barrier provides about eight orders of magnitude, while the weak interaction that occurs in the post-tunneling process contributes the remaining 25 orders of magnitude. The weak interaction truly dominates (by $\approx 17$ orders of magnitude) in regulating the slowness of the thermonuclear processes in the Sun.

It is the low value of the probability associated with the weak interaction that causes reaction (a) of the $pp$-cycle to be so much slower than reactions (b) or (c). We recall (Section 11.3) that, on average, a proton participates in reaction (a) once in (about) 10 Gy. In reaction (b), since the Coulomb barrier is similar to that in reaction (a), the tunneling probability is comparable to that for reaction (a). However, the post-tunneling process in reaction (b) involves the interaction between two nuclei so as to form a third stable nucleus. The interaction in reaction (b) therefore operates by way of the *strong* force, in sharp contrast to reaction (a), where the *weak* force is at work. In the nature of things, we expect that the strong force operates on much shorter time-scales than the weak force. In support of this expectation, we note that measurements of the cross-section for reaction (b) indicate that the post-tunneling process in (b) operates almost 18 orders of magnitude more rapidly than in reaction (a). As a result, instead of a time-scale of order $10^{18}$ sec between occurrences of reaction (a), reaction (b) occurs on time-scales of order seconds. We shall return to discuss the time-scale for reaction (c) in Section 11.7, after we quantify how the tunneling probability depends on charge and mass.

## 11.6 TEMPERATURE DEPENDENCE OF THERMONUCLEAR REACTION RATES

A significant characteristic of the Gamow tunneling probability $P_G(v)$ (Equation 11.10) is the occurrence of the particle speed in the *denominator* of the argument of the exponential term. This has the effect that $P_G(V)$ falls off exponentially rapidly to zero as the speed decreases below a value that is related to $V_c$ (see Equation 11.6). In the opposite limit, for speeds $V > V_c$, $P_G(V)$ increases at first, but in the limit $V \gg V_c$, the value of $P_G(V)$ eventually saturates at a value of unity.

In contrast to this behavior of the tunneling factor as a function of velocity $V$, there is a very different behavior for the velocity distribution $f(V)$ as a function of $V$. A significant property of a thermal velocity distribution $f(V) \sim V^2 \exp(-V^2 / V_{th}^2)$ is that as the speed increases from $V = 0$, the value of $f(V)$ at first increases because of the $V^2$ term in front of the exponential. However, as $V$ increases, the exponential term eventually dominates over the $V^2$ term. As a result, the number of available particles falls off exponentially rapidly at high speed.

In order to quantify the overall rate of thermonuclear reactions, we must perform an integral of the product $\Pi(V) = P_G(V)f(V)$ over all velocities. Because of the contrasting behavior of each of the terms as a function of velocity, the integral receives essentially zero contribution from particles with low speeds or from particles with high speeds. The integrand peaks at an intermediate velocity $V_o$, corresponding to energy $E_o$. The particles that contribute most to the rate of thermonuclear reactions are those that lie within a range of velocities $\Delta V$ in the neighborhood of $V_o$. As a result, when we integrate over all velocities, the thermonuclear reaction rate $r_{tn}$ is proportional to $f(V_o)$ (the number of particles in the thermal distribution at $V = V_o$) times $\Delta V$.

Converting from velocity to energy, we note that the exponential term in $f(V)$ converts to $f(E) \sim \exp(-E/kT)$, while $P_G(V)$ converts to $P_G(E) = \exp(-\beta/\sqrt{E})$. In the expression for $P_G(E)$, $\beta = CZ_1 Z_2 \sqrt{A}$ and $C' = 2\pi^2 e^2 \sqrt{(2m_p)}/h = 1.23 \times 10^{-3}$ c.g.s. units. Since the mean thermal energy $kT$ ($\approx 1.9 \times 10^{-9}$ ergs) in the core of the Sun is of order 1 keV (= $1.6 \times 10^{-9}$ ergs), it is convenient (Clayton 1968) to express energy in units of keV: $E_k = E/(1 \text{ keV})$. In these units, $C'$ is replaced by $C'_k = 31 \text{keV}^{0.5}$.

In terms of energy, the product $\Pi(E) = P_G(E)f(E)$ has a maximum value at an energy $E_o$ where the sum of the two terms $\beta/\sqrt{E} + E/kT$ in the exponent is a minimum. Taking the derivative with respect to energy, we find that this minimum occurs when

$$\frac{1}{kT} - \frac{\beta}{2E_0^{3/2}} = 0 \tag{11.11}$$

This leads to $E_o = (\beta kT/2)^{2/3}$: this is the energy at which the particles in the thermal distribution participate with maximum effectiveness in quantum tunneling and, therefore, also in thermonuclear reactions. For example, in the case of reaction (a) in the $pp$-cycle in the core of the Sun, we have $Z_1 = Z_2 = 1$ and $A = 0.5$. These lead to $\beta = 22$ keV$^{0.5}$. Since $kT \approx 1.2$ keV in the core of the Sun, we find $E_o \approx 5.6$ keV, i.e., $\approx 4.7$ times larger than the mean thermal energy. The velocity of particles with energy $E_o$ is therefore $\approx \sqrt{4.7}$ times the mean thermal speed, i.e., $\approx 2.2V_{th}$. This confirms our discussion in Section 11.5.1 (in the fourth paragraph from the end of the section).

Using the estimate of $E_o$, i.e., and rearranging Equation 11.11 as follows $\beta/\sqrt{E_o} = 2E_o/kT$, we find that the rate of thermonuclear reactions $f_r$ is proportional to $\Pi(E_o)$, i.e., $f_r \sim \exp(-3E_o/kT)$. Because of the exponential factor, the rate $f_r$ is quite sensitive to temperature. To quantify this, let us insert the expression derived earlier for $E_o$ and take the natural logarithm. We find $\ln(f_r) = -3(\beta/2)^{2/3}/(kT)^{1/3}$. It is often convenient to write the reaction rate in terms of a power law of the temperature, $f_r \sim T^\delta$. This leads to

$$\delta \equiv \frac{d \ln f_r}{d \ln T} = +\left(\frac{\beta^2}{4k}\right)^{1/3} \frac{1}{T^{1/3}} \tag{11.12}$$

Inserting c.g.s. values for $\beta$ (for the $pp$-reaction) and $k$, and expressing the temperature in units of $10^6$ K (i.e., $T_6 \equiv T/10^6$ K) we find

$$\delta = +\frac{11.1}{T_6^{1/3}} \tag{11.13}$$

# Energy Generation in the Sun

In the core of the Sun, where $T_6 \approx 15$–$16$, Equation 11.13 indicates that $\delta \approx 4$–$5$. Thus, the rate of $pp$-reaction increases rather rapidly as temperature increases.

## 11.7 RATE OF REACTION (C) IN THE PP-I CYCLE

Reaction (c) (Section 11.1) involves a larger Coulomb barrier than do reactions (a) or (b). It is interesting to see quantitatively how sensitive the tunneling barrier is to the reacting nuclei.

In calculating the quantity $\beta$ for reaction (c), using $Z_1 = Z_2 = 2$ and $A = 1.5$, we find $\beta = 152$ keV$^{0.5}$. Setting $kT = 1.2$ keV, this leads to $E_o = 20.3$ keV, which is much larger than the 5.6 keV value for reaction (a). As a result, the reaction rate $f_r$, which is proportional to $\exp(-3E_o/kT)$ (see paragraph leading up to Equation 11.12), is reduced in reaction (c) compared to reaction (a) by $\exp(-3[20.3-5.6]/1.2) \approx 10^{-16}$ in the core of the Sun. (Note, $kT \approx 1.2$ keV in the solar core.) However, the post-tunneling process in reaction (c) depends on the strong force: the rate of this process therefore greatly exceeds that for (weak-force) reaction (a). Empirically, the excess in rates is found to be of order $10^{25}$ (Clayton 1968, p. 380).

Combining the factors $10^{-16}$ and $10^{25}$, we see that reaction (c) has a rate that is $10^9$ times more frequent than each proton-proton reaction (a). However, there is one more factor that must be included in the argument, namely, the abundance of He$^3$ nuclei in the Sun. The equilibrium abundance of He$^3$ nuclei is much smaller than the proton abundance: specifically, in equilibrium, for every proton, there are only $10^{-5}$ He$^3$ nuclei in the Sun. This has the effect that the mean free time interval that a particular He$^3$ nucleus must wait between collisions with another He$^3$ nucleus is $10^5$ times longer than the mean free time for collisions with a proton.

The combination of the enhancement factor of $10^9$ (due to tunneling and post-tunneling processes) and the decrease of $10^5$ (due to abundances) has the net effect that an individual He$^3$ nucleus has a collision leading to reaction (c) in a time-scale which is some $10^4$ times shorter than the time-scale for a proton to undergo reaction (a). As a result, whereas a time-scale of order $10^{10}$ years is characteristic of reaction (a), the time-scale for reaction (c) is of order $10^6$ years.

For reactions other than reaction (a), the numerical coefficient 11.1 in the expression for $\delta$ must be replaced by $11.1(Z_1Z_2)^{2/3}(A/0.5)^{1/3}$, where the 0.5 refers to the reduced mass that enters into reaction (a). Following the argument used in deriving Equation 11.13, we find that reaction (c) in the core of the Sun has a rate that increases as $T^{16}$. The great sensitivity to temperature arises from the sensitivity of tunneling to the strength of the Coulomb barrier.

However, because reaction (a) is so much slower than either (b) or (c), the overall time-scale for the $pp$-I chain is determined essentially entirely by reaction (a). Therefore, the temperature sensitivity of nuclear reactions in the Sun is controlled by reaction (a), i.e., rate of the $pp$-I chain is proportional to $T^{4-5}$. If something were (somehow) to cause the core of the Sun to increase its temperature by a factor of (say) two (without changing any other physical parameters), then the $pp$-I chain would occur at a rate that exceeds the rate in the present Sun by a factor of 16–32.

## EXERCISES

11.1 From Exercise 1.5 in Chapter 1, you already know the values of $V_{esc}$ for main sequence stars with masses of 0.1, 0.3, 1, 3, and 10 $M_\odot$. Assuming thermal speeds in the core $V_{th} \approx V_{esc}$, evaluate the ratio $r_c/\lambda_p = V_c/V_{th}$ in the core of each star (where $V_c$ is given by Equation 11.6). Show that on the main sequence, the ratio $r_c/\lambda_p$ does not vary by more than a factor of roughly two.

11.2 Using the tunneling probability formula $P(V) \approx \exp(-2\pi^2 r_c/\lambda_p)$, calculate $P(V)$ for the five stars in Exercise 11.1. Show that $P(V)$ for the $10M_\odot$ star is two to three orders of magnitude larger than for the $1M_\odot$ star, while $P(V)$ for the $1M_\odot$ star is three to four orders of magnitude larger than for the $0.1M_\odot$ star. Show how these results help us to explain the empirical results that the luminosity of a $10M_\odot$ star exceeds $L_\odot$ by about 1000, while $L_\odot$ exceeds the luminosity of a $0.1M_\odot$ star by about 1000.

## REFERENCES

Audi, G., & Wapstra, A. H., 1993. "The 1993 atomic mass evaluation: (I) Atomic mass table", *Nuclear Phys.* A. 565, 22.
Bahcall, J. N., Serenelli, A. N., & Basu, S., 2006. "10,000 standard solar models: a Monte Carlo simulation", *Astrophys. J. Suppl.* 165, 400.
Bethe, H. A., 1939. "Energy production in stars", *Phys. Rev.* 55, 434.
Bethe, H. A., & Critchfield, C. L., 1938. "The formation of deuterons by proton combination", *Phys. Rev.* 54, 248.
Bonanno, A., & Frohlich, H.-E., 2015. "A Bayesian estimation of the helioseismic solar age", *Astron. & Astrophys.* 580, A130.
The Borexino Collaboration, 2020. "Experimental evidence of neutrinos produced in the CNO fusion cycle in the Sun", *Nature*, 587, 577.
Clayton, D. D., 1968. "Thermonuclear reaction rates", *Principles of Stellar Evolution and Nucleosynthesis*, McGraw-Hill, New York, pp. 283–361.
Connelly, J. N., Bizzarro, M., Krot, A. N., et al., 2012. "The absolute chronology and thermal processing of solids in the solar protoplanetary disk", *Sci.* 338, 651.
De Broglie, L., 1924. "Recherches sur la théorie des quanta". PhD Thesis, Sorbonne University, Paris.
Fermi, E., 1934. "Versuch einer theorie der β-strahlen", *Zeits. f. Physik* 88, 161.
Gamow, G., 1928. "Zur quantentheorie des atomkernes", *Zeitschrift fur Physik* 51, 204.
Gamow, G., & Teller, E., 1936. "Selection rules for β-disintegration", *Phys. Rev.* 49, 895.
Hofstadter, R., 1956. "Electron scattering and nuclear structure", *Rev. Modern Phys.* 28, 214.
Hofstadter, R., Fechter, H. R., & McIntyre, J. A., 1953. "High energy electron scattering and nuclear structure determinations", *Phys. Rev.* 92, 978.
Mullan, D. J., 2006. "Why is the Sun so large?" *Amer. J. Phys.* 74, 10.
Schiff, L. I., 1955. *Quantum Mechanics*, McGraw Hill, New York, pp. 309–310.
Schrödinger, E., 1926. "An undulatory theory of the mechanics of atoms and molecules", *Phys. Rev.* 28, 1049.
Spitzer, L., 1962. *Physics of Fully Ionized Gases*, Interscience, New York, pp. 127–128.
Squires, G., 1998. "Francis Aston and the mass spectrograph", *J. Chem. Soc., Dalton Trans.* 3893.
Thomson, G. P., 1946. "Dr F. W. Aston, F. R. S.", *Nature* 157, 290.

# 12 Neutrinos from the Sun

As a result of the calculations in Chapters 5, 7, and 9, we have obtained a model for the interior of the Sun. The question now is: is there any way to check our model to see if the calculations are consistent with reality?

In the 1960s, there was only one answer to this question: we need to detect a certain kind of energetic particle (neutrino) that emerges from nuclear reactions in the core of the Sun. The goal of such experiments would be to check that the numbers of neutrinos that reach the Earth, as well as their energies, are consistent with the properties we calculated for nuclear reactions in the solar core.

The existence of neutrinos was first postulated by Wolfgang Pauli in 1930 in order to preserve the laws of conservation of momentum and energy in certain radioactive decays. Pauli's approach was a bold one: he had to postulate that in these decays, a hitherto unseen particle with zero electric charge must emerge with a finite energy and momentum, but with a mass that must be so small as to be almost zero, certainly much smaller than the mass of an electron. The term "neutrino" was subsequently coined by Fermi for Pauli's unseen particle. The absence of electric charge means that the neutrino does *not* interact with its surroundings by means of electromagnetic processes. The fact that the neutrino is associated with radioactive decay (a process that is driven by the weak interaction) means that the neutrino interacts with other particles via the weak force.

We have already seen (Chapter 11, Section 11.5.3) how the weak force in the Sun makes for very long time-scales in certain reactions, whereas the strong force makes reactions occur much more rapidly. There is also a significant difference in strength between the weak force and the electromagnetic force. Because of this difference, photons (signatures of the electromagnetic force) and neutrinos (signatures of the weak force) behave very differently as they propagate inside the Sun. In Chapter 9, Section 9.3, we determined that photons originating in the core of the Sun take *millions of years* to escape from the Sun. In contrast to the photons, we shall see that neutrinos from nuclear reactions in the solar core can reach the surface of the Sun in a matter of no more than *a few seconds*. This is a striking illustration of how much less effective the weak force is in comparison with the force we are more familiar with in our macroscopic world, i.e., the electromagnetic force.

## 12.1 GENERATION AND PROPAGATION OF SOLAR NEUTRINOS

Every time the *pp*-I chain occurs, a neutrino emerges from the first step of the chain (Section 11.1, reaction (a)). A complete *pp*-I chain requires this step to occur twice. As a result, since the *pp*-chain occurs some $10^{38}$ times per second in the Sun (see Section 11.2), we expect that there are roughly $2 \times 10^{38}$ neutrinos generated per second in the Sun's core.

Are these neutrinos likely to be absorbed as they pass through the Sun? Or can they escape more or less freely? To answer this, we return to the same sort of calculation we did in Section 9.3 when we were considering how *photons* propagate inside the Sun. The relevant physical quantity that we now need to evaluate is the mean free path $\lambda_m$ that a *neutrino* can travel between collisions with the atoms/ions/nuclei/electrons it encounters as it propagates inside the Sun.

In general, when a projectile moves through a medium containing $n$ "target objects" per cc, each with a cross-section of $\sigma$, the mean free path is given by $\lambda_m = 1/n\sigma$. In the case of photons, where the opacity $\kappa$ is conventionally expressed in units of cm$^2$ gm$^{-1}$, the product $n\sigma$ can be replaced by the product $\kappa\rho$ (Section 3.3). As a result, as we have already seen (Section 9.3), the mean free path of a photon $1/\kappa\rho$ is found to be of order $\approx 0.001$ cm in the Sun's core.

Turning now to neutrinos, we revert to the general formula $\lambda_m = 1/n\sigma$. The "target objects" that a neutrino from the core of the Sun encounters on its way to the surface are mainly protons and electrons. In the core of the Sun, where the mass density $\rho$ is $\approx 140$ gm cm$^{-3}$ (see Chapter 9, Table 9.1), the number density of nuclei is roughly $\rho/m_H$ (where $m_H = 1.67 \times 10^{-24}$ gm). Thus, $n \approx 10^{26}$ cm$^{-3}$, mainly protons, but including He and a few "metal" ions. For each proton, there is also roughly one electron present in this (almost completely ionized) gas.

At this point, we encounter the key difference between photons and neutrinos. On the one hand, photons that try to propagate through the ions in the core of the Sun "see" the ions as having, on average, effective areas of order $\sigma(\text{phot}) \approx 10^{-23}$ cm$^2$. The reason that the cross-section has a value larger than the Thomson cross-section (see Equation 3.1) is that the photon interacts via *electromagnetism* with some bound electrons that still remain attached to certain heavy nuclei even at temperatures as large as 15–16 MK. On the other hand, for neutrinos, electromagnetism is not important: neutrinos interact with the nuclei in the Sun by means of the *weak* force. For this, the cross-section is *much* smaller than the Thomson value. In order to determine the neutrino cross-section, Cowan et al. (1956) searched for neutrinos emerging from a fission reactor: when a uranium or plutonium nucleus undergoes fission, some of the by-products (e.g., strontium-90, cesium-137, iodine-131) are unstable to beta decay, and these decays are accompanied by neutrinos with energies of up to a few MeV. Although these beta decays involve elements that are not significant in the Sun, the *energies* of the neutrinos that emerge are comparable to those that occur in the Sun (see Figure 12.1). Using a specially designed detector, Cowan et al. found a mean neutrino reaction rate in the detector between 0.6 and 2.9 events per hour, depending on the power level of the reactor. Running for almost 1400 hours in order to build up a significant sample, Cowan et al. determined that for the neutrinos emerging from the fission reactor, the cross-section for interactions with the nuclei in their detector had a value $\sigma \approx 6 \times 10^{-44}$ cm$^2$. In subsequent more refined experiments, the cross-section was found to be larger by a factor of about two. To a good approximation, we may take $\sigma(\text{neut}) \approx 10^{-43}$ cm$^2$ for the neutrino cross-section with energies of order 1 MeV.

The contrast between the photon cross-section $\sigma(\text{phot})$ and the neutrino cross-section $\sigma(\text{neut})$ is noteworthy: the difference amounts to some *20 orders of magnitude*. "Weak" (compared to electromagnetism) is indeed an appropriate adjective to describe the interaction that MeV neutrinos have with matter.

For neutrinos with energies of a few MeV, the cross-section is insensitive to energy, and so we can, without serious error, apply the cross-section determined from fission reactor neutrinos to the conditions in the core of the Sun. Combining the value of $\sigma(\text{neut})$ with the value of $n \approx 10^{26}$ cm$^{-3}$ in the core of the Sun, we see that even in the densest region of the Sun, the neutrino mean free path $\lambda_m = 1/n\sigma \approx 10^{17}$ cm. In terms of a unit of length that is more familiar to astronomers, this equals one-tenth of a *light-year*: a block of lead (with a density about 0.1 times the density at the center of the Sun) could be as much as one light-year thick and a neutrino would have a good chance of passing right through the entire block! Up to this point in the discussion of solar neutrinos, we have been assuming that the density of the material through which the neutrino passes is equal to the density at the center of the Sun. But of course this will not be appropriate as the neutrinos move outward from the core of the Sun: as they do so, they will pass through gas of increasingly lower density. As a result, the value of $\lambda_m$ becomes even *larger* than the value of $10^{17}$ cm mentioned earlier. But even at the center of the Sun, the value of $\lambda_m$ exceeds the solar radius ($\approx 7 \times 10^{10}$ cm) by more than six orders of magnitude.

As a result, the neutrinos from the *pp*-reaction in the core of the Sun barely "feel" the material of the solar interior at all. Less than one neutrino in a million will undergo a scattering between the core of the Sun and its surface. For the rest, the Sun is essentially "transparent", and from this perspective, the neutrinos simply stream freely out of the Sun. With essentially zero rest mass, a neutrino travels at the speed of light: once a neutrino is generated in the core, it reaches the surface in a time of $R_\odot/c = 2.3$ sec. Some 500 sec later, the neutrino passes the Earth's orbit and continues its journey into deep space: in the interstellar gas, electron/proton densities are at most 1 cm$^{-3}$, and as a

Neutrinos from the Sun

result, the mean free path for a neutrino is at least $10^{43}$ cm. Such a neutrino can in principle traverse the entire visible universe without interacting.

The lack of scattering in solar material does not mean that the neutrinos feel *no* effects whatsoever from passing through the Sun. In fact a certain type of effect does occur, one that causes the neutrino to change into another type of neutrino. We will return to this later, after we describe the experiments that have been built to detect solar neutrinos.

## 12.2 FLUXES OF *PP*-I SOLAR NEUTRINOS AT THE EARTH'S ORBIT

As a result of the *pp*-I chain, we predict that the Sun generates roughly $N_n \approx 2 \times 10^{38}$ neutrinos per second. When the neutrinos pass by the Earth, at a distance of $D = 1$ AU from the Sun, the flux of solar neutrinos should therefore be roughly $F_n = N_n/4\pi D^2 \approx 6 \times 10^{10}$ cm$^{-2}$ sec$^{-1}$. How well does this prediction match up with observations? Analysis of multiple experimental measurements of the actual flux at Earth orbit of neutrinos from the *pp*-chain (Bergstrom et al. 2016) have led to the following value: $5.97^{+0.04}_{-0.03} \times 10^{10}$ cm$^{-2}$ sec$^{-1}$. This range overlaps with our rough estimate of the flux.

The neutrinos emerging from the *pp*-I chain have a range of energies; all of the energies are less than 0.42 MeV. Other channels of the *pp*-chain, as well as contributions from the CNO cycle (see Section 12.3.2), ensure that the Sun generates other neutrinos with a range of energies. The spectrum of all known types of solar neutrinos, calculated from a detailed model of the Sun and evaluated at the mean distance (1 AU) of the Earth's orbit, is shown in Figure 12.1. Neutrinos that emerge from reactions involving only two outgoing particles are emitted at unique energies: these appear as vertical "lines" in the figure. Reactions in which more than two outgoing particles are present (including $pp \rightarrow De^+\nu$) give rise to a "continuum" of energies for the neutrinos, up to a well-defined maximum "cut-off" energy, which is determined by the difference in energy between initial and final state. The "continuum" in which we are most interested in this subsection is the one with the largest amplitude, i.e., the one labeled *pp* in Figure 12.1.

For the "lines" occurring in the solar neutrino spectrum, the ordinate in Figure 12.1 refers to the flux in units of particles cm$^{-2}$ sec$^{-1}$ at Earth. For the "continua", the ordinate in Figure 12.1 refers to a differential energy flux, in units of particles cm$^{-2}$ sec$^{-1}$ MeV$^{-1}$ at Earth.

In terms of overall flux, the neutrinos from the Sun are predominantly those that emerge from reaction (a) of the *pp*-I chain. As Figure 12.1 illustrates, the differential flux of *pp* neutrinos has a maximum amplitude of $F(max) = (2–3) \times 10^{11}$ cm$^{-2}$ sec$^{-1}$ MeV$^{-1}$, while the cut-off energy is $E(cut-off) = 0.42$ MeV. To calculate the total flux of these solar neutrinos at Earth orbit, we need to determine the area under the *pp* curve in Figure 12.1. This area is certainly no larger than the area we would obtain if we were to represent the *pp* curve as a rectangle with height $F(max)$ and width $E(cut-off)$. That is, the total flux of *pp* neutrinos at Earth cannot be any larger than the product of this height and this width, i.e., no larger than $(8–13) \times 10^{10}$ neutrinos cm$^{-2}$ sec$^{-1}$. This upper limit is consistent with our rough estimate of $6 \times 10^{10}$ neutrinos cm$^{-2}$ sec$^{-1}$ given earlier.

## 12.3 NEUTRINOS FROM REACTIONS OTHER THAN *PP*-I

In the Sun, most of the energy generation occurs via the *pp*-I chain of reactions that were discussed in Chapter 11. Now, in order to provide a more complete discussion of the neutrinos that come from the Sun, we need to look at certain less frequent reactions that also occur in the solar core, in particular the so-called *pp*-II and *pp*-III chains. These reactions do not contribute much to the energy output of the Sun, but for historical reasons, they are important: they contribute significantly to the solar neutrino fluxes that were first actually *detected* on Earth in the late 1960s. In fact, for the first 25 years (or so) of solar neutrino experiments, the only neutrinos that could be detected were those from the *pp*-III chain. The reason has to do with the fact that, as far as a neutrino detector is concerned, the incoming neutrinos must have a certain minimum "threshold" energy before the

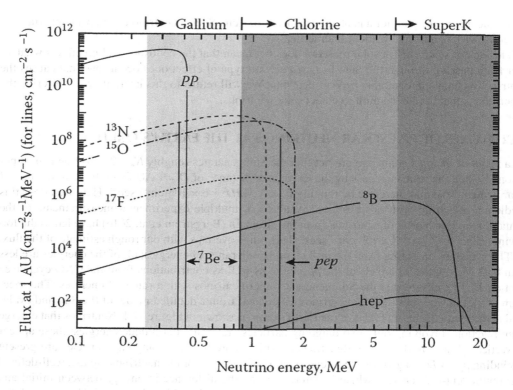

**FIGURE 12.1** Fluxes of solar neutrinos as a function of energy at a distance of 1 AU from the Sun. Different ordinates are used depending on whether one is dealing with "lines" or with "continua". (From Nakamura K. 2000. *Euro. Phys. J. C*, 15, 366. With permission.)

detector can respond. Each type of detector has its own particular threshold, depending on the specific reaction that is involved in the process of detection. An important aspect of Figure 12.1 can be seen by inspecting the upper edge of the figure: there, one sees the threshold energies of three different types of detectors that have been used in attempts to detect solar neutrinos. Information pertaining to these different detectors will be discussed in Section 12.4.

### 12.3.1 *pp*-II AND *pp*-III CHAINS

Both of these chains start off with the reactions *pp*-I (a) and (b) (see Section 11.1) to produce He$^3$. Then, instead of interacting with another He$^3$ nucleus (as happens in reaction (c) in Section 11.1), both *pp*-II and *pp*-III at first rely on the following reaction:

$$He^3 + He^4 \rightarrow Be^7 + \gamma \tag{d}$$

The question that we need to ask is this: how fast does reaction (d) go compared with reaction (c) of the *pp*-I chain? To address this, we first consider the difference in Coulomb barrier tunneling by proceeding analogously to the discussion in Section 11.7. For reaction (d), we use $Z_1 = Z_2 = 2$ and $A = 1.71$: this leads to $\beta = 162$ keV$^{0.5}$. Using $kT = 1.2$ keV in the solar core, we find the energy at which the reaction is most effective $E_o = (\beta kT/2)^{2/3}$ has a value of 21.1 keV. This is somewhat larger than the 20.3 keV value for reaction (c) of the *pp*-I chain. As a result, the tunneling rate, which

# Neutrinos from the Sun

is proportional to $\exp(-3E_c/kT)$ (see discussion prior to Equation 11.12), is reduced in reaction (d) compared to reaction (c) by a factor of almost 10 in the core of the Sun. Thus, quantum tunneling reduces the reaction rate of (d) compared to (c) by about one order of magnitude.

On the other hand, a significant factor that strongly *favors* the occurrence of reaction (d) over reaction (c) has to do with the fact that $He^4$ is much more abundant in the Sun than $He^3$ is: the excess is some four orders of magnitude. As a result, any nucleus of $He^3$ finds itself likely to collide, in a given time interval, with $10^4$ times more $He^4$ nuclei than with $He^3$ nuclei.

The final determination of how rapidly reaction (d) occurs compared to reaction (c) has to do with what happens in the post-tunneling process, when the strong force comes into operation. There is no easy way to see what differences should be expected when the strong force comes into play: one must rely on detailed quantum mechanical calculations. These indicate (Clayton 1968) that for (d), the reaction rate has a numerical value that is $10^4$ times *smaller* than for reaction (c).

Combining the reduction in tunneling rate ($10^{-1}$) with the increase in abundance ($10^4$) and the decrease ($10^{-4}$) in the post-tunneling rate, we find that the net effect is that reaction (d) occurs about 10 times less frequently in the solar core than reaction (c). As a result, the *pp*-I chain occurs about 90% of the time in the solar core, while the *pp*-II and *pp*-III chains, in combination, occur about 10% of the time.

Now we turn to separate considerations of the *pp*-II and *pp*-III chains.

First, we consider the *pp*-II chain. Following reaction (d), the *pp*-II chain proceeds according to the reactions:

$$Be^7 + e^- \rightarrow Li^7 + \nu \tag{e}$$

$$Li^7 + p \rightarrow 2He^4 \tag{f}$$

Reaction (e), which involves electron capture, leads to two (and only two) particles in the exit channel. As a result, the neutrinos have a unique energy. The energy difference between the ground states of the nuclei is 0.86 MeV. A neutrino "line" corresponding to reaction (e) appears in Figure 12.1 at an energy of 0.86 MeV.

As it happens, the $Li^7$ nucleus emerging from reaction (e) also has an excited state at an energy of 0.48 MeV above ground: this lies low enough that it also lies *below* the energy of the ground state of $Be^7$. The transition from the ground state of $Be^7$ to this excited state is allowed energetically, and it produces a neutrino with an energy of $0.86 - 0.48 = 0.38$ MeV. A neutrino "line" also is plotted in Figure 12.1 at this energy. Laboratory measurements indicate that the 0.86 MeV transition occurs about 90% of the time. This accounts for the fact that in Figure 12.1, the vertical line representing the 0.86 MeV neutrinos has an amplitude that is about 10 times larger than the amplitude of the line that represents the 0.38 MeV neutrinos.

Now we consider the *pp*-III chain. Following reaction (d), the *pp*-III chain proceeds as follows:

$$Be^7 + p \rightarrow B^8 + \gamma \tag{g}$$

Reaction (g) differs from reaction (e) in the qualitative sense that in (g), the repulsive force between two positively charged particles has to be penetrated, whereas in (e), there is an attractive force between the $Be^7$ nucleus and the electron. For these reasons, the *pp*-II chain gets off to a faster start than the *pp*-III chain. Detailed calculations show that the *pp*-II chain occurs about 100 times more frequently in the Sun than the *pp*-III chain.

The next step in the *pp*-III chain is historically the most important:

$$B^8 \rightarrow Be^8 + e^+ + \nu \tag{h}$$

Reaction (h) is the reaction that was the first to allow solar neutrinos to be detected on Earth. Three particles emerge from the decay, and as a result, the neutrino energies are spread across a continuum. Significantly, the cut-off energy of the continuum is quite large, some 14 MeV (see Figure 12.1). This large value arises mainly from the difference between the rest masses of the parent and the daughter nuclei (see Audi and Wapstra 1993), namely, 8.021864 a.m.u. ($B^8$) and 8.003111 a.m.u. ($Be^8$). If the decay in reaction (h) were to occur between the *ground states* of parent and daughter, then the cut-off energy would be the energy corresponding to the total mass difference (0.0188 a.m.u.), i.e., 17.5 MeV. However, although the decay starts in the ground state of $B^8$, quantum mechanical considerations indicate that a transition to the ground state of $Be^8$ is forbidden: instead, the decay is constrained to go to an excited state of $Be^8$ that lies 2.9 MeV above ground. The emergence of a positron on the r.h.s. of reaction (h) also requires the system to supply a "creation energy" of 0.511 MeV. As a result, the cut-off energy of the neutrino spectrum emerging from reaction (h) is 17.5 − 2.9 − 0.5 = 14.1 MeV.

The final step in the *pp*-III chain is as follows:

$$Be^8 \rightarrow 2He^4 \qquad (i)$$

The net effect of the *pp*-III chain is that four protons plus an alpha nucleus combined into two alpha particles. In other words, starting with four protons as intake, the outcome is one alpha nucleus.

### 12.3.2 OTHER REACTIONS THAT OCCUR IN THE SUN

The *pep* reaction involves an electron capture by a proton and a subsequent collision with another proton:

$$p + e^- + p \rightarrow D + \nu \qquad (j)$$

The energetics are the same as reaction (a) of the *pp*-I chain, except that the electron appears on the left-hand side. But in the case of reaction (j), (in contrast to reaction (a)), only *two* particles emerge on the right-hand side. As a result, the emergent neutrino in the *pep* reaction has a unique energy: 1.442 MeV. In Figure 12.1, a vertical line labeled *pep* can be seen at this energy. Reaction (j) in the Sun occurs only once for every 400 *pp* interactions as described by reaction (a).

The *Hep* reaction leads to a neutrino continuum with a cut-off at 18.8 MeV:

$$He^3 + p \rightarrow He^4 + e^+ + \nu \qquad (k)$$

The neutrinos emerging from reaction (k) are certainly the most energetic neutrinos generated by solar nuclear reactions (see Figure 12.1). However, the number of neutrinos generated by reaction (k) in the Sun are so rare that, as of 2021, there are only reports in the literature of *upper limits* on the flux of neutrinos $F(k)$ from reaction (k) compared to the flux of neutrinos $F(a)$ which originate in reaction (a): $F(k)/F(a)$ is found to be less than one part in $3 \times 10^5$ (The Borexino Collaboration 2018).

In stars hotter than the Sun, energy is generated preferentially by a "bi-cycle" of reactions in which carbon nuclei act as catalysts for fusing four protons into one helium. In this "bi-cycle", three beta decays occur (from $N^{13}$, $O^{15}$, and $F^{17}$), each with the emission of a neutrino having a continuous energy spectrum extending up to 1–2 MeV. These continua are shown in Figure 12.1. In the Sun, observations indicate that the CNO cycle contributes on the order of 1% to the Sun's energy output (The Borexino Collaboration 2020). In view of the smallness of this contribution, we will not consider CNO neutrinos any further in this "first course".

## 12.4 DETECTING SOLAR NEUTRINOS ON EARTH

The very smallness of the interaction cross-section that allows neutrinos to escape from the center of the Sun has the inevitable corollary that detection of neutrinos on Earth requires efforts that are nothing short of Herculean.

There are two general classes of experiment for the detection of neutrinos. In one class, we rely on the properties of certain nuclei to absorb a neutrino, thereby transforming the initial nucleus into the nucleus of a new element: the goal is then to identify the amount of the new element produced in a given time interval. In the second class of experiments, we do not use nuclear physics at all: instead, we detect events in which a fast neutrino "smashes into" an electron in a certain medium (e.g., water), giving the electron a speed exceeding that of light in the medium. When that happens, a burst of Cherenkov radiation is emitted and is detected by light-sensitive phototubes.

### 12.4.1 CHLORINE DETECTOR

The first neutrino detector, built by Raymond Davis Jr. in the 1960s, used a large tank of cleaning fluid ($C_2Cl_4$) containing 520 tons of $Cl^{37}$. The goal was to have solar neutrinos interact with $Cl^{37}$ nuclei to produce nuclei of $Ar^{37}$, and then count how many argons were in the tank after a certain length of running time. In order to avoid contamination from backgrounds, the detector was buried deep, almost one mile, underground in a mine in South Dakota (Davis et al. 1968).

The (forward) decay reaction

$$Ar^{37} \rightarrow Cl^{37} + e^+ + \nu \tag{l}$$

is driven by the mass difference between the ground states of $Ar^{37}$ and $Cl^{37}$, corresponding to an energy of 0.814 MeV (Audi and Wapstra 1993).

As a result, the Davis detector (which records events driving the prior reaction backward), responds only to neutrinos with energies in excess of 0.814 MeV. In principle, this means that if neutrino capture were to occur mainly via a transition from the $Cl^{37}$ ground state to the $Ar^{37}$ ground state, then Davis should be able to detect the line neutrinos from the *pep* reaction and from the higher energy $Be^7$ decay, as well as continuum neutrinos from $B^8$ decay, the *Hep* reaction, and three decays in the CNO bi-cycle.

However, the nuclear physics is such that transitions from the $Cl^{37}$ ground state to excited states of the $Ar^{37}$ nucleus are preferred, especially to a level at an energy of about 5 MeV above the ground state. As a result, neutrinos in the $B^8$ and *Hep* continua dominate the signal in the Davis detector. Of these, the B neutrinos are dominant by far.

Now that we know which continua will be dominant, we turn to the experimental results. In principle, we are seeking a measurement of the neutrino flux. A convenient unit can be devised that incorporates the likelihood of a neutrino being detected. The common unit for discussing solar neutrino experiments is the solar neutrino unit (SNU): this is defined to be one neutrino capture per second in a detector that contains $10^{36}$ target nuclei.

Why is this unit useful? Because the interaction cross-section between a neutrino and one of the target atoms is expected to be of order $10^{-43}$ cm$^2$, while the input flux from the Sun in the $B^8$ continuum is expected to be a few times $10^7$ neutrinos cm$^{-2}$ sec$^{-1}$. The product of these numbers yields an expected capture rate in a detector of a few times $10^{-36}$ per second. In view of this, a detector containing $10^{36}$ targets should yield a detection rate of a few per second. By definition, a detection of one per second per $10^{36}$ targets equals 1 SNU.

The Davis detector contained roughly $10^{31}$ chlorine target nuclei. The standard solar model predicted that in one day of running (i.e., about $10^5$ sec), the Davis detector should record about two neutrino captures. When the exact calibration was done, it was found that two captures per day would correspond to a solar neutrino rate of 8.1 SNUs (see Figure 12.2). Even allowing the

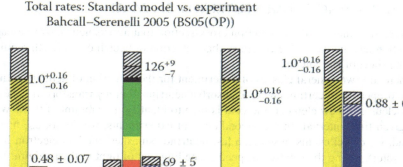

**FIGURE 12.2** Neutrino counting rates: comparison between theory and experiments. The rates are given in units of SNUs (see text). (Taken from de Gouvea 2006). Original permission to use this figure in 2009 was granted for a version which appeared on the website of the late John N. Bahcall.

experiment to run for several months at a time, the total yield of argon atoms in the tank at the end of the run was expected to be no more than a few hundred, out of a tank containing some $10^{31}$ chlorine atoms: the chemical expertise required to flush out those few argon atoms from an "ocean of chlorine" was truly impressive. For his work, Davis was awarded the Nobel Prize in Physics in 2002.

The experimental results obtained by Davis were a surprise. The observed count rates, when averaged over 20 years and more, yielded a rate of only about 0.6–0.7 captures per day. The corresponding average solar neutrino rate is 2.6 ± 0.2 SNUs.

Davis' experiment led to the startling conclusion that the experimental capture rates of solar neutrinos were smaller than predicted (8.1 SNUs) by a factor of about three. This shortfall became known as the "solar neutrino problem". We shall return to this later.

### 12.4.2 Cherenkov Emission

If a neutrino collides with an electron in a medium, the electron, called the "knock-on electron", picks up some of the neutrino energy. Solar neutrinos, with energies of up to 14 MeV, can create knock-on electrons that also have energies measured in MeV. Such electrons travel with speeds that are close to the speed of light *in vacuo*. If such an electron travels through a medium where the speed of light is reduced to (say) $0.7c$ (such as water), the knock-on electron will be moving faster than light in the medium. This causes emission of light in a Cherenkov cone, with an opening angle

Neutrinos from the Sun

determined by the electron's energy. To make the electron fast enough for the Cherenkov process to be possible, the initial neutrino must have a minimum energy.

The Kamiokande detector in Japan, containing some 2000 tons of water, was instrumented with a spherical shell of phototubes to track the Cherenkov cones from solar neutrinos: this detector came online in 1983. Subsequently, the super-Kamiokande detector, with 50,000 tons of water and with more than 10,000 phototubes, came online in 1996. In both cases, the minimum energy required to create knock-on electrons with significant Cherenkov emission is $\geq 5$ MeV (e.g., Rothstein 1992; Takeuchi 2005). As a result, neither detector could record the main ($pp$) neutrinos from the Sun: the detectors could respond only to the upper end of the spectrum of $B^8$ neutrinos.

The standard solar model predicted that the neutrino detection rates should be $1 \pm 0.2$ SNUs. But the experimental results yielded no more than 0.4–0.5 SNU.

A major advantage of the Kamiokande detectors compared to the chlorine experiment of Davis is that they provide information as to the *direction* of the incoming neutrino. The data confirmed that the neutrinos are indeed coming from the Sun. The leader of the Kamiokande experiments, M. Koshiba, shared with Davis the 2002 Nobel Prize award.

### 12.4.3 Gallium Detectors

In order to detect the most abundant neutrinos from the Sun (i.e., those from reaction (a) of the $pp$-I chain), it is necessary to devise a detector in which the threshold energy lies well below the cut-off energy (0.42 MeV) of the $pp$ neutrinos. As it happens, an isotope of gallium satisfies this criterion.

The relevant scheme on which this neutrino detector is based is the decay

$$Ge^{71} \rightarrow Ga^{71} + e^+ + \nu \qquad \text{(m)}$$

The mass difference between $Ge^{71}$ and $Ga^{71}$ corresponds to an energy difference of only 0.23 MeV. Consequently, in the inverse reaction, neutrinos can be captured by $Ga^{71}$ if the neutrino energy exceeds 0.23 MeV. Most of the $pp$ neutrinos emerging from the Sun satisfy this criterion.

With a detector sensitive to the most abundant solar neutrinos, the count rate is predicted to be much larger than in the chlorine detector or in the Cherenkov detectors: the gallium detectors were predicted to respond at the rate of 126 SNUs.

Two experiments were built, one in Russia (SAGE, using 50 tons of liquid gallium, with operations starting in 1990: see Abdurashitov et al. 1999) and one in Italy (GALLEX, using a solution containing 30 tons of gallium, with operations starting in 1991: see Kirsten 2008).

The detection results from both experiments were in agreement with each other, some 67–69 SNUs: both detection rates were definitely *lower* than the predictions.

### 12.4.4 Heavy Water Detector

A detector containing 1000 tonnes of heavy water ($D_2O$), surrounded by an even larger volume of clean "ordinary water" ($H_2O$), was buried 2 km below ground at the Sudbury Neutrino Observatory (SNO) in Ontario, Canada (see Figure 12.3). The container of heavy water plus "ordinary water" was viewed by almost 10,000 photomultiplier tubes, arranged on a geodesic dome framework, in order to detect the faint flashes of radiation emitted by particle interactions inside the heavy water. The size of the instrument can be estimated by comparison with the two workers who are visible near the bottom of the image. The entire container plus geodesic dome was immersed in a 30-meter barrel of ordinary water ($H_2O$): the barrel is as tall as a 10-story building.

The presence of deuterium allowed three distinct classes of reactions to occur involving neutrinos (Ahmad et al. 2002).

**FIGURE 12.3** The inner part of the SNO detector. Notice the scale of this detector: the scale can be estimated from the size of the human beings near the bottom of the image. (Image courtesy of SNO.)

    a. Knock-on electrons are created as in Kamiokande; these gave results similar to those in Section 12.4.2.
    b. Neutrinos associated with neutron/proton decays (called *electron neutrinos*) interact with D to cause the neutron to decay into a proton. The nucleus then becomes a "di-proton", which is unstable (see Section 11.5.2). There is a rapid decay into two free protons plus an electron. If the electron is fast enough, a Cherenkov pulse can be detected. The standard solar model predicted 30 of these events per day: the experiment actually recorded only about 10 per day.
    c. Neutrinos associated with decays *other* than neutron/proton decays belong to distinct families: they are referred to as μ-neutrinos and τ-neutrinos, to indicate the decays with which

Neutrinos from the Sun

they are associated. All three neutrino families can interact with deuterium by a process known as the neutral current reaction. This splits the deuterium nucleus, and a free neutron emerges. In the presence of a suitable contaminant nucleus (such as $Cl^{35}$, added to the water tank in the form of table salt), neutron capture can occur and gamma rays are emitted.

The standard solar model predicted about 30 neutrinos per day: the experiment recorded essentially that rate. For the first time, a neutrino detector responded in the way that was predicted by the standard solar model.

## 12.5 SOLUTION OF THE SOLAR NEUTRINO PROBLEM

For a decade or more after Davis announced his first results, the commonest explanation for the solar neutrino problem was that there must be something wrong with the solar model. Attempts were made by solar modelers to add extra effects in the Sun (strong magnetic fields, fast rotation, atypical metal abundances), but these were mostly *ad hoc*. However, as helioseismology (see Chapters 13 and 14) came into its own in the 1980s and 1990s, it became clear that there was very little wrong with the profile of physical parameters inside the solar model. This led to the conclusion that the solution of the solar neutrino problem must lie in the physics of the elementary particles.

According to the standard model of particle physics, the fundamental constituents of matter consist of six "flavors" of quarks (two of which exist in protons and neutrons) and six leptons. The latter consist of electrons, μ-mesons, and τ-mesons, plus the "flavors" of corresponding neutrinos (electron neutrinos, μ-neutrinos, and τ-neutrinos). Leptons interact only through the weak force, and also (if they are electrically charged) through the electromagnetic force. Although in the standard model all neutrinos have zero mass, experimental evidence from cosmic rays indicates that this is not exactly true. It turns out that neutrinos have nonzero masses, although the masses are orders of magnitude less than the next lightest lepton (the electron). The existence of finite mass has the effect that neutrinos in different flavors can "mix" among themselves.

The Sun generates in its nuclear reactions electron neutrinos only: the reason for this is that all of the decays that generate neutrinos in the Sun involve electrons only (see Section 11.1, reaction (a), and Chapter 12, Section 12.3.1, reactions (e) and (h), and Section 12.3.2, reactions (j) and (k)). As a result, the Sun does *not* generate either μ- or τ-neutrinos directly in any of the nuclear reactions that occur in the core. However, as the electron neutrinos propagate outward from the core of the Sun, passing through the radiative core plus the convective envelope, and then (after leaving the Sun) passing through the interplanetary plasma that lies between the Sun and Earth, the electron neutrinos undergo a mixing process, thereby producing neutrinos in the other two flavors. If enough mixing occurs so as to populate equally all three flavors, then roughly equal numbers of neutrinos are produced in all three flavors.

As a result, only about one-third of the (electron) neutrinos generated at the Sun survive to reach the Earth as electron neutrinos. The remaining two-thirds reach the Earth as roughly equal mixtures of μ-neutrinos and τ-neutrinos.

The chlorine and gallium detectors are sensitive only to electron neutrinos. Their count rates are smaller than expected because the detectors do not "see" the μ- or τ-neutrinos. The Cherenkov pure-water experiments are in principle sensitive to all three neutrinos, but in practice the cross-section for scattering off electrons in the water favors the electron neutrinos. Therefore, the pure-water detectors respond best to the one-third electron neutrinos, with a weaker response to the other two-thirds. But when the SNO experiment (c) (see Section 12.4.4, paragraph (c)) was performed using salty water, all three flavors of neutrinos could participate in the reactions and were therefore detectable.

The history of the solar neutrino "problem" reads like an exciting detective story. It took some 35 years of "big science" in multiple countries to identify the "culprit". The case was solved in 2002 by the SNO experiment (see Section 12.4.4c). The first director of the SNO Observatory, A. B. McDonald, was awarded the 2015 Nobel Prize in Physics.

Two significant results emerged from the neutrino detective story. First, as regards the physics of the internal structure of the Sun, the solar models survived a stringent test. Second, in the field of particle physics, a new window "beyond the standard model" was opened up. Both areas of research, solar physics and particle physics, benefited from the long and hard process of attending to the details of solving the solar neutrino problem.

## EXERCISES

12.1 Use the isotope masses in Table 11.1 to show that the cut-off energy in reaction (k) (Section 12.3) is 18.8 MeV.

12.2 For reaction (k) (Section 12.3), show that the energy $E_o$ at which quantum tunneling has maximum effectiveness (see Chapter 11, Section 11.6) is equal to 10.1 keV. Using this value of $E_o$, show that, due to Coulomb effects alone, reaction (k) occurs almost $10^5$ times less frequently than reaction (a) (Chapter 11, Section 11.1).

12.3 Using tabulated values of atomic weights for C and Cl, show that a detector that contains 520 tons of $C_2Cl_4$ contains close to $10^{31}$ atoms of chlorine.

## REFERENCES

Abdurashitov, J. N., Bowles, T. J., Cherry, M. L., et al., 1999. "Measurement of the solar neutrino capture rate by SAGE and implications for neutrino oscillations in vacuum", *Phys Rev. Lett.* 83, 4686.

Ahmad, Q., Allen, R. C., Andersen, T. C., et al., 2002. "Direct evidence for neutrino flavor transformation from neutral-current interactions in the Sudbury Neutrino Observatory", *Phys. Rev. Lett.* 89, 011301.

Audi, G., & Wapstra, A. H., 1993. "The 1993 atomic mass evaluation: (I) Atomic mass table", *Nuclear Phys. A.* 565, 22.

Bergstrom, J., Gonzalez-Garcia, M. C., Maltoni, M., et al., 2016. "Updated determination of the solar neutrino fluxes from solar neutrino data", *J. High Energy Phys.* 132.

The Borexino Collaboration, 2018. "Comprehensive measurement of $pp$-chain solar neutrinos", *Nature* 562, 505.

The Borexino Collaboration, 2020. "Experimental evidence of neutrinos produced in the CNO fusion cycle in the Sun", *Nature* 587, 577.

Clayton, D. D., 1968. "Major nuclear burning stages in stellar evolution", *Principles of Stellar Structure and Nucleosynthesis*, McGraw-Hill, New York, pp. 362–435.

Cowan, C. L., Reines, F., Harrison, F. B., et al., 1956. "Detection of the free neutrino: A confirmation", *Sci.* 124, 103, 1956.

Davis, R., Harmer, D. S., & Hoffman, K. C., 1968. "Search for neutrinos from the Sun", *Phys. Rev. Lett.* 20, 1205.

de Gouvea, A. 2006. "TASI lectures on neutrino physics", *Physics in D≥4 TASI 2004*, ed. J. Terning et al., World Scien. Publ. Co., Singapore, pp. 197–258.

Kirsten, T., 2008. "Retrospect of GALLEX/GNO", *J. Phys. Conf. Ser.* 120, 052013 (IOP Publishing).

Nakamura, K., 2000. "Solar neutrinos", *Euro. Phys. J. C* 15, 366.

Rothstein, I. Z., 1992. "Solar $v_e$ production does not enhance event rates in the Kamiokande detector", *Phys. Rev. D* 45, R2583.

Takeuchi, Y., 2005. "Solar neutrino measurements in Super-Kamiokande", *Nucl. Phys. B (Proc. Suppl.)* 149, 125.

# 13 Oscillations in the Sun
## *The Observations*

We have already (in Chapter 12) raised the important question: how can we possibly check on our models of the internal structure of the Sun? After all, the interior of the Sun is surely one of the most inaccessible parts of the world we live in. So it is natural to raise the question: how do we know we are on the right track? Could it be that our calculations are far from reality, or maybe are just plain wrong? Is it possible to check on these calculations?

One way to address this issue is by studying neutrinos, which come from the hottest parts of the solar interior, where nuclear reactions occur. The neutrinos allow us to check on our calculations in the very core of the Sun (see Chapter 12).

But starting in the 1970s, and especially following a landmark experiment in 1980 at the South Pole (where the Sun was observed without interruption for more than 100 hours), a second method of testing the solar models became available. In terms of physics, the properties of the entire solar interior at (almost) all radial locations from center to surface can be checked by studying the properties of certain classes of waves that propagate inside the Sun. There are three classes of waves that are useful in the present study: they are referred to as "*p*-modes", "*g*-modes", and "*r*-modes", where the letters refer to the dominant restoring force of the waves (pressure, gravity, and rotation, respectively). A special subclass of modes that are confined to the surface layers of the Sun are referred to by the special title of "*f*-modes", where the "*f*" stands for fundamental. (We have already encountered the *f*-modes in Section 1.5 in connection with the determination of a precise value of the solar radius.) These various classes of waves provide us (in principle) with a "wave window into the Sun" extending from the surface (where the radial coordinate $r$ has values close to $R_\odot$) deep into the interior (down to values of $r$ as small as $0.1$–$0.2 R_\odot$ or even smaller). In this chapter and in Chapter 14, we turn to a study of these waves and describe how they can help us to check our calculations of the internal structure of the Sun at radial locations that lie within the "wave window". The fact that the "wave window" extends (almost) all the way into the core of the Sun, while neutrinos probe the core itself, means that, roughly speaking, neutrinos and waves provide complementary approaches to testing the models of the internal structure of the Sun.

The Sun, although appearing to the unaided eye as being constant in its output, nevertheless is not absolutely unchanging. The most obvious forms of solar variability are sunspots: dark regions on the surface that appear and disappear on semiregular time-scales of *days* to *years*. In Chapter 16, we will describe in detail the physical properties of sunspots. But in the context of the present chapter, the most important facet of spots is that vertical magnetic fields are present in the darkest center of the spot (the umbra), while in the surrounding less dark region (the penumbra), the field lines become progressively more inclined away from the vertical (forming a "ramp") as we move outward. An observational physical consequence of "ramps" in the Sun will be discussed in Section 13.6.

However, apart from sunspots, when one observes the Sun with sufficiently high resolution, one finds that there are some highly regular variations in velocity and intensity that occur on time-scales of *minutes*. In this case, the periodicities of the variations are not at all semiregular: on the contrary, they occur at highly precise frequencies, which are (in the best cases) reproducible within a few parts per 100,000 every time one observes the Sun. These extremely periodic variations provide a scientific means for us to study the solar interior in a way that is similar to the way in which seismologists obtain information about the Earth's interior by studying earthquakes. "Helioseismology" is the term now used to describe this scientific study.

The purpose of this chapter is to describe the *observations* that allow us to determine the properties of the Sun's periodic variations. (Theoretical discussions will be presented in Chapter 14.)

The variations can be studied from the point of view of *temporal* variations alone (with no regard for spatial resolution). They can also be studied in data that have been *spatially* resolved across the disk of the Sun. We turn first to the purely temporal variations.

## 13.1 VARIABILITY IN *TIME* ONLY

When the Sun is observed "as a star", data are gathered without regard to spatial location on the surface of the Sun. The detector integrates over the entire disk of the Sun. Using high *spectral* resolution, small variations in *velocity* can be detected in the radiation that reaches Earth from the Sun. When these variations are analyzed as a *time series*, a power spectrum is obtained, showing how much power occurs (in units of velocity-squared per unit frequency) as a function of frequency. An example is shown in Figure 13.1 (Fossat et al. 1981). The data were obtained by observing the Sun continuously for a time interval $T_o$ lasting roughly $0.5 \times 10^6$ sec: this feat of observing continuously for about 5 days was made possible by making the observations of the Sun from a site at the geographic South Pole (where the Sun does not suffer diurnal setting or rising for time intervals as long as 6 months). Along the abscissa in Figure 13.1 is plotted the frequency in units of millihertz (mHz), while the ordinate shows the power in velocity ($\sim V^2$ per unit frequency). The frequency resolution, of order $1/T_o$, is $\approx 2$ microhertz ($\mu$Hz).

The striking result in Figure 13.1 is that large amounts of power are observed at certain sharply defined frequencies, while at other frequencies, there is so little power that it can hardly be distinguished from noise. The impression is that the Sun emits at a number of discrete frequencies that are so sharp that the spectrum seems to consist of a number of "spikes" (or, in more mathematical parlance, "delta-functions"). Each spike has a width in frequency of order 1 $\mu$Hz.

Inspection of Figure 13.1 illustrates that the "delta-functions" contain significant quantities of power at certain sharply defined frequencies $v$ that extend between (roughly) 2.5 mHz and (roughly) 4.5 mHz. The corresponding periods ($= 1/v$) range from as long as (about) 400 sec down to as

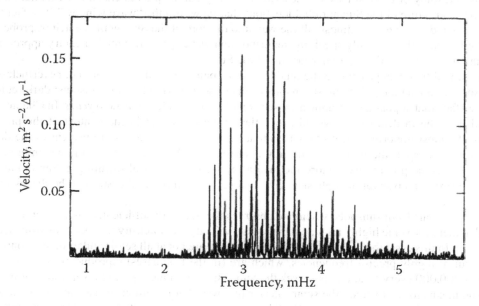

**FIGURE 13.1** Power spectrum of solar oscillations in *velocity*. (From Fossat et al. 1981; used with permission from Springer.)

short as (about) 220 sec. Earlier observations of this type, obtained at lower resolution (because they were taken at mid-latitude ground-based sites where the Sun rose and set every 24 hours), had detected only a broad peak of power centered at frequencies of about 3.3 mHz, i.e., periods of about 300 sec: for this reason, the early observers referred to the oscillations as "5-minute oscillations". Subsequently, when theoretical work showed that the peaks arise from acoustic waves (i.e., pressure waves) in the Sun, the peaks were referred to as "*p-modes*".

The plot in Figure 13.1 uses a *linear* axis for the power scale. This allows us to quantify readily the *largest* peaks in the power spectrum. But a linear plot makes it difficult to identify the *smallest-amplitude* oscillations in the spectrum. In order to enhance our ability to see the smaller oscillations, a logarithmic plot (taken from the website of the Solar and Heliospheric Observatory [SOHO]) is presented in Figure 13.2. Space-borne detectors are not subject to limitations of rising/setting of the Sun: the observing interval used for Figure 13.2 was 800 days. The range in frequency $v$ extends from less than 0.5 mHz to 8 mHz. The oscillations with the largest power levels (having amplitudes of 4000–5000 m$^2$ sec$^{-2}$ Hz$^{-1}$) exist at $v = 3$–$3.5$ mHz, i.e., in the 5-minute range.

In an independent study of *p*-mode amplitudes from that shown in Figure 13.2, Kiefer et al. (2018) used 22 years of data from the GONG network (which can resolve modes with $l = 0$–$150$) in order to identify the frequency at which the *p*-mode velocity amplitude is maximum: it was found to lie at $3079.76 \pm 0.17$ µHz. But when one considers the maximum in squared velocity amplitude (which depends on the product of amplitude and width of the mode), a slightly different maximum emerges: ≈3200 µHz. The corresponding periods of these two maxima in frequency are 325 and 312 second, i.e., well within the limit of the so-called 5-minute oscillations. The maximum mean velocity amplitude of an individual *p*-mode in the GONG data set was found to be 37 cm sec$^{-1}$ (Kiefer et al. 2018). Although the amplitude of individual *p*-modes is small, there are (rare) events in the Sun when coherent patches of *p*-mode oscillations can arise in which the maximum velocity amplitude

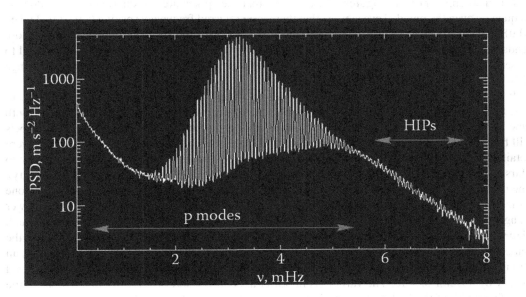

**FIGURE 13.2** Logarithmic plot of *p*-mode power in the Sun. Abscissa: frequency $v$ in units of mHz. Ordinate: power spectral density (PSD) of *velocity* oscillations in the Sun derived from 800 days of measurements. The *p*-modes are the dominant signal at low frequencies (< 5 mHz). At high frequencies, a different phenomenon appears: high interference peaks (HIPs), arising from partial wave reflection in the solar atmosphere. (Courtesy of SOHO/GOLF consortium. SOHO is a project of international cooperation between ESA and NASA.)

in the coherent patch can rise to as large as ±1 km sec$^{-1}$ (McClure et al. 2019): such events give rise to short-lived "flashes" of Doppler shift at localized regions of the solar surface.

The plot in Figure 13.2 allows us to identify some (weak) oscillatory power (rising above the background noise) out to a frequency as large as $\nu \approx 5$ mHz (i.e., out to periods no less than 200 sec). We shall see later (Section 13.5.5) that there is a physical reason why $p$-modes do not exist in the Sun at periods less than ~200 sec. The identifiable modes with the smallest identifiable amplitudes at large $\nu$ in Figure 13.2 have amplitudes of less than 100 m$^2$ sec$^{-2}$ Hz$^{-1}$ (after subtracting the background), i.e., almost 100 times weaker than the peak power that is emitted at periods of about 5 minutes.

On the low-frequency side of the peak, oscillatory power can be identified (after subtracting the background) down to frequencies as low as (roughly) $\nu \approx 1.5$–1.6 mHz (i.e., periods as long as roughly 10 minutes). The last identifiable modes at low $\nu$ contain power which, after subtracting the background (which rises steeply between $\nu = 1.5$ mHz and $\nu = 0.2$ mHz), have numerical values of power of perhaps 10 m$^2$ s$^{-2}$ Hz$^{-1}$, i.e., almost three orders of magnitude smaller than the peak power in the 5-minute range.

The data in Figures 13.1 and 13.2 illustrate that the Sun produces power at a multitude of remarkably "spiky" peaks in frequency, i.e., the Sun is oscillating ("ringing") in many *very specific* tones. The tones are caused by acoustic modes (i.e., pressure waves) trapped inside the Sun (see Chapter 14). The narrowness of the "spiky" peaks in Figure 13.1 is striking. The ratio of line frequency ($\nu \approx 3$ mHz) to line width ($\delta\nu \approx 1$ μHz) can be of order $Q \approx \nu/\delta\nu \approx 3000$. And if even better frequency resolution is available (e.g., from spacecraft observations), the line width may be found to be even narrower than 1 μHz. Therefore, when the Sun "rings" in a particular mode, the "quality factor" $Q$ of the resonant cavity as regards that mode can be even larger than 3000, perhaps of order 10 thousand. Such a large $Q$ value indicates that once a $p$-mode is generated, that mode may persist for as long as 10 thousand periods. Given that the periods of the strongest $p$-modes are about 5 minutes, any particular mode may persist for as long as 50,000 minutes, i.e., $\approx 1$ month.

It is also apparent from Figures 13.1 and 13.2 that the spikes are not distributed at random in frequency: even the unaided eye can see that there is a preferred frequency spacing ($\approx 0.07$ mHz, i.e., 70 μHz) between adjacent peaks. Actually, careful analysis of the spectrum indicates that a more fundamental frequency spacing turns out to have about twice this value: many modes are found to be separated by a so-called large spacing $\Delta\nu$ that is found to have a numerical value in the range 135–138 μHz (Appourchaux et al. 1998).

It is important to note that although the *frequency* of any given $p$-mode remains the same at all times, this is not true as regards the *peak power* in the mode: this power fluctuates significantly with time. An example of the variability in the power of a particular mode ($l = 0$, $n_r = 21$: these labels will be defined later) over a time interval of several months is shown in Figure 13.3. The data were obtained by an instrument on board the USSR PHOBOS mission during its 160-day trajectory to Mars in 1988: the instrument was used to measure variations in *intensity* (rather than velocity). The mode amplitude is plotted in a perspective 3-D diagram with time (in units of days) along one horizontal axis, frequency of the mode (in units of μHz) along another horizontal axis, and power along the vertical axis (in units of parts per million squared per Hz). The particular mode to which Figure 13.3 refers is a mode with $\nu \approx 3034$ μHz: this mode lies very close to the frequency where the velocity amplitude of $p$-modes reaches its maximum value (Kiefer et al. 2018). In the course of an observing run that extends over an interval of at least 120 days, the *power* of the mode is observed to vary by a factor of $\approx 10$. The variations in power appear as a sort of "ridge of mountains" in the plot. The width of the highest "mountains" can be seen to be as long as about 1 month, consistent with lifetimes estimated earlier based on the $Q$ value. Although the peak of the ridge certainly changes in "height" as time goes on, the data indicate that the peak of the "ridge" does *not* shift significantly in *frequency*. The reason why peaks come and go in the "mountain range" has to do with how the mode is generated: we shall see (see Section 14.8) that individual $p$-modes are generated by convective flows at various depths below the surface in the Sun, and these flows are highly

# Oscillations in the Sun

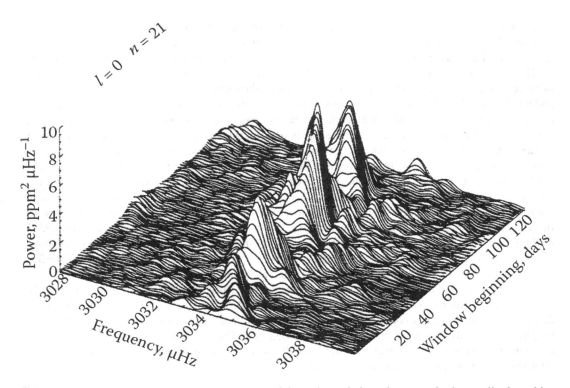

**FIGURE 13.3** A 3-D plot showing, as a function of time, the variations that occur in the amplitude and in the frequency of one particular mode of oscillation in the Sun. The mode in this figure has degree $l = 0$ and radial order $n_r = 21$. (The label $n = 21$ in the upper left corner of the figure corresponds to $n_r = 21$ in our notation.) Of the two axes that lie in the "horizontal plane" in the plot, the one on the right-hand side of the figure indicates the passage of time (in units of days) since the beginning of the observing window. This particular observing run lasted more than 4 months. The second "horizontal axis" indicates the frequency of the oscillation (in units of μHz). The particular mode in the figure has a frequency lying between 3028 and 3038 μHz. Rising above the "horizontal plane" in the figure, the "vertical axis" shows the power in the oscillation mode at each instant and at each frequency: units of power are ppm² μHz⁻¹ (where ppm = parts per million). (From Gavryusev and Gavryuseva 1997; used with permission from Springer.)

variable, forming and decaying on time-scales of order 10–20 minutes. But despite the obvious (large) changes in *amplitude* that occur in Figure 13.3, the *frequency* of the mode remains constant within a fraction of 1 μHz.

## 13.2 VARIABILITY IN *SPACE* AND *TIME*

The data in Figures 13.1 and 13.2 refer to variability in *time* only: such data are obtained when the Sun is observed *as a star*, with no attempt to resolve the Sun's disk spatially. However, valuable information about the Sun can also be extracted from the *spatial* properties of the variations. To do that, the Sun must be observed with data that are not only well resolved in time, but also resolved *spatially*. The higher the angular resolution used to obtain the data, the smaller the patches on the Sun's surface that can be examined for oscillation.

From a mathematical perspective, when one analyzes the properties of spatial variations on a spherical surface, it is natural to use "spherical harmonic functions" $Y_{lm} = P_l^m(\cos\theta)e^{im\phi}$ to describe the surface structure. Here, $\theta$ is the colatitude and $\phi$ is the longitude. The index $l$ refers to structure in the latitudinal direction, between the north pole ($\theta = 0$) and the south pole ($\theta = \pi$).

The index $m$ refers to structure in longitude. Initially, we neglect longitudinal variations, and consider $m = 0$. This allows us to reduce $Y_{lm}$ to the Legendre functions $P_l(\cos \theta)$. For $l = 0, 1, 2$, and 3, the first four Legendre functions are as follows: $P_l(x) = 1, x, (3x^2 - 1)/2$, and $(5x^3 - 3x)/2$, respectively.

The parameter $l$ is referred to as the "angular degree" of the mode: the value of $l$ is the number of nodes (i.e., regions of zero amplitude) that exist in the oscillatory structure between the north and south poles. For modes with $l = 0$ (no nodes in latitude), gas motions are synchronized over the entire surface: in such a mode, at a given instant, the gas is moving outward (at all points of the surface), and then one half-cycle later, the gas is moving inward (at all points of the surface). For modes with $l = 1$, there is one node in latitude, at $\cos(\theta) = 0$, i.e., the equator: this means that, at a given instant, when the gas in the northern hemisphere is moving outward, the gas in the southern hemisphere is moving inward. One half-cycle later, the northern gas moves inward, while the southern gas moves outward. For modes with $l = 2$, at a given instant, gas moves outward between the north pole and colatitude $\cos^{-1}(1/\sqrt{3}) = 55°$ (corresponding to latitude 35° N), gas moves inward in the equatorial regions (at latitudes between 35° N and 35°S), and gas moves outward from 35°S to the south pole. One half-cycle later, the outward motion is confined to the equatorial regions, while the polar "caps" move inward. For modes with $l = 3$, nodes occur at colatitudes $\cos^{-1}(\sqrt{3/5}) = 39°$ (i.e., at latitude 51° N), at latitude 0 (the equator), and at latitude 51° S.

The larger the $l$ value, the more nodes can be "squeezed" into the range of latitudes from +90° to −90°, and the closer the nodes approach each other on the stellar surface. When $l$ is large, the linear distance between adjacent nodes along a great semicircle from the north pole to the south pole is roughly equal to the length of that semicircle divided by $l$. The distance between adjacent nodes is equivalent to one-half of one wavelength. Formally, a linear distance that can be regarded as the "horizontal wavelength" $\lambda_h$ of a mode is given by

$$\lambda_h = \frac{2\pi R_\odot}{\sqrt{l(l+1)}} \tag{13.1}$$

This tells us that a mode with a degree of (say) $l = 250$ has $\lambda_h \approx 17,500$ km. The angular scale of such a length on the Sun's surface, as observed from Earth, is about 20–25 arcsec: therefore, in order to obtain meaningful information about modes with $l > 250$, we need to make observations of the Sun with angular resolutions that are at least as good as 5–10 arcsec: if we can manage to achieve such spatial resolution, then we will be able to "fit in" a few "pixels" across one wavelength of the $l = 250$ modes. But if an instrument were to have an angular resolution that is no better than (say) 30 arcsec, then we could not expect to resolve the modes with $l = 250$: in such a case, we cannot hope to obtain any reliable information about modes that have $l$ values any larger than perhaps 50–100. In this context, it is relevant to mention data for two specific instruments that have played (or are playing) significant roles in helioseismic studies: (i) SOHO/MDI has an angular resolution of about 3 arcsec: this is good enough to resolve modes with $l$ values up to about 300; (ii) SDO/HMI has angular resolution of about 1 arcsec: this allows resolution of modes with $l$ values up to about 1000.

So, given an observing scheme that allows us to measure *velocities* across the disk of the Sun with resolutions that are as good as *a few arcseconds*, we can analyze the data not only in terms of the *temporal* properties, but also in terms of its *spatial* properties (up to $l \approx 300$). To extract a power spectrum corresponding to a given $l$ value, a data set (which has been averaged over longitude) has to be numerically convolved with the particular spherical harmonic $Y_l$. For each $l$, the resulting series is subjected to time-series analysis, and a power spectrum is obtained for that $l$ value: the power spectrum will consist of "spikes" at a number of discrete frequencies (reminiscent of Figure 13.1). Repeating the analysis for many different $l$ values, the resulting power spectrum can conveniently be plotted in 2-D, with spatial information (the degree of the mode, $l$) along one axis, and temporal information (the frequency of the mode, $\nu$) along the other axis.

In Figure 13.4, the 2-D plot was extracted from 16 years of data from SOHO/MDI. The data are plotted (for modes with degree $l$ up to 300) with error bars on the $\nu$ values. With 16 years of data, i.e., with a data string extending over a period $T \approx 0.5 \times 10^9$ sec, we expect to be able to determine $\nu$ values with a precision no better than $\sim 1/T \approx 10^{-9}$ Hz. Note that in order to make the error bars in $\nu$ ($\pm \sigma$) visible on the scale of the plot, the authors have plotted *not* the value of $\sigma$ itself, but $\sigma$ multiplied by a number that is at least one thousand, and in the case of the lowest-lying curve, has a value of 100 thousand. Thus, the lowest $\nu$ for a mode plotted in Figure 13.4 is about 0.9 mHz (at $l \approx 80$), and the plotted "error" is about ±0.1 mHz. Dividing this "error" by the enhancement factor of 100,000, which is associated with the lowest line in Figure 13.4, the corresponding $\sigma$ is $10^{-6}$ mHz, i.e., of order $10^{-9}$ Hz, as expected from the value of $1/T$.

What do the results in Figure 13.4 tell us about the Sun? First of all, they indicate how many modes can be detected with high precision in the Sun: e.g., in the case of "fundamental modes" (i.e., *f*-modes: the lowest-lying curve in the figure), one can see that there is a plotted value of frequency at every $l$ value between 91 and 300, i.e., a frequency has been measured for at least 200 *f*-modes. The authors of Figure 13.4 state that in each year of their 16-year observing run, they were able to obtain reliable fits to between 1900 and 2300 modes. (Many more modes may be present in the Sun, but the amplitudes are not large enough to be detected reliably by SOHO/MDI.) This means that on average, each ridge in Figure 13.4 contains about 100 individual modes.

Moreover, Figure 13.4 tells us that oscillations in the Sun occur in locations that are *not* spread uniformly over the entire plane of $\nu$ vs. $l$. Instead, the observed dots occur preferentially in groups, with each group appearing as a more or less narrow well-defined "ridge" going from lower left to

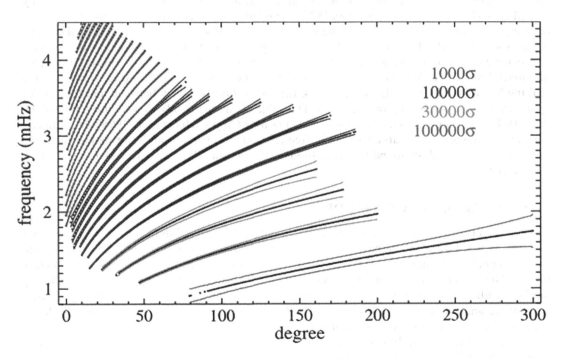

**FIGURE 13.4** Spatial-temporal power spectrum of solar oscillations. Abscissa: angular degree $l$ of the mode. Ordinate: frequency $\nu$ of the mode in units of mHz. Each data point is plotted as a black dot. Errors in frequency ($\sigma$) have been amplified by factors of at least 1000 in order to render them visible on the scale of this plot. Observations were obtained during 16 years of observation with SOHO/MDI from 1996 to 2011. (From Larson and Schou 2015; used with permission from Springer.)

upper right. To determine which frequencies occur at a given $l$, (say $l = 100$), the procedure is to start at the location labeled $l = 100$ on the abscissa, and then draw an imaginary line vertically upwards. This line intersects the lowest "ridge" at $v \approx 1$ mHz. Thus, the longest period mode with $l = 100$ has a period of about 1000 sec. Extending the imaginary line further, the next mode occurs on the second lowest ridge, with a $v \approx 1.5$ mHz, i.e., $P \approx 670$ sec. Although it might seem like a trivial point, it is worth noting that there are *no* solar $p$-modes with $l = 100$ at periods between 670 and 1000 sec: the Sun really does select only certain well-defined periods for its $p$-modes. Extending the vertical line further, we find that the highest frequency mode that can be reliably detected by SOHO/MDI with $l = 100$ occurs when the vertical line intersects the ninth-lowest ridge at $v \approx 3.5$ mHz, i.e., $P \approx 290$ sec.

At higher frequencies, modes with $l = 100$ are not plotted in Figure 13.4: this indicates that the observers could not identify confidently any $l = 100$ modes in the 16-year data set. If we had chosen for our search a value $l = 250$, we would find that only one mode (the $f$-mode) is detectable in the data set, with $v \approx 1.5$ mHz. Thus, the upper right-hand corner of the plot is "empty" in the sense that if the Sun does indeed have modes there, the SOHO/MDI detectors could not detect such modes confidently even after observing continuously for 16 years: either the amplitude of such modes was too small or the background noise on length-scales of order 17,500 km (see Equation 13.1) for $l = 250$ was too large.

We can also choose other values of $l$ and run our eye vertically through Figure 13.4 to identify mode frequencies $v$ where intersections with the ridges lie. In this way, frequencies can in principle be derived for modes with all values of $l$ down to $l = 0$. For each value of $l$, the modes are found to form a series of "spikes" in frequency. Each such list corresponds to a vertical "cut" through Figure 13.4. By way of example, we note that, for $l = 0$, some 20 peaks were listed by Duvall et al. (1988), with $v$ ranging from 1824 to 4669 µHz. Of interest to us here, we note that the list happens to contain three entries at the following values of $v$: 2899, 3034, and 3169 µHz. (The mode at $v = 3034$ µHz happens to be the one that appears in Figure 13.3.) Notice that the intervals between these three modes are 135 and 135 µHz. These differences are specific examples of a much broader trend in the data. If we examine modes with higher $v$ for *a fixed value of l*, a striking feature of solar oscillations emerges: the interval in frequency between adjacent modes approaches a *constant* asymptotic value $\Delta v$. The value of $\Delta v$ varies only slowly with $l$: for $l$ in the range 0–10, $\Delta v$ is found to lie in the range 135–138 µHz (Appourchaux et al. 1998). The numerical value of this so-called large spacing in frequency contains important information about the time interval required for sound to travel through the interior of the Sun from center to surface (see Section 14.6.1).

## 13.3 RADIAL ORDER OF A MODE

The question is: for a fixed value of $l$, what do the different "spikes" in frequency in Figure 13.4 correspond to? Why are there only certain frequencies in the list of "spikes"? Empirically, the answer is not immediately obvious. We need to turn to theory (Chapter 14) for assistance in this regard. Theory tells us that each mode of the Sun follows a certain pattern *on the surface of the Sun* (with nodes at well-defined latitudes, determined by $l$). However, when we depart from the surface and probe into the interior of the Sun, theory indicates that each mode also has a certain functional form *in the radial coordinate between the center of the Sun and the surface*. The functional form in radius is called the "radial eigenfunction", and it contains a series of "ups and downs" (see, e.g., Chapter 14, Figures 14.2 and 14.3). Between each "up" and the next "down", the eigenfunction passes through zero at a specific radial location. Each such zero is a node of the radial eigenfunction. The number of nodes $n_r$ that exist in the eigenfunction as we follow a radial line between the center of the Sun and the surface is called the *"radial order"* of the mode.

Now we are in a position to interpret the "spikes" in frequency for a given $l$: each "spike" in $\nu$ corresponds to a particular integer $n_r$. For a fixed $l$, the value of $n_r$ increases as the $\nu$ increases towards the upper boundary of Figure 13.4.

In contrast to the angular *degree* $l$, which can be determined *empirically* by examining how oscillations with a particular period are distributed across the *surface* of the Sun (which is clearly observable from Earth), there is no *purely empirical* way to determine the value of $n_r$. To determine the value of $n_r$, we need to have some way to "see" what is happening below the surface of the Sun. How is such a thing possible? The answer is: it has to be done using a theory of some kind. The radial order can be determined only by comparing the observed frequency with calculations of the interior structure of the Sun and seeing which frequency fits best. Such an exercise leads to the conclusion that the three $l = 0$ modes mentioned earlier with $\nu = 2899$, 3034, and 3169 µHz correspond to $n_r = 20$, 21, and 22. In terms of this notation, the mode in Figure 13.3 corresponds to $l = 0$ and $n_r = 21$. (Note that the authors of Figure 13.3 use the slightly different notation $l = 0$, $n = 21$: the reason we prefer to use the label $n_r$ for the radial order rather than $n$ is to avoid confusion with the parameter $n$, which we have already used as the polytropic index in Chapter 10.)

Now that we have introduced the radial order $n_r$, we are in a position to interpret an essential feature of Figure 13.4, specifically, the "ridges" that can be seen extending from lower left to upper right. Model fitting indicates that along each of these ridges, *the radial order $n_r$ retains a unique value*. The value of $n_r$ is smallest for the ridges that lie closest to the lower edge of Figure 13.4. Model fitting suggests that the lowest-lying visible ridge in Figure 13.4 has $n_r = 0$: this mode is referred to as the "fundamental" or $f$-mode where the amplitude of the eigenfunction rises monotonically from small values near the center to large values very close to the surface. The highest lying ridge plotted in Figure 13.4 (which contains only a few modes) has $n_r = 27$. However, the upper boundary in Figure 13.4 is by no means an upper limit on the *largest* values of $n_r$ detected in the Sun: e.g., Duvall et al. (1988), in their spatial-temporal study, have listed one mode having $n_r$ as large as 35. In the opposite limit of small $n_r$, the *lowest* value of $n_r$ reported by Duvall et al. (1988) is $n_r = 5$. Modes with even lower values of $n_r$ (down to $n_r = 1$) have been identified in GONG data at frequencies below 1 mHz by using special averaging techniques (Salabert et al. 2009).

## 13.4 WHICH $p$-MODES HAVE THE LARGEST AMPLITUDES?

Inspection of Figures 13.1 and 13.2 shows that the Sun appears to pump power preferentially into modes with periods of order 5 minutes, i.e., $\nu \approx 3300$ µHz. Such modes correspond to relatively low $l$ and/or relatively large $n_r$ values. For example, in tables of modes that were reliably identified in the Sun in the earliest data sets (e.g., Duvall et al. 1988), modes with $l = 0$ and those with $l = 1$ have been identified with $n_r$ values ranging from $n_r = 12$ to $n_r = 33$. Modes with $l = 2$ and $l = 3$ have been reliably identified with $n_r$ values ranging from $n_r = 6$ to $n_r = 35$. Outside these ranges, e.g., at $n_r < 6$, it is difficult to detect modes reliably in most ground-based data sets. In terms of frequency, the highest and lowest $\nu$ that are reliably detected are close to 5 and 1.0 mHz, respectively (Figure 13.2).

It is natural to ask: why does the Sun preferentially pump power into acoustic modes with just this range of frequencies? Why do we not detect modes with $\nu$ as high as (say) 10 or 20 mHz or as low as (say) 0.1–0.5 mHz? As regards the low $\nu$ behavior, we postpone the discussion until we have a mathematical description of the eigenfunctions (Chapter 14, Section 14.8). But we do not need to understand eigenfunctions in order to determine why modes with $\nu > 5$ mHz are absent. So let us turn now to that piece of the puzzle.

## 13.5 TRAPPED AND UNTRAPPED MODES

The reason why $p$-modes are detectable in the Sun has to do with the fact that certain acoustic waves are able to build up to amplitudes that are so large that we can detect them in observations by instruments located on (or near) Earth. And the reason for the large amplitude is that certain waves are *not allowed* to propagate freely out of the Sun from the place where they are generated. Instead, once those waves are generated, there is something about the Sun's internal structure that causes the waves to be *trapped*: given favorable conditions, trapped waves that bounce back and forth inside the Sun can continue (over the course of a lifetime that may be as long as a month) to gather in energy from the surrounding granules with their significant kinetic energy associated with convective motions. In this way, trapped modes have a chance to build up to large amplitudes. It is precisely such waves that give rise to the high levels of power in the discrete peaks in Figures 13.1 and 13.2.

On the other hand, certain other acoustic waves *can* propagate freely from their place of origin, and these waves may ultimately reach into the upper atmosphere of the Sun. Those propagating waves, which are not subject to trapping (and therefore not of interest in a study of *p*-modes), are of great interest in the context of the heating of the solar chromosphere (see Section 15.9).

In this section, we ask: can we identify a dividing line between trapped and untrapped acoustic waves in the Sun? We shall find that, for waves that are propagating *vertically*, the dividing line occurs at a certain critical wave period $P_c$: vertically propagating waves with periods *longer* than $P_c$ are trapped, while vertically propagating waves with periods *shorter* than $P_c$ are *not* trapped. We shall refer to $P_c$ as the acoustic "cut-off" period.

In this regard, we recall the empirical result that the *p*-modes with significant amplitudes are observed to have $\nu$ = 2.5–4.5 millihertz (mHz): such frequencies correspond to wave periods of 220–400 sec. This tells us that there is something about the solar atmosphere that causes waves with such periods to be trapped: the "something" has to do with the fact that the atmosphere is stratified, with a characteristic length-scale (the "scale height" $H$) and a characteristic speed (the sound speed $c_s$). As a result, there is a characteristic time-scale, of order $t_s = H/c_s$, for sound waves traveling through the atmosphere: we shall find that the cut-off period is proportional to $t_s$.

### 13.5.1 Vertically Propagating Waves in a Stratified Atmosphere

Consider the propagation of a sound wave *vertically* in an atmosphere that is in HSE. As a result of HSE, the pressure in the unperturbed atmosphere has a vertical distribution $p_o(h)$ that obeys the equation

$$\frac{dp_o}{dh} = -g\rho_o \qquad (13.2)$$

In this equation, the vertical coordinate $h$ increases in the *upward* direction. The right-hand side represents the weight of 1 cm³ of gas acting as a downward force on each square centimeter of the atmosphere at height $h$. This downward force is balanced by (the gradient of) the pressure acting upward on that square centimeter.

We now imagine that we impose a small vertical displacement $\xi$ on a parcel of gas caused by a local perturbation in pressure. We assume that the displacement raises or lowers the parcel in such a way that initially there is no change in density. Denoting the new pressure by $p(h) = p_o(h) + \Delta p$, we see that the vertical force acting on 1 cm³ of gas is

$$\Delta F = \frac{dp}{dh} + g\rho_o \qquad (13.3)$$

This force is no longer equal to zero: HSE is not satisfied. In view of the conservation of momentum (i.e., Newton's second law of motion), the imbalance in vertical forces $\Delta F$ acting on 1 cm³ of gas (with mass $\rho_o$) leads to a vertical acceleration $d^2\xi/dt^2$ such that

$$\rho_o \ddot{\xi} = -\frac{dp}{dh} - g\rho_o = -\frac{d}{dh}(\Delta p) \qquad (13.4)$$

# Oscillations in the Sun

In Equation 13.4, double dots denote the second derivative with respect to *time*. This is the equation of motion for the parcel of gas as it responds to the change in ambient pressure gradient. (In Section 7.1, we also applied Newton's law of motion to gas in the convection zone, where the breakdown of HSE also leads to an imbalance of forces, with consequent vertical acceleration.)

As the parcel of gas moves, its density does not remain constant: the internal density changes by a finite amount $\Delta\rho$ if the magnitude of the displacement $\xi$ varies with $h$. To quantify this, we consider the conservation of mass associated with the displacement. Suppose that the initial parcel of gas spanned an interval of height $\Delta h$, but when it is displaced, that parcel spreads out over a different interval of height $\Delta\xi$. There are two possibilities. On the one hand, if $\Delta\xi$ *exceeds* $\Delta h$, the parcel has been "stretched", and the internal density of the gas *decreases* as a result of the displacement. On the other hand, if $\Delta\xi$ is *less than* $\Delta h$, the parcel has been compressed, and the internal density of the gas *increases* as a result of the displacement. In both cases, the fractional change in density is related to the ratio of $\Delta\xi$ to $\Delta h$. In the limit, the fractional change in density occurring in the parcel of gas can be written as

$$\frac{\Delta\rho}{\rho} = -\frac{d\xi}{dh} \tag{13.5}$$

Note that if $\xi$ has the same value at all $h$, i.e., if the atmosphere is displaced as a whole by a constant amount, the derivative $d\xi/dh$ equals zero and there is no change in density at any location.

Now that we have taken into account the conservation of momentum and the conservation of mass, it remains to address the conservation of energy. In the present case, we will incorporate this by assuming that the sound waves propagate *in an adiabatic manner*, i.e., $p \to \rho^\gamma$. Thus, $\Delta p/p = \gamma\Delta\rho/\rho = -\gamma d\xi/dh$. To first order, we therefore replace $\Delta p$ in Equation 13.4 by $-\gamma p_o d\xi/dh$.

Using this in Equation 13.4, we find

$$\rho_o \ddot{\xi} = \frac{d}{dh}\left(\gamma p_o \frac{d\xi}{dh}\right) \equiv \gamma p_o \xi'' + \gamma \xi' \frac{dp_o}{dh} \tag{13.6}$$

On the right-hand side, primes denote *spatial* derivatives with respect to $h$. Dividing through by $\rho_o$ and noting that the adiabatic sound speed $c_s$ is given by $c_s^2 = \gamma p_o / \rho_o$, we can rewrite Equation 13.6 in the form

$$\ddot{\xi} = c_s^2 \xi'' - \gamma g \xi' \tag{13.7}$$

Equation 13.7 is the equation that describes how *a sound wave propagates* vertically *through a stratified atmosphere*, i.e., in an atmosphere where finite gravity is present.

In the special case where gravity is absent, we can set $g = 0$. In this case, the background medium is unstratified, i.e., it is homogeneous (with equal density) in all directions. In such a case, Equation 13.7 reduces to the standard wave equation $\ddot{\xi} = c_s^2 \xi''$ describing waves propagating through a homogeneous medium with speed $c_s$. In this case, there are no limitations on the direction of travel or on the frequency of the waves that may propagate: the wave properties are determined solely by the properties of sound in the ambient medium.

However, in the presence of gravity, the second term on the right-hand side of Equation 13.7 comes into play. This is the term that makes a significant difference to the properties of sound wave propagation if the ambient medium is no longer homogeneous but now stratified.

## 13.5.2 Simplest Case: The Isothermal Atmosphere

Let us consider the case of an isothermal atmosphere with temperature $T$. In this case, we have already (Section 5.1) seen that the density is stratified as a function of height according to an exponential law: $\rho(h) = \rho_o \exp(-h/H)$ where the scale height $H$ is given by the expression $H = R_g T / g\mu_a$ where $\mu_a$ is the mean atomic weight of the atmospheric gas. For a perfect gas, the sound speed $c_s$ can be written as $c_s^2 = \gamma R_g T / \mu_a = \gamma g H$.

In order to proceed with the solution in this case, it is convenient to transform to dimensionless variables. We introduce a new dimensionless length coordinate: $h' = h/2H$. Note that in order to convert the dimensional length $h$ to dimensionless form, we normalize to a length that is *not* equal to the scale height, but is equal to *twice* the scale height.

We also introduce a new dimensionless time-scale by normalizing the time variable to a time-scale related to the characteristic time-scale $t_s = H/c_s$ mentioned earlier in Section 13.5. But again, rather than using $t_s$ as the normalizing factor, we instead use *twice* the value of $t_s$. That is, the dimensionless time-scale is defined by $t' = t/(2H/c_s)$.

Converting now the temporal and spatial derivatives to the new dimensionless variables of time and height, and making use of the relation $c_s^2 = \gamma g H$, we find that Equation 13.7 takes on the form

$$\ddot{\xi} = \xi'' - 2\xi' \qquad (13.8)$$

In Equation 13.8, dots now denote differentiation with respect to the dimensionless time $t'$, while primes now denote differentiation with respect to the dimensionless length $h'$. Equation 13.8 describes, in terms of dimensionless length and time variables, how a stratified atmosphere responds to a sound wave propagating vertically.

To solve Equation 13.8, one further change of variables is helpful: we replace $\xi$ with the auxiliary variable $u = \xi \exp(-h')$. This leads to the equation

$$\ddot{u} = u'' - u \qquad (13.9)$$

In order to describe wave motion, we seek a periodic solution to this equation: $u = u_o e^{i\omega t'}$ where $u_o$ is a function of the spatial coordinate $h'$ only, $\omega$ is the angular frequency associated with the dimensionless time $t'$, and $i = \sqrt{-1}$. Substituting this in Equation 13.9, we find

$$u_o'' = -(\omega^2 - 1)u_o \equiv -A u_o \qquad (13.10)$$

where $A$ is defined as $\omega^2 - 1$.

Mathematically, Equation 13.10 has two well-known classes of solutions, depending on the algebraic sign of $A$.

Class (i) $A > 0$. The parcel of gas undergoes a displacement $u$ that is simple harmonic motion (i.e., proportional to an exponential with an *imaginary* argument) in the (dimensionless) height coordinate $h'$:

$$u_o = \exp(\pm i h' \sqrt{A}) \qquad (13.11)$$

When combined with the sinusoidal time factor $e^{i\omega t'}$, Equation 13.11 represents a *sound wave freely propagating* in a vertical direction through the atmosphere.

The condition that the value of the parameter $A$ must exceed zero for the solution of Equation 13.11 to be valid means (see Equation 13.10) that $\omega^2$ must exceed unity. That is, the wave *frequency must* exceed *a certain value in order that the wave may propagate freely in the vertical direction*.

Class (ii) $A < 0$. In this case, the frequency $\omega$ is so small that $\omega^2 - 1$ takes on a negative value. In this case, Equation 13.10 contains a positive number on the right-hand-side. The solution is an exponential with a *real* argument:

$$u_o = \exp(\pm h' \sqrt{A}) \qquad (13.12)$$

In order to avoid divergence at infinity, only the *damped* solution in Equation 13.12 is physically meaningful. The nature of the solution in Class (ii) is very different from the solution in Class (i). The damped solution indicates that *vertical* waves *do not propagate through the stratified atmosphere at low frequencies*.

### 13.5.3 Critical Frequency and the "Cut-Off" Period

The transition between the *oscillatory* (*propagating*) solutions in Class (i) and the *damped* (*non-propagating*) solutions in Class (ii) occurs at $A = 0$, i.e., at $\omega^2 = 1$. Reverting to dimensional variables, the corresponding transition occurs at the critical (angular) frequency

$$\omega = \omega_{ac} \equiv \frac{c_s}{2H} \tag{13.13}$$

This critical (angular) frequency, identified by subscript "ac", is referred to as *the acoustic cut-off (angular) frequency for vertically propagating waves*.

The corresponding cut-off *period* $P_{ac}$ for *vertically propagating waves* is given by

$$P_{ac} = \frac{2\pi}{\omega_{ac}} = \frac{4\pi H}{c_s} = \frac{4\pi}{g}\sqrt{\frac{R_g T}{\gamma \mu_a}} \tag{13.14}$$

### 13.5.4 Physical Basis for a Cut-Off Period

Why, in physical terms, does a cut-off period exist in the Sun's atmosphere? Why is it that waves with periods longer than $P_{ac}$ cannot propagate vertically? To answer this, consider what happens if one tries to launch a wave vertically with a certain period into the atmosphere. If the wave period is longer than $P_{ac}$, the spatial extent of one wavelength ($\lambda = c_s P_{ac}$) of such a wave extends over many ($4\pi$, i.e., $> 10$) scale heights of the atmosphere: at one end of the wavelength, the local density differs from the density at the other end of the wavelength by a factor of $e^{10}$, i.e., by a factor more than 10,000. With such a large density contrast, during the time that the wave is propagating vertically across one of its own wavelengths, the atmosphere has time to "adjust itself" to the perturbation: the effects of the adjustment are to cancel out the wave, thereby preventing it from propagating upward. The wave is said to be "trapped", i.e., confined to the lower layers of the atmosphere where the wave originated. The stratified atmosphere in effect can "short out" the long-period waves.

On the other hand, a vertical wave with a short period ($< P_{ac}$) can propagate vertically across one of its own wavelengths before the ambient atmosphere has time to adjust: in the absence of this adjustment, the wave is no longer "shorted out" by the atmosphere. Instead, the short-period wave is free to continue its vertical propagation upward into the overlying atmosphere. Such waves have the ability to travel vertically upward into the overlying chromosphere and corona.

### 13.5.5 Numerical Value of the Cut-Off Period

Now we come to a key question: what is the numerical value of the critical period in the Sun? In the upper photosphere of the Sun, where $T = 4860$ K and $\mu_a \approx 1.3$, we find

$$P_{ac} \approx 195 - 200 \text{ seconds} \tag{13.15}$$

The corresponding cut-off (linear) frequencies ($\nu = \omega/2\pi$) are $\nu_{ac} \approx 5.0 - 5.1$ mHz.

Recall that the *p*-modes in the Sun have detectable amplitudes for frequencies that are no greater than (about) 5 mHz (see Figure 13.2). Now that we have derived the concept of the acoustic cut-off of a stratified atmosphere, we can understand why *p*-modes are not detectable with periods shorter than (roughly) 200 sec: such waves, with frequencies $\nu > 5$ mHz, have $\nu > \nu_{ac}$. Such waves are free to propagate vertically through the solar atmosphere in accordance with Equation 13.11. Such waves are therefore *not* trapped: they escape easily from the location where they are generated. They do not "stick around" long enough to build up their amplitude to the large values associated with *p*-modes.

On the other hand, acoustic waves with periods longer than 200 sec *cannot* propagate vertically through the upper solar photosphere: mathematically, their "wave form" is described by the negative sign in Equation 13.12. When such waves encounter the upper photosphere, they are not permitted to propagate further in a vertical direction: *instead, they are reflected back down into the Sun*. This sets up the possibility of those waves becoming trapped. And if they are trapped, then they can "stick around" long enough to have energy pumped into them by the convective motions. The more energy is pumped in, the larger their amplitudes become, and the easier it is for us to detect them with our near-Earth instruments at a distance of 150 million km from the Sun.

The principal conclusion of this section is the following: the effects of atmospheric stratification explain why *p*-modes are *not* detectable at $\nu$ *larger* than (roughly) 5 mHz.

In order for waves to be trapped, the Sun needs to provide a "cavity" of some kind to contain the waves. We have now identified wave reflection as a reason why there exists an *upper* boundary to such a cavity in the Sun. In Section 14.7, we shall discuss a mechanism that gives rise to the existence of a *lower boundary* to the cavity.

## 13.6 WAVES PROPAGATING IN A *NON-VERTICAL* DIRECTION

The existence of a cut-off in wave frequencies as regards propagation of waves in the solar atmosphere has been discussed in Sections 13.5.1–13.5.5 in the context of *vertically propagating waves*. Since the gravitational acceleration *g* is directed exactly vertically in any given location on the Sun, vertical wave propagation is subject to the entire magnitude of solar gravity, $g = 27,420$ cm sec$^{-2}$ (see Equation 1.13). As we have found (see Equation 13.14), the acoustic cut-off period $P_{ac}$ depends on $1/g$. For vertically propagating waves, we have found that $P_{ac}$ has a numerical value of 195–200 sec (see Equation 13.15). This means that the Sun should not contain waves propagating vertically in its atmosphere with periods any longer than 200 sec.

However, in certain regions of the Sun, waves can be seen propagating upwards with periods $P$ that are definitely *longer* than the critical value $P_{ac} \approx 200$ sec. In one case, Kobanov et al. (2013) show images of the location of high levels of acoustic power at $P = 300$ sec. In another case, the wave periods are observed to be as large as 500 sec (e.g., Rajaguru et al. 2019).

How are we to understand the propagation of such long-period waves? Since $P_{ac}$ scales as $1/g$, one way to *increase* the value of $P_{ac}$ would be (in principle) to *decrease* the effective value of *g*. How might we achieve that goal? It is helpful here to recall the work of Galileo from the early 1600s when he was trying to measure how fast a falling object moves and/or accelerates under the influence of gravity. To measure speed/acceleration, it is essential to have a reliable clock. Unfortunately, Galileo's clocks (either water clocks or his own heartbeats) were not good enough to measure the speed/acceleration of a falling object if that object were released *vertically*: in such a case, the object moved too quickly for Galileo's clocks to measure the fall time reliably. What was he to do? He decided to use an inclined plane (i.e., a "ramp"): if the ramp were tilted at an angle θ relative to the vertical, the effective gravity $g_{eff}$ would be reduced from *g* to $g \cos\theta$. Thus, if θ = 60 degrees, then $g_{eff}$ has a value of 0.5*g*. It is key to note that the regions where sunspot waves with $P = 300$ sec are observed (Kobanov et al. 2013) are in a very particular part of a sunspot known as the penumbra, where magnetic field lines are tilted relative to the vertical by angles that can be as large as 60 degrees or more. Magnetic fields have the effect that gas that is significantly ionized is forced to move along the direction of the field lines (see Sections 16.6.1 and 16.6.2.2).

Therefore, in the penumbra, gas is constrained to move along ramps where $g_{eff}$ may be as small as $0.5g$ (or even less). As a result, the local $P_{ac}(loc)$ is no longer confined to values of 195–200 sec (as in Equation 13.15). Instead, $P_{ac}(loc)$ in the penumbra may increase to as large as 400 sec (or larger). This easily allows the 5-minute oscillations (where $p$-mode power is observed to be maximum), with $P = 300$ sec, to satisfy the local propagation condition $P < P_{ac}(loc)$. The results of Kobanov et al. (2013) clearly show that 5-minute oscillations *can* and *do* propagate up into the chromosphere and even into the low corona in regions where a suitable ramp is available. Such a ramp provides a "magneto-acoustic portal" for long-period waves (which are normally cut-off from propagating up into the chromosphere) to reach the chromosphere (Kontogiannis et al. 2010). Galileo would be interested to know that after an interval of 400 years, solar physicists are also putting ramps to good use in their studies.

In view of the fact that acoustic waves with periods longer than 200 seconds should, in principle, remain trapped beneath the surface of the Sun and should never reach the chromosphere, the fact that ramps in the magnetic field allow otherwise "trapped" waves to reach the chromosphere has led to a dramatic statement by Cally and Moradi (2013): "Active regions are open wounds in the Sun's surface". One of the meanings of this statement is that "internal material" such as acoustic waves with $P > 200$ sec, which should ordinarily remain hidden beneath the solar surface, may find a way to break up through the surface of an active region if that region contains an appropriate "ramp". Another meaning of Cally and Moradi's phrase is that a wave of one mode (which would ordinarily be trapped beneath the surface) can undergo a conversion into another mode (see Section 16.7.7) that is not subject to trapping, thereby gaining access to the upper atmosphere.

## 13.7 LONG-PERIOD OSCILLATIONS IN THE SUN

The presence of oscillations in the Sun at periods of a few minutes has been known for several decades. These are the well-studied $p$-modes.

But we can also ask: might there be oscillations in the Sun with much longer periods (e.g., hours)? We shall see in Chapter 14 that two distinct classes of oscillation can exist (under certain conditions) in a compressible sphere, one at high frequency (the $p$-modes, where gas pressure is the main restoring force), the other at low frequency (the $g$-modes, where gravity is the main restoring force). Since $p$-modes have certainly been detected in the Sun, it is natural to ask: have $g$-modes also been detected in the Sun? The answer to this question is controversial at the time of writing (2021): although claims of detection of hour-long periodicities have been reported by some observers, other observers have not yet confirmed the detections.

One reason for the difficulty of detecting $g$-modes in the Sun has to do with the extensive convective envelope occupying the outermost one-quarter to one-third of the solar radius (see Chapter 7). As it turns out, $g$-modes can propagate freely only in a medium that is *stable* against convection: a vertical displacement of an element of gas in such a medium is acted upon by gravity to restore the element to its starting position. Such a response allows a small-amplitude wave (i.e., a $g$-mode) to propagate in the medium. The gas in the deep radiative interior of the Sun (which was modeled in Chapters 8 and 9) is convectively stable and is therefore expected to support $g$-modes. But in a convectively *unstable* medium, the gravity force cannot restore the element to its original position: instead, the gravity force continues to drive an upward (or downward) element up (or down) farther and farther away from its starting point (see Section 6.8). Convective motions, with vertical velocities of up to a few km sec$^{-1}$, are the preferred manner for gas to move in response to the gravity force in unstable gas (such as the solar convection zone). Even though there might well be $g$-modes in the radiative interior of the Sun, the existence of a thick convective envelope could effectively hide such waves from our sight.

Fortunately, there are certain stars where (unlike the situation in the Sun) the surface layers are convectively stable. In Chapter 14, once we have identified an important characteristic of the periods of $g$-modes, we shall cite unambiguous evidence for $g$-modes in a particular category of such stars.

Although the possible existence of *p*-modes and *g*-modes in a compressible (nonrotating) sphere has been known theoretically for many decades (since the work of Cowling [1941]), there can also be a third class of oscillation modes (referred to as *r*-modes or *Rossby waves*) if the star is rotating. The periods of *r*-modes in a star are related to the rotation period of the star: in the case of the (slowly rotating) Sun, this means that *r*-modes are expected to have periods of order months, i.e., orders of magnitude longer than the periods of *p*- or *g*-modes. The first reported detection of *r*-modes in the Sun appeared in 2018, when Löptien et al. analyzed 6 years of data from the Solar Dynamics Observatory (SDO): the latter has enough spatial resolution to allow detection of photospheric granules, and the vorticities of granule motions were used to identify *r*-modes. Further details will be given in Chapter 14. Prior to the solar detections, *r*-modes had also been detected in certain rapidly rotating hot stars (Van Reeth et al. 2016).

## 13.8 *p*-MODE FREQUENCIES AND THE SUNSPOT CYCLE

We shall see in Section 16.1.4 that the Sun undergoes a cycle of magnetic activity during which the number of sunspots on the surface waxes and wanes on a time-scale of 11 years (or so). It is of interest to ask: does the magnetic activity have an effect on the *p*-modes as the Sun changes from solar minimum to solar maximum? The answer is yes: the frequency of any given *p*-mode is observed to *increase* when the Sun has *more* sunspots. The amount of the frequency shift Δν(mx–mn) between solar maximum and solar minimum is smaller for modes with lower frequencies. Using data from SOHO/GOLF over an interval of 18 years spanning cycle 23 and part of cycle 24, Salabert et al. (2015) reported the following shifts: for modes with frequencies between 1.8 and 2.75 mHz, Δν(mx–mn) ≈ +0.1 μHz, for modes with frequencies between 2.4 and 3.1 mHz, Δν(mx–mn) ≈ +0.3 μHz, and for modes with the highest frequencies (3.1–3.8 mHz), Δν(mx–mn) ≈ +0.6 μHz. In an independent analysis of SOHO/MDI data, Jain et al. (2012) found that the frequency shifts of *p*-modes tracked the sunspot number so well during cycle 23 that the double peak in sunspot numbers (known as the Gnevyshev gap: see Section 16.1.4) that occurred in the years 2000–2002 could also be identified in the values of Δν. This suggests that the Gnevyshev gap involves real physical changes inside the Sun and is not simply a quirk of statistics.

So sensitive are the current measures of *p*-mode frequencies that it has become possible to detect a small shift in frequency even between one solar *minimum* and the next solar *minimum*. Broomhall (2017), in an analysis of GONG data from two solar minima (1996–1997 and 2008–2009), found that the frequency shifts in 2008–2009 (a long and deep minimum) were definitely *smaller* than in the 1996–1997 minimum. The differences, averaged over all modes, amounted to Δν(mn1–mn2) ≈ +0.01μHz, corresponding to a magnetic field in the 2008–2009 minimum being *weaker* than in the 1996–1997 minimum by about 1 G. It is remarkable that the average frequencies of solar *p*-modes can now be measured with a precision as good as 10 *nano*hertz.

## EXERCISES

13.1 In Chapter 1, Exercise 1.5, you have already calculated surface gravities for five "main sequence" stars with masses of $0.1–10 M_\odot$. Calculate the cut-off period $P_{ac}$ for each star, assuming $\mu_a = 1.3$, and setting $T = T_{\text{eff}}$ as calculated in Chapter 1, Exercise 1.6.

13.2 Using the properties of the same five "main sequence" stars as in Exercise 13.1, calculate the critical gravity period $P_g$ (Chapter 1, Equation 1.24) for each star.

13.3 How large must the degree $l$ of a mode in the Sun be in order to have a horizontal wavelength equal to (a) a supergranule diameter (≈ 30 thousand km; see Chapter 15), (b) a granule diameter?

## REFERENCES

Appourchaux, T., Rabello-Soares, M.-C., & Gizon, L., 1998. "The art of fitting $p$-mode spectra. II. leakage and noise-covariance matrices", *Astron. Astrophys. Suppl.* 132, 131.

Broomhall, A. M., 2017. "A helioseismic perspective on the depth of the minimum between solar cycles 23 and 24", *Solar Phys.* 292, 67.

Cally, P. S., & Moradi, H., 2013. "Seismology of the wounded Sun", *Mon. Not. Roy. Astron. Soc.* 435, 2589.

Cowling, T. G., 1941. "The non-radial oscillations of polytropic stars", *Mon. Not. Roy. Astron. Soc.* 101, 367.

Duvall, T. L., Harvey, J. W., Libbrecht, K. G., et al., 1988. "Frequencies of solar $p$-mode oscillations", *Astrophys. J.* 324, 1158.

Fossat, E., Grec, G., & Pomerantz, M. A., 1981. "Solar pulsations observed from the geographic south pole", *Solar Phys.* 74, 59.

Gavryusev, V. G., & Gavryuseva, E. A., 1997. "Statistical properties of the pulses in solar $p$-mode power", *Solar Phys.* 172, 27.

Jain, R., Tripathy, S. C., Watson, F. T., et al., 2012. "Variation of solar oscillation frequencies in solar cycle 23", *Astron. Astrophys.* 545, A73.

Kiefer, R., Komm, R., Hill, F., et al., 2018. "GONG $p$-mode parameters through two solar cycles", *Solar Phys.* 293, 151.

Kobanov, N. I., Chelpanov, A. A., & Kolobov, D. Y., 2013. "Oscillations above sunspots from the temperature minimum to the corona", *Astron. Astrophys.* 554, A116.

Kontogiannis, I., Tsiropoula, G., & Tziotziou, K., 2010. "Power halo and magnetic shadow in a solar quiet region", *Astron. Astrophys.* 510, A41.

Larson, T. P., & Schou, J., 2015. "Improved helioseismic analysis of medium-$l$ data from MDI", *Solar Phys.* 290, 3221

Löptien, B., Gizon, L. Birch, A. C., et al., 2018. "Global-scale equatorial Rossby waves as an essential component of solar internal dynamics", *Nature Astron.* 2, 568.

McClure, R. L., Rast, M. P., & Pillet, V. M., 2019. "Doppler events in the solar photosphere: Superposition of fast granular flows and $p$-mode coherence patches", *Solar Phys.* 294, 18.

Rajaguru, S. P., Sangeetha, C. R., & Tripathi, D., 2019. "Magnetic fields and supply of low-frequency acoustic wave energy to the solar chromosphere", *Astrophys. J.* 871, 155.

Salabert, D., Garcia, R. A., & Turck-Chieze, S., 2015. "Seismic sensitivity to sub-surface solar activity", *Astron. Astrophys.* 578, A137.

Salabert, D., Leibacher, J., Appourchaux, T., et al., 2009. "Low signal-to-noise solar $p$-modes", *Astrophys. J.* 696, 653.

Van Reeth, T., Tkachenko, A., & Aerts, C., 2016. "Interior rotation of a sample of γ Doradus stars from ensemble modelling of their gravity-mode period spacings", *Astron. Astrophys.* 593, A120.

# 14 Oscillations in the Sun
## *Theory*

In order to understand in physical terms why the Sun exhibits oscillations at precisely defined frequencies, we consider in this chapter the oscillations in an idealized "star". Specifically, we revert to the topic of polytropes (see Chapter 10), since these provide in certain cases an analytic form for the radial profile of pressure and density inside a star: in the presence of such radial profiles, oscillations of various kinds ($p$-modes, $g$-modes) can occur. Of course, if we were undertaking a detailed examination of the Sun, we would have to make use of the full numerical radial profiles of pressure and density: but unfortunately, those numerical solutions make it more complicated to derive the properties of the oscillations. So in this first course in solar physics, we simplify the problem by considering the oscillation modes of a polytrope. Results from the polytropic case contain many of the important characteristics of oscillations in the "real Sun": in particular, we will discover (see Section 14.6) that $p$-modes have the property (at high frequency) of being separated from adjacent modes by a constant interval of *frequency*, while $g$-modes have the property (at long periods) of being separated from adjacent modes by a constant interval in *period*.

In this chapter, we derive a pair of first-order differential equations (Equations 14.17 and 14.18) that describe the properties of oscillations in a polytrope. The pair of equations we shall derive represent a simplification of the full oscillation problem (which actually requires four equations to specify completely). The simplification from four equations to two equations is known as the "Cowling approximation". Moreover, in line with our choice of polytropes, Cowling (1941) also relied on polytropes to provide the "background" ambient medium in which the oscillations would occur. Students will have an easier time exploring the computational properties of the simpler system.

The idealized "star" that we will use for our discussion of $p$-modes and $g$-modes is assumed to be nonrotating. Cowling (1941) also confined his attention to nonrotating stars because he writes that, in the rotating case, "the mathematical difficulties are much greater". But the "real Sun" does actually rotate (see Section 1.11). In the presence of rotation, a third class of modes of oscillation, the $r$-modes, can be generated. Because the forces involved in driving $r$-modes are distinct from those in $p$-modes and $g$-modes, the description of $r$-modes requires a different approach from the Cowling approximation. In Section 14.10, we will address the physics of $r$-modes.

## 14.1 SMALL OSCILLATIONS: DERIVING THE EQUATIONS

In order to derive the equations that govern oscillations in a (nonrotating) polytrope, we follow the discussion first given (in the midst of World War II) by the British mathematician T. G. Cowling (1941). Let the material at any point in the polytrope undergo a vector displacement **h** (where boldface type denotes a vector quantity). This vector displacement is in a general direction, but the component of **h** along the *radial* direction is of particular interest: we refer to this radial component as $R$. The displacement is accompanied by localized perturbations in density, pressure, and gravitational acceleration: we refer to these as $\delta\rho$, $\delta p$, and $\delta g$.

Our goal in this section is to derive three equations that will allow us to solve for the three physical quantities that describe the oscillations: $R$, $\delta\rho$, and $\delta p$.

The perturbations induced in the star by the displacement **h** have the effect that the equation of HSE is no longer satisfied. The imbalance of pressure and gravity forces leads to an acceleration that is given by the equation for *conservation of momentum*

$$\rho \frac{d^2 \mathbf{h}}{dt^2} = -\nabla p - \rho \mathbf{g} \tag{14.1}$$

where the spatial gradient (vector) operator $\nabla$ includes three components, including one that points along the radial (outward) direction. Equation 14.1 is a more general (3-D) form of the 1-D Equation 7.1. (Note that in Equation 7.1, where we were considering convective motions, our goal was to estimate only *vertical* motions.) Instead of containing only the vertical velocity $V$, Equation 14.1 includes the 3-D vector velocity $\mathbf{v} = d\mathbf{h}/dt$. Moreover, in Equation 7.1, the pressure gradient involved the derivative $d/dz$, where the depth coordinate $z$ increases *inward*, whereas in Equation 14.1, the vector operator $\nabla$ involves a component $d/dr$, where the radial coordinate $r$ increases *outward*: this accounts for the difference in sign in the first term on the right-hand side. Why do we need to consider more than 1-D (radial) motions in the present chapter? Because the oscillations in the Sun are *not* confined to the radial direction: most of the oscillations are actually *non-radial* in nature. This point was made explicit in the title of Cowling's (1941) article.

The essential feature of any oscillation is that some physical quantity, responding to two types of forces (one that drives the system away from equilibrium, the second that attempts to restore equilibrium) undergoes periodic motion. Suppose that the displacement **h** is periodic, with a time dependence $e^{i\omega t}$, where $i = \sqrt{(-1)}$ and the angular frequency $\omega$ is related to the frequency $\nu$ (used in Chapter 13) by $\omega = 2\pi\nu$. Then the left-hand side of Equation 14.1 can be written as $-\rho\omega^2 \mathbf{h}$. Retaining only terms that are of first order in the perturbation amplitude, the right-hand side of Equation 14.3 becomes

$$-\nabla \delta p - g \delta \rho - \rho \delta g \tag{14.2}$$

In order to keep the discussion as simple as possible but retain the essential physics of oscillation, we now invoke what is called the "Cowling approximation": we neglect changes in the gravity, i.e., we set $\delta g = 0$. Why is it plausible to neglect changes in the gravitational acceleration? Because the mass in a star is concentrated toward the center: the central density is much larger than the density in the outer regions: e.g., in the Sun the central density exceeds the density in the photosphere by a factor of more than $10^8$. Now, the oscillations we consider here consist of motions that have maximum amplitudes in the outer regions of the star, where the eigenfunctions reach their maximum amplitudes (see Figures 14.2 and 14.3). As a result, the mass interior to a certain point remains almost unchanged by the slight changes associated with oscillations. This allows us to assume $\delta g = 0$ as a reasonable simplifying approximation.

The small oscillations occur as perturbations in a medium (the polytrope) that is in HSE. This allows us to write $g = -\nabla p/\rho$, and so we can rewrite Equation 14.2 in the form

$$-\rho\omega^2 \mathbf{h} = -\nabla \delta p + (\delta\rho/\rho)\nabla p \tag{14.3}$$

This equation has three components. Let us consider one of those components, namely the radial component. This leads to the following expression for the quantity $R$:

$$\rho\omega^2 R = \frac{\partial}{\partial r}(\delta p) - \frac{\delta \rho}{\rho} \frac{\partial p}{\partial r} \tag{14.4}$$

This gives us the first equation (of three) that relate our three unknowns $R$, $\delta p$, and $\delta \rho$.

# Oscillations in the Sun

Turning now to the *conservation of mass*, the equation of continuity $\partial\rho/\partial t + \nabla \cdot (\rho v) = 0$ can be written, to first order in the perturbations, as

$$\delta\rho = -\nabla \cdot (\rho \mathbf{h}) \tag{14.5}$$

We can now eliminate $\mathbf{h}$ from Equation 14.3 by taking the divergence of both sides:

$$\omega^2 \delta\rho = -\nabla^2 (\delta p) + \nabla \cdot \left[ \frac{\delta\rho}{\rho} \nabla p \right] \tag{14.6}$$

In a spherical object, it is natural to separate the oscillations into two components: one depends only on the angular coordinates, and the second is a function of the radial coordinate $r$. The Laplacian operator in Equation 14.6 contains a radial component and an angular component. The latter can be described by a spherical harmonic, $Y_{lm}$. Here, we ignore the $m$ (longitudinal) subscript and consider only the latitudinal variations, which are characterized by $l$, the degree of the mode. In this case, the angular (latitudinal) part of the Laplacian in Equation 14.6 can be written as $-l(l+1)\delta p/r^2$.

In spherical coordinates, the expressions for the radial components of Laplacian and divergence lead to the following form for Equation 14.6:

$$\omega^2 \delta\rho = \frac{l(l+1)}{r^2} \delta p - \frac{1}{r^2} \frac{\partial}{\partial r} \left( r^2 \frac{\partial \delta p}{\partial r} - r^2 \frac{\delta\rho}{\rho} \frac{\partial p}{\partial r} \right) \tag{14.7}$$

We note that on the right-hand side of Equation 14.7, there are terms inside the large parentheses that are reminiscent of terms on the right-hand side of Equation 14.4. Using the latter equation in Equation 14.7 leads to a second equation that relates our three unknowns:

$$\omega^2 \delta\rho = \frac{l(l+1)}{r^2} \delta p - \frac{1}{r^2} \frac{\partial}{\partial r} (\rho \omega^2 r^2 R) \tag{14.8}$$

Now that we have eliminated the angular dependences, there is only one remaining independent variable: the radial coordinate. As a result, we can safely replace the partial derivative $(\partial/\partial r)$ in Equations 14.4 and 14.8 by the total derivative $(d/dr)$.

Equations 14.4 and 14.8 provide us with two equations that relate the displacement of the fluid element $R$ in the radial direction to the perturbations in pressure $\delta p$ and density $\delta \rho$. A third equation is needed if we are to solve for the three unknowns. Having already used the equations that describe conservation of *mass* (Equation 14.5) and conservation of *momentum* (Equation 14.1), we now turn to the equation for conservation of *energy* in order to derive a third equation relating the three unknowns $R$, $\delta p$, and $\delta \rho$.

The simplified form of the energy equation that we use in this first course in solar physics is the following: the oscillations are assumed to be *adiabatic*. That is, when the oscillations occur, the total pressure variation $\Delta p$ is related to the total density variation $\Delta \rho$ by an adiabatic relation: $\Delta p/p = \gamma \Delta\rho/\rho$. Here, $\gamma$ is referred to as the "adiabatic exponent". In a notation analogous to the definition of the polytropic index (see Chapter 10, Equation 10.1), Cowling (1941) writes $\gamma$ in the form $\gamma = 1 + (1/N)$. For a typical adiabatic index $\gamma = 5/3$ (such as that which occurs in a monatomic nonionizing gas), the value of $N$ takes on the particular value 1.5.

What are the total changes in pressure that occur as a result of the oscillation? First, there is $\delta p$ itself. However, since the oscillating element of fluid has also moved a radial distance $R$, the fluid element finds itself at a radial location where the ambient pressure is *different* from the value it had in the unperturbed location. Thus, the total change in pressure $\Delta p$ associated with the oscillation is the sum of two terms: $\Delta p = \delta p + R(dp/dr)$. An equivalent sum of terms applies also to the total change in density. Then the adiabatic version of the energy equation can be written:

$$\frac{\delta p + R(dp/dr)}{p} = \left(1 + \frac{1}{N}\right)\left(\frac{\delta \rho + R(d\rho/dr)}{\rho}\right) \quad (14.9)$$

The Cowling approximation has therefore provided us with three equations, Equations 14.4, 14.8, and 14.9, for three unknowns.

## 14.2 CONVERSION TO DIMENSIONLESS VARIABLES

To help cast the equations into more convenient form, we introduce some dimensionless variables. In this process, we are guided by the choices that were made in Chapter 10 in connection with dimensionless variables in a polytrope of order $n$.

First, we change from the dimensional frequency $\omega$ to a dimensionless quantity $\alpha$ according to the definition:

$$\alpha = \frac{\omega^2(1+n)}{4\pi G \rho_c} \quad (14.10)$$

The fact that $\alpha$ is dimensionless can be verified by recalling the definition of the Emden unit of length $r_o$ (see Equation 10.9): the combination $1/\sqrt{(G\rho)}$ (which occurs in Equation 1.24 and also in Equation 14.10) has the dimensions of time-squared, and therefore when multiplied by $\omega^2$ (with dimensions 1/time-squared) as in Equation 14.10 leads to a dimensionless quantity. Note that the (dimensional) frequency $\omega$ scales as $\sqrt{\alpha}$.

Second, we reduce the radial displacement $R$ to dimensionless form $X$ by normalizing to the Emden unit of length: $X = R/r_o$. Analogously, we express the radial coordinate $r$ as a new dimensionless variable $x = r/r_o$. We reduce the pressure and density perturbations to dimensionless forms by normalizing to their respective values at the center of the "star": $\theta = \delta p/p_c$ and $\eta = \delta \rho/\rho_c$.

In terms of these dimensionless variables, we could (if we wished) convert Equations 14.4, 14.7, and 14.9 into dimensionless form. So far, the derivation is quite general and could be applied to any particular star in order to solve for the three unknowns.

But now, following Cowling (1941) and in the spirit of Chapter 10, we restrict our attention to the case of a "polytropic star". Specifically, we now apply our three equations (in dimensionless form) to a spherical object where the density profile is given by the particular functional form appropriate for a polytrope. That is, referring to Equation 10.6, we are considering an object where the density $\rho(x)$ at radial location $x$ is related to the central density by the specific form $\rho(x)/\rho_c = y(x)^n$ where $y(x)$ is the local value of the Emden solution for polytrope $n$ at radial location $x$.

Then we find that Equation 14.8 becomes

$$\alpha \eta = \frac{l(l+1)}{x^2} \theta - \frac{1}{x^2} \frac{d}{dx}(\alpha x^2 y^n X) \quad (14.11)$$

Equation 14.11 describes how the density perturbation $\eta$ is related to the pressure perturbation $\theta$ and to (the radial gradient of) the radial displacement $X$.

In dimensionless form, Equation 14.4 becomes

$$\alpha y^n X = \frac{d\theta}{dx} - (1+n)\eta y' \quad (14.12)$$

where $y'$ denotes the spatial gradient $dy/dx$. Equation 14.12 describes how the radial displacement $X$ is related to the (radial gradient of the) pressure perturbation $\theta$ and the density perturbation $\eta$.

Oscillations in the Sun

Finally, in order to get our third equation in dimensionless form, Equation 14.9 can be written as

$$y\eta(1+N) = N\theta - (n-N)Xy^n y' \tag{14.13}$$

Equation 14.13 is the third equation we have been seeking in order to describing how changes in (dimensionless) density are related to changes in (dimensionless) pressure and to changes in (dimensionless) radial displacement. Since Equation 14.13 is an algebraic relationship between $\eta$, $\theta$, and $X$, we can use Equation 14.13 to eliminate $\eta$ from Equations 14.11 and 14.12. This leads to two differential equations for two unknowns, the radial displacement $X$ and the pressure perturbation $\theta$.

Substituting the expression for $\eta$ into Equation 14.12, and gathering terms in $\theta$ on the left-hand side, we find

$$\frac{d\theta}{dx} - \frac{(1+n)Ny'}{(1+N)y}\theta = X\left[y^n\alpha - \frac{(1+n)(n-N)}{1+N}y^{n-1}(y')^2\right] \tag{14.14}$$

Substituting the expression for $\eta$ into Equation 14.11, and gathering terms in $X$ on the left-hand side, we find

$$\frac{1}{x^2}\frac{d}{dx}(x^2 y^n X) - \frac{(n-N)}{1+N}y^{n-1}y'X = \theta\left[\frac{l(l+1)}{\alpha x^2} - \frac{N}{(1+N)y}\right] \tag{14.15}$$

Equations 14.14 and 14.15 are coupled differential equations: the radial gradient of one variable (on the l.h.s.) is expressed in terms of the value of the other variable (on the r.h.s.). These are the two equations we need in order to calculate the oscillations occurring inside a polytrope.

In order to put the equations into a more convenient form for numerical integration, we define auxiliary variables, one for the pressure fluctuation $\theta$, the other for the radial displacement $X$. In order to see which auxiliary variable is most helpful, we note that the left-hand side of Equation 14.14 can be written in the form $\theta' - (Ey'/y)\theta$ where $E = N(1+n)/(1+N)$ is a numerical constant. This form of a differential equation suggests an integrating factor $y^{-E}$. This leads us to convert the pressure perturbation variable $\theta$ and the radial displacement variable $X$ to new auxiliary variables:

$$w = \theta y^{-E}, z = Xy^E \tag{14.16}$$

The two new variables $w = w(x)$ and $z = z(x)$ describe how the pressure perturbation and radial displacement vary as a function of radial location in the polytrope. In terms of these two variables, and also introducing the constant $Q = 2E - n$, we finally arrive at two ordinary differential equations for $w$ and $z$ as functions of the radial coordinate $x$:

$$\frac{dw}{dx} = zy^{-Q}\left(\alpha - \frac{(1+n)(n-N)}{1+N}\frac{(y')^2}{y}\right) \tag{14.17}$$

$$\frac{d}{dx}(x^2 z) = wy^Q\left(\frac{l(l+1)}{\alpha} - \frac{Nx^2}{(1+N)y}\right) \tag{14.18}$$

## 14.3 OVERVIEW OF THE EQUATIONS

Let us summarize what we have done up to this point. Equations 14.17 and 14.18 describe the profiles of radial displacements (~$z(x)$) and fluctuations in pressure (~$w(x)$) that occur when a "star" oscillates with a particular frequency (~$\sqrt{\alpha}$). The "star" extends in radial coordinates from $x = 0$ (the center) to $x = x_1$ (the surface, where $x_1$ is different for each value of the polytropic index $n$). Because the entire star from center to surface is involved in determining the solutions of Equation 14.17 and 14.18, the oscillations we are considering are *truly global in nature*. Inside the star, the function $y(\sim T)$ varies from $y = 1$ at the center to $y = 0$ at the surface. The radial gradient $y'$ is zero at the center, and then takes on negative values throughout the rest of the star. The star obeys a polytrope equation of state: $p \sim \rho^{1+1/n}$ and the oscillations are adiabatic: $\Delta p/p = (1 + 1/N)\Delta\rho/\rho$. The oscillations vary in latitude such that $l$ nodes exist between the north pole and the south pole of the star.

For any chosen polytrope (specified by the index $n$), the Lane–Emden equation can be integrated (either analytically or numerically) to obtain a table of values of $y$ and $y'$ as a function of $x$ between 0 and $x_1$. A value is assigned to the adiabatic index $N$ (typically $N = 1.5$), and this then fixes the value of $Q$. Then the second term inside the large brackets on the right-hand side of Equations 14.17 and 14.18, as well as the $yQ$ terms, can be evaluated at all tabulated values of $x$ between 0 and $x_1$. This provides the "background information" required to undertake a numerical integration for the two unknowns $w(x)$ and $z(x)$.

The final quantity that must be assigned in order to integrate Equations 14.17 and 14.18 is the frequency parameter $\alpha$. In our search for solutions to the equations, we shall start by assigning a small value to $\alpha$ and integrate the equations to determine what value the pressure parameter $w(surf)$ will have at the surface of the polytrope. Then we shall step through a series of values of $\alpha$ to find out the special frequencies (the "eigenfrequencies") for which $w(surf)$ is found to have the particular value of zero.

The properties of oscillations in a polytrope can be determined by integrating Equations 14.17 and 14.18 numerically for $w(x)$ and $z(x)$ with appropriate boundary conditions at the center and at the surface. Numerical integration can be performed either by programming a Runge-Kutta subroutine or by using one of the widely available software packages such as MATLAB® or Mathematica.

In order to begin the integration at the center of the star, asymptotic functional forms for $w$ and $z$ must be specified: see step 6 in Section 14.4.1. The boundary conditions at the surface of the "star" for an eigenmode are that $w \to 0$ and $z \to 0$ as $x \to x_1$.

By experimenting with different choices of polytropic index $n$, one can learn a great deal about the properties of oscillations in stars.

The properties of Equations 14.17 and 14.18 are such that, in asymptotic terms, there are two distinct classes of eigenmodes: one is relevant in the limit where the frequency parameter $\alpha \to \infty$, the second is relevant in the limit where the frequency parameter $\alpha \to 0$ (see Section 14.6). In the limit of high frequency ($\alpha \to \infty$), pressure dominates as the restoring force: such modes are referred to as *p*-modes. The *p*-modes exist in all polytropes. In the limit of long period ($\alpha \to 0$), gravity (or more specifically, buoyancy) dominates as the restoring force: these are *g*-modes. Unlike the *p*-modes, *g*-modes do *not* exist in all polytropes: because of the presence of the term $(n - N)$ in Equation 14.17, *g*-modes with finite periods do not exist if the polytropic index $n$ is less than the value that one has assigned to $N$ (typically $N = 1.5$).

## 14.4 THE SIMPLEST EXERCISE: *p*-MODE SOLUTIONS FOR THE POLYTROPE *n* = 1

In order to get a feel for how the oscillation equations work and how they lead to eigenfrequencies, it is instructive to integrate Equations 14.17 and 14.18 numerically for the case of a particularly simple polytrope, namely, $n = 1$. Although that polytrope makes no claim to describe any actual star, it still retains the overall structure of high pressure and density at the center and much lower pressure and density at the surface. For present purposes, the outstanding advantage of the polytrope $n = 1$ is that the functions $y(x)$ and $y'(x)$ are known analytically: $y(x) = \sin(x)/x$, $y'(x) = \cos(x)/x - \sin(x)/x^2$. As a result, we do not need to prepare a table of values of $y$ and $y'$: the local values can be calculated analytically. The surface of the star occurs at $x_1 = 3.14159$. We assume $N = 1.5$. This leads to $E = 1.2$ and $Q = 1.4$.

Oscillations in the Sun

To perform the calculation, one must first specify a certain value of $\alpha$: this remains fixed throughout the "star" as one integrates Equations 14.17 and 14.18 for a particular value of the frequency of the mode. Once a value has been chosen for $\alpha$, we start at the center of the star ($x = 0$, $y = 1$, $y' = 0$), and increase the value of $x$ by some chosen increment $\Delta x$. Integrate the coupled Equations 14.17 and 14.18 for the two unknowns $w$ and $z$ at each step. Because the two equations are coupled, one integrates outward first in (say) $w$ using the current value of $z$: given the current value of $z$, the right-hand side of Equation 14.17 can be evaluated, and this therefore allows one to evaluate $dw/dx$. Knowing this, one can take a step $\Delta x$ and use one's numerical scheme (e.g., Runge-Kutta) to calculate an updated value of $w$. This is then inserted in the right-hand side of Equation 14.18 in order to evaluate $dz/dx$. This then allows the numerical scheme to calculate an updated value of $z$ across the step $\Delta x$. This process is repeated for all tabulated values of $x$ between 0 and 3.14159. The result is a table of values of $w$ and $z$ as a function of $x$.

Of special interest is the value of $w(x_1)$ at the star's surface (where the parameter $x$ has the particular value $x = x_1 = 3.14159$). In most cases, for arbitrary values of $\alpha$, the value of $w(x_1)$ will be found to be nonzero. But as one repeats the exercise with increasing values of $\alpha$, one will find that, in certain special cases (with $\alpha$ taking on certain discrete values), when the calculation reaches the surface of the polytrope, the computed value of $w(x_1)$ turns out to be equal to zero. Those special cases are the eigenmodes ($p$-modes) of the polytrope.

### 14.4.1 PROCEDURE FOR COMPUTATION

1. Pick a value of $l$ among the set 0, 1, 2, and 3.
2. Pick a starting guess for $\alpha$, the (dimensionless) frequency. Because of the choice of normalizations, the starting guess for $\alpha$ should not be too far from unity. A recommended starting guess is $\alpha = 0.1$. The reason for this choice is as follows: when extensive calculations are performed for $l = 0, 1, 2$, and 3, the lowest eigenfrequency in the $n = 1$ polytrope is found to have the numerical value $\alpha \approx 1, 0.3, 0.55$, and 0.7, respectively.
3. Start the integration near the center of the polytrope by setting $x = 0.01$ (or $x = 0.1$ if you are confident about your equation solver routine).
4. At that value of $x$, evaluate the local values of $y(x) = \sin(x)/x$ and $y'(x) = \cos(x)/x - \sin(x)/x^2$.
5. With choices now made for $l$, $\alpha$, $n (= 1)$, and $N (= 1.5)$, you have all the information you need to compute the local numerical values of the expressions in large brackets on the right-hand side of Equations 14.17 and 14.18.
6. You will have to choose initial values for $w$ and $z$. What initial values should you use for $w$ and $z$? When one examines the asymptotic properties of Equations 14.17 and 14.18, it turns out that in the limit $x \to 0$, the functional form of $z$ is as follows:

$$z(x) = x^{l-1} \tag{14.19}$$

for all values of $l$. Also in the limit $x \to 0$, the functional form of $w$ is

$$w(x) = \frac{\alpha x^l}{l} \tag{14.20}$$

for $l > 0$, and $w = 1$ for $l = 0$. Using Equations 14.19 and 14.20, evaluate $w$ and $z$ at whatever (small) value of $x$ you have chosen as the starting point.
7. Increase $x$ by $\Delta x = 0.01$ (if that is your choice of step size). Using the values of $w$ and $z$ from step 6 and recalculating the local values of $y$ and $y'$ at the new value of $x$, evaluate the right-hand side of Equations 14.17 and 14.18. With the new numerical values for the derivatives, step forward to calculate the new values of $w$ and $z$.

212                                                                                                      Physics of the Sun

8.  Repeat step 7 until $x$ has a value that is slightly smaller than $x_1 = 3.14159$. Where should you stop the integration in $x$? That is, how close should you approach the limiting value $x_1 = 3.14159$? You cannot go too close, because then the terms in $1/y$ on the right-hand side of both equations will become infinitely large. One possible approach is to stop the integration in $x$ when the value of $y(x)$ has decreased to a "small" value, such as 0.001. (Recall that starting at the center, $y$ has a value of 1.0.) For purposes of the oscillation calculation, this stopping point may be considered to be "the surface" of the polytrope.
9.  Once you reach this "surface", your code will give you a certain value for the pressure fluctuation variable $w$. Call this $w(surf)$, and enter this value into a table alongside the value you specified for $\alpha$ (in step 2).
10. Now, pick a new, *larger* value for $\alpha$. How large should the new value of $\alpha$ be? Recommended increases are 0.1 up to $\alpha = 5$. That is $\alpha = 0.2, 0.3, 0.4, \ldots, 4.9, 5.0$. Then increase the increment to 0.5 for values of $\alpha$ between 5 and 20. Then use increments of three for $\alpha$ up to (about) 300. Alternatively, you could (if you like) choose constant steps in $\log(\alpha)$.
11. For each value of $\alpha$, repeat steps 3–9. For each $\alpha$, tabulate the value you compute for $w(surf)$. Since $\alpha$ is proportional to the square of the frequency (see Equation 14.10), it is more convenient to convert from $\alpha$ to a dimensional frequency using the unit $v_g \approx$ 100 µHz (Chapter 1, Section 1.12). For a polytrope with index $n$, the conversion factor for a "star" with mass and radius equal to the solar values is (Mullan and Ulrich 1988) $v = v_g \sqrt{(3C_c \alpha/(n + 1))}$ where $C_c$ is the central condensation of the polytrope (Chapter 10, Section 10.9). In the present case, $n = 1$, this leads to $v = 222.0\sqrt{\alpha}$ µHz.
12. Once you have computed results for all values of $\alpha$ from 0.1 to about 300, plot $w(surf)$ versus $\alpha$, or (more conveniently) $w(surf)$ versus $v$. Two such plots are shown in Figure 14.1, one for $l = 2$ (solid curve) and the other for $l = 0$ (dotted curve). One sees that, as the

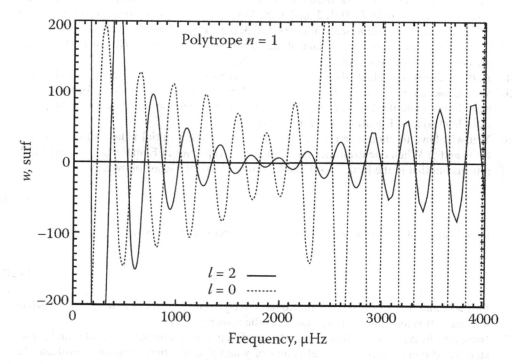

**FIGURE 14.1**  $w(surf)$ is the surface value of the oscillation variable $w$ (the pressure fluctuation) as a function of frequency for polytrope $n = 1$ and for two values of the degree $l$. Crossings of the horizontal line $w(surf) = 0$ identify the eigenfrequencies of $p$-modes in the $n = 1$ polytrope.

# Oscillations in the Sun

frequency increases, $w(surf)$ swings back and forth between positive and negative values. For each value of $l$, there exists a discrete set of values of $\nu$ (i.e., $\nu_1$, $\nu_2$, $\nu_3$ . . .) at which the computed curve crosses the horizontal axis, i.e., $w(surf)$ passes through a value of zero. At such values of $\nu$, the pressure fluctuation in the oscillation takes on the particular value of zero at the surface of the polytrope. Such an oscillation is a $p$-eigenmode of the polytrope for the particular value of $l$ that was chosen for the calculation. The sequence of values $\nu_i$ are eigenfrequencies of the degree-$l$ $p$-modes for that polytrope.

For your chosen value of $l$, you will now have a series of eigenfrequencies for the $p$-modes that occur in the $n = 1$ polytrope.

13. Pick a new value of $l$ and repeat steps 2–12. For each value of $l$, you will find a different series of eigenfrequencies.

## 14.4.2 COMMENTS ON THE $p$-MODE RESULTS: PATTERNS IN THE EIGENFREQUENCIES

Our results for the Cowling approximation in a polytrope, although greatly simplified, nevertheless allow us to draw valuable conclusions that would remain valid if we had applied the (more complicated) full oscillation equations to a detailed solar model. Because of this, it is instructive to examine certain properties of the output of the polytrope oscillation program.

To begin with, we note that at the highest frequencies plotted in Figure 14.1, the last four eigenfrequencies for $l = 0$ are found to be at $\nu = 3490, 3653, 3816,$ and $3980$ µHz. For $l = 2$, the highest four eigenfrequencies are found to be at $\nu = 3474, 3637, 3801,$ and $3965$ µHz. Two patterns are striking here.

First, for $p$-modes with a given degree $l$, the intervals between adjacent eigenfrequencies are $\Delta\nu = 163, 163,$ and $164$ µHz for $l = 0$ and $\Delta\nu = 163, 164,$ and $164$ µHz for $l = 2$. That is, both sequences of $p$-modes show a striking asymptotic behavior: there is a (roughly) constant *frequency* interval between adjacent eigenmodes. The asymptotic frequency separation we have found here is slightly larger than $\Delta\nu$ ($= 153$ µHz) obtained for the $n = 1$ polytrope from a more precise calculation (Mullan and Ulrich 1988): this difference can be ascribed to inadequacies in treating the surface boundary conditions in the solutions of the equations presented in the prior figure.

Although the present results pertain only to the $n = 1$ polytrope, it will be shown later (Section 14.6) that a constant *frequency* interval between adjacent $p$-modes is predicted to be a *general property* of the oscillation equations *in the limit of high frequencies*. For reasons that will soon become clear, we refer to $\Delta\nu$ as the "large separation" between adjacent modes. In the Sun, observations indicate that the "large separation" $\Delta\nu$ has values of 135–136 µHz for modes with $l = 0$–3 (Appourchaux et al. 1998). Clearly, this observed separation is smaller than the 153–163 µHz that we have found for the $n = 1$ polytrope. But this is not a matter for any great concern: we have never claimed that the $n = 1$ polytrope is supposed to be an accurate representation of *the Sun itself*. The point is, there *does exist* a "large separation" for the $p$-modes in the "real Sun", just as we have discovered for $p$-modes in our model of the $n=1$ polytrope.

Second, we note that the sequences of eigenfrequencies for $l = 0$ and $l = 2$ pair up with each other such that the two curves in Figure 14.1 cross the horizontal axis at *almost* the same frequencies. This indicates that the eigenfrequencies for certain $p$-modes with $l = 0$ and $l = 2$ differ from each other by an amount that is small compared to the "large separation" of either sequence. The frequency separations between corresponding $l = 0$ and $l = 2$ $p$-modes are $\delta\nu(0-2) = 16, 16, 15,$ and $15$ µHz, i.e., about one order of magnitude smaller than the numerical values of $\Delta\nu$. The frequency differences $\delta\nu(0-2)$ are referred to as "small separations", to distinguish them from the "large separations" ($\Delta\nu$) between adjacent $p$-modes at constant $l$.

Although we do not present the results graphically here, we note that when the analog of Figure 14.1 is plotted for $l = 1$ and $l = 3$, results similar to those in Figure 14.1 emerge. There is again an asymptotic "large separation" $\Delta\nu$ of about 163 µHz in frequency between adjacent modes with the same $l$ value, and a "small separation" between corresponding $l = 1$ and $l = 3$ modes. In this case, the "small separations" turn out to be $\delta v(1-3) = 10-12$ µHz.

In the Sun, empirically it is found that the "large separations" are 135–136 µHz, while the "small separations" for $p$-modes with frequencies of 3500–4000 µHz are $\delta v(0-2) \approx 10$ µHz and $\delta v(1-3) \approx 12$ µHz (Appourchaux et al. 1998). We notice that the numerical values of the "small separations" are about one order of magnitude smaller than the numerical values of the "large separations". Of course there is no reason to expect the $n = 1$ polytrope to reproduce the structure of the Sun in detail. Nevertheless, it is encouraging that even with the simplification of using the $n = 1$ polytrope, and also using the simplification of the Cowling approximation, we recover some important empirical features of the Sun's $p$-mode eigenfrequencies.

An important property of the small and large separations is that they can be used (in conjunction with computer codes that follow the structural changes in the Sun as it becomes older) to determine the age of the Sun. Bonanno and Frohlich (2015) have used the empirical ratio of small/large separations extracted from more than 25 years of helioseismic data obtained by the Birmingham Solar Oscillations Group to derive an age of the Sun of $4569 \pm 6$ Myr. This compares remarkably well with the age of the oldest inclusions in meteorites: $4567.3 \pm 0.16$ Myr (Connelly et al. 2012).

Finally, there is one further piece of information that we need in order to interpret Figures 13.1 and 13.2. It is this: given the frequencies of two adjacent $p$-modes with $l = 0$ (say the modes observed at $v = 3034$ and $3169$ µHz: see Chapter 13, Section 13.2), it is observed that there exists a mode with $l = 1$ with a frequency which is *almost exactly half-way* between the two adjacent $l = 0$ modes. Thus, the ($l = 1$, $n_r = 21$) mode is observed at $v = 3099$ µHz, only 0.1% away from the midpoint frequency of the two surrounding $l = 0$ modes. Now, modes with $l = 0$ are excited in the Sun with almost equal power to those with $l = 1$. Therefore, rather than seeing in Figures 13.1 and 13.2 separations between peaks of 135 µHz (the "large separation"), one sees separations of only about one-half that value, i.e., about 70 µHz (i.e., 0.07 mHz).

### 14.4.3 EIGENFUNCTIONS

Now that frequencies of the $p$-modes have been identified, it is also important to consider the structure of the radial eigenfunctions. Two examples are shown in Figure 14.2, where we plot the radial profile of the function $w$ (corresponding to the pressure perturbation) in Equations 14.17 and 14.18. The abscissa in Figure 14.2 is the radial coordinate $x$ in Equations 14.17 and 14.18: it runs from 0 (at the left-hand side) corresponding to the center of the "star" to the boundary value $x = x_1 = \pi$ (at the right-hand side) appropriate for the $n = 1$ polytrope.

Two features are noticeable about the eigenfunctions in Figure 14.2. First, the numerical values have excursions on both sides of the $w = 0$ (horizontal) axis. As a result, there exist a finite number of "nodes" where the eigenfunction passes through the value of zero. The number $n_r$ of times that an eigenmode crosses the $w = 0$ axis between center and surface is used to label the mode as being of "radial order $n_r$".

Second, as we approach the surface, the excursions of the eigenfunction increase to larger (absolute) values. The peaks in the eigenfunction can be considered as "antinodes" where the pressure fluctuation has a local maximum. The antinode that occurs nearest to the surface has a larger amplitude than those lying somewhat deeper. (This is not always true for some of the very lowest order $l$ modes, such as $l = 0$, but for moderate and high $l$ values, the antinode nearest the surface has the largest amplitude.)

A question that is of particular interest in solar physics is the following: at what radial location is the largest antinode of any given $p$-mode eigenfunction situated? The answer to this question has a bearing on the basic question: why are certain $p$-modes excited to large amplitude in the Sun while other $p$-modes are not? For the two examples in Figure 14.2, the radial locations at which the

# Oscillations in the Sun

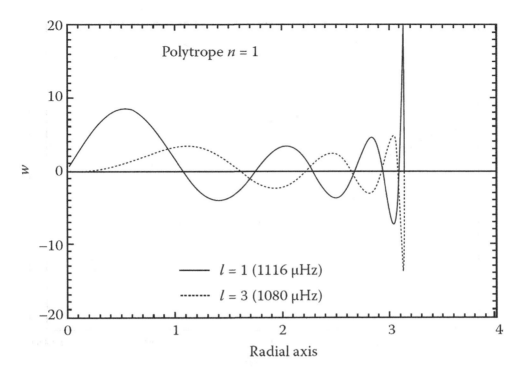

**FIGURE 14.2** Eigenfunctions in $w$ plotted as a function of radial position for two $p$-modes with similar frequencies in the polytrope $n = 1$. The two $p$-modes differ in degree $l$ by $+2$, and they differ in radial order by $-1$. The radial coordinate in the abscissa extends from center to surface.

largest antinode occurs are located at about 0.99 times the radius of the $n = 1$ polytrope. That is, the largest antinodes occur at depths $z_{an}$ that are close to the surface, no more than 1% of the stellar radius below the surface. The larger the numerical value of $n_r$, the closer the last (and largest) antinode lies to the surface (for a given $l$ value). In order to demonstrate this result in more detail, we show in Figure 14.3 some details of eigenfunctions that do *not* refer to a polytropic model but that instead were obtained from *a realistic solar model*. For the three modes shown, with $n_r = 10$, 15, and 25 (and $\nu = 1610$, 2290, and 3650 µHz, respectively), we see that the largest antinodes lie at fractional depths of 0.5%, 0.2%, and < 0.1% of $R_\odot$, respectively. These depths will be important subsequently when we consider why certain $p$-modes are excited in the Sun more effectively than others are.

The eigenfunctions in Figure 14.3 were computed theoretically using a more sophisticated oscillation code than the one we have used to generate Figures 14.1 and 14.2. What is *not* shown in Figure 14.3 is how much power the Sun actually pumps into the various modes. In fact, there are striking differences in the levels of power that are observed to occur in the modes whose eigenfunctions are plotted in Figure 14.3. The frequency of the $n_r = 25$ mode in Figure 14.3 ($\nu = 3650$ µHz) is such that the mode lies near the peak of power for solar $p$-modes (see Figure 13.2): the power level is observed to be almost 5000 m$^2$ s$^{-2}$ Hz$^{-1}$. On the other hand, the frequency of the $n_r = 10$ mode in Figure 14.3 ($\nu = 1610$ µHz) is such that the mode lies in the barely detectable regime in Figure 13.2: the observed power level is perhaps 10 m$^2$ s$^{-2}$ Hz$^{-1}$ above background. Thus, the observations indicate that the ($l = 1$, $n_r = 10$) mode is present in the Sun at a power level which is almost three orders of magnitude *smaller* than the power in the ($l = 1$, $n_r = 25$) mode. The intermediate mode ($l = 1$, $n_r = 15$), at $\nu = 2290$ µHz, is present in Figure 13.2 at a power level which is about 10 times smaller than

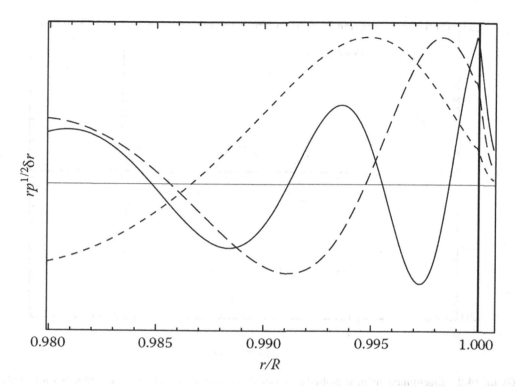

**FIGURE 14.3** Eigenfunctions of (scaled) radial displacement $\delta r$ (analogous to $z$ in Equation 14.18) for $l = 1$ $p$-modes in a realistic model of the Sun. The abscissa is radial location expressed in terms of the solar radius $R$. Notice that the horizontal scale is greatly expanded: only the outermost 2% of the radial coordinate (near the surface) is plotted. Dotted, dashed, and solid curves refer to $p$-modes with radial orders $n_r$ = 10, 15, and 25, respectively. (From the website of Jorgen Christensen-Dalsgaard at https://phys.au.dk/~jcd/oscilnotes/Lecture_Notes_on_Stellar_Oscillations.pdf. The above figure occurs on page 88 of the jcd notes, in Figure 5.9. Used with permission of J. C-D.)

the peak power. We will return to these power levels later when we consider the excitation of the modes in Section 14.8.

The eigenfunctions in Figure 14.2 belong to two modes separated by the "small separation" $\delta v$(1–3). As can be seen from Figure 14.2, each of the two eigenfunctions crosses the $w = 0$ axis several times between the center and the surface. Inspection shows that, of the two eigenfunctions in Figure 14.2, the $l = 1$ curve has *one more node in the radial direction* than the $l = 3$ curve has. It is a general relation that modes differing by *two* units in $l$ are separated by the "small separation" $\delta v$ if they also differ by *one* unit in $n_r$. That is, $v(l + 2, n_r) \approx v(l, n_r + 1)$.

Although not plotted in Figure 14.2, when we plot the radial profile of the *radial displacement z of the oscillations*, the maximum excursions of $z$ are found to be smaller by an order of magnitude or more than the excursions of the pressure fluctuation variable $w$. The fact that pressure fluctuations are dominant confirms that we are dealing with $p$-modes.

## 14.5 WHAT ABOUT $g$-MODES?

Our choice of polytrope $n = 1$ (chosen for the simplicity of its analytic solution) prevents us from discussing $g$-modes: because of our choice $N = 1.5$, no $g$-modes with finite periods exist in any

# Oscillations in the Sun

polytrope with $n < 1.5$. If we wished to numerically calculate g-modes in a polytrope, we must use $n > 1.5$ (assuming $N = 1.5$). As we have seen (Chapter 10), for polytropes with $n > 1.5$, no analytic formulas exist for the polytrope structure (apart from the uninteresting case of $n = 5$ for an infinitely distended star). Therefore, a study of g-modes in polytropes requires us first to determine numerically a table of values of $y$ and $y'$ as a function of $x$, and then interpolate in this table to obtain local values of $y$ and $y'$ at each value of $x$ in the right-hand side of Equations 14.17 and 14.18. Such a study has been reported for the polytropes $n = 2, 2.5, 3, 3.5$, and $4$ by Mullan (1989).

Since the case $n = 3.25$ is relevant for the radiative interior of the Sun (see Section 10.2), we focus on that case here. An abbreviated table of $y$ and $y'$ values as a function of $x$ in the $n = 3.25$ polytrope has already been given (see Table 10.1). A more extended version of that table, including more than 800 rows, was prepared so that it could be used for interpolation in the right-hand side of Equations 14.17 and 14.18. Those equations were then numerically integrated from center to surface, using a series of frequencies appropriate for g-modes, using the steps outlined earlier in Section 14.4.1.

There are four alterations to the steps outlined in Section 14.4.1 when we discuss g-modes. First, the range of permissible $l$ does not include $l = 0$: so step (1) should read: "Pick a value of $l$ among the set 1, 2, 3". Second, in step (4), the local values of $y$ and $y'$ for each value of $x$ cannot be obtained analytically: instead, they must be obtained by interpolating into Table 10.1 (or an extended version thereof). Third, in step (10), one must choose a new, *smaller* value of $\alpha$: the study of g-modes requires going to smaller and smaller frequencies, i.e., towards longer and longer *periods*. Fourth, in step (11), the conversion factor from $\alpha$ to (dimensional) frequency for $n = 3.25$ for an object with the mass and radius of the Sun is $\nu = 789.0\sqrt{\alpha}$ µHz.

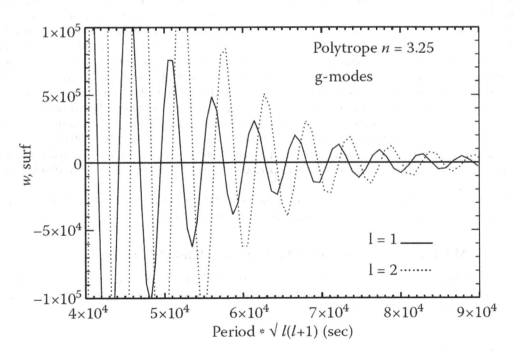

**FIGURE 14.4** Calculation of g-modes in the polytrope $n = 3.25$. Plotted is $w(surf)$, the surface value of $w$, one of the two oscillation variables, as a function of the mode *period* times $\sqrt{l(l+1)}$ for $l = 1$ and 2. Eigenmodes occur where $w(surf) = 0$. Note that the separation between adjacent eigenmodes approaches a constant value in *period*. The asymptotic separation in the quantity $P\sqrt{l(l+1)}$ is about 2500 sec. Contrast this plot with Figure 14.1, where the abscissa was given in terms of *frequency*.

Apart from these alterations, the numerical integration proceeds step by step at each frequency as described in Section 14.4.1. The result is again that, at each frequency, the code provides a value for the oscillation variable, $w(surf)$, at the surface of the polytrope.

Analogous to the plot in Figure 14.1 for the $p$-modes, we present in Figure 14.4 the surface values of $w$ in the polytrope $n = 3.25$ for $g$-modes with $l = 1$ and $l = 2$. In contrast to Figure 14.1, where we plotted $w(surf)$ as a function of *frequency*, in Figure 14.4, it makes more sense to plot $w(surf)$ for the $g$-modes as a function of *the period*. The modes we plot have periods between about 8 and 16 hours. Once again, the zero points of $w(surf)$ define the locations of eigenmodes. Inspection of Figure 14.4 shows that, for the $g$-modes, the separation between adjacent eigenmodes approaches an asymptotic limit *that is constant in period*. In the case of the particular examples that are plotted in Figure 14.4, the period separation between adjacent eigenmodes times $\sqrt{l(l+1)}$ is found to be about 2500 sec. For $l = 1$, this corresponds to a period separation of 1770 sec, i.e., about 30 minutes between adjacent modes. For $l = 2$, this corresponds to a period separation of 1020 sec, i.e., about 17 minutes between adjacent modes.

Thus, whereas $p$-modes exhibit constant asymptotic separation in *frequency* (with a "large separation" $\Delta \nu$ between the *frequencies* of adjacent modes at high frequency), $g$-modes exhibit constant asymptotic separation $\Delta P$ in *period*. This dramatic distinction in *asymptotic behavior* between $p$-modes and $g$-modes is a striking feature of the oscillation equations derived by Cowling (1941) in Equations 14.17 and 14.18.

As already noted (Section 13.7), $g$-modes have not yet been reliably detected in the Sun. Therefore, unfortunately the Sun cannot yet be used to test the asymptotic prediction of constant $\Delta P$ between adjacent modes. In principle, in this textbook about the Sun, we should say nothing more about $g$-modes. However, in a broader framework, we consider it worthwhile to note that in stars where no extensive convective envelope is present, $g$-modes demonstrating the asymptotic behavior of constant separation in *period* have been reliably detected. As an example, we show in Figure 14.5 results obtained by Zhang et al. (2020) for a particular star where the effective temperature (6947 K) is sufficiently hot that the convective envelope is too shallow to suppress the $g$-modes in the outer regions of the star. Analyzing 4 years of Kepler satellite photometric data, Zhang et al. identified 17 low-frequency peaks with periods in the range from 0.7 days to 1.2 days: the mean interval between adjacent *periods* was found to be $2756.2 \pm 0.8$ seconds. The constancy of the *period* interval between adjacent modes is remarkable: it is precisely what is predicted asymptotically for $g$-modes in a nonrotating medium. Based on a lack of rotational splitting of the $g$-modes, Zhang et al. concluded that the rotation period of the core of the star must be longer than 550 days, i.e., the star is rotating so slowly that it can be well described by Cowling's equations for a nonrotating star. Zhang et al. (2020) identified the modes with periods of 0.7–1.2 days as belonging to dipole ($l=1$) gravity modes.

We turn now to a discussion as to why the $p$-modes and the $g$-modes display these distinctly different asymptotic behaviors, one in *frequency* and the other in *period*.

## 14.6 ASYMPTOTIC BEHAVIOR OF THE OSCILLATION EQUATIONS

We can use the mathematical properties of Equations 14.17 and 14.18 to see why $p$-modes have asymptotically equal spacing in *frequency* between adjacent modes, whereas $g$-modes have asymptotically equal spacing in *period*.

### 14.6.1 P-MODES

As regards $p$-modes, asymptotic behavior emerges in the limit of high frequencies, $\alpha \to \infty$. In this limit, Equation 14.17 reduces to $dw/dx = \alpha z y^{-Q}$ while Equation 14.18 reduces to $d(x^2 z)/dx = -w\psi(x,$

# Oscillations in the Sun

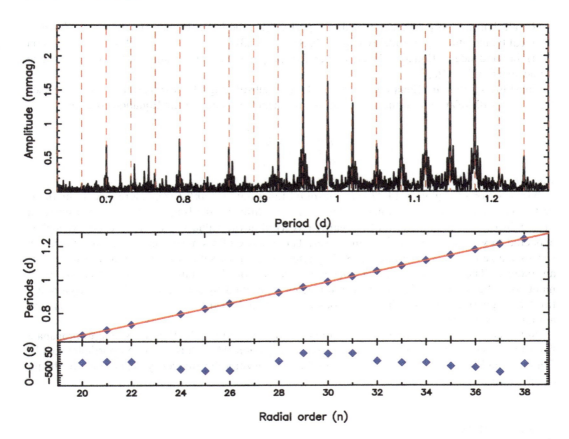

**FIGURE 14.5** Top panel: amplitude of oscillations, plotted as a function of *period*, in a star hotter than the Sun. Each spike represents a *g*-mode with a different radial order. Lower panel: period of each spike is plotted against the radial order of the best-fitting model mode. The red line passing through the observed periods is (very close to) a straight line, indicating that a (nearly) constant interval in *period* exists between adjacent modes, as predicted for *g*-modes in the asymptotic limit of long periods in a nonrotating star. (From Zhang et al. 2020; used with permission of X. Zhang.)

$y$) where $\psi = Nx^2 y^{Q-1}/(N+1)$. Let us concentrate on the values of the quantities $w$ and $z$ close to the "surface" of the star, i.e., at a fixed value of $x$ (close to $x_1$). In essence, this is a rough method of concentrating on the value of $w$ which we referred to as $w(surf)$ when we plotted Figure 14.1. Although this is not a mathematically rigorous procedure, it helps us to see (roughly) that we can write $dz/dx \approx -Aw$ where $A = y^{Q-1}/\Gamma$ is a constant at a fixed value of $x$. (We use the quantity $\Gamma$ to replace $(N+1)/N$. The numerical value of $A$ depends on the (almost zero) value of $y$ close to $x_1$. Again treating $y^{-Q}$ as essentially constant (because we are treating a point at a fixed value of $x$ close to the surface), we take the second derivative of $w$ and find $d^2w/dx^2 \approx ay^{-Q}dz/dx$. Substituting the prior expression for $dz/dx$, we find

$$\frac{d^2w}{dx^2} \approx -w\alpha A y^{-Q} \approx -w\left(\frac{\alpha}{y\Gamma}\right) \qquad (14.21)$$

Recall that in a polytrope composed of a perfect gas, the Lane–Emden function $y$ is related to the local temperature by $y = T/T_c$ (Chapter 10, Section 10.3). The combination $y\Gamma$ (which occurs on the right-hand side of Equation 14.21) is therefore proportional to the square of the local adiabatic sound speed: $y\Gamma = A'(c_s)^2$.

Substituting this in Equation 14.21, and also substituting the expression $\nu = A''\sqrt{\alpha}$ (see Equation 14.10) relating $\alpha$ to the (dimensional) frequency, we can rewrite Equation 14.21 as

$$\frac{d^2w}{dx^2} \approx -w\left(\frac{B\nu}{c_s}\right)^2 \qquad (14.22)$$

where $B^2 = 1/AA''^2$.

The solution of Equation 14.22 is a sinusoidal function: $w \sim \sin(xB\nu/c_s)$. Here, $w$ is to be interpreted as $w(surf)$, (i.e., the ordinate in Figure 14.1), while $\nu$ is the frequency (i.e., the abscissa in Figure 14.1). The quantity $x$ can be regarded as fixed at the value $x = x_1$. As a result, $w(surf)$ is expected to vary sinusoidally as $\nu$ increases. The essence of a sinusoid is that it passes through successive zeroes as the argument $x_1 B\nu/c_s$ passes through the set of discrete values $j\pi$ where $j$ is an integer. The interval between successive zeroes of the sinusoid corresponds to increments of $\pi$ between successive eigenfrequencies multiplied by $x_1 B/c_s$. Thus, in this asymptotic limit of high frequencies, adjacent modes are predicted to differ in *frequency* by a constant amount $\Delta\nu = \pi c_s/Bx_1$.

Note that $\Delta\nu \sim c_s/x_1$. Now, the combination $x_1/c_s$ is related to $t_s$, the time for sound to propagate from the Sun's center to the surface (Chapter 9, Equation 9.6). Thus, $\Delta\nu$ is proportional to $1/t_s$. Detailed mathematical work shows that, in fact, the asymptotic frequency separation $\Delta\nu$ should equal $1/(2t_s)$.

### 14.6.2 g-MODES

As regards g-modes, the treatment of asymptotic behavior follows the prior discussion for p-modes, except that now we consider the asymptotic limit of *low* frequencies, i.e., $\alpha \to 0$. In this limit, the dominant term in $dz/dx$ is $l(l+1)/\alpha$, while the dominant term in $dw/dx$ is proportional to $(n - N)$. Repeating the prior steps, we again find a sinusoidal solution, with zeroes separated by a constant interval $\pi$ in the argument. In this case, however, adjacent modes differ by a constant value of the argument $A_g = (n - N)\sqrt{l(l+1)}/\nu$.

There are two features to be noted about $A_g$. First, the presence of $n - N$ has the effect that sinusoidal solutions exist only for $n > N$: in a polytrope where $n < N$, the solutions are no longer propagating waves, but are instead damped exponentials.

Second, $A_g$ includes the frequency in the *denominator*. Thus, $A_g$ is proportional to the *period* of the mode. As a result, the constant interval between neighboring g-modes is proportional to the *period*. Thus adjacent g-modes (with fixed $l$) are separated by a constant interval $P_o/\sqrt{l(l+1)}$ in the *period*. The quantity $P_o$ is predicted to have a well-defined value for a given value of the polytropic index: e.g., for $n = 3$, $P_o = 3497$ sec, while for $n = 3.5$, $P_o = 1927$ sec (see Mullan 1989). Thus, for modes with degree $l = 1$, the asymptotic separation in period between adjacent modes is predicted to be 41 minutes for $n = 3$, and 23 minutes for $n = 3.5$. As can be seen from Figure 14.4, for the intermediate case $n = 3.25$, the asymptotic separation for $l = 1$ modes is about 30 minutes.

In contrast to the asymptotic behavior of p-modes (where the asymptotic interval in *frequency* is determined by the radial profile of the *sound speed*), it is not surprising that for g-modes, the asymptotic interval in *period* between adjacent g-modes is *not* determined by the sound speed: instead, it is determined by the radial profile of a very different physical quantity known as the Brunt–Vaisala frequency ($\nu_{BV} \sim n - N$).

In the solar convection zone, where $n = 1.5$, and therefore $n = N$, g-modes are exponentially damped.

# 14.7 DEPTH OF PENETRATION OF p-MODES BENEATH THE SURFACE OF THE SUN

Now that we know that *p*-modes are associated with the propagation of *sound* waves, we can obtain a valuable piece of information in the context of the following question: how deeply into the Sun do *p*-modes penetrate? The answer depends on the value of the wavelength of the *p*-mode.

We have already introduced (see Equation 13.1) the concept of "horizontal wavelength" $\lambda_h = 2\pi R_\odot/\sqrt{[l(l+1)]}$ in connection with modes of angular degree *l*: this wavelength is a measure of how many nodes exist along the meridian from north to south pole. Associated with $\lambda_h$, we introduce the "horizontal wave number" $k_h$ defined by $k_h = 2\pi/\lambda_h = \sqrt{[l(l+1)]}/R_\odot$.

A second component of the wave number that enters into the solar *p*-modes is associated with the wavelength in the *radial* direction $\lambda_r$. The "radial wave number" is defined by $k_r = 2\pi/\lambda_r$, such that the "total wave number" *k* is defined by $k^2 = k_h^2 + k_r^2$.

The propagation of sound waves in a medium occurs in such a way that there is a well-defined relationship between the (angular) *frequency* $\omega$ of the wave, its *wavelength* (or wave number *k*), and the speed of sound $c_s$ in the medium. In the simplest case of a uniform medium, this so-called dispersion relation is especially simple: $c_s = \omega/k$. But the Sun is not a uniform medium: the speed of sound increases with depth as we move farther inward below the surface. As a result, the dispersion relation for sound waves propagating through the Sun at any depth has a more complicated form:

$$k_h^2 + k_r^2 = \frac{\omega^2}{c_s^2} \quad (14.23)$$

Let us use Equation 14.23 to consider what happens to a sound wave as it penetrates deeper and deeper below the surface of the Sun. At the surface, the degree of the mode is identified in terms of the angular degree *l*: this fixes the value of $k_h$, and the sound wave associated with the mode of degree *l* retains that value of $k_h$ at all depths. Also at all depths, $\omega$ retains a constant value.

The essential aspect of the Sun is that, with increasing depth below the surface, the ambient temperature *T* increases (as we have computed in Chapters 5, 7, and 9). This increase in *T* leads to an *increase* in $c_s^2 = \gamma R_g T/\mu_a$ as we move deeper into the Sun. As a result, the right-hand side of Equation 14.23 *decreases* as depth increases. At a certain depth $z_r$, the right-hand side falls to such a small value that is becomes equal to (the constant quantity) $k_h^2$. At that depth, the only way to satisfy Equation 14.23 is for $k_r$ to become zero. When that happens, propagation in the *radial* direction is no longer permitted. The wave number becomes entirely horizontal. When the wave was near the surface, it had a finite value for both $k_h$ and $k_r$: such a wave would propagate at a well-defined angle relative to the radial direction. But at depth $z_r$, the wave propagates horizontally and can therefore penetrate no deeper into the Sun. Thus, although the wave starts off its journey into the Sun by propagating into deeper layers at a finite angle to the radial direction, the wavefront gradually becomes more and more bent (refracted) away from the radial direction, until at depth $z_r$, the wave becomes horizontal. After that happens, the wave then begins to refract back toward the surface. The layer with depth $z_r$ serves as the lower boundary of the acoustic cavity for that wave mode.

The depth $z_r$ is indicative of the maximum depth to which a wave with a given degree *l* penetrates into the Sun. We cannot expect that such a wave will be able to provide much (or any) information about what is happening in the deeper interior of the Sun, at depths in excess of $z_r$. If we wish to study conditions at radial locations of (say) *r* in the deep interior of the Sun, we must make sure to study the properties of waves that *can* propagate into depths $z_r$ which are at least as great as the depth $R_\odot - r$.

Setting $k_r^2 = 0$ in Equation 14.23 and setting $c_s^2 = \gamma R_g T/\mu$, we find that the depth $z_r$ occurs when the local temperature *T* has the value $T_r = \omega^2 \mu/\gamma R_g k_h^2$. At what depth does the temperature have such a value? Well, as long as we are considering depths that are not too far beneath the surface of the Sun, specifically no more than about 200,000 km (roughly 0.3 $R_\odot$), we know that the solar structure is determined by convective heat transport. In such conditions, the temperature gradient is (roughly) equal to the adiabatic gradient, $g/C_p$. This means that the temperature as a function of depth is $T(z) = T(z_o) + g(z - z_o)/C_p$ (see Chapter 7, Equation 7.5). We set $z_o = 0$ where $T(z_o) \approx 6000$ K. Throughout most of the convection zone, $T(z) \gg 6000$ K. Therefore, the depth $z_r$ is essentially equal to $T_r C_p/g$. Recalling that $C_p$ can be set equal to $\gamma R_g/\mu(\gamma - 1)$ (see Chapter 6, Equation 6.5), we find

$$z_r = \frac{\omega^2}{k_h^2} \frac{1}{g(\gamma - 1)} \tag{14.24}$$

In the limit of large $l$, and for modes with (linear) frequency $\nu$ (= $\omega/2\pi$), the depth of penetration can be expressed as a fraction of the solar radius roughly as follows:

$$\frac{z_r}{R_\odot} = \frac{4\pi^2 \nu^2 R_\odot}{l^2} \frac{1}{g(\gamma - 1)} \tag{14.25}$$

Thus, the *larger* the degree $l$, the *shallower* is the penetration of the mode beneath the surface of the Sun.

We have already seen (Chapter 7) that there is one particular depth in the Sun that is of special interest, namely, the depth at which the base of the convection zone lies. As an illustration of Equation 14.25, it is therefore instructive to ask: how large must $l$ be in order to have the depth of penetration no deeper than the convection zone? In such a case, $z_r/R_\odot \leq 0.3$. Inserting this in Equation 14.25, and using $g = 27,420$ cm sec$^{-2}$ and $\gamma = 5/3$, we find that the degree $l$ must exceed a value equal to $l_c \approx 22,400\nu$. The modes most commonly excited in the Sun have $\nu \approx 0.003$ Hz. This leads to $l_c \approx 60$–70. This indicates that if we want to use $p$-modes to study the convection zone in the Sun, it will be best to concentrate on the properties of modes with degree $l$ in excess of 60–70. Referring to Figure 13.4 in Chapter 13, we see that modes lying on the prominent ridges toward the right-hand side of the figure (with $l > 70$) consist of modes that are all confined within the solar convection zone.

For modes with large $l$, i.e., for modes that do not penetrate deeply beneath the surface, the eigenfunctions are effectively "squeezed" into a shell in the outer parts of the Sun between a depth of $z_r$ and the surface. For example, observed modes with the largest $l$ in Figure 13.4, with $l \approx 200$, and $\nu = 3$ mHz, are confined to a shell that penetrates beneath the surface to a distance of only about $0.03 R_\odot$, i.e., to a depth of only about 21,000 km. And yet there still exist a series of modes, each with its own radial order $n_r$, which must be "squeezed" into this thin shell. It is obvious that, in such a case, even relatively small values of $n_r$ will result in having the last antinode situated very close to the surface.

What about modes with $l < 60$–70? Such modes have the ability to penetrate into the radiative interior of the Sun before they are refracted back towards the surface. In principle, modes with the lowest values of $l$ (e.g., $l = 1, 2$, or 3) are capable of reaching in almost to the nuclear-generating core regions of the Sun. Therefore, if we are interested in processes occurring in the Sun below the base of the convection zone, our main focus should be on the low-$l$ modes.

Another aspect of $p$-mode propagation inside the Sun is that, given the speed of the acoustic waves, one can calculate how much *time* it will take for any given $p$-mode to propagate downward to its reflection depth $z_r$ and then propagate back up to the surface. This timing can be altered if, along the propagation path, the wave happens to encounter a localized active region: in such a region, acoustic waves may convert to a different wave mode with a different propagation speed (see Section 16.7.7). This conversion to another mode may alter the propagation time. If this alteration in timing can be quantified, this can give astronomers on Earth the ability to actually estimate the location of an active region on the "far side" of the Sun before it ever rotates onto the visible disk. Verification of this ability has been provided by observations with the STEREO spacecraft (Section 16.7.7).

# Oscillations in the Sun

## 14.8 WHY ARE CERTAIN p-MODES EXCITED MORE THAN OTHERS IN THE SUN?

In order to understand why some $p$-modes are excited to large amplitude in the Sun, while other $p$-modes are hardly excited at all, let us recall an important feature of the observed power spectrum in Chapter 13, Figure 13.2. We have seen (Chapter 14, Section 14.4.3) that the solar mode with $l = 1$ and $n_r = 10$ is present at a power level that is 500 times smaller than the mode with $l = 1$ and $n_r = 25$. And for the mode with $l = 1$ and $n_r = 15$, the power level is some 10 times smaller than the mode with $l = 1$ and $n_r = 25$. How can we understand this preference of the Sun to excite the mode with $n_r = 25$ almost one thousand times more effectively than the mode with $n_r = 10$?

### 14.8.1 Depths Where p-Modes Are Excited

To address this, consider the eigenfunctions plotted in Figure 14.3. We note that the largest antinode of the mode with $l = 1$ and $n_r = 10$ lies at a depth of 0.5% of the solar radius, i.e., at a depth of 3500 km below the photosphere. On the other hand, the mode with $l = 1$ and $n_r = 25$ has its largest antinode at depths of <700 km. The ($l = 1$, $n_r = 15$) mode has its largest antinode at an intermediate depth, 1400 km.

These results lead us to ask the question: is there some physical quantity in the Sun that can provide power to the $p$-modes and is favorable for excitation of $p$-modes at depths of <700 km but is less favorable (by a factor of 10) at depths of 1400 km, and is even less favorable (by factors of 500) at depths of 3500 km?

### 14.8.2 Properties of Convection at the Excitation Depth

The most obvious characteristic of depths between $z$ <700 km and $z = 3500$ km is that they lie within the solar convection zone. The principal characteristic of that zone is that the gas is driven effectively to finite velocities by means of convective instability. In the turbulent motions characteristic of solar convection, granules come into existence and subsequently go out of existence on time-scales of a few minutes. When a compressible medium is in motion, it is inevitably a source of pressure fluctuations, i.e., sound waves. As a result, the solar convection motions are effective generators of sound waves. This raises the possibility that such waves may serve as a source of $p$-modes, provided that conditions are favorable to allow transfer of energy into the modes.

At what spatial location inside the Sun is energy likely to be transferred most effectively into a $p$-mode? The answer is: at the location where the mode's eigenfunction has its largest antinode. That is where the mode "likes" to have a large pressure fluctuation. This suggests that $p$-modes can be excited in the Sun most effectively if the largest antinode (i.e., in general, the antinode lying closest to the photosphere) lies at a depth where convection generates sound waves effectively.

How much power do the convective flows in the Sun emit as sound waves? The maximum available power can be computed by noting that an individual granule survives for only about one turnover time, i.e., for a time interval $t_c \approx D/V$, where $D$ is a length associated with circulation around the convection cell, and $V$ is the convective velocity (see Section 6.5). When a granule reaches the end of its lifetime and loses its identity by dissolving back into the ambient medium, it is as if the energy density of the convective flow ($E_d \sim \rho V^2$ ergs cm$^{-3}$), equivalent to a ram pressure, is made available (over a time-scale of order $t_c$) as a pressure pulse in the ambient medium. The maximum available power $P_p$ emerging from each cm$^3$ in the pressure pulse is of order

$$P_p \approx \frac{E_d}{t_c} \approx \frac{\rho V^3}{D} \cdot \text{ergs cm}^{-3} \text{sec}^{-1} \tag{14.26}$$

By integrating over the linear extent of the granule ($\sim D$), we find that the maximum available *flux* of pressure $F_p$ from the dissolving granule is $\sim \rho V^3$ ergs cm$^{-2}$ sec$^{-1}$.

Only a fraction of $F_p$ is converted into a flux of sound waves, $F_s$, with periods in the 5-minute range. In the photosphere, where the sound speed is $\approx 10$ km sec$^{-1}$, a 5-minute sound wave has a wavelength $\lambda \approx 3000$ km, i.e., larger than the linear extent of the granule $D$. The dissolving granule acts in essence as a "short antenna" for radiating sound waves with wavelength $\lambda$. Antenna theory indicates that an antenna of length $D$ is quite inefficient at emitting waves with $\lambda > D$. Specifically, by considering the details of a multipole expansion, it can be shown that the efficiency of emission from a short antenna is proportional to $(D/\lambda)^{2m+1}$, where $m = 1$ for dipole emission and $m = 2$ for quadrupole emission. It turns out that the sound emitted by solar convection is generated mainly by quadrupole terms: as a result, the efficiency of sound emission by solar granules scales as $(D/\lambda)^5$.

The periods $P_s$ ($=\lambda/c_s$) of the sound waves emerging from a cell with lifetime $t_c$ are comparable to $t_c$. As a result, we can write $D/\lambda \approx V/c_s$. This leads to the following estimate for the flux of sound $F_s$ emitted by granular gas motions:

$$F_s \sim F_p \left(\frac{V}{c_s}\right)^5 \sim \rho V^3 M^5 \tag{14.27}$$

where $M = V/c_s$ is the Mach number associated with the convective flows.

At this point in the argument, and in order to proceed with a quantitative discussion, it is essential that we (somehow) determine how the convective velocity $V$ varies with depth beneath the solar surface. Unfortunately, the model of the convection that we computed in Chapter 7 does not contain this information: we made no attempt to compute $V$ at each depth because we did not attempt to apply "mixing-length theory" in detail. Instead, we "skipped over" the superadiabatic and ionizing layer and went right to the limit of setting the temperature gradient equal to the adiabatic temperature gradient. By referring to more detailed models, we pointed out (Section 7.7) that the layer that we "skipped over" has a linear extent of a few Mm. These are precisely the range of depths that we now need to know about if we are to be successful in discussing the excitation of $p$-modes. Therefore, with the approach we have adopted in this "first course in solar physics", we have to admit that we are not really in a position to provide a self-contained reliable quantitative answer to the question "why are certain $p$-modes excited more than others?"

Rather than leave this important question unanswered, however, it is worthwhile to refer briefly to one particular solar model in which the depth dependence of the convective velocity was explicitly tabulated. Inspection of that model (Baker and Temesvary 1966) indicates that at depths of 700, 1400, and 3500 km below the photosphere, the combination of parameters $\rho V^3 M^5$ (which is proportional to the flux of sound energy, see Equation 14.27) takes on numerical values of $3.4 \times 10^4$, $2.3 \times 10^3$, and 24 ergs cm$^{-2}$ sec$^{-1}$, respectively. That is, at a depth of 1400 km, $F_s$ is reduced by a factor of $\approx 10$ compared to $F_s$ at 700 km: this could explain why the power observed in the ($l = 1, n_r = 15$) $p$-mode is 10 times smaller than the power observed in the ($l = 1, n_r = 25$) $p$-mode. Note also that at 3500 km, $F_s$ is reduced by a factor of $\approx 1000$ compared to $F_s$ at 700 km: this could explain why the power observed in the ($l = 1, n_r = 10$) $p$-mode is almost 1000 times smaller than the power observed in the ($l = 1, n_r = 25$) $p$-mode. These estimates of the depths at which various $p$-modes are excited are admittedly simplistic: nevertheless, there is good overlap between our estimates and the depths of the acoustic sources that have been extracted by other methods: e.g., 140–550 km (Kumar 1994), and $2000 \pm 500$ km (Shelyag et al. 2006).

Finally, note that if we consider modes with large $l$, where the eigenfunctions are "squeezed" into a thin shell close to the solar surface (see Equation 14.25), even rather small values of $n_r$ may result in the last antinode lying quite close to the surface, i.e., right in the zone where acoustic generation by convection is highly efficient. This explains why, at large values of $l$, modes with small values of $n_r$ (e.g., $n_r = 4$) can be excited to detectable amplitudes (see Figure 13.4).

## 14.9 USING p-MODES TO TEST A SOLAR MODEL

Now that we have computed a solar model, we have obtained tables of values of various physical parameters as a function of the radial distance from center to surface inside the Sun. The question arises: how can we test the validity of the results we have obtained? After all, they are numbers in a table, and their values are only as good as the assumptions and approximations that went into their calculation. It would be good to have an independent means of checking. This is where helioseismology comes into its own: it allows us to "peer into" the interior of the Sun and check some of the physical variables we have calculated. In particular, the fact that p-modes with different values of $l$ can probe into the Sun to different depths (see Equation 14.25) means that, by studying modes with judiciously chosen $l$ values, we have in principle access to a tool to explore the conditions that exist at different radial locations inside the Sun.

### 14.9.1 GLOBAL SOUND PROPAGATION

We have already seen (Chapter 14, Section 14.6.1) that the asymptotic frequency separation $\Delta v$ should equal $1/(2t_s)$, where $t_s$ is the time required for sound to propagate from the center of the Sun to the surface. In view of this, we can now see the significance of a calculation we did in Chapter 9, Section 9.4. There, we computed the value of $t_s$ for our complete solar model, and found $t_s = 3804$ sec. Using that, we find $\Delta v = 1/2t_s = 131.5$ µHz. This is within 2%–4% of the observational values of $\Delta v$: 135–136 µHz (Appourchaux et al. 1998). This tells us that our model for the Sun is doing a good job of reproducing a key global property of the "real Sun".

### 14.9.2 RADIAL PROFILE OF THE SOUND SPEED

A solar model provides a radial profile of (among other things) the sound speed from center to surface. Once this is available, it is in principle possible to calculate a table of the eigenfrequencies of p-modes with various values of $l$ and $n_r$.

The modifier "in principle" in the previous sentence is meant to emphasize that Equations 14.17 and 14.18 refer only to the case of a polytrope: in the case of a realistic solar model, no single value of the polytropic index exists throughout the entire model. Therefore, new (non-polytropic) versions of Equations 14.17 and 14.18 must be derived in which the radial profile of sound speed is incorporated explicitly. Also, for maximum precision, the Cowling approximation would have to be replaced with a more complete set of (four) equations.

Once those changes have been made, a table of mode frequencies can be calculated. These can be checked against the measured frequencies in order to determine how good the model is. In general, the calculated frequencies will not reproduce the observed values. The discrepancies can be used to determine what numerical changes need to be made to the model sound speeds in order to achieve better fits. An example is shown in Figure 14.6, where relative discrepancies $\delta c_s^2/c_s^2$ between the model values of $c_s^2$ and the values of $c_s^2$ required by the p-mode frequencies are plotted as a function of radial location between the center of the Sun and the surface.

The first thing to notice about Figure 14.6 is that although discrepancies between model and data *are* present, the discrepancies are relatively small: nowhere inside the Sun is $\delta c_s^2/c_s^2 \approx \delta T/T$ larger than 0.4%. This is a striking endorsement of the reliability of current solar models: the run of temperature inside a solar model from center to surface, ranging from > 10 million K to a few thousand K, reproduces what happens at all radial locations inside the Sun to better than a few parts per thousand. It was this success in testing models of the solar interior that forced neutrino physicists (who were trying to solve the solar neutrino problem: see Section 12.5) to switch attention away from the solar models (now known to be reliable at the parts per thousand level) and begin to concentrate on the physics of neutrinos themselves: as it turned out, the "standard model" of particle physics needed to be updated to allow neutrinos to have a nonzero mass.

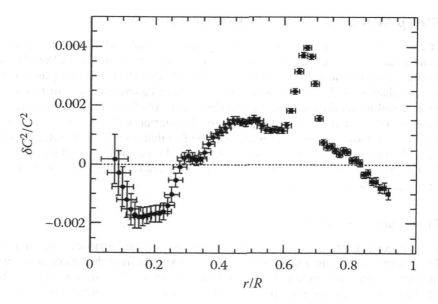

**FIGURE 14.6** Radial profile of discrepancies between sound speeds (squared) in a solar model and the sound speeds (squared) required to reproduce the observed eigenfrequencies. (Courtesy of SOHO/MDI consortium. SOHO is a project of international cooperation between ESA and NASA.)

Because modes of different degree $l$ penetrate into the Sun by different amounts (see Equation 14.25), discrepancies between modes with the largest $l$ values contain information about errors in the model in the outermost layers of the Sun. Modes with intermediate $l$ values ($\approx 60$–$70$) may be used to probe conditions down to the base of the convection zone (at radial locations $r \approx 0.7 R_\odot$). As can be seen from Figure 14.6, the base of the convection zone is the site of the largest discrepancies in sound speed: this suggests that certain physical phenomena occurring at the interface between the convection zone and the radiative core may not yet be properly incorporated in the solar model. Among these phenomena might be rotational shear, or magnetic fields, or overshooting of convection: however, allowance for such complications would take us far beyond the limits of a first course in solar physics.

In the deepest regions of the radiative interior, information about the model is contained in $p$-modes with the lowest $l$ values. But even then, the innermost part of the Sun, at radial locations within (say) $0.2 R_\odot$ of the center of the Sun, cannot be probed with great reliability by $p$-mode data. This explains why the error bars in Figure 14.5 become considerably larger in the innermost regions of the Sun.

### 14.9.3 The Sun's Rotation

We do not need helioseismology to study the rotation on the *surface* of the Sun: that rotation can be observed directly. The observations (Section 1.11) show that the Sun rotates faster at the equator than at high latitudes, with a difference of almost 30% between the equator and the poles.

When it comes to studying the Sun's rotational properties *beneath* the surface, we must rely on helioseismology. In describing the modes that exist inside the Sun, we have concentrated on only two of the integers that specify a mode: $l$, and $n_r$. These are related to properties of the modes in the *latitudinal* and *radial* directions respectively.

# Oscillations in the Sun

However, in order to study rotation, we would also need to include, in our spherical harmonic analysis, an index $m$ to describe properties of modes in a third direction, namely, in *longitude*. For a spherical harmonic mode with any given value of the degree $l$, there exist in principle $2l +1$ submodes with $m$ values varying from $m = -l$ to $m = +l$. The algebraic sign of $m$ indicates the longitudinal direction in which the mode propagates. Now, the Sun is rotating in the longitudinal (azimuthal) direction (parallel to the equator) with a speed $V_r$ that, at the equator and on the surface, has a magnitude of about 2 km sec$^{-1}$. A *p*-mode (sound wave) that propagates in the *same* longitudinal direction as rotation propagates with speed $c_s + V_r$ relative to a nonrotating frame. On the other hand, a *p*-mode that propagates *opposite* to the direction of rotation propagates with speed $c_s - V_r$ relative to a nonrotating frame. These differences in propagation speed lead to differences in the frequencies of eigenmodes with positive and negative values of $m$. As a result, a mode with a particular value of $l$ has a power spectrum that no longer contains only a single "spike" at frequency $v_o$: instead, in the case $m = \pm 1$, there are now extra spikes shifted by $\delta v$ on either side of the central peak. The relative shift in frequency, $\delta v/v_o$, is essentially equal to $V_r/c_s$. Since $c_s$ is reliably known at any particular radial location (see Figure 14.6), a value for $V_r$ can be obtained. And by making use of the depth dependence of mode penetration as a function of degree $l$, one can obtain the radial profile of rotation. The *p*-modes allow us to probe rotation in the regions of the Sun that extend from the surface in to radial locations of a few tenths of $R_\odot$.

Figure 14.7 shows the profile of rotational angular velocity that has been determined by analysis of SOHO/MDI data between 1996 and 2011 by Larson and Schou (2015). (The same data set was used to construct Figure 13.4.) The results in Figure 14.7 are presented as a function of two variables: (i) the fractional radial coordinate ($r/R$) inside the Sun and (ii) the latitude.

Examining first the *surface* of the Sun, i.e., at $r/R = 1$, the curves in Figure 14.7 show that at low latitudes (0 deg), the surface rotates relatively rapidly: $\Omega/2\pi \approx 450$–$460$ nanohertz (nHz), i.e., a rotation period of 25.2–25.7 days. (This range overlaps the equatorial rotation period of 25.37 days obtained by averaging surface Doppler data over 14 years: see Section 1.11.) At higher

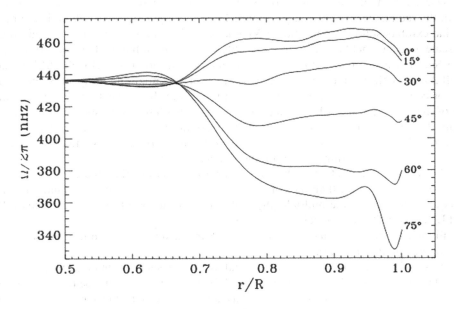

**FIGURE 14.7** Angular velocity versus fractional radius inside the Sun at six different latitudes. (From Larson and Schou 2015; used with permission of Springer.)

latitudes, the surface rotates more slowly: at 75 deg latitudes, $\Omega/2\pi \approx 340$–$350$ nHz, i.e., a rotation period of 33.1–34.0 days. Although the surface Doppler data allow in principle the study of rotation all the way to the poles (see Chapter 1, Equations 1.22 and 1.23), the helioseismological data are not sensitive enough at latitudes above 75° to allow Figure 14.7 to be extended reliably to the poles.

The fact that the Sun's *surface* does not rotate as a solid body has been known for a long time (see Section 1.11). What Figure 14.7 demonstrates is that it is not merely the *surface* of the Sun that departs from solid-body rotation: this feature also exists in the *interior* of the Sun.

The long-known departure from solid body rotation at the *surface* is called latitudinal differential rotation (LDR). As we examine the *subsurface gas* in Figure 14.7, we see that LDR persists (with values similar to the surface) down to depths of at least 0.2 solar radii, i.e., to radial locations as small as $0.8R_\odot$. At greater depths, as we approach the base of the convection zone, at radial location $r \approx 0.7R_\odot$, there is a remarkable convergence of the angular velocities to essentially a unique value. Below the convection zone, in the radiative interior, i.e., at $r < 0.65R_\odot$, and in as far as a radial coordinate of order $0.5R_\odot$, the Sun rotates at essentially the same rate at all latitudes: $\Omega/2\pi \approx 435$ nHz. That is, the radiative interior of the Sun, at least in its outer regions, rotates *as a solid body* with a period of 26.6 days. The rotation of the Sun in the inner regions ($r < 0.5R_\odot$) are not known as reliably as they are in the outer regions, because very few $p$-modes penetrate into those deep regions of the interior. A narrow boundary layer exists between the base of the convection zone and the outer edge of the radiative interior: this layer is known as the "tachocline".

The fact that the radial profiles in the convection zone in Figure 14.7 have nonzero slopes means that the convection zone of the Sun, at fixed latitude, has a variation in angular velocity as a function of *radius*. Thus, the Sun exhibits *radial* differential rotation (RDR), in addition to the LDR evident at the surface.

Why exactly the Sun shows the rotational properties shown in Figure 14.7 is not readily explainable in terms of the physics of a "first course". For example, from the simplest perspective, one might expect that the convection zone, with its turbulent stresses (which produce a highly effective viscosity), should be able to enforce solid body rotation in the convection zone more easily than in the radiative interior. And yet Figure 14.7 shows quite the opposite: it is the radiative interior that exhibits solid body rotation. (What keeps the radiative zone in solid body rotation? A large-scale magnetic field could do this, possibly even a primordial field left over from the time when the Sun first condensed out of a magnetic interstellar cloud: see Gough 2017; Wood and Brummell 2018.) With the interior in solid body rotation, one might then expect (since the radiative interior contains 98% of the Sun's mass) that the gas in the convection zone (amounting to only 2% of the mass) could easily be "kept in line" and also forced into solid body rotation. But once again, this expectation is contrary to what occurs in the real Sun. Some powerful internal dynamics must be at work to drive the convection zone into differential rotation: apparently, the forces at work in the convection zone (rotation, gravity, thermal buoyancy, viscous stresses) have the overall effect that the rotation of the convection zone is slower than the core (by about 10%) at high latitudes, but is faster than the core (by about 10%) at low latitudes. Computational models are required to incorporate multiple physical effects if they are to replicate successfully the observed LDR and RDR (e.g., Kitchatinov 2005).

We shall return to the rotational properties of the Sun when we discuss how magnetic fields are generated in the Sun (Chapter 16).

The results in Figure 14.7 rely on measuring small frequency splittings in mode frequencies due to rotational motions. We have already seen (Section 13.8) that solar $p$-mode frequencies are observed to vary in a systematic way during the solar cycle. This raises the question: do the *rotational splittings* in solar $p$-modes also show variations during the solar cycle? The answer appears to be "No" (Broomhall et al. 2012): the 11-year cycle does not significantly alter the rotation of the solar interior.

## 14.10 *r*-MODES IN THE SUN

In a rotating star, the existence of Coriolis force gives rise to the possibility of a new class of wave modes that propagate in longitude, and in which the restoring force is not dominated by pressure (as in *p*-modes) or by gravity (as in *g*-modes) (Papaloizou and Pringle 1978: hereafter PP78). Waves in this class (called *r*-modes) are globally coherent horizontal oscillations dominated by the Coriolis force: the waves do *not* propagate in the radial direction. In PP78, the theory of *r*-modes was developed in the context of a class of white dwarf stars known as dwarf novae, where rotation periods can be as short as 10 seconds (!). In the present book, we are interested in the Sun, where the rotation period is five to six orders of magnitude longer than in dwarf novae (see Section 1.11). With such slow rotations, it is expected that *r*-modes in the Sun might have amplitudes that are much more difficult to detect than in dwarf novae. In fact, it was not until 2018 that the first unambiguous detection of *r*-modes *in the Sun* was published (Löptien et al. 2018).

However, suggestions that *r*-modes might be present in the Sun had been made more than three decades earlier (e.g., Wolff and Hickey 1987: hereafter WH87) based on the observation (at certain times) of solar activity at preferred longitudes. An *r*-mode is defined by WH87 as a "toroidal oscillation of swirling horizontal motions" with periods no shorter than the star's rotation period. The restoring Coriolis force conserves angular momentum. WH87 pointed out that *r*-modes could have detectable effects on observed solar properties for the following reason: in models of the deep convection zone, the convection cells have sizes and vorticities that are comparable to those of *r*-modes. Moreover, Wolff (1992) pointed out that the beating of various pairs of *r*-modes in the Sun could be relevant for understanding the presence of periodicities in various solar features in the range of 100–1000 days.

On small length-scales (Chowdhury et al. 2010), the *r*-modes reduce to a type of wave that was first discussed by Rossby (1939) in the context of the Earth's atmosphere, where they play a major role in determining the weather on large scales (in the so-called semi-permanent centers of action). As a result, in the term *r*-modes, the *r*- can be interpreted as standing for either rotation or Rossby. The modes are also referred to as "inertial oscillations", depending on the historical context or on the geometric approximations that enter into the derivation (see WH87). In a spherical object, *r*-modes are characterized by parameters analogous to those we have already encountered for *p*-modes and *g*-modes, namely, $l$ is the degree of the mode (= number of nodes between N and S pole), and $m$ is a longitudinal mode number.

Following on from PP78, the properties of *r*-modes in main sequence stars and in polytropic spheres (polytropes are useful in the real world!) have been analyzed by Saio (1982). To a first-order approximation, Saio confirmed a result from PP78 to the effect that the angular frequency $\omega$ of the *r*-modes (as observed by someone who is co-rotating with the star) is as follows:

$$\omega = \frac{2m\Omega}{l(l+1)} \qquad (14.28)$$

The effects of the Coriolis force are such that, if an *r*-mode is observed in a frame of reference that is rotating *with the star*, the flow pattern of each *r*-mode will be seen to drift *retrograde* (opposite to the direction of rotation), with each $l$ value having its own rotation period that is *slower by a well-defined amount* than the star's rotation period. Modes with the largest $l$ and $m$ values have periods that are close to the rotation period of the star: in the Sun, this is a period of almost 1 month. The solar *r*-modes with the smallest $l$ and $m$ values are predicted to have periods that are *longer* than 1 month by factors of a few.

The work of Saio (1982) showed strikingly that the calculation of eigenfunctions for *r*-modes is not as simple as we described earlier for *p*-modes and *g*-modes. In the latter cases, we have already mentioned that the modes can be described adequately by two first-order differential equations

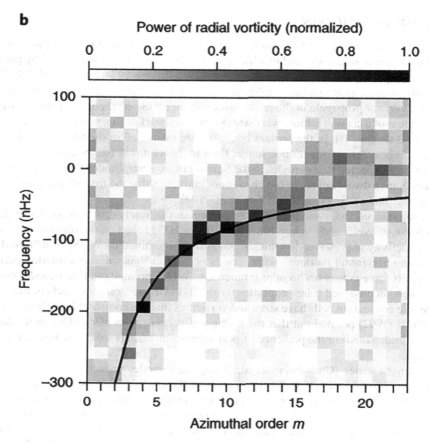

**FIGURE 14.8** *r*-modes in the Sun. Power spectrum of surface radial vorticity determined from granulation tracking (Löptien et al. 2018). Abscissa: longitudinal order *m*, with only sectoral modes (with $l = |m|$) plotted. Ordinate: frequency (in nanohertz) of mode relative to the rotation frequency of the equatorial regions of the Sun. (Used with permission from Springer.)

(Equations 14.17 and 14.18). But Saio showed that for *r*-modes, the number of differential equations increases to six, thereby reinforcing the statement of Cowling (1941) to the effect that, when rotation is included, "the mathematical difficulties are much greater". As a result, in the present "first course", we will not further consider Saio's six equations.

Reliable observational detection of individual *r*-modes with particular values of longitudinal order (*m*) in the Sun was first reported by Löptien et al. (2018). They determined the values of radial vorticity in near-surface layers by tracking the surface motions of granules: observations with enough angular resolution to identify individual granules were provided by images from the HMI instrument on SDO (with an angular resolution of 1"). Analyzing vorticity maps in terms of $Y_l^m$, and isolating the sectoral modes (i.e., modes with $l = m$, with power concentrated in equatorial regions), a power spectrum was obtained as a function of the azimuthal order *m*. By averaging together 6 years of SDO/HMI data, significant power was found to occur in modes with *m* ranging from 3 to 15. These modes were predicted to lie at (angular) frequencies given by Equation 14.28 with $l=m$. The corresponding (linear) frequencies of these modes are $\nu(m) = \omega/2\pi = 2\Omega/2\pi(m+1)$. Since the solar rotational period in the equatorial regions is of order 25–35 days, i.e., $2\pi/\Omega \approx 2 \times 10^6$ sec, we expect to find $\nu(m) \approx 1000$ nHz/$(m+1)$. Thus, modes with (e.g.) $m = 3, 9$, and 15 are predicted to occur close to $\nu \approx 250$, 100, and 60 nHz. In fact, Löptien et al. (2018) reported a total of 13 modes, including the prior three modes at 259, 86±6, and 47±7 nHz. The functional form of $\nu(m) \sim 1/(m+1)$ was found to fit the peaks in the power spectrum well (see Figure 14.8).

In a quantitative sense, a significant conclusion of Löptien et al. (2018) is the following: the amplitudes of the vorticity of the sectoral $r$-modes are found to be nearly as large as the vorticity associated with convection cells on similar length-scales. This conclusion confirms the reasoning of WH87 that $r$-modes can serve as an essential component of solar dynamics. Moreover, Löptien et al. (2018) report that the $r$-modes have amplitudes that do not vary much with depth, at least down to depths of 21 Mm.

Confirmation of the observational results of Löptien et al. (2018) has been reported by Liang et al. (2019) in spacecraft data sets spanning 21 years of observations: modes with azimuthal order $m$ ranging from 3 to 15 were again reported as having measurable amplitudes, with $m = 10$ having the largest amplitude. Further confirmation of $r$-modes with $m = 3$–15 in the Sun has been provided by analysis of ground-based GONG data spanning 17 years (Hanson et al. 2020): the signals are most reliably detected for $m$ values in the range 8–11. It is notable that $r$-modes with $m = 10$ have rotation periods of 150–158 days (Chowdhury et al. 2010): we shall encounter periodicities of this order when we consider certain aspects of solar magnetic activity (Section 16.9).

In view of the slowness of solar rotation compared to the rotation in some early-type stars, it is perhaps not surprising that $r$-modes were detected with confidence in fast rotating stars prior to their discovery in the Sun (Van Reeth et al. 2016). Specifically, a group of stars with spectral type ranging from mid-A to late-B stars had been found to have the following characteristic in their power spectra: a sharp "spike" occurs at a certain frequency, accompanied by a broad "hump" of power at *lower* frequencies (Balona 2013). Saio et al. (2018) suggested that this "hump and spike" property could be interpreted in the context of $r$-modes as follows: the "spike" corresponds to the stellar rotation frequency itself, while the "hump", with multiple frequencies, all of which move more *slowly* than the stellar rotation, could be ascribed to $r$-modes moving in the retrograde direction.

## EXERCISES

14.1 Perform the step-by-step procedure described in Section 14.4.1 for $p$-modes with $l = 1$ and 3. Plot the equivalent of Figure 14.1 and obtain a table of the eigenfrequencies for $l = 1$ and $l = 3$ $p$-modes in the $n = 1$ polytrope for an object with solar mass and radius. For each $l$ value, determine the "large separations" $\Delta\nu$ (in μHz) between adjacent modes. And for appropriate pairs of modes, determine the "small separations" $\delta\nu(1$–$3)$ between modes with $l = 1$ and $l = 3$.

14.2 You have already (Chapter 10, Exercise 1) calculated a table of values of $y$ and $y'$ for the polytrope $n = 3.25$. Use your tabulated values (including interpolation if necessary) and the step-by-step procedure in Section 14.5 to integrate Equations 14.17 and 14.18 in the $n = 3.25$ polytrope for $g$-modes with $l = 3$. Plot the results in the form shown in Figure 14.4. Compare your eigenperiods for $l = 3$ with those for $l = 1$ in Figure 14.4: each $l = 3$ period should lie close to the period of an $l = 1$ mode.

## REFERENCES

Appourchaux, T., Rabello-Soares, M.-C., & Gizon, L., 1998. "The art of fitting $p$-mode spectra: II. Leakage and noise-covariance matrices", *Astron. Astrophys. Suppl.* 132, 131.

Baker, N. H., & Temesvary, S., 1966. "A solar model", *Table of Convective Stellar Envelope Models*, 2nd ed. NASA Institute for Space Studies, New York, pp. 18–28.

Balona, L., 2013. "Activity in A-type stars", *Mon. Not. Roy. Astron. Soc.* 431, 2240.

Bonanno, A., & Frohlich, H.-E., 2015. "A Bayesian estimation of the helioseismic solar age", *Astron. Astrophys.* 580, A130.

Broomhall, A.-M., Salabert, D., Chaplin, W. J., et al., 2012. "Misleading variations in estimated rotational frequency splittings of solar $p$-modes", *Mon. Not. Roy. Astron. Soc.* 422, 3564.

Chowdhury, P., Khan, M., & Ray, P. C., 2010. "Short-term periodicities in sunspot activities during the descending phase of solar cycle 23", *Solar Phys.* 261, 173.

Christensen-Dalsgaard, J., 2003. "Properties of solar and stellar oscillations", p. 88 of www.phys.au.dk/~jcd/oscilnotes/chap-5.pdf

Connelly, J. N., Bizzarro, M., Krot, A. N., et al., 2012. "The absolute chronology in the solar protoplanetary disk", *Sci.* 338, 651.

Cowling, T. G., 1941. "The non-radial oscillations of polytropic stars", *Monthly Not. Royal Astron. Soc.* 101, 367.

Gough, D. O., 2017. "Is the Sun a magnet?" *Solar Phys.* 292, 70.

Hanson, C. S., Gizon, L., & Liang, Z.-C., 2020. Rossby waves observed in GONG+ ring-diagram flow maps", *Astron. Astrophys.* 635, A109.

Kitchatinov, L. L., 2005. "Reviews of topical problems: The differential rotation of stars", *Physics Uspekhi* 48, 449.

Kumar, P., 1994. "Properties of acoustic sources in the Sun", *Astrophys. J.* 428, 827.

Larson, T. P., & Schou, J., 2015. "Improved helioseismic analysis of medium-$l$ data from SOHO/MDI", *Solar Phys.* 290, 3221.

Liang, Z.-C., Gizon, L., Birch, A. C., et al., 2019. "Time-distance helioseismology of solar Rossby waves", *Astron. Astrophys.* 626, A3.

Löptien, B., Gizon, L., Birch, A. C., et al., 2018. "Global-scale equatorial Rossby waves as an essential component of solar internal dynamics", *Nature Astron.* 2, 568.

Mullan, D. J., 1989. "$g$-mode pulsations in polytropes: High-precision eigenvalues and the approach to asymptotic behavior", *Astrophys. J.* 337, 1017.

Mullan, D. J., & Ulrich, R. K., 1988. "Radial and non-radial pulsations of polytropes: High-precision eigenvalues and the approach of $p$-modes to asymptotic behavior", *Astrophys. J.* 331, 1013.

Papaloizou, J., & Pringle, J. E., 1978. "Non-radial oscillations of rotating stars", *Mon. Not. Roy. Astron. Soc.* 182, 423.

Rossby, C.-G., 1939. "Relations between variations in the intensity of the zonal circulation of the atmosphere and the displacements of the semi-permanent centers of action", *J. Marine Res.* 2, 38.

Saio, H., 1982. "$r$-mode oscillations in uniformly rotating stars", *Astrophys. J.* 256, 717.

Saio, H., Kurtz, D. W., Murphy, S. J., et al., 2018. "Theory and evidence of global Rossby waves in upper main-sequence stars: $r$-mode oscillations in many Kepler stars", *Mon. Not. Roy. Astron. Soc.* 474, 2774.

Shelyag, S., Erdelyi, R., & Thompson, M. J., 2006. "Acoustic wave propagation in the quiet solar subphotosphere", *Astrophys. J.* 651, 576.

Van Reeth, T., Tkachenko, A., & Aerts, C., 2016. "Interior rotation of a sample of $\gamma$ Doradus stars from ensemble modelling of their gravity-mode period spacings", *Astron. Astrophys.* 593, A120.

Wolff, C. L., 1992. "Intermittent solar periodicities", *Solar Phys.* 142, 187.

Wolff, C. L., & Hickey J. R., 1987. "Solar irradiance changes and special longitudes due to $r$-modes", *Sci.* 235, 1631.

Wood, T. S., & Brummell, N. H., 2018. "A self-consistent model of the solar tachocline", *Astrophys. J.* 853, 97.

Zhang, X., Chen, X., Zhang, H., et al., 2020. "A pre-main-sequence $\gamma$ Dor-$\delta$ Sct hybrid with extremely slow internal rotation in a short-period eclipsing binary KIC9850387 revealed by asteroseismology", *Astrophys. J.* 895, 124.

# 15 The Chromosphere

So far, when we have discussed the Sun, we have been interested in the material that extends from the visible surface *downward* into the *interior* of the Sun. The visible surface, the "photosphere" (the "light sphere"), provides the light that dominates human vision. Our model of the interior of the Sun, extending over the entire radial extent from center to photosphere, spanned a radial distance of some 700,000 km. When we computed the model, we did so in three segments, focusing on distinct laws of physics that play dominant roles in each segment. As it turned out, the three segments were found to be of unequal radial depth. The model of the deep interior (Chapter 9) extended over some 500,000 km. The model of the convection zone (Chapter 7) had a depth of some 200,000 km. And the model of the photosphere (Chapter 5) spanned no more than a few hundred kilometers in linear extent.

Now, we turn our attention in the opposite direction. Instead of starting at the photosphere and moving *inward* towards the *center* of the Sun, we now start at the photosphere and move *upward* and *outward*. This brings us into the more rarefied gas that forms the outer atmosphere of the Sun. And just as we did for the interior, it will be convenient to recognize that different laws of physics are dominant in different segments of the outer atmosphere. We shall find it convenient to again discuss three segments of the outer atmosphere: the chromosphere (Chapter 15), the corona (Chapter 17), and the solar wind (Chapter 18). Of these, the linear extents are again very different: the solar wind is by far the largest, extending over vast distances of interplanetary space, with linear scales up to 10 *billion* km; the corona can be detected with optical equipment out to distances of 1–2 *million* km; and the chromosphere has a thickness of no more than a few *thousand* km.

Thus, as we move outward from the surface, the chromosphere is by far the thinnest of the three segments, by analogy with the thinness of the photosphere as regards the interior of the Sun. This raises the question: why should we spend time on such a narrow region? What is there for us to learn about solar physics by paying attention to such a thin shell of gas? The answer is: the chromosphere allows us to study the effects of certain waves with the important property that they can effectively transport energy from one region of the Sun to another. The simplest type of wave that propagates through a nonmagnetic gas is a sound wave, in which regions of high and low pressure contain molecules that "push each other" through the gas at a well-defined speed (the "sound speed", $c_s$). The value of $c_s$ is determined by how fast the molecules are moving randomly due to the finite temperature of the gas. We have already seen (Chapters 13 and 14) that sound waves provide a mechanism for us to probe the *interior* of the Sun: in those cases, the sound waves do not appreciably alter the physical conditions as the waves propagate. Now we turn to another location in the Sun where sound waves play quite a different role: in the chromosphere, the propagating waves can have a significant effect on the physical properties of the gas. We shall see that the waves can deposit energy, thereby leading to significant local heating of the gas. The goal of the present chapter is to study the following specific issues: How are the waves generated? How much energy do the waves carry? How is the wave energy actually deposited as heat in the surrounding gas?

We shall also eventually see (Chapter 16) that in certain parts of the Sun, the gas is permeated by magnetic fields: in the presence of such fields, sound waves are not the only wave modes propagating through the medium. In addition to sound waves, there can also be waves in the magnetic field itself (Alfven waves), and there can also be magnetosonic waves, where gas pressure and magnetic fields intertwine in ways that lead to "fast" waves and "slow" waves. These waves can also deposit heat in the gas in addition to what is deposited by the sound waves. Careful observations of different regions in the chromosphere may reveal the presence of waves of different kinds.

DOI: 10.1201/9781003153115-15

## 15.1 DEFINITION OF THE CHROMOSPHERE

The word "chromosphere" is derived from a Greek word meaning "color sphere". Why is the word "color" used to describe this structure? The reason has to do with the phenomena that are visible to the human eye during an eclipse of the Sun.

Two distinct phenomena can be seen during an eclipse as the Moon blocks out the brilliant light of the photosphere. One of these phenomena lasts for a relatively long time (minutes), while the other is over "in a flash" (in seconds). But both tell us something valuable about the Sun's atmosphere.

i. The long-lasting phenomenon, which can be seen as long as the total phase of the eclipse lasts (up to a maximum duration of about 7 *minutes*), is an extended whitish region (see Figure 15.1) that extends above the surface to radial distances of a few solar radii: this is the corona (which will be discussed in Chapter 17).

ii. The short-lived phenomenon is visible only for a few (4–8) *seconds* at the start and end of totality (see innermost rings of Figure 15.1). Brightly colored "patches" of a ring can be seen, confined to a narrow region very close to the solar limb. The predominant color of the patchy ring is "rose-colored". The fact that the patches are obviously colored (in

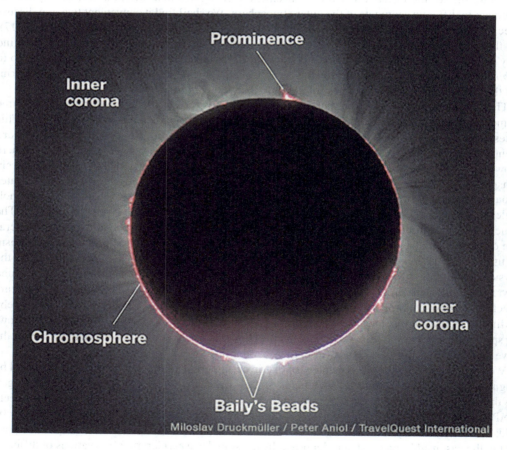

**FIGURE 15.1** The chromosphere (and inner corona) of the Sun as seen during a total eclipse of the Sun. The chromosphere is confined to a narrow region close to the Moon's edge, and it has a pronounced reddish (rose-colored) hue (Used with permission of M.Druckmuller).

# The Chromosphere

contrast to the white corona) gives rise to the term "chromosphere". We shall see that the fact that the dominant color is red tells us something significant about the physics of the chromosphere (see Section 15.10.3).

When a spectrum is obtained, the chromosphere reveals a large number of *emission* lines in the spectrum, of which the strongest is the line known as $H\alpha$ in the red portion of the spectrum. The dominance of $H\alpha$ in the chromospheric emission gives rise to the reddish hue in Figure 15.1. The presence of emission lines in the flash spectrum provides a remarkable contrast to the photosphere, where most of the spectrum (emitted in light visible to the unaided eye) contains *absorption* lines (see Chapter 3, Figure 3.5). The fact that the chromosphere lasts for only a few seconds gives rise to the phrase "flash spectrum" for the chromospheric emission lines. The strongest lines originate in hydrogen (including the lines in the Balmer series known as $H\alpha$, $H\beta$, $H\gamma$, and $H\delta$ at wavelengths of 6563, 4861, 4340, and 4102 Å respectively). There are also prominent chromospheric lines due to helium (5876 Å), and ionized calcium (Ca II $H$ and $K$ at 3968 and 3934 Å).

The fact that the chromosphere does *not* extend as a uniform ring around the Moon's edge in Figure 15.1 indicates that the chromosphere is *not* homogeneous but "patchy". Images obtained in the Ca II $H$ line (see Figure 15.2) reveal the presence of short-lived (with lifetimes of at most a few minutes) "spiky" features ("spicules") rising up from the photosphere to heights of a few thousand km. Similar images of spicules can also be obtained by observing in the core of the $H\alpha$ line: Pereira et al. (2016) state, "spicule shapes, extent, and lifetimes are essentially identical whether observed in $H\alpha$ or in the Ca II $H$ line". Inside each spicule, upward gas motions are confined within linear structures that are defined by local magnetic fields. Variations in magnetic field properties from one

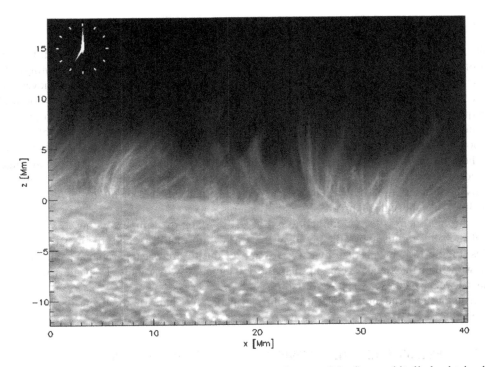

**FIGURE 15.2** Snapshot from a time series of Ca II images of a part of the Sun, and its limb, obtained with the 3 Å wide filter of the broadband filter imager (BFI) attached to the Solar Optical Telescope (SOT) on the *Hinode* spacecraft. The solar disk occupies the lower half of the image. Spicules can be seen against the dark background sky in the upper half. The height scale has been carefully determined, relative to the vertical continuum (5000 Å) at optical depth unity. (From Judge and Carlsson 2010; used with permission of P. Judge.)

part of the solar surface to another causes the chromospheric emission to appear "patchy". As can be seen in Figure 15.2, the strongest emission from spicules occurs at altitudes between the disk of the Sun (the photosphere) and heights of order 2–3 Mm, with fainter parts of some spicules extending up to as high as 5–10 Mm. What about the width of a spicule? The image in Figure 15.2 suggests that the width of a spicule is ≤ 0.1 times the length, i.e., at most several hundred km, i.e., smaller than granule widths. Upward speeds of gas in spicules are found to be tens of km s$^{-1}$ for long-lived (referred to as "Type I") spicules and up to 150 km s$^{-1}$ for short-lived ("Type II") spicules.

The presence of emission lines in the flash spectrum indicates that in the chromospheric gas, bound electrons in atoms and ions are able to cascade down from upper energy levels to lower ones: this process must start by significant numbers of free electrons being present in the gas, from which they can be captured by the ions. The fact that free electrons are abundant indicates that the local temperature in the chromosphere is greater than the temperature in the photosphere. Something has *heated up the gas in the chromosphere to temperatures in excess of those in the photosphere*. It will be our primary goal in this chapter to quantify the heating process.

Historically, the discovery of the emission line at 5876 Å in the flash spectrum is noteworthy. This line was observed for the first time during a solar eclipse in 1868 when a spectroscope was used to view the Sun during the few seconds of the flash spectrum. The line could not be identified with any known material on Earth at that time: the name "helium" was given to the material, after the Greek work "helios" for the Sun. It would take 30 years before helium was identified as a trace element that is also present in Earth's atmosphere, with a percentage abundance of less than 0.001%. In the Sun, the abundance of helium is much larger than on Earth: next to hydrogen, helium is the most abundant element in the solar atmosphere, accounting for some 10% of the atoms in any given volume. In Earth's atmosphere, helium contributes only about 5 parts per million.

## 15.2 LINEAR THICKNESS OF THE CHROMOSPHERE

The fact that the flash spectrum lasts only for a few seconds contains information on the linear extent of the chromosphere along the radial direction. To see this, we note that the timings of the various phenomena occurring during an eclipse are determined by how fast the Moon moves across our line of sight to the Sun. Now, the Moon is in orbit around the Earth such that one orbit (360 deg) requires about 30 days. This corresponds to an angular velocity of 0.5 deg per hour, i.e., 1800 arcsec per hour, or 0.5 arcsec per second of time. As a result, in a time interval of roughly 4–8 sec, the Moon traverses an arc having an angular extent of roughly 2–4 arcsec. At the distance of the Sun, where the conversion factor is 725.3 km per arcsec (Chapter 1, Section 1.2), such an angular extent corresponds to a linear extent of roughly 1500–3000 km. This (very rough) estimate of the chromospheric thickness is consistent with the brightest parts of the spicule bases in Figure 15.2: see the linear height scale along the left-hand border.

The chromosphere is truly confined to a thin shell around the Sun, extending to no more than 0.5% of the solar radius above the photosphere. But within that thin shell, the existence of many spicules (Figure 15.2) means that if we try to assign a value to a physical parameter at any particular height, we will be in essence averaging spatially and temporally over multiple spicules and multiple regions between spicules.

In terms of independent variable in the chromosphere, we shall use the linear height above the photosphere.

## 15.3 OBSERVING THE CHROMOSPHERE ON THE SOLAR DISK

Observations during a total solar eclipse were the first to allow human beings to see the "rosy-hued" chromosphere *at the limb* of the Sun (see Figure 15.1). With specially constructed telescopes in good locations (e.g., in space), there is no need to wait for an eclipse: images of the *limb* allow us to detect

# The Chromosphere

spicules (see Figure 15.2). However, it is also possible to observe the chromosphere by judicious choice of observing conditions on the *disk* of the Sun.

To see why this is the case, let us recall the results plotted in Figure 3.7. By observing in a wavelength range that lies close to the center of strong lines (such as Ca II $H$ and $K$, or $H\alpha$, or the Si I line at 1256 Å), we are essentially probing the solar atmosphere at vertical heights of 1500–2000 km. Recalling the linear thickness of the chromosphere (1500–3000 km), which has been revealed by the flash spectrum, we see that, if we observe at a wavelength that allows our line of sight to penetrate no deeper than 1000–2000 km above the photosphere, this will put us right in the midst of the chromosphere.

When the disk of the Sun is observed in the center of the Ca II $K$ line, the chromosphere is seen to be nonuniform in brightness, especially around sunspots (see Figure 15.3). In Figure 15.3, the things that first catch the eye are localized enhancements in brightness lying in certain regions of the surface. These bright features lie within a range of latitudes, roughly between 10 and 30 deg, in both the northern and southern hemispheres. Some of these bright features are spatially associated with localized dark features ("sunspots"). The bright features in the Ca K images of the chromosphere are larger in area than the sunspots: the bright features are referred to as "plages" (Latin for "wounds": see also Cally and Moradi 2013) or "active regions".

In view of the eye-catching nature of plages and sunspots (which are definitely sites of stronger than average magnetic fields on the Sun), we may ask: does the chromosphere exist *outside* magnetic regions? The answer is a definite "Yes". Apart from the plages and spot regions, the rest of the solar chromosphere has a characteristic appearance, which is illustrated in the expanded view in Figure 15.4: this is an image taken in the Si I line at 1256 Å, which is formed at an altitude comparable to that of Ca II $K$. A characteristic topological aspect of Figure 15.4 can be identified by inspection: one can trace a thread of bright elements distributed over essentially all regions of the image, with the thread helping to serve as a border surrounding darker patches. The bright elements in the image are connected to one another across the surface of the Sun: this connectivity gives rise to the descriptive term "network" for the bright rims. The darker centers are referred to as "cells".

This topology of cells surrounded by borders is reminiscent of what we observe in the photospheric features called granules (see Chapter 6, Figure 6.1). However, whereas the photospheric granules consist of *bright* centers surrounded by *dark* rims, in the case of the chromosphere, we

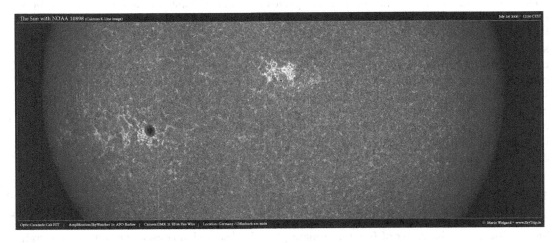

**FIGURE 15.3** Image of the Sun in the light of the K line of ionized calcium. Taken in SHAMS Observatory, Karachi, Pakistan. (Used with permission of Sajjad Ahmed.)

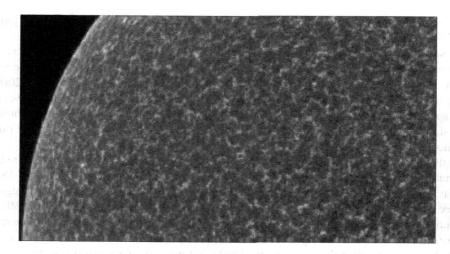

**FIGURE 15.4** A quiet region of the solar chromosphere viewed in the line Si I 1265 Å. Chains of brighter elements can be identified as a "network" that "snakes" across much of the surface. The bright network surrounds darker regions ("cells"). The "cells" have horizontal dimensions similar to those of the "supergranules" that were first discovered on dopplergrams (cf. Section 6.12) (Judge and Peter 1998; used with permission of Springer).

have the opposite topology: *dark* cell centers surrounded by *bright* rims. (The analogy to "[dark] fields" separated by "[bright] hedgerows" comes to mind here.) The topological similarities to the granules gives rise to the term "supergranules" for the features that appear in the Si I images. We have already mentioned supergranules in the context of *velocity* data (see Section 6.12). Now, in the context of chromospheric data, we have another definition: a supergranule can be defined as a darker cell surrounded by a rim of bright network. The sizes of the supergranules in Figure 15.4 are entirely consistent with those discussed in Section 6.12.

The fact that chromospheric features (supergranules, network) can be identified at *all* regions of the solar surface implies that whatever source of energy is creating the chromosphere is present at *all* locations of the Sun's surface. Although it is obviously true that strong magnetic fields contribute to enhancing the chromosphere in plages and near spots, it does not necessarily follow that magnetic fields are *essential* for the chromosphere to exist. Rather, it is important to note that even when magnetic fields are *not* present in strength, the Sun has access to a nonmagnetic source of energy that suffices to power the chromosphere with its characteristics of supergranules and network *at all locations on the solar surface*. One such source of energy is sound waves generated by the turbulent convective flows of compressible gas: anyone who has been close to a jet engine when it is operating knows that the fast (turbulent) airflow emerging from the jet is an effective generator of sound waves. And because turbulent convective flows exist at all locations on the Sun's surface (granules are present all over the Sun), sound wave emissions also emerge into the solar atmosphere from all locations on the surface. We shall return to this topic when we discuss the corona.

## 15.4 SUPERGRANULES OBSERVED IN THE $H\alpha$ LINE

Another strong spectral line useful for observing the chromosphere is $H\alpha$. When the limb of the Sun is observed at the center of $H\alpha$, the chromosphere is observed to consist not only of an overall region of emission, but also of multiple discrete bright linear structures similar to the spicules that have already been described (Figure 15.2).

The Chromosphere

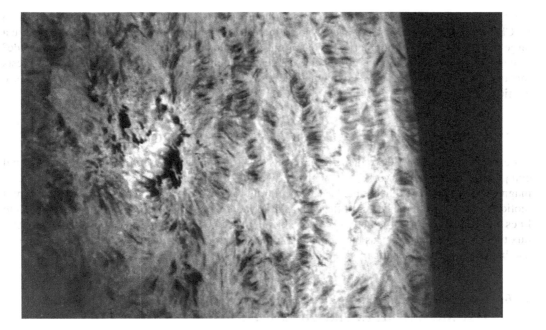

**FIGURE 15.5** Image of a portion of the solar disk in $H\alpha$. The dark features that look like "brushstrokes" are composed of individual "spikes", referred to as "dark mottles". Each dark mottle is probably a spicule. (Image taken at the National Solar Observatory, which is operated by the Association of Universities for Research in Astronomy, under a cooperative agreement with the National Science Foundation.)

When the disk is observed at the center of $H\alpha$ (see Figure 15.5), supergranules can also be identified, but the bright/dark topology is inverted relative to the Ca K/Si I images. After one stares at the image for some time, one may discern a pattern reminiscent of "fields surrounded by hedgerows": but in $H\alpha$, the "fields" are brighter than average, while the "hedgerows" are darker than average. Within the hedgerow, individual short dark straight streaks (or "brushstrokes") can be identified: these are referred to by solar observers as "dark mottles". It seems likely that each dark mottle is actually a spicule seen in projection against the bright disk. The image indicates that the "dark mottles" seen in $H\alpha$ (i.e., the spicules) are not distributed uniformly over the disk: instead, they are confined to the "hedgerows". With practice, the eye can identify, in each $H\alpha$ image of the Sun, brighter patches of surface surrounded by groups of dark spicules. The brighter patches correspond to (relatively dark) supergranule cells in Ca K/Si I images. The (dark) spicules in $H\alpha$ lie along the (bright) network which is observed in Ca K/Si I images. Because each spicule is a well-defined linear feature where gas seems to be confined to a (more or less) cylindrical shape, it is believed that each spicule owes its existence to a locally stronger magnetic field that helps to hold the gas in the cylinder together during the lifetime of the spicule (see Section 16.6.1). Why should stronger magnetic fields be confined to the edges of supergranules? Because the mainly horizontal flow from the center of the supergranule outwards sweeps up magnetic fields and deposits them in the edges, i.e., in the network, where one supergranule "runs into" its neighbor.

Why would a spicule appear bright when seen at the limb and dark when seen on the disk? The answer has to do with the background. At the limb, where the line of sight eventually passes out into empty space, there is no background light to absorb, and as a result, the spicule appears as an emitting structure. But on the disk, the spicule material has plenty of H atoms in the $n = 2$ energy level that are capable of absorbing photons with wavelengths near 6563 Å coming from the photosphere.

Why are there large populations of H atoms in the $n = 2$ state in a spicule? An answer is provided by Chapter 3, Section 3.3.2: even a small local increase in temperature inside a spicule can cause a large increase in the $n = 2$ population. What might cause a local increase in temperature in a spicule? The fact that spicular material is confined to a restricted width transverse to an upward axis suggests that a magnetic field is present, and magnetic fields can give rise to localized energy deposition (see Section 16.10).

## 15.5 THE TWO PRINCIPAL COMPONENTS OF THE CHROMOSPHERE

Observations in both Ca K and in $H\alpha$ indicate that the chromosphere consists of two principal components, cell and network, each with its distinct properties. High-resolution observations of magnetic fields indicate that the fields are concentrated in the network. It appears that the horizontal motions outward from the center of the cell (as described in Section 6.12) "sweep up" magnetic field lines and deposit them in the network. Each spicule in the network represents a localized magnetic flux tube in which the local gas has become energized (probably by magnetic energy in some form) and has reached up to heights of several thousand km.

## 15.6 TEMPERATURE INCREASE FROM PHOTOSPHERE TO CHROMOSPHERE: EMPIRICAL RESULTS

Images of the Sun in Ca K or $H\alpha$ exhibit one important and noticeable difference from an image in the white light continuum. The latter shows a pronounced limb *darkening*: the intensity at the limb in visible light is a very significant 60% fainter than the center of the disk (see Chapter 2, Equation 2.5 and discussion). The observed limb darkening $I(\mu) = a + b\mu$ (where $b$ is a positive number) can be ascribed to the source function $S(\tau) = a + b\tau$ (Section 2.5.3), indicating that the temperature is *increasing* as the optical depth in the continuum increases. However, when the Sun is observed in certain lines that emphasizes the chromosphere, the eye is struck by the fact that limb darkening is not immediately evident (e.g., Figure 15.4). On the contrary, there are more or less extended regions where the limb may actually be somewhat *brighter* than the center of the disk. This is a significant difference from what occurs in the visible continuum. This difference suggests that the temperature in the chromosphere is *not* increasing as the optical depth (in the chromospheric line) increases. In fact, the behavior is precisely the opposite: analysis of the chromospheric emission lines that are seen in the "flash spectrum" indicates that the temperature of the gas *increases* as the optical depth *decreases*. That is, the temperature in the chromosphere *rises* as the height above the photosphere increases. This is in striking contrast to the behavior of temperature in the photosphere: there, as the height increases, the temperature in the photosphere decreases or approaches a constant value (see Table 5.3).

How much does the temperature rise in the chromosphere? An empirical determination of an answer to this question requires detailed study of various lines, both absorption and emission, as well as continua, in the solar spectrum. This leads to a profile of temperature versus height. It is found that the profile of the chromospheric temperature rise in a cell differs somewhat, and in a systematic manner, from the profile of the chromospheric temperature rise in the network. In fact different pieces of the network, some of which are observed to be brighter than others, also have somewhat different profiles. And in the cell, some areas are darker than others, and these also yield somewhat different profiles. In Figure 15.6, we illustrate the profiles that have been obtained by analyzing the spectral data for the average cell center (B), a dark point within a cell (A), the average network (D), and a bright network element (E). (The notation is that of Vernazza et al. 1981.) Note that the data in Figure 15.6 refer to one-dimensional models: i.e., at each height above the photosphere, a single value of temperature is assigned. In view of the presence of spicules (Figure 15.2), the temperature at any given height in Figure 15.6 represents an average over many spicules.

# The Chromosphere

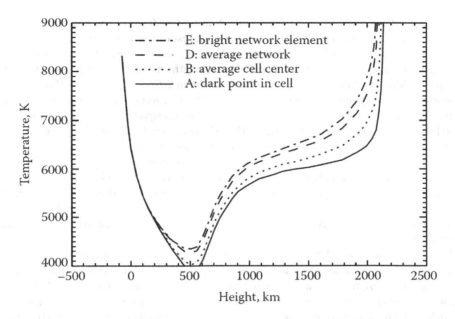

**FIGURE 15.6** Temperature profiles as a function of height in a one-dimensional model from the photosphere up into the chromosphere. The height scale is set to zero in the photosphere ($\tau = 2/3$), which is plotted near the *left*-hand side of this figure. (This is *opposite* to the convention that was used to plot an "average" 1-D model in Figure 4.2.) In Figure 15.6, negative heights refer to layers of the Sun that lie *below* the photosphere. Four distinct regions on the solar surface that differ from each other in observed brightness (labeled A, B, D, and E) are illustrated. Above the photosphere, notice (i) the temperature minimum at $h \approx 500$ km, (ii) the rising temperatures between $h = 500$ and 1000 km, (iii) the temperature "plateau" between $h \approx 1000$ and 2000 km, and (iv) the steep increase in temperature above $\approx 2100$ km. (The four profiles were constructed by plotting data from Vernazza et al. 1981.)

Let us consider some important features of the profiles in Figure 15.6.

First of all, all four profiles merge into a single profile in and below the photosphere: whatever differences exist between the regions A, B, D, and E must originate *above the photosphere*. In fact, it is not until we examine the solar atmosphere at heights of a few hundred kilometers (or more) above the photosphere that the temperature profiles begin to display measurable differences.

Second, in all four cases, the temperature passes through a minimum value at heights close to 500 km above the photosphere. We refer to this feature in each curve as the "temperature minimum". Temperatures at the minimum range from about 4400 K in the brightest element (E), to less than 4000 K in the darkest element (A): all of these temperatures lie significantly below the "boundary temperature" of 4858 K predicted by the Eddington model (Section 2.10). Thus, although the Eddington model works reasonably well in and below the photosphere (in particular in its quantitative explanation of limb darkening), it becomes less reliable in the optically thin regions of the atmosphere. In particular, there is nothing whatsoever in the Eddington model that can explain the *increase* in temperature that sets in above heights of about 500 km. We regard the temperature increases that become prominent at heights above 500 km as an indication that we have entered into "the lower chromosphere". At these heights, some physical process has started to deposit heat into the gas, and the amount of heat deposited is larger in features observed to be brighter (compare feature E with feature A in Figure 15.6).

It should be noted that 4000 K is not the lowest temperature that has been reported in a chromospheric model: when 2-D MHD calculations were performed in conjunction with radiative transfer, the temperature in a magnetically quiet region was predicted to decrease to as low as 1660 K

(Leenaarts et al. 2011) over a height range from 1 Mm up to 3.6 Mm. The best hope of obtaining observational confirmation of this very-low-temperature result was at one time expected to be provided by the Atacama Large Millimeter Array (ALMA), operating in the Chilean Andes mountains at an altitude of 5 km, and with 10 bandpasses at wavelengths as long as 6–8.5 mm and as short as 0.3–0.4 mm. ALMA was first used for solar observations in 2016, achieving angular resolutions of order 0.01 arcsec (corresponding to linear scales of 7 km, i.e., some three times smaller than even DKIST can resolve): the data revealed the presence of "chromospheric holes" where the local temperature fell to 3000 K (da Silva Santos et al. 2020). However, evidence for gas as cool as 1660 K in the chromosphere does not yet (as of November 2021) seem to be available in the literature.

Third, above 500 km, the temperature at first begins to increase rather steeply, but at heights of about 1000 km, the temperature profile flattens out to some extent. This gives rise to a sort of "plateau" in the temperature at heights between (roughly) 1000 and 2000 km. The temperatures in the plateau are mainly in the range 6000–6500 K. This plateau can be regarded as defining the "middle chromosphere". In the plateau, the local temperature exceeds the temperature minimum by roughly 2000 K.

Fourth, at heights of about 2100 km, there is a steep increase in temperature in all four features: this is referred to as the "upper chromosphere", and is the lowest-lying portion start of a narrow "transition region" where temperatures rise up rapidly to coronal values (of order 1 MK).

Fifth, within the plateau, the four distinct features differ from one another in temperature by several hundred degrees. For example, at $h = 1500$ km, the average cell center has $T \approx 6200$ K, while the bright network element has $T \approx 6600$ K. Although the difference in temperature of 400 K is only some 20% of the 2000 K increase in temperature above the temperature minimum, we shall find (Section 15.10.2) that this relatively small difference in temperature actually requires a much large difference in energy deposition rate in the network than in the cell.

Sixth, the overall thickness of the chromosphere, from "lower" to "upper", is some 1600–1700 km. This overlaps with the range of thicknesses reported in Section 15.2 from eclipse timings.

Seventh, we have mentioned (Figure 15.2) that discrete structures (spicules) exist at certain locations in the network: these may extend upward in height to a few thousand kilometers. Some spicules therefore have heights that exceed the thickness of the chromospheric profiles in Figure 15.6. The "real chromosphere" includes spicules that are not well described by the results in Figure 15.6. The latter should be regarded as representing some sort of average over "spicular" and "non-spicular" regions of the solar atmosphere.

## 15.7 TEMPERATURE INCREASE INTO THE CHROMOSPHERE: MECHANICAL WORK

The most striking result in Figure 15.6 is that the local temperature *increases* as the height increases above the photosphere. This result indicates that it is no longer useful to think in terms of the Eddington atmosphere, where radiative equilibrium was operative. In the latter conditions, we have seen (Chapter 2, Equation 2.40) that the temperature should vary as $T^4 \sim \tau +$ const, i.e., $T$ should approach a constant value as we go higher up in the atmosphere, where $\tau \to 0$. In the Eddington atmosphere, therefore, there should certainly be no tendency for the temperature to *increase* as we move toward smaller values of $\tau$.

Clearly, something quite different from radiative equilibrium is operating in the chromosphere. What could it be?

Up to this point in our calculations of the solar model (Chapters 5, 7, and 9), we have always been dealing with material where the temperature falls off monotonically as radial distance from the center of the Sun increases. In the presence of such a negative radial gradient of temperature, it is natural to think in terms of the heat that flows *down* the temperature gradient.

Now, as we enter into the chromosphere, $T$ starts to *increase* as the radial distance increases. The radial gradient of temperature is now *positive*. Such a positive gradient cannot be a consequence of

# The Chromosphere

classical heat flow from the inner portions of the Sun. There must be a different physical process in operation in order to raise the temperature in the chromosphere: this process is believed to be *mechanical work*. One possible source of such work can be identified with the thermodynamic term *PdV*, which occurs when a suitable pressure *P* compresses the volume *V* of 1 gm of gas. Where can we find suitable pressures to perform such work in the solar atmosphere? One such source is sound waves: we have already seen that the Sun supports multiple *p*-modes, each of which is a sound wave. Sound waves are longitudinal modes in which pressure compresses and rarefies the local gas as the wave propagates past any point.

Thus, sound waves, by their very nature as a propagating series of compressions and rarefactions, *can* do mechanical work on the gas in the Sun. However, the fact that the chromosphere extends *above* the photosphere by linear distances of a few thousand kilometers indicates that the waves responsible for chromospheric heating are *not* identical to the trapped *p*-modes: the latter are *trapped below the photosphere*, whereas now we need to have waves that are capable of *propagating above the photosphere*. As we have seen (Equation 13.15), sound waves that are capable of freely propagating vertically above the photosphere must have periods shorter than 195–200 sec, i.e., they must have frequencies larger than 5 mHz.

The increase in chromospheric temperature is found to occur in *both* the cell and the network. However, material in the network increases in temperature faster than material in the cell (see Figure 15.6: compare curve E to curve A). Over the same range of heights, the average network gas is hotter than the average cell gas by an extra few hundred degrees. This suggests that the supply of mechanical energy is greater in the network than in the cell. Although the excess temperature in the network *seems* relatively small (about 20%), we shall see that it actually requires a larger difference in mechanical energy deposition.

## 15.8 MODELING THE CHROMOSPHERE: THE INPUT ENERGY FLUX

The aim of any attempt to model the chromosphere is to calculate how the temperature varies as a function of height. Specifically, how fast does the temperature increase above the boundary value that is predicted by the photospheric model? Does it increase by (say) 1000 K over a height interval of 10 km? or 100 km? or 1000 km?

In order to calculate the temperature rise, let us consider how sound waves could provide mechanical energy to the gas. Let us start in the photosphere with sound waves that are propagating upwards, transporting a certain flux of acoustic energy. Let us calculate what happens as these waves propagate upward. A sound wave can be characterized by an amplitude in velocity $\delta V$. In a medium of density $\rho$, the mass contained in 1 cm³ is $\rho$ grams: therefore, the wave energy in 1 cm³ (i.e., the kinetic energy density of the wave) is $0.5\,\rho\delta V^2$ ergs cm⁻³. The waves propagate at the speed of sound $c_s$. As a result, the acoustic energy flux carried upwards by the sound waves is given by

$$F(ac) = 0.5\rho\delta V^2 c_s \tag{15.1}$$

in units of ergs cm⁻² sec⁻¹.

In the photosphere of the sun, our solar model (see Table 5.3) informs us that $\rho \approx 3 \times 10^{-7}$ gm cm⁻³. We also know that the local (adiabatic) speed of sound $c_s$ is given by the formula $\sqrt{(1.67 R_g T/\mu)}$, where $\mu$ is the mean molecular weight. In the photosphere, the numerical value of $c_s$ is $\approx 7$–8 km sec⁻¹. This leads to $F(ac) \approx 0.11\,\delta V^2$.

What are we to use for $\delta V$, the amplitude of the sound waves in the photosphere? We have seen (Section 3.8.1) that line profiles in the solar spectrum have excess widths over and above what the lines would have in the presence of purely thermal motions. The excess widths, of order 0.75–1.5 km sec⁻¹ and ascribed to "microturbulence", in all likelihood include contributions from convective flows and sound waves in the photosphere. The observed amplitude of the turbulence may therefore be regarded as an upper limit on the amplitude of sound waves in the photosphere.

Of the observed microturbulence of 0.75–1.5 km sec$^{-1}$, let us suppose that sound waves contribute no more than 50%: i.e., we assume that the amplitude of sound waves in the photosphere is no more than $\delta V$ (photo) $\approx$ 1 km sec$^{-1}$. If this is a reliable assumption, then it would set a limit on $F$(ac) of $\leq 1.1 \times 10^9$ ergs cm$^{-2}$ sec$^{-1}$. Compared to the energy flux passing through the photosphere in the form of radiation ($F_\odot = 6.2939 \times 10^{10}$ ergs cm$^{-2}$ sec$^{-1}$: see Section 1.9), we see that $F$(ac) in the photosphere is less than 2% of the overall flux of energy propagating upwards through the solar atmosphere. In view of this, we are certainly not discussing a major channel for the transport of energy through the photosphere: radiation is still by far the dominant channel for energy transport in the visible layers of the Sun.

Any acoustic energy that *is* present in the photosphere and that contributes to microturbulent line broadening certainly includes some *p*-modes with periods in excess of (about) 200 sec. However, such long-period waves cannot propagate up into the (nonmagnetic) chromosphere. (The "ramps" that allow such waves to reach the chromosphere occur only in certain magnetic areas, namely sunspot penumbrae.) The only segment of the acoustic flux that is of interest as far as the heating of the (nonmagnetic) chromosphere is concerned is the segment where the waves have periods that are short enough to allow vertical propagation. This segment contains only those waves with periods *shorter* than 200 sec. In order to estimate the flux of sound waves that can actually reach the chromosphere (thereby contributing to heating of the gas up there), we need to reduce the above upper limit on $F$(ac).

What fraction of $F$(ac) reaches the chromosphere? The answer depends on the spectrum of the acoustic power that is generated by the convective motions. Most of $F$(ac) is expected to be generated at periods corresponding to granule turnover times, or lifetimes (see Section 14.8.2), i.e., at periods of 300–600 sec (Section 6.2). Waves with periods of less than 200 sec are therefore expected to contribute only a fraction to the overall spectrum. According to one theoretical estimate (Musielak et al. 1994), the acoustic energy flux that reaches the chromosphere $F$(chr) is no more than $5 \times 10^7$ ergs cm$^{-2}$ sec$^{-1}$. That is, only a few percent of $F$(ac) is in the form of waves that are free to propagate vertically in the solar atmosphere: as expected, the great majority of $F$(ac) created by granules that survive for a time interval of 300–600 sec is in the form of waves with periods that are longer than 200 sec. Musielak et al. (1994) find that their theoretical estimate of $F$(chr) is quite sensitive to various assumptions about the properties of turbulence. It is entirely possible that the prior estimate of $F$(chr) could be in error by a factor of two or more.

In view of the uncertainties, a conservative range of theoretical estimates of the flux of acoustic energy that is available as the input for chromospheric heating in the Sun may be

$$F(chr) = 10^{7-8} \text{ergs cm}^{-2} \text{ sec}^{-1} \qquad (15.2)$$

Since the two basic components of the chromosphere (cells, network) are observed to differ in brightness, it seems plausible that the lower limit of the range of $F$(chr) in Equation 15.2 might be suitable to apply to the cell, while the upper limit in Equation 15.2 might apply to the network. We shall return to this when we discuss the heating in quantitative terms.

Is there any observational evidence to support the hypothesis of acoustic waves as an important element in the heating of the chromosphere? Yes: in the comprehensive review by Carlsson et al. (2019), the following statement occurs with regard to the *lower* chromosphere: "The signatures of acoustic waves can be seen in all chromospheric diagnostics as 'sawtooth' behavior in the CaII lines (because of the temporal variation of velocity and thus Doppler shift associated with shocks)". The presence of shock waves is especially detectable in time-resolved observations of "bright grains" in the H and K lines of Ca II: these observations indicate that the minimum size of structures in the solar chromosphere is in the tens of km range (Kalkofen 2012). However, by way of contrast,

Carlsson et al. (2019) point out that in the *upper* chromosphere, there are no clear shock signatures: this may occur because acoustic waves may be converted to other wave modes when they reach the higher levels (see Section 15.12.1).

A consistency check on the theoretical acoustic fluxes in Equation 15.2 can be found in an early attempt to estimate the total radiant energy flux $F(rad)$ emitted by the chromosphere. Based on the strongest emission lines that were then observable, Withbroe and Noyes (1977: WN) estimated $F(rad)$ to have a value of $4 \times 10^6$ ergs cm$^{-2}$ sec$^{-1}$ in the quiet Sun and in coronal holes, and to have a value of $2 \times 10^7$ ergs cm$^{-2}$ sec$^{-1}$ in active regions. It is actually quite difficult to obtain the total radiant flux from the chromosphere because large numbers of weak absorption lines in the spectrum of a star with a chromosphere can actually contribute significantly to the chromospheric radiant flux (e.g., Houdebine 2010). As a result, the fluxes quoted by WN may be considered as lower limits to the true chromospheric radiant energy losses. Thus, the fact that the WN estimates of radiant losses in quiet (active) Sun are lower by a factor of 2.5 (5) than the lower (upper) limits in Equation 15.2 does not necessarily mean that there is inconsistency with the values in Equation 15.2.

## 15.9 MODELING THE CHROMOSPHERE: THE ENERGY DEPOSITION RATE

What happens to the flux of acoustic energy $F(chr)$ as it propagates upward in the Sun's atmosphere? At first, the amplitudes of the waves are small enough that the waves simply "ride" through the gas, dissipating no energy. In this regime, the energy flux of the waves remains constant. In the upper photosphere, where the temperature is almost constant with height (Section 2.10), $c_s$ is also almost constant with height. However, the *density* $\rho$ is by no means constant with height: it is actually decreasing as height increases, following an exponential law (see Equation 5.4). As a result, in order to keep $F(chr) \sim \rho \delta V^2 c_s$ constant with height, the wave amplitude $\delta V$ must vary as $1/\sqrt{\rho}$. This means that $\delta V$ must *increase* exponentially with increasing height according to $\delta V(h) \sim \exp(+h/2H_p)$. As we have already seen (Section 5.1), in the solar photosphere $H_p = 114-140$ km.

We have seen that the amplitude of sound waves in the photospheric layers $\delta V$ (photo) may be of order 1 km s$^{-1}$. Compared to the local sound speed, the sound wave amplitudes in the photosphere are $\approx 0.1 c_s$. Applying the exponential growth formula, we see that when the waves reach a height $h_s$ where $\exp(h_s/2H_p) \approx 10$, then the amplitude of the sound waves will have grown to a value $\delta V(h_s)$, which approaches $c_s$. This occurs at a height $h_s \approx 4.6 H_p$ above the photosphere, i.e., at a linear altitude of $h_s \approx 520-640$ km above the photosphere.

What happens to an acoustic wave when its amplitude becomes comparable to the local sound speed? To see what happens, we note that a sound wave consists of a crest and a trough: the wave is moving forward relative to the background medium at speed $c_s$. However, the meaning of the term "amplitude of the wave" means that the matter in the crest is moving with a speed of $\delta V$ *relative to the wave*. That is, the matter in the crest of the wave is moving relative to the background medium at a speed $\delta V + c_s$, while the material in the trough of the wave is moving relative to a stationary observer at speed $-\delta V + c_s$. When $\delta V$ approaches $c_s$, the material in the crest overtakes the material in the trough. Then the wave profile becomes so steep that a vertical step in pressure develops: in this condition, the sound wave has become a shock front. This behavior is reminiscent of water waves on the surface of the ocean as those waves approach a shelving beach (i.e., a region where the depth of the water is decreasing as the wave moves closer to the shore): when the wave becomes vertical, the wave can no longer continue to be a sinusoidal motion. At that point, the wave "breaks" and deposits its energy in the form of a churning whitecap. Analogously, when an acoustic wave evolves to the condition of a shock front, the pressure jump across the wave "breaks", leading to local churning and compression of the gas. As a result, the *PdV* work leads to a conversion of the original wave energy into localized heat.

This leads us to an important conclusion about a certain region in the solar atmosphere, particularly the region $h_s \approx 520\text{–}640$ km above the photosphere. At such heights, we expect that sound waves from the photosphere will begin to "break" and, as a result, acoustic energy will begin to be deposited effectively in the solar atmosphere. In this regard, it is important to note from Figure 15.6 that this height range is precisely where all four of the empirical models of the chromosphere shown in the figure indicate that the temperature reaches a minimum, and starts to increase upward. In view of what we have said about acoustic waves undergoing steepening and forming shock waves, it is natural to attribute the empirical increase in temperature at heights of 500 km or so above the solar photosphere to the onset of shock heating.

In contrast to the Eddington model, where $T$ was predicted to *fall off* slowly and approach a nonzero limiting value as optical depth $\tau \to 0$ (and height increases), now the dissipation of acoustic power, i.e., the addition of extra energy to the ambient gas, has the effect that the temperature should *start to increase* above a certain height, $h_s$. In fact, the empirical models (Figure 15.7) do indeed show that the temperature in the solar atmosphere reaches a minimum value, $T(\min)$, in the vicinity of the height $h_s$.

The region of the "temperature minimum" may be thought of as a boundary between the upper photosphere (below) and the lower chromosphere (above).

Although we expect that shock heating will *set in* at heights of order $h_s$, we do not expect the acoustic energy to be deposited *in its entirety* at a single location. For one thing, any local heating causes the local scale height to increase, and this helps to postpone further steepening of the wave to greater heights. Instead of instantaneous local dissipation, the process is spread out in the vertical direction such that the acoustic flux falls off roughly as $\exp(-h/\lambda_d)$, where $\lambda_d$ is referred to as a "dissipation length-scale". Since dissipation is associated with steepening of the waves, and the steepening is associated with the falling off in density (which occurs on an $e$-folding scale of $H_p$), we expect that $\lambda_d$ might be of order a few times $H_p$. For purposes of rough estimation, $\lambda_d \approx 300$ km might be plausible. With this choice, we expect that fully developed shock dissipation should occur at heights of order $h_s + \lambda_d \approx 820\text{–}940$ km above the photosphere: this is consistent with detailed shock modeling (Carlsson and Stein 1992) where the shocks are observed to form at altitudes of about 1 Mm.

Once the dissipation length-scale is known, we can estimate the average volumetric rate $E(\text{chr})$ at which acoustic energy is deposited into each cubic cm of the atmosphere: $E(\text{chr}) \approx F(\text{chr})/\lambda_d$. Inserting the value of $E(\text{chr})$ given earlier in Equation 15.2, we find that acoustic energy is deposited into the chromosphere at a volumetric rate that is, at least as to order of magnitude, given by

$$E(chr) \approx 0.3 - 3 \text{ ergs cm}^{-3}\text{sec}^{-1} \qquad (15.3)$$

This is a (rough) estimate of how rapidly acoustic energy is being deposited every second into each cubic cm of the Sun's chromosphere. Note that we have arrived at a finite range of possible energy deposition rates: this is appropriate because it is well known empirically that different regions of the chromosphere emit radiant energy at different rates. As proof of this statement, note that in Figure 15.6, feature A is a dark point in the interior of a supergranule cell, where the intensity of chromospheric emission is at a minimum. On the other hand, feature E, a bright element in the network surrounding a supergranule cell, can be considered as a region where the intensity of chromospheric radiation is a maximum. From a physics perspective, it is plausible to consider feature A as containing gas in which energy is being deposited (by whatever means) at a *lower* rate than is the case for feature E. Our estimates of the volumetric energy deposition rate in Equation 15.3 span a range of one order of magnitude: it would not be inconsistent for us to consider that the lower limit in Equation 15.3 could be appropriate for feature A, while the upper limit in Equation 15.3

# The Chromosphere

could be appropriate for feature E. When we discuss magnetic fields in the Sun (Chapter 16), we shall find that the network is a region where magnetic fields are stronger, whereas in the interior of a supergranule cell, the fields are weaker: in this context, it would not be surprising if the energy deposition rate is larger in regions of stronger field (where acoustic waves are not the only wave modes which exist: see Section 15.12.1).

## 15.10 MODELING THE EQUILIBRIUM CHROMOSPHERE: RADIATING THE ENERGY AWAY

When mechanical energy is deposited into a cubic cm of gas, the gas attempts to get rid of the energy by whatever means are available. One of the most efficient means available to gas at the temperature minimum is to increase the local temperature by a finite amount and then use the increased efficiency of radiative losses at the higher temperature to radiate the energy away. If the gas is successful in finding a way to radiate energy at a rate of 0.3–3 ergs cm$^{-3}$ sec$^{-1}$, then an equilibrium can be reached: the local temperature can achieve a more or less steady state.

Let us turn now to a calculation of the excess temperature that could allow the solar atmosphere to reach such an equilibrium.

### 15.10.1 RADIATIVE COOLING TIME-SCALE

We first need to estimate how long it takes for gas to cool by means of radiation. Suppose a parcel of gas is heated (for whatever reason) to a temperature $T$ that is hotter than its surroundings: the latter are at temperature $T_o$. How long would it take for the heated parcel to radiate away its excess heat energy? The gas (with density $\rho$) in a volume element $dV$ has excess internal energy $E(exc)$ = $\rho C_v(T - T_o)dV$ ergs, where $C_v$ is the specific heat per gram.

How quickly can this excess energy be radiated from this volume element? It depends on what form of radiation is available to the gas. Suppose the radiation is predominantly in the continuum. Let the surface area of the volume element be $dA$. If the volume element lies deep enough in the atmosphere, it will be optically thick, i.e., $\tau \gg 1$. In such a case, the energy will be radiated from the surface $dA$ with maximum effectiveness, namely at a rate given by the Planck function: the emergent intensity is such that the rate at which energy is radiated out of each square centimeter into a background medium with temperature $T_o$ (integrated over $4\pi$ solid angle) is given by the difference in source functions: $S_{bb}(T) = 4\sigma_B(T^4 - T_o^4)$ (see Chapter 2, Equation 2.37). In this limit, the rate $(dE/dt)_{rad}$ at which the excess energy in the volume element can be radiated away in the continuum would simply by $S_{bb}(T)dA$, i.e., $4\sigma_B\left(T^4 - T_o^4\right)dA$ ergs sec$^{-1}$.

However, as we move upward in the solar atmosphere and encounter gas with smaller and smaller densities, the volume element will *not* always turn out to be optically thick. Eventually, the line of sight through the element will become optically thin, i.e., $\tau$ will become <1. In such a case, rather than emitting radiation at the rate given by the Planck function, the emergent intensity is reduced by the factor $\tau$ (Chapter 2, Equation 2.18): $S = \tau S_{bb}(T)$. This leads to a cooling rate of

$$\left(\frac{dE}{dt}\right)_{rad} = 4\tau\sigma_B(T^4 - T_o^4)dA \tag{15.4}$$

With this rate of energy loss, how long will it take for radiation to cause the parcel to cool down to $T = T_o$? This "cooling time" $t_{cool}$ is given roughly by $E(exc)/(dE/dt)_{rad}$. This leads to

$$t_{cool} \approx \frac{E(exc)}{(dE/dt)_{rad}} = \frac{\rho C_v(T - T_o)}{4\sigma_B(T^4 - T_o^4)}\frac{1}{\tau}\frac{dV}{dA} \tag{15.5}$$

This estimate of the cooling time depends on the local conditions and also on the optical depth of the parcel. In general, the ratio of the volume of the element $dV$ to its surface area $dA$ is associated with the linear scale $ds$ of the element: $dV/dA \approx ds$. Moreover, according to the definition of optical depth, we can also write the optical depth of the element in terms of the linear scale: $\tau = \kappa \rho\, ds$. Substituting this in Equation 15.5, we obtain an expression that is independent of the size of the element (as long as it is optically thin):

$$t_{cool} = \frac{C_v(T-T_o)}{4\sigma_B \kappa (T^4 - T_o^4)} \tag{15.6}$$

This expression is valid for the regions in the solar atmosphere where continuum radiation is efficient. In higher layers, where emission lines become more efficient radiators, we do not expect to find Equation 15.6 as useful.

### 15.10.2 Magnitude of the Temperature Increase: The Low Chromosphere

Now that we know how rapidly energy can be radiated away from a volume element near the temperature minimum, we can estimate the equilibrium value of the local increase in temperature $\Delta T = T - T_o$ that occurs as a result of deposition of mechanical energy at a volumetric rate $E(chr)$.

An increase in the local temperature by an amount $\Delta T$ causes the local thermal energy density to increase by $\Delta E = C_v \rho \Delta T$ ergs cm$^{-3}$. This excess energy can be radiated away at a rate that is determined by the cooling time-scale $t_{cool}$:

$$\left(\frac{dE}{dt}\right)_{rad} \approx \frac{\Delta E}{t_{cool}} = 4\sigma_B \kappa \rho (T^4 - T_o^4) \tag{15.7}$$

The units of the right- and left-hand sides of Equation 15.7 are ergs cm$^{-3}$ sec$^{-1}$. Equilibrium is possible if the rate at which energy is being deposited into a unit volume $E(chr)$ (see Equation 15.3) is equal to the rate at which energy is radiated out of that unit volume $(dE/dt)_{rad}$. Equilibrium therefore occurs when

$$4\sigma_B \kappa \rho (T^4 - T_o^4) = 0.3 - 3 \tag{15.8}$$

Inserting the value of the Stefan–Boltzmann constant $\sigma_B$, we find that the gas in the solar atmosphere *can* reach equilibrium if the following relationship is satisfied:

$$\kappa (T^4 - T_o^4) \approx (1-10) \times 10^3 / \rho \tag{15.9}$$

Can we find a solution to this equation? In order to answer this, we need to know how the opacity $\kappa$ depends on temperature and density in the upper parts of the solar photosphere. We have already seen (Section 3.7) that $\kappa$ can be fitted in certain regimes of temperature with power laws in density and temperature. In the present case, we are interested in gas such as that which exists in the low chromosphere: in such gas, we know that the temperature lies below $10^4$ K. At such temperatures, we have already seen (Section 3.7) that the opacity can be approximated by the expression $\kappa \approx 10^{-32} \rho^{0.3} T^9$. The steep dependence on temperature is noteworthy: it arises mainly because an increase in temperature (in the temperature range $T < 10^4$ K) leads to rapid increases in the populations of the upper levels of hydrogen atoms. Inserting this in Equation 15.9, we find

$$T^9 (T^4 - T_o^4) \approx (1-10) \times 10^{35} / \rho^{1.3} \tag{15.10}$$

# The Chromosphere

Solutions of this equation, for a given density $\rho$, indicate the equilibrium temperature to which gas in the solar atmosphere of density $\rho$ would be heated if (i) energy were deposited in that gas at a rate given by Equation 15.3 and (ii) continuum opacity from an optically thin medium determines the radiative losses.

What value of density should we use in Equation 15.10? The answer depends on where exactly in the solar atmosphere the mechanical energy is being deposited. Densities in the solar atmosphere vary over a wide range. In the photosphere, our solar model (Table 5.3) suggests $\rho \approx 3 \times 10^{-7}$ gm cm$^{-3}$. Densities at the temperature minimum, i.e., some 4.6 scale heights above the photosphere, are lower than the photospheric densities by factors of $e^{-4.6} = 0.01$. Thus, local densities in the low chromosphere have values that are no larger than roughly $3 \times 10^{-9}$ gm cm$^{-3}$. In the upper chromosphere, at heights of 2000 km, i.e., at least $14 H_p$ above the photosphere, the densities are smaller than photospheric values by $e^{-14} \approx 10^{-6}$. As a result, when we consider conditions in the solar chromosphere, we are interested in the solutions of Equation 15.10 over a range of densities from (roughly) $3 \times 10^{-9}$ gm cm$^{-3}$ to $3 \times 10^{-13}$ gm cm$^{-3}$. (For future reference, we note that the latter *mass* density corresponds to a *number* density in the upper chromosphere of order $2 \times 10^{11}$ protons cm$^{-3}$.) Using the condition of hydrostatic equilibrium, i.e., $\rho(h) = \rho_o \exp(-h/H_p)$, we can associate (roughly) each value of density with a corresponding height above the photosphere. (We use $H_p = 140$ km and $\rho_o = 3 \times 10^{-7}$ gm cm$^{-3}$.)

Let us assume that the background atmosphere (before acoustic waves are present) has $T_o = T(\min) \approx 4000$ K (see Figure 15.6). Using this, we can obtain solutions to Equation 15.10 for any choice of density throughout the aforementioned range. For clarity, we consider two distinct components of the chromosphere: in one, the deposition of acoustic flux occurs at a low rate (we use the number 1 in brackets on the right-hand side of Equation 15.10), while in the other, the acoustic flux is deposited at a 10 times higher rate (as given by the number 10 in brackets on the right-hand in Equation 15.10). The corresponding solutions to Equation 15.10 are presented in Figure 15.7. We see that, over a range of heights from about 500 km to about 1000 km, the temperature is predicted to rise steeply to a value that is at least 2000 K above the temperature minimum. Thus, acoustic dissipation, in combination with continuum radiative losses, appears to account quite well for the initial rise in temperature in the low chromosphere.

The high-flux solution agrees best with empirical curve E, the bright network element. The low-flux solution lies closer to empirical curve A, the dark point in the supergranular cell. At heights in the low chromosphere, the high-flux solution gives rise to temperatures that, at any particular height, are hotter than on the low-flux solution by several hundred degrees. Thus, even though the empirical curves appear to be separated in temperature by a relatively small amount (a few hundred degrees), that temperature difference corresponds to input rates of mechanical energy that differ by a factor of 10.

Why are the rates of mechanical energy deposition in the bright network elements 10 times larger than in the dark point in the cell? One obvious difference between such locations is the magnetic field strength: supergranule flows cause magnetic fields to be strong in the network but weak in the cell. The strong fields in the network provide channels for spicules to exist. As a result, it seems likely that the presence of 10 times enhanced mechanical energy deposition rates in the network may be related to magnetic fields.

### 15.10.3 Magnitude of the Temperature Increase: The Middle Chromosphere

The results in Figure 15.7 show clearly that although we have been fairly successful in fitting the temperatures in the *low* chromosphere (i.e., $h \approx 500$–1000 km), using Equation 15.10, the fit definitely breaks down in the middle chromosphere. The reason for the breakdown is related to the choice of source function that was used for the radiative losses in the low chromosphere: we chose

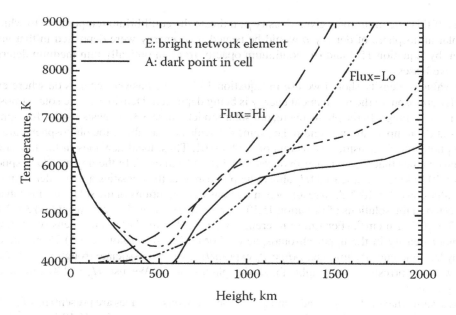

**FIGURE 15.7** Theoretical fits to chromospheric temperature increases using low and high fluxes of acoustic waves. Curves A and E: portions of two of the empirical curves in Figure 15.7. Lines labeled Flux: solutions of Equation 15.10 in the limits of low and high flux.

the blackbody relation $S_{bb} \sim T^4$, which is valid only as long as the continuum photons and the gas are tightly coupled in local thermodynamic equilibrium (LTE).

However, as we rise to greater altitudes above the solar photosphere, and the density falls off exponentially, the coupling between continuum photons and gas diminishes. The gas becomes less and less efficient as a radiator of the Planck function.

At the same time, bound levels in the dominant atoms and ions are being increasingly well populated by the rising temperatures. As a result, emission lines from certain bound levels become increasingly effective coolants of the gas in the middle chromosphere. Among these lines, the $h$ and $k$ lines of Mg II are the maximally effective contributors to radiative cooling at heights of about 1.1 Mm in the lower regions of the middle chromosphere (Carlsson and Leenaarts 2012). In the upper regions of the middle chromosphere, the temperature begins to reach values close to 7000 K: at such temperatures, we have already seen (Section 4.3), hydrogen atoms in the chromosphere are approaching 50% ionization. And if ionization of hydrogen is occurring, an accompanying effect will be an increase in the populations of bound states of hydrogen atoms. These increased populations help to strengthen the bound-bound transitions, and hydrogen lines can become strong enough that they dominate the radiative cooling at heights around 2–2.3 Mm above the photosphere (Carlsson and Leenaarts 2012). The radiative cooling rates in these various lines have temperature and density dependences that depart significantly from those of Planck curves.

In order to understand the plateau in temperature in the middle chromosphere, it is important to recall that the flux of acoustic power responsible for chromospheric heating originates in the convective turbulence below the photosphere. As a result, the acoustic flux $F(ac)$ is maximum near the photosphere, and it diminishes with increasing height. Moreover, the mechanical energy deposited in the middle chromosphere does not go simply into increasing the local temperature. Instead, the energy is diverted in increasingly large amounts to internal degrees of freedom: population of bound levels and (ultimately) the ionization of hydrogen (and helium).

# The Chromosphere

In the upper part of the middle chromosphere, hydrogen ionization rises above the 50% level.

The combination of reduced rates of input of mechanical energy, the onset of strong radiative cooling that occurs predominantly in emission lines, and the siphoning off of energy into bound levels and ionization leads to a plateau in the temperature. The bound levels of hydrogen act, in effect, as a kind of thermostat for the middle chromosphere.

From the plateau in the middle chromosphere, where hydrogen is roughly 50% ionized, strong emission of lines in the Balmer series occurs. The first member of the Balmer series, $H\alpha$ (at a wavelength of 6563 Å), is the strongest emitter from the chromosphere in the visible spectrum. The red color of this strong line accounts for the "rose-colored hue" that is a common feature of the flash spectrum that can be seen by the unaided eye briefly during an eclipse of the Sun (see Figure 15.1).

### 15.10.4 MAGNITUDE OF THE TEMPERATURE INCREASE: THE UPPER CHROMOSPHERE

In the upper chromosphere, where the temperature increases above 7000 K, rapidly approaching $10^4$ K and higher, hydrogen approaches complete ionization. No longer are there internal degrees of freedom (bound levels, ionization) available to absorb mechanical energy. No longer are there strong continua or lines available to radiate away the mechanical energy. Equilibrium is not possible: there is a "runaway" of the temperature to high values.

With only thermal energy available, and with the low density of the gas (approaching $10^{-13}$ gm cm$^{-3}$), the deposition of energy even at a rate $E$(chr) that is much lower than in Equation 15.3 leads to rapid local heating. To see this, note that the thermal energy density $e$, which is comparable to the local pressure $R_g \rho T/\mu$, obeys the equation $de/dt = E$(chr) if there are no longer any effective (radiative) channels to carry away the energy. Thus, even if the deposition rate is as low as (say) 0.1% of the lowest value in Equation 15.3, the rate of temperature increase in gas with a density of $10^{-13}$ gm cm$^{-3}$ is $dT/dt = 0.001\mu E$(chr)$/R_g \rho$. With $\mu = 0.5$ in ionized hydrogen, and $R_g = 8.31 \times 10^7$ ergs gm$^{-1}$ deg$^{-1}$, we find $dT/dt \approx 20$ K sec$^{-1}$. This provides an upper limit on the heating: the gas will also tend to expand (thereby cooling) because of local pressure enhancement. Before the expansion cooling becomes effective, the local temperature can increase in 1–2 minutes by several thousand K.

This can help us to understand why the temperature rises steeply in the upper chromosphere (see Figure 15.6). This region of steep temperature gradient is referred to as the transition region (or interface region) between chromosphere (at temperatures up to $10^4$ K) and corona (at temperatures of order $10^6$ K).

For future reference, we note that at the top of the chromosphere, where $T \approx 10^4$ K, and number densities are of order $2 \times 10^{11}$ cm$^{-3}$, the gas pressures ($p = 2N_e kT$) are of order 0.6 dyn cm$^{-2}$. When we discuss the corona (Chapter 17), it will be valuable to compare this pressure near the *top of the chromosphere* with the pressure near the *base of the corona*.

## 15.11 THE IRIS SATELLITE

Now that we have introduced the terms transition region (TR) and interface region (IR), it is timely to describe a satellite that was launched in 2013 specifically to study how the physical properties of the gas in the chromosphere and in the IR vary as a function of height. The satellite is called IRIS: Interface Region Imaging Spectrograph. It contains a 19-cm UV telescope orbiting in a Sun-synchronous orbit, and observing the Sun through three filters strategically located at UV wavelengths. The filter labeled far ultraviolet (FUVshort) observes at wavelengths 1332–1358 Å, where certain spectral lines are sensitive to gas at temperatures of 25,000 K (25 kK). The second filter (FUVlong) observes at wavelengths 1389–1407 Å where some lines are sensitive to temperatures between 80 kK and 150 kK. The third filter (near UV: NUV) observes at wavelengths 2782–2835 Å, which contain the strongest lines of Mg II: these lines are formed at temperatures of order 10 kK. The angular resolution is 0.33 arcsec; the velocity resolution is 1 km s$^{-1}$, and the temporal resolution is 2 seconds. The field of view (FOV) is 175 × 175 arcsec$^2$, i.e., a square with sides of length 130,000 km on the Sun, large enough to include a dozen supergranules or more: the angular resolution is good enough to distinguish easily between network and internetwork features within an individual supergranule. Images of the area of the Sun that are under observation at any instant can be obtained by forming images of the slit-jaws in various lines.

Carlsson et al. (2019) have described various telescopes on the ground and in space that have been useful over the years in studying the solar chromosphere and the IR. However, Carlsson et al. (2019) specifically refer to IRIS as a "game changer" in this area of research in part "because of the public availability of large amounts of high-quality chromospheric data". In conjunction with improvements in numerical modeling, the improvements in data provided by IRIS have "led to progress" in our understanding of chromospheric observational properties. An example of the vast data sets that have been accumulated by IRIS is provided by the Mg II $h$ resonance line at 2803.52 Å: by observing the profile of this line in all regions of the solar disk, astronomers now have access to a data set containing 4 million profiles in internetwork gas, 0.2 million profiles in network, 0.1 million profiles in plage, and 0.038 million profiles in sunspots (Schmit et al. 2015). Many of the profiles contain a broad absorption line with double emission peaks near line center: in most cases, the violet-shifted emission peak is observed to be stronger than the redshifted emission peak. However, in certain locations (especially in plage), the emission peak is seen to be a single line. Realistic modeling of these profiles requires the use of 3D-radiative-MHD codes. These codes indicate that the strength of the emission core is not a reliable indicator of local temperatures: however, shifts in the line core are reliable indicators of line-of-sight velocities in the chromospheric gas where the local optical depth in the line core is of order unity. Quantitative interpretation of the IRIS Mg II $h$ line data sets will in principle provide information on how the physical conditions in the chromosphere vary as we move from network to internetwork to plage to sunspots. To probe the velocity fields at levels in the chromosphere that lie above and below the regions where the Mg II $h$ line is formed, IRIS data on the strongest lines of C II (at 1334–1336 Å) can be used, depending on the local structure in density and/or temperature.

By following wave motions in IRIS lines formed at different heights in the solar atmosphere, it has been possible to identify the presence of slow-mode MHD shock waves (see Section 15.12.1) in regions of strong magnetic field (network, plage) but also in less magnetic regions such as the internetwork quiet Sun (Carlsson et al. 2019). Moreover, magnetic waves of a transverse nature (e.g., Alfvenic modes) are observed to be "ubiquitous in the chromosphere", and these waves carry "a large energy flux upward" into the Sun's outer atmosphere. In this regard, it is noteworthy that by combining IRIS data with observations from the Solar Dynamics Observatory (SDO), Kayshap et al. (2020) have identified waves propagating upward in plage regions from the photosphere to the transition region (at the level where the Si IV 1400 Å line forms). Remarkably, the periods of these waves range from as short as 2 minutes to as long as 9 minutes. Now, the 2-minute waves certainly have shorter periods than the acoustic cut-off of the atmosphere (i.e., 200 seconds: Section 13.5.5), so there is no contradiction as regards their ability to propagate upwards through the atmosphere. But 9 minutes is definitely much longer than the nominal cut-off acoustic period: something other than the idealized treatment of Section 13.5.5 must be at work. The answer is that magnetic fields are present in the plage regions that Kayshap et al. are studying: in the presence of such fields, two effects may come into play. First, the field lines, if properly inclined, can serve as "ramps" for long-period waves to access the chromosphere; second, the sound waves can undergo mode transformation into slow-mode MHD waves. The latter waves have very different vertical transmission properties from those of acoustic waves.

In view of the presence of shock waves in the chromosphere, it is noteworthy that IRIS data have sufficiently high angular resolution that they can be used to determine separately the physical conditions on both sides of a shock, i.e., in the upstream and downstream flows. This is a truly impressive technical achievement. Ruan et al. (2018) have demonstrated quantitatively that the data on the observed jumps in density and temperature across shocks in the solar chromosphere are consistent with the Rankine-Hugionot relations (which describe how physical quantities in post-shock gas are related theoretically to the same quantities in pre-shock gas).

## 15.12 A VARIETY OF WAVE MODES IN THE CHROMOSPHERE?

The goal of the discussions in Sections 15.9 and 15.10 was to determine if it might be possible to interpret quantitatively the heating of the solar chromosphere in terms of one particular wave mode, namely, *acoustic* waves emitted by the turbulent convection. The fact that we chose to concentrate on one particular wave mode (the acoustic mode, with its intrinsic aspect of compressibility) in our discussion was not random: rather, it was based on the wealth of observational evidence that provides unambiguous support for the presence of a large array of acoustic waves ("*p*-modes") propagating throughout the inner layers of the Sun (see Chapter 13). These acoustic modes are generated in source regions that in some cases lie within $0.1\%R_\odot$ of the photosphere (see Section 14.4.3): with sources so close to the surface, it would not be surprising if some fraction of the acoustic modes might find it possible to make their way upwards into chromospheric gas which lies some $(0.1-0.2)\%R_\odot$ *above* the photosphere. This raises the question: is there any observational evidence for the presence of a flux of acoustic waves in the chromosphere that is sufficient to replenish the energy flux lost by the chromosphere?

Data pertaining to this question have been reported by means of high-resolution observations of the Sun obtained by instruments carried by long-duration balloon flights. A balloon experiment labeled "Sunrise" observed the Sun using a 1-meter telescope that was transported across the Atlantic Ocean from Sweden to Canada at stratospheric altitudes (36–37 km, above the ozone layer, to permit UV observations) in two flights during the years 2009 (when the Sun was magnetically inactive) and 2013 (when the Sun contained active regions) (Solanki et al. 2017). Each flight lasted about 5 days. During the first flight, Bello Gonzalez et al. (2010) observed a patch of Sun extending over 45" × 45" 'with a spatial resolution of order 70–100 km on the Sun: this resolution is better than that which was available in any previous search for acoustic power in the Sun. Doppler shifts of the center of the Fe I line at 5250.2 Å were measured in order to obtain line-of-sight velocities in granules and intergranular lanes at effective heights of 200–300 km above the photosphere. Bello Gonzalez et al. interpreted their data by applying transmission factors for pure acoustic waves propagating upwards in the solar atmosphere. The waves they detected were found to have periods ranging from 100 seconds to 190 seconds: all of these periods are certainly short enough that they are able freely to propagate vertically through the solar atmosphere (see Section 13.5.5). Bello Gonzalez et al. concluded that the lower chromosphere in the patch of quiet Sun they observed contains vertically propagating acoustic waves with energy fluxes in the range $(6.4-7.7) \times 10^6$ ergs cm$^{-2}$ sec$^{-1}$, originating mainly in intergranular lanes. These fluxes are about 50% larger than the earlier estimates of Withbroe and Noyes (1977) for the quiet Sun, and they are not far from our lower limit estimate of $F(chr)$ in Equation 15.2.

It is noteworthy that the observations used by Bello Gonzalez et al. (2010) were obtained in the year 2009, when the Sun was experiencing one of its deepest and longest-lasting minima in magnetic activity: as a result, observational conditions in 2009 were actually optimized for searching for chromospheric heating of an *acoustic* (i.e., nonmagnetic) nature.

Independent data obtained from ground-based observations of a quiet area at the center of the solar disk reported (Abbasvand et al. 2020) that *acoustic* waves do supply enough power to the middle chromosphere (at heights of 1000–1400 km above the photosphere) to compensate for the observed radiative losses. Thus, at least in the height range 1000–1400 km, Abbasvand et al. confirmed that the chromosphere *in the quiet Sun* is adequately heated by acoustic waves. However, in the upper chromosphere (heights of 1400–1800 km), acoustic waves were found to be *insufficient* to compensate for the observed radiative losses. A source of energy other than acoustic waves is required to account for the heating of the upper chromosphere.

When the Sunrise balloon flew for a second time in 2013, active regions were present in abundance on the Sun. As a result, an acoustic interpretation of the 2013 wave data could be inadequate. In fact, we note that the chromospheric heating flux that is required (according to Withbroe and Noyes 1977) in active regions exceeds by factors of two to three the acoustic wave flux reported by Bello Gonzalez during the 2009 flight. This suggests that in active regions, the acoustic energy flux may need to be supplemented by factors of two to three in active regions. What might the nature of the supplementary energy flux in active regions be? The fact that active regions are areas of the Sun where the local magnetic fields are stronger than average suggest that magnetic effects might play a role.

What role(s) might the magnetic field play in heating the chromosphere? One possible role could be to supply extra wave modes that do not exist in nonmagnetic gas. We discuss this possibility in Section 15.12.1.

### 15.12.1 The "Plasma Beta" Parameter and Conversions between Wave Modes

How might we quantify the magnetic effects present in active regions compared to quiet Sun? If the solar atmosphere were composed solely of a hydrodynamic fluid, the fluid could be characterized by a single characteristic speed (the speed of sound $c_s$), with well-defined phase correlations between density and velocity. But the solar atmosphere, consisting as it does of partially ionized gas (capable of carrying electric currents), inevitably also contains magnetic fields that vary in strength $B$ from one location to another. The presence of a field leads to significant changes in the properties of the waves that are able to propagate in the solar chromosphere: for a comprehensive discussion, see Jess et al. (2015). Associated with the field, a second characteristic wave speed enters the problem, namely, the Alfven speed $v_A = B/\sqrt{(4\pi\rho)}$. The ratio $\beta = c_s^2/v_A^2$ (referred to by the term "plasma beta") is a measure of the extent to which gas pressure or magnetic pressure dominates in any element of gas. The magnetohydrodynamic (MHD) interaction between gas and field in any location in the chromosphere now gives rise to two characteristic "magnetosonic" speeds, referred to as the *fast mode*, with speed $v_f$, and the *slow mode*, with speed $v_s$. If we happen to be observing a region of the chromosphere where the field is weak, i.e., $\beta \gg 1$, it turns out that $v_f \approx c_s$ (with magnetic and kinetic pressures in phase), while $v_s \leq v_A$ (with magnetic and kinetic pressures out of phase). In the opposite limit, i.e., in a region of strong field where $\beta \ll 1$, it turns out that $v_f \approx v_A$, while $v_s \leq c_s$.

The variation of the parameter $\beta$ with altitude in the Sun plays an important role on the properties of wave propagation. Specifically, the gas in the deep photosphere (and below) exists in the regime $\beta \gg 1$, while in the corona, the ambient medium is in the regime $\beta \ll 1$. Therefore, at any given position on the Sun, as we move vertically upward into the chromosphere, we eventually pass through a critical layer where $\beta = 1$: this layer, known as the "magnetic canopy" (Kontogiannis et al. 2014), is a critical layer where local conditions actually enable waves to undergo transformations ("conversions") from one mode into another. In regions of strong field (e.g., in the bright network element E in Figure 15.6), the $\beta = 1$ layer lies at a relatively lower altitude (of order 1500 km: Kontogiannis et al. 2011). Therefore the upper chromosphere and part of the mid-chromosphere over network exists mainly in the regime $\beta \ll 1$, where the slow magnetosonic mode propagates at a speed $\leq c_s$. (This does *not* mean that the slow-mode wave *becomes* a sound wave: on the contrary, a sound wave can propagate in all directions from a given starting point, whereas the slow-mode wave *cannot* propagate at all in directions that are perpendicular to the field.) If, therefore, we were to discover that a fluctuation were propagating upwards in network element E at the local sound speed, we could not immediately conclude that we were observing an acoustic wave in the chromosphere: we might (if the local conditions were appropriate) actually be observing a slow-mode wave. In order to distinguish between the slow-mode wave and an acoustic mode, we would need to obtain further observational information regarding the relative phases between magnetic and kinetic pressure. Similarly, if we are observing a region of weak field (e.g., the dark internetwork element A in Figure 15.6), the $\beta = 1$ layer lies at a relatively higher altitude, as high as 2.4 Mm in one case (Kontogiannis et al. 2011): therefore, the chromosphere in such a region exists mainly in the regime $\beta \gg 1$. In such a case, if we were to detect a fluctuation propagating upwards in element A at speed $c_s$, we could not necessarily conclude that we were observing a *bona fide* acoustic wave: we might instead be observing a fast-mode wave. As a result, while the interpretation of Doppler data in terms of acoustic waves (as reported by Bello Gonzalez et al. 2010) may be acceptable in regions where the Sun is very quiet, the interpretation becomes less straightforward if magnetic fields are present in the observed patch of Sun.

The fact that one wave mode can "convert" into another wave mode at certain locations in the Sun's atmosphere can be put to good use when we wish to "locate" active regions on the "far side" of the Sun, up to 14 days before they can be seen from Earth. We will address this topic in Section 16.7.7.

Is there a way to observe gas that lies above the canopy, as well as gas that lies below the canopy, using a single line in the spectrum? Yes, there is: the $H\alpha$ line is helpful in this regard. Following the discussion in Section 3.8.2, if observer A uses a filter to observe close to the center of $H\alpha$ (within, say, ¼ Å), then the image obtained by A will show gas *above* the canopy. But if observer B uses a filter that is tuned away from line center (by, say, 0.8 Å), then the image obtained by B will show gas that lies *below* the canopy. As a result, images A and B will look quite different from each other: one image (below the canopy) contains granulation, the other (above the canopy) does not.

The fact that waves of various modes can exist in a magnetized gas has the effect that it is not a simple matter to calculate how the different modes will propagate in the solar chromosphere. Due to spatial variations in the ambient properties, each wave mode is subject to different laws of refraction, reflection, and transmission. At the magnetic canopy, the Alfven speed and the sound speed are nearly equal. As a result, when one considers the two-dimensional problem of an originally purely acoustic wave propagating vertically from the photosphere and encountering the canopy, the effect is that the incoming wave undergoes "conversion" into slow and fast magnetosonic waves (Kontogiannis et al. 2014). The emerging slow wave can continue to propagate into the upper chromosphere along a magnetic "ramp" (with field lines inclined to the vertical by an angle θ) above the canopy provided that the period of the wave is shorter than the local cut-off value of 200 (seconds)/cosθ (see Section 13.6). But the emerging fast wave suffers a different fate: it is refracted when it encounters a region where the vertical gradient of $v_f$ becomes large (such as in a region where the density falls off rapidly as height increases): as a result, the fast mode is unlikely to reach the upper chromosphere but is more likely to be directed downward into the photosphere.

Jess et al. (2015) do not confine attention only to the contributions of *compressible* waves in the process of heating the outer solar atmosphere: they also discuss at length the presence of *incompressible* waves. Such waves can be generated by the horizontal components of convective motions, giving rise to torsional Alfven waves and kink modes. Because of the absence of compressibility, these modes are subject in general to less efficient damping than the compressible waves: however, in certain conditions, a code that includes 1.5 dimensions suggests that Alfven waves can "convert" into slow magnetosonic waves: these can form shocks that can heat the chromosphere (Arber et al. 2016). However, in order to analyze properly the combined effects of Alfven, acoustic, and magnetosonic waves, the "conversion" problem in the solar atmosphere must be analyzed with a fully three-dimensional approach (e.g., Cally and Khomenko 2019). The complications of such an approach are far beyond the limits of this "first course".

Ultimately, if we are to solve the physics problem of how exactly the energy present in a magnetosonic wave is actually converted into heating of the chromosphere, we need in principle to observe how the waves are dissipated. Jess et al. (2015) state that this goal of discovering evidence for dissipation has not yet been achieved: the dissipation of magnetosonic waves probably occurs on length-scales that are so short in the Sun as to be well beyond the current limits of observational resolution. For example, if MHD shock formation is at work, then the dissipation is expected to occur over length-scales comparable to the thickness of the shock layer, i.e., 10–20 km (Goodman and Kazeminezhad 2010). Even smaller length-scales are suggested for dissipation if it is permissible to consider an analogy with a particular solar environment in which *in situ* measurements have actually resolved the dissipation scale: we refer to the solar wind as it flows past Earth's orbit (Smith et al. 2001). These data indicate that magnetosonic wave dissipation (in a turbulent medium with electron density $n_e$ cm$^{-3}$) occurs on a length-scale known as the ion-inertial scale $L_{ii}$ (in cm) $\approx 2 \times 10^7/\sqrt{n_e}$. If the same dissipation process is at work in the solar chromosphere, where $n_e$ ranges from $10^{11}$ cm$^{-3}$ to $10^{10}$ cm$^{-3}$ (Vernazza et al. 1981), the $L_{ii}$ dissipation scale in the chromosphere would be no larger than a few *meters*. Even with the resolution that DKIST provides, such a small length cannot be resolved.

## EXERCISE

15.1 The estimates of chromospheric heating given in Equation 15.10 are obtained by picking a particular fitting formula for the opacity in Equation 15.9, $\kappa \approx 10^{-32}\rho^{0.3}T^9$. Other choices of fits to the opacities are possible, using a different coefficient and different exponents. Choose values of 5, 7, and 11 for the temperature exponent, and values of 0 and 0.5 for the density exponent. For each pair of exponents, recalculate the fitting formula such that, in all cases, log $\kappa$ = 4 when log $\rho$ = 0 and log $T$ = 4, and then recalculate the curves labeled Flux=Hi and Flux=Lo in Figure 15.7.

## REFERENCES

Abbasvand, V., Sobotka, M., Svanda, M., et al., 2020. "Observational study of chromospheric heating by acoustic waves", *Astron. Astrophys.* 642, A52.

Arber, T. D., Brady, C. S., & Shelyag, S., 2016. "Alfven wave heating of the solar chromosphere: 1.5D models", *Astrophys. J.* 817, 94.

Bello Gonzalez, N., Franz, M., Martinez Pillet, V., et al., 2010. "Detection of large acoustic energy flux in the solar atmosphere", *Astrophys. J. Lett.* 723, L134.

Cally, P. S., & Khomenko, E., 2019. "Fast to Alfven mode conversion and ambipolar heating in structured media. I. Simplified cold plasma model", *Astrophys. J.* 885, 58.

Carlsson, M., De Pontieu, B., & Hansteen, V. H., 2019. "New view of the solar chromosphere", *Ann. Rev. Astron. Astrophys.* 57, 189.

Carlsson, M., & Leenaarts, J., 2012. "Approximations for radiative cooling and heating in the solar chromosphere", *Astron. Astrophys.* 539, A39.

Carlsson, M., & Stein, R. F., 1992. "Non-LTE radiating acoustic shocks and CaII K2V bright points", *Astrophys. J. Lett.* 397, L59.

Da Silva Santos, J. M., de la Cruz Rodriguez, J., Leenaarts, J., et al., 2020. "The multithermal chromosphere: inversions of ALMA and IRIS data", *Astron. Astrophys.* 634, A56.

Goodman, M. L., & Kazeminezhad, F., 2010. "Simulation of MHD shock generation, propagation, and heating in the chromosphere", *Astrophys. J.* 708, 268.

Houdebine, E. R., 2010. "Observation and modelling of main-sequence star chromospheres", *Mon. Not. Roy. Astron. Soc.* 403, 2157.

Jess, D. B., Morton, R. J., Verth, G., et al., 2015. "Multiwavelength studies of MHD waves in the solar chromosphere", *Space Sci. Rev.* 190, 103.

Judge, P. G., & Carlsson, M., 2010. "On the solar chromosphere observed at the limb with HINODE", *Astrophys. J.* 719, 469.

Judge, P. G., & Peter, H., 1998. "The structure of the chromosphere", *Space Sci. Rev.* 85, 187.

Kalkofen, W., 2012. "The validity of dynamical models of the solar chromosphere", *Solar Phys.* 276, 75.

Kayshap, P., Srivastava, A. K., Tiwari, S. K., et al., 2020. "Propagation of waves above a plage as observed by IRIS and SDO", *Astron. Astrophys.* 634, A63.

Kontogiannis, I., Tsiropoula, G., & Tziotziou, K., 2011. "HINODE and SOHO/MDI quiet sun magnetic field. Implications for the height of the magnetic canopy", *Astron. Astrophys.* 531, A66.

Kontogiannis, I., Tsiropoula, G., & Tziotziou, K., 2014. "Transmission and conversion of magneto-acoustic waves on the magnetic canopy in a quiet Sun region", *Astron. Astrophys.* 567, A62.

Leenaarts, J., Carlsson, M., Hansteen, V., et al., 2011. "On the minimum temperature in the quiet solar chromosphere", *Astron. Astrophys* 530, A124.

Musielak, Z., Rosner, R., Stein, R. F., et al., 1994. "On sound generation by turbulent convection: A new look at old results", *Astrophys. J.* 423, 474.

Pereira, T. M. D., Rouppe van der Voort, L., & Carlsson, M., 2016. "The appearance of spicules in high resolution observations of CaII H and H$\alpha$", *Astrophys. J.* 824, 65.

Ruan, W., Yan, L., He, J., et al., 2018. "A new method to comprehensively diagnose shock waves in the solar atmosphere based on simultaneous spectroscopic and imaging observations", *Astrophys. J.* 860, 99.

Schmit, D. F., Bryans, P., De Pontieu, B., et al., 2015. "Observed variability of the solar MgII h spectral line", *Astrophys. J.* 811, 127.

Smith, C. W., Mullan, D. J., Ness, N. F., et al., 2001. "The day the solar wind (almost) disappeared: Magnetic field fluctuations, wave refraction, and dissipation", *J. Geophys. Res.* 106, 18625.

Solanki, S., Riethmuller, T. L., Barthol, P., et al., 2017. "The second flight of the SUNRISE balloon-borne solar observatory: Overview of instrument updates, the flight, the data, and first results", *Astrophys. J. Suppl.* 229, 2.

Vernazza, J. E., Avrett, E. H., & Loeser, R., 1981. "Structure of the solar chromosphere. III. Models of the EUV brightness components of the quiet Sun", *Astrophys. J. Suppl.* 45, 635.

Withbroe, G. L., & Noyes, R. W., 1977. "Mass and energy flow in the solar chromosphere and corona", *Ann. Rev. Astron. Astrophys.* 15, 363 (WN).

# 16 Magnetic Fields in the Sun

Up to this point, we have been considering the Sun in terms of material that can be described reliably by the laws of "ordinary" gas dynamics and radiative transfer. This has been sufficient to allow us to describe in some detail the overall structure of the Sun, including radial profiles of pressure, temperature, and density from the center all the way to the surface. But the very concept of a "radial" profile incorporates the assumption that the profile is the same in all directions, i.e., the material is spherically symmetric. This is certainly an adequate assumption deep in the interior of the Sun.

However, as we approach the surface, certain features become apparent in the Sun where departures from spherical symmetry are more or less severe. One such effect is introduced by rotation: the Sun departs from a spherical shape by having a slightly oblate figure. But the effect is so small that the unaided eye cannot see the effect. In fact, reliable measurements of the oblateness are quite difficult to make (Chapter 1, Section 1.10). In this chapter, we consider the departures from spherical symmetry that arise from the presence of magnetic fields.

## 16.1 SUNSPOTS

In the context of observations using photons in the visible part of the spectrum, the most dramatic departures from spherical symmetry on the Sun's surface are sunspots (see Figure 16.1). These are darker areas of the surface that are sometimes large enough to be seen by the unaided eye. Occasional reports of naked-eye sunspots by Chinese observers are on record for the past two millennia. To be sure, the advent of telescopes has greatly increased the observability of sunspots. But even during the time period 1600–1650, when telescopes first became available, there are records of as many as 33 naked-eye sunspots, including one by Galileo himself (Vaquero 2004).

Sunspots spanning a wide range of sizes appear from time to time, in an unpredictable way, as dark spots somewhere on the surface of the Sun, usually as a pair of spots or in groups (see Figure 2.3). As solar rotation carries a pair of spots across the disk of the Sun, one spot is in the lead, and the other follows. This gives rise to the notation "leader" spot and "follower" spot. Thus, when a pair of spots first rotates onto the disk, appearing at the east limb, we see first the leader. And when the pair eventually (about two weeks later) reaches the west limb, it is the follower spot that disappears from view last. Although the spot pair lies mainly in the east-west direction, observers find that the follower spot lies in general slightly *farther* from the equator than the leader. That is, the line joining leader and follower does not lie exactly east-west, but is tilted slightly: the tilt angle is small for pairs of spots lying near the equator, and the tilt increases with increasing latitude, reaching a maximum of 10–12 degrees at latitudes of 30–35 degrees (Hale et al. 1919). This property of sunspot tilt angles is referred to as "Joy's law", named for the astronomer who analyzed almost 3000 sunspots over three to four sunspot cycles. As we shall see later (Section 16.9), Joy's law will play a role in explaining why the Sun undergoes a cycle of magnetic activity every 11 years or so.

As can be seen in the expanded view of a spot in Figure 16.1, a large spot (either leader or follower) consists of a darker central core (the "umbra", the Latin word for "shadow") surrounded by a "penumbra". The penumbra has radial (more or less horizontal) striations that alternate between bright and dark, giving an overall impression that the penumbra is on average intermediate in brightness between the umbra and the photosphere (see Figures 2.3 and 16.1).

Why are there two distinct regions in a spot, one being (more or less) uniformly dark and the other having horizontal striations? The answer, at least in part (see Section 16.5), has to do with the fact that in the umbra, there exists a magnetic field that is observed to be nearly *vertical*, while in the penumbra, the magnetic fields are observed to be more nearly *horizontal*. Quantifying this statement, Jurcak et al. (2018) used Hinode spectropolarimetric data to determine the vector magnetic field in 79 active regions that occurred during the years 2006–2015. They found that, at the umbral-penumbral (UP) boundary of all the sunspots they studied, there exists a preferred value for *one* component of the field, namely, the *vertical component* $B_{ver}(UP)$. With a likelihood of 99%, the value of $B_{ver}(UP)$ was found to lie in the range 1849–1885 G. The most probable value of $B_{ver}(UP)$ was found to be 1867 G. It is remarkable that such a narrowly confined range of $B_{ver}(UP)$ persists in sunspots over a 10-year period: this period spans the end of cycle 23 and the rise to maximum of cycle 24. In a follow-up 3-D MHD numerical study, Schmassmann et al. (2021) have found that a criterion derived by Gough and Tayler (1966) for the onset of convection in the presence of a (specifically) *vertical* field (see Section 6.9) may help to explain the observed existence of a preferred value for $B_{ver}(UP)$.

### 16.1.1 Spot Temperatures

To determine the temperature of the gas in a spot, we start with the question: how dark is the umbra relative to the photosphere? The answer depends on the wavelength: the shorter the wavelength, the darker the intensity of the umbra $I_\lambda(*, \mu)$ compared to the intensity of the undisturbed photosphere at the same distance from disk center $I_\lambda(\mu)$. At wavelengths of 4000 Å, large spots may have $I_\lambda(*, \mu)$ values that are 10–20 times smaller than $I_\lambda(\mu)$ (Bray and Loughhead 1979; their Table 4.1). The contrast between umbra and photosphere becomes less pronounced as the spot is observed at longer and longer wavelengths: around 1 micron, the spot intensity is almost half as bright as the photosphere. Model atmospheric fits to umbral radiation allow one to obtain the profile of temperature versus optical depth in the umbra. Expressing temperatures in terms of the variable that appears in the Saha equation (Section 4.2), $\theta = 5040/T$, the differences between the temperature variable $\theta(*, \tau)$ inside the spot at optical depth $\tau$ and the temperature variable $\theta(\tau)$ in the undisturbed photosphere at the same value of $\tau$ (but not at the same physical depth), are observed to be as large as $\Delta\theta \approx 0.3$–$0.4$ at $\tau \approx 1$. Since the local temperature in the photosphere at $\tau \approx 1$ is close to 6000 K, i.e., $\theta(1) \approx 0.84$, this leads to $\theta(*, 1) \approx 1.14$–$1.24$. This corresponds to a temperature in the spot of 4100–4400 K at $\tau \approx 1$. That is, the gas in the "photosphere" of the spot (i.e., around $\tau \approx 1$) is some 1600–1900 K *cooler* than the gas at equal optical depth in the photosphere. Estimates of the *effective* temperature of a spot are 4100–4200 K, i.e., almost 2000 K cooler than the photosphere. The fractional deficit in effective temperature in the umbra is about 30% compared to the undisturbed photosphere. Thus, the bolometric flux ($(\sim T_{eff}^4)$) of radiation emerging from the umbra is only $0.7^4$ times the photospheric flux, i.e., the flux emerging from the umbra is only about 25% of the flux emerging from the photosphere.

Apart from the darkness of umbrae relative to the photosphere, is there any observational evidence for cool temperatures in sunspots? Indeed there is: spectra of umbrae contain molecules that are essentially nonexistent (due to dissociation at higher temperatures) in the photosphere. The commonest molecules in umbrae are hydrides, including those of the elements C, Mg, Fe, Al, B, and Be. Next in abundance are oxides, including those of the elements Ti, Zr, La, Sc, V, and Ba. The list also includes AlF and BF. Rotational temperatures have been derived for some of the umbral molecules and are in some cases found to be as low as 1100 K. There can be no doubt: the gas in sunspots is definitely colder than the gas in the photosphere.

### 16.1.2 Why Are Sunspots Cooler than the Rest of the Photosphere?

The fact that some 75% of the photospheric energy flux is blocked in a sunspot umbra indicates that some physical mechanism is at work to cause a severe perturbation of energy flow in the umbra.

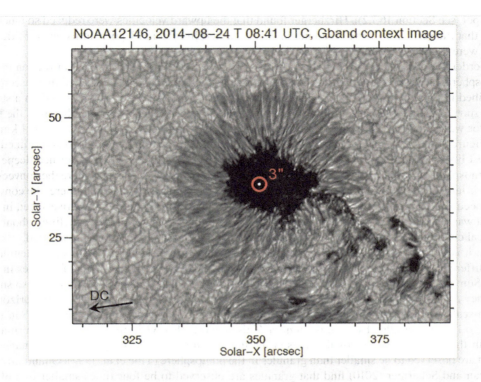

**FIGURE 16.1** A sunspot, showing the dark umbra (center) plus striated penumbra, surrounded by undisturbed photosphere (containing granules). The darkest part of the umbra is denoted by a 3"-wide red circle: spectra were obtained in the portion of the umbra that lies inside this circle. The arrow labeled DC points towards the center of the solar disk. (From Lohner-Bottcher et al. 2018; used with permission of ESO.)

What might that mechanism be? To address that question, let us first consider the unperturbed photosphere of the Sun. Recall that in Section 6.3 and Figure 6.3, we discussed the use of "dopplergrams" to measure vertical velocities over a finite area of the Sun's photosphere. The data indicate that in the solar granulation, the bright rising cores contain upward flows with speeds of $\geq 2$ km s$^{-1}$, while the dark intergranular lanes contain downward flows with speeds of $\geq 3$ km s$^{-1}$. These vertical flows (in conjunction with temperature differences $\Delta T$ between bright and dark gas) are responsible for the normal convective transport of energy in the upward direction through the Sun's photosphere (see Equation 6.2). In a sunspot umbra, it is clear that the normal upward transport of energy is severely interrupted. Since convection carries a significant fraction of upward energy flux in the photosphere and an even larger fraction beneath the photosphere (see Section 6.7.3), we need to address the following question: could it be that something significant is impeding convection in the umbra? Since we have identified the key properties of convection in the photosphere in terms of vertical velocities and temperature differences $\Delta T$ between rising and falling gas, the following question is relevant: how do the vertical velocities and $\Delta T$ in the umbra of a sunspot compare with those in the photosphere?

In principle, dopplergrams could be used to perform a differential analysis between umbra and photosphere. Hirzberger (2003) used this approach to examine the vertical velocities in small pores in the Sun. (Pores are the smallest regions of cool sunspot-like material on the Sun: they differ from spots in that pores consist only of "umbra", without any evidence for "penumbra" surrounding the cool spot (see Section 16.7.2)). Hirzberger found that the upward velocities were reduced so much in pores that the vertical velocity could not be distinguished from noise. Upper limits on velocities in pores were found to be of order 0.1 km s$^{-1}$.

In order to undertake a differential study of convection in an umbra relative to convection in the photosphere, more precise results can be obtained by using the method of "C-shaped line bisectors" described in Section 3.8.3. Löhner-Böttcher et al. (2018: hereafter LB18) have reported on a study of 13 spots in which they used a line of Ti I: in the photosphere near one particular spot, the line bisector was found to be C-shaped with a maximum mean convective speed of almost 0.4 km s$^{-1}$ (see Figure 16.2). Such a speed is about 30% higher than the maximum mean speed determined for the Fe I line used in Figure 3.8: the Ti I line, being weaker than the Fe I line, is formed deeper in the atmosphere than the Fe I line, thereby sampling more effectively the layers where the convective speed is larger. But in the umbra, the results are found to be remarkably different: there, the convective speed was found to be no larger than 0.02 km s$^{-1}$. That is, the average convective speed in the umbra was found to be smaller than in the nearby quiet Sun by a factor of 20. Something about the physical conditions in the umbra is leading to a dramatic reduction in the convective speed.

But it is not only the *speed* of convection that is reduced in an umbra. Values of the temperature difference $\Delta T$ between rising and falling gas (which can be many hundreds of degrees in the quiet Sun: see Section 6.6) are also affected. In fact, the $\Delta T$ values in sunspot umbrae are so small that they are not easily measurable. As a result, the prominent bright/dark pattern (with horizontal length-scales of order 1000 km) that is such a characteristic of convection in the quiet Sun (see Figure 6.1) is essentially absent in the umbra. This suggests that $\Delta T$ in the umbra is much smaller than in the photosphere. A possible reason for this feature may be that granules in a magnetic region are observed to be smaller than granules in the photosphere (Title et al. 1992). Quantitatively, Narayan and Scharmer (2010) find that granules are observed to be four times smaller in a plage with mean field 800 G than in the photosphere. In 3-D MHD models, Tian and Petrovay (2013) show that, whereas in the presence of weak fields the pattern of surface cells looks very similar to observed solar granulation, in strong fields things look quite different: the surface now contains a configuration of structures that look nothing like solar granulation, with horizontal length-scales much smaller than those of granules in the photosphere. In one particular sunspot model, the numerical value of the horizontal length-scale of convection cells was found to be of order 80 km, i.e., more than 10 times smaller than the diameters of granules in the quiet Sun (Mullan 1974). Tian and Petrovay use the descriptive biological term "brain-pattern" to describe the configuration of convection cells with small horizontal dimensions in the presence of strong magnetic fields. In the presence of cells that have small horizontal scales, photon energy exchange between rising and falling gas will be more effective than in normal granulation, thereby reducing the value of $\Delta T$. A large number (>2000) of small transient bright features in a single umbra were studied by Hinode (Watanabe 2009): these "umbral dots" were found to have mean diameters of 150–200 km (i.e., some 10 times smaller than the mean diameter of photospheric granules) and lifetimes of <15 minutes. Magnetic field strengths in umbral dots have been reported in the range 1000–2700 G (Falco et al. 2017). Yadav et al. (2018), also using Hinode data for multiple (42) umbrae, suggested that umbral dots are manifestations of magneto-convection.

As a result, when we try to calculate the upward convective flux in an umbra using Equation 6.2, the combination of reduced velocities and reduced $\Delta T$ have the effect that the resulting upward flux in an umbra is small compared to the flux being carried up in the quiet Sun. This provides a physical explanation for why the umbra of a sunspot appears dark.

But that still leaves unanswered the question: what is the *physical cause* for small convective velocities in an umbra? LB18 sought an answer to this by searching for a correlation between the (small) umbral speeds and various physical parameters: they found that the most significant (anti-)correlation is between the umbral speed and the strength of the magnetic field in the spot. In their sample of spots, they found that the (anti-)correlation coefficient between umbral speed and field strength was −0.79. This strong correlation indicated that the umbral speed would fall formally to

# Magnetic Fields in the Sun

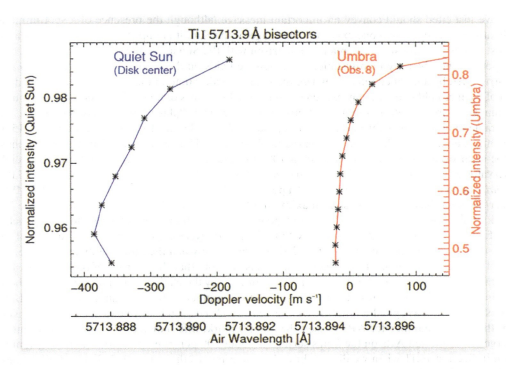

**FIGURE 16.2** Bisector analysis of the Ti I 5713.9 Å line in the quiet Sun photosphere (blue curve) and in an umbra (red curve). (From Lohner-Bottcher et al. 2018; used with permission of ESO.)

zero if the field strength were to be as large as 2.78 kiloGauss (kG). This is a very interesting result: it shows that, if umbral magnetic fields can be as large as (roughly) 3 kG, then the primary cause of small convective speeds in an umbra has something *directly to do with the magnetic field*. In Section 16.4.2, we shall discuss what the observations actually tell us regarding the strength of the magnetic fields occurring in the umbrae of sunspots: we shall find that the fields present in sunspot umbrae are indeed large enough to reduce the convective speed, in some cases to essentially zero.

In Section 16.7.1, we shall consider the physical process at work in "magneto-convection", i.e., in situations where magnetic fields interfere effectively with convection.

### 16.1.3 Areas of Spots and Plages

How large are sunspots? Their areas can be readily measured in a white-light image of the solar disk. The smallest spots, consisting of umbra only (without any noticeable penumbra), are called "pores" and have angular diameters of 2–5 arcsec: that is, the smallest pores are comparable in size to the sizes of individual granules. (We shall see in Section 16.7.2 that this is not a coincidence.) The largest pores have diameters of no more than 10 arcsec: once a pore grows to a diameter of 10″ or more, a penumbra appears, and the feature becomes a *bona fide* sunspot, consisting of a dark umbra in the center and a less dark penumbra around the outer edge of the umbra. Large spots have areas that are often cited in units of "millionths of the solar hemisphere [MSH]". The largest spot ever recorded had an area of $A \approx 6300$ MSH, i.e., it occupied about 0.6% of the visible hemisphere. More commonly, sunspots have areas of up to a few hundred MSH: 95% of spots have $A < 500$ MSH (Bray and Loughhead 1979, p. 229, their Table 6.1). Although spots are indeed striking visual phenomena when seen against the background of the solar surface, the numerical values of areal coverage that we have just cited should convey an important message: although the presence of a spot immediately attracts the eye when one examines an image of the Sun, it is important to note that even the largest spot is actually small in area in comparison with the Sun as a whole.

Spots are not the only manifestations of magnetic fields on the solar surface. In fact, each spot is surrounded by a magnetized area where the field is weaker than in the umbra of the spot, but the area of weaker field strength may exceed the spot area by factors of up to 10 or more. The surrounding magnetic area is referred to as an "active region" (AR) or "plage". To observe the plages, it is preferable to observe in a chromospheric line, such as $H\alpha$ or the Ca II $K$ line (e.g., see Figure 15.5). How much area do the plages occupy on the Sun's surface? The answer is: it depends on the phase of the solar cycle. In a study of more than 100,000 images of the Sun in Ca II $K$, taken at eight observatories around the world in the years 1893–2018, Chatzistergos et al. (2019) have found that the *daily* areal coverage by plages reached a peak of 12% of the solar disk area (i.e., 120,000 MSH) on certain days during the 1957 solar maximum. (The *annual* median value of areal coverage had a peak of about 8%, also in 1957.) The corresponding peak in total areal coverage for sunspots has been found (in a study of 32,223 spots by SOHO/MDI in cycle 23) to be no more than 1% of the solar surface on any one day (Valio et al. 2020). Thus, even when the Sun is at its most active, the area of the surface occupied by magnetic features is no more than 10% (or so) of the visible disk.

A numbering convention for ARs was adopted in 1972 by the National Oceanic and Atmospheric Administration (NOAA): an AR is assigned a NOAA number if there are spots in the AR. But a spotless AR is not assigned a number. By the year 2002, i.e., after 30 years of using the convention, the NOAA number for ARs passed the 10,000 mark: this indicates that the average rate of emergence of spotted ARs on the Sun's surface is about 300 per year. We will refer to this number again in Section 16.11 when we discuss how the Sun removes the magnetic helicity of all of these newly emerging ARs.

### 16.1.4 Spot Numbers: The "11-Year" Cycle

Observers of sunspots who collect data over long intervals of time (decades and centuries) have discovered that the number of spots on the surface of the Sun varies with time. The numbers increase and decrease in a nearly cyclical manner: sometimes there are many spots on the surface, while at other times there are few (or even no) spots visible (see Figure 16.3).

In order to quantify this variability, observers have devised certain rules to count the number of spots visible on the Sun on any given day. Since the year 1848, the commonest system in use has been the Zurich sunspot number, $R_Z$, which counts both single spots and groups of spots. Each day, observers use an image of the Sun to calculate a (daily) value of $R_Z$. In order to make plots over extended periods of time, the daily values are typically averaged over monthly or annual intervals: whereas individual daily values may exceed 500, the monthly averages do not exceed 400 (Acero et al. 2017), and the annual averages (plotted in Figure 16.3) are even smaller, never exceeding 200.

When there are a large number of spots visible on the surface of the Sun, and yearly averages of $R_Z$ rise to values of 100–200 or more, the Sun is said to be at "maximum activity". When average values of $R_Z$ are small (e.g., <10–20), the Sun is said to be in a stage of "minimum activity". These phases are also known as "solar maximum" and "solar minimum" respectively. On the scale of the plot in Figure 16.3, it looks as if each cycle has only one well-defined maximum: however, when a finer time-scale is used for the plot, some cycles are found to contain two maxima close together in time. Among recent solar cycles, the ones with maximum around 1980 and around 2001 were found to have a single maximum, but the cycles with maxima around 1990 and 2012 had double maxima, separated by an interval of 1–2 years. The double peaks are separated by the "Gnevyshev gap", first discussed in the literature by Gnevyshev (1967).

The interval of time between one "solar minimum" and the next minimum is not constant: the interval can be as short as nine years and as long as 12+ years. The average length of the sunspot cycle (based on $R_Z$ values) is about 11 years. In Figure 16.3, we see that in the second half of the

Magnetic Fields in the Sun

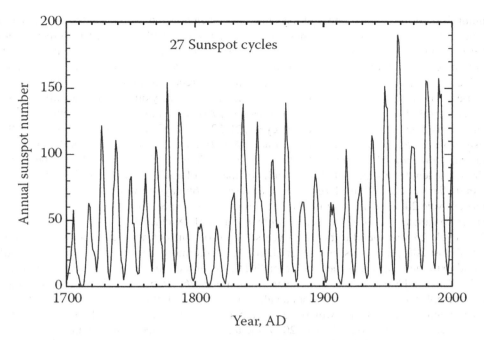

**FIGURE 16.3** Number of sunspots, averaged over one year, plotted as a function of time over a span of three centuries. (The data were obtained from the Solar Influences Data Center at www.sidc.be/sunspot-data/)

20th century (when the "Space Age" began), "solar maxima" were observed in the years 1957, 1970, 1980, 1990, and 2000–2001. As of the time of writing (September 2021), the most recent "solar maximum" was observed in 2014. For ease of reference, solar observers have devised a running number to label each solar cycle: by convention, the cycle that is assigned the number 1 began at the minimum of activity in 1755, when detailed spot records began to be kept on a regular basis. (Some earlier cycles are included in Figure 16.3, but the data for those cycles are less reliable, and no cycle number has been assigned to them.) When telescopes allowed the first modern records of sunspots in the early 1610s, observers such as Galileo (in Italy) and Thomas Herriott (in England) would make drawings of spots on some days, but then weeks or months or years would elapse without any records being made. From 1650 to 1720, the Sun was almost entirely devoid of spots, a period now known as the "Maunder minimum". After 1720, interest in spots grew more widespread, but regular observations did not begin until the 1750s. Cycle 3 was the largest cycle (in terms of sunspot number) in the 18th century. In the 19th century, cycles 4 and 5 were the smallest cycles, while cycles 8 and 11 were the largest. In the 20th century, cycle 14, with a peak in 1907, was the smallest cycle, while cycle 19, with a peak in 1957 (at the dawn of the Space Age), was the largest cycle. The most recent solar maximum (cycle 24, in 2014) had a maximum sunspot number that was smaller than almost all of the 20th-century cycles. As of the time of writing (December 2021), the Sun is in cycle 25: the first active region of this cycle emerged on the Sun on July 6, 2019 (Chelpanov and Kobanov 2020). If it takes 5–8 years for the cycle to reach maximum (see Section 16.9), then the maximum of cycle 25 should occur somewhere between 2024 and 2027: during that interval, the Parker Solar Probe (see Section 18.7.2) is scheduled to achieve its closest approach to the Sun.

When one inspects results such as those shown in Figure 16.3, the human eye is typically drawn to the peaks (solar maxima) in the curve. But it is also worthwhile to examine the "dips" (solar minima) in the curve. How small can the $R_Z$ value become in a solar minimum? The answer depends on the length of time over which one averages the individual (daily) values. A different way to quantify the spottedness of the Sun, or more precisely, the *lack of spottedness*, especially near solar minimum, is in terms of the number of days in a year when *no* spots were detected on the Sun, i.e., when the daily value of $R_Z$ was zero. These "spotless" days (reported in the website https://wwwbis.sidc.be/silso/spotless) vary in number from cycle to cycle. Since the year 1849, the year 1913 had the largest number ($N(s\text{-}d)$) of spotless days (310). In the years 2008 and 2009, $N(s\text{-}d) = 270$ and 260 respectively. And for years that occurred during the most recent minimum of activity, there were 275 spotless days in 2019. Thus, we may say that the Sun was "very quiet" (magnetically speaking) in the years 1913, 2008, 2009, and 2019. The value of $N(s\text{-}d)$ can also be used to characterize the level of activity over an entire solar cycle: in cycles 21, 22, and 23, $N(s\text{-}d)$ had values of 273, 309, and 813 respectively. Clearly, the much larger value of $N(s\text{-}d)$ in cycle 23 indicates that this was a cycle of unusually *low magnetic activity*. Another measure of low activity in cycle 23 can be found by examining the length of time that elapsed with no "medium-large" flares: the GOES data contained not a single flare of class C2 (or larger) for an interval of 466 days between the end of cycle 23 and the start of cycle 24. (For the definition of classes of flares, see Section 17.19.1.) And in the 3-year interval from January 2007 to December 2009, not a single X-class flare occurred. Moreover, the length of cycle 23 was found to be 12.83 years (based on sunspot numbers): this was the longest cycle in almost 200 years (Kossobokov et al. 2012). Furthermore, the end of cycle 23 was marked by several extreme solar properties (de Toma 2011): the annual mean sunspot number, which had been 12.6, 13.4, and 8.6 at the previous solar minima of 1976, 1986, and 1996, fell to as low as 2.9 in 2008; the polar field strength fell from a value of 5–10 G (in 1996) to 3–4 G (in 2008); the interplanetary magnetic field (IMF), which had been 5.48, 5.76, and 5.11 nanotesla (nT; $1 = 10^{-5}$G) in 1976, 1986, and 1996, fell to 3.93 nT in 2009; the annual mean value of solar wind speed in 1976, 1986, and 1996 had been 455, 453, and 423 km s$^{-1}$, whereas it decreased to 364 km s$^{-1}$ in 2009; the polar coronal holes had smaller areas (4%–5% of the solar surface) in 2008 than in 1996 (7%–8% of the solar surface).

Not only do the *numbers of sunspots* increase and decrease with a period of ~11 years: there is also an 11-year cycle in the mean *latitudinal positions* of spots. At solar maximum, the spots are observed to lie at higher latitudes (up to 30–40 degrees), whereas at solar minimum, the spots are observed to lie at lower latitudes (10–20 degrees). When sunspot positions are plotted as a function of time, the systematic displacement towards lower latitudes as time goes on gives rise, in each 11-year cycle, to an easily recognized pattern that is referred to as the "butterfly diagram" (see Figure 16.10). Also displaying a "butterfly diagram" as the solar cycle progresses is the *toroidal* component of the solar magnetic field (Mordvinov et al. 2012), indicating that sunspots are generated by the toroidal component of the Sun's field. This conclusion will be important when we consider why the Sun has an 11-year cycle (Section 16.9).

During the years 2011–2014, several workshops were held by solar astronomers with the goal of developing a new approach to quantifying the sunspot number (Clette et al. 2016). The new sunspot number is referred to as the group sunspot number (GSN). When earlier sunspot records are revisited, the yearly averages of the GSN are found to be in some cases different from the Zurich numbers, with maximum values between 250 and 300, i.e., somewhat larger than the values of $R_Z$ in Figure 16.4. Acero et al. (2017) conclude that using the newer GSN approach, the daily sunspot number may at times reach numerical values in excess of 550.

Notice that the sunspot number (as plotted in Figure 16.3) indicates changes that occur during each cycle as measured by *optical images of the photosphere*. But even with the large changes in $R_Z$ that are apparent in Figure 16.3 between solar minimum and solar maximum, the overall changes in the total *flux* of optical radiation from the Sun does not change by more than about 0.1% (see Figure 1.1). That is, the solar cycle causes only miniscule changes in the radiation *from the photosphere*. Other *photospheric* properties that also showed no significant alterations (in the quiet Sun) over a 6-year interval spanning the maximum of solar cycle 24 include vertical flows, horizontal flows, continuum contrast, and network magnetic fields (Roudier et al. 2017). In stark contrast, when we discuss the Sun's corona later (Chapter 17), we shall see that in the course of a solar cycle, the radiation emitted by coronal gas at temperatures $>2.5 \times 10^6$ K is observed to vary by a factor of a few *orders of magnitude* (Takeda et al. 2019). Apparently, the coronal material is much more sensitive than the photospheric material is to the magnetic changes occurring in the course of a solar cycle.

The data in Figure 16.3 are based mainly on telescopic sightings of spots. Is there any way to extend the study of solar cycle to times *before* the invention of the telescope? One possibility is based on the fact that the solar wind (see Chapter 18) makes it difficult for energetic charged particles ("galactic cosmic rays" or GCR) to reach the Earth. When sunspots are plentiful, the GCR flux at Earth is reduced (see Figure 18.6) by as much as 20%. When a GCR reaches Earth, it causes a "shower" of energetic particles in the atmosphere, and these can generate the radioactive nuclides $Be^{10}$ and $C^{14}$ (Beer et al. 2018). Ice cores can be drilled in Greenland to study the contents of ice that was "laid down" (in the form of snow and rain) centuries or millennia ago: the abundance of $Be^{10}$ in such cores shows variations that (in recent centuries) are found to be correlated significantly with the 11-year sunspot cycles. And the $C^{14}$ abundances can be extracted from tree ring records extending back some 10 millennia: the records indicate a periodicity of 208 years and one of 2300 years. The interference between the 11-year cycle and these longer cycles gives rise at certain times to "grand maxima" and "grand minima" of solar activity on time-scales of a century or more. The most famous "grand minimum" occurred in the years 1650–1720 when sunspots were observed to be very rare in the Sun: this period is called the "Maunder Minimum", during which Europe experienced a "little ice age". Interestingly, the $Be^{10}$ record demonstrates that although spots were rare in the years 1650–1720, nevertheless, the access of GCR to Earth continued to be blocked every 11 years, indicating that the Sun's magnetic fields were still undergoing an 11-year cycle. It is possible that the large solar maximum in 1957 (see Figure 16.3) might correspond to one of the "grand maxima". The reader might find it interesting to speculate whether, based on the observed weakness of solar cycle 24 (see Figure 18.4), the Sun might be heading into another grand minimum (and another "little ice age") in the 21st or 22nd century.

### 16.1.5 Spot Lifetimes

How long do spots live? Small spots may appear and disappear in a matter of hours. Larger spots require days or weeks to reach maximum size, and days or weeks to decay. A formula that has been used to estimate spot lifetime in terms of area is the following: the lifetime $T$ (in units of days) is (roughly) proportional to the maximum area of the spot $A$ (in units of MSH), with a proportionality constant of 0.1, i.e., $T(\text{days}) \approx 0.1A(\text{MSH})$ (Bray and Loughhead 1979, p. 229).

In her study of sunspot groups recorded in Greenwich by photography during the years 1874–1906, Annie Maunder (1909) reported that among 624 spot groups that were seen in at least two solar rotations during those 32 years, the 12 longest lived groups that were reliably identified survived for five solar rotations.

Most spots decay by breaking up into smaller units, and these are then eroded over time by the turbulent erosion by convective cells around the periphery: this turbulence shreds the spot by tearing off pieces of magnetic field.

### 16.1.6 Energy Deficits and Excesses

Something unexpected happens to the power output from the Sun in the course of the sunspot cycle (see Figure 1.1). You might expect that when sunspots are *most* abundant, the Sun would emit *less* power. But this is *not* what is observed. Instead, exactly the opposite is observed: the solar power output is observed to have *maximum* values in or around the years 1980, 1990, and 2000, when there are *most* sunspots on the surface. This surprising discovery emerged from spacecraft data during the last decades of the 20th century: it was only in those decades that the precision of measurements of the total (bolometric) luminosity of the Sun, integrated over all wavelengths, became at least as good as 0.1%. (In previous years, observations from the ground were plagued by uncertainties arising from Earth's atmosphere, which blocks some 2% of the luminosity.) When such precision became available in instruments that also remained stable enough over an entire 11-year cycle, the data indicated that the solar luminosity has a *maximum* value when the number of spots is largest. The excess power output at solar maximum compared to solar minimum is of order 0.1% (see Figure 1.1).

This is counterintuitive: when there are lots of spots, each dark umbra emits less power than the undisturbed photosphere, and therefore, one would expect the solar output to be a minimum.

However, sunspots are not the only contributors to perturbations in the solar energy output. Careful photometry of the photosphere *in the vicinity of sunspots* reveals the presence of multiple small features that are slightly *brighter* (by at most a few percent) than the undisturbed photosphere when viewed in white light. These bright point-like features (called "faculae") are much less obvious to the human eye than sunspots. In fact, even with telescopes, faculae are almost impossible to pick out near the center of the solar disk: the easiest place to observe them is in the vicinity of sunspots as the latter approach the limb (see Figure 16.4).

The excess of facular flux above the photospheric value helps to offset *some* of the flux deficit of a large sunspot in the vicinity. It has been reported (Hempelmann and Weber 2012) that facular emission actually *over*compensates for the spot deficit when the sunspot number is not too large ($R_Z = 100 \pm 100$), but that spot deficits "win out" at larger values of $R_Z$. This might explain why solar power output is slightly larger (by 0.1%) at solar maxima, when "typical" values of $R_Z$ lie in the range $100 \pm 100$.

In this regard, it is relevant to note that in certain stars which have Sun-like spectra, Radick et al. (1998) and Shapiro et al. (2014) have reported that some stars exhibit variability that is dominated by faculae, while in other stars, the variability is dominated by starspots. The transition between facular-dominated and spot-dominated variability is found to occur at a certain level of magnetic activity: variability in stars with lower activity levels (such as the Sun) are facular dominated, while in stars with higher activity levels, the variability is spot dominated.

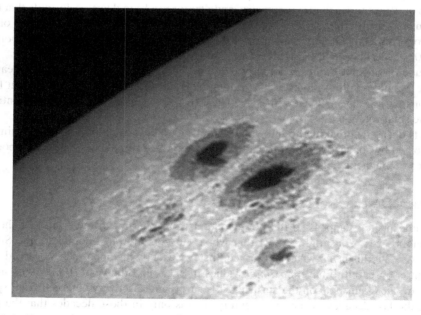

**FIGURE 16.4** Faculae in a white-light image of sunspots near the limb of the Sun. Faculae, located in the vicinity of sunspots, appear slightly *brighter* than the undisturbed photosphere. (Photo by Damian Peach; used with permission.)

Magnetic Fields in the Sun 269

## 16.2 CHROMOSPHERIC EMISSION

Another departure from spherical symmetry in the solar atmosphere, which we came across in the preceding chapter, appears when we observe the chromosphere. When the Sun is viewed in the core of a chromospheric line (such as the Ca II $K$ line, or $H\alpha$: see Figure 3.7), the Sun is not spherically symmetric. Quantitatively, as we have seen, the differences in brightness between network (along the edges of supergranules) and cell centers (in the central locations of supergranules) suggest that mechanical energy is being deposited in the network at a rate that exceeds the rate in the cell centers by a significant factor, possibly by as much as an order of magnitude (see Equation 15.12). The excess brightness in the network is known to be correlated with locally stronger magnetic fields. Why should that be so? Because vertical fields that happened (at an earlier time) to be situated near the cell center are swept towards the edge of the supergranule as time goes on due to horizontal flows from the center of a supergranule towards its edge (see Section 6.12). We may infer that the enhanced brightness of network (relative to the cell center) is associated with the presence of locally stronger magnetic fields (see Section 16.7.6) that have been swept by horizontal flows into the network. Each mottle (spicule) seen in $H\alpha$ along the edge of a supergranule (i.e., in the network) (see Section 15.4) owes its existence to a local patch of enhanced vertical magnetic field.

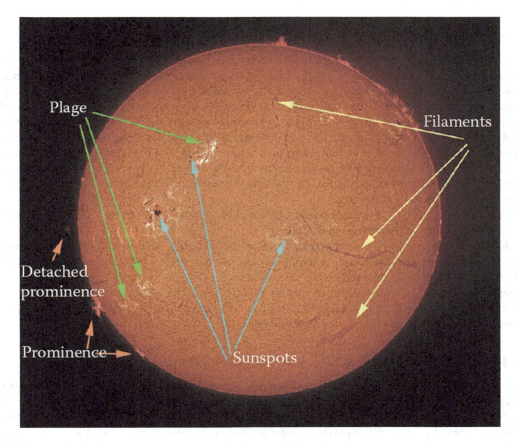

**FIGURE 16.5** The chromosphere: departures from spherical symmetry may include sunspots, plage, filaments, and prominences. (Copyright Peter Ward/Advanced Telescope Supplies; used with permission.)

Also, when we observe the chromosphere, we find that there are features that have a close connection with sunspots in the photosphere (see Figure 16.5). In the chromosphere, the locations of sunspots (as determined from photospheric images) are found to be surrounded by regions of enhanced emission ("plages": see Chapter 15, Figure 15.4). Thus, sunspots are in fact only one component of a more extended physical structure, a "plage", in which chromospheric emission is enhanced. The (white-light) faculae are co-located with plage. The overall feature, including sunspots, plages, and faculae, is called an "active region".

Active regions do not appear randomly at all locations of the solar surface: from centuries of observation, it has been found that there are more favored areas (near the equator, at latitudes of no more than ±35 degrees) and less favored (or even forbidden) areas (near the poles, at latitudes in excess of ±35 degrees).

Also present in chromospheric lines are features called "prominences". These are structures that were first observed in emission standing above the limb of the Sun: they consist of material that appears to be suspended "in midair". We shall see (in Section 16.7.4) that magnetic fields play a key role in suspending the prominence material.

Prominences can also be viewed on the disk of the Sun when the latter is observed in $H\alpha$: in such cases, they appear as dark more-or-less ribbon-shaped features ("filaments": see Figure 16.6) located preferentially at positions where the surface magnetic fields change polarity. The filaments can be "quiescent", i.e., stationary for hours or days, but they can also, at the end of their lifetime, reveal rapid evolution as the prominence material "erupts" rapidly, either because of heating or expulsion of the material into the upper atmosphere.

## 16.3 MAGNETIC FIELDS: THE SOURCE OF SOLAR ACTIVITY

Why does the Sun depart from spherical symmetry? The answer is: because of the presence of magnetic fields. These fields give rise to a variety of observational phenomena. The umbrella term "magnetic activity" is used to cover the magnetically driven phenomena that are such a striking characteristic of the Sun at times. Under the term "magnetic activity", we include the presence of sunspots, faculae, the chromospheric network, and prominences, all of which are more-or-less long-lived phenomena that can be regarded as quasi-stationary in nature. On the other hand, "magnetic activity" also includes phenomena that are by no means stationary, such as flares and coronal mass ejections (CMEs). Both of the latter involve highly time-dependent processes that disturb the solar atmosphere in striking ways, giving rise (at times) to "fireworks displays" involving the most energetic phenomena in the solar system.

Our aim in this chapter is to describe, in terms of physical processes, how magnetic effects give rise to a rich variety of phenomena in the Sun.

Before discussing the general properties of magnetic fields and their interactions with plasma, and in order to keep the discussion rooted in the Sun, we start with what the observations tell us about the magnetic fields themselves. We need first to understand how astronomers measure the strength of the fields in solar features of various kinds. Once we have a feel for the orders of magnitude of the field strengths in various features, we will turn to the physics to determine which processes are most relevant in the various phenomena.

## 16.4 MEASUREMENTS OF SOLAR MAGNETIC FIELDS

There are three observational approaches to measuring solar fields. One involves remote sensing using optical photons (Section 16.4.1): we observe certain photons coming from a certain feature on the surface of the Sun, examine the spectral and polarimetric properties of the photons, and infer the strength and direction of the field on the solar surface in the feature under observation. A second approach uses remote sensing using radio photons: polarization data again contains information on the strength and direction of local magnetic fields: these data give information about field strengths

Magnetic Fields in the Sun

in "magnetic loops" in the solar corona. A third approach involves direct measurements of the field *in situ* in the plasma that streams out of the Sun (the "solar wind": see Chapter 18) into interplanetary space, and then extrapolate back (almost all the way to the surface of the Sun) to infer the magnetic field strengths near the surface of the Sun.

Let us consider the three observational approaches to measuring fields in the Sun.

### 16.4.1 MEASUREMENT OF MAGNETIC FIELDS ON THE SUN: OPTICAL DATA

To measure solar magnetic fields, we use techniques that seek to identify changes in the shape of a spectral line when a field is present. Two different physical effects have been used by solar astronomers to measure the magnetic fields in the Sun. One is the Zeeman effect, which results in shifts to atomic energy levels in the presence of a field: these shifts are amenable to a simple physical interpretation and will be described later. The second effect is the Hanle effect, which leads to changes in how polarized radiation is scattered in the presence of a field. The Hanle effect in the Sun is sensitive to very weak fields, in which the Zeeman splitting is comparable to the natural width of the spectral line: this occurs in fields with strengths ranging from a few G to some tens of G. If future polarimeters can achieve sensitivities that are good enough to detect degrees of polarization as small as a few parts in $10^6$, it may be possible to measure the global field of the Sun (i.e., 6–12 G: see Section 16.4.6) using the Hanle effect (Vieu et al. 2016). We are interested mainly in fields in active regions on the Sun: in such areas, the regions with the smallest fields (referred to as "dead calm" areas: see Section 16.4.1.5) have field strengths of 10–20 G. Even in these dead calm regions, the fields are already for the most part strong enough for the Zeeman effect to dominate the Hanle effect. As a result, we will not discuss the Hanle effect further in this "first course".

#### 16.4.1.1 Zeeman Splitting

How is a spectral line altered in the presence of a magnetic field? To answer this, let us recall what happens in the *absence* of the field. Each spectral line involves the transition of an electron from one atomic energy level $E_1$ to another level $E_2$. In the absence of external magnetic fields, the energy levels are determined by atomic structure. Radiation from an atom is spherically symmetric: there is no preferred direction in the problem. When the atom is observed from any direction, what is observed is a single line with frequency $v_o = (E_2 - E_1)/h$, i.e., a single line with a wavelength $\lambda_o = c/v_o$.

Now introduce an external magnetic field. Two aspects of the situation change. First, the energies of the atomic levels are altered: this will cause the lines to shift in wavelength. Second, the photons that emerge have properties that are no longer spherically symmetric: observers who are situated in different viewing positions see different spectra.

To understand how magnetic processes affect atoms, we first refer to a basic result of magnetostatics: what happens when one places an object with a magnetic moment in a field? Recall that when iron filings are sprinkled on paper near a bar magnet, a clear pattern is seen: each iron filing, which is a small magnet in itself with its own magnetic moment, aligns itself with the local magnetic field lines. Now, every electron has an intrinsic "spin" (with angular momentum $\hbar/2$), and associated with this spin is a magnetic moment that is referred to as the Bohr magneton: $\mu_B = e\hbar/2m_ec$. In a magnetic field $B$, an electron can settle into one of two states: one with $\mu_B$ *parallel* to the external field and another with $\mu_B$ *antiparallel* to the external field. In one of these states, the electron *gains* an energy $+\mu_B B$, while in the other state, the electron energy is *reduced* by $\mu_B B$. Thus, an electron which initially was in an atomic level with a particular energy $E_1$ now finds that the level "splits" into two levels, with energies $E_1 + \mu_B B$ and $E_1 - \mu_B B$.

What will we observe if we detect the photons that emerge from the aforementioned atom? The answer depends on the direction from which we mark the observation. Suppose we choose to make the observations parallel or antiparallel to the external magnetic field: that is, we choose to "look straight along the field". Let us also suppose for simplicity that the energy level $E_2$ has the property that the level does not undergo any splitting in a magnetic field (atomic levels with this property do exist). In that case, what we see is the following: the original single line at a frequency $v_o$, i.e., at wavelength $\lambda_o = c/v_o$, is now seen to consist of two lines (a "doublet"), at frequencies $v_o \pm \Delta v$, where $\Delta v = \mu_B B/h$.

This conversion of a single line into a doublet (when you observe parallel or antiparallel to the field lines) as a result of a magnetic field is called Zeeman splitting, after the discoverer (see Figure 16.6).

The wavelengths of the two components of the doublet are $\lambda_o + \Delta\lambda$ and $\lambda_o - \Delta\lambda$, where $\Delta\lambda = \lambda_o^2 \Delta v / c$. Inserting the value for $\mu_B$, we find that in the presence of a magnetic field, the wavelength shift is $\Delta\lambda = \text{const} \times B\lambda_o^2$ where the constant, equal to $e/4\pi m_e c^2$, has the numerical value of $4.9 \times 10^{-5}$ cm Gauss$^{-1}$ cm$^{-2}$. For convenience, if we express wavelengths in units of Å (= $10^{-8}$ cm), we find

$$\Delta\lambda = 4.9 \times 10^{-13} B(G) \lambda_o^2 \qquad (16.1)$$

Equation 16.1 is valid for the simplest case, when only the electron's magnetic moment is responding to the external magnetic field. This is referred to as the "normal" Zeeman effect.

The actual Zeeman effect in "real atoms" differs slightly from the prior formula because there are other sources of magnetic moments in addition to the intrinsic magnetic moment of the electron itself. For example, when an electron orbit has a finite angular momentum, that orbit may also have an associated magnetic moment: the vector describing the magnetic moment points in a direction which is perpendicular to the plane of the orbit, and has a magnitude proportional to the angular momentum of the electron in its orbit. The process by which an external field interacts with an orbital magnetic moment is quantitatively different from the process by which the external field

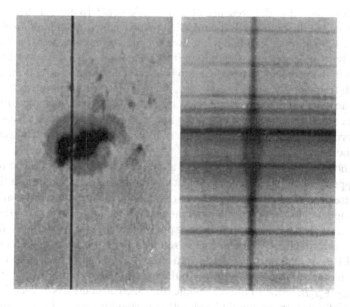

**FIGURE 16.6** Zeeman splitting of spectral lines in a magnetic region on the Sun. The vertical slit of the instrument is located as shown in the left-hand image: the slit overlaps (about halfway down the image) with the umbra of a sunspot. On the right-hand side, a (vertical) spectral line, which is single in the undisturbed Sun (at top and bottom), becomes multiple when the slit overlaps the umbra, where magnetic fields are strong. (Image downloaded from the website of High Altitude Observatory, a division of the National Center for Atmospheric Research, funded by the National Science Foundation.)

# Magnetic Fields in the Sun

interacts with the electron spin. As a result, the prior expression is only part of the story of the wavelength shift for any given transition. Each "real" transition has a factor $g_L$ associated with it (the so-called Landé g-factor), and the right-hand side of the prior expression must be multiplied by $g_L$. For transitions of various kinds, the numerical value of $g_L$ may be as small as zero (if orbital and spin magnetic effects cancel) or as large as (roughly) three. Moreover, depending on the atomic structure, rather than splitting into two components, a line may split into multiple components. When multiple components are present, this is referred to as the "anomalous" Zeeman effect.

To give a numerical example, consider an atom that has a line in the visible part of the spectrum, at (say) $\lambda_o = 5000$ Å. In a field of 3000 G (typical for a sunspot umbra), and assuming $g_L \approx 1$, we find that $\Delta\lambda \approx 0.037$ Å. Therefore, in order to detect a clean splitting of the lines in a sunspot, observers are required to use instruments with a resolving power (defined by the ratio of $\lambda_o/\Delta\lambda$) of more than 100,000. Achieving such a resolving power requires careful attention to instrumental design. However, in solar physics, such resolving powers are available: in fact, we have already cited (see Figure 16.2) results from the Sun that were obtained by a spectrometer with a resolving power of >700,000.

Note that in Equation 16.1, the Zeeman splitting increases with the *square* of the wavelength: as a result, one can improve the chances of observing Zeeman splittings by observing in the infrared, where wavelengths are longer than those in visible light. As an extreme example, we may cite the use of a pair of Mg I emission lines at wavelengths near 12 μm (Hong et al. 2020). We shall refer to observations that take advantage of infrared lines in Section 16.4.2.

### 16.4.1.2 Zeeman Polarization: The Longitudinal Case

The splitting of a single line into two (and only two) components occurs when we make observations *along* the magnetic field. This is referred to as *"longitudinal* Zeeman splitting".

Observations show that the two lines of a doublet are not merely different in their wavelength; they also differ in polarization: the two lines are circularly polarized in opposite senses. To see why this is so, consider a spectral line in which the upper level is not affected by the field (i.e., its Landé g-factor = 0), but the lower level undergoes normal Zeeman splitting. In the lower level, the angular momentum (spin) of the electron has a component along the field of $\pm\hbar/2$, depending on whether the electron spin is in the "up" or "down" position relative to the field.

To understand how a passing photon interacts with an electron in the split level, we note that the angular momentum (spin) of a circularly polarized photon is $\pm\hbar$, depending on whether the photon has right- or left-hand polarization. Consider an electron that is sitting in the "down" position, with spin $-\hbar/2$. If a photon with left-hand polarization (i.e., spin $-\hbar$) passes by that electron, the electron cannot interact with the photon, because if the interaction occurred, then the electron would have to absorb the photon's spin, add it to its own, and enter a state with spin of $-3\hbar/2$. Such a state is not available to the electron: the only available states have spins of $\pm\hbar/2$. As a result, the electron simply ignores the left-hand polarized photon, and the photon passes through unperturbed. But now consider the case where the passing photon is right-hand polarized, i.e., the photon has spin $+\hbar$. Now, the electron *can* interact with the photon, adding the photon's spin $+\hbar$ to its own ($-\hbar/2$), and ending up with spin $+\hbar/2$. Such a spin *is* allowed: the electron simply transitions to the "up" position. Thus, an electron in the "down" position preferentially absorbs right-hand polarized photons out of the beam. The remaining photons, i.e., the left-hand circularly polarized photons, pass through and reach the observer: the observer therefore sees left-hand circular polarization as the dominant component of the absorption feature.

How do we know that circularly polarized photons interact with electrons in this way? Because experimental confirmation is available in the laboratory. Specifically, in a thin sheet of iron, the magnetization can be chosen so that the elementary magnets in the iron all tend to be aligned in one particular direction. Then if a circularly polarized photon passes through the sheet, the photon will be scattered preferentially if its polarization has the correct sign to flip an aligned magnet. Of course, such a test does not work with *optical* photons: the iron sheet prevents them from passing through. But if one uses high-energy photons (gamma rays), then these *can* pass through the iron sheet, and the aligned magnets can be flipped. In fact, this property of photon-electron interactions played a role in establishing the existence of a physical phenomenon known as parity violation in weak interactions: a key experiment was performed by Goldhaber et al. (1958).

Returning now to our Zeeman doublet, we recall that a "down" electron has a specific energy shift, depending on the direction of the field. Let us consider a case in which the field points *toward* the observer. Recall that the magnetic moment of the electron is proportional to the electron spin, and the proportionality factor depends on the (negative) charge of the electron. As a result, when the electron spin is sitting in the "down" position, the magnetic moment is sitting "up" relative to the field. Therefore, the electron has an excess energy $+\mu_B B$ relative to the unperturbed state. This means that the lower energy level is no longer the unperturbed value $E_1$, but a larger value: $E_1 + \mu_B B$. The upper energy level is (by assumption) still at the unperturbed value $E_2$. As a result, the frequency of the transition is no longer equal to $(E_2 - E_1)/h$, but takes on a lower value. A lower value for the frequency means a longer wavelength for the photon: therefore, we are discussing the *redward*-shifted component of the Zeeman doublet, at wavelength $\lambda_o + \Delta\lambda$.

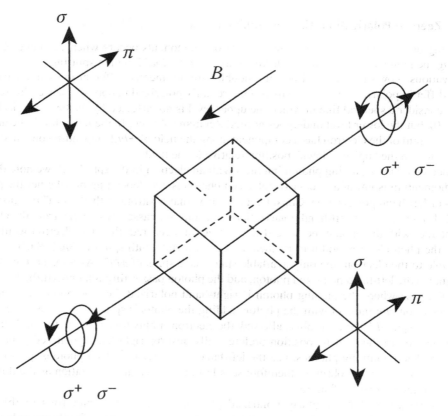

**FIGURE 16.7** When observations are made either parallel to, or antiparallel to, the direction of the magnetic field (vector $B$), Zeeman splitting leads to components which are *circularly* polarized. When observations are made perpendicular to the field, the split components are *linearly* polarized. (From PHYWE Systeme GmbH and Co, Gottingen, Germany; used with permission.)

We conclude that when the field in a certain location of the Sun is pointing *toward the observer* and a Zeeman line in absorption in observed, the *redward* component will be observed to be circularly polarized in the left-handed sense. Analogous arguments can be applied to the "up" electron, showing that the *blueward* component, when seen in absorption, will be right-hand polarized.

Conversely, if we observe a region where the solar field is pointing *away from the observer*, the blueward component of the Zeeman doublet will be left-hand polarized.

Thus, measurements of the polarization of a spectral line provides a powerful diagnostic of the *direction* of magnetic fields in the Sun. If I point my telescope at a particular umbra on the Sun, observe a Zeeman doublet in absorption, and find that the blueward component, at wavelength $\lambda_o - \Delta\lambda$, is (say) right-hand polarized, then I know that the field lines are pointing toward me. In order to point towards me (on Earth), such field lines are directed outward from the Sun. And I also know how strong the field is along the line of sight, by measuring the shift $\Delta\lambda$ and inserting this into Equation 16.1.

### 16.4.1.3 Zeeman Polarization: The Transverse Case

Magnetic fields in the solar atmosphere are complicated in spatial structure. As a result, although there are certainly possibilities for observing "straight down the field" in favorable situations, this is not always the case. In other locations, the line of sight will turn out to be perpendicular to the field lines.

If I observe the photons that propagate in directions perpendicular to the external magnetic field, the results are different from the longitudinal case (see Figure 16.7). In the perpendicular case, known as the "transverse Zeeman effect", we observe not two, but three components. Two components are still found at wavelengths $\lambda_o + \Delta\lambda$ and $\lambda_o - \Delta\lambda$ (as before), but now there is also a third component at the undisplaced wavelength $\lambda_o$. In the case of the transverse Zeeman effect, the polarizations of the components are observed to be *linear* (rather than circular): the undisplaced line is linearly polarized *parallel* to the field (this line is referred as the π component), while the two shifted components are polarized *perpendicular* to the field (these lines are referred to as the σ components, where Greek sigma is assigned because the word perpendicular in German "*senkrecht*" starts with the letter *s*).

Because of the polarization properties, by (i) measuring the splitting $\Delta\lambda$, (ii) counting components of the split line, and (iii) measuring polarizations, we can determine the strength of the field and also the angle of the magnetic field relative to our line of sight. This is a powerful diagnostic for the properties of magnetic fields, especially in the Sun where the fields in an active region can be very complicated, with many changes from one location to another.

### 16.4.1.4 Babcock Magnetograph: Longitudinal Fields

The polarization properties of the Zeeman effect provide a practical technique for measuring weak solar magnetic fields. No longer do we have to build an instrument that can cleanly separate the components of the Zeeman doublet. Instead, solar observers (starting with Babcock 1953) use the trick of observing in one polarization at a certain wavelength that is shifted by $+\Delta\lambda$ on one side of line center, and in the opposite polarization at a wavelength which is shifted by $-\Delta\lambda$ on the other side of line center. By carefully choosing $\Delta\lambda$ so that the observations are made on the steepest part of the absorption line profile, even a slight amount of Zeeman splitting can then be detected. This allows detection of fields as weak as tens of Gauss on the Sun.

An instrument that takes advantage of the circular polarization properties of the Zeeman doublet is referred to as a Babcock magnetograph. It is useful for measuring the *longitudinal* field at all points of the solar disk. Such instruments have been in widespread use for decades for daily monitoring of solar magnetic fields.

### 16.4.1.5 Vector Magnetograph

Instruments that measure both linear and circular polarization are called "vector magnetographs". These instruments measure the four "Stokes parameters" ($I$, $Q$, $U$, $V$) that determine the shape of a spectral line in the presence of a magnetic field (see Section 3.8.4). The parameter $I$ measures the total strength of the line; $V$ measures the circular polarization of the line, which is controlled by the component $B_{LOS}$ of the magnetic field strength lying *along* the line of sight (*LOS*); $Q$ and $U$ measure the linear polarization of the line, and these are controlled by the component $B_T$ of the magnetic field lying *transverse* to the line of sight. It can be shown (Asensio Ramos and Martinez Gonzalez 2014) that, in observational terms, $V$ depends on the *first* derivative of $I$ as a function of wavelength $\lambda$: as a result, the value of $V$ is *linearly* proportional to the value of $B_{LOS}$. However, $Q$ and $U$ depend on the *second* derivative of $I$ as a function of wavelength $\lambda$: as a result, the values of $Q$ and $U$ are proportional to the *square* of $B_T$. As a result, in the presence of noisy data, the value of $B_{LOS}$ can be determined more reliably than the value of $B_T$. Thus, in data obtained with the SDO/HMI vector magnetograph, the noise in the value determined for $B_{LOS}$ is about 10 G, but the noise in the $B_T$ value is about 100 G (Virtanen et al. 2019).

Early in 2010, observers at the National Solar Observatory in Arizona started to make daily observations of the *entire solar disk* using a vector magnetograph named SOLIS (Synoptic Optical Long-term Investigations of the Sun). Results of vector fields extracted from the daily images can be "stitched" together into "synoptic maps" that allow the reader to view the results over the course of one complete Carrington rotation (CR: lasting 27.2753 days by definition: see Section 1.11). To be sure, individual active regions on the Sun had previously been observed from time to time with other vector magnetographs, but SOLIS was the first instrument to obtain *daily* measurements of the *entire* visible disk over an *entire* CR (#2092). Also in 2010, with the launch of SDO/HMI, vector magnetograph data became available from space as well. The vector data yield the three components of the magnetic field in the solar photosphere: radial, latitudinal, and longitudinal. Virtanen et al. (2019) have analyzed the results from SOLIS over the time interval 2010–2017, from CR 2092 to CR 2190, i.e., for almost 100 Carrington rotations of the Sun. By "stitching" together the maps for each CR, Virtanen et al. obtained a "super-synoptic" map extending over the entire 7–8 year period.

The Hinode spacecraft has provided vector magnetograph data to be obtained in large areas of quiet Sun, especially in the cell centers (also referred to as internetwork [IN]). The angular resolution of the measurements is very high: 0.16 arcsec. The results (Bellot Rubio and Orozco Suarez 2019) indicate that the IN magnetic field strength is dominated by fields with strengths of order 100–200 G, and these fields are found to be predominantly *horizontal*, i.e., parallel to the Sun's surface. As regards the azimuth of the horizontal fields, they are distributed randomly. On the other hand, observations of the quiet Sun indicate that stronger fields (> 600 G) are mainly vertical (Pastor Yabar et al. 2018).

In the case of the very quiet Sun, the balloon-borne instrument called SUNRISE, flying in June 2009 during a deep solar minimum, examined a weak field region at the center of the Sun's disk in search of emerging magnetic loops (Martinez Gonzalez et al. 2012: hereafter MG12). With angular resolutions comparable to that of Hinode (of order 100 km on the Sun), each newly emerging loop was identified as a bipolar pair of regions with + and − values of $V$ (representing the two footpoints of the loop, with one footpoint containing *vertical* field lines pointing towards the observer, while the second footpoint contains vertical field lines pointing away from the observer), separated by a region of *horizontal* field (with finite values of $Q$ and $U$) between the pair. This combination of vertical fields at footpoints and horizontal fields between the footpoints is what would be expected if magnetic field were in the form of magnetic loops emerging into the atmosphere from *beneath* the surface. As time goes by, the bipolar pair separates from each other spatially as the magnetic loop rises higher in the solar atmosphere. Over the course of 40–60 minutes of observations, some 500 loops were observed to emerge in the field of view: the advantage of the very low level of solar activity during the SUNRISE flight was that individual loops could be identified and

# Magnetic Fields in the Sun

followed in their evolution as they emerged into the solar atmosphere from beneath the surface. The observations showed clearly that a loop of magnetic field, which is observed to be present in the solar atmosphere at any instant $T_o$, actually emerged from beneath the surface at an earlier time $T_i < T_o$. And prior to $T_i$, no sign of that loop was present. The emergence of magnetic loops through the surface suggests that some process (probably a dynamo) is at work beneath the surface, actively generating new magnetic loops.

An amazingly detailed observational study of the way in which one particular loop emerged first in the photosphere, then moved up into the chromosphere, and finally reached the corona has been recorded using a very different observing program by Kontogiannis et al. (2020). Seven separate instruments in space and on the ground were combined to follow a quiet region of the Sun (close to solar minimum) for a 2-hour interval during which a small loop emerged. The progression of loop emergence in the data is such that one can see clearly (in their Figure 8) how the footpoints of the loop, with opposite magnetic polarity, separate monotonically from each other in the photosphere during the course of the 2-hour sequence of observations: initially, the footpoints were very close to each other (no more than 2–3 arcsec apart), but after 2 hours, they are separated by almost 10 arcsec. And when the loop reaches the corona, the authors derive the differential emission measure (DEM: see Section 17.5) during two intervals of time, one at early times in the event, and the second some 90 minutes later: in the first interval, the DEM peaks at $\log T = 6.0$, while in the second interval, the DEM peaks at $\log T = 6.1$. These peak temperatures are entirely appropriate for the "cooler" component of the corona ($\log T \leq 6.1$), which is known to persist more or less unchanged at all phases of the solar cycle (see Figure 17.9 and Section 17.6).

Significantly, in their study of SUNRISE data, MG12 found that loops do *not* emerge in all locations equally. There were certain locations where (during certain observing sessions) loops preferred to emerge, while in other areas, loops emerged only rarely. Specifically, MG12 identified regions with areas of order 50 Mm² (equal to the area of about 50 granules) in which *no* loops emerged during their 40–60 minutes of observations. MG12 referred to these areas as "dead calm": such regions are good candidates for the weakest field regions on the surface of the Sun, with local fields of perhaps no more than 10–20 G.

## 16.4.2 Magnetic Field Strengths in Sunspot Umbrae

Because the behavior of gas in sunspots is controlled by magnetic fields, our understanding of the physics of sunspots will be more precise the better we can measure the strength of the magnetic fields that exist in the spot. The strongest fields are observed to occur in the umbra of a spot, and those umbral fields are observed to be more or less vertical, i.e., perpendicular to the surface of the Sun.

Space-based magnetic field data from SOHO/MDI (Watson et al. 2011) were obtained for cycle 23 (1996–2010). Using an automated search technique, more than 30,000 spots were identified in the data (including repeated observations of spots that lived for more than one day). The SOHO/MDI measures the line-of-sight ("longitudinal") magnetic field on the Sun (in contrast to SDO/HMI that measures vector fields). Magnetic field strengths in the umbrae of the spots in the Watson et al. sample are shown in Figure 16.8. Watson et al. state, "the majority of measurements fall between 1500 and 3500 Gauss". Measurements of sunspot magnetic fields extending partially into solar cycle 24 have subsequently been reported by Watson et al. (2014) using data from SOHO/MDI (which operated up until 2011), from SDO/HMI (starting in 2010), and from ground-based data (using Zeeman splitting in an infrared line) from National Solar Observatory. In all three data sets, it was found that the *darker* the umbra, the *stronger* the field. The ranges of field strengths in the three data sets were found to overlap well with the range reported by Watson et al. (2011) for cycle 23.

It might be wondered: is there a *lower* limit on the strength of a field in a sunspot umbra? The lower limit of 1500 G mentioned earlier coincides with the smallest value of field strengths in pores with diameters of 1.5–3.5 Mm (Verma and Denker 2014). Maybe such a lower limit

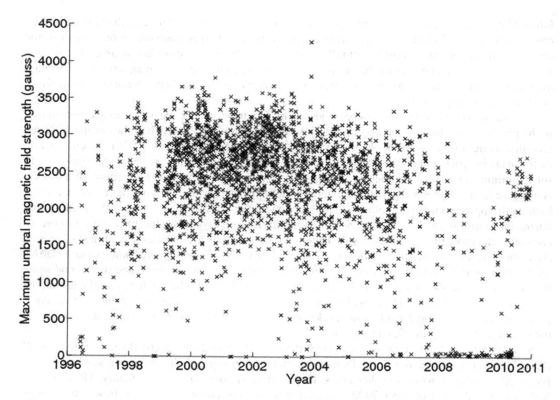

**FIGURE 16.8** Maximum magnetic field strength in the umbrae of sunspots during the years 1996–2010, i.e., during solar cycle 23. The measurements were made by the MDI instrument on board SOHO (Watson et al. 2011; used with permission of ESO).

does exist in sunspot umbrae, but if it does, it is not immediately obvious from the results in Figure 16.8, in part because SOHO/MDI measures only the component of the magnetic field *along the line of sight*.

In the opposite limit, we may also ask: is there an *upper* limit on umbral field strength? In a survey of 32,000 sunspot groups observed at four different ground-based observatories between 1917 and 2004, Livingston (2006) found fields in excess of 4 kG in 55 groups, and fields in excess of 5 kG in five groups: one single example (out of 32,000) was found of a spot with a record field of 6.1 kG. Livingston concluded that for fields in excess of 3 kG, the frequency of spots with field strength $B$ falls off roughly as $B^{-9}$. The strongest fields reported so far in sunspots include data obtained with the vector magnetograph on the Hinode satellite: Okamoto and Sakurai (2018) detected a field of 6.25 kG, and Castellanos Duran et al. (2020) detected a field of 8.1 kG.

Another feature of the umbral field strengths in Figure 16.8 is that the fields tend to be stronger at solar maximum (in the years 2001–2002) than at solar minima (1996, 2009).

We note that, as reported by LB18, fields as large as 2.8 kG in umbrae have an important physical property: they are capable of reducing the convection speeds in umbrae to essentially zero. As can be seen by visual inspection of Figure 16.8, fields of 2.8 kG were among the most frequently occurring in cycle 23. Therefore, small (or zero) convection speeds are expected to be common in the umbrae of spots during cycle 23. For umbrae with fields >3 kG, this conclusion would be even stronger.

# Magnetic Fields in the Sun

## 16.4.3 Orderly Properties of Sunspot Fields

Results from Babcock magnetographs reveal a high degree of order in the fields on the Sun. We have already mentioned that spots typically appear in pairs, a leader and a follower. When a magnetograph is applied to each spot, a highly ordered behavior emerges. At any given time, almost all leader spots in (say) the Sun's northern hemisphere are observed to have the same magnetic polarity. That is, leader spots in the northern hemisphere are found mainly to have fields which (say) point *toward* the observer. At the same time, almost all follower spots in the northern hemisphere will exhibit fields that point *away from* the observer. And simultaneously, in the southern hemisphere, the situation will be precisely reversed: the fields in leader spots will point mainly *away from* the observer. When the same observations are repeated 11 years later, the fields are found to be reversed: leader spots in the northern hemisphere will now be found mainly to have fields pointing *away from* the observer. Thus, the true solar magnetic cycle is actually (roughly) 22 years long.

These rules are referred to as Hale's polarity law, from the observer who discovered the effect in the early 1900s. In the earlier sentences, we have been careful to include the modifiers "almost" and "mainly", because Hale's law is not absolute: in the 30-year span 1/1/1989–12/31/2018, Zhukova et al. (2020) examined a total of 8606 bipolar active regions observed by SOHO, SDO, and several ground-based stations and found that Hale's law was violated in only ≈3.0% of cases. Nevertheless, in the highly turbulent medium that is the Sun's convective envelope, it is amazing to find that *any* regularity in behavior can withstand the constant pummeling by the ever-present eddies: in fact, the Sun manages to have 97% of its spot groups obey Hale's law.

## 16.4.4 Remote Sensing of Solar Magnetic Fields: Radio Observations

The Sun's corona (Chapter 17) emits radiation over a broad band of radio wavelengths. The source of the radio emission depends on the local conditions and on the frequency. At frequencies in the microwave band, between (roughly) 1 and 20 GHz, two principal emission mechanisms contribute to the radio flux. One is a free-free process, where electrons are accelerated when they pass close to ions in the coronal plasma (see Section 3.3.1). The second has to do with electrons "gyrating" in a magnetic field: circular motion involves acceleration of a centrifugal nature, and when a charged particle accelerates, this acceleration leads to the emission of electromagnetic radiation. This "gyro-emission" has a preference to be emitted at certain frequencies, namely, at the "gyrofrequency" $v_B$ (see Section 16.6.1) and its harmonics.

As we shall see (Section 16.6.1), the value of $v_B$ depends only on the field strength in the plasma emitting the radiation. Therefore, if $v_B$ can be derived from observations, we can determine the field strength in the coronal plasma.

When radio data are obtained, polarization once again (just as in the case of optical photons; see Sections 16.4.1.2 and 16.4.1.3) plays an important role in determining magnetic properties of the emitting regions on the Sun. In some active regions, the free-free emission is observed to be circularly polarized. That is, when the radio flux is measured at a certain frequency $v$ in right-hand polarization $F_R$, this flux differs from the flux at the same frequency in left-hand polarization $F_L$. The degree of polarization $d_p = (F_R - F_L)/(F_R + F_L)$ is observed to have values that may be as large as tens of percent. It can be shown theoretically (e.g., Lee 2007) that $d_p$ is simply proportional to the ratio of $v_B/v$. Therefore, a measurement of $d_p$ at frequency $v$ can be converted to $v_B \approx d_p v$, and thence to the field strength $B$ in the coronal emitting region.

Values of $B$ in the corona span a wide range: in a survey of coronal loops in 10 different active regions, Schmelz et al. (1994) reported $B$ ranging from as low as 30 G to almost 600 G. Over certain sunspots, the coronal field strength has been reported to be as large as 1800–2000 G (Lee 2007).

### 16.4.5 How Are Coronal Fields Related to Fields in the Photosphere?

The optical approach to measuring magnetic fields in the Sun typically provides information about the fields lying in (or near) the *photosphere*. The radio approach typically provides information about fields in the plasma *above* the solar surface, mainly in the corona (see Chapter 17). In order to understand the connection between magnetic results at different levels of the solar atmosphere, astronomers typically use computing approaches such that, given information on the field in the photosphere, one can use the physical properties of magnetic fields to calculate what the fields should be like at increasing altitudes upward into the corona.

The simplest computing approach to this problem is to assume that the field that is computed up in the corona is a "potential field" $B_{pot}$, i.e., one that can be defined as the gradient of a scalar quantity. With the latter definition, $\nabla \times B_{pot}$ is identically zero, and hence, in the potential field (PF), no currents flow in the corona. However, in order to make the coronal model more realistic and allow solar wind to escape, an extra refinement is to add a "source surface" (SS) at a certain radial distance $r_{SS}$ such that, at $r > r_{SS}$, the field lines are assumed to be strictly radial. The combination of potential field and source surface is referred to as the PFSS model of coronal fields. Can the PFSS predictions of coronal fields be tested by observations? Yes: when stereoscopic views of coronal streamers became available by means of simultaneous observations by STEREO A and B (separated by ~50 degrees in orbital position), the PFSS closed fields were found to overlap well with the observed locations of helmet streamers (Telloni et al. 2014).

A more realistic approach is to assume that *some* currents are actually flowing in the corona, but, in order to allow tractable computations, a specific assumption is made as follows: the currents present in the corona are assumed to flow parallel/antiparallel to the field. In such a case, the Lorentz force is identically zero, and the fields $B_{ff}$ are referred to as "force-free fields". An example of the computationally intensive calculation that is needed in order to calculate a force-free field in the corona is demonstrated by Tadesse et al. (2014).

A great advantage of knowing the values of both $B_{pot}$ and $B_{ff}$ is that the energy $E_p$ of the potential field in many circumstances represents the lowest energy state of the coronal magnetic field, without any currents to be dissipated. The energy in the force-free field, $E_{ff}$, is larger than $E_p$ because $E_{ff}$ includes energy associated with the coronal currents. The *difference* in energy between $E_{ff}$ and $E_p$ represents what is called the "magnetic free energy" $E_{mfe}$ associated with currents flowing in the corona. If the corona is to release magnetic energy (as it does in flares, by dissipating some of the coronal currents), there is a maximum amount of energy that the release can have: that maximum is $E_{mfe}$. It is important to note that not *all* of $E_{mfe}$ is released in a flare: the reason is that the field is not permitted to release its energy freely. Instead, the process of field evolution is subject to a constraint due to the conservation of a physical quantity known as "magnetic helicity", which is related to twisting of the field (see Section 16.11). In a magnetized plasma where dissipation occurs, it turns out (e.g., Priest et al. 2016) that although magnetic energy can be released quite rapidly, magnetic helicity is a more "rugged" quantity that is (almost) conserved even when magnetic energy is decaying. A flare provides vivid evidence that some magnetic energy has been dissipated on a short timesale, but the fact that *some* magnetic twist survives for a longer time than magnetic energy means that not *all* of the currents in the corona are dissipated in the time-scale of the flare. As a result, the endpoint of the magnetic field after a flare has occurred is not strictly potential: some currents remain.

### 16.4.6 Direct Magnetic Measurements in Space: The Global Field of the Sun

Clearly, when we use the term "direct measurement" of magnetic fields, it is not a question of measuring the fields *in* the atmosphere of the Sun itself. Instead, we perform the measurements in interplanetary space where spacecraft instruments can be in contact with the interplanetary fields. The importance of such measurements is that magnetic fields that were at one time situated in the

Sun's atmosphere are actually transported out into interplanetary space by the expanding solar wind. (This is a special case of a phenomenon we shall discuss later [Section 16.6.2.2], namely, that fields and plasma can in certain circumstances be effectively "frozen together".) Measurements of the magnetic field strengths *in situ* in interplanetary space are the nearest we can get to "direct measurements" of solar fields. If we can measure the field strength in space, at a certain distance from the Sun (e.g., near Earth's orbit at 1 AU), we may be able to calculate how strong the fields are back at the surface of the Sun.

Measurements of the magnetic field strength in interplanetary space have been made since the 1960s, when spacecraft first escaped beyond the confines of the Earth's magnetic field and were free to sample the true interplanetary magnetic field itself. At first, when there was no clear knowledge as to how strong the IMF might be, some of the early magnetometers were so swamped by the background of magnetic fields caused by the spacecraft itself that they could not reliably identify the IMF. It was soon realized that the detectors had to be sensitive enough to measure IMF fields with strengths of order $10^{-5}$ Gauss, i.e., 1 nanotesla. (For convenience in discussing IMFs, the nanotesla [i.e., $10^{-5}$ G] is referred to for brevity as 1 gamma [1$\gamma$]. For comparison, the magnetic field at the Earth's magnetic north pole is roughly 60,000$\gamma$.) Detection of fields by spacecraft which are as weak as 1$\gamma$ is possible only if the fabricators of the spacecraft make sure that electric currents in the spacecraft itself, or materials used in the construction of the various components of the spacecraft, do not generate fields that are as large as, or larger than, the IMF.

Once the satellite makes a reliable measurement of the IMF, the goal is to extrapolate the IMF back to the Sun: to do this with confidence, one needs to make allowance for certain properties of the solar wind (Chapter 18). Allowing for these, it is found that much of the IMF emerges from the polar regions of the Sun. The fact that the north and south poles of the Sun contain magnetic fields is strongly suggested by certain images of the solar corona, especially those that are taken close to solar minimum. On August 1, 2008, an eclipse of the Sun occurred (see Figure 17.3), and on the day of the eclipse, it happened that there was not a single sunspot visible on the solar surface. (For evidence of the lack of sunspots on that day, see http://sidc.oma.be/news/105/welcome.html.) Thus, the Sun contained none of the strong fields associated with active regions and sunspots on the day of the eclipse. This gives the best opportunity to detect the weaker fields associated with the Sun as a whole. Inspection of Figure 17.3 reveals that the north and south polar caps of the Sun exhibit bright and dark streaks that are reminiscent of how iron filings line up when they are scattered on a piece of paper located near the north and south poles of a bar magnet.

Extrapolation of IMF data indicate that the radial component of the magnetic field near the north and south poles of the Sun is in the range from 6 to 12 G (Hundhausen 1977). These numbers are subject to revision if the solar wind properties have not been incorporated correctly. But to the extent that Hundhausen's estimates are valid, we see that the solar polar fields are stronger than the fields at the Earth's magnetic poles by up to one order of magnitude.

The magnetometers that have been flown on spacecraft are such that not only the *magnitude* of the IMF can be measured, but also its *direction*. Such measurements indicate that at any instant, there is a preferred direction for the field at the Sun's north pole, and simultaneously the field at the south pole has the opposite direction. The preferred directions at the two poles remain constant for about a decade. And then, at intervals of time that range from as short as (about) nine years to as long as (about) 12 years, the field directions at the solar poles reverse sign. The polar reversals do not always occur in coincidence: they may be separated by periods of months or a year. During such periods, both poles of the Sun have the same magnetic polarity. However, for 90% of the cycle, there is a clearly defined polarity for the global field of the Sun. The fact that the global polarity switches in (about) 11 years indicates that the true magnetic cycle of the Sun has a period of (about) 22 years.

The fields of 6–12 G that exist at the poles of the Sun and extend far out into interplanetary space represent the *global* magnetic field of the Sun. The fact that the field reverses sign every 9–12 years indicates that the global magnetic field of the Sun is subject to periodic behavior. This is quite different from the Earth's magnetic field, which, although not strictly constant, nevertheless retains a more-or-less constant value over time-scales of many thousands of years.

Can the polar fields of the Sun be recorded by Zeeman techniques? Fields with strengths of at most 12 G are so weak that they are close to the limit of observability for Babcock magnetographs. Moreover, fields in the polar regions of the Sun tend to be radially directed. As a result, observations from Earth see these fields as mainly transverse to the line of sight. Therefore, Babcock magnetographs, which are sensitive to the field components along the line of sight, are not ideally suited to detecting the global polar fields of the Sun. Nevertheless, estimates of mean polar fields are reported as part of the output from certain solar magnetic observatories, e.g., the Wilcox Solar Observatory (see http://wso.stanford.edu/gifs/Polar.gif). The polar fields recorded in this way are found to have strengths of at most 2–3 G. These are certainly weaker than the 6–12 G estimates mentioned earlier: perhaps the solar wind corrections should be reexamined, or perhaps the Wilcox data are averaging over polar field strengths in different ways. Whatever the source of the discrepancy, the Wilcox data confirm the space-based discovery that the polarities of the Sun's north and south poles reverse every 11 years (or so).

## 16.5 EMPIRICAL PROPERTIES OF GLOBAL AND LOCAL SOLAR MAGNETIC FIELDS

We now have information about two apparently distinct components of magnetic fields in the Sun. One is global and quite weak (6–12 G), while the others (especially in spots) are highly localized and very strong (several kG). The localized fields are stronger than the global field by factors of at least 100, and in some cases by almost 1000. In the umbra of a sunspot, the direction of the field is found to be essentially vertical, i.e., perpendicular to the solar surface. As we move from the umbra outward into the penumbra, the field lines are observed to tilt more and more toward a horizontal direction: the (nearly) horizontal penumbral fields give rise to dark and bright "striations" (or "zebra stripes") that are the hallmark of the penumbra when it is observed at high angular resolution (Figure 16.1). As a result of the high inclination of penumbral fields at an angle of $\theta$ relative to the vertical, the effective gravity acting on gas in the penumbra is reduced below the "true" value of solar gravity $g_\odot = 27{,}420$ cm s$^{-2}$ to an effective value of only $g_\odot \cos\theta$. As we have already seen (Section 13.6), this reduction in effective gravity in the penumbra *increases* the acoustic cut-off period ($P_{ac} \sim 1/g$: see Equation 13.14) from a normal value of 200 seconds (see Equation 13.15) to periods as long as 300–500 seconds. This allows waves that would normally remain trapped below the photosphere to have access to the chromosphere.

The active regions surrounding sunspots have field strengths ranging from 30 G to several hundred G (e.g., Schmelz et al. 1994). In the quiet Sun, average magnetic fields (averaged over a field of view of, say, 10 arcsec$^2$) are weaker, and those fields are not uniformly distributed: the field is highly clumped into compact flux ropes. In the center of each flux rope, the field may reach values of 1–2 kG. This leads to the graphic term "pin-cushion" that is sometimes used to describe the nature of the magnetic field structure in the quiet Sun (Parker 1974). Given a field of view of any particular instrument, the average magnetic field strength that will be measured by that instrument for that field of view depends on how many "magnetic pins" happen to lie within the field of view. Active regions are locations where there are enhanced areal densities of the "magnetic pins". Sunspot umbrae are locations where the areal density of "magnetic pins" reaches a maximum: a very efficient process sweeps in, and holds together, the vertical magnetic flux ropes that are the defining characteristic of an umbra.

Both for the global field and for the localized strong fields, there is a 22-year cycle in the direction of the magnetic fields. For several years prior to solar maximum, there is found to be a close correlation between the direction of the global field in (say) the northern hemisphere and the preferred polarity of leader spots in the northern hemisphere at the same time. These empirical results indicate that there is a close physical connection between the Sun's weak global fields and the strong fields that appear from time to time on the surface in highly localized structures.

In order to understand why such a connection exists, we need to understand how magnetic fields and the gas in the Sun act and react upon each other. To achieve this understanding, we need to take a long step back from the Sun and come down to the level of individual charged particles moving through a magnetic field. We need to spend considerable effort (in Section 16.6) on the physical processes governing the interactions of magnetic fields and charged particles before we apply those processes to specific phenomena on the Sun (Section 16.7). Investment of effort at this stage will pay off well when we return to the data pertaining to the rich variety of solar magnetic phenomena.

## 16.6 INTERACTIONS BETWEEN MAGNETIC FIELDS AND IONIZED GAS

To understand the physical process whereby magnetic fields and gas interact with each other, we need to understand the forces that magnetic fields exert on charged particles.

In a gas that is electrically neutral, such as the air that we all breathe on Earth to stay alive, magnetic fields have no significant dynamical effects. The motions of the (neutral) gas in the near-ground atmosphere on Earth (i.e., the winds) are not at all affected by the Earth's magnetic field. But things are very different in the Sun's atmosphere. It is the fact that the gas in the Sun's atmosphere is electrically charged that opens up the possibility of interesting interactions between the field and the gas. As we shall see, the interactions go both ways: in certain circumstances, the field forces the gas to behave in a certain way, while in other circumstances, the gas forces the field to behave in a certain way. Interactions with such widely different properties lead to a great variety of interesting behaviors on the Sun.

### 16.6.1 MOTION OF A SINGLE PARTICLE

When we deal with electrically charged particles, we need to recognize a fundamental distinction between the ways in which such particles respond to electric and magnetic fields. For example, a particle with electric charge $e$, when placed in an *electric* field **E** is acted on by a force $F = e\mathbf{E}$: this force is *parallel* to the electric field if the charge is positive, and this force is *antiparallel* to the electric field if the charge is negative. In response to this force, Newton's second law of motion $F = ma$ (where $m$ is the mass of the particle and $a$ is the acceleration), a charged particle, when placed in an electric field, will experience an acceleration of $eE/m$ in the direction of the field (if the charge is positive) or opposite to the direction of the field (if the charge is negative). As a result, the motion of an electric charge with a certain mass in the presence of an *electric* field is determined solely by the direction and magnitude of the field.

But when a charged particle moves in a *magnetic* field, the motion is no longer along the field. Instead, the motion is more complicated. Suppose that a charged particle moves with (vector) velocity **V** in a (vector) magnetic field **B**: such a particle experiences a motional electric field $\mathbf{E}_m = (1/c)\mathbf{V} \times \mathbf{B}$. As a result, the charged particle is subject to the Lorentz (vector) force $(e/c)\mathbf{V} \times \mathbf{B}$. Here, we use **boldface** to denote that a quantity is a vector, with magnitude and direction. The magnitude of **B** or **V** (boldface) is written as $B$ or $V$ (without bolding). The symbol "×" between two vectors denotes the cross product of the two vectors: the magnitude of the cross product $\mathbf{V} \times \mathbf{B}$ is equal to $|\mathbf{V}||\mathbf{B}| \sin \theta$, where $\theta$ is the angle between the vectors. We see that the Lorentz force acts in a direction that does not depend *only* on the direction and magnitude of the magnetic field: instead, the Lorentz force also depends on the direction and magnitude of the *velocity* vector. The Lorentz force acts in a direction perpendicular to the velocity and perpendicular also to the magnetic field.

As regards the units, if we express the charge $e$ in electrostatic units, and **B** in Gauss, then the Lorentz force is in units of dynes.

For a particle of mass $m$ and charge $e$, Newton's second law of motion says that in a magnetic field, the equation of motion is

$$m\frac{d\mathbf{V}}{dt} = \frac{e}{c}\mathbf{V} \times \mathbf{B} \tag{16.2}$$

The solutions of Equation 16.2 have certain distinct properties. First, if a charged particle moves in a direction that is parallel, or antiparallel, to the magnetic field, then the right-hand side of Equation 16.2 has a numerical value that is identically zero. This means that the magnetic field exerts *no* force on such a particle. As a result, there is no acceleration if a charged particle moves *along* the magnetic field: whatever speed the particle had to start with, it will retain that speed as long as it keeps moving parallel (or antiparallel) to the field. Thus, if a magnetic field line stretches from point A to point B (and if no other complicating factors are present), an electron or a proton can propagate freely at constant speed between A and B.

Second, in all other cases, when there is a component of particle velocity that is *perpendicular* to the field, then the Lorentz force on the particle is finite in magnitude and acts in a specific direction. The specific direction of the Lorentz force is *not* along the magnetic vector *nor* is it along the velocity vector: instead, the force is perpendicular to *both the magnetic field vector and to the velocity vector*. Suppose the field is in the +z-direction, and the particle starts to move exactly perpendicular to the magnetic field, in (say) the +y-direction. Then taking the vector product $\mathbf{V} \times \mathbf{B}$, we find that the Lorentz force initially acts in the +x-direction. If the electric charge is positive, then the particle motion will not remain in the +y-direction but will instead be forced (by the Lorentz force) to be accelerated in the x-direction. Once an x-component of velocity occurs, the Lorentz force $\mathbf{V} \times \mathbf{B}$ will develop a new component in the −y-direction. This will eventually reduce the y-velocity to zero, at which point the x-velocity will have its maximum magnitude. However, the y-velocity will not stop there. Instead, the y-component of the velocity will increase in the −y-direction, causing the x-component of the force to become negative. This will cause the x-velocity to decrease, eventually falling to zero, at which point we are back to the initial condition.

The net effect of the Lorentz force is that the particle (in the presence of a field pointing in the z-direction) moves in a *circular path in the x–y plane*. Because the electric charge enters into the Lorentz force, a positively charged particle moves along the circular path in one direction, while a negatively charged particle moves along the circular path in the opposite direction. In both cases, the Lorentz force is directed *towards the center* of the circle. It is helpful at this point to refer to the circles in Figure 16.7: although the figure was originally designed to illustrate the circular polarization of certain photons propagating parallel (or antiparallel) to the magnetic field, we can use it (in part) also as an illustration of the circular orbits in the x–y plane when a charged particle moves in the presence of a field line pointing in the +z-direction. Note that there are arrows on the circles in Figure 16.7: positively charged particles rotate around the field lines in one direction while negatively charged particles rotate around the field lines in the opposite direction.

Now that we know that charged particles describe circular paths in magnetic fields, we are ready to take an important step and obtain quantitative insight into the effects of magnetic fields in the solar atmosphere. To do this, the essential question to raise is the following: *how large (in linear measure is the circular path of a charged particle (such as a proton or electron) in the magnetic fields that exist in the solar atmosphere?* We can answer that question by balancing the forces that act on a particle moving in a circular orbit: in this case, the two forces to be balanced are the centrifugal force and the Lorentz force.

These forces can be balanced in a circular orbit of radius $r_g$ provided that

$$\frac{eVB}{c} = \frac{mV^2}{r_g} \tag{16.3}$$

Implicitly, we recall that opposite charges circulate in opposite directions: but the magnitudes of the forces are our main interest here. Solving the prior equation leads to an expression for $r_g$:

$$r_g = \frac{mcV}{eB} \qquad (16.4)$$

This formula indicates that when a charged particle describes its circular motion in a magnetic field, the particle cannot depart from the field line *in the direction perpendicular to B* by a distance of more than $r_g$. The subscript $g$ denotes that we are dealing with a length-scale that is referred to as the "gyroradius": this is the linear size of the orbit on which the particle "gyrates" around the field.

Recalling that motion of the particle parallel or antiparallel to the field is unconstrained, the true "orbit" of a charged particle in a magnetic field is a helix: the particle is free to move longitudinally (i.e., up and down the field line) at will, but the particle cannot depart from the field line in the transverse direction by more than $r_g$.

A key aspect of Equation 16.4 is that $r_g$ is proportional to the particle's speed $V$. As a result, the *time* required for a charged particle to traverse one gyration circle is determined by the ratio of orbital circumference to particle speed: $t_g = 2\pi r_g/V$. Notice that this time-scale is *independent* of the speed of the particle. The associated frequency $v_B = 1/t_g = eB/2\pi mc$, called the "gyro-frequency", depends only on $B$. If a value can be estimated for $v_B$ in any locality on the Sun (such as an active region), that can provide information about the field strength in that active region (see Section 16.4.4).

Now we come to the question *at the heart of understanding why magnetic fields in the Sun interfere with convection*: what is the numerical value of a typical $r_g$ in the solar atmosphere?

Consider a thermal proton in the photosphere, where $T = 6000$ K. The mean thermal speed of the proton is $V \approx 10^6$ cm sec$^{-1}$. Inserting proton mass and charge, we find that in a field of $B$ Gauss, $r_g$ (cm) $\approx 100/B$. Thus, in a very quiet region of the photosphere where the field has a strength of (say) 10 G (see Section 16.4.1.5), protons are constrained to gyrate no more than 10 cm away from the field line. In a region where the field is (say) 1000 G (e.g., in a sunspot), protons can move only 1 mm (!) away from the field line. Electrons are even more tightly constrained: if we consider thermal electrons, the gyroradii mentioned earlier for protons must be reduced by factors of 43 (to as small as 0.02 mm (!!) for electrons in a sunspot).

In the corona, the temperatures are larger, of order $10^6$ K (see Chapter 17). This leads to gyroradii that are about 10 times larger than in the photosphere. Proton gyroradii in the corona, in regions with $B = 10$ G, are therefore of order 100 cm.

The most striking aspect of these gyroradii is how small they are compared to *any* of the relevant length-scales in the Sun, such as the solar radius ($\approx 10^{11}$ cm), the scale height in the atmosphere ($\approx 10^7$ cm in the photosphere), or the size of a granule ($\approx 10^8$ cm). The gyroradii of thermal protons and electrons are miniscule compared to most of the relevant length-scales in the Sun. (However, when we discuss flares in Section 17.19, it is possible that in such events, some of the relevant processes of energy release may be occurring on scales comparable to the gyroradius.)

The conclusion is clear: *charged particles in the Sun cannot move very far in directions that lie perpendicular to the magnetic field*. A popular way of saying this is to state that, in the solar atmosphere, *the ionized gas is "tied" tightly (or "frozen") to the field lines.*

At first glance, one might expect that neutral gas (where particles have electric charge $e = 0$) should *not* be affected by the Lorentz force: this might be taken to mean that in the photosphere (where hydrogen is at least 99.9% neutral), the fields might have little effect on the gas. But, as it turns out, this is not the case. The presence of even a few ions and electrons gives the magnetic field "something to hold on to": and then the charged particles communicate the magnetic forces to neutrals by means of collisions (e.g., see Mullan 1971). If, in a given volume element, there is only one ion for every 1000 neutrals, then that ion has to collide eventually with 1000 neutrals in order to pass on the magnetic forces which the ion is responding to. A slight local spatial separation may open up between neutrals and ions as these 1000 collisions do their work, but the separations (referred to as "ambipolar diffusion") occur over length-scales that are small compared to the length-scales of features in the solar photosphere, such as granules. In the low chromosphere, where the number densities of ions are of order $10^{11}$ cm$^{-3}$, and neutrals have densities of order $10^{16}$ cm$^{-3}$, ion-neutral separations are expected to be no more than a few cm. Essentially, even in the photosphere, where the gas is more neutral (in an electrical sense) than anywhere else in the Sun, the gas is still effectively "frozen" to the field.

### 16.6.2 Motion of a Conducting Fluid

So far, we have considered the interaction between a magnetic field and a single charged particle. Now we move to the macroscopic case, where we consider a fluid (plasma) composed of many individual charged particles. In such a fluid, electrons and ions can move in different directions: the result is that, in the plasma, a finite current can flow. The current density $j$ is given by $e(N_i \mathbf{V}_i - N_e \mathbf{V}_e)$ where $\mathbf{V}_i$ and $\mathbf{V}_e$ are velocities of ions and electrons, and $N_i$ and $N_e$ are number densities of ions and electrons in the plasma. For example, when an electric field $\mathbf{E}$ is present pointing along the $+x$ direction, ions move with velocity $\mathbf{V}_i$ in the $+x$ direction, and electrons move with speed $\mathbf{V}_e$ in the $-x$ direction: *both* of these motions contribute to the flow of current in the $+x$ direction.

Each cubic cm of the solar atmosphere contains $N_i$ ions, each of which is acted on by a Lorentz force of $+(e/c)\mathbf{V}_i \times \mathbf{B}$. Each cubic cm of the solar atmosphere also contains $N_e$ electrons: on each electron, the Lorentz force equals $-(e/c)\mathbf{V}_e \times \mathbf{B}$. The equation of motion for 1 cm$^3$ of solar material, with total mass $\rho = N_i m_i + N_e m_e \approx N_i m_i$ and bulk velocity $\mathbf{V}$, now includes not only the terms with which we are familiar from hydrodynamics (pressure gradient and gravity), but also a term that describes the Lorentz force acting on that cubic cm:

$$\rho \frac{d\mathbf{V}}{dt} = -\nabla p - \rho g + \frac{1}{c} \mathbf{j} \times \mathbf{B} \qquad (16.5)$$

This equation describes the dynamical effects that a magnetic field exerts on the fluid. Let us look in detail at the nature of the magnetic forces: they have interesting properties that will help us to understand why the solar atmosphere contains a variety of magnetic phenomena.

#### 16.6.2.1 Magnetic Pressure and Tension

The gas pressure $p$ enters into Equation 16.5 because the gradient of $p$ exerts a well-known force on the gas. This is true even in the absence of magnetic effects, such as in the Earth's atmosphere. (Localized winds start their motion by leaving a region where gas pressure is high and moving towards a region where the pressure is lower.) On small length-scales, such as those that occur in solar granules, the force due to $\nabla p$ can be considered isotropic without serious error. However, this is not true of the Lorentz force. The term in $\mathbf{j} \times \mathbf{B}$ in Equation 16.5 can be rewritten in a way that brings out the fact that a magnetic field in fact exerts a force that is certainly *not* the same in all directions.

To see this, we use one of Maxwell's equations, namely, curl $\mathbf{B} = (4\pi/c)\mathbf{j}$, to replace $\mathbf{j}$ in the Lorentz force in Equation 16.5. Then the Lorentz force becomes $(1/4\pi)$ curl $\mathbf{B} \times \mathbf{B}$. This can be rewritten, using vector identities, as the sum of two components $\mathbf{L}_1 + \mathbf{L}_2$, where $\mathbf{L}_1 = -\nabla(B^2/8\pi)$, and $\mathbf{L}_2 = (\mathbf{B} \cdot \nabla)\mathbf{B}/4\pi$.

Comparing with Equation 16.5, $\mathbf{L}_1$ has the same form as the term $-\nabla p$. This suggests (at first sight) that the magnetic field gives rise to a pressure analogous to gas pressure. The magnitude of the "magnetic pressure" is $p_{mag} = B^2/8\pi$. (If we express $B$ in units of Gauss, $p_{mag}$ is in units of dyn cm$^{-2}$.) If $\mathbf{L}_1$ were the only term we needed to consider in the Lorentz force, we could be tempted to think that the magnetic pressure at any position might behave just like the gas pressure. But such a conclusion would be incorrect.

Because the Lorentz force also includes $L_2$. The expression for $L_2$ can also be written as the sum of two components, $L_{2a} + L_{2b}$. Let us define a unit vector ê along the direction of the magnetic field. Then $L_{2a}$ can be written as $ê(ê \cdot \nabla)B^2/8\pi$: this is a force that acts along the vector ê, i.e., *along* the field lines. As regards the magnitude of this component, the magnitude is equal and opposite to $-\nabla(p_{mag})$. As a result, $L_{2a}$ *cancels* the component of $L_1$ that lies *along the field direction*. The net effect is that although $p_{mag}$ at first sight appears to be analogous to the gas pressure, with equal pressures in all directions, this is not the complete picture of the Lorentz force. In a more complete picture, we find that the gradient of magnetic pressure *along the field direction* is actually zero. With the removal of the effects of the gradient *along* the field, it turns out that the magnetic field gradient exerts a force only in the direction *perpendicular* to the field direction. This is a striking indication of anisotropy in the presence of a magnetic field.

The component $L_{2b}$ can be written $(B^2/4\pi)(ê \cdot \nabla)ê$. By considering the unit vector and its gradient, it can be shown that $L_{2b}$ is a vector lying in a direction *perpendicular* to the field lines. The magnitude of $L_{2b}$ is equal to $B^2/4\pi R_{curv}$, where $R_{curv}$ is the radius of curvature of the field lines. The vector $L_{2b}$ points *toward the center of curvature* of the field lines. The term $B^2/4\pi$ represents a *magnetic tension* $T_m$ along the field lines.

The fact that magnetic fields give rise to both tension and pressure (although in *different* directions) should alert us to the fact that magnetic fields will have effects that may have no analogs in the simpler world of gas dynamics (such as in the atmosphere we live in on Earth). The fact that the Lorentz force is highly anisotropic is important for understanding magnetic activity in the Sun.

From a dimensional point of view, the units of pressure are equivalent to the units of energy density. Therefore, the energy density of the magnetic field is equal to $W_{mag} = B^2/8\pi$. If $B$ is in units of Gauss, $W_{mag}$ is in units of ergs cm$^{-3}$. For example, in the umbra if a sunspot, with a field strength of order 3 kG (see Figure 16.9), the magnetic energy density is of order $4 \times 10^5$ ergs cm$^{-3}$. How does that compare with the thermal energy density ($E(th) = NkT$) in 1 cm$^3$ of photosphere material? The number density of protons in the photosphere $N$ is roughly $2 \times 10^{17}$ cm$^{-3}$ (Section 5.6). Therefore, with $T \approx 6000$ K, we find $E(th) \approx 2 \times 10^5$ ergs cm$^{-3}$. The fact that a 3 kG field (such as occurs in a sunspot umbra) has an energy density greater than the energy density contained by the photospheric gas in thermal motions suggests that the magnetic field in an umbra contains enough energy to have a significant effect on the thermal properties of the photospheric plasma. This physical conclusion contributes to having gas in the umbra *cooler* than gas in the photosphere.

### 16.6.2.2 The Equations of Magnetohydrodynamics (MHD)

The equations that describe the macroscopic interaction between a moving fluid and a magnetic field are those of magnetohydrodynamics (MHD) (also referred to as "hydromagnetics"). We shall find that there exists in MHD an analog to the result (Equation 16.4) that an individual charged particle in the solar atmosphere is confined close to the field lines.

To obtain the MHD equations, we start with two of Maxwell's electrodynamic equations:

$$\nabla \bullet \mathbf{B} = 0 \tag{16.6}$$

$$\frac{\partial \mathbf{B}}{\partial t} = -c\nabla \times \mathbf{E} \tag{16.7}$$

In a resistive medium, the current **j** that flows in response to an electric field **E** is proportional to **E** (according to Ohm's law). The constant of proportionality is determined by a physical quantity known as the electrical conductivity $\sigma_e$ of the plasma: this quantity is a measure of how effective the plasma is at conducting current. When we express the charge on the electron in electrostatic units ($|e| = 4.8 \times 10^{-10}$), the conductivity $\sigma_e$ has units of sec$^{-1}$. In a fully ionized plasma with temperature $T$, the value of $\sigma_e$ in electrostatic units can be shown to have a value of order $10^7 \, T^{3/2}$ sec$^{-1}$ (Spitzer 1962).

In terms of **j**, Ohm's law can be written as follows:

$$\mathbf{j} = \sigma_e(\mathbf{E} + \mathbf{E}_m) = \sigma_e\left(\mathbf{E} + \frac{\mathbf{V} \times \mathbf{B}}{c}\right) \tag{16.8}$$

where $\mathbf{E}_m$ is the motional electric field mentioned in Section 16.6.1. In Equation 16.8, the external electric field can be written as $\mathbf{E} = \mathbf{j}/\sigma_e - (1/c)\mathbf{V} \times \mathbf{B}$. Inserting this in Equation 16.7, we find $\partial \mathbf{B}/\partial t = -c\nabla \times (\mathbf{j}/\sigma_e) + \nabla \times (\mathbf{V} \times \mathbf{B})$. Replacing **j** with $(c/4\pi)\nabla \times \mathbf{B}$, and using Equation 16.6, we finally have the following equation describing how the magnetic field evolves with time:

$$\frac{\partial \mathbf{B}}{\partial t} = \nabla \times (\mathbf{V} \times \mathbf{B}) + \eta_e \nabla^2 \mathbf{B} \tag{16.9}$$

The quantity $\eta_e$ in Equation 16.9, defined by $\eta_e = c^2/4\pi\sigma_e$, is called the magnetic diffusivity. If the magnetic diffusivity in a plasma has a nonzero value, then the medium allows the magnetic field to diffuse away on a finite time-scale. The larger $\eta_e$, the faster the field diffuses away.

Equation 16.9 describes how magnetic fields and fluid motions interact with each other. There are two distinct terms on the right-hand side of Equation 16.9 that lead to very different physical behaviors. It is important to consider the effects of these two terms separately.

First, suppose the velocity **V** of the fluid is zero. Then the first term on the right-hand side of Equation 16.9 vanishes. We are left with an equation in which the time derivative of the field is related to the second spatial derivative of the field. This is a diffusion equation: it describes how the magnetic field diffuses (i.e., *decays*) as time goes on. If the spatial properties of the field are such that significant changes in field strength occur over length-scales of $L$, we can approximate $\nabla^2$ in Equation 16.9 as to order of magnitude by $1/L^2$. Defining the time-scale $\tau_d = L^2/\eta_e$, we see that Equation 16.9 can be written $\partial \mathbf{B}(t)/\partial t = -\mathbf{B}(t)/\tau_d$. The solution of this equation is straightforward: $\mathbf{B}(t) = \mathbf{B}(0)\exp(-t/\tau_d)$, i.e., the strength of the field diffuses away on an *e*-folding time-scale $\tau_d$. That is, a magnetic field in a stationary medium (with $\mathbf{V} = 0$) does not remain constant with time, but decays on a characteristic time-scale $\tau_d$ given by

$$\tau_d = \frac{4\pi\sigma_e L^2}{c^2} \tag{16.10}$$

What does this decay time-scale signify? The energy in the electric current is dissipated by resistive effects at a rate $\mathbf{j} \cdot \mathbf{j}/\sigma_e$ such that in the time-scale $\tau_d$, resistive dissipation within 1 cm$^3$ leads to a reduction in the magnetic energy $W_{\text{mag}}$ in that cubic cm by an amount of order $B^2/8\pi$. Thus, resistive dissipation causes the field strength to decay on a time-scale of $\tau_d$. That is, the second term on the right-hand side of Equation 16.9 has to do with the *reduction* in strength of the initial field: given the fact that $\sigma_e$ in the Sun has a finite (nonzero) value, this process of magnetic dissipation is *always* occurring in the Sun to some extent.

Second, suppose that the conductivity of the medium is so large that $\eta_e \to 0$. Then the second term on the right-hand side of Equation 16.9 becomes negligible compared to the first. That is, diffusion of the field is negligible. The surviving term, which includes the velocity of the medium, can be shown to have the following property: if you choose a particular parcel of fluid that contains a magnetic field **B** within its area **A** and follow that parcel as it moves around in response to local forces, the amount of magnetic flux ($= \int \mathbf{B}d\mathbf{A}$) enclosed by that parcel of fluid *remains constant* as time goes on. That is, magnetic flux neither enters nor leaves the parcel of fluid as it moves. The phrase commonly used to describe this behavior is that the field and the fluid are "frozen together".

This is the equivalent, in the fluid limit, of the tightly bound nature of single particle motion: the limit of infinite conductivity is formally equivalent to the limit in which the radius of gyration (see Equation 16.4) is so small (compared to other lengths in the problem) that the gyroradius can be taken to be effectively zero.

Whereas the second term on the right-hand side of Equation 16.9 always has to do with the *decay* of an initial field, the first term on the right-hand side of Equation 16.9 can, in the right circumstances, lead to the opposite process, i.e., a *strengthening* of the field. In view of the universal process of field dissipation operating in the Sun, a competing strengthening process of some kind must be occurring continually inside the Sun in order to understand why *new* magnetic field loops are continually observed to be rising up through the solar surface, even when solar activity is at its quietest (see Section 16.4.1.5). The generic name "dynamo operation" is used for the process that produces new magnetic fields inside the Sun, even if all the physical details of such an operation may not yet be fully understood. Different suggestions for the possibly relevant physical processes are referred to by titles such as $\alpha\Omega$, $\alpha^2$, $\alpha^2\Omega$, and flux transport, among others (Charbonneau 2014). Exploring the details of these processes would take us far beyond the scope of the present first course.

### 16.6.2.3 Time-Scales for Magnetic Diffusion in the Sun

The electrical conductivity of the gas in the solar atmosphere is determined by the rate at which electrons (and ions) undergo collisions with the ambient medium when the electrons (and ions) attempt to carry the current. In the limit of complete ionization (e.g., in the corona, or deep below the surface), the collisions are determined by Coulomb effects. In such a case, the conductivity is given by the Spitzer formula $\sigma_e \approx 10^7 T^{3/2}$ sec$^{-1}$ (Spitzer 1962: note that we have converted Spitzer's formula from electromagnetic units [e.m.u.] to electrostatic units [e.s.u.] using the conversion factor $c^2$.) In the upper chromosphere ($T = 10^4$ K) and in the corona ($T = 10^6$ K), typical values of $\sigma_e$ are $10^{13}$ sec$^{-1}$ and $10^{16}$ sec$^{-1}$, respectively.

In the photosphere and low chromosphere, where the degree of ionization may be much less than unity, $\sigma_e$ is definitely not as large as the Spitzer value. In a partially ionized gas, $\sigma_e$ is proportional to the ratio of the number densities of electrons to neutrals. According to Bray and Loughhead (1979, Table 4.7), at optical depth $\tau = 1$ in the photosphere, and at $\tau = 1$ in the umbra of a sunspot (where the degree of ionization is lower than in the nonmagnetic photosphere), $\sigma_e$ has numerical values of order $10^{12}$ sec$^{-1}$ and $10^{11}$ sec$^{-1}$, respectively.

Knowing these realistic numerical values of $\sigma_e$, we can evaluate the time-scale for a field to decay in the Sun. Suppose we consider one of the smallest identifiable magnetic units on the Sun: a pore. The characteristic length-scale is comparable to granule diameters, i.e., $L = 10^8$ cm. Using this in Equation 16.10, along with the photospheric value of $\sigma_e$ ($10^{12}$ sec$^{-1}$), we find that the pore is expected to decay (by ohmic dissipation) on a time-scale of order $\tau_d \approx 10^8$ sec, i.e., about 3 years. How does this compare with the observed lifetimes of pores? Not at all well: the latter are observed to disappear after at most a few hours, i.e., some $10^4$ times shorter than ohmic dissipation predicts. At the other extreme, consider the largest spot ever observed: with an area $A = 6300$ millionths of the visible hemisphere, the associated linear scale $L$ is about $8 \times 10^9$ cm. According to Equation 16.10, the decay time in the photosphere would be $10^{12}$ sec, i.e., some 30,000 years. However, the observed lifetime was less than one year. Once again, the observed lifetime is shorter by about $10^4$ compared to the ohmic decay time-scale.

The conclusion is that when we consider a pore (or a larger spot), resistive dissipation is not a significant contributor to the decay of the structure. As a result, the second term on the right-hand side of Equation 16.9 is very small compared to the first term: it is as if $\eta_e \to 0$, i.e., it is effectively as if there is no diffusion at all. Therefore, the magnetic field in the pore, and in other magnetic structures with length-scales as large as (or larger than) granules, can be considered as "frozen" into the gas. Physical processes of erosion due to gas motions around the outer limit of the pore (spot) are what alter the pore (spot) so that it survives for only a "short" time.

One of the interesting features about MHD in the context of solar physics is that, although field and gas are "frozen" together, sometimes the field dominates the gas, and at other times the gas dominates the field. Which of the two is dominant in any given situation depends on the relative energy densities. In both cases, however, the field and gas are effectively frozen together.

## 16.7 UNDERSTANDING MAGNETIC STRUCTURES IN THE SUN

Now let us see how the effects of MHD operate in a variety of solar features. The solar atmosphere provides a number of interesting situations where we may profitably study the effects of MHD in different limiting conditions.

### 16.7.1 SUNSPOT UMBRAE: INHIBITION OF CONVECTION

In an "ordinary" convection cell (i.e., granule), when no magnetic field is present, the circulation of the gas (which is responsible for upward transport of heat) occurs in several stages. (1) Hot matter starts its upward circulation at depth H and rises vertically to the photosphere in the bright center of the granule. (2) As the matter approaches the photosphere, it expands (due to reduced ambient density) and spreads out horizontally. In this phase, the material cools off, mainly by radiative losses into space. (3) The cooled material finds a location where it can sink vertically: this occurs in the dark intergranular lanes, and the gas returns eventually to depth $H$. (4) The material eventually, as a result of fluctuations, absorbs some excess heat, and this begins the circulation of a new cell. Each granule lives long enough to allow roughly one complete circulation to occur.

In the photosphere, with densities of $(2-3) \times 10^{-7}$ gm cm$^{-3}$ and granulation flow velocities $V \approx (1-3)$ km sec$^{-1}$, the energy densities of the flows ($\approx 0.5\rho V^2$) are $10^{3-4}$ ergs cm$^{-3}$. These convective properties are ultimately determined by the requirement that the gas must transport outward the flux of energy that is provided by nuclear reactions deep inside the Sun.

Now we ask the question that is relevant to understanding a sunspot: what happens to the circulation in a granule when a magnetic field is present? In an umbra, the magnetic field lines are mainly vertical and have strengths as large as 3000 G. Since ionized gas can move freely along field lines, stages (1) and (3) of the granule (vertical) motions are unaffected by the field. However, stage (2) is severely impeded: matter that contains even a small degree of ionization of photospheric gas is effectively frozen to the field lines. Since the latter are vertical, they impede the gas from flowing freely in a horizontal direction. The energy density of the field ($W_{mag} = B^2/8\pi$) is $\approx 4 \times 10^5$ ergs cm$^{-3}$, i.e., 40–400 times greater than the energy densities of the granular flows. Because of the excessive magnetic energy density, the "frozen fields" are capable of preventing the horizontal flows in stage (2) of granule flow.

The overall effect is that granule circulation (at least in its regions of horizontal flows) is inhibited by the (vertical) umbral field. But it is precisely that circulation which allows convection to transport heat to the solar surface in the nonmagnetic Sun. If some mechanism has the effect that the circulation in a granule (i.e., in a convection cell) is shut down, then convection cannot function properly. In the case of an umbra, the shutting down of circulation by vertical magnetic fields is so severe that it would not be an exaggeration to state that convection can apparently not function at all in an umbra. To be sure, radiation is available to carry some heat upward, but this is not very effective. As a result, the upward heat flux decreases below the normal value by a significant factor. The umbra becomes darker (by many tens of percent) than the photosphere.

Although gas motions in the umbra are still permitted in the vertical direction, the restriction on the horizontal dimensions of "cells" has the effect that the horizontal size of convection "cells" in an umbra become considerably smaller than the horizontal dimensions (~1 Mm) of granules in nonmagnetic Sun. (This leads to the descriptive term "brain-pattern" used by Tian and Petrovay 2013: see Section 16.1.2.) Even if the up and down flow speeds may be individually relatively large, an observer who is limited in spatial resolution of 100–200 km will be averaging over several small cells: this will lead to a significantly reduced upward mean flow speed. This could account for the greatly reduced C-shaped bisector speeds reported by LB18 in the umbrae of spots (Section 16.1.2).

### 16.7.2 PORES: THE SMALLEST SUNSPOTS

Sunspots are dark because a vertical magnetic flux tube inhibits convection *in one or more* granules. But what happens if a magnetic flux tube emerges onto the solar surface with a diameter that is *smaller* than a typical granule diameter, i.e., smaller than about 1 Mm? In such a case, the flux tube is not wide enough to impose control over the complete circulation of the granule. The gas flows can "shift over" into a nonmagnetic area, and convection can proceed more or less uninhibited. Thus, the smallest sunspots (i.e., pores that consist of umbra only, without any penumbra) must have diameters at least as large as a single granule, i.e., ≥ 1–2 Mm. In fact, in a Hinode study of almost 10,000 pores (Verma and Denker 2014), the pore areas were found to be of order 5 Mm² or less. Pores tend to be formed in supergranule boundaries where increased concentrations of magnetic elements are located. In order for a pore to be formed, the observations indicate that the magnetic flux (= field strength times area) must exceed a critical value of $(4-5) \times 10^{19}$ Maxwells. (The unit of 1 Maxwell [or 1 Mx] is defined by a field of 1 G times 1 cm².) In terms of this criterion, pores with diameters of 1.5–3.5 Mm (i.e., with areas of $1.8-10 \times 10^{16}$ cm²) must contain fields with strengths in excess of 0.45–2.5 kG. In fact, the observed fields in such pores are always in excess of 1.5 kG (Verma and Denker 2014). If the magnetic flux exceeds $1-1.5 \times 10^{20}$ Mx, the pore develops a penumbra and therefore makes a transition to a full-fledged sunspot (Leka and Skumanich 1998).

### 16.7.3 SUNSPOTS: THE WILSON DEPRESSION

Magnetic fields exert a pressure *perpendicular* to the field lines (Section 16.6.2.1). In an umbra, where the magnetic field lines are *vertical*, the field therefore exerts a pressure of $B^2/8\pi$ in the *horizontal* direction. The gas inside the flux tube also exerts a pressure $p_{in}$. In order for the flux tube to be a stable structure, the sum of these forces must be balanced by the gas pressure $p_{ext}$ in the external (nonmagnetic) medium:

$$p_{ext} = p_{in} + \frac{B^2}{8\pi} \tag{16.11}$$

Let us consider some typical numerical values for the photosphere. At $\tau \approx 1$ in the undisturbed photosphere, $p_{ext} \approx 10^5$ dyn cm$^{-2}$ (see Table 5.3). A sunspot in which the vertical field has $B =$ (say) 1 kG requires, for stability, that $p_{in} = p_{ext} - B^2/8\pi$ have the numerical value $0.6 \times 10^5$ dyn cm$^{-2}$. If the sunspot has $B = 1.5$ kG, then $p_{in}$ must be $\approx 0.1 \times 10^5$ dyn cm$^{-2}$, i.e., an order of magnitude *smaller* than the pressure in the undisturbed photosphere. Truly, the umbral flux tube is a region that has been almost "evacuated" of gas.

Recalling that in a sunspot umbra, the field can be as strong as 3 kilogauss, we see that in order to contain such a field, $p_{ext}$ must be at least as large as $3.6 \times 10^5$ dyn cm$^{-2}$. In fact, to allow for the presence of *any* finite gas pressure inside the spot, $p_{ext}$ must be even larger, perhaps $(4-5) \times 10^5$ dyn cm$^{-2}$. Such high pressures are simply not available in the vicinity of the level $\tau \approx 1$ in the photosphere. Therefore, in order to contain the sunspot fields by horizontal pressure from the surrounding (nonmagnetic) gas, we must rely upon gas pressures that lie *deeper than* the photosphere. The fact that material in an umbra lies deeper than in the photosphere gives rise to the term "Wilson depression" $W_D$, named in honor of an 18th-century English observer who first observed the effect in sunspots near the limb.

How large is $W_D$? Observations of 12 sunspots with Hinode have been reported by Löptien et al. (2020). Since the field lines are not necessarily precisely vertical in all umbrae, but instead the field lines tend to "fan out" above the surface, the field lines are to some extent curved. The presence of curved field lines adds an extra term to Equation 16.11, the "curvature force". Using this modified form of Equation 16.11, Löptien et al. (2020) reported that, in their sample of spots, $W_D$ ranged from 500 to 700 km. It was also found that the stronger the umbral field, the deeper the Wilson depression.

If we imagine that we could (somehow) stand in the undisturbed photosphere at the level $\tau = 1$, and look horizontally into a sunspot, what would we see? We would see a medium where the gas pressure is greatly reduced compared to the gas in which we are "standing". In other words, the gas in the spot would be "missing". Where did the missing gas go? The answer is: the cooler conditions have caused the gas to have a smaller scale height, causing the density/pressure to fall off more rapidly as height increases. This fall-off can be interpreted as having caused the gas in the umbra to "slump" to greater depths.

### 16.7.4 Sunspots: What Determines Their Lifetimes?

The answer to this question depends on a long-standing argument about sunspots, concerning their subsurface structure: is a spot a "monolithic" structure extending deep into the convection zone? Or is it a "cluster" of smaller flux tubes that have been swept together by subsurface flows?

In the cluster model (Parker 1979), the cooling effects of magnetic fields are confined to a "shallow" layer close to the solar surface. In principle, one might analyze the subsurface structure by measuring how $p$-modes are slowed down as they propagate across a sunspot (where the local cooling reduces the local sound speed). In a comprehensive review, Moradi et al. (2010) argue that the evidence from wave travel times favors a sunspot as being a shallow feature, with a depth of order 2–2.5 Mm. An illustration of a model of this type is shown in Figure 16.9 (Solov'ev and Kirichek 2014), where the spot depth is suggested to be of order 4 Mm, i.e., somewhat larger than Moradi

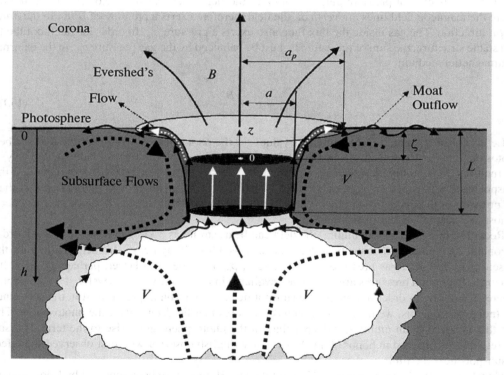

**FIGURE 16.9** Cross-section of a shallow sunspot. $\zeta$ is the Wilson depression, L is the lower sunspot boundary, dotted lines with arrows denote subsurface flows which hold the magnetic cluster together within a range of depths from zero (at the top) down to L (at the bottom). Note the two-layer structure: cooler (darker) material (above depth = L) where convection is suppressed and hotter (brighter) material at depths below L. The spot acts as an insulated "plug" that stops heat from penetrating upwards through the umbra, thereby "damming" up heat in the lower layers. (Solov'ev and Kirichek 2014).

et al. (2010) suggest, but still "shallow" compared to the solar radius. An essential aspect of the cluster model is that it assumes there is a circulation of material around the spot, with inflow near the surface and outflow at deeper layers. The near-surface inflow acts as a "collar" that sweeps near-surface field lines together when the spot first appears on the Sun. The lifetime of a sunspot according to this model depends on how long the "collar" lasts: once the collar fades away (for whatever reason), the spot is pulled apart by convective turbulence, with individual flux tubes being shredded from the outer regions first.

However, probing the properties of layers that lie several Mm below the surface by analyzing helioseismic data is not a simple process, and the quantitative details of the depth dependence are still not definitive. As regards the "monolithic" versus "cluster" models, based on model simulations of sunspots, it is possible that some spots (especially the longest lived ones) are actually monolithic, with roots that extend deeply into the convection zone (Rempel 2011).

### 16.7.5 PROMINENCES

Near the topmost portions of a prominence, the magnetic field lines are mainly horizontal. Material on such field lines can be supported against gravity. Ionized material on a horizontal field line is not permitted to move vertically because that would involve motions perpendicular to field lines. Motion along the field lines is permitted, but this can be impeded if there is a dip in the field lines.

### 16.7.6 FACULAE

Faculae (a plural noun: singular = facula) are flux tubes with diameters that are *too small* to create pores. Without the ability to impede the circulation in a complete granule, there is no reason why convection should be inhibited. As a result, a facula does *not* appear as a dark feature in the photosphere. However, the presence of vertical fields has the effect (see Equation 16.11) that the pressure inside faculae is lower than in the external gas. The reduced internal pressure allows us to see deeper inside faculae, i.e., faculae also exhibit Wilson depressions, although not as large as in the largest spots. Depressions in faculae may be of order 100–200 km. The deepest gas that we can see in a facula is surrounded by walls of hotter gas extending upward by 100–200 km.

What effect does this have on what we see when we observe faculae? There is almost no observable effect when a facula is observed near the center of the solar disk. However, when a facula is near the limb of the Sun, conditions are different. Now, our line of sight enters the facula at an angle to the vertical. This allows us to observe granules lying beyond (i.e., outside) the walls of the tube. It is as if we were permitted to insert a slender glass tube into a furnace: if we were to look into the tube from a point of view that is not too far off-axis, we could get a glimpse (through the glass wall of the tube) of deep regions of the furnace that are hotter than the surface regions. This deeper, hotter material emits more brightly than the material at the surface. As a result, each facula is seen as a localized feature that is brighter than the surrounding photosphere. The excess in brightness can be up to a few percent relative to the nonmagnetic photosphere.

Facular excesses play a significant role in explaining the surprising observation that the "solar irradiance" is observed to be *larger* at solar maximum (i.e., when sunspot numbers are at a maximum) than at solar minimum (see Section 1.4). For each sunspot, there are many faculae in the surrounding active region. Even though the power deficit of a sunspot (many tens of percent) is much larger than the power excess in any individual facula, the area occupied by the multiple faculae in the vicinity of the sunspot is large enough that the facular excesses in the Sun can more than compensate for the spot deficit.

### 16.7.7 EXCESS CHROMOSPHERIC HEATING: NETWORK AND PLAGES

In Chapter 15, we saw that the chromosphere in cell centers is heated by acoustic waves emerging from turbulent convection. Excess heating of the chromosphere in the network, and in plages, can be understood in terms of an extra source of wave energy in those regions. Since the network and plages can be identified with confidence as locations where the magnetic field strengths are larger than in quiet Sun, it is natural to look to the magnetic field as the source of extra wave energy.

The fact that a magnetic field is associated with a tension $T_m = B^2/4\pi$ along the field lines suggests that we consider the analog to transverse waves on a stretched string. Classical mechanics tells us that if a string is under a tension $T_m$, and the string has a mass density $\rho$, the speed of transverse waves on the string is $\sqrt{(T_m/\rho)}$. Now, even though a magnetic field line in and of itself has no mass, nevertheless in the solar atmosphere, where field and gas are "frozen together", the gas is tightly tied to the field. Therefore, when the field lines move, the gas also moves along, thereby conferring (in effect) a mass density on the field equal to that of the ambient gas. Analogous to the stretched string, therefore, a magnetic field line can support a transverse wave mode that propagates at a speed

$$V_A = \sqrt{(T_m/\rho)} = B/\sqrt{4\pi\rho} \qquad (16.12)$$

This propagation speed is referred to as the Alfven speed, after the Swedish physicist who, in the midst of World War II, discovered this wave mode in a magnetized gas (Alfven 1942).

Let us look at numerical values for the Alfven speed in the Sun. In the photosphere, where $\rho \approx 3 \times 10^{-7}$ gm cm$^{-3}$, the numerical value of $V_A$ is given by 515 $B$ cm sec$^{-1}$, if $B$ is in Gauss. In the upper chromosphere, where $\rho \approx 3 \times 10^{-13}$ gm cm$^{-3}$, $V_A = 5.15\ B$ km sec$^{-1}$. In the low corona, where number densities are $10^{8-9}$ cm$^{-3}$, i.e., $\rho \approx 2 \times 10^{-(15-16)}$ gm cm$^{-3}$, $V_A = 60-200\ B$ km sec$^{-1}$. Thus, in the photosphere, even in the umbra of a sunspot, Alfven speeds are no more than 10–20 km sec$^{-1}$. These speeds are not greatly different from the local sound speed. But in the corona, in active regions where the fields can be as large as 1000 G or more (Lee 2007), Alfven speeds may be as large as tens of thousands of km sec$^{-1}$ (e.g., Schmelz et al. 1994).

The *flux* of Alfven waves $F_A$ in the presence of transverse velocity fluctuations of $\delta V$ is of order $\rho(\delta V)^2 V_A$. In the photosphere, turbulent velocities of order 1 km sec$^{-1}$ can distort local magnetic fields such that $\delta V$ may also be of order 1 km sec$^{-1}$: this leads to $\rho(\delta V)^2 \approx 3 \times 10^3$ ergs cm$^{-3}$. If $B$ = 10–100 G, we find $F_A$ in the photosphere can be of order $10^{7-8}$ ergs cm$^{-2}$ sec$^{-1}$. This is comparable to the acoustic flux that might have access to the chromosphere (see Equation 15.2). But Alfven waves are not subject to the same dissipation as sound waves: therefore, it is possible that Alfven waves originating in the photosphere might supply a significant flux of energy to the chromosphere or even to the corona.

The fields in the Sun exist in the presence of gas in which sound travels as a succession of expansions and contractions moving at speed $c_s$. The combined effects of fields and gas lead to the existence of wave modes called fast-mode and slow-mode MHD waves: these are hybrid waves in which gas pressure and magnetic fields combine in different ways. In a medium where $V_A \ll c_s$ (such as deep in the photosphere), the fast mode behaves as a sort of sound wave. On the other hand, in a medium where $V_A \gg c_s$ (such as in the corona), the slow mode behaves as a sort of sound wave guided by the field.

A remarkable property of the existence of several different modes of waves has already been mentioned in Section 15.12.1 in the context of vertically propagating waves, namely, in favorable conditions, one wave mode can *convert* into another (e.g., Cally and Moradi 2013). However, the conversion process is not limited to vertical propagating waves: it can also happen to waves that, as they propagate around the Sun, go from quiet Sun into an active region, or *vice versa*. As a result, astronomers can determine the position of an active region on the hidden hemisphere of the Sun. The reasoning starts by noting that a wave that was originally propagating as a pure sound wave (e.g., a $p$-mode moving at speed $c_s$) can, upon encountering an active region, suddenly be converted to an MHD wave. This leads to two important consequences as regards our study of the Sun.

First, the wave propagation speed changes from $c_s$ to a value of order $V_A$. This change in propagation speed has the effect of altering the travel time of the wave between any two points in the Sun (see Section 14.7): the alteration in travel time can be as much as 40 seconds in the presence of a 1 kG field (Cally and Moradi 2013). This travel time alteration can be used to analyze helioseismic data in order to predict if there are active regions present on the "far side" of the Sun *and where they are located* (Lindsey and Braun 2000). At first, and specifically prior to 2011, the only way to check these predictions was to wait for solar rotation to carry the active regions eventually onto the visible disk. But during the halcyon years 2011–2015, the STEREO mission, with its two satellites observing opposite hemispheres of the Sun from very different vantage points, made it possible for the first time to view the *entire* Sun (back *and* front) *simultaneously*. The analytical predictions of far-side features could now be checked in real time. Liewer et al. (2017) reported that during a nine-month interval in 2011 and 2012, analysis of the helioseismic data from the HMI instrument on SDO led to the prediction of 22 far-side active regions *in all 22 cases*, the STEREO EUV detectors (which could "see" the far side directly) did indeed find EUV emission at the predicted sites. (After 2015, one of the STEREO satellites no longer transmitted data, so the whole-Sun feature was no longer available. But STEREO A still provides valuable information.) Thus, helioseismic data can indeed be used to "see behind" the Sun reliably, even when we no longer have the advantage of making observations from two STEREO spacecraft.

Second, MHD waves can supply mechanical energy to the solar gas at rates that are different from those of sound waves: the flux of energy can be larger (if $V_A$ is larger than $c_s$), and furthermore, the physical process of dissipating the wave energy is different from that which occurs in sound waves. For example, MHD waves can be dissipated by resistance to current flow, but this mode of dissipation is not accessible to sound waves. The combination of increased flux and extra dissipation has the effect that chromospheric heating in a magnetic region can be significantly larger than in a nonmagnetic region. This can explain why the chromosphere in a network region (where fields are stronger) is observed to be brighter than the chromosphere in the centers of supergranule cells (where fields are weaker) (see Section 16.2).

### 16.7.8 Magnetic Field and Gas Motion: Which Is Dominant?

In the photosphere, we have now seen two distinct and interesting behaviors of the magnetic field. In some cases (sunspots, pores), the field causes the solar surface to be *darker* than normal, while in other cases (network, faculae), the field causes enhanced *brightening*.

Can we identify a transition between these behaviors? We have already identified one such transition in the case of pores. There, the critical parameter was the diameter of the flux tube: in order for darkening to occur, the pore must be larger than the diameter of a typical granule.

We can also consider the matter from the perspective of frozen flux. The essence of frozen flux is that gas and field are forced to move together. This raises the question: which one dominates? Does the gas dominate the field or does the field dominate the gas? The answer is: the Sun provides us with the luxury of a "yes and no" answer. Examples of both situations can be found in different features.

The question of dominance can be discussed in terms of the energy densities available to gas and field in the photosphere. Moving gas has kinetic energy density $E_d = 0.5\rho V^2$ ergs cm$^{-3}$. Inserting typical values of density ($2-3 \times 10^{-7}$ gm cm$^{-3}$) and convective velocity (1–3 km sec$^{-1}$) in the photosphere, we find $E_d \approx (0.1-1.4) \times 10^4$ ergs cm$^{-3}$. What strength of magnetic field is comparable to the value of $E_d$? Answer: a field with energy density $E_{mag} = B^2/8\pi$ is comparable in energy density to $E_d$ if $B \approx 150-600$ G.

Therefore, in any magnetic structures in the photosphere where the local field strengths are in excess of (roughly) 600 G, we expect to find that the gas flows are *not* sufficiently energetic to "push the field around". In such situations, the field will "win out" and impose changes on the gas. Sunspots, where the field suppresses convection and darkness ensues, are a prominent example of this behavior.

On the other hand, in photospheric structures where the field strength is less than (roughly) 150 G, the field is *not* sufficiently energetic to "push the gas around". In such situations, the gas "wins out" and imposes changes on the field. For example, in the network, moving gas induces wave modes on the field, thereby heating the overlying chromosphere and causing local brightening.

## 16.8 AMPLIFICATION OF STRONG SOLAR MAGNETIC FIELDS

Where do the strongest fields in the Sun come from? Part of the answer is that the weak polar fields can be amplified by the differential rotation observed on the solar surface. The amplification occurs because the field is "frozen" into the solar material (see Section 16.6.2.2), and the latter rotates differentially in latitude. As we saw in Section 14.9.3, the rotation period at the Sun's surface is 25.7 days near the equator and 31.3 days at 60° latitude, i.e., a difference of about 20%. We have seen (Section 16.4.5) that the solar polar field has a strength of 6–12 G. Assuming that the polar field is a dipole (i.e., its field lines run in the north-south direction), the equatorial field strengths are expected to be of order 3–6 G. Because of field freezing, the equatorial field (which is originally in the north-south direction) will be sheared (i.e., stretched) by differential rotation. Both components of differential rotation (latitudinal and radial) can come into play. But for simplicity, let us consider only LDR. Then a particular field line, after one rotation (i.e., after 25.7 days) will return to the same longitude on the equator, but the high-latitude section of the same field line will lag behind by about 20% of a rotation. After five rotations, i.e., after 4.2 months, the high-latitude section of the field line will have fallen behind by about one full rotation, i.e., the equatorial will have "lapped" the polar section. In one year, the polar portion will have lost 2.8 full rotations on the equatorial portion of the field line.

The excess stretching of the field leads to field lines that become more and more stretched out in longitude. That is, although the initial (polar) fields essentially were directed from north to south (i.e., they were *poloidal* fields), the stretching due to differential rotation leads to increasingly strong fields in the east-west direction (i.e., *toroidal* fields). It is this tendency for stretched fields to be mainly toroidal that causes most pairs of sunspots (which originate in the strong fields stretching mainly in the east-west direction) to lie almost east-west (see Section 16.1): each pair of spots originates in a strong toroidal (almost) east-west magnetic flux tube.

How strong do the toroidal fields become? The answer depends on how the area of a flux tube is distorted by the stretching motion. As the field lines are stretched, the area of a flux tube will likely be "squeezed". How much will the squeezing be? This is a complicated problem and it is not easy to give a simple answer. But suppose, for the sake of numerics, that an increase in length by 20% (after one rotation) leads to a reduction in area by (say) $\varphi = 10\%$. To conserve magnetic flux, the reduction in area by 0.1 in 25.7 days would mean that the toroidal field would be larger than the initial value (3–6 G) by ≈1.1 after one rotation. After one year, i.e., after 14.2 equatorial rotations, the toroidal field strength near the equator would be increased by $1.1^{14.2}$, i.e., by a factor of 3.9. After 2, 3, 4, and 5 years, the initial equatorial field of 3–6 G would be amplified by factors of 15, 58, 220, and 870. Thus, the toroidal field strength at the equator would be 2600–5200 G after 5 years. These values are comparable to the field strengths observed in sunspot umbrae (see Figure 16.8).

Suppose our estimate of $\varphi$ is too large: suppose a more realistic value is $\varphi \approx 0.05$. Then the toroidal field strength would require about 8 years to reach a strength of 1 kG. Thus, depending on the actual value of $\varphi$, the continuous operation of LDR *could* result in fields as strong as sunspot fields in time-scales of 5–8 years.

So far, we have considered only LDR as we see it at the surface. But the amount of LDR varies as we examine different depths beneath the surface. In Figure 14.6, we saw that the angular velocity difference between gas at 0 deg latitude and 60 deg latitude is maximum at radial locations of 0.9–0.95 solar radii. As a result, the stretching of poloidal fields will build up faster at depths of 35–70 thousand km below the surface. Another region of strong shear occurs at the interface between convection zone and radiative interior: there, a strong shear occurs over a relatively short

# Magnetic Fields in the Sun

interval in the radial coordinate. Magnetic fields that are frozen into such a highly sheared medium may also generate fields of strength ≥1 kG in relatively short time-scales.

However, in the radiative interior, where the gas rotates almost as a solid body, there is little or no tendency for the poloidal field to undergo stretching.

In view of these processes, toroidal fields of order 1 kG can be built up in the course of a few years not just in the surface layers, but at all depths throughout the convection zone. Now, the sunspot cycle is observed to last, in fact, some 11 years on average. So the stretching time-scales estimated earlier are in the right ballpark to allow surface fields to build up to kG strength in the course of (roughly) one-half of the sunspot cycle.

Since stretching of field lines is an inherent process in a medium with frozen fields and differential rotation, the question arises: what eventually stops the process of stretching? Why do the surface fields reach strengths of a few kilogauss and not much more? One reason has to do with buoyancy forces. To see how this operates, consider the application of Equation 16.11 to a stretched flux tube. The internal pressure $p_{in}$ is lower than the ambient pressure $p_{ext}$ by the amount $B^2/8\pi$. How does the temperature inside the flux tube compare with the temperature outside? To answer this, we note that, deep in the interior of the Sun, where radiative transport dominates (see Chapters 8 and 9), photons can carry heat efficiently back and forth between neighboring parcels of gas. These photons are not impeded in any way by the magnetic field. As a result, the *temperatures* inside and outside the flux tube *can* remain essentially equal even though the pressures inside and outside are *not* equal. Therefore, reduced pressure $p_{in}$ corresponds to *reduced* density $\rho_{in}$ inside the flux tube: $\rho_{in}/\rho_{ext} = p_{in}/p_{ext} = 1 - (B^2/8\pi p_{ext})$. Notice that, in order to avoid negative densities inside the flux tube, the maximum value which the field strength can have is $B_{max} = \sqrt{(8\pi p_{ext})}$. Given the pressure at the base of the convection zone ($4.3 \times 10^{13}$ dyn cm$^{-2}$: see Section 7.9), a firm upper limit on $B_{max}$ is of order 33 MG.

Because the flux tube contains gas with lower density than in the ambient (nonmagnetic) medium, buoyancy forces come into play and push the flux tube upward. How strong are the buoyancy forces? In the presence of gravity $g$ and a density difference $\Delta\rho = \rho_{ext} - p_{in}$, buoyancy creates an upward acceleration $a_b$ which is given by (see Equation 7.2):

$$a_b = g\frac{\Delta\rho}{\rho} \approx \frac{gB^2}{8\pi p_{ext}} \qquad (16.13)$$

In the presence of this acceleration, how long does it take for a parcel of gas to be buoyed up to the surface? To make the time as long as possible, let us consider a parcel of gas starting from as deep as we can reasonably assume, i.e., near the base of the convection zone. Such a parcel starts at a depth $D \approx 2 \times 10^{10}$ cm (see Section 7.9). What field strength should we use? Well, in the interest of making the rise as fast as possible, let us consider $B$ to have its largest permissible value, $B_{max}$. In that case, and assuming that the flux tube is free to rise, Equation 16.13 indicates that the full acceleration of gravity ($a_b \approx g \approx 2 \times 10^4$ cm sec$^{-2}$) would come into play. In such conditions, the parcel could rise to the surface in a time $\tau_r = \sqrt{(2D/a_b)} \approx 1400$ sec if the parcel met no resistance along the way. This is less than one-half hour, a very short time indeed in the context of the solar 11-year cycle! Of course, the upper limit of 33 MG for $B_{max}$ is extreme: estimates of the maximum field strengths that may exist in the convection zone and still be consistent with the observed properties of sunspots suggests that $B_{max}$ may be no larger than $10^5$ G (Choudhuri and Gilman 1987). At the base of the convection zone, this would lead to values of $a_b$ that are no larger than 0.2 cm sec$^{-2}$. With this slower buoyant acceleration, the rise time to the surface is found to be a few days. Even allowing for the presence of resistance due to convective turbulence before reaching the surface, the buoyant time-scales are likely to be no longer than ≈ 1 year.

The effects of buoyancy have a well-defined effect in the Sun: they cause flux tubes to move "up and out". And the stronger the field, the faster the buoyancy forces bring it up to the surface. As a result, when we try to impose the condition that the Sun must make a strong (toroidal) field by amplifying its (weak) poloidal field, there is a race against time. On the one hand, differential rotation takes a finite time (5–8 years) to stretch the field and amplify it. On the other hand, as the field becomes stronger, the stronger are the buoyancy forces that want to drive the flux tube "up and out".

## 16.9 WHY DOES THE SUN HAVE A MAGNETIC CYCLE WITH $P \approx 10$ YEARS?

The Sun's magnetic cycle occurs over a time interval $P$ whose value is observed to span a range from about 9 years to about 12 years, as the largest solar magnetic features (sunspots and their accompanying active regions) increase and decrease in numbers. The sunspots are the regions where the field grows to its largest values (3 kG), and the poles are the locations where the weak global fields (6–12 G) are easiest to identify.

In order to understand why the large magnetic features in the Sun have a cycle, Babcock (1961) argued as follows. Let us start by considering the global field of the Sun at time $t_o$. Let the global field at $t_o$ be directed in such a way that the Sun's north pole has field lines that point *outward* from the Sun. Also at $t_o$, suppose for simplicity that there are no sunspots or active regions on the surface, i.e., the Sun is at sunspot minimum and ready to start a new cycle. Let us see if we can understand the directions of the fields that occur in sunspots in this new cycle.

Differential rotation operates on the poloidal field, and in the course of 5–8 years, the poloidal field lines are stretched out so as to form strong toroidal fields (≥1 kG) beneath the surface. This can be considered as the growth phase of the cycle. At certain locations and at certain (unpredictable) times, when something causes the local toroidal field to become unstable, a section of toroidal field rises up and breaks through the solar surface: thus sunspots owe their existence to the toroidal field (Mordvinov et al. 2012). The time-scale for buoyancy to bring up a field of strength 3 kG to the surface is short, perhaps as short as a few days. The breakthrough forms an active region containing (typically) a pair of sunspots (leader plus follower) with a definite polarity. The active region surrounding the sunspots also has a leading portion and a following portion. Given the *outward* direction of the Sun's north pole field in the northern hemisphere at time $t_o$, the leader spot (and the nearby portion of the surrounding active region) during the growth phase will have a magnetic field $B$(lead) whose lines point *outward* from the Sun. But the follower will have a magnetic field $B$(follow) whose lines point *inward*. (Conversely, in the southern hemisphere, at the same time, pairs of spots will exhibit leaders with *inward* field lines and followers with *outward* field lines.) That is, during the growth phase of the cycle, the leader spots in a given hemisphere retain the same sense of the magnetic field as exists at the pole in that hemisphere at time $t_o$. This helps us to understand Hale's polarity law.

Since the sign of $B$(lead) is the same as the sign of $B$(pole) in its hemisphere, it follows that the sign of $B$(follow) must be *opposite* to the sign of $B$(pole) in its hemisphere. This is an important feature in helping us understand the solar cycle. Each spot pair is surrounded by an active region, which retains the overall polarity of the leader and follower spots. Because of Joy's law (see Section 16.1), the follower spot lies slightly *farther* from the equator than the leading spot. Therefore, if any mechanism is at work to cause spot pairs to migrate closer to the poles as time goes on, the *following* spot will reach the pole first. Since the sign of $B$(follow) is *opposite* to the sign of $B$(pole), as more and more follower spots reach the pole, they eventually overwhelm the local polarity and replace it with the opposite polarity. See Figure 16.10 where following (leading) and leading spots in the northern hemisphere are distinguished by red (blue) colors respectively in cycles 21 (~1980) and 23 (~2001), but by the opposite colors in cycles 22 (~1990) and 24 (~2014), in accordance with Hale's law. The build-up of following polarity at the poles leads to a reversal of the field (from red to blue, or blue to red) at the Sun's poles every 11 years (or so). This process is the essence of the Babcock (1961) theory of the solar cycle.

What mechanism might be at work to force spots to migrate systematically towards the poles? One possibility is the meridional circulation that exists in the near-surface layers of the Sun (Section 1.11). The speed of this near-surface flow has a maximum amplitude (on average) at solar

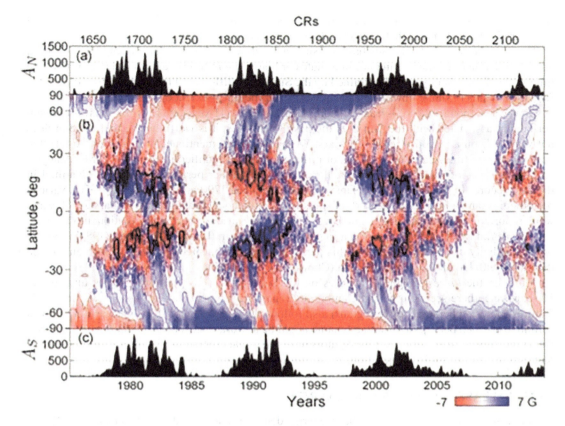

**FIGURE 16.10** (a) Changes in total sunspot area in the northern hemisphere in solar cycles 21–24. (b) Time-latitude diagram of the averaged magnetic fields (blue: negative fields; red: positive fields). Sunspot areas are shown by black contours corresponding to 100 MSH. (c) Changes in total sunspot area in the southern hemisphere in solar cycles 21–24. (From Mordvinov and Yazev 2014.)

minimum of 10–20 m sec$^{-1}$ at latitudes of order 30–50 degrees (Komm et al. 2015). The meridional flow speed is slower at solar maximum. Gas flowing at a mean speed of ≈10 m sec$^{-1}$ can drag material from equator to pole (a distance of $\pi R_\odot/2 \approx 10^{11}$ cm) in a time of order $10^8$ sec ≈ 3 years.

This systematic flow to the poles occurs simultaneously with a diffusive (random walk) process due to the horizontal motions of gas in supergranules: with diameters of order $d = 30{,}000$ km and horizontal velocities of order 0.3 km sec$^{-1}$, the associated diffusivity $D \approx dv$ is of order $10^{14}$ cm$^2$ sec$^{-1}$. In the presence of such a random walk, the time-scale required to cover a distance $L$ is $\tau_d \approx L^2/D$. Therefore, in order for a flux tube to be transported from equator to pole, i.e., across a distance of $L \approx 10^{11}$ cm, the time required is of order $10^8$ sec, i.e., ≈ 3 years. Because of Joy's law, this random walk favors the arrival of follower polarity at the pole, thereby leading again to reversal of the polar field. Combining meridional circulation and supergranule diffusion, we expect that the time-scale for reversal of the polar fields is probably no longer than 3 years. (For an instructive illustration of these processes, see Sanchez et al. [2014].)

With this information, we can now address the question: why does the sunspot cycle occur on a $P$ of order 10 years? The physical properties of the Sun itself set the various time-scales that go into determining $P$. First, there is a time-scale on which fields can be amplified by differential rotation (5–8 years); second, there is a time-scale for the fields to be buoyed up by gravity to the surface (<1 year); third, there is a time-scale for the fields to be transported up to the polar regions

(<3 years). Combining these time-scales, we see that a time interval of 9–12 years *could* encompass many or all of the elements that contribute to a solar cycle.

This helps us to see why the Sun has a sunspot (and active region) cycle whose length is measured *not* in millennia or centuries, and *not* in minutes or hours, but in time-scales of order 10 years. The time-scale of the solar cycle is determined by the Sun's own differential rotation, its gravity, the diffusivities of supergranule flows, meridional flows, and the linear extent of its surface.

However, although the ≈10-year (or more accurately, the 11-year) cycle is the most prominent periodicity associated with the Sun over the past 3 centuries, it is worth noting that the Sun does not vary *only* on a period of order 10 years. We have already mentioned (see Section 16.1.4) that periodicities on time-scales of centuries or millennia can be identified in certain records. Analysis of various types of solar activity (spots, flares) show that other periods can also be identified in the data record. In one study of the sunspot record between 1700 and 1969 (Wolff 1976), various periodicities (and with various degrees of statistical significance) ranging from 3 years to 180 years were reported, possibly arising from beating between various inertial oscillation modes (i.e., $r$-modes). And in an analysis of gamma-ray observations from flares, Rieger et al. (1984) reported on a periodicity of 154 days: recently, this period has been connected specifically with an $r$-mode (Section 14.10) having $m = 10$ in the Sun (Chowdhury et al. 2010).

Although the *largest* magnetic features on the Sun (spots, active regions) clearly undergo an 11-year cycle, there seems to be some ambiguity regarding the question: do the *smallest* magnetic features (pores, ephemeral active regions) also follow the 11-year cycle? If the answer were to be a definitive no, then perhaps a different dynamo *might* be at work to generate the small-scale features. But at the present time, different investigators have arrived at contradictory answers: future work is required in order to settle the question.

## 16.10 RELEASES OF MAGNETIC ENERGY

We have seen that magnetic fields have energy densities equal to their pressures, i.e., $W_{mag} = B^2/8\pi$ ergs cm$^{-3}$. In favorable circumstances, most of the "free" energy in the field (i.e., the energy of coronal magnetic field in excess of the energy of the potential field [see Section 16.4.5]) can be converted into other forms. The two most prominent classes of events in the Sun that owe their existence to release of magnetic energy are flares and coronal mass ejections. We will discuss flares in Chapter 17, in the context of the solar corona, and CMEs in Chapter 18, in the context of the solar wind. However, in order to set the stage for a discussion of CMEs, we need to consider one more physical property of magnetic fields. We discuss this in the next subsection.

## 16.11 MAGNETIC HELICITY

We are already familiar with the fact that magnetic fields are vector quantities, which have magnitude and direction. The total energy $E$ in a magnetized plasma (including kinetic and magnetic energy) is a positive definite scalar quantity that, in an ideal plasma (i.e., one with no dissipation), remains constant as time goes on. In such a case, $E$ is said to be an "invariant" of the system. However, if some process of dissipation can operate in the plasma (e.g., due to viscosity or resistivity), $E$ will become progressively smaller as time goes on. The dissipation occurs at small length-scales, and in order for the energy to get down to such scales, the energy must undergo a "cascade" from large to small scales. Such a cascade of energy is a natural feature of a medium where turbulence is well developed.

However, there is a second scalar parameter that describes a different aspect of a magnetized plasma: magnetic helicity, $H$, which is a measure of how *twisted* the field lines are. Given that the field $B$ in a region is defined as the curl of a "vector potential" $A$, the value of $H_m$ per unit volume is defined as the dot product $A.B$ of the two vectors. The units of $H_m$ are Mx$^2$. (For definition of Mx, see Section 16.7.2.) A coronal loop with two footpoints rooted in photospheric granules can have its field lines twisted by the granular motions. The twisting leads to the flow of current along the loop. If the loop is twisted through an angle of $2\pi$ radians between one footpoint and the other, the loop is said to be subject to a twist $T$ of 1 turn. If the loop becomes twisted too much, the entire loop starts to erupt and change its shape into coils and "super-coils": this coiling is described by the term "writhe". A magnetic field in a region of the Sun can be assigned a helicity value of $H_m$, which is formally a measure of twist plus writhe. Because current can flow along two opposite directions in a loop, the value of $H_m$ is always accompanied by an algebraic sign: + or − refers to the handedness of a screw that follows the twist in any particular loop. In an ideal plasma (i.e., one without any dissipation), $H_m$ is an invariant. However, if there is finite dissipation, the value of $H_m$ will undergo changes at a *slower* rate than those occurring in $E$: in fact, $H_m$ may retain an essentially constant value for a time that is almost as long as the global diffusion time-scale (Equation 16.10) (Priest et al. 2016). As a result, the decay of $E$ cannot occur in an arbitrary manner but must satisfy the additional constraint $H_m \approx$ constant (see Section 16.4.5). (An analogy occurs in frictionless mechanics when an object slides down an inclined plane. If the object has the capacity to rotate as it slides, then total energy is not the only quantity to be conserved: one must also ensure that the angular momentum is conserved.) The reason for the very different temporal behavior of $E$ and $H_m$ is that in 3-D MHD, whereas $E$ cascades towards *smaller* length-scales (and is therefore readily dissipated at atomic scales), $H_m$ undergoes an "inverse cascade" towards *larger* scales (where dissipation is less effective). As a result, helicity persists in a plasma for longer times than $E$ persists. $H_m$ is said to be a more "rugged" invariant than $E$.

As regards the determination of $H_m$ in the Sun, magnetic fields measured in active regions on the Sun's surface are frequently extended up into the corona by assuming (see Section 16.4.5) that the fields in the corona are force-free, i.e., $\nabla \times \mathbf{B} = \alpha \mathbf{B}$. In this equation the parameter $\alpha$ has a different sign depending on the handedness of the current: in loops where writhe is not important, the sign of $\alpha$ can be taken as a proxy for the sign of $H_m$. Pevtsov et al. (1995) extracted values of $\alpha$ for 60 ARs that were observed in the years 1988–1994 (solar cycle 22). They also used data in the literature to determine values of $\alpha$ during cycles 20 and 21. Remarkably, they discovered that, for ARs in the northern hemisphere (NH), the algebraic sign of $\alpha$ is negative in the majority (76%) of their sample, while in the SH, $\alpha$ is positive in the majority (69%) of their sample. Apparently, the Sun is capable of imposing systematically opposite helicities in ARs which lie N and S of the equator, with maximum numerical values for $H_m$ at latitudes of 15–25 degrees. Also remarkably, this pattern of $\alpha > 0$ (<0) in SH (NH) was observed to persist in cycles 20, 21, and 22, even though the global field of the Sun switched sign between each pair of cycles: apparently, a change in the sign of $\mathbf{B}$ (between one cycle and the next) does *not* necessarily lead to a change in the sign of $\nabla \times \mathbf{B} / \mathbf{B}$.

In their work, Pevtsov et al. (1995) pointed out that earlier research had already indicated that a very different aspect of the Sun (its wind: see Chapter 18) also shows signs of opposite helicities in NH and SH. Specifically, Bieber et al. (1987) analyzed the spiral magnetic field predicted by Parker (1958) in the solar wind (see Section 18.5): it emerged that $H_m$ is negative in the NH wind and positive in the SH wind. Moreover, this feature was found to remain unchanged in different solar cycles (despite the change in polarity of the solar field). These features are clearly analogous to the properties displayed by ARs on the solar surface.

Pevtsov et al. (1995) suggested that differential rotation is probably *not* the primary cause of the hemispheric pattern behavior for $\alpha$: instead, convective turbulence may be responsible. Why would that be? Because it has been shown computationally, by Moreno-Insertis and Emonet (1996), that magnetic flux tubes rising up from the solar interior will *not* survive their upward passage through the convection zone turbulence *unless* the fields are twisted by some minimum amount. In other words, the presence of helicity acts as a powerful antidote against the shredding process that would otherwise occur as the flux tube traverses the convection zone. The inherent twist of flux ropes emerging at the surface of the Sun naturally supplies helicity into an emerging AR. This helicity associated with each AR is the dominant contributor to the global magnetic helicity of the Sun (Zhang 2012).

In order for the Sun to make a transition from one solar cycle to the next, the magnetic flux (especially the strong toroidal flux) of the old cycle has to be removed to make room for the flux belonging to the new cycle. This requires removal of some $10^{24}$ Mx of toroidal flux per solar cycle: this can be achieved by a combination of solar wind outflow, eruption of filaments (see Section 16.2), and CMEs (Bieber and Rust 1995). However, the CMEs do not remove only magnetic *flux* from the Sun: Rust (1994) and Low (1996) have suggested that CMEs also remove magnetic *helicity* from the Sun. Since the hemispheric helicity has the same sign in all cycles, if the helicity of one cycle were not removed from the Sun (somehow), then the amount of helicity would simply keep accumulating in the solar atmosphere, but this is not observed to happen. The amount of helicity that CMEs remove from the Sun in the course of a solar cycle is estimated to be $2.5 \times 10^{46}$ Mx$^2$ (Demoulin et al. 2016), enough to deal with the emergence of 200–300 ARs per year. We have already noted (Section 16.1.3) that in the Sun, some 300 ARs emerge on the surface on average each year: it seems that CMEs are capable of removing essentially all of the helicity that the ARs introduce into the solar atmosphere.

Is there observational evidence indicating that CMEs are in fact associated with removal of helicity? Nindos and Andrews (2004) analyzed 133 large flares in cases where the value of $H_m$ in the local AR could be measured before each flare: they found that some flares were accompanied by CMEs while others (with comparable flare energy) were not. A clear distinction between these two groups of flares as regards helicity was found: ARs with large $H_m$ were more likely to generate a CME than ARs with small $H_m$. We shall return to this aspect of observational evidence, in one particular active region, in Section 18.9.7.

## EXERCISES

16.1 Calculate the Zeeman splitting of a line at $\lambda = 6000$ Å in fields of 1, 100, and $10^4$ G.

16.2 Consider an electron, a proton, and a lead nucleus gyrating in the Earth's magnetic field ($B = 1$ G) with a variety of energies. Calculate the radius of gyration for each particle in cases where the kinetic energy is (a) 1 eV, (b) 1 MeV, and (c) 1 GeV.

16.3 In the space between the stars (the interstellar medium: ISM), energetic particles (galactic cosmic rays) gyrate about a field with a strength of about $3 \times 10^{-6}$ G. Determine the relativistic $\gamma$ factor for an ultrarelativistic proton that has a radius of gyration of 10 AU in this field.

16.4 Calculate the Alfven speed in the ISM, where the gas number density is 1 proton cm$^{-3}$. (The mass of a proton is $1.67 \times 10^{-24}$ gm.)

16.5 Calculate the range of Alfven speeds in the interplanetary medium near Earth: the field strengths range from one to $10 \times 10^{-5}$ G and the number densities range from one to 10 protons cm$^{-3}$.

## REFERENCES

Acero, F. J., Carrasco, V., Gallego, M., et al., 2017. "Extreme value theory and the new sunspot number series", *Astrophys. J.* 839, 98.

Alfven, H., 1942. "Existence of electromagnetic-hydrodynamic waves", *Nature* 150, 405.

Asensio Ramos, A., & Martinez Gonzalez, M. J., 2014. "Hierarchical analysis of quiet Sun magnetism", *Astron. Astrophys.* 572, A98.

Babcock, H. W., 1953. "The solar magnetograph", *Astrophys. J.* 118, 387.

Babcock, H. W., 1961. "The topology of the Sun's magnetic field and the 22-year cycle", *Astrophys. J.* 133, 572.

Beer, J., Tobias, S. M., & Weiss, N. O., 2018. "On long-term modulation of the Sun's magnetic cycle", *Mon. Not. Roy. Astron. Soc.* 473, 1596.

Bellot Rubio, L., & Orozco Suarez, D., 2019. "Quiet Sun magnetic fields: an observational view", *Living Rev. Sol. Phys.* 16, 1.

Bieber, J. W., Evenson, P. A., & Matthaeus, W. H., 1987. "Magnetic helicity of the Parker field", *Astrophys. J.* 315, 700.
Bieber, J. W., & Rust, D. M., 1995. "The escape of magnetic flux from the Sun", *Astrophys. J.* 453, 911.
Bray, R. J., & Loughhead, R. E., 1979. "The properties of sunspot groups", *Sunspots*, Dover, New York, pp. 225–246.
Cally, P. S., & Moradi, H., 2013. "Seismology of the wounded Sun", *Mon. Not. Roy. Astron. Soc.* 435, 2589.
Castellanos Duran, J. S., Lagg, A., Solanki, S. K., et al., 2020. "Detection of the strongest magnetic field in a sunspot light bridge", *Astrophys. J.* 895, 129.
Charbonneau, P., 2014. "Solar dynamo theory", *Ann. Rev. Astron. Astrophys.* 52, 251.
Chatzistergos, T., Ermolli, I., Krivova, N. A., et al., 2019. "Analysis of full-disk Ca II K spectroheliograms", *Astron. Astrophys.* 625, A69.
Chelpanov, A. A., & Kobanov, N. I., 2020. "The first active region of the new cycle", *Solar Phys.* 295, 94.
Choudhuri, A. R., & Gilman, P. A., 1987. "The influence of the Coriolis force on flux tubes rising through the solar convection zone", *Astrophys. J.* 316, 788.
Chowdhury, P., Khan, M., & Ray, P. C., 2010. "Short-term periodicities in sunspot activities during the descending phase of solar cycle 23", *Solar Phys.* 261, 173.
Clette, F., Cliver, E. W., Lefevre, L., et al., 2016. "Preface to topical issue: Recalibration of the sunspot number", *Solar Phys.* 291, 2479.
Demoulin, P., Janvier, M., & Dasso, S., 2016. "Magnetic flux and helicity of magnetic clouds", *Solar Phys.* 291, 531.
De Toma, G., 2011. "Evolution of SH's during the minimum between cycles 23 and 24", *Solar Phys.* 274, 195.
Falco, M., Puglisi, G., Guglielmino, S. L., et al., 2017. "Comparison of different populations of granular features in the solar photosphere", *Astron. Astrophys.* 605, A87.
Gnevyshev, M. N., 1967. "On the 11-year cycle of solar activity", *Solar Phys.* 1, 107.
Goldhaber, M. Grodzins, L., & Sunyar, A. W., 1958. "Helicity of neutrinos", *Phys. Rev.* 109, 1015.
Gough, D. O., & Tayler, R. J., 1966. "The influence of a magnetic field on Schwarzschild's criterion for convective instability", *Mon. Not. Roy. Astron. Soc.* 133, 85.
Hale, G. E., Ellerman, F., Nicholson, S. B., et al., 1919. "The magnetic polarity of sunspots", *Astrophys. J.* 49, 153.
Hempelmann, A., & Weber, W., 2012. "Correlation between the sunspot number, total solar irradiance, and the terrestrial insolation", *Solar Phys.* 277, 417.
Hirzberger, J., 2003. "Imaging spectroscopy of solar pores", *Astron. Astrophys.* 405, 331.
Hong, J., Bai, X., Lee, Y., et al., 2020. "MgI 12.32 μm line in a flaring atmosphere", *Astrophys. J.* 898, 134.
Hundhausen, A. J., 1977. "An interplanetary view of coronal holes", *Coronal Holes and High- Speed Wind Streams*, ed. J. B. Zirker, Colorado Associated University Press, Boulder, CO, p. 301.
Jurcak, J., Rezaei, R., Bello Gonzalez, N., et al., 2018. "The magnetic nature of the UP boundary in sunspots", *Astron. Astrophys.* 611, L4.
Komm, R., Hernandez, I. G., Howe, R., et al., 2015. "Solar-cycle variations of sub-surface meridional flow", *Solar Phys.* 290, 3113.
Kontogiannis, I., Tsiropoula, G., Tziotziou, K., et al., 2020. "Emergence of small-scale magnetic flux in the quiet Sun", *Astron. Astrophys.* 633, A67.
Kossobokov, V., Le Mouël, J.-L., & Courtillot, V., 2012. "On solar flares and cycle 23", *Solar Phys.* 276, 383.
Lee, J., 2007. "Radio emissions from solar active regions", *Space Sci. Rev.* 133, 73.
Leka, K. D., & Skumanich, A., 1998. "Evolution of pores and the development of penumbrae", *Astrophys. J.* 507, 454.
Liewer, P. C., Qiu, J., & Lindsey, C., 2017. "Comparison of helioseismic far-side active region detections with STEREO", *Solar Phys.* 292, 146.
Lindsey, C., & Braun, D. C., 2000. "Seismic images of the far side of the Sun", *Science*, 287, 1799.
Livingston, W., Harvey, J. W., Malanushenko, O. V., et al., 2006. "Sunspots with the strongest magnetic fields", *Solar Phys.* 239, 41.
Löhner-Böttcher, J., Schmidt, W., Schlichenmaier, R., et al., 2018. "Absolute velocity measurements in sunspot umbrae", *Astron. Astrophys.* 617, A19 (LB18).
Löptien, B., Lagg, A., van Noort, M., et al., 2020. "Connecting the Wilson depression to the magnetic field of sunspots", *Astron. Astrophys.* 635, A202.
Low, B. C., 1996. "Solar activity and the corona", *Solar Phys.* 167, 217.
Martinez Gonzalez, M. J., Manso Sainz, R., Asensio Ramos, A., et al., 2012. "Dead calm areas in the very quiet Sun", *Astrophys. J.* 755, 175.

Maunder, A. S. D., 1909. *Catalogue of Recurrent Groups of Sunspots for the Years 1874 to 1906*, H.M. Stationery Office, Edinburgh.

Moradi, H., Baldner, C., Birch, A. C., et al., 2010. "Modeling the subsurface structure of sunspots", *Solar Phys.* 267, 1.

Mordvinov, A. V., Grigoryev, V. M., & Peshcherov, V. S., 2012. "Large scale magnetic field of the Sun and evolution of magnetic activity", *Solar Phys.* 280, 379.

Mordvinov, A. V., & Yazev, S. A., 2014. "Reversals of the Sun's polar magnetic fields", *Solar Phys.* 289, 1971.

Moreno-Insertis, F., & Emonet, T., 1996. "The rise of twisted magnetic tubes in a stratified medium", *Astrophys. J. Lett.* 472, L53.

Mullan, D. J., 1971. "The structure of transverse hydromagnetic shocks in regions of low ionization", *Mon. Not. Roy. Astron. Soc.* 153, 145.

Mullan, D. J., 1974. "Sunspot models with Alfven wave emission", *Astrophys. J.* 187, 621.

Narayan, G., & Scharmer, G. B., 2010. "Small-scale convection signatures associated with a strong plage solar magnetic field", *Astron. Astrophys.* 524, A3.

Nindos, A., & Andrews, M. D., 2004. "The association of big flares and CME's", *Astrophys. J. Lett.* 616, L175.

Okamoto, T. J., & Sakurai, T., 2018. "Super-strong magnetic fields in sunspots", *Astrophys. J. Lett.* 852, L16.

Parker, E. N. 1958. "Dynamics of the interplanetary gas and magnetic fields", *Astrophys. J.* 128, 664.

Parker, E. N., 1974. "Magnetic fields in the Sun", *Bull. Amer. Astron. Soc.* 6, 18.

Parker, E. N., 1979. "Sunspots and the physics of magnetic flux tubes. I. The general nature of the sunspot", *Astrophys. J.* 230, 905.

Pastor Yabar, A., Martinez Gonzalez, M. J., & Collados, M., 2018. "Magnetic topology of the north solar pole", *Astron. Astrophys.* 616, A46.

Pevtsov, A. A., Canfield, R. C., & Metcalf, T. R., 1995. "Latitudinal variation of helicity of photospheric magnetic fields", *Astrophys. J. Lett.* 440, L109.

Priest, E. R., Longcope, D. W., & Janvier, M., 2016. "Evolution of magnetic helicity during flares and CME's", *Solar Phys.* 291, 2017.

Radick, R. R., Lockwood, G. W., Skiff, B. A., et al., 1998. "Patterns of variation among Sun-like stars", *Astrophys. J. Suppl.* 118, 239.

Rempel, M., 2011. "Subsurface magnetic field and flow structure of simulated sunspots", *Astrophys. J.* 740, 15.

Rieger, E., Share, G. H., Forrest, D, J., et al., 1984. "A 154-day periodicity in the occurrence of hard solar flares?" *Nature* 312, 623.

Roudier, Th., Malherbe, J. M., & Mirouh, G. M., 2017. "Dynamics of the photosphere along the solar cycle from SDO/HMI", *Astron. Astrophys.* 598, A99.

Rust, D. M., 1994. "Spawning and shedding helical magnetic fields in the solar atmosphere", *Geophys. Res. Lett.* 21, 241.

Sanchez, S., Fournier, A., & Aubert, J., 2014. "Advection-dominated flux-transport solar dynamo models", *Astrophys. J.* 781, 8.

Schmassmann, M., Rempel, M., Bello Gonzalez, N., et al., 2021. "Characterization of magneto-convection in sunspots: The GT stability criterion", *Astron. Astrophys.* 656, 92.

Schmelz, J. T., Holman, G. D., Brosius, J. W., et al., 1994. "Coronal magnetic structures observing campaign: 3. Coronal and magnetic field diagnostics derived from multiwaveband active region observations", *Astrophys. J.* 434, 786.

Shapiro, A. I., Solanki, S. K., Krivova, N. A., et al., 2014. "Variability of Sun-like stars: reproducing observed photometric trends", *Astron. Astrophys.* 569, A38.

Solov'ev, A. A., & Kirichek, E. A., 2014. "The sunspot: Shallow or deep?" *Geomagn. Aeron.* 54, 915.

Spitzer, L., 1962. "Encounters between charged particles", *Physics of Fully Ionized Gases*, 2nd ed., Interscience, New York, pp. 120–154.

Tadesse, T., Pevtsov, A. A., Weigelmann, T., et al., 2014. "Global solar free magnetic energy and electric current density distribution of Carrington rotation 2124", *Solar Phys.* 289, 4031.

Takeda, A., Acton, L., & Albanese, N., 2019. "Solar cycle variation of coronal irradiance observed with YOHKOH", *Astrophys. J.* 887, 225.

Telloni, D., Antonucci, E., Dolei, S., et al., 2014. "Stereoscopic investigations on fluctuations in the outer solar corona", *Astron. Astrophys.* 565, A22.

Tian, C., & Petrovay, K., 2013. "Structures in compressible magnetoconvection and the nature of umbral dots", *Astron. Astrophys.* 551, A92.

Title, A. M., Topka, K. P., Tarbell, T. D., et al., 1992. "On the differences between plage and quiet Sun in the solar photosphere", *Astrophys. J.* 393, 782.

Valio, A., Spagiari, E., & Marengoni, M., et al., 2020. "Correlations of sunspot physical characteristics during solar cycle 23", *Solar Phys.* 295, 120.
Vaquero J. M., 2004. "A forgotten naked-eye sunspot recorded by Galileo", *Solar Phys.* 223, 283.
Verma, M., & Denker, C., 2014. "Horizontal flow fields observed in Hinode images", *Astron. Astrophys.* 563, A112.
Vieu, T., Gonzalez, J. M., Yabar, A. P., et al., 2016. "How to infer the Sun's global magnetic field using the Hanle effect", *Mon. Not. Roy. Astron. Soc.* 465, 4414.
Virtanen, I. I., Pevtsov, A. A., & Mursula, K., 2019. "Structure and evolution of the photospheric magnetic field in 2010–2017", *Astron. Astrophys.* 624, A73.
Watanabe, H., Kitai, R., & Ichimoto, K., 2009. "Characteristic dependence of umbral dots on their magnetic structure", *Astrophys. J.* 702, 1048.
Watson, F. T., Fletcher, L., & Marshall, S., 2011. "Evolution of sunspot properties during solar cycle 23", *Astron. Astrophys.* 533, A14.
Watson, F. T., Penn, M. J., & Livingston, W., 2014. "A multi-instrumental analysis of sunspot umbrae", *Astrophys. J.* 787, 22.
Wolff, C. L., 1976. "Timing of solar cycles by rigid internal rotations", *Astrophys. J.* 205, 612.
Yadav, R., Louis, R. E., & Mathew, S. K., 2018. "Investigating the relation between sunspots and umbral dots", *Astrophys. J.* 855, 8.
Zhang, H., 2012. "Reversal magnetic chirality of solar active regions", *Mon. Not. Roy. Astron. Soc.* 419, 799.
Zhukova, A., Khlystova, A., Abramenko, V., et al., 2020. "A catalog of bipolar AR's violating the Hale polarity law, 1989–2018", *Solar Phys.* 295, 165.

# 17 The Corona

Since ancient times, people fortunate enough to witness a total eclipse of the Sun have been able to see, for a few minutes, a remarkable phenomenon with the unaided eye: this feature becomes visible during the short interval of time (no more than 7–8 minutes) when the brilliant light from the photosphere of the Sun is blocked totally by the Moon. When the total eclipse begins, the disappearance of the solar photosphere does not mark the onset of the complete darkness of night. Instead, witnesses see a faint residual brightness in an extended region surrounding the dark side of the Moon. At its brightest, the faint light has an intensity that is no more than several millionths of the brightness at the center of the solar disk (van de Hulst 1950). The faint light is called the corona, the Latin word for "crown", because it appears that the Sun is "wearing" a (faint) covering on top of its brilliant (but hidden-by-the-Moon) photosphere.

Coronal radiation in the visible part of the spectrum has several components. The first, the $K$-corona, is continuum radiation (due to electron scattering) that dominates close to the Sun. Inside radial locations of $1.1R_\odot$, the $K$-corona has an intensity of a few times $10^{-6}$ in units of the intensity at the center of the solar disk (van de Hulst 1950). The $F$-corona is due to scattering of sunlight off interplanetary dust particles: the letter "$F$" indicates that the spectrum of this part of the corona contains many of the Fraunhofer lines that occur in the spectrum of the Sun itself. The $F$-corona dominates the coronal emission at radial locations $r > 2.0$–$2.2\ R_\odot$. Van de Hulst (1950) suggests that the $K$-corona is the "real corona" (varying during the solar cycle), while the $F$-corona is a "spurious corona" (no variation with the solar cycle). A third component, the $E$-corona (not mentioned by van de Hulst) gives rise to multiple emission lines in the visible and infrared portions of the solar spectrum. An example of a spectrum of the $E$-corona, obtained during an eclipse of the Sun by Voulgaris et al. (2012), is shown in Figure 17.1. Observations obtained during eclipses that occur at various stages of the 11-year solar cycle indicate that the emission lines in the $E$-corona vary in their relative strengths during the solar cycle.

The corona is observed to have shapes that differ systematically from one eclipse to another. These different behaviors can be seen in Figures 17.2 and 17.3, which were obtained by means of sophisticated image processing (Druckmüller et al. 2014). On the one hand, in Figure 17.2, obtained on July 22, 2009 (when the Sun was in the deepest observed minimum of solar activity), the corona at high latitudes (close to the N and S poles) is characterized by striations radiating out from the Sun, looking like iron filings scattered on a piece of paper close to a bar magnet. These striations trace the global (poloidal) magnetic field of the Sun. At low latitudes in Figure 17.2, the global field is obscured by features called "streamers": some of these are labeled "helmet streamers" because they resemble a certain type of World War I military helmet at lower heights, with a narrowing "spike" at greater altitudes. On the other hand, when the image in Figure 17.3 was obtained in 2001, the Sun was close to activity maximum: the "iron filings" around the poles are not readily apparent, and now there are "helmet" streamers at essentially all latitudes. The size of helmet streamers, as measured by the separation between their footpoints on the surface of the Sun, can be estimated roughly from Figures 17.2 and 17.3: the linear sizes are of order several tenths of a solar radius, i.e., of order a few times $10^{10}$ cm.

It is not immediately obvious from Figures 17.2 or 17.3, but the *motions of gas in the solar atmosphere* may be distinctly different inside and outside the helmet: Dolei et al. (2015) report that, inside one particular helmet, the gas remains essentially *stationary* at radial distances $r \leq 3.5R_\odot$, whereas outside the helmet, the gas is *expanding* radially outward at speeds of 40 km s$^{-1}$ already at $r = 2.5R_\odot$ and at speeds of 140 km s$^{-1}$ at $r = 5R_\odot$. Thus, the "solar wind" (see Chapter 18) escapes freely from the *outside* of a helmet streamer but not from the *inside*. Is there any physical reason

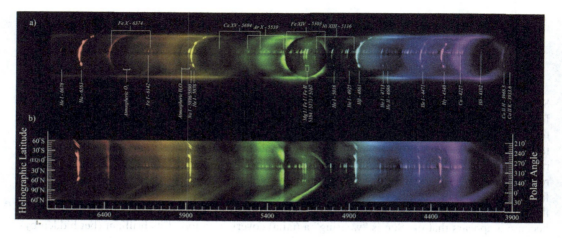

**FIGURE 17.1** Spectrum of the solar corona obtained during an eclipse. The coronal lines (which constitute the $E$-corona) due to highly ionized stages of Ar, Ca, Fe, and Ni are labeled by wavelength and ionization stage along the top of the upper spectrum. Along the bottom of the upper spectrum, chromospheric lines in the flash spectrum are labeled; four lines in the Balmer series are present. (Voulgaris et al. 2012; used with permission.)

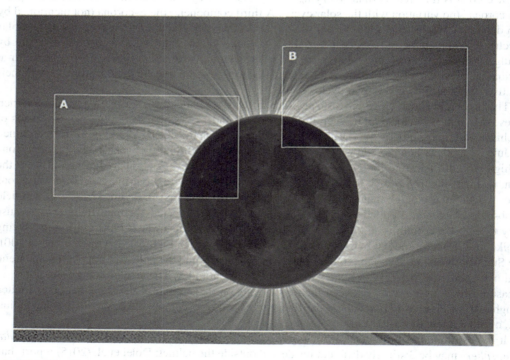

**FIGURE 17.2** The corona as photographed during a total eclipse on July 22, 2009, when the Sun was in a very deep activity *minimum*. North pole is at the top. Notice that at both north and south poles, there are striations reminiscent of iron filings near a magnet: the striations indicate the global (poloidal) magnetic field of the Sun. Boxes A and B were inserted by Druckmüller et al. (2014), but we do not discuss them here. (Used with permission of S. Habbal.)

The Corona

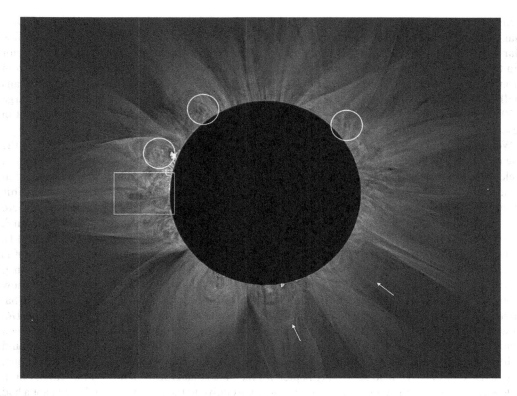

**FIGURE 17.3** The corona as photographed during a total eclipse on June 21, 2001, when the Sun was close to *maximum* activity. North pole is at the top. The global poloidal field is for the most part obscured, due to multiple streamers that are present at essentially all latitudes. Certain features in the image have been marked by Druckmüller et al. (2014), but we do not discuss them here. (Used with permission of S. Habbal.)

why the region of stationary gas in a streamer should extend out to $r \approx 3.5 R_\odot$? We will return to this question in Section 18.3.

To the unaided eye, the corona has a pearly white color. For this reason, the corona as seen in Figures 17.2 and 17.3 is referred to as the "white-light corona". Subsequently, we will compare this corona with what is observable in other spectral regions, especially in X-rays. Instrumental measurements show that the spectrum of the $F$ corona is more or less identical to sunlight, apart from the presence of several emission lines that have no counterpart in the photosphere. The corona appears brightest at locations close to the surface of the Sun, and the brightness decreases with increasing distance from the photosphere

The origin of the corona, i.e., the identification of the physical process(es) that heat(s) the corona, is a long-standing problem in solar physics. Magnetic fields play a key role in structuring the corona and in heating it. In this chapter, we summarize the physical parameters that have been determined for the corona in various locations on the Sun. These provide boundary conditions that have to be explained by solar researchers.

## 17.1 ELECTRON DENSITIES

Quantitative measurements of the brightness of the corona during eclipses indicate that the brightest regions of the corona, near the edge of the Moon's disk, i.e., at the base of the corona, have intensities of no more than a fraction $f \approx 5 \times 10^{-6}$ times the brightness at the center of the (uneclipsed) solar disk. A remarkable feature of the white light corona (at least out to radial distances of a few solar radii) is that the light is observed to be *polarized*. Coronal light in which the electric vector is in the radial direction (pointing directly away from, or toward, the Sun) differs in intensity from coronal light in which the electric vector is in the tangential direction. The difference is by no means small: at radial locations between 1 and 3 solar radii ($R_\odot$), the degree of polarization can be as large as 50% (van de Hulst 1950). The presence of significant polarization indicates that coronal light in these regions of the corona arises as a result of scattering of radiation off free electrons.

We can use this information to obtain a first estimate of the density of free electrons $n_e$ at the base of the corona. When we view the corona from Earth, our line of sight passes through a column of electrons that has a number density of $n_e$ at its densest (where our line of sight reaches a radial distance closest to the Sun), combined with an effective transverse length $L$ through the corona. This column of electrons is capable of scattering a fraction of $5 \times 10^{-6}$ of the light from the photosphere into our line of sight. The column density of electrons along our line of sight is $N_e = n_e L$ cm$^{-2}$. Combining this column density of electrons with the Thomson cross-section $\sigma_T$ (Equation 3.1), the fraction of light from the photosphere that is expected to be scattered into our line of sight is $f = n_e L \sigma_T$. Inserting $f = 5 \times 10^{-6}$, we find that $n_e L$ must have a numerical value of about $10^{19}$ cm$^{-2}$. What is a reasonable estimate for the transverse length of our line of sight? It is difficult to see how it could be larger than one solar radius: therefore, an upper limit of $10^{11}$ cm is plausible for $L$. What about a lower limit? The inner corona is seen to fall off in intensity with a scale height of order $0.1 R_\odot$ (Newkirk 1967). What could give rise to such a scale height? A possible answer has already been mentioned in Section 5.1: if a gas is in hydrostatic equilibrium (HSE), the density/pressure should decline with increasing height according to a scale-height $H_p = R_g T/g\mu$. Now, in the upper corona, the existence of outflow of solar wind (Chapter 18) indicates that HSE is definitely *not* valid, but in the low corona, where the wind is still moving slowly (relative to the sound speed), HSE is not a bad approximation. We shall see (Section 17.3) that the low corona has $T \approx 10^6$ K: in such conditions, hydrogen is fully ionized, and so $\mu \approx 0.5$. This leads to $H \approx 7 \times 10^9$ cm $\approx 0.1 R_\odot$. Thus, a lower limit on $L \approx 10^{10}$ cm is plausible near the base of the corona. Using these limits, the prior estimate of $n_e L \approx 10^{19}$ cm$^{-2}$ leads to the conclusion that the coronal base density $n_e$ is of order $10^8$–$10^9$ cm$^{-3}$.

Detailed analysis of the coronal brightness and how it falls off with increasing distance from the Sun have been analyzed by (e.g.) Newkirk (1967): he finds that, indeed, the densities at the base of the corona are in the range $10^{8-9}$ cm$^{-3}$. The radial profile of density is found to vary between polar regions of the Sun (where the corona has lower densities) and the equatorial regions (where higher densities are present). The density also varies between solar minimum and solar maximum. However, in all cases the density profile decreases monotonically as one observes farther from the Sun.

How can we understand why the densities at the base of the corona have the values estimated earlier? Recall that the gas density in the photosphere of the Sun, i.e., about $3 \times 10^{-7}$ gm cm$^{-3}$ (see Table 5.3), corresponds to number densities (assuming pure hydrogen) of order $2 \times 10^{17}$ cm$^{-3}$. Thus, between photosphere and coronal base, the number density of particles decreases by a factor of $\sim 10^{8-9}$. We shall see that the temperature in the corona is higher than in the photosphere by a factor of $\sim 200$. Thus, the pressure at the coronal base is lower by $\sim 2 \times 10^{-(6-7)}$ than the photospheric pressure. In the presence of hydrostatic equilibrium, this would correspond to traversing about 13–15 scale heights. The scale height $H_p$ in the photosphere is known to be 114–140 km (see Equation 5.5). In the chromosphere, where temperatures rise to perhaps 1.5 times the photospheric value, along with a reduction in the mean molecular weight by a factor of almost two, the value of $H_p$ in the chromosphere is expected to be of order 300 km. Thus, the linear height corresponding to 13–15 scale heights should be between 1500 km and 4500 km. Even though this is a simple argument, it is noteworthy that the linear height up to the base of the corona overlaps well with the observed (very rough) thickness of the chromosphere (Section 15.2): that is, it is reasonable to assume that the base of the corona occurs at a height that is equal (essentially) to the height of the gas at the top of the chromosphere.

# 17.2 ELECTRON TEMPERATURES

The physical parameters of the corona, specifically its density and temperature, cannot be derived by the techniques that were used to study the photosphere. In the latter, the theory of radiative transfer was the tool that provided information as to the variation of physical quantities as a function of optical depth $\tau$. But in the corona, the gas is so rarefied that $\tau$ is always $\ll 1$, i.e., optically thin conditions prevail. So we have to rely on different techniques if we wish to determine numerical values for the key physical parameters in the corona. We shall typically use the temperature as independent variable in the corona.

## 17.2.1 Optical Photons

The first reliable estimates of temperatures in the coronal gas were obtained by Edlen (1945), based on his study of fine structure atomic energy levels in isoelectronic sequences, i.e., a series of different elements in different stages of ionization such that all members of the series contain the *same number* of bound electrons. (An example of such a sequence can be found in Edlen (1936): this sequence, each with 12 bound electrons, contains the ions Mg I, Al II, Si III, ... Ca IX, ... Fe XV, the last of which will play a role in Figure 17.8.) By searching for patterns in the energy intervals that separate certain fine-structure levels in each of the ions in particular isoelectronic series, Edlen was able to identify the upper and lower energy levels that give rise to certain emission lines in the *visible spectrum* of the corona.

Edlen showed that the strongest coronal emission line in the *red* part of the visible spectrum (at $\lambda = 6375$ Å) originates as a (forbidden) transition between two fine-structure energy levels that exist in the Fe X ion (see Figure 17.1), i.e., iron with nine electrons removed. Edlen also showed that the strongest coronal emission line in the *green* part of the visible spectrum (at $\lambda = 5303$ Å) originates in a (forbidden) transition between two fine-structure levels in the Fe XIV ion (see Figure 17.1), with 13 electrons removed. The removal of 9 or 13 electrons from iron atoms requires a source of energy: assuming that the electron temperature in the plasma $T$ supplies the necessary energy, it is found that $T$ must be of order 1–2 MK, i.e., some 200 times hotter than the photosphere. (In what follows, we use the abbreviation MK for million degrees Kelvin.) More precisely, the red Fe X line is formed at about 1.2 MK, while the green Fe XIV line is formed at about 1.8 MK. A line at 5694 Å due to Ca XV (see Figure 17.1) is formed at about 2.3 MK: this line has never been observed at solar minimum (Voulgaris et al. 2012), indicating that at solar minimum the hotter coronal gas above 2 MK more or less disappears (see Figure 17.10).

For an image of the corona during two eclipses in which regions of the corona that emit the 6374 Å line appear in red, while regions of the corona that emit the 5303 Å line appear in green, see Figure 17.4. In the top two panels, the image contains only the emissions in the *lines* themselves. In the lower two panels in Figure 17.4, as well as the red and green line emissions, the image is overlaid with a white-light *continuum* image.

Since Edlen's discovery of these large temperatures, the key question about the solar corona has been: how does the Sun manage to heat up electrons in its atmosphere to temperatures that are some 200 times hotter than the photosphere? Already when we discussed the chromosphere (Chapter 15), we raised the question as to how *chromospheric* gas could become heated above the photospheric temperature. In the chromosphere, the problem was relatively mild: we "only" had to explain why the temperature should increase above the photospheric value by a factor of about two. In that case, mechanical heating due to acoustic waves was found to be adequate to provide much of the chromospheric heating, at least in the low-to-mid chromosphere. When we come to the corona, we are faced with an analogous problem, except that now we have to account for an increase in temperature by a factor of at least 200. To be sure, we are dealing with gas that has a $10^9$ times *lower density* than the gas in the photosphere: as a result, even a relatively small flux of mechanical energy may be all that is needed to boost the temperature to the MK mark. Nevertheless, the questions remain: what

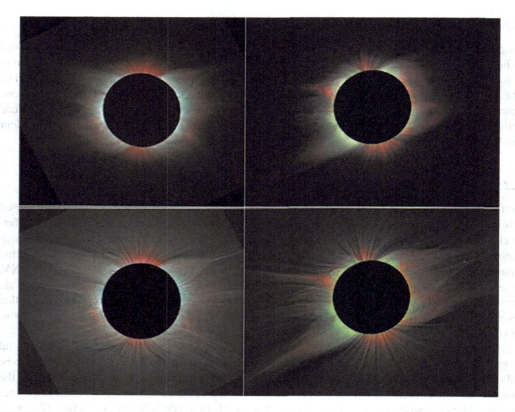

**FIGURE 17.4** Images of two eclipses of the Sun: March 29, 2006 (left), and August 1, 2008 (right). Upper panels: the *E*-corona, showing emission in red and green lines only; lower panels: white light image of the *K+F* corona is superposed on the red/green image in the upper panel. (From Habbal et al. 2010; used with permission of S. Habbal.)

is the source of the mechanical energy? and how much flux of that energy is needed to account for the observed temperature?

While Edlen's major achievement was to help to determine the *temperature* in the corona, his study also helped, interestingly enough, to set an upper limit on the electron *density* in the corona. To see why this is so, we note that the two coronal emission lines analyzed by Edlen were found to be "forbidden lines" (see Section 3.1): the transitions occurred in both cases between the $P_{3/2}$ ground level and a fine-structure $P_{1/2}$ level that lies about 2 electron-volts (eV) above the ground level. Electric dipole transitions are not allowed between such levels according to the common selection criteria that apply to *LS* coupling in an atom. (The selection rules indicate that an electron starting in a *P* orbit should be allowed to make a transition only to either an *S* orbit or a *D* orbit, but *not* to another *P* orbit.) However, the much rarer magnetic dipole transitions *can* allow transitions from a *P* state to another *P* state. Edlen estimated that their radiative probabilities are in the range 10–500 sec$^{-1}$, i.e., the lifetimes of the upper level are in the range 2–100 millisec. (Laboratory measurements now indicate that the lifetime of the upper level of the red Fe X line at 6374 Å is $14.2 \pm 0.2$ millisec [Brenner et al. 2009], well within the range estimated by Edlen.) This requires that the Fe X and Fe XIV ions must be preserved free from collisions for time-scales that may have to be as long as 0.1 sec. This sets an upper limit on the local electron density. The mean free time between collisions $\tau_c$ is given by the formula $1/(n_e \sigma V)$, where $\sigma$ is the collision cross-section and $V$ is the mean electron speed. For Coulomb collisions, $\sigma \approx 2 \times 10^{-4}/T^2 \approx 2 \times 10^{-16}$ cm$^2$ in the corona (with $T = 1$ MK). The thermal velocity of coronal electrons has a mean value of $7 \times 10^5 \sqrt{T}$ cm sec$^{-1}$, i.e., $V \approx 7 \times 10^8$ cm sec$^{-1}$. Therefore, in order to ensure that the mean free time between collisions is longer than (say) 0.1 sec, $n_e$ should not exceed the limiting value of order $10^9$ cm$^{-3}$. As we have already seen (Section 17.1), the intensity of scattered light at the base of the corona is observed to be such that $n_e$ is indeed in the range $10^{8-9}$ cm$^{-3}$, consistent with Edlen's limit.

In passing, we note that the fine-structure splitting of about 2 eV between the $P_{3/2}$ and the $P_{1/2}$ levels in Edlen's ions is very large compared to the fine-structure splitting that we normally see in optical spectra. A famous pair of lines in the yellow part of the solar optical spectrum, the lines that were labeled the *D* lines by Fraunhofer when he discovered absorption lines in the solar spectrum in 1814, lie at wavelengths of 5890 and 5896 Å. (These can be seen as a single [blended] yellow emission line in the flash spectrum in Figure 17.1.) The separation of the two *D* lines in wavelength occurs because of the fine-structure splitting between $P_{3/2}$ and $P_{1/2}$ levels in neutral sodium, analogous to the lines that Edlen studied in the corona. However, in the case of the yellow *D* lines, the fine-structure splitting is only 0.002 eV, i.e., three orders of magnitude smaller than in Fe X and Fe XIV. The large difference in splitting arises because of the highly stripped nature of the ions in Edlen's study: on the one hand, the electron in sodium responsible for the *D* lines moves in a Coulomb field due to a net nuclear charge of $Z = 1$, but on the other hand, the electron that produces the Fe X or Fe XIV coronal lines moves in a Coulomb due to a net nuclear charge of 9 or 13. These large charges lead to significant increases in the energies of the bound levels (see Section 3.2.1), as well as in the fine-structure splittings in Fe X and Fe XIV, compared to those in sodium.

### 17.2.2 X-RAY PHOTONS

At temperatures of order 1 MK, the mean thermal energies of the particles in the coronal plasma are of order 0.1 keV. In such a plasma, much of the radiation emerges in spectral lines with energies of order 0.1 keV, extending in energy up to a few times this value.

Photons with energies of 0.1–1 keV have wavelengths of roughly 100–10 Å. Such photons are referred to by astronomers as "soft" X-rays. ("Hard" X-rays are those with energies of $\geq 10$ keV.) Soft X-rays are strongly absorbed in the Earth's atmosphere, and therefore cannot be observed from the ground. Direct detection of even the strongest lines in the coronal X-ray spectrum had to await the launching of rockets and spacecraft that would carry instruments into regions of space above the Earth's atmosphere. Such instruments were first launched in the late 1940s. Solar X-ray astronomy came into its own in the 1960s with the launch of a series of satellites called Orbiting Solar Observatories (OSO), and also with the flight of the Skylab space station (in orbit during the years 1973–1974). The last in the OSO series, OSO-8, was launched in 1975.

Examples of X-ray spectra of the Sun in two different wavelength ranges are shown in Figures 17.5 and 17.6. The wavelength range in Figure 17.5 (13–20 Å) corresponds to photon energies between (roughly) 0.5 keV and 1 keV. The wavelength range in Figure 17.6 (3–7 Å) corresponds to energies extending to higher values, (roughly) 2–4 keV.

It is striking how completely different the spectrum of the Sun is at X-ray wavelengths compared to what we see when we view the Sun in visible light. In the latter, there is a strong continuum (which we can see with our own eyes when a rainbow is visible), from which many *absorption* lines remove light (see Figure 3.4): this is characteristic of the radiation that emerges from the *optically thick* photosphere. But in the coronal spectrum in Figure 17.5, we see that the continuum is relatively weak, and the spectrum is dominated by a multitude of strong *emission* lines.

By comparing the wavelengths of the lines in the spectrum with tables of lines observed in laboratory plasmas, many of the emission lines from the solar corona in Figures 17.5 and 17.6 have been identified (see labels of the various lines in both figures). Interestingly, and in corroboration of Edlen's pioneering work on the interpretation of optical photons, many of the observed X-ray lines

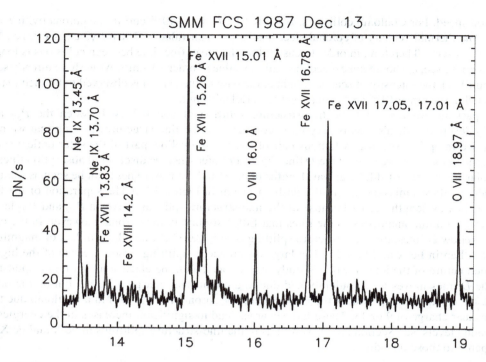

**FIGURE 17.5** X-ray spectrum of a quiescent active region on the Sun between wavelengths of 13 and 20 Å obtained by the flat crystal spectrometer on board Solar Maximum Mission (SMM) on December 13, 1987. (Del Zanna and Mason 2014; used with permission of ESO.)

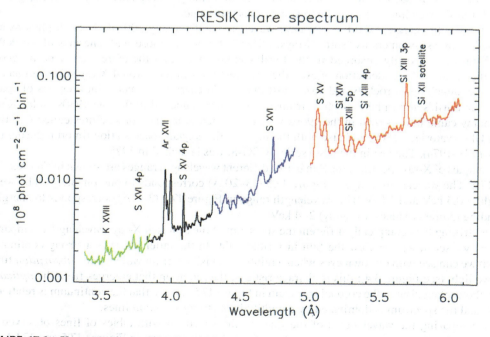

**FIGURE 17.6** X-ray spectrum of the Sun during a flare obtained by the Polish-led RESIK instrument on board the Russian spacecraft CORONAS-F on February 22, 2003. Different colors indicate different channels of the instrument. (Del Zanna and Mason 2018; used with permission of Springer.)

can be assigned to highly stripped stages of ionization of some of the more abundant elements in the Sun, including oxygen, neon, and iron. For the lines present in Figure 17.5, the emitting element with the highest levels of ionization is Fe, including Fe XVII and Fe XVIII, i.e., iron which has lost three and four more of its electrons than the most highly ionized iron (Fe XIV) that was discussed by Edlen.

Prominent in the solar X-ray spectrum are the Lyman-series lines of the hydrogenic ions of several elements. The Lyman lines occur when an electron makes a transition into the ground state (the $n = 1$ level) from levels with $n = 2, 3, 4, 5. \ldots$ . A hydrogenic ion is one in which only one electron remains in bound orbit around the nucleus. In an element of atomic number Z, the Bohr model of the hydrogen atom indicates that Lyman lines are predicted to lie at wavelengths proportional to $1/Z^2$. (See Exercises 3.4–3.7.) Spectra of the Sun in X-rays in the quiet Sun or in flares have been obtained by 16 different X-ray satellite detectors during the years 1977–2007 (Doschek and Feldman 2010). These spectra have revealed Lyman-$\alpha$ lines of the hydrogenic ions of O ($Z = 8$) at 19.0 Å (see Figure 17.5), Ne ($Z = 10$) at 12.16 Å, Mg ($Z = 12$) at 8.4 Å, Al ($Z = 13$) at 7.2 Å, Si ($Z = 14$) at 6.2 Å, S ($Z = 16$) at 4.75 Å (see Figure 17.5), Ca ($Z = 20$) at 3.0 Å, and Fe ($Z = 26$) at 1.8 Å.

The presence of highly stripped ions is a clear indication that electron temperatures $T_e$ in the plasma are high. Can we determine how high $T_e$ actually is? To answer this, we can do the following thought experiment: suppose we were to strip the last remaining electron off O, Ne, and Mg, how much energy would that require? To answer this, we note that, according to the Bohr theory of the atom (see Section 3.2.1), the ionization potentials $I(Z)$ required to strip *all* Z electrons off an element with atomic number Z are larger by factors of $Z^2$ than the ionization potential of hydrogen (13.6 eV).

Now when we applied the Saha equation to a medium with low electron pressure (such as occur in the chromosphere and corona), we found that hydrogen begins to undergo significant (50%) ionization at temperatures of 7100–7200 K (see Section 4.3). Analogously, in order to generate significant populations (50%) of the hydrogenic ions of O, Ne, and Mg, we need to solve the Saha equation $\theta I - 2.5\log T = -\log p_e$ (see Equation 4.6) for cases with $I = I(Z)$. In the low corona, the electron pressure does not differ greatly from that in the upper chromosphere (see Section 17.10). Therefore, if we set $\log p_e = 0$ (as in Section 4.3, for the upper chromosphere), we shall not make a serious error. The logarithmic temperature term is slowly varying, and so the solution for the temperature of 50% ionization in each ion is roughly $T \sim Z^2$. Since the appropriate T for 50% ionization is about 7000 K for hydrogen ($Z = 1$), when we set $Z = 8, 10$, and 12, we find that 50% of O, Ne, and Mg are in the hydrogenic state when the temperature has values of roughly at $T = 0.5$–1 MK.

Thus, the observational detection of Lyman-$\alpha$ lines of hydrogenic O and S (and also other elements) in the solar X-ray spectrum provided significant corroboration that Edlen (1945) was prescient in his identification of the red and green emission lines in the visible spectrum of the corona as arising in highly stripped iron in gas with temperatures of order 1 MK.

## 17.3 "THE" TEMPERATURE OF LINE FORMATION

Each emission line in the X-ray spectrum of the Sun, arising as it does from a specific element (say, Fe) and from a specific stage of ionization of that element (say, Fe XIV), is emitted from gas in which the temperature is not strictly uniform but spans a finite range. However, if we observe a strong line from Fe XIV emitted by the coronal gas, this tells us that the range of electron temperatures $T_e$ in that gas cannot be arbitrarily broad. If $T_e$ were too low, it would be impossible for the Fe atoms to be stripped of 13 electrons: therefore, all lines originating in the Fe XIV ion would necessarily be weak. If, on the other hand, $T_e$ were too high, even more than 13 electrons would be stripped from the ion, forming Fe XV or higher stages of ionization. Once again, in such a case, lines from Fe XIV would no longer be emitted in significant quantities. Detailed atomic structure calculations show that any given ion (say, Fe XIV) is present in maximum abundance when the

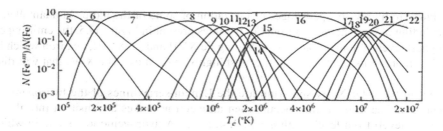

**FIGURE 17.7** Fractional abundances of different ionization stages of iron as a function of electron temperature. The ordinate is the ratio of the density of Fe in ionization stage m+ to the total density of Fe. Each curve in the figure is labeled with an integer $m$, where $m = 0$ corresponds to neutral iron (i.e., Fe I), while $m = 9$ and 13 correspond to Fe X and Fe XIV respectively. (From Jordan 1969; used with permission of Blackwell Publishing.)

temperature of the gas has a certain value: according to the calculations of Jordan (1969), the peak abundance of Fe XIV occurs at about 2 MK (see Figure 17.7). For Fe X, the peak abundance occurs at $T = 1$ MK. A full calculation of what fraction of Fe is in any one of the charge states from 4+ to 22+ when the temperature ranges from $10^5$ K to $2 \times 10^7$ K is shown in Figure 17.7 (Jordan 1969). As the figure shows, some ions (e.g., 7+, 16+) are dominant over a rather broad range of temperatures. On the other hand, other ions are present in large fractions (>0.3, say) only within rather narrow ranges of temperatures (e.g., 10, 11, 12, 13+). It is the latter charge states that are of most interest in studying the solar corona.

Now it is true that at $T \approx 2$ MK, gas containing iron will contain *some* iron ions that have lost "only" 8, 10, or 12 electrons, while other iron ions will be present that have lost 14 or 16 electrons. However, the dominant ion of iron in that gas is (according to Figure 17.7) Fe XIV. As a result, spectral lines originating in transitions between energy levels of the Fe XIV ion (such as Edlen's "green line" in the visible spectrum) will be maximally strong in gas with $T \approx 2$ MK. Given the existence of such a peak, it is reasonable to refer to "the temperature of formation" $T_i(f)$ of each line (labeled by $i$) in the X-ray spectrum.

Note that the temperatures we refer to in discussing ionization processes are *electron* temperatures: they are measures of the mean thermal speeds of electrons in the plasma. Why do the temperatures refer to electrons? Because it is the fast motion of passing (free) electrons that determines whether an iron ion in that gas, in the presence of those free electrons, can retain its own bound electrons, or whether it would be energetically favorable to move to a higher ionization stage, or to a lower one. And it is also the passing (free) electrons, with their characteristic thermal speeds of order $kT$, that are responsible for inducing transitions of bound electrons in (say) Fe XIV from a lower to an upper energy level, thereby emitting a particular spectral line.

## 17.4 EMISSION LINES THAT ARE POPULAR FOR IMAGING THE CORONA

### 17.4.1 SOHO/EIT

The SOHO spacecraft, launched in 1996, contained an instrument called EIT (Extreme Ultraviolet Imaging Telescope) that obtained images of the entire Sun in four narrow-band filters centered at certain wavelengths: (i) 171 Å, including lines of Fe IX/X, maximally sensitive to gas with temperature $T_p \approx 1$ MK; (ii) 195 Å, including a line of Fe XII, $T_p \approx 1.5$ MK; (iii) 284 Å, including a line of Fe XV, $T_p \approx 2$ MK; and (iv) He II 304 Å, $T_p \approx 80{,}000$ K. An example of images of the Sun obtained in these four filters on a random day are shown in Figure 17.8, with each panel from left to right corresponding to (i) blue, (ii) green, (iii) yellow, and (iv) red. The bright spots in all four images are due to active regions that happened to occupy certain positions on the visible disk on the day when the

The Corona    317

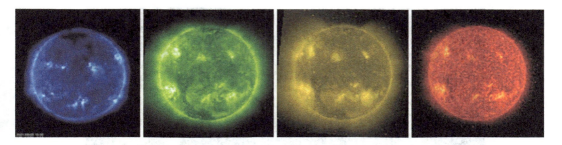

**FIGURE 17.8** Images of the Sun obtained on September 29, 2021, by SOHO/EIT in narrow bandpasses of wavelength corresponding to temperatures (from left to right) of $T_p$ = 1 MK, 1.5 MK, 2 MK, and 0.08 MK. (Courtesy of SOHO/EIT consortium. SOHO is a project of international cooperation between ESA and NASA.)

images were obtained. Extended dark regions around the north pole (at the top of all four images) are coronal holes where densities of the coronal gas are locally low.

It is natural for the human eye, when examining the images in Figure 17.8, to have its attention drawn to the brightest parts of the images, i.e., the active regions. In the 284 Å (yellow) image, it is obvious that most of the radiation in the image does in fact come from active regions: in the spaces between the active regions, there are extended dark patches over the surface, indicating that material with temperatures of 2 MK is essentially not detectable outside active regions. In other words, emission from 2 MK gas is essentially not detectable in areas of quiet Sun.

However, the 195 Å (green) image (caused by 1.5 MK gas) has a different texture: the active regions are *not* the only locations that are the sites of detectable X-ray emission. In addition to the active regions, one also sees a "fuzzy" emission extending over almost the entire disk of the Sun. An enlarged 195 Å image, obtained on a day when the Sun was very quiet (with no sunspots at all on the visible hemisphere), is shown in Figure 17.9: this image indicates that the 1.5 MK gas is distributed over (more or less) the entire surface of the *quiet* Sun. In this regard, we note that in a study of the eclipse of August 17, 2017, even when the sunspot number was $N_s$ = 45 (i.e., moderately active), Boe et al. (2020) concluded that "the vast majority of the plasma in the corona" has $T_e$ < 1.5 MK. A small minority of hotter plasma is confined to active regions. Although some of the hotter plasma may at times be ejected in the solar wind, Boe et al. stress that such hotter material will "only compose a small fraction of the total solar wind plasma". Thus, the green images in Figures 17.8 and 17.9 are indeed representative of *most* of the gas in the corona that exists in the *quiet Sun*. This indicates that, whatever the physical process is that heats the coronal gas to 1.5 MK, the process is by no means confined to active regions but is operating essentially everywhere on the Sun. On the other hand, the data in Figure 17.8 indicate that, in order to heat the coronal gas to 2 MK or hotter, the Sun needs to have active regions. This conclusion will become even more convincing when we discuss results from the YOHKOH spacecraft (Figure 17.13).

## 17.4.2 SDO/AIA

The SDO spacecraft, launched in 2010, contained an instrument called AIA (Atmospheric Imaging Assembly) that obtained images of the entire Sun in 10 bandpasses, each centered at a certain wavelength but also spanning a finite range of wavelengths: inside the finite range, each bandpass contains lines that, depending on the stage of ionization, are strongest in quiet Sun or coronal holes, in active regions, or in flares. The lines in the 10 AIA bandpasses originate in gas with a range of "peak temperatures" $T_p$ from as low as a few thousand K to more than 10 million K. Bandpasses (i)–(iii) observe the photosphere in filters peaked at 1700 Å, 4500 Å, and 1600 Å, sensitive to gas with $T_p \approx$

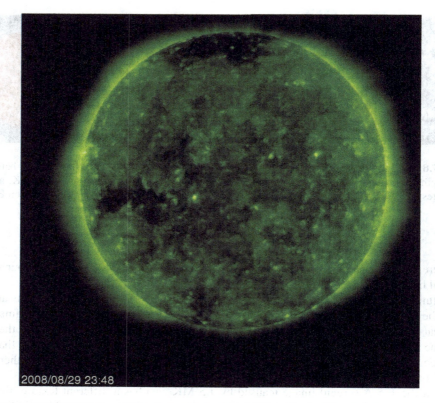

**FIGURE 17.9** SOHO/EIT image of the Sun on August 29, 2008, in the Fe XII 195 Å line, with $T_p = 1.5$ MK. On this day, the number of sunspots $N_s$ on the surface was observed to be precisely zero: in fact, $N_s$ had been observed to have the value zero on 27 days in the month of August 2008, i.e., the Sun was very quiet. But 195 Å emission can be seen at essentially all regions of the Sun's surface, i.e., in the *quiet* Sun. (Courtesy of SOHO/EIT consortium. SOHO is a project of international cooperation between ESA and NASA.)

4500 K, 6000 K, and $10^4$ K respectively; (iv) He II 304Å, $T_p \approx 80{,}000$ K; (v) 171 Å, including an Fe IX line, $T_p \approx 0.6$ MK; (vi) 193 Å, including an Fe XII line, $T_p \approx 1.2$ MK (but if the Sun flares, this channel also detects a line from Fe XXIV, with $T_p \approx 20$ MK); (vii) 211 Å, including an Fe XIV line, $T_p \approx 2$ MK; (viii) 335 Å, including an Fe XVI line, $T_p \approx 2.5$ MK; (ix) 94 Å, including an Fe XVIII line, $T_p \approx 6.3$ MK; and (x) 131 Å, including an Fe VIII line, $T_p \approx 0.4$ MK (but if the Sun flares, this channel also detects a line from Fe XXI, with $T_p \approx 10$–16 MK).

### 17.4.3 Hinode/EIS

Raster scans of smaller portions of the Sun's surface can be made in a large number of lines with the EIS instrument on Hinode: the wavelength range spanned by EIS is 170–290 Å, and this range includes more than 200 lines that are potentially detectable by EIS (Landi and Young 2009). An example of the advantages of using EIS lines is given by Hannah and Kontar (2012), who used 48 of these lines that are sensitive to temperatures ranging from a few tenths of 1 MK up to several MK in order to construct a quantity known as the "differential emission measure" (DEM: see Section 17.5) for various features in the Sun. The larger number of lines makes for a more reliable determination of the DEMs in quiet Sun (with a peak in DEM at $\log T \approx 6.0$), active regions (with a peak in DEM at $\log T \approx 6.25$, and flares (with a peak in DEM at $\log T \approx 7.0$). Using a smaller number of lines, namely, the six SDO/AIA channels, Aschwanden et al. (2011) reported that DEM peak temperatures in different regions of the Sun were found to be lowest in the coronal hole around the south pole: $\log T \approx 5.7$.

## 17.5 QUANTITATIVE ESTIMATES OF THE "EMISSION MEASURE" OF CORONAL GAS

Because many lines in the X-ray spectrum are associated with fairly narrow ranges of $T_e$, it seems reasonable to use $T_e$ as the independent variable when we wish to interpret (in physical terms) the amount of radiation emitted by coronal gas in a particular line. That is, although we have used optical depth and linear distances as independent variables for the various regions of the Sun we have discussed so far, those variables are not helpful when discussing the corona.

Once $T_e$ is chosen as independent variable, we need to ask: what are we to use as the physical parameter to specify *the amount of the gas in the solar atmosphere* that contributes to the intensity of a particular spectral line characterized by a particular value of $T_e$? The answer is: in a region of the solar atmosphere with volume $V(T)$ (containing gas with a temperature within a narrow range $\Delta T$ centered on $T$), we will use a quantity known as the differential emission measure, DEM = $n_e^2(T) V(T)$. Here, $n_e(T)$ is the electron density of the gas with temperature inside the range $\Delta T$ centered on $T$.

Why is DEM an appropriate choice? Because the strength of photons in any coronal line is determined by the number of collisions that occur each second per unit volume between free electrons (with density $n_e$ cm$^{-3}$) and a particular ion of a particular element (with density $n_i$ cm$^{-3}$). Each particular ion is most abundant in a temperature range $\Delta T$ centered on $T$. To form a particular coronal line, a collision between a passing electron and an ion containing one or more bound electrons pumps one of the bound electrons to an upper level: a photon emerges when that electron returns to a lower level. Consider 1 cm$^3$ of plasma in which there exists one ion and one electron. Knowing the quantum properties of the ion and how fast the electron moves (i.e., temperature $T$), it is possible to calculate quantum mechanically the rate at which a particular line would be emitted. Multiplying by the photon energy $h\nu_L$ of the spectral line, this yields an energy emission rate $\Phi_L(T)$ (ergs cm$^3$ sec$^{-1}$) for that line from that 1 cm$^3$ volume at temperature $T$. As $T \to 0$, it is hard to excite any atomic levels, and so $\Phi_L(T) \to 0$ as $T \lozenge 0$. And at the highest $T$, above (say) 10 MK, all the elements are ionized, and there are few (or no) bound levels to radiate lines: only the (weak) free-free continua remain, and $\Phi_L(T)$ is once again small. As a result, for each line, $\Phi_L(T)$ has a peak value at an intermediate temperature. In the solar atmosphere, where a broad mixture of elements exists, each element contributes somewhat differently to an overall $\Phi(T)$ function. The result is a function that peaks at $T = 1-3 \times 10^5$ K (see Figure 17.13). This function is referred to as the "radiative loss function" for the optically thin gas in the corona.

If the 1 cm$^3$ volume contains $n_e(T)$ electrons at temperature $T$ and one ion, the energy emission rate in one line will be $n_e(T)\Phi_L(T)$. If the 1 cm$^3$ volume contains $n_e(T)$ electrons and $n_i(T)$ relevant ions of the appropriate element and charge state, and if all the photons can escape without being blocked by intervening material, then the energy emission rate will be $n_e(T) n_i(T) \Phi_L(T)$ ergs cm$^{-3}$ sec$^{-1}$. Electrons are supplied mainly by ionization of the dominant element (hydrogen, with $n_H$ nuclei cm$^{-3}$), i.e., $n_e \approx n_H$. The total number of ions of any element is determined by the abundance $\zeta$ of the element relative to hydrogen: $n_i = \zeta n_H$. As a result, $n_e(T) n_i(T) \sim n_e(T)^2$. Finally, if the coronal source we are observing has a volume of $V_c(T)$ for gas at temperature $T$, the source emits energy at a rate $\sim n_e(T)^2 V_c(T)\Phi_L(T)$ ergs sec$^{-1}$. The combination of $n_e(T)^2 V_c(T)$ is called the differential emission measure at temperature $T$.

When we observe on Earth a coronal line emitted by an active region with an energy flux of $F_E$ (in units of ergs cm$^{-2}$ sec$^{-1}$), and we wish to interpret the amount of that emission in terms of local physical quantities in the solar atmosphere, we first need to address the following question. How large is the energy flux emitted in that line back at the solar surface? Because radiant energy falls off as the distance squared, and our (near-)Earth observations are made at radial distances of $D = 1$

AU, we know that the radiant energy of any emission line at the solar surface will be more intense than at 1 AU by the square of the ratio of $D$ to the radius of the Sun $R_\odot$. We have already seen (see last sentence in Section 1.5) that $D/R_\odot \approx 215$. Therefore, the radiant energy $F_S$ of the emission line at the surface of the Sun will exceed the radiant energy $F_E$ of that line as measured at Earth by a factor of $215^2$. That is, $F_S = 215^2 F_E$. Multiplying by the area of the coronal source in units of cm$^2$, we find the rate at which energy emerges from the coronal source in that line $E_S$ (ergs sec$^{-1}$). Equating this to $n_e(T)^2 V_c(T) \Phi_L(T)$ and knowing (from quantum mechanics) the value of $\Phi_L(T)$, we can estimate the value of DEM for temperature $T$. Strategically choosing lines at various values of $T$ and summing over values of DEM at all available temperatures gives a total EM (also in units of cm$^{-3}$) for the corona as a whole.

Some coronal detectors measure the X-rays from the entire Sun (e.g., the EVE instrument on SDO). What value of EM would be appropriate in such conditions? The total volume $V$ of the corona visible from Earth is essentially the visible hemisphere area $A_{ss} = 2\pi R_\odot^2 = 3 \times 10^{22}$ cm$^2$ *times* the scale height $H \approx 7 \times 10^9$ cm, i.e., $V \approx 2 \times 10^{32}$ cm$^3$. (In a stratified atmosphere, this is the volume of an equivalent *uniform* atmosphere with a density equal to that at the base of the corona.) Given $n_e = 10^{8-9}$ cm$^{-3}$ at the coronal base, the EM for the entire visible hemisphere of the Sun is expected to be in the range from $2 \times 10^{48}$ cm$^{-3}$ to $2 \times 10^{50}$ cm$^{-3}$. Is there any evidence that these estimates are realistic? To answer this, we show in Figure 17.10 results obtained by the EVE-MEGS-A instrument on SDO, which observed the Sun over a broad range of the XUV/EUV/UV continuum during an interval of over 4 years between the date of first light (April 30, 2010) and the time when the charge-coupled device (CCD) instrument which was used to make the observations failed (May 26, 2014): this period included the maximum of solar cycle 24.

In the top panel, EM is plotted as a function of time from just after solar minimum (of cycle 23) through solar maximum (of cycle 24). Separate curves refer to gas with three different temperature ranges (cooler, medium, and hotter). We see that the vertical scale covers a range from $1 \times 10^{48}$ cm$^{-3}$ to $2 \times 10^{49}$ cm$^{-3}$. Adding up the contributions of the three ranges of temperature, the maximum

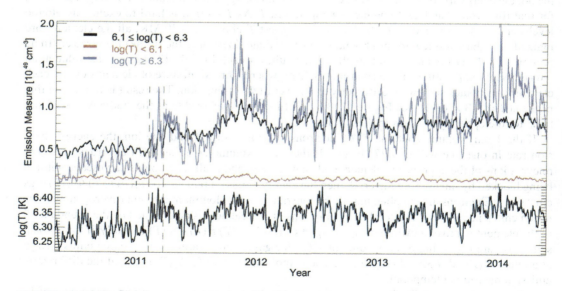

**FIGURE 17.10** Top panel: Values of EM from the SDO/EVE instrument over a 4-year interval. SDO/EVE records radiation from the entire solar disk. Different curves refer to the EM of gas that lies in different ranges of temperature: cooler (log $T$ < 6.1), medium (6.1<log $T$ < 6.3), and hotter (log $T$ > 6.3). Bottom panel: emission-measure-weighted mean temperature for the complete EVE data set. (Schonfeld et al. 2017; used with permission of S. Schonfeld.)

values of EM are of order $3 \times 10^{49}$ cm$^{-3}$. This range overlaps well with our estimated range of EM values. It seems that our order of magnitude estimates of density and volume are not unrealistic.

Independent X-ray data on the quiet Sun were obtained in February–November 2009 during the long and exceptionally low solar minimum by a Polish-led instrument (SphinX) on a Russian satellite (Sylwester et al. 2019). Analyzing a series of several hundred intervals of time when *no activity* was visible, Sylwester et al. found that a 1-temperature fit indicated a quiet Sun temperature of 1.69 MK and EM = $1.17 \times 10^{48}$ cm$^{-3}$. Sylwester et al. note, "a cool component of DEM similar in temperature to the one found from SphinX data is always present" in the results of Schonfeld et al. (2017): the data in the latter paper were obtained starting in April 2010, i.e., some 5 months *after* the end of the SphinX data were gathered. Moreover, the EM of the cool component reported by Sylwester et al. agrees well with the results reported from EVE for the cooler component (Figure 17.10).

It is of interest to inquire: when did the deepest part of the 2008–2009 solar minimum actually occur? Using three different indicators of activity, White et al. (2011) conclude that the minimum can best be identified with the time interval October 15–December 2 in the year 2008, although data from STEREO/EUVI suggest that the minimum may have extended into the first few months of the year 2009 (Nitta et al. 2014). In view of this, most of the SphinX data and all of the earliest SDO/EVE data were both obtained *after the minimum had occurred*, i.e., when solar activity was starting to increase in the earliest phase of cycle 24.

In a separate experiment on a different Russian satellite, Reva et al. (2018) set an upper limit on the presence of hot material *in the quiet Sun* by searching for emission in a particular spectral line (the Lyman-alpha line of hydrogenic Mg XII at $\lambda$ = 8.42 Å): this line requires a plasma temperature of at least 4 MK to excite it. Reva et al. (2018) reported that, in a search of data obtained during a 10–11-day interval (February 18–28, 2002) with 105 sec cadence, *not a single* reliable detection of Mg XII emission was obtained in roughly 10,000 images of the quiet Sun. Even in non-flaring active regions, no Mg XII emission was detected. The only sites where Mg XII emission was reliably detected by Reva et al. were in active regions where flares had actually occurred. These observations indicate that the *quiet Sun* does *not* contain detectable amounts of material with $T \geq 4$ MK. In support of this conclusion, Hannah et al. (2010) used data obtained during the years 2005–2009 by the RHESSI spacecraft to search for gas in the quiet Sun with temperatures of ≥5 MK: they reported that the DEM of such gas must be less than $10^{-6}$ of the peak DEM that occurs in the quiet corona.

## 17.6 THE SOLAR CYCLE IN X-RAYS

It is clear from Figure 17.10 that the EM varies significantly as the Sun goes from minimum (in early 2010) to maximum (in 2014). In particular, the EM of the "hot" material (with $T > 2$ MK) increases by at least one order of magnitude at solar maximum (2014) compared to its EM value in 2010, shortly after solar minimum. This behavior is associated with the emergence of many active regions during solar maximum, each containing gas that is hotter than 2 MK. Is this behavior unique to solar cycle 24? Or do other cycles also show significant changes in X-rays during a cycle? To answer that question, we illustrate 25 years of data from SOHO/EIT images that were obtained in a spectral line of Fe XV at 284 Å, corresponding to a temperature of about 2 MK, in Figure 17.11. Visual inspection shows clearly that at times of solar minima (1996, 2009, 2019), the Sun in Fe XV is significantly fainter than at times of solar maximum (2001–2002, 2014).

When the Sun is observed in images corresponding to even hotter temperatures, the magnitude of the change in solar emission is even greater than for observations at 2 MK. For example, the YOHKOH spacecraft had sensors that were sensitive to radiations from gas with temperatures up to 10 MK (Takeda et al. 2019): during the course of a solar cycle (1991–2001), the EM was found to vary by a factor of order 30.

On the other hand, when we examine the cooler coronal gas (log T < 6.1, i.e., T<1.3 MK), we see in Figure 17.10 the following striking feature: this "cooler" component of the corona changes remarkably little (if at all) as the Sun goes from solar minimum to maximum. As Schonfeld et al.

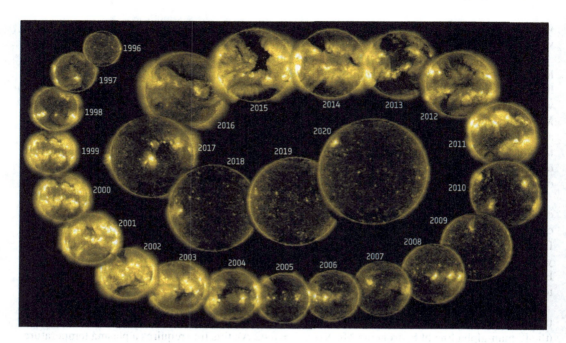

**FIGURE 17.11** Montage of images of the Sun obtained by SOHO/EIT in the Fe XV line at 284 Å in the course of 25 years. The 284 Å line is formed at a temperature of 2 MK. (Courtesy of SOHO/EIT consortium. SOHO is a project of international cooperation between ESA and NASA.)

(2017) have stated, "the quiet-Sun corona does not respond strongly to the solar cycle". Since the solar cycle is fundamentally a magnetic phenomenon, we might rephrase Schonfeld et al. as follows: the continuum emission at XUV/EUV/UV wavelengths from the quiet-Sun corona apparently does *not* respond strongly (if at all) to the large-scale magnetic fields that come and go on an 11-year cycle. Data providing independent support for this conclusion had been reported previously by Brooks et al. (2009) based on observations of several spectral lines at two solar minima (1998, 2006): although separated in time by 9 years and using different instruments (SOHO/CDS and Hinode/EIS), the DEMs for the quiet Sun in the range $\log T = 5.4$–$6.4$ were found to be essentially identical in both cases, with a peak at $\log T = 6.1$. The latter temperature is identical to the upper limit of temperature in the "cooler" component reported by Schonfeld in Figure 17.10. The constancy in time of the corona's *cooler* component is a noteworthy conclusion to which we shall return when we discuss the source of coronal heating.

Confirmation of the dichotomy in the temporal behavior of "hotter" and "cooler" coronal emission lines during the solar cycle is provided by independent data obtained by a different instrument (SOHO/CDS) (Del Zanna 2015). These data show that emission in a hotter (2.5 MK) line disappears at solar minimum, whereas in a cooler (1 MK) line, the active regions disappear at minimum, but the quiet Sun does not: instead, the quiet Sun remains visible as a "fuzzy" glow that persists even during the weakest solar minimum (see Figure 17.9, when there were no sunspots detected on the visible hemisphere).

## 17.7 THE SOLAR CYCLE IN MICROWAVE RADIO EMISSION

The solar cycles recorded in X-rays in Figure 17.11 refer to a time interval that lasts for 25 years. A much longer data series (70+ years) of nonoptical emission from the Sun is provided by observations

The Corona                                                                                                                                                                          323

of the radio flux $F_{10.7}$ emitted by all sources present on the solar disk at a wavelength of 10.7 cm, i.e., a frequency of 2.8 GHz. (Why was 10.7 cm chosen? The choice of wavelength lies "in the history of radar development during the Second World War" [Tapping 2013].) The observing equipment measures the radio flux from the Sun every day for an interval of one hour around noon in a bandpass with a width of 0.1 GHz. The measurements are cited in "solar flux units": 1 s.f.u. = $10^{-22}$ W m$^{-2}$ Hz$^{-1}$. In the 2.8±0.1 GHz bandpass, the solar radio emission is sensitive to physical conditions in the upper chromosphere and at the base of the corona. In regions where magnetic fields are weak (quiet Sun, and even in many active regions), the 10.7 cm flux is largely emission which occurs due to the thermal free-free process (*bremsstrahlung*). In sunspot regions, the dominant emission is thermal gyroresonance. These observations have been performed on a daily basis at a number of locations in Canada since 1947 (Tapping 2013). At solar maximum, $F_{10.7}$ is observed to reach (*monthly averaged*) values as large as 280 s.f.u., with individual *daily* values as large as 458 s.f.u. in 1947. At solar minimum, $F_{10.7}$ certainly has a smaller value than at solar maximum, but even in the lowest solar minima, $F_{10.7}$ does *not* fall to zero: instead, *there is a firm minimum below which the radio flux does not go*. Inspection of the daily values of $F_{10.7}$ indicates minimum values (in s.f.u.) in the low 60s: e.g., 61–62 in 1953, 64–65 in 1964, 65–66 in 1985, 65–66 in 1996, 64–65 in 2008, and 63–64 in 2019. Notice that the solar minima are not all identical in the radio flux: for example, the minimum in 2008 emits a weaker radio flux than the minimum in 1996. We have already seen (Section 13.8) that the frequency shifts in $p$-modes were observed to be smaller in 2008 than in 1996, consistent with the mean solar magnetic field being weaker in 2008: the fact that the 10.7 cm flux was also smaller in 2008 than in 1996 suggests that $F_{10.7}$ includes a detectable contribution from the magnetic field even when the latter is at its weakest.

Independent corroboration of the results from 10.7 cm data is provided over a range of frequencies by Shimojo et al. (2017), who have observed the Sun since 1950 at four frequencies between 1 and 9.4 GHz. Although the fluxes (at all four frequencies) can vary by a factor of two or so at the six solar maxima in the database, the behavior at solar minima is different: the observed fluxes are almost exactly the same at all five solar minima in the database.

The (near) constancy of the radio fluxes at several distinct solar minima, when gyro effects are minimized, probably represent free-free emission from the upper chromosphere and the base of the corona.

How are we to understand why free-free emission from the inactive Sun $F_{10.7}$ never falls below a certain value ($\approx$ 60 s.f.u.)? We can obtain a simplified answer to this question by proceeding in the following manner. According to Figure 17.10, the inactive solar corona (i.e., material with log $T$ < 6.1) has $EM \approx$ a few $\times 10^{48}$ cm$^{-3}$. The radiative loss function (Figure 17.13) includes free-free emission that can be seen as the rising curves at log $T$ > 7. Extrapolating back to a temperature appropriate for the quiet corona (log $T \approx$ 6.1), we estimate that free-free emission has $\Phi(T)$ perhaps of order $10^{-23}$ erg cm$^3$ sec$^{-1}$. Combining this with the measured $EM \approx$ a few $\times 10^{48}$ cm$^{-3}$ for the inactive corona, we expect that the inactive solar corona should radiate free-free emission at a rate of a few $\times 10^{25}$ erg sec$^{-1}$. On a sphere of radius 1 AU, this emission will be spread out at Earth orbit over an area of order $10^{27}$ cm$^2$ to generate a flux of order $10^{-2}$ erg cm$^{-2}$ sec$^{-1}$. Converting to the units used for 1 s.f.u., this flux corresponds to a value of order $10^{-5}$ W m$^{-2}$. The spectral shape of free-free emission is almost flat at all frequencies less than $h\nu_c \approx kT$ (and the spectrum falls off exponentially at higher frequencies). With $T \approx 10^6$ K, this means that the free-free emission is spread almost uniformly over a frequency range of $\nu_c \approx$ a few times$10^{16}$ Hz, indicating that the flux per unit Hz is of order a few times $10^{-21}$ W m$^{-2}$ Hz$^{-1}$ = a few tens of s.f.u. It is worth noting that this very crude estimate is not too far from the observed firm lower limit on the observed flux: 60 s.f.u. The upper chromosphere gas (with its higher densities than the coronal density) also contributes to the lower flux limit, and there is also a small contribution (amounting to probably no more than a few s.f.u. at solar minimum) due to magnetic effects.

## 17.8 ION TEMPERATURES

How can we determine the temperature $T_i$ of ions in the corona? Is $T_i$ equal to $T_e$? In principle, if there are sufficient collisions between ions and electrons, the thermal energy should be equilibrated, and $T_i$ should equal $T_e$. This is more likely to happen in denser gas, such as occurs in the densest streamers. In the less dense gas of "coronal holes", equilibration of ion and electron temperatures is more difficult to achieve.

In the event that equilibration occurs, the thermal velocities of (say) iron ions in a coronal region where the electron temperature is 1–2 MK should have mean values $V_{th} = \sqrt{(2kT_e/Am_H)}$ (where $A = 56$ for iron), i.e., $V_{th} = 17$–$24$ km sec$^{-1}$. Therefore, if we measure the line width of a coronal iron line, the half-width of the line $\Delta\lambda$ should in principle have a value that is related to $\lambda$ by the relationship $\Delta\lambda = \lambda V_{th}/c$. However, empirical data reveal that the observed line widths are *larger* than the thermal predictions. What causes the excess line widths? Could it be that the temperature of ions is enhanced relative to the temperature of electrons? Possibly: in fact, some extremely high temperatures (>100 MK) have been reported for certain ions in fast solar wind (e.g., Cranmer et al. 2008). But there is another possibility: there might be nonthermal motions ("turbulence") in the corona, and the ion lines might be broadened by the Doppler effect caused by those motions. We have already come across the idea of "microturbulence" in a very different context: "microturbulence" plays a role in the photospheric spectrum of the Sun when that spectrum is analyzed in terms of a one-dimensional model (Section 3.8.1). Quantitatively, however, there is a large difference in the amplitude of microturbulence between photosphere and corona: whereas in the photosphere the amplitude of the microturbulence is 1–2 km sec$^{-1}$, the amplitude in the corona is much larger, up to 60 km sec$^{-1}$ at altitudes of order $0.3R_\odot$ above the surface (Wilhelm et al. 1998). Waves on the magnetic field might explain this coronal "microturbulence" (Section 17.18.1): indeed, some models of coronal wind acceleration by magnetic waves predict that the wave amplitudes could be as large as 60 km sec$^{-1}$ (or more) in fast wind (Tu and Marsch 1997). The presence of such large turbulent motions makes it difficult to determine with confidence how much of the observed broadening of coronal lines can be attributed specifically to thermal motions. But if it could be shown that indeed the heavy ions are definitely much hotter than protons or electrons, that might contain important information about the physical process that heats the heavy ions (Cranmer et al. 2008).

## 17.9 DENSITIES AND TEMPERATURES: QUIET SUN VERSUS ACTIVE REGIONS

In the quiet Sun (QS), Brosius et al. (1996) used a rocket experiment to determine that $DEM(T)$ has a maximum value at $T = 1$–$2$ MK. This confirms the early work of Edlen (1945).

In contrast to the results obtained in the QS, Brosius et al. found that, in active regions (ARs), $DEM(T)$ remains significantly large at temperatures up to $T = 4$–$5$ MK. Thus, ARs contain material that is *definitely hotter*, by factors of up to two to three in temperature, than that which is present in the quiet Sun. (This has already been mentioned in our discussion of the EIT images in Figure 17.8 and also in Section 17.5.) Can we say anything about the relative spatial locations of hotter and cooler gas inside an AR? Yes, we can: stereoscopic triangulation of 70 loops in a single AR using simultaneous observations by the strategically located spacecraft STEREO A and B has shown (Aschwanden et al. 2009) that two classes of AR loops can be identified: "the hottest loops are found in the core of the AR, while the coolest are preferentially found in the peripheral plage region". In more recent work, Ghosh et al. (2017) added further details about the AR corona: (1) hot core loops ($T = 3$–$5$ MK), (2) warm loops ($T = 1$–$2$ MK), and (3) fan loops ($T = 0.6$–$1$ MK) located on the AR periphery and with longer lifetimes than other loops. However, in addition to the extra class of loops, Ghosh also mentioned, "a significant amount of diffuse plasma spread over a large area [of the AR] at coronal temperatures without any well-defined visible structures". We have already mentioned the presence of diffuse ("fuzzy") plasma distributed widely in the QS (Section 17.4.1 and Figure 17.9): the work of Ghosh et al. (2017) indicates that such "fuzzy" plasma is also present in ARs.

The reliability of DEM values from any region on the Sun can be improved by using as many lines as possible, distributed over as wide a range of temperatures as possible, such as those provided by Hinode/EIS (see Section 17.4.3). One particularly long loop observed by Hinode/EIS (Gupta et al. 2019) has allowed temperature to be determined by two independent methods: (i) density variations as a function of height in the loop were found to follow an exponential with a scale height of $59 \pm 3$ Mm, corresponding to $T = 1.37$ MK; (ii) EMs constructed for various lines intersected at a unique temperature at heights ranging from 1.1 to 1.25 $R_\odot$ and the unique temperature was found to be 1.37 MK. The agreement between two independent methods of obtaining temperatures in this coronal loop is remarkable.

As regards densities, in the quiet Sun, Brosius et al. (1996) (based on ratios of certain spectral line intensities) report densities of $10^9$ cm$^{-3}$: this is at the upper end of the range of densities that we reported in Sections 17.1 (based on the intensity of coronal light) and in Section 17.2.1 (based on properties of Edlen's forbidden lines). Moreover, in active regions, Brosius et al. report densities that are larger by factors of four to five relative to the quiet Sun.

In a more recent study of ratios of various pairs of lines in EIS/Hinode data, Gupta (2017) has also reported densities at the base of the corona in quiet Sun that lie in the range between $3 \times 10^8$ cm$^{-3}$ and $10^9$ cm$^{-3}$, while in active regions, the densities at the base of the corona are found to range from $10^9$ to $10^{10}$ cm$^{-3}$.

Thus, coronal material in active regions is observed to be *denser* than coronal material in the quiet Sun by factors of at least three and possibly 10.

## 17.10 GAS PRESSURES IN THE CORONA

Let us notice an important point about the pressure at the base of the corona. Now that we have information about temperatures and densities, we can evaluate empirically the gas pressures at the base of the corona ($p_{cb} = 2N_e kT$). Inserting $T = 1$–2 MK and $N_e = 1$–$4 \times 10^9$ cm$^{-3}$ (to include quiet Sun and active regions), we find coronal base pressures in the range 0.3–2 dyn cm$^{-2}$.

The physical significance of these pressures becomes apparent when we compare them with the pressure at the top of the chromosphere, $p_{tc}$. In Section 15.10 (last paragraph), we noted that $p_{tc}$ is of order 0.6 dyn cm$^{-2}$. It is important to notice that the range of $p_{cb}$ *overlaps* with $p_{tc}$. It seems that the pressure at the base of the corona may be essentially identical to the pressure at the top of the chromosphere.

This discussion of pressure reminds us that when hydrostatic equilibrium applies, it is useful to think in terms of pressure scale heights, $H_p$. The scale height is defined to be the vertical distance across which the density (or pressure) falls off by a factor of $e$. In terms of this definition, the number $n_p$ of scale heights that separates the top of the chromosphere from the base of the corona is given by $n_p = \ln(p_{tc}/p_{cb})$. Inserting the prior ranges of values of $p_{tc}$ and $p_{cb}$ and noting that $n_p$ cannot be negative, we find that $n_p$ ranges from at most 0.7 to a value approaching zero. In chromospheric gas, $H_p$ is a few hundred km. As a result, the transition from the chromosphere to the corona (we refer to this as "the transition region") occurs across a height range that is no more than 100–300 km: it may in fact be close to zero.

On a scale of one solar radius, even a transition over 300 km qualifies as relatively abrupt. If we perform the thought experiment of starting in the photosphere and moving up in altitude through the solar atmosphere, we will find that, after we pass through the chromosphere, the onset of the corona occurs over a height scale of no more than 300 km. Such a "transition region" is essentially discontinuous.

We shall return in Section 17.16, once we discuss thermal conduction, to discuss a physical reason how such a discontinuity might arise in the solar atmosphere.

## 17.11 SPATIAL STRUCTURE IN THE X-RAY CORONA

Hot material at $T = 4$–5 MK in the corona is highly localized to active regions: it is not found in the quiet Sun (Reva et al. 2018). On the other hand, the cooler coronal material, at temperatures of 1–2 MK, can be found essentially everywhere in the quiet Sun, which extends over large portions of the Sun's surface. As an illustration, we have already shown in Figure 17.9 an image of the Sun obtained in the 195 Å line of Fe XII, a line which is formed at $T \approx 1.5$ MK. Visual inspection of Figure 17.9 indicates that the 1.5 MK coronal material is spread more or less everywhere throughout the corona: one can see a "fuzzy glow" permeating most of the field of view (apart from coronal holes). However, when the Sun is imaged with an instrument sensitive to hotter gas ($T \geq 3$ MK), the result is a much "patchier" picture: see Figure 17.12.

Apparently, as long as we exclude coronal holes, the Sun finds a way to heat material *almost everywhere in the corona* to temperatures of order 1.5 MK, but heating the gas to temperatures of > 3 MK is a less common phenomenon that occurs only in active regions.

The contrast between the low-to-mid chromosphere and the hotter coronal regions is worth noting. No matter where one looks on the surface of the Sun, one finds chromospheric material at temperatures of 6–7 thousand K. But the hotter corona is far from spherically symmetric. This suggests that whatever is heating the hottest parts of the corona is distinct from whatever is heating the low-to-mid chromosphere. The spherical symmetry of the latter led us (in Chapter 15) naturally to the conclusion that acoustic waves from the ubiquitous convection may be heating the low-to-mid chromosphere. In Chapter 15, we provided quantitative evidence in favor of this hypothesis.

But in the hotter coronal regions, there must be additional localized sources of mechanical energy. Since the hotter coronal regions coincide with active regions, with their locally strong magnetic fields, it is natural to conclude that the heating of coronal material to > 3 MK is associated intrinsically with magnetic processes. Magnetic fields can provide mechanical energy over and above what is supplied by the acoustic waves from convection. One source of this extra energy is a variety of wave modes that exist in a magnetized plasma and that have no counterparts in a nonmagnetic medium (Alfven waves: Chapter 16, Equation 16.12). But there are also physical processes in a magnetized medium that do not occur in a nonmagnetic medium. The most striking of these ("magnetic reconnection") will be discussed in Section 17.19.11.

## 17.12 MAGNETIC STRUCTURES: LOOPS IN ACTIVE REGIONS

The fact that localized magnetic fields are associated with coronal heating receives strong confirmation from images such as Figure 17.12. This image was obtained with the YOHKOH spacecraft, where the imager responds mainly to gas at "hotter" temperatures above 2.5 MK (Yoshida et al. 1995). Even a casual inspection of the image shows that the strongest emission comes from features that appear to have shapes of a particular type, namely, the emission looks like it is coming from features that are best described by the terms "arcs", "arches", or "loops". Each arch has well-defined "footpoints" rooted in the solar surface, while the central part of the arch reaches up to a greater or lesser height above the surface.

One particularly clear example of a loop appears in Figure 17.12 about halfway between the center of the Sun and the right-hand limb. That loop has a length that can be estimated visually from Figure 17.12: the footpoints are separated by a distance $D_p$ of about one-half a solar radius, i.e., $\approx 3 \times 10^{10}$ cm. Other loops in X-ray images of the corona can be identified as having smaller footpoint separations. In some images, loops can be identified with footpoints that are separated by not much more than a couple of times $10^9$ cm. How high does a loop with footpoint separation $D_p$ extend above the surface of the Sun? At one extreme, magnetic field properties are such that it is unlikely that a stable loop would extent to heights exceeding $D_p$ by a large factor: such elongated loops would tend to "pinch off" near their base. At the other extreme, it is also unlikely that the loop would reach up only a very low height, much less than $D_p$: such a squat loop would need to have a shape that was essentially flat on top. In general, a loop with footpoint separation $D_p$ is expected to have a maximum height above the surface that is also of order $D_p$.

The largest loops in Figure 17.12 have lengths comparable to the spatial scales mentioned earlier in connection with streamers in the white-light corona (Figures 17.2 and 17.3). In fact, some helmet streamers in the white-light corona are spatially correlated with some of the loops seen in X-rays.

# The Corona

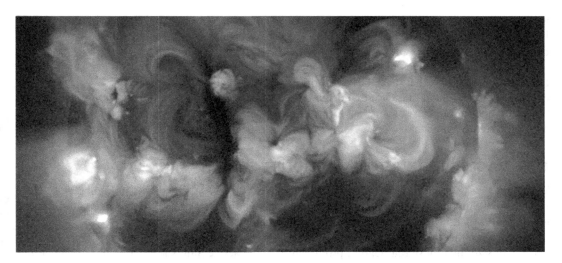

**FIGURE 17.12** Image of the Sun in X-rays emitted by gas at temperatures hotter than (roughly) 2.5 MK (YOHKOH/SXT). (Image is from the YOHKOH mission of ISAS, Japan. The X-ray telescope was prepared by the Lockheed-Martin Solar and Astrophysics Laboratory, the National Astronomical Observatory of Japan, and the University of Tokyo with the support of NASA and ISAS.)

The difference is that X-ray observations allow the loops to be observed *on the disk* of the Sun, and also in regions where the loops extend beyond the limb. In contrast, the white-light data can record loops only when these are extended above the limb, where the brilliant photosphere does not overwhelm the (faint) coronal light. In X-rays, the photosphere is dark, and as a result, loops of coronal emission are easily detected even when viewed against the background of the solar disk.

When the loop shapes in Figure 17.12 are compared with photospheric magnetic data, it is found that each arch (or loop) is associated with an active region. Each loop follows the location of a magnetic flux rope that emerges from the surface of the Sun at a location ("footpoint") where the magnetic polarity has a particular sign, and reenters the Sun at another footpoint (some distance away), where the magnetic polarity has the opposite sign.

Such a loop is said to be "closed", because both footpoints are rooted in the denser material of the solar surface. In the presence of a closed loop, solar material is constrained to follow the loop field lines (Section 16.6.1): the material is forbidden from flowing across the field lines. In this sense, it can be said that coronal gas is "trapped", and density builds up on the loop. As a result, if there are magnetic waves or other processes that supply energy to the loop, the energy supply has a "captive audience", i.e., the trapped gas in the loop. This material can be subjected to prolonged heating, which will be greater or less, depending on how much mechanical energy the magnetic field can deposit in the gas. The trapping of plasma by magnetic fields is the ultimate source of the hotter and denser gas observed in active regions (see Section 17.4.1).

## 17.13 MAGNETIC STRUCTURES: CORONAL HOLES

Coronal holes (CHs) differ from the remainder of the Sun not merely in having lower than average density and lower than average electron temperature. They also differ in the sense that they exhibit a different magnetic topology. In the holes, the magnetic data indicate that the magnetic field lines, rather than containing closed loops, are in fact *open* to space. As a result, coronal material is not trapped but can stream out freely from the Sun. In this situation, there is no "captive audience", and the density and temperature do not build up as much as on the closed loops of active regions. Whatever supply of magnetic energy is available in the CH goes into accelerating the plasma away from the Sun. So efficient is this acceleration that the fastest solar wind is observed to emerge from CHs.

As regards the lifetimes of CHs, those located at N and S poles are present at essentially all times: there is no sign that these CHs ever disappear. Using data from solar cycles 21–23, Hewins et al. (2020) have found that CHs that occur at low latitudes generally last for about one solar rotation, although in rare cases, some last for as long as 3 years. They also found that coronal holes do not rotate rigidly, in general.

## 17.14 MAGNETIC STRUCTURES: THE QUIET SUN

What can we say about the solar image of the cooler coronal gas that appears in Figure 17.9? This image of the Sun on a very quiet day (with no sunspots on the visible disk) shows that material with temperatures of 1.5 MK is found essentially at all locations in the quiet corona.

Is it possible that this ubiquitous 1.5 MK material might be attributable to magnetic loops, maybe smaller in size than those that feature prominently in the active Sun (such as the loops that can be easily identified in Figure 17.12)? In some cases, the answer appears to be "Yes". There are many small bright "points" that catch one's eye in Figure 17.9: it is easy to imagine that these *might* be due to small magnetic loops, each with its own localized source of magnetic energy.

But it is important to notice that there also exists in Figure 17.9 a much more extensive and diffuse, almost spherically symmetric, "fuzzy glow" that permeates essentially the entire image. What could account for this? Could acoustic waves from the convective turbulence explain this? The advantage of acoustic waves from convection is that they are emitted uniformly over the entire surface of the Sun, thereby possibly explaining the (near) spherical symmetry on the "fuzzy glow". We shall visit this possibility in Section 17.18.1. Wave modes other than acoustic may also be at work. For example, waves that start their existence as purely acoustic modes emerging from convective turbulence might be able to find a way to reach the corona by means of mode conversion in the presence of magnetic fields (Section 16.7.7). If that happens, then a combination of acoustic waves and magnetic fields might be responsible for this "fuzzy glow". In this regard, it is relevant to note that weak magnetic fields can be detected at essentially all locations of the quiet Sun (see Section 16.4.1.5): to be sure, the strength of the fields is not as large as in active regions, and the fields tend to be mainly horizontal (Bellot Rubio and Orozco Suarez 2019), presumably rooted in vertical fields that create small loops. However, from visual inspection of Figure 17.9, it is not obvious that any loop-like structures can be identified with certainty in the "fuzzy glow". We shall revisit this in Section 17.18.2.

## 17.15 WHY ARE QUIET CORONAL TEMPERATURES OF ORDER 1–2 MK?

At the end of Chapter 15, we noted that the material in the upper chromosphere is less and less efficient at disposing of any mechanical energy deposited therein. The cooling mechanism that provided such an effective thermostatic effect in the low-to-mid chromosphere diminishes greatly in efficiency in the upper chromosphere. The reason for the decreased cooling efficiency is that hydrogen is becoming almost completely ionized, and free protons plus free electrons are much less effective radiators than the bound electrons in hydrogen atoms. (Recall that bound electrons are much better *absorbers* of photons than are free electrons: see Section 3.3.1.) As a result, if any mechanical energy is deposited at the top of the chromosphere, the temperature in the gas increases rapidly ("runs away") to much higher values.

What will stop the temperature runaway above the upper chromosphere? The answer is: the runaway will stop when an additional source of cooling (over and above radiation) comes into play in order to dispose of the deposited energy. If and when that happens, a new equilibrium condition becomes possible.

# The Corona

Two additional cooling options are available in the solar corona. One has already been mentioned in the context of coronal holes (Section 17.13): the material begins to flow outward from the Sun, carrying away thermal energy. This option is available in coronal holes because the magnetic field lines there are mainly vertical and open to interplanetary space. The openness of the field lines is a characteristic signature of coronal holes.

## 17.15.1 Thermal Conduction by Electrons

But in the quiet Sun, the outflow option is not always available as a cooling mechanism: the magnetic fields in the quiet Sun's corona are observed to be for the most part horizontal (Section 16.4.1.5), indicative of closed loops. Although these loops in effect shut down the possibility of easy outflow, that does not mean that another form of cooling is excluded. An option that becomes available to assist in cooling is the mechanism of *thermal conduction*.

We have already come across this mechanism when we discussed (in Chapter 8) how heat is transported deep inside the Sun. According to Equation 8.1, in the presence of a temperature gradient $dT/dr$, the flux of heat down the temperature gradient is $F = -k_{th}(dT/dr)$, where $k_{th} = (1/3) \lambda V_t \rho C_v$ is the thermal conductivity (Equation 8.2). In Chapter 8, we calculated the value of $k_{th}$ in a manner appropriate for the deep interior of the Sun: there, we noted that the principal agent for the transport of heat was *photons*. Now we need to consider conduction in the corona, where photons are no longer the principal agents of heat transport. In the corona, it is *thermal electrons* that transport the heat most effectively.

The prior formula for $k_{th}$ remains valid, but now we need to reexamine the four parameters relevant in the case of thermal electrons in the corona, where electron densities are $n_e$ cm$^{-3}$. (i) The mean thermal speed of the electrons is given by $V_t = \sqrt{(2kT/m_e)}$. (ii) The mass density of the corona is given by $\rho = n_e \mu m_H$. (iii) The specific heat per gram at constant volume (Equation 6.3) is given by $C_v = (3/2)k/\mu m_H$. (iv) The mean free path $\lambda$ is given by $1/n_e \sigma$, where the electron-ion collision cross-section is $\sigma$.

To calculate $\sigma$ for electrons in the corona, we note that collisions occur because of the Coulomb forces. We have already seen (Section 11.3) that the Coulomb cross-section is given by $\sigma_c \approx \pi e^4 \Lambda / (kT)^2$ where the (slowly varying) Coulomb logarithm is $\Lambda$. In coronal conditions, the numerical value of $\Lambda$ is typically 10–20 (Spitzer 1962).

Combining the factors in Equation 8.2, we find that the thermal conductivity in the solar corona is given by

$$k_{th} = \frac{1}{3} \frac{(kT)^2}{n_e \pi e^4 \Lambda} \sqrt{\frac{2kT}{m_e}} \frac{3k}{2\mu m_H} n_e \mu m_H \qquad (17.1)$$

Collecting terms, we see that the electron density $n_e$ cancels out, and the conductivity depends only on a power law of the temperature, according to $k_{th} = k_o T^{2.5}$. In the corona, inserting appropriate values for the physical constants, we find that the numerical value of the coefficient $k_o$ is about $10^{-6}$ in c.g.s. units.

The high power of the temperature dependence in $k_{th}$ (index = 2.5) is noteworthy. It has the effect that although thermal conduction by electrons is negligible in the (cold) photosphere (where photons dominate the transport of energy), this is no longer the case in coronal conditions. In the corona, with temperatures that are 200–400 times larger than in the photosphere, the electron thermal conductivity is of order one *million* times more effective than in the photosphere. As a result, thermal conduction in the corona is a physical process that cannot be neglected.

Now that we have an expression for the thermal conductivity, we revert to a discussion of energy balance, such as that which was given earlier (Section 15.12) in the context of the chromosphere. Given that energy is deposited at a certain rate every second into 1 cm$^3$ of the corona, we need to know the rate $E_{cond}$ at which the energy is conducted out of that cubic cm per second by conduction.

This rate, i.e., the rate of energy loss per unit volume, is given by the spatial divergence of the heat flux $F$.

To proceed, we note that in the presence of a closed loop of half-length $L$, the coronal part of the loop (near the apex of the loop) has a temperature $T$ that is much larger than the footpoints. As a result, the temperature gradient $dT/dr$ along the loop is given more or less reliably by the ratio $T/L$. This leads to a heat flux downward toward the surface $F = k_{th}dT/dr \approx k_oT^{2.5}T/L$. The spatial divergence of this flux, div $F$ can be written roughly as $F/L$. Therefore, if we express everything in c.g.s. units, we find that the conductive contribution to the rate of cooling is $E_{cond} \approx k_oT^{3.5}/L^2$ ergs cm$^{-3}$ sec$^{-1}$. Once again, we note that the rate of conductive cooling due to thermal electrons in the corona has an even steeper dependence on temperature, with an index of 3.5.

### 17.15.2 Radiative Losses

Although radiative cooling operates with reduced effectiveness in the corona compared to the chromosphere, we may not conclude that radiative losses from the corona should be neglected altogether. On the contrary, they *do* contribute to removing energy that has been deposited in the coronal gas.

In the chromosphere, our discussion of radiative losses was cast in terms of the local opacity (see Equation 15.4). Now that we are in the corona, it is more convenient (see Section 17.5) to express the radiative losses in terms of an optically thin total loss function $\Phi(T)$ that includes lines (each with loss function $\Phi_L(T)$) plus continua. (Unfortunately, this total $\Phi(T)$ is sometimes written as $\Lambda(T)$, where $\Lambda$ is the Greek form of the initial letter of the word "loss": but because we have already used the symbol $\Lambda$ for the Coulomb logarithm, we prefer on the whole to use $\Phi(T)$ here for the total radiative loss function. However, as the sole exception to this choice, we note that

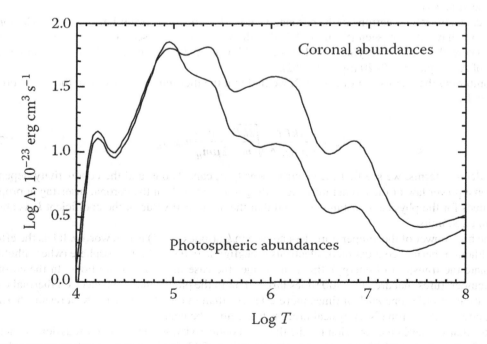

**FIGURE 17.13** Radiative loss function (labeled $\Lambda$ in this figure, rather than $\Phi(T)$) as a function of temperature, for two different sets of elemental abundances. (Figure kindly provided by J. Raymond, Harvard-Smithsonian Center for Astrophysics [2008].)

the vertical axis in Figure 17.13 uses the label $\Lambda$ to denote $\Phi(T)$.) The function $\Phi(T)$ is defined in such a way that for an optically thin medium with electron density $n_e$ and temperature $T$, the rate of energy loss from 1 cm$^3$ per second due to radiation is given by $n_e^2 \Phi(T)$. The numerical value of $\Phi(T)$ at any given $T$ depends on the particular mixture of chemical elements present in the gas: since certain "metals" in certain stages of ionization emit preferentially at certain temperatures, if those metals are more abundant in the mixture, then the value of $\Phi(T)$ will be larger at those temperatures. Results obtained by Raymond (2008) for $\Phi(T)$ for two different mixtures of elements are plotted in Figure 17.13. The curve labeled "coronal abundances" differs from the curve labeled "photospheric abundances" in the sense that hot gas (>2 MK) in active regions is observed to have relatively larger fractions of certain elements: the more abundant elements (enhanced in abundance by factors of two to four relative to photospheric values) are observed to be those that have *lower* values of the first ionization potential (FIP). In the quiet Sun and in coronal holes, the coronal gas has abundances equal to those of the photosphere: the presence of enhanced abundances of low-FIP elements (referred to as "FIP bias") is observed to be confined to active regions (Del Zanna 2019). The cause of FIP bias must occur in the chromosphere, where mixtures of ionized and neutral versions of different elements exist co-spatially: if a local source of electromotive force (e.g., due to MHD waves) is available, the neutrals and ions will respond differently to the force (Rakowski and Laming 2012).

For both mixtures of chemical elements in Figure 17.13, we see that $\Phi(T)$ has a maximum value at temperatures of 0.1–0.2 MK: at the maximum, the numerical value of $\Phi(T)$ reaches a value that is close to $10^{-21}$ erg cm$^3$ sec$^{-1}$. There is a sharp decrease in $\Phi(T)$ at $T < 10^4$ K and a more gradual decline (almost monotonic) in $\Phi(T)$ at temperatures between 1 and 10 MK. This behavior is reminiscent of the shapes of the opacity curves (Figure 3.3). This is no coincidence: the opacity due to lines and continua arises from the same atomic levels as are responsible for effective radiative losses in an optically thin plasma.

It is important to note that, in the conditions that are appropriate to the quiet solar corona, i.e., at temperatures of at most a few MK, the radiative loss function *decreases* as the temperature *increases*. We can approximate the slope of the curve in Figure 17.13 at "coronal" temperatures (i.e., between 1 and 10 MK) by the function $\Phi(T) \sim 1/T^{0.5}$. Visual inspection of Figure 17.13 suggests that, for a gas with photospheric abundances, the numerical value of $\Phi(T)$ in the "coronal" temperature range (i.e., 1–10 MK) can be approximated fairly well by the expression $\Phi(T) \approx 10^{-19}/T^{0.5}$ ergs cm$^3$ sec$^{-1}$.

Using this approximation, we see that the volumetric loss of energy by optically thin radiation in coronal conditions is roughly $E_{rad} = n_e^2 \Phi(T) \approx 10^{-19} n_e^2 / T^{0.5}$ erg cm$^{-3}$ sec$^{-1}$.

### 17.15.3 Combination of Radiative and Conductive Losses

Now we can return to the energy balance in the corona. If mechanical energy is deposited in unit volume at a rate of $E_{mech}$ ergs cm$^{-3}$ sec$^{-1}$ (by a mechanism that we leave unspecified at this point), this deposited energy can be carried off from that unit volume by the combined effects of radiation and conduction according to

$$E_{mech} = E_{rad} + E_{cond} \approx 10^{-19} n_e^2 T^{-0.5} + 10^{-6} T^{3.5} / L^2 \qquad (17.2)$$

In order to arrive at the lowest energy state, the corona tends towards a temperature $T_c$ such that the rates of radiative and conductive energy loss are comparable: this makes the cooling time-scale as long as possible. At temperatures that are either higher or lower than $T_c$, the cooling time-scale would be shorter, and the Sun would have to provide larger fluxes of mechanical energy to compensate for the more rapid cooling.

Equating the magnitudes of the two terms on the right-hand side of the prior equation at temperature $T_c$, we find that

$$T_c^4 \approx 10^{-13} n_e^2 L^2 \qquad (17.3)$$

It is convenient at this point to convert from electron density $n_e$ to coronal pressure: $p = 2n_e k T_c$ (where the factor 2 indicates that electrons and ions both contribute equally to the gas pressure in the corona). This allows us to rearrange Equation 17.3 as follows:

$$T_c^6 \approx \frac{10^{-13}}{4k^2}(pL)^2 \qquad (17.4)$$

Inserting the numerical value of Boltzmann's constant $k = 1.38 \times 10^{-16}$ ergs K$^{-1}$, we find

$$T_c^6 \approx 1.3 \times 10^{18}(pL)^2 \qquad (17.5)$$

Taking the sixth root of each side, we find

$$T_c \approx 1050(pL)^{1/3} \qquad (17.6)$$

In Equation 17.6, if $p$ is in units of dyn cm$^{-2}$ and $L$ is in units of cm, the units of $T_c$ are degrees K.

What should we use for the pressure $p$? In the low corona, $p$ is essentially equal to the pressure at the top of the chromosphere (see Section 17.10). At the top of the chromosphere (see Section 15.10.4), $p \approx 0.6$ dyn cm$^{-2}$.

What values are appropriate for loop lengths in the corona? The earlier discussion (Section 17.12) about loops in the corona indicates that $L$ values range from a few times $10^9$ to a few times $10^{10}$ cm. Substituting these values, we find that at the lower limit of $L$, we can set $pL \approx 10^9$: this leads to $T_c \approx 1$ MK. At the upper limit of $L$, we can set $pL \approx 10^{10}$, leading to $T_c \approx 2$ MK.

It is remarkable that simply by considering the lowest energy configuration, with equal loss rates in radiation and conduction, we have arrived at estimates of coronal temperatures that are consistent with the empirical results in the quiet Sun, where $T_c$ spans the range 1–2 MK (Section 17.4.1).

To be sure, we have not solved the problem of coronal heating completely: Equation 17.6 depends on knowing the values of the two parameters $p$ and $L$: we have not described how we might go about estimating values for these parameters, which are determined by certain physical processes in the Sun. It may be presumed that the value of $p$ is ultimately determined by the flux of mechanical energy that is generated by the convection zone. And the values of $L$ are controlled by the processes that generate magnetic fields in the Sun. So in calculating $T_c$ by means of Equation 17.6, let us not read more than we should into the result: we *can* obtain a numerical result for $T_c$ but *only if* we first turn to *empirical results* in order to identify the most appropriate values to insert for the quantities $p$ and $L$.

## 17.16 ABRUPT TRANSITION FROM CHROMOSPHERE TO CORONA

We have already noted (Section 17.10) that the transition between chromosphere and corona in the Sun occupies a spatial width that is quite narrow. The thickness of the transition is no more than 300 km, and may be much less. Now that we know about thermal conduction in the corona, we return to the question: why is the transition so abrupt?

The answer has to do with the fact that with the hot corona lying above the chromosphere, heat is conducted from the corona down into the chromosphere. The corona adopts a thermal structure such that a certain heat flux $F = -k_{th} dT/dr$ flows downward, supplying energy *from above* to the upper chromosphere. This is distinct from the supply of (acoustic) energy that emerges from the convection zone and enters the chromosphere *from below*.

# The Corona

At coronal temperatures, $k_{th} \sim T^{2.5}$ is so large that a given heat flux $F$ can be transported by a small temperature gradient, $dT/dr$. But as the temperature decreases toward chromospheric values, the value of $k_{th}$ decreases rapidly. In the upper chromosphere, where temperatures are lower than coronal values by factors of 100, $k_{th}$ is a mere $10^{-5}$ times its coronal value. Therefore, in order to transport the same heat flux $F$ downward, $dT/dr$ must become $10^5$ times *larger* in the chromosphere than in the corona. That is, the temperature gradient must become much steeper in the chromosphere than in the corona.

For numerical purposes, we note that in a coronal loop with half-length $L$, $dT/dr \approx T/L$. Inserting coronal values (T = $10^6$ K, $L = 10^{9-10}$ cm), we see that in the coronal portion of a loop, $dT/dr$ has a numerical value of typically $10^{-(3-4)}$ K cm$^{-1}$. In order to transport the same flux $F$ in the much less conductive chromosphere, we need to increase $dT/dr$ by $10^5$. This leads to $dT/dr \approx$ 10–100 K cm$^{-1}$. Of course, the heat flux may not remain strictly constant all the way down from the corona into the chromosphere: some of the energy may be dissipated along the way by radiative losses. But even if only 1%–10% of the coronal heat flux survives into the chromosphere, a temperature gradient of order 1 K cm$^{-1}$ would be required to transport that flux in the upper chromosphere. In the presence of such a gradient, the transition from the upper chromosphere (at $T = 10^4$ K) to a region where the temperature is (say) $10^5$ K, would occur across a spatial distance of no more than $10^5$ cm, i.e., 1 km.

Our previous estimate of the thickness of the transition (≤300 km), based on a comparison of pressures (see Section 17.10), can easily accommodate such a steep conductive structure. On the scale of the solar radius, the transition region between chromosphere and corona is essentially a discontinuity.

The discussion in the present section is based on a highly idealized treatment of conditions in the corona and chromosphere, as if there are a unique temperature in the chromosphere at all locations and a unique temperature in the corona. Actually, as we have already seen, there are spatial inhomogeneities in the chromosphere (spicules) and spatial structures in the corona (loops). In places where spicules exist, their lengths (up to several thousand km) allow them to extend well into the corona. As a result, the localized *roughness* of the solar surface is such that it is not accurate to think of a uniform spherically symmetric thin shell of thickness 1 km separating the chromosphere from the corona at all points on the solar surface. Instead, the transition region occurs at different heights above the photosphere in different locations, depending on local conditions. Nevertheless, wherever the transition from chromosphere to corona does occur, it is abrupt, occurring across spatial scales that may be as short as 1–300 km.

## 17.17 RATE OF MECHANICAL ENERGY DEPOSITION IN THE CORONA

We have seen that the corona in the quiet Sun is observed to be essentially always at temperatures of 1–2 MK (see Figure 17.10: especially the curve labeled "log $T$ < 6.1"). In view of this, it seems reasonable to conclude that the quiet corona has reached an equilibrium: as fast as the mechanical energy is deposited in unit volume, the material in that volume disposes of the energy through radiation and conduction losses (Section 17.15), leading to a stable temperature between 1 and 2 MK.

Now that we know the temperature and density of the solar corona, we can evaluate the volumetric radiation loss rate $E_{rad} = n_e^2 \Phi(T) \approx 10^{-19} n_e^2 / T^{0.5}$. In the quiet Sun, where $n_e \approx 3 \times 10^8$ cm$^{-3}$ (Section 17.9) and $T$ = 1–2 MK, this formula leads to $E_{rad}$(QS)$\approx 10^{-5}$ ergs cm$^{-3}$ sec$^{-1}$. In active regions, where $n_e$ is enhanced over quiet Sun by as much as 4–5 and $T$ is enhanced by 2–5, $E_{rad}$ (AR) is enhanced over quiet Sun values by a factor of perhaps 5–10. Thus, $E_{rad}$ (AR) ≥ $5 \times 10^{-5}$ ergs cm$^{-3}$ sec$^{-1}$. Since conductive losses are essentially equal to the radiative losses, we see that the rate of volumetric input of mechanical energy into the corona $E_{mech}$ must be roughly as follows:

$$E_{mech}(QS) \approx 2 \times 10^{-5} \cdot \text{ergs cm}^{-3}\text{sec}^{-1} \qquad (17.7)$$

$$E_{mech}(AR) \approx 10^{-4} \cdot \text{ergs cm}^{-3}\text{sec}^{-1} \tag{17.8}$$

It is worthwhile to compare these rates of volumetric energy deposition in the *corona* with the analogous rates in the *chromosphere* (see Equation 15.3): $E_{mech}(chr) = 0.3-3$ ergs cm$^{-3}$ sec$^{-1}$. In each cubic centimeter of the corona, mechanical energy is apparently being deposited at a rate that is three to five orders of magnitude *smaller* than in the chromosphere. That is, the demands of the chromosphere for mechanical energy are much greater than the demands of the corona. Despite the large reduction in mechanical energy deposition rate in the corona, the greatly reduced density of coronal material (compared to chromospheric densities) allows even a low rate of energy input to heat the coronal gas to temperatures of 1–2 MK.

Now that we know the rate of emission $E_{rad}$ (QS) from unit volume of the quiet corona, we can ask: how much power does the entire quiet corona emit in the form of radiation? The quiet corona is distributed over essentially the entire solar surface, i.e., it has an area $A(QS) \approx 4\pi R_\odot^2 = 6 \times 10^{22}$ cm$^2$. With an exponential scale height $H_c \approx 7 \times 10^9$ cm in the low corona (Section 17.5), the volume of the quiet corona over the entire Sun is $V_c(QS) \approx A(QS)H_c \approx 4 \times 10^{32}$ cm$^3$. Multiplying this by $E_{rad}$ (QS), we find that the radiative power of the quiet corona is $L_{cor} \approx 4 \times 10^{27}$ ergs sec$^{-1}$. Comparing to the total power output from the Sun (see Equation 1.11), we see that the quiet corona radiates in X-rays at a rate that is only one millionth of the photospheric radiation rate. The corona is truly a faint accessory of the Sun.

## 17.18 WHAT HEATS THE CORONA?

The problem of coronal heating is an active topic of research interest. The heating hypotheses can conveniently be divided into two major groups: waves and non-waves.

### 17.18.1 Wave Heating

One possibility for heating the corona is that a flux of waves (of some sort) is entering the corona and dissipating their energy there. According to this viewpoint, by analogy with our discussion of the chromosphere (Chapter 15), the local volumetric rate of energy deposition would be given by the divergence of the flux of energy $F(w)$ in the waves. Also by analogy with our discussion of the chromospheric heating, the divergence can be approximated by the ratio $F(w)/\lambda_d$, where $\lambda_d$ is a dissipation length (see Section 15.11). In a stratified medium, if dissipation is driven by nonlinear processes (such as shocks), $\lambda_d$ might be a few times the local scale height $H_p$. If this line of reasoning applies to the low corona (where $H_p \approx 7 \times 10^9$ cm: Section 17.5), this suggests that $\lambda_d$ may be $(2-3) \times 10^{10}$ cm. In the context of this "wave model of coronal heating", the necessary wave flux $F(w) = E_{mech}\lambda_d$ would be of order

$$F(QS) \approx 5 \times 10^5 \text{ergs cm}^{-2}\text{sec}^{-1} \tag{17.9}$$

in the quiet Sun, and

$$F(AR) \approx 3 \times 10^6 \text{ ergs cm}^{-2}\text{sec}^{-1} \tag{17.10}$$

in active regions.

Recalling that in the chromosphere, the wave flux $F$(chr) has previously been estimated to be of order $10^{7-8}$ ergs cm$^{-2}$ sec$^{-1}$ (Equation 15.2), we see that the coronal flux of waves (if waves are indeed the source of coronal heating) is smaller than the chromospheric wave flux by a factor of at least three, and maybe by as much as 200.

In the chromosphere, the nature of the waves that perform the heating (at least in the low-to-mid chromosphere) can be identified with a fair degree of confidence: acoustic waves emitted by the

# The Corona

turbulent convection beneath the photosphere are a good candidate. As a quantitative confirmation of this hypothesis, the theory of sound emission from convective turbulence predicts enough wave flux to perform the required heating (Section 15.10.2). Moreover, the fact that convection is always present, and occurs at all locations of the solar surface, means that the low-to-mid chromosphere is present essentially spherically symmetric on the solar surface.

But what might be the source of waves that could be responsible for *coronal* heating? Could some acoustic waves be responsible? Since coronal heating is stronger in active regions, it is natural to suspect that waves of a magnetic nature might serve the purpose. Let us consider two candidates.

### 17.18.1.1 Acoustic Waves?

What about acoustic waves? It is true that a large fraction of the acoustic flux coming up from the convection zone is dissipated in the chromosphere. But a "large fraction" does not necessarily mean "all". Might there be some acoustic wave flux "left over" at the top of the chromosphere? After all, even if as much as 99.5% of a flux of $10^8$ ergs cm$^{-2}$ sec$^{-1}$ were dissipated in the chromosphere, the surviving 0.5%, i.e., $5 \times 10^5$ ergs cm$^{-2}$ sec$^{-1}$, would suffice to supply the necessary flux of energy to heat the corona, at least in the quiet Sun. The spherical symmetry of the acoustic flux could help to explain why the "fuzzy glow" of 1.5 MK gas in the quiet Sun can be found almost everywhere on the surface (Figure 17.9).

However, there are empirical reasons that make it difficult to accept the acoustic wave heating possibility: the last of the OSO missions (OSO-8, launched in 1975) was used to search for acoustic waves coming up into the corona. The flux was found to amount to no more than $7 \times 10^4$ ergs cm$^{-2}$ sec$^{-1}$ (Bruner 1978). This is almost an order of magnitude smaller than what is required to heat even the quiet Sun corona (Equation 17.9). The data suggest that there are simply not enough acoustic waves reaching the corona to supply the heating required even for the quiet Sun.

### 17.18.1.2 Alfven Waves?

We have seen (Section 16.7.7) that the flux of Alfven waves in the photosphere $F_A$ could be of order $10^{7-8}$ ergs cm$^{-2}$ sec$^{-1}$ in regions where $B = 10$–100 G (assuming a velocity amplitude of 1 km sec$^{-1}$: see Section 16.7.7). Now, we have also seen (Equation 17.9) that a wave flux of $5 \times 10^5$ ergs cm$^{-2}$ sec$^{-1}$ would suffice to heat the quiet Sun's corona. Comparison with $F_A$ indicates that a field strength of even 0.5 G in the photosphere (quite weak by solar standards) *could* be sufficient to provide enough Alfven wave flux to be considered as a candidate for heating the quiet Sun corona. An advantage of Alfven waves is that, other things being equal, they do not dissipate as quickly as acoustic waves: the main reason is that there are no compressions and rarefactions associated with Alfven waves. As we have already seen (Section 15.9), a wave with constant energy flux will have a velocity amplitude growing as $1/\sqrt{\rho}$ as the wave propagates upwards: as a result, the amplitude of the Alfven wave may grow much larger than the amplitude assumed in the photosphere. For example, when the wave reaches an altitude where the density has fallen off by a factor (say) $10^{3-4}$ relative to the photosphere, the Alfven waves may have amplitudes of some tens of km. Such amplitudes could explain the observed line widths of as much as 60 km sec$^{-1}$ that have been reported in the low corona (see Section 17.8).

However, we must remember some important provisos. First, waves generated in the photosphere must survive into the corona, and second, the waves must be dissipated in the corona, i.e., on length-scales of a few times $10^{10}$ cm. The first of these may be difficult to satisfy because Alfven waves tend to be reflected from a jump in density. Now, between the photosphere (where, with mass densities of $2$–$3 \times 10^{-7}$ [Chapter 5, Table 5.3], the number densities are of order $10^{17}$ cm$^{-3}$) and the coronal base (where number densities are of order $\approx 10^9$ cm$^{-3}$ [Section 17.10]), there is a reduction in density by $\varphi_d \approx 10^{-8}$. This jump in density occurs mainly across a length-scale of 1–2 thousand km (the chromosphere), but it also includes a nearly discontinuous (smaller) jump between chromosphere and corona (see Section 17.16). Waves with periods of a few minutes (such as those emitted by gas circulating in granules), in regions where the Alfven speed $V_A$ is at least a few km sec$^{-1}$, will have wavelengths $\lambda_w = PV_A$ that exceed 1000 km. In such a case, the Alfven waves will "sense" the change in density from photosphere to corona as essentially a discontinuity. Across a discontinuity where the density jumps by a factor of $\phi_d$, the flux of the transmitted wave is only $4\sqrt{\phi_d}$ of the incident flux (Alfven and Falthammer 1963). Using the earlier estimate of $\phi_d \approx 10^{-8}$, the fraction of the incident waves that survives into the corona is only $4 \times 10^{-4}$. Thus, even if we were to allow the field in the photosphere to be as large as 100 G, in which case the photospheric flux is as large as $F_A \approx 10^8$ ergs cm$^{-2}$ sec$^{-1}$, the flux transmitted into the corona might be no more than $4 \times 10^4$ ergs cm$^{-2}$ sec$^{-1}$. This is too small to supply the wave flux even for the quiet corona (Equation 17.9).

It is possible that, rather than restricting attention to Alfven waves generated in the *photosphere*, we should consider the possibility that the Sun may generate Alfven waves elsewhere. Magnetic reconnection (see Section 17.19.11) events in the chromosphere or in the corona might provide localized sources of Alfven waves, but it is hard to obtain quantitative estimates of wave fluxes from such transient events.

### 17.18.2 Non-Wave Heating: The Magnetic Carpet

A hypothesis that relies on magnetic fields, but which may also explain the near-spherical symmetry of the quiet Sun corona, is called the "magnetic carpet" (Figure 17.14).

In Figure 17.14, the background (with local hot spots in white) represents a segment of an image of the Sun taken in the same Fe XII line as in Figure 17.8. The background reminds us that the "fuzzy glow" of material at 1–2 MK is present at most locations in the solar corona. Superposed on the image are magnetic field lines. Each field line is rooted in a pixelated feature on the Sun's surface that is either white or dark. White and dark pixels represent opposite magnetic polarity: therefore, each field line begins and ends in opposite polarity spots, looping upward between one

**FIGURE 17.14** Illustration of the "magnetic carpet" in the solar atmosphere. (Image available for public use at the NASA/Goddard website http://stargazers.gsfc.nasa.gov/images/sun_images/sunspots_cmes_occur/aerialcarpt.gif)

white footpoint and one dark footpoint. As regards linear scale, each pixelated feature is some $10^4$ km across. (On such a scale, the solar diameter spans 140 pixels.)

Any given pixelated feature may have multiple field lines emerging from it, each connecting to a separate pixel of opposite polarity at another location. Thus, although field lines fill up most of the available space at high altitudes, when one approaches the surface, the field lines become clumped, or concentrated into specific "threads" that emerge from specific locations. The analogy to a domestic carpet, with its multiple clumps of thread emerging from a substrate, in which all clumps are "rooted", leads to the nomenclature "magnetic carpet" for the structure that dominates in the low solar corona.

However, the analogy with a domestic carpet should not be taken too literally. The Sun's magnetic carpet is *not* a static structure: far from it. Temporal variability is *of the essence* because new magnetic fields are continually emerging from beneath the solar surface. The image in Figure 17.14 is only a single "still frame" taken from a movie showing dramatic variability. Any particular one of the many white and dark pixelated features that appear in the "still frame" in Figure 17.14 actually emerges at a certain point in time, splits apart or coalesces with neighbors, drifts around the surface because of granule motions, and eventually disappears. The time-scale during which any given feature survives from emergence to disappearance has been found to be 1–2 days. As the features on the surface evolve, driven by continual emergence of new magnetic fields, the magnetic field loops in the solar atmosphere also evolve, expanding, contracting, distorting in a multitude of ways (Schrijver et al. 1997).

One possible outcome of this complex process is that field lines in the corona can find themselves in situations where they undergo a process known as "magnetic reconnection" (see Section 17.19.11). Magnetic energy can be released in each reconnection event and converted into heat. It is possible that the energy released as a result of the multitude of small localized reconnection events that occur in the magnetic carpet every second may be an important source of energy that heats the upper solar chromosphere in quiet regions up to altitudes of order 2 Mm (Chitta et al. 2014). It is less obvious that the carpet will also definitely heat the corona, although Chitta et al. (2014) suggest that perhaps better observations will in the future alter this conclusion. An advantage of the magnetic carpet as a coronal heating mechanism is that, since there exists at least some field present in most parts of the Sun's surface at any given time (see Section 16.4.1.5), coronal heating due to the magnetic carpet *might* account for the (almost spherically symmetric) "fuzzy glow" permeating most of the solar surface (see Figure 17.9). However, a puzzle still remains: if in fact magnetic loops are present in the carpet, why is it that in Figure 17.9, we can see no evidence for definitive loops (in contrast to the clear evidence for loops in Figure 17.12)? One possible answer may be that the loops in the magnetic carpet might be so small that they are not clearly identifiable in images such as Figure 17.9. Future higher resolution observations may help to resolve this puzzle.

## 17.19 SOLAR FLARES

The most spectacular and energetic phenomena on the Sun are associated with explosive events called "flares". These are transient short-lived brightenings that can make their appearance in multiple regions of the electromagnetic spectrum, ranging (in some flares) over 12–14 orders of magnitude in energy, from γ-rays with energies up to hundreds of MeV to meter-wave radio photons with energies of $10^{-6}$ eV. Here, we first describe the general properties of flares, and then examine the physical processes at work. The flare-generating physical process in which we shall be most interested is referred to as "magnetic reconnection" (Sections 17.19.10).

In the broadest sense, two general classes of flares have been identified: confined and eruptive. The former are limited to a small region of the Sun, whereas the latter eject material in the form of a "coronal mass ejection" (see Section 18.9) that can propagate out into interplanetary space.

### 17.19.1 GENERAL

Flares are difficult to see in the optical continuum ("white light flares") because the surface of the Sun is so bright: in a 2.5-year period around solar maximum in 1980 (when flares occur with maximum frequency: see Section 17.19.2), only 12 white light flares were detected (Neidig and Cliver 1983).

Instead of searching in "white light", it has been found that flares are much easier to detect from the ground when the Sun is observed in chromospheric lines: e.g., in a two-year period 1978–1979, 15,500 flares were recorded in H$\alpha$ (Kurochka 1987), roughly 1000 times more frequent than white-light flares. A striking feature of eruptive flares, when they are observed in chromospheric radiation, is that the regions of strongest emission in optical/UV are often confined to two (more or less) linear features called "ribbons": an example of such ribbons is shown in Figure 17.15 (from Milligan et al. 2014). The ribbons are interpreted to be areas where energy has been magnetically channeled (by thermal conduction or by electron beams) down to the chromosphere from the corona: in fact, the ribbons are found to coincide well with the regions in the solar photosphere where the vertical electric current density is largest (Musset et al. 2015). The two ribbons are observed to lie closest to each other in the early stages of the flare, but as the flare develops, the ribbons systematically separate spatially from each other. By quantifying the changes in the magnetic flux $\varphi$ intercepted by the ribbons as they separate spatially, the time derivative of $d\varphi/dt$ can be evaluated. Then, using Faraday's law of induction, one can infer the magnitude of the voltage drop that exists along a particular line (the "separator", which connects two magnetic "null points": see Section 17.19.11) in the flare region (Tschernitz et al. 2018). The potential drops in certain flares are found to be as large as a few million volts. In the presence of such voltages, electrons can be accelerated to energies of 100 keV in

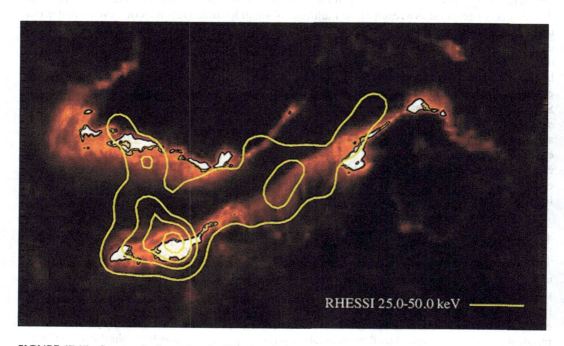

**FIGURE 17.15** Image of a flare taken by Hinode in the Ca II *H* line on February 11, 2011. Orange areas with white inclusions are referred to as flare "ribbons", where the Ca II *H* emission is most intense. The brightest patches of each ribbon are shown in white, surrounded by black contours marking the 80% level of peak intensity. One ribbon (almost forming a straight line) is situated mainly along the diagonal from lower left to upper right. The second ribbon, in the upper left quadrant, looks like a distorted letter "U". Superposed on the Ca II *H* image are contours in a hard X-ray image from the RHESSI satellite taken around the same time as the Ca II *H* image: the yellow contours denote the 30%, 50%, and 80% levels of the RHESSI X-ray fluxes. Note that the highest peaks in X-ray flux (where electron beams are magnetically channeled downwards to strike the chromosphere) coincide spatially with the ribbons. (Milligan et al. 2014; used with permission of R. Milligan.)

fractions of a second: these fast electrons are channeled downwards along local magnetic field lines to the photosphere, where they emit hard X-rays and dump their energy in the optical ribbons. The strength of the associated electric field is related to the rate of magnetic reconnection occurring in the overlying corona.

Flares are also easy to detect in X-rays, where the brightness of the entire Sun can increase by an order of magnitude or more in a matter of minutes or even seconds, and then decay to the previous level of brightness. A plot of intensity of radiation from a flare as a function of time is referred to as a "light curve". Examples of flare light curves, as recorded by one particular series of X-ray satellites (GOES 10, 12) are shown in Figure 17.16. Along the abscissa is plotted the time, in this case covering an interval of three days in 2005, when the Sun was moderately active. The ordinate shows the soft X-ray flux measured by GOES in two different wavelength (energy) ranges: 1–8 Å (1.5–12 keV) (upper curves), and 0.5–4 Å (3–25) keV (lower curves). Flares emit more energy at lower energies (upper curves) than at higher energies (lower curves). The labels A, B, C, M, and X (each spanning one decade in flux) along the right-hand side are a lettering system that has become commonplace in classifying flares according to the magnitude of the *peak flux* that is measured during that flare in the GOES 1–8 Å channel: the smallest A-class flare is defined to have a class of A1, with a peak flux of $10^{-8}$ W m$^{-2}$, while the smallest X-class flare (X1) has a peak flux of $10^{-4}$ W m$^{-2}$. Within each class, the peak fluxes are further subclassified by a number from 1 to 10: thus, an M6 flare has a peak flux of $6 \times 10^{-5}$ W m$^{-2}$. The largest flares detected by GOES have peak fluxes (as measured by the GOES 1–8 Å channel at 1 AU) of order $(2-4) \times 10^{-3}$ W m$^{-2}$, corresponding to class X20-X40. Compared to the total radiant flux received from the Sun at 1 AU (~1361 W m$^{-2}$; see Figure 1.1), we see that the largest solar flare emits (at its peak) a flux in X-rays that is of order one millionth of the Sun's total radiant output. Flares too faint to be detected by GOES (i.e., flares with fluxes lower than GOES A1-class) were detected by the SphinX detector (see Section 17.5), which was some 100 times more sensitive than GOES. The weakest SphinX flares were assigned to two new classes below A-class: S-class (for "small"), corresponding to a flux of $10^{-9}$ to $10^{-8}$ W m$^{-2}$, and Q-class (for "quiet"), corresponding to $10^{-10}$ to $10^{-9}$ W m$^{-2}$ (Sylwester et al. 2019).

Inspection of the curves plotted in Figure 17.16 indicates that each flare is indeed a transient event, characterized (typically) by a fast rise in radiative flux, followed (typically) by a slower decline with a shape resembling that of a damped exponential. (The acronym FRED is sometimes used to describe the flare curve: Fast Rise, Exponential Decay.) When data are obtained simultaneously in other wavelength ranges (radio, optical, and UV spectral lines), flares are also apparent as rapid increases in those spectral regions, followed by slower declines. The various time-scales may be different in different wavelength ranges. In the hardest X-rays (with energies of several hundred keV), the rise time-scale of a flare may be very abrupt, in some cases as short as 0.05 seconds (Altyntsev et al. 2019): this short time-scale places significant demands on any flare model. As can be seen in Figure 17.16, the largest flare events observed in the (softer) X-rays detected by GOES may take the better part of 1 day to return to the "quiet Sun level", whereas the smaller events are complete in less than (sometimes much less than) 1 hour.

Although flares are difficult to detect in "white light" but easy to detect in X-rays, this does not mean that flares emit more energy in X-rays than in optical light. On the contrary, careful analysis suggests (Kretzschmar 2011) that flares emit some 70% of their energy in *optical* photons.

### 17.19.2 How Many Solar Flares Have Been Detected?

The first solar flare to be reported was seen independently by two English astronomers Carrington (1859) and Hodgson (1859) in white light. Because white light flares are not reported very often (because of the brilliant background of the solar photosphere), a wide range of detectors, spanning radiation with wavelengths as short as 1 Å (or less) and as long as meters, have been used over the intervening years to facilitate the observation of flares. Because of different sensitivities in the various detectors, and depending also on how bright the Sun is at that particular wavelength, the number

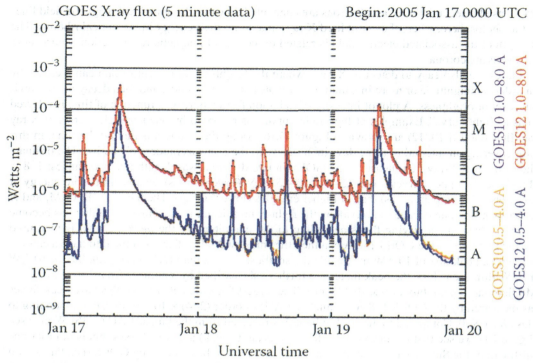

**FIGURE 17.16** Time profiles of soft X-ray flux emitted by the entire Sun in two ranges of wavelengths (ranges of energies) during a three-day interval in 2005, when the Sun was moderately active. Data are in the public domain. (From Space Weather Prediction Center, Boulder, CO; National Oceanic and Atmospheric Administration; U.S. Department of Commerce. Used with permission.)

of detectable flares varies from one wavelength range to another. We have already noted that in an analysis of 2 years of Hα observations, Kurochka (1987) reported 15,500 flares.

Using the GOES 1–8 Å X-ray data set over the course of a 37-year interval (1975–2011, i.e., three solar cycles), Aschwanden and Freeland (2012) used an automated analysis routine to identify a (remarkable!) total of 338,661 X-ray flares. The average rate of flares per year is almost 10 thousand. However, the annual rates vary greatly (by factors of ~100): the smallest number of flares in any single year was 186 in 2008 (when the Sun was passing through a very deep minimum in magnetic activity), while the greatest annual number was 18,797 in 1979 (when the Sun was close to the maximum of cycle 21). The existence of a pronounced correlation between numbers of flares and the Sun's (magnetic) activity level leads to a significant conclusion: the energies released during flares owe their existence (somehow) to magnetic fields.

As regards the numbers of flares with different peak fluxes, the (cumulative) numbers of flares with peak fluxes >X1, >M1, and >C1 were found to be 248, 3986, and 35,221 respectively. Thus, (small) C-class flares are more frequent than (large) X-class flares by a factor of >100. Quantitatively, the (differential) number of flares with maximum fluxes $x$ in the range from $x$ to $x + dx$ can be fitted with the function $N(x) = N_1 (1+ x/x_1)^{-\alpha}$: in large flares ($x \gg x_1$), this function behaves as a power law $N(x) \sim x^{-\alpha}$, while in small flares ($x \ll x_1$), $N(x)$ approaches a constant value $N_1$. For each year in the 37-year interval, Aschwanden and Freeland (2012) extracted a value for α, the power-law index: the values were found to range from 1.7 to 2.3, with a mean value of $\langle\alpha\rangle = 1.98$ and a standard deviation of ±0.11. Interestingly, the power-law index is (within ~5% error bars) essentially invariant across

The Corona                                                                                                              341

three solar cycles. This is noticeably different behavior from the *number* of flares per year: that number varies by a factor of ~100 between years of minimum and maximum activity.

The empirical fact that the maximum fluxes of flares obey a power-law distribution contains important information about flare physics. First, because there is no evidence for a peak in the distribution, this means that there is no "characteristic (or preferred) maximum flux" associated with solar flares: instead, there is simply a broad range of maximum fluxes available to the solar atmosphere when local conditions lead the Sun to "decide" to create a solar flare. Second, Aschwanden and Freeland (2012) point out that their empirical finding of $<\alpha> = 1.98 \pm 0.11$ is consistent with the prediction of a flare theory called "self-organized criticality" (SOC) (Lu and Hamilton 1991): in SOC theory, a flare occurs as a result of an "avalanche" of many small magnetic reconnection events (Section 17.19.10). An avalanche of this kind in the solar atmosphere is driven by the fact that convective turbulence causes the local magnetic fields to undergo continual stretching and stressing as time goes on. The eventual occurrence, in this scenario, of a "reconnection avalanche" is analogous to avalanches in a sandpile as extra grains of sand are added slowly to the pile. In the case of flares, the amplitude of any particular flare is determined by the number of small reconnection events that contribute to the flare.

In a discussion of SOC theory, Aschwanden (2011) shows that if the distribution of *peak fluxes* in GOES flares follows a power law with an index $\alpha$, then the total *energy* of the flare (proportional to maximum flux *times* the duration of the flare) will be found to follow a distribution that is also a power law. Significantly, however, the power-law index $\alpha_E$ of the *energy* distribution in SOC theory is *not* the same as the index $\alpha$ of the peak flux distribution. Instead, $\alpha_E = (\alpha+1)/2$. Inserting $\alpha \approx 1.98$ (as found earlier), the flare energy distribution is predicted to have a slope $\alpha_E \approx 1.5$. The significance of the latter number is that, when one integrates over all flare energies, the *largest* flares dominate in the integral. On the other hand, in some theories of coronal heating, it is claimed that the *smallest* flares ("nanoflares": Parker [1972]) dominate the heating: if SOC is a realistic flare theory, the "nanoflare" claim is not supported by the GOES data. (But note that GOES detects only relatively large flares with peak fluxes larger than A1-class). However, when flares are observed with hard X-ray instruments such as RHESSI, flares that are too small to be detected by GOES may become detectable. Among such smaller flares, the slope of the energy distribution is sometimes found to be steeper than 2.0 (Aschwanden et al. 2016): in the presence of such steep slopes, nanoflares *might* dominate coronal heating. In the nanoflare model of Parker (1972), the major focus of the discussion was on active regions, where the gas in local "hot spots" might be heated to temperatures of 4–5 MK or larger: indeed, such hot spots have been detected in flaring active regions (Reva et al. 2018). On the other hand, in the quiet Sun, nanoflares are known to contribute very little to the DEM, no more than one part per million (Hannah et al. 2010). In order to draw reliable conclusions about the overall role of nanoflares in coronal heating, Aschwanden et al. (2016) stress that more homogeneous data sets are required: specifically, an instrument is required that will allow the same method and the same time intervals to be used for *all* flares from the largest (with energies of $10^{32}$ ergs or more) to the smallest (with energies of order $10^{23}$ ergs). Only in the case where such a homogeneous data set becomes available can we expect to construct a flare energy distribution that is reliably self-consistent on all scales.

Moreover, from a purely mathematical perspective, it is not a trivial matter to determine precisely the slope of a power-law distribution of flare-related quantities. Fitting a power law is not always a stable process, because the value of the power law can depend on how one defines the onset of each flare (Ryan et al. 2016). As a result, depending on the definition of the flare onset, the slope of the energy distribution could change from steeper than −2.0 to shallower than −2.0: such a change would make all the difference between, on the one hand, a conclusion that nanoflares dominate coronal heating, and, on the other hand, a conclusion that large flares dominate coronal heating.

### 17.19.3 FLARE TEMPERATURES AND DENSITIES

In Figure 17.16, the flare X-rays have been measured by GOES in two different energy bands: one band spans a lower range (1.5–12 keV), and the other band spans a higher range (3–25 keV). By comparing the relative fluxes of flare X-rays in the two bands when the flare is at peak intensity, an estimate of a "color temperature" $T_c$ can (in principle) be estimated if the flare is assumed to be isothermal. The numerical values of $T_c$ in samples of many flares are found to range from 5 to 25 MK (Feldman et al. 1996). Independently, a "color temperature" for flares can also be obtained using data in somewhat higher energy ranges from the RHESSI spacecraft: these $T_c$ estimates are found to be systematically larger (by a few MK) than those obtained from GOES (Warmuth and Mann 2016). These systematic differences in $T_c$ values derived from GOES and RHESSI data indicate that the assumption that flare plasma is isothermal is incorrect. A more reliable approach is to construct the differential emission measure in order to identify the temperature $T_p$ at which DEM is a maximum (see Section 17.5). Ryan et al. (2014) have derived the DEM for a sample of 149 M- and X-class flares observed by SDO/AIA: this analysis leads to $T_p = 12.0 \pm 2.9$ MK. For the same flares, the $T_c$ derived from GOES and from RHESSI data are found to be larger than $T_p$ by factors of $1.4 \pm 0.4$ and $1.9 \pm 1.0$ respectively. Because of the (incorrect) assumption of isothermality that enters into the determination of the "color temperature" $T_c$ from GOES or RHESSI, the value of $T_p$ derived from DEM is a more physically reliable indicator of "the" temperature in solar coronal gas.

At the peak of a flare (in X-ray emission), the local gas in the flare volume typically reaches its highest temperature. At later times, the temperature of the plasma cools. Based on the strengths of various X-ray lines emitted by Fe ions in ionization stages ranging from XXIV to XIV, the average rate of cooling has been estimated (in a sample of 72 M-class and X-class flares) to be $3.5 \times 10^4$ K sec$^{-1}$ (Ryan et al. 2013): that is, flare plasma cools from (about) 20 MK to "coronal" temperatures (1–2 MK) in a time interval of 5–10 minutes, on average. If there is no source of prolonged heating after the flare reaches maximum intensity (due, e.g., to another flare occurring nearby), the cooling time-scale is determined by the effectiveness of radiative and conductive cooling.

Although a flare temperature of order 12 MK (as mentioned earlier) already indicates a 10-fold increase in temperature compared to the quiet corona, even higher temperatures have been identified in certain flares if the detector can respond to hotter gas. The very hottest gas in flares has been detected by observing in the lines of the most highly ionized elements. The Japanese satellite Hinotori, launched in 1981, observed 13 large flares (mainly X-class) in the 1.8Å Lyman-α line of the hydrogenic ion Fe XXVI. In about 50% of these flares, a "super-hot" component was observed, with maximum temperatures of 30–40 MK (Tanaka 1986).

The coolest gas recorded in flares has been recorded by the SphinX instrument, with its X-ray sensitivity some 10–100 times better than GOES. During the 9 months from February to November 2009, when the Sun was in a deep activity minimum, SphinX nevertheless detected 1604 flares, the largest being a C2.1 class flare (Gryciuk et al. 2017). The hottest flare temperature was found to be 7.94 MK, but for many of the flares, the temperature did not exceed 2 MK. In fact, one flare (#890 in the list of Gryciuk et al.) was listed as having a temperature of only 1.24 MK, i.e., log $T = 6.1$, equal to the temperature of the cooler component of the quiet corona (see Figure 17.9): in this flare, the energy release was apparently so small that, although X-rays were detected (probably due to an increase in the local gas density), the local gas apparently did not heat up measurably compared to the quiet Sun.

In comparison with the quiet (i.e., non-flaring) corona, where $T = 1-2$ MK, we see that the gas in large flares at the peak of the DEM is hotter than the quiet Sun corona by a factor that may be as large as 10 or so. It is an empirical fact that whatever causes a large flare to occur can lead to a great deal of heating (with a temperature increase of ~10-fold or more) in the coronal gas.

Electron *densities* $n_f$ in coronal flare plasma are also larger than the densities in the quiet Sun. For example, in a sample of flares analyzed by Moore and Datlowe (1975), the mean $n_f$ was found to be $0.2-0.7 \times 10^{11}$ cm$^{-3}$, depending on certain assumptions about the flare volume. Even higher densities (>$10^{13}$ cm$^{-3}$) have been derived in flare plasma by analyzing the Lyman continuum (Machado et al. 2018). Compared with the quiet Sun coronal densities of $1-4 \times 10^9$ cm$^{-3}$ (see Section 17.9), the conclusion is that the densities in flaring gas are greater than in the quiet Sun by factors which may

The Corona 343

be as large as 3–4 orders of magnitude. Something about the physical processes occurring in a flare leads to significant compression and significant heating of gas in the solar atmosphere.

### 17.19.4 SPATIAL LOCATION AND EXTENT

The X-ray detector on the GOES spacecraft (Figure 17.16) *cannot* make an image of the Sun: it can only measure the flux of X-rays from the entire solar disk. As a result, there is no way to tell from GOES data alone *on which part of the Sun* any given flare occurred. But other instruments (e.g., SDO/AIA), which can make images in optical or EUV photons, demonstrate unambiguously that flares occur in active regions, especially in active regions where sunspots have complex umbrae with opposite polarities in close spatial contact. Based on a study of almost 4000 flares and their associated spots, Greatrix (1963) showed that the intensity of a flare is related to changes in the *magnetic flux* of the associated spots (see a discussion of flare ribbons and magnetic flux in Section 17.19.1).

When the largest flares are imaged in H$\alpha$, they are found to spread out spatially to cover an area that is a significant fraction of the area of the active region. The linear extent of the largest flares on the surface of the Sun can be several times $10^9$ cm, with areal coverage of order $10^{19}$ cm$^2$. In the vertical direction, flare plasma may extend up to altitudes that are not greatly different from the extent on the surface. As a result, the total volume $V_f$ of a large solar flare can be $10^{28-29}$ cm$^3$.

Although most of the attention in solar flare research is paid to the rapid (and spectacular!) energy release process, there is actually another less dramatic process that also deserves attention: the build-up and storage of energy in the magnetic field prior to a flare. The latter process is much less dramatic than the flare itself. Build-up and storage occurs slowly over time intervals of hours and days. It occurs in active regions because magnetic field lines are rooted in the photosphere, where convective motions are ubiquitous. At the footpoint of a magnetic loop, gas that is frozen into the field and subjected to convective motions pushes the field lines around in complicated ways, with twistings and stretchings and braidings. These complex motions cause stresses to build up in the field lines, and the stressed fields serve as a reservoir in which free magnetic energy may be stored. At certain times and places, a trigger comes into operation, releasing the free energy on a short time-scale as a flare. The trigger causes a transition from a long-drawn-out storage process (on time-scales of $10^5$ sec) to a rapid release process (on time-scales of seconds): the two processes occur on time-scales differing by as much as five orders of magnitude.

### 17.19.5 ENERGY IN NONTHERMAL ELECTRONS

In order to appreciate the physics of flares, it is important to estimate the amount of energy released in one such event. Solar flares come in a very broad range of energies. Some are so small that the only evidence for an event is a slight short-lived increase in brightness in the chromospheric line H$\alpha$. In such cases, the flare energy emerges purely in the form of photons. Measurements provide the excess luminosity in H$\alpha$ over and above the luminosity in the quiet Sun. Integrating the excess luminosity over the lifetime of the flare yields an energy $E_\alpha$ for the flare in H$\alpha$ photons. However, if we wish to evaluate the total energy of a flare, we have to take account of the energy that emerges in other photons and in other forms.

In large flares, hard X-rays emerge with photon energies of tens or hundreds of keV. Such X-rays indicate the presence of fast electrons, also with energies of tens or hundreds of keV, i.e., much higher than the typical thermal energies in the corona. (In 1 MK plasma, thermal energies $kT$ are of order 0.1 keV.) Large flares are quite efficient at accelerating "nonthermal electrons". From the shape of the hard X-ray spectrum, information can be extracted about the number $N_e$ of nonthermal electrons as a function of energy: typically, the electron spectrum is found to be a power law in energy:

$$\frac{dN_e}{dE} \sim E^{-\delta} \tag{17.11}$$

where the spectral index $\delta$ is found to range between 2 and 5, with most values between 3 and 4.5 (Brown et al. 1981). To evaluate the total energy $E_{nt}$ contained in the nonthermal electrons in any particular flare, we need to multiply $dN_e/dE$ for that flare by $E$ and integrate over the energy spectrum from a minimum energy $E_{min}$ to a maximum energy $E_{max}$. The result is

$$E_{nt} \sim \frac{1}{E_{min}^{\delta-2}} - \frac{1}{E_{max}^{\delta-2}} \tag{17.12}$$

Since $\delta$ typically exceeds 2, the value of $E_{nt}$ is determined mainly by the term that depends on $E_{min}$. The smaller $E_{min}$ is, the larger $E_{nt}$. It is not easy to extract the value of $E_{min}$ reliably from observational data, but it is probably no more than 20 keV. If we set $E_{min}$ = 10–20 keV, then $E_{nt}$ can reach values approaching $10^{32}$ ergs. In a study of some of the largest solar flares ever observed, Lin and Hudson (1976) report that nonthermal electrons with energies in the range 10–100 keV "constitute the bulk of the flare energy": these electrons are found to "contain 10–50% of the total energy output". Apparently, a flare site is very effective (up to 50%) at accelerating many electrons to energies of tens of keV: the number of fast electrons that are accelerated every second in a sample of 18 flares observed in hard X-rays by the RHESSI spacecraft has been estimated to be $10^{32-36}$ sec$^{-1}$ (Mann and Warmuth 2011). These "primary" particles emerge from the flare site and propagate into other regions of the solar atmosphere, depositing various fractions of their energy into the ambient corona, chromosphere, and (in some cases) the photosphere.

Nonthermal electrons generate not only hard X-rays, but they also, in regions of magnetic field, emit radio emission at microwave wavelengths by mean of gyrosynchrotron radiation. Do the same electrons give rise to both types of electromagnetic radiation? Krucker et al. (2020) have reported on a comparative study of hard X-ray data from the RHESSI spacecraft and radio data from the Nobeyama Radio Heliograph (NoRH). RHESSI operated from 2002 to 2018. The Nobeyama detectors measured solar radio data at two microwave frequencies, 17 and 34 GHz. In a sample of 82 individual burst components observed simultaneously by NoRH and RHESSI during 40 flares, Krucker et al. found a linear correlation between hard X-ray flux and microwave flux, with a correlation coefficient of 92%. Krucker et al. concluded, "the same population of accelerated non-thermal electrons . . . produce both the . . . microwave emission and the . . . hard X-ray emission".

### 17.19.6 Where Are Flare Electrons Accelerated?

To answer this, we note that hard X-rays are emitted when the fast electrons from a flare site propagate downwards (guided by the magnetic fields of a flaring loop) into denser gas in the chromosphere (or maybe even in the photosphere) and, colliding there with ambient ions, radiate free-free photons with energies no larger than (but comparable to) the electron energy. In order to find out where the electrons originated (up in the corona), one approach is to measure how long it takes for the electrons to travel from their source down to the denser gas: i.e., the goal is to measure the *time of flight* of electrons. Faster electrons should reach the denser gas *before the slower electrons arrive*. As a result, the onset times of more energetic X-rays (created by faster electrons) occur slightly *earlier* than the onset times of less energetic X-rays (created by slower electrons). Aschwanden et al. (1996) reported on a sample of flares in which the times of flight were measured for X-ray energies as low as 30 keV and as high as 250 keV: the onset times of the high energy photons were found to occur *earlier* than those at lower energy, with differences ranging from 0.059 sec to 0.140 sec. Clearly, to make reliable estimates of these time differences, the detectors must be able to measure onset times with a precision measured in milliseconds. Knowing how fast electrons with energies of 250 and 30 keV travel, the differences in times of flight from the acceleration site to the dense gas where X-rays are emitted can be converted into altitudes of the site of acceleration. Aschwanden et al. (1996) found that the altitudes at which the fast electrons originate ranged from 14 to 31 Mm. Subsequent analysis (Aschwanden 2002) of some 100 flare events observed by both YOHKOH and the Compton Gamma-Ray Observatory (CGRO) found that the altitudes of the flare acceleration site ranged from 5 to 60 Mm. These altitudes lie definitely higher than the top of the loops from which flare radiation was observed. These results suggest that the instability that triggers the flare (by accelerating nonthermal electrons) does *not* occur *in* the loop: whatever the process is that does the triggering occurs *above the top of the loop.*

### 17.19.7 OTHER CHANNELS OF FLARE ENERGY

As well as the energy contained in fast electrons, flares also manifest energy in other forms (see, e.g., Sturrock 1980). The amount of thermal energy in flare plasma ($\approx N_f k T_f V_f$) over and above the thermal energy in an equal volume of "quiet" coronal plasma can be $10^{30}$ ergs and more.

Some flares, especially those with the largest energies, may lead to the acceleration of "solar energetic particles" (SEP). These are protons, neutrons, and other nuclei with energies up to thousands of MeV. Where do these energetic particles originate? To address this, the charge states of the SEP can be used to indicate the temperature of the source material (Reames 2016): the results indicate that the SEP are typically accelerated from the quiet Sun and from active regions, but not directly from flare plasma itself. The acceleration probably occurs in shock fronts launched by the flare itself or by a CME (see Chapter 18).

Whatever it is that accelerates the SEP, it is an observational fact that some of the SEP are directed downwards, and some are directed upwards. Consider first those that move *downward* towards the photosphere. These can generate gamma rays when they encounter the denser gas of the photosphere: in a 4-year time span, 18 flares were observed by the Fermi satellite to emit detectable gamma-ray continuum emissions at energies > 100 MeV (Ackermann et al. 2014). Less energetic SEP, when they strike the photosphere, can excite gamma-ray lines, including the 0.511 MeV line from positronium, the 2.223 MeV line from deuterium (when a flare neutron merges with a photospheric proton), and de-excitation lines from nuclei of $C^{12}$ and $O^{16}$ at 4.4 and 6.3 MeV respectively (Chupp et al. 1973).

Second, let us consider the SEP that are directed *upwards*. In suitable conditions, they may escape into the solar wind and can be detected by spacecraft lying outside the shielding of Earth's magnetic field. The GOES spacecraft has a particle detector that records a "solar proton event" (SPE) if the flux of protons with energies >10 MeV exceeds 10 particle flux units (p.f.u.), where 1 p.f.u. = 1 particle cm$^{-2}$ sec$^{-1}$ ster$^{-1}$ at geostationary orbit. A catalog of 261 SPEs detected by GOES in a 41-year interval (1976–2017) can be found at the website www.ngdc.noaa.gov/stp/satellite/goes/doc/SPE.txt. The number of recorded SPE events varies greatly from one year to another. In solar minimum years, an entire year can elapse without a single SPE event (e.g., in 1996, 2007, 2008, and 2009), while in solar maximum years, as many as 23 SPE events have been recorded. As regards the magnitude of the events, the fluxes of >10 MeV protons was found to exceed $10^4$ p.f.u. in nine (3%) of the 261 SPEs; in most events, the peak flux was in the range 10–100 p.f.u. For an earlier time interval (1955–1986), Shea and Smart (1990) compiled data for SPEs (using the same definition as earlier) from a variety of detectors: the compilation contains 218 events, with an average annual rate of about seven SPEs per year: this rate is close to the average annual rate reported by the GOES catalog.

The most energetic protons in SPEs can have energies up to several GeV: such particles, although rare, have the ability to "break through" the Earth's magnetic shielding and be detected by instruments on the ground called neutron monitors (NMs). These detections are called "ground-level events" (GLEs), of which 73 have been recorded. At the location of each NM, the Earth's field and overlying atmosphere shields the NM from particles with energies that are lower than a certain threshold, i.e., only particles with energies in excess of that threshold can reach that particular NM. An NM that is near the geomagnetic equator may have an energy cut-off as high as 17 GeV (in Thailand), while in Antarctica, the energy cut-off is <1 GeV. Information on all of the 73 GLEs identified so far can be found at the official site in Oulo, Finland, at https://gle.oulu.fi/#/. There is a striking difference between solar cycles as regards the numbers of GLEs: cycle 23 had 16, cycle 24 had only one, and cycle 25 has already had one even though the cycle is only about 2 years old at the time of writing (December 2021).

The fact that fast neutrons are associated with certain flares is clearly illustrated by the emission of the 2.223 MeV line of deuterium due to SEP directed *downwards* to the photosphere. But what about the neutrons that are directed *upwards*? Can they be detected by near-Earth detectors? With a half-life of only 15 minutes for free neutrons, the traversal of the distance from Sun to Earth (requiring at least 8–9 minutes, even for the fastest particles) must be accompanied by *some* decay (even allowing for relativistic time dilation). Neutrons have the advantage that, since they have zero electric charge, the Earth's magnetic field poses no obstacle to them: the neutrons can penetrate unimpeded through the field to reach even satellites in low-Earth orbit, such as the International Space Station (ISS). Koga et al. (2017) used a neutron detector attached to the ISS during the years 2010–2015 (i.e., spanning the maximum of solar cycle 24) and reported that neutrons with energies in the range 30–120 MeV were reliably detected in association with 28 flares.

Finally, some flares are observed to eject bulk material from the corona into the solar wind: these events, called "coronal mass ejections" (CMEs), will be discussed in Chapter 18. Their kinetic energies can reach values of order $10^{32}$ ergs, comparable to the total energies in nonthermal electrons in large flares. The effects of CMEs can in exceptional cases lead to disturbances on the Earth. For example, following the large flares of August 1972, the Earth's magnetic field was perturbed so much by one (or more) CMEs moving past the Earth at high speed that dozens of magnetically sensitive sea mines that had been dropped into a harbor in North Vietnam were unintentionally detonated within an interval of some 30 seconds (Knipp et al. 2018).

Adding up the various channels among which flare energy is distributed, estimates indicate that the largest solar flares are found to have energies of a few times $10^{32}$ ergs. Thus, one of the largest flares reported by a space-borne total irradiance monitor (Kopp et al. 2005) was observed to have a total radiated energy of $6 \pm 3 \times 10^{32}$ ergs. From a theoretical point of view, Aulanier et al. (2013) have argued, based on a 3-D MHD simulation of a large sunspot with the strongest magnetic field ever recorded in a spot, that the maximum energy that might be released in a solar flare (under optimal conditions) could be $\sim 6 \times 10^{33}$ ergs.

At the opposite extreme of energies, might one wish to specify "the smallest solar flare"? Probably not: depending on the sensitivity of the detector and on how quiet the Sun is in the wavelength being used to make the observations, one could in principle pick out smaller and smaller events that might qualify as "flares". In hard X-rays (where the "quiet corona" emits at a very low level), events can be identified as flares with $E$(tot) as small as $\approx 10^{26}$ ergs (Lin et al. 1984). These small events are sometimes referred to as "microflares": the prefix "micro" indicates they have energies that are of order $10^6$ times smaller than the largest flares. If it could ever be demonstrated with confidence that there are events in the Sun with $E$(tot) as small as (roughly) $10^{23}$ or $10^{20}$ ergs, they might justifiably be referred to as "nanoflares" or "picoflares".

Is there anything on Earth to which we can compare solar flares in terms of energy release? The closest event may be a nuclear explosion whose energy release is quoted in terms of the equivalent tonnage (or megatonnage: MT) of TNT. In a 1 MT explosion, the energy released is $4 \times 10^{22}$ ergs. Each "nanoflare" in the Sun (if such events occur) would generate the equivalent of a 2.5 MT explosion on Earth.

### 17.19.8 Do Flares Perturb Solar Structure Significantly?

The largest flares, with energy releases of a few times $10^{32}$ ergs, certainly involve important disruptions of the active region in which they occur: field lines connecting different spots in the active region are rearranged, and some matter may be ejected into space. However, in the larger context of the Sun as a whole, flares represent only a small perturbation.

To see this, we note that even the largest flares have durations that are no more than a fraction of a day (see Figure 17.16). During such an interval (≈ 50,000 sec), the nuclear reactions in the core of the Sun continue to pour out energy at the standard rate, i.e., $4 \times 10^{33}$ ergs sec$^{-1}$. Therefore, over the duration of the largest flares, the Sun puts out $2 \times 10^{38}$ ergs of nuclear energy. Compared to this, the energy released in even the largest flares is only a small perturbation (of order one-millionth) of the solar energy budget.

### 17.19.9 Energy Densities in Flares

Now that we know, for large flares, that $E_t(\text{fl}) \approx 10^{32}$ ergs, while the flare volume is estimated to be $V_f \approx 10^{28-29}$ cm$^3$ (see Section 17.19.4), we can estimate the mean energy density in large flares. The result is $E_t(\text{fl})/V_f \approx 10^{3-4}$ ergs cm$^{-3}$.

What is the origin of this energy density? Since flares are observed to occur in active regions, it is reasonable to expect that the magnetic field is somehow responsible. In quantitative terms, we have already seen (Section 16.6.2.1) that magnetic fields have an energy density $W_{\text{mag}} = B^2/8\pi$ ergs cm$^{-3}$ if $B$ is expressed in Gauss.

Magnetic fields have energy densities of $W_{\text{mag}} = 10^{3-4}$ ergs cm$^{-3}$ in regions where the field strengths are 160–500 G. Now, we have already seen that active regions in the Sun have fields of hundreds of Gauss (Section 16.5). Thus, the energetics suggest that the energies contained in magnetic fields in active regions are sufficiently strong that they *could* (at least in principle) supply the energy that is released in a flare. The problem is, first, to identify a mechanism that has the ability to convert magnetic energy into fast electrons and heat. But there is a more pressing problem: this conversion must occur *rapidly* enough to be consistent with the fastest time-scale observed in flares, namely, on a time-scale that may be as short as a few seconds or even less. Let us now turn to one such mechanism.

### 17.19.10 Physics of Flares (Simplified): Magnetic Reconnection in 2-D

Magnetic reconnection is a process that leads to the conversion of magnetic energy into plasma kinetic energy, and (eventually) into thermal energy. According to Longcope and Forbes (2014), the process is not completely understood, but the process essentially involves an electric current flowing in a current sheet (see Figure 17.17a and b). An electric field $E$ causes magnetic flux to transfer from domains that lie on top of, and below, the current sheet to domains that lie at the right- and left-hand tips of the current sheet. As a result of flux transfer, the field reaches a state of lower energy, with free magnetic energy being released. Information about the electric field involved in reconnection can be derived from observations of the motions of chromospheric features known as flare ribbons (Section 17.19.1) as these features move across the surface of the Sun and intersect more and more magnetic flux in the chromosphere (Tschernitz et al. 2018).

Ten lines of observational evidence have been listed by Aschwanden (2020) in support of the hypothesis that *magnetic reconnection* is the basic physical process at work in solar flares. In this section and the next, we summarize certain aspects of reconnection.

According to the Sweet–Parker mechanism (named after the two researchers who first proposed a model, see Parker 1957), the simplest model of reconnection occurs in a two-dimensional (2-D) environment when two regions of magnetic field, containing field lines with opposite polarities approach each other in the vicinity of an X-point, where $B = 0$ (see Figure 17.17a). In the limit of MHD, where fields and plasma are "frozen" together in the distant inflow, plasma plus field flow in towards the X-point from two directions, and plasma plus field flow outward away from the X-point in two different directions. The inflow and outflow are overall perpendicular to each other. In a further development of the process, Petschek (1964) proposed a model that includes the presence of MHD shock fronts as a part of the model (see Figure 17.17b).

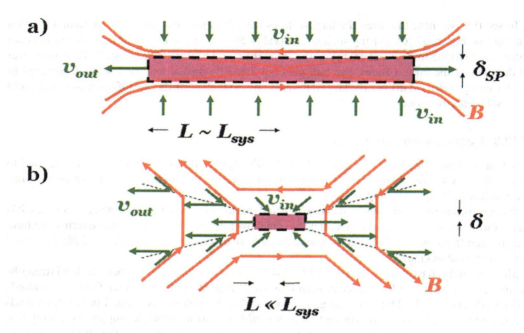

**FIGURE 17.17** Idealized schematics of magnetic reconnection in 2-D. (a) Sweet–Parker model. (b) Petschek model. Field lines are shown in red, with arrowheads indicating direction of the field. The field strength $|B|$ goes to zero at an X-point between the upper and lower red lines. Purple area: current sheet, i.e., a region of enhanced plasma density surrounding the X-point. Plasma flows slowly inwards with speed $v_{in}$ *towards* the current sheet from top and bottom. Plasma flows rapidly outwards with speed $v_{out}$ *away from* the current sheet towards left and right. In (b), dotted lines indicate MHD shock fronts. (Courtesy of P. Cassak.)

The emergent fields are *weaker* than the incoming fields. Since magnetic energy is proportional to $B^2$, the magnetic energy in the outflow is *reduced* compared to the inflow. To conserve energy, some of the original magnetic energy is converted into another form: kinetic.

The essential physical process at work is that an ion that is initially "frozen to" (i.e., gyrating around) an incoming field line (in the regions above and below the current sheet in Figure 17.17a) eventually finds itself swept into a region near the center of the current sheet (the neutral line) where $B = 0$. As the ion approaches the neutral line, the gyroradius, which is proportional to $1/B$ (Equation 16.4), becomes so large that it exceeds the distance between the particle and the sheet. In such conditions, the ion is sometimes moving in regions where it is still associated with the incoming field, and at other times, it finds itself moving in regions where oppositely directed fields exist. As a result, the ion is able to sample two distinct regimes of magnetic flux in Figure 17.17a, and is therefore no longer "frozen" to one particular field. In this situation, the ion essentially "breaks loose" from the field to which it was at first attached: the ion is now free to flow in a different direction, namely, in the outflow direction (towards the right and left sides of Figure 17.17a and b).

However, the idealized scheme in Figure 17.17, with its smooth large-scale flows of material inwards and outwards, can become significantly distorted in the presence of plasma instabilities and turbulence. Specifically, as a result of a plasma process known as a "tearing instability", as the current sheet becomes thinner, it is expected to break up into multiple small "magnetic islands" (or "plasmoids"). These multiple islands can generate large numbers of tiny reconnection sites where length-scales are so small that the time-scale for the plasma to diffuse across field lines (thereby converting magnetic energy into kinetic energy) is reduced greatly (e.g., Cassak et al. 2017). This aspect of greatly reduced time-scales is important if a flare model is to be successful in replicating the shortest observed time-scales on which solar flares can release energy (~0.05 sec in some cases: Altyntsev et al. 2019). Evidence for the presence of multiple plasmoids in a particular solar flare (presumably emerging from the reconnection site of the flare) has been reported in YOHKOH data (Nishizuka et al. 2010).

In the context of the SOC model of flares (Section 17.19.2), the magnetic analog of individual grains of sand that eventually lead to an avalanche in a sandpile is the occurrence of a lot of individual "grains" (i.e., small reconnection events) leading to a multitude of small energy releases by means of reconnections. Perhaps each small reconnection event is due to an encounter between neighboring magnetic islands.

## 17.19.11 Physics of Flares (More Realistic): Magnetic Reconnection in 3-D

In the solar corona, reconnection models that rely on strictly 2-D processes (such as Figure 17.17) are of limited applicability. In order to obtain a realistic physics appreciation of what happens in reconnection, it is necessary to extend the discussion to 3-D. Complexities arise in 3-D that have no analog in the 2-D models, and we will not attempt, in this first course, to describe fully the processes coming into play. But it is worthwhile to introduce two fundamental aspects of 3-D reconnection: the "spine" and the "fan", and how these have been identified in recent observational data.

3-D reconnection occurs in the vicinity of a magnetic "null point" where the strength of all three components of the field go to zero. Such a null point can occur above an active region when multiple loops are emerging with varying orientation through the surface as time goes on: most of the null points are created when two bipoles interact with each other. For example, Cook et al. (2009) identified the presence of almost 3000 magnetic nulls in the solar corona over the course of two solar cycles: near solar maximum, their search found an average of 15–17 nulls per day, mainly at low latitudes (where active regions are commonest) and mainly at low altitudes (<0.25 $R_\odot$ above the photosphere).

Priest and Titov (1996) described what is perhaps the simplest possible representation of a 3-D null point (see Figure 17.18): in terms of Cartesian coordinates, the $z$-axis defines a "spine" along which a particular magnetic field line approaches the null point from one side, and an oppositely directed field lines approaches the null point from the other side. Bundles of field lines close to the $z$-axis also approach the null point but, before reaching $z = 0$, the bundles spread out parallel to the $xy$-plane. In the $xy$-plane, a continuum of field lines emerges from (or converges on) the null point: these lines are said to form a "fan" in the $xy$-plane. In special cases, the fan consists of straight field lines emerging or converging; in other cases, the fan consists of curved field lines, although still confined to the $xy$-plane.

In the "real" solar corona, the field lines are more complicated than those in the simple model of Figure 17.18. However, it may still be possible to identify the "spine" and the "fan" in a particular flare. As an example, we show in Figure 17.19 a model of the force-free fields that occur in the corona in a region where two bipoles (P1/N1 and P2/N2, P for positive, N for negative) are located close to each other in the photosphere (Sun et al. 2013). A magnetic 3-D null can be seen situated at a certain altitude above the pole N1. From the null, two spines (red lines) emerge, one "inner" (heading down towards N1 on the nearby solar surface), the other "outer" (leading to a remote brightening at N2, where that spine encounters the solar surface, sometimes as far as 200 Mm away from the null point [Liu et al. 2011]). Also emerging from the null point is a fan: but in this case (unlike the simplest case shown in Figure 17.18), the fan does *not* lie in a single plane. Instead, the field lines in the fan form a 3-D "drape" or "dome" extending from the null point down to the chromosphere. Any nonthermal electrons that are accelerated at the null point (or at other locations on the dome) are guided downwards by the "dome", and these electrons strike the chromosphere at the footpoints of the "dome", forming a (more or less) circular "ribbon" of light in the chromosphere (see Figure 17.19(b)).

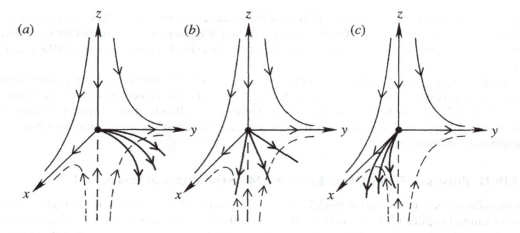

**FIGURE 17.18** Idealized perspective view of magnetic reconnection at a 3-D magnetic null point located at the origin. At the origin of the $xyz$ coordinate system, there exists a null point, where the field strength $B = 0$. Field lines lying exactly on the $z$-axis are referred to as the "spine". Field lines lying in the $xy$-plane are described as the "fan" (from Priest and Titov 1996).

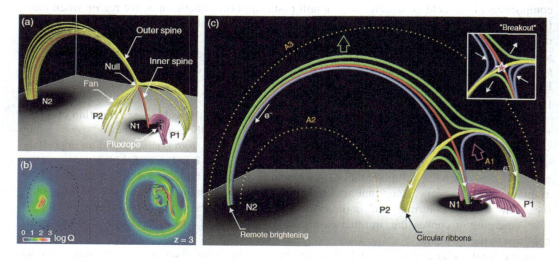

**FIGURE 17.19** Model of coronal fields in fan-spine topology associated with an M2-class solar flare on November 15, 2011. (a) Two bipolar pairs exist in the photosphere. The region contains a null point above N1. The spines (red, one "inner", the other "outer") and the fan (actually a "dome" in 3-D) (yellow) are marked. (b) Circle on the right contains the chromospheric "footpoints" of the dome. Red region on the left is the footpoint of the outer spine. (c) 2-D slice through the magnetic system illustrating possible physical processes pertinent to the observed event. (From Sun et al. 2013; used with permission of X. Sun.)

Clearly, a large number of complicated variants of the scenario shown in Figure 17.19 are possible depending on the number of bipoles (and their field strengths and their relative locations and orientations) that contribute to the coronal field in any particular active region. It requires a very knowledgeable eye to interpret observations correctly in terms of the topology of a fan (dome) plus spine. For further examples of such interpretations, see (e.g.) Masson et al. (2009) and Liu et al. (2011).

### 17.19.12 Consequences of Magnetic Reconnection

What consequences does reconnection have in the solar corona? Using typical coronal values of field and density, we find that the coronal $V_A$ has values that are typically several hundreds of km sec$^{-1}$ and more. Thus, the gas emerging from a reconnection site forms a high-speed jet that flows out into the ambient corona.

Collisional processes ensure that the jet energy will eventually be deposited in a finite volume around the reconnection site, leading to local heating. How hot will the heated plasma be? If all of the kinetic energy can be converted into heat, then the temperature $T_f$ of the heated plasma will be such that the mean thermal speed $\sqrt{2R_g T/\mu}$ is comparable to $V_A$. For $V_A = 500$–$1000$ km sec$^{-1}$ and $\mu = 0.5$, this corresponds to $T_f \approx 10$–$30$ MK, consistent with temperatures deduced from X-ray data (Section 17.19.3). Thus, reconnection is a mechanism whereby magnetic energy is converted (ultimately) into the thermal energy of hot flare plasma.

Another process that occurs at a reconnection site is particle acceleration. The motional electric field $E_f = (1/c)\mathbf{u} \times \mathbf{B}$ at the reconnection site can accelerate electrons. In conditions where $E_f$ exceeds a certain limit $E_D$ (known as the Dreicer field: see Dreicer 1960), the electrons experience "runaway" to high energies, up to 100 keV or more. In a large flare, as we have seen, the fast electrons may actually contain total energies of $10^{32}$ ergs, a large fraction of the overall flare energy.

Finally, reconnection results in changes in field line connectivity: magnetic field lines that were previously closed, i.e., that previously looped back to the Sun's surface, can become open to interplanetary space. This can allow a volume of plasma that was previously trapped (on closed field lines) to escape from the Sun. This process may occur during events known as coronal mass ejections (CMEs: see Section 18.8).

As a specific example of how magnetic energy release can explain the observations of a single flare, consider an X2.1-class flare that was studied in detail by Feng et al. (2013). Vector magnetograms were obtained by SDO/HMI for the surface of the Sun in the active region before and after the flare. Using these data, two types of magnetic field were computed for the corona above the active region: one is the potential field $B_p$ (containing no currents), and the other is a force-free field $B_{ff}$ (containing nonzero currents) (see Section 16.4.5). The total energy $E_p$ in $B_p$ above the active region was calculated to be in the range $5$–$5.5 \times 10^{32}$ ergs, and this energy remained essentially constant throughout the flare. The total energy in the force-free field, $E_{ff}$, above the active region was also calculated. The magnetic free energy that is the maximum energy available for this active region to release in a flare, $E_{mfe}$, is equal to $E_{ff} - E_p$. Feng et al. (2013) demonstrated that $E_{mfe}$ underwent a clear *decrease* $\Delta E_{mfe}$ precisely at the time when the GOES satellite detected the maximum X-ray flux from the flare. That is, during the flare, $E_{mfe}$ was found to drop from $1.1 \times 10^{32}$ ergs before to $(0.55$–$0.6) \times 10^{32}$ ergs afterwards. The reduction in $E_{mfe}$ was found to be nominally $\Delta E_{mfe} \approx 6.4 \times 10^{31}$ ergs: however, because of assumptions involved in the force-free calculation, the numerical value of $\Delta E_{mfe}$ is subject to uncertainties, possibly by as much as a factor of three. Thus, the magnetic free energy available to power the flare $\Delta E_{mfe}$ (max) *may* be as large as $2 \times 10^{32}$ ergs.

The essential physics question now becomes: is the energy difference $\Delta E_{mfe}$ (or $\Delta E_{mfe}$ (max)) sufficient to provide the *observed* release of different forms of energy in the flare? To address that issue, Feng et al. (2013) analyzed the energies associated with three channels of flared output: photons, nonthermal electrons, and coronal mass ejection. First, as regards photons, Feng et al. found that results from SDO/EVE indicate that at wavelengths from 1 to 370 Å, the flare radiated photons with a total energy of $2.2 \times 10^{31}$ ergs: if other wavelengths are included, this photon energy is expected to increase by "a few". Even the nominal value of $\Delta E_{mfe}$ suffices to explain this bolometric radiated energy. Second, as regards nonthermal electrons, Feng et al. (2013) used RHESSI data to determine that the energy contained in nonthermal electrons in this flare was $U_e \approx 7.9 \times 10^{30}$ ergs, with an uncertainty of perhaps a few: once again, the nominal value of $\Delta E_{mfe}$ easily suffices to explain the energy in nonthermal electrons. (Clearly, the energy in nonthermal electrons in this flare did not contribute anywhere near the fraction of 50% reported by Lin and Hudson [1976] for the large flares that occurred in August 1972.) Third, the flare was accompanied by a CME (see Section 18.8): the sum of kinetic and potential energies for the CME were estimated to be $6.5 \times 10^{31}$ ergs. Combining the radiated energy plus the nonthermal electron energy plus the CME energy, we arrive at a total flare energy of about $1.3 \times 10^{32}$ ergs. Comparing this energy with $\Delta E_{mfe}$ (max) $\approx 2 \times 10^{32}$ ergs, Feng et al. (2013) concluded that, quantitatively, the magnetic "free energy is able to power the flare and the CME in AR 11283".

We note that if we use the nominal value of $\Delta E_{mfe}$, the change in $E_{mfe}$ during the flare amounted to about 10% of the energy in the potential field. These magnetic energy estimates are consistent with independent estimates for 173 flares by Aschwanden et al. (2014), who reported that $E_{mfe}/E_p$ was found to lie in the range 1%–25%.

In summary, in a solar flare, the Sun apparently uses magnetic reconnection to convert the available magnetic free energy into various channels, including photons, accelerated particles, and bulk motion of material that is ejected from the Sun. Different flares divide up the magnetic free energy in different proportions among these channels, depending on details of the physical conditions in the active region where the flare occurred.

## 17.19.13 Can Flares Be Predicted?

In view of the negative effects that certain flares can have on Earth, there is widespread interest in identifying methods that might allow the onset time of a flare to be forecast some hours or days before it actually occurs. One of the favored approaches relies on applying neural networks to a certain data set of flares (the "learning set") that have already occurred and identifying the "true positives", "true negatives", "false positives", and "false negatives". The goal is to learn if the network can then identify a small number of variables that are found to be more helpful in making reliable predictions of which active regions will produce a flare within a certain number of hours: statistically, some of these methods claim success rates of 70%–80% (e.g., Al-Ghraibah et al. 2015; Bhattacharjee et al. 2020).

## REFERENCES

Ackermann, M., Ajello, M., Albert, A., et al., 2014. "High-energy gamma-ray emission from solar flares", *Astrophys. J.* 787, 15.
Alfven, H., & Falthammer, C.-G., 1963. "Magnetohydrodynamics", *Cosmical Electrodynamics: Fundamental Principles*, 2nd ed., Clarendon Press, Oxford, pp. 73–133.
Al-Ghraibah, A., Boucheron, L., E., & McAteer, R. T. J., 2015. "An automated classification approach to magnetic energy build-up", *Astron. Astrophys.* 579, A64.
Altyntsev, A. T., Meshalkina, N. S., Lysenko, A. L., et al., 2019. "Rapid variability in the SOL-2011-08-04 flare: Implications for electron acceleration", *Astrophys. J.* 883, 38.
Aschwanden, M. J., 2002. "Particle acceleration and kinematics in solar flares", *Space Sci. Rev.* 101, 1.
Aschwanden, M. J., 2011. "The state of SOC of the Sun during the last three solar cycles", *Solar Phys.* 274, 119.
Aschwanden, M. J., 2020. "Petschek reconnection rate and Alfven Mach number of magnetic reconnection outflows", *Astrophys. J.* 895, 134.
Aschwanden, M. J., Boerner, P., Schrijver, C. J., et al., 2011. "Automated temperature and EM analysis of coronal loops and AR's using SDO/AIA", *Solar Phys.* 283, 5.
Aschwanden, M. J., Crosby, N. B., Dimitripoulou, M., et al., 2016. "25 years of SOC: Solar and astrophysics", *Space Sci. Rev.* 198, 47.
Aschwanden, M. J., & Freeland, S. L., 2012. "Automated solar flare statistics in soft X-rays over 37 years of GOES observations", *Astrophys. J.* 754, 112.
Aschwanden, M. J., Wills, M. J., Hudson, H. S., et al., 1996. "Electron time-of-flight distances and flare-loop geometries", *Astrophys. J.* 468, 398.
Aschwanden, M. J., Wuelser, J.-P., Nitta, N. V., et al., 2009. "Instant stereoscopic tomography of active regions", *Astrophys. J.* 695, 12.

Aschwanden, M. J., Xu, Y., & Jing, J., 2014. "Global energetics of solar flares. I. magnetic energies", *Astrophys. J.* 797, 50.

Aulanier, G., Demoulin, P., Schrijver, C. J., et al., 2013. "The standard flare model in 3-D: II. Upper limit on solar flare energy", *Astron. Astrophys.* 549, A66.

Bellot Rubio, L., & Orozco Suarez, D., 2019. "Quiet Sun magnetic fields: An observational view", *Living Rev. Sol. Phys.* 16, 1.

Bhattacharjee, S., Alshehhi, R., Dhuri, D. B., et al., 2020. "Supervised convolutional neural networks", *Astrophys. J.* 898, 98.

Boe, B., Habbal, S., Druckmuller, M., et al. 2020. "CME-induced thermodynamic changes in the corona inferred from Fe XI and Fe XIV emission observations during the 2017 August 21 total solar eclipse", *Astrophys. J.* 888, 100.

Brenner, G., Lopez-Urrutia, J. R. C., Bernitt, S., et al., 2009. "On the transition rate of the Fe X red coronal line", *Astrophys. J.* 703, 68.

Brooks, D. H., Warren, H. P., Williams, D. R., et al., 2009. "HINODE/EIS observations of the temperature structure of the quiet corona", *Astrophys. J.* 705, 1522.

Brosius, J. W., Davila, J. M., Thomas, R. J., et al., 1996. "Measuring active and quiet Sun coronal plasma properties with extreme-ultraviolet spectra from SERTS", *Astrophys. J. Suppl.* 106, 143.

Brown, J. C., Smith, D. F., & Spicer, D. S., 1981. "Solar flares observations and their interpretations", *The Sun as a Star*, ed. S. Jordan, NASA Sci. Tech. Inf. Branch, Washington, DC, NASA SP-450, p. 200.

Bruner, E. C., 1978. "Dynamics of the solar transition region", *Astrophys. J.* 226, 1140.

Carrington, R. C., 1859. "Description of a singular appearance seen in the Sun", *Mon. Not. Roy. Astron. Soc.* 20, 13.

Cassak, P. A., Liu, Y.-H., & Shay, M. A., 2017. "A review of the 0.1 reconnection rate problem", *J. Plasma Phys.* 83, 715830501.

Chitta, L. P., Kariyappa, R., van Ballegooilen, A. A., et al. 2014. "Non-linear force-free field modelling of the solar magnetic carpet", *Astrophys. J.*, 793, 112

Chupp, E. L., Forrest, D. J., Higbie, P. R., et al., 1973. "Solar γ-ray lines observed during solar activity August 2–11, 1972", *Nature* 241, 333.

Cook, G. R., Mackay, D. H., & Nandy, D., 2009. "Solar cycle variations of coronal null points", *Astrophys. J.* 704, 1021.

Cranmer, S. R., Panasyuk, A. V., & Kohl, J. L., 2008. "Improved constraints on the preferential heating and acceleration of oxygen ions in the extended solar corona", *Astrophys. J.* 678, 1480.

Del Zanna, G., 2019. "The EUV spectrum of the Sun: chemical abundances", *Astron. Astrophys.* 624, A36.

Del Zanna, G., & Andretta, V., 2015. "The EUV spectrum of the Sun: irradiances during 1998–2014", *Astron. Astrophys.* 584, A29.

Del Zanna, G., & Mason, H., 2014. "Elemental abundances and temperatures of quiescent solar AR cores", *Astron. Astrophys.* 565, A14.

Del Zanna, G., & Mason, H., 2018. "Solar UV and X-ray spectral diagnostics", *Living Rev. Solar Phys.* 15, 5.

Dolei, S., Spadaro, D., & Ventura, R., 2015. "Visible light and UV observations of coronal structures", *Astron. Astrophys.* 577, A34.

Doschek, G., & Feldman, U., 2010. "The solar UV-X-ray spectrum from 1.5 to 2000 Å", *J Phys. B. At. Mol. Op. Phys.* 43, 232001.

Dreicer, H., 1960. "Electron and ion runaway in a fully ionized gas", *Phys. Rev.* 117, 329.

Druckmüller, M., Habbal, S. R., & Morgan, H., 2014. "Discovery of a new class of coronal structures in white light eclipse images", *Astrophys. J.* 785, 14.

Edlen, B., 1936. "Mg I isoelectronic sequence from Ti to Co", *Zeits. f. Astrophys.* 103, 536.

Edlen, B., 1945. "The identification of the coronal lines", *Monthly Not. Royal Astron. Soc.* 105, 323.

Feldman, U., Doschek, G. A., Behring, W. E., et al., 1996. "Electron temperature, emission measure, and X-ray flux in A2 to X2 X-ray class solar flares", *Astrophys. J.* 460, 1034.

Feng, L., Wiegelmann, T., Su, Y., et al., 2013. "Magnetic energy partition between the CME and the flare from AR 11283", *Astrophys. J.* 765, 37.

Ghosh, A., Tripathi, D., Gupta, G. R., et al., 2017. "Fan loops observed by IRIS, EIS, and AIA", *Astrophys. J.* 835, 244.

Greatrix, G. M., 1963. "On the statistical relations between flare intensity and sunspots", *Mon. Not. Roy. Astron. Soc.* 126, 123.

Gryciuk, M., Siarkowski, M., Sylwester, J, et al., 2017. "Flare characteristics from X-ray light curves", *Solar Phys.* 292, 77.

Gupta, G. R., 2017. "Spectroscopic evidence of Alfven wave damping in the off-limb solar corona", *Astrophys. J.* 836, 4.

Gupta, G. R., Del Zanna, G., & Mason, H. E., 2019. "Exploring Alfven wave damping in a long off-limb coronal loop up to $1.4R_\odot$", *Astron. Astrophys.* 627, A62.

Habbal, S. R., Druckmüller, M., Morgan, H., et al., 2010. "Mapping the distribution of $T_e$ and Fe charge states in the corona", *Astrophys. J.* 708, 1650.

Hannah, I. G., Hudson, H. S., Hurford, G. J., et al., 2010. "Constraining the hard X-ray properties of the quiet Sun with new RHESSI observations", *Astrophys. J.* 724, 487.

Hannah, I. G., & Kontar, E. P., 2012. "DEM from regularized inversion of Hinode and SDO data", *Astron. Astrophys.* 539, A146.

Hewins, I. M., Gibson, S. E., Webb, D. F., et al., 2020. "The evolution of CH's over three solar cycles", *Solar Phys.* 295, 161.

Hodgson, R., 1859. "On a curious appearance seen in the Sun", *Mon. Not. Roy. Astron. Soc.* 20, 15.

Jordan, C., 1969. "The ionization equilibrium of elements between carbon and nickel", *Mon. Not. Roy. Astron. Soc.* 142, 501.

Knipp, D. J., Fraser, B. J., Shea, M. A., et al., 2018. "On the little-known consequences of the August 4 1972 ultra-fast coronal mass ejecta", *Space Weather*, 16, 1635.

Koga, K., Muraki, Y., Masuda, S., et al., 2017. "Measurement of solar neutrons on 05 March 2012", *Solar Phys.* 292, 115.

Kopp, G., Lawrence, G., & Rottman, G., 2005. "The total irradiance monitor (TIM): Science results", *Solar Phys.* 230, 129.

Kretzschmar, M., 2011. "The Sun as a star: observations of white-light flares", *Astron. Astrophys.* 530, A84.

Krucker, S., Masuda, S., & White, S. M., 2020. "Microwave and hard X-ray observations by NORH and RHESSI", *Astrophys. J.* 894, 158.

Kurochka, L. N., 1987. "Energy distribution of 15000 solar flares", *Soviet Astron.* 31, 231.

Landi, E., & Young, P. R., 2009. "CHIANTI: An atomic data base for emission lines. X", *Astrophys. J.* 706, 1.

Lin, R. P., & Hudson, H. S., 1976. "Non-thermal processes in large solar flares", *Solar Phys.* 50, 153.

Lin, R. P., Schwartz, R. A., Kane, S. R., et al., 1984. "Solar hard X-ray microflares", *Astrophys. J. Lett.* 28, 421.

Liu, W., Berger, T. E., Title, A. M., et al., 2011. "New evidence of fan-spine magnetic topology", *Astrophys. J.* 728, 103.

Longcope, D. W., & Forbes, T. G., 2014. "Breakout and tether-cutting eruptions models are both catastrophic (sometimes)", *Solar Phys.* 289, 2091.

Lu, E. T., & Hamilton, R. J., 1991. "Avalanches and the distribution of solar flares", *Astrophys. J. Lett.* 380, L89.

Machado, M. E., Milligan, R. O., & Simoes, P. J. A., 2018. "Lyman continuum observations of solar flares using SDO/EVE", *Astrophys. J.* 869, 63.

Mann, G., & Warmuth, A., 2011. "Budget of energetic electrons during solar flares", *Astron. Astrophys.* 528, A104.

Masson, S., Pariat, E., Aulanier, G., et al., 2009. "The nature of flare ribbons in coronal null-point topology", *Astrophys. J.* 700, 559.

Milligan, R. O., Kerr, G. S., Dennis, B. R., et al., 2014. "The radiated energy budget of chromospheric plasma in a major solar flare", *Astrophys. J.* 793, 70.

Moore, R. L., & Datlowe, D. W., 1975. "Heating and cooling of the thermal X-ray in solar flares", *Solar Phys.* 43, 189.

Musset, S., Vilmer, N., & Bommier, V., 2015. "Hard Xray emitting energetic electrons and photospheric electric currents", *Astron. Astrophys.* 580, A106.

Neidig, D. F., & Cliver, E. W., 1983. "The occurrence frequency of white-light flares", *Solar Phys.* 88, 275.

Newkirk, G., 1967. "Structure of the solar corona", *Annual Rev. Astron. Astrophys.* 5, 213.

Nishizuka, N., Takasaki, H., Asai, A., et al., 2010. "Multiple plasmoid ejections in the 2000/11/24 flare", *Astrophys. J.* 711, 1062.

Nitta, N. V., Aschwanden, M. J., Freeland, S. L., et al., 2014. "The association of solar flares and CME's during the extended solar minimum", *Solar Phys.* 289, 1257.

Parker, E. N., 1957. "Sweet's mechanism for merging magnetic fields in conducting fluids", *J. Geophys. Res.* 62, 509.

Parker, E. N., 1972. "Topological dissipation and the small-scale fields in turbulent gases", *Astrophys. J.* 174, 499.

Petschek, H. E. 1964. "Magnetic field annihilation", *The Physics of Solar Flares*, ed. W. N. Hess, NASA publication SP-50, Washington DC, p. 425.

Priest, E. R., & Titov, V. S., 1996. "Magnetic reconnection at 3-dimensional null points", *Phil. Trans. Roy. Soc.* 354, 2951.

Rakowski, C. E., & Laming, J. M., 2012. "On the origin of slow solar wind: helium abundance variations", *Astrophys. J.* 754, 65.
Raymond, J. C., 2008. Harvard-Smithsonian Center for Astrophysics. Personal communication.
Reames, D. V., 2016. "Temperature of the source plasma in gradual SEP events", *Solar Phys.* 291, 911.
Reva, A., Ulyanov, A., Kirichenko, A., et al., 2018. "Estimate of the upper limit on hot plasma DEM in non-flaring active regions based on Mg XII spectroheliograph data from CORONAS-F/SPIRIT", *Solar Phys.* 293, 140.
Ryan, D. F., Chamberlin, P. C., Milligan, R. O., et al., 2013. "Decay-phase cooling and inferred heating of M- and X-class solar flares", *Astrophys. J.* 778, 68.
Ryan, D. F., Dominique, M., Seaton, D., et al., 2016. "Effects of flare definitions on the statistics of derived flare distributions", *Astron. Astrophys.* 592, A133.
Ryan, D. F., O'Flannagain, A. M., Aschwanden, M. J., et al., 2014. "The compatibility of flare temperatures observed with AIA, GOES, and RHESSI", *Solar Phys.* 289, 2547.
Schonfeld, S. J., White, S. M., Hock-Mysliwiec, R. A., et al., 2017. "Daily differential emission measure distributions derived from EVE spectra", *Astrophys. J.* 844, 163.
Schrijver, C. J., Title, A. M., Van Ballegooijen, et al., 1997. "Sustaining the quiet photospheric network: The balance of flux emergence, fragmentation, merging, and cancellation", *Astrophys. J.* 487, 424.
Shea, M. A., & Smart, D. F., 1990. "A summary of major solar proton events", *Solar Phys.* 127, 297.
Shimojo, M., Iwal, K., Asai, A., et al., 2017. "Variation of the solar microwave spectrum in the last half century", *Astrophys. J.* 848, 62.
Spitzer, L., 1962. *Physics of Fully Ionized Gases*, Interscience, New York.
Sturrock, P. A., 1980. *Solar Flares: A Monograph from Skylab Solar Workshop II*, Colorado Associated University Press, Boulder.
Sun, X., Hoeksema, J. T., Liu, Y., et al., 2013. "Hot spine loops and the nature of a late-phase solar flare", *Astrophys. J.* 778, 139.
Sylwester, B., Sylwester, J., Siarkowski, M., et al., 2019. "Analysis of quiescent corona X-ray spectra from Sphinx during the 2009 solar minimum", *Solar Phys.* 294, 176.
Takeda, A., Acton, L., & Albanese, N., 2019. "Solar cycle variation of coronal irradiance observed with YOHKOH", *Astrophys. J.* 887, 225.
Tanaka, K., 1986. "Solar flare X-ray spectra of Fe XXVI and Fe XXV from the Hinotori satellite", *Publ. Astron. Soc. Japan* 38, 225.
Tapping, K. F., 2013. "The 10.7 cm solar radio flux", *Space Weather* 11, 394.
Tschernitz, J., Veronig, A. M., Thalmann, J. K., et al., 2018. "Reconnection fluxes in eruptive and confined flares", *Astrophys. J.* 853, 41.
Tu, C. Y., & Marsch, E., 1997. "Two-fluid model for heating of the solar corona and acceleration of the solar wind by high-frequency Alfven waves", *Solar Phys.* 171, 363.
van de Hulst, H. C., 1950. "The electron density of the solar corona", *Bulletin Astron. Inst. Netherlands* 11, 135.
Voulgaris, A. G., Gaintatzis, A. S., Seiradakis, J. H., et al., 2012. "Spectroscopic coronal observations during the total solar eclipse of 11 July 2010", *Sol. Phys.* 278, 187.
Warmuth, A., & Mann, G., 2016. "Constraints on energy release in solar flares from RHESSI and GOES observations", *Astron. Astrophys.* 588, A115.
White, O. R., Kopp, G., Snow, M., et al., 2011. "The solar cycle 23–24 minimum", *Solar Phys.* 274, 159.
Wilhelm, K., Marsch, E., Dwivedi, B. N., et al., 1998. "The solar corona above polar coronal holes as seen by SUMER on SOHO", *Astrophys. J.* 500, 1023.
Yoshida, T., Tsuneta, S., Golub, L., et al., 1995. "Temperature structure of the solar corona", *Publ. Astron. Soc. Japan*, 47, L15.

# 18 The Solar Wind

We have already seen that the solar corona, with its temperature in excess of 1 MK, is the site of some events related to several distinct physical processes: high levels of ionization, effective thermal conduction, and emission of spectral lines over a broad range of the electromagnetic spectrum. Now we turn to a different physical property of the corona that also depends on the fact that the temperature is of order 1 MK. We shall find that, given the global properties of the Sun (specifically, the numerical values of its mass and radius), the gas that comprises the corona cannot (in certain circumstances) "stand still" but must undergo expansion. The expanding material is called the "solar wind", and in its earliest stages (close to the Sun), the wind is intertwined with the corona and therefore is subject to guidance by the coronal magnetic fields. Eventually, as we shall see (Section 18.10.1), the wind at some point "breaks free" from the coronal field, and from that point on, the wind becomes an independent entity, heading outward for an eventual encounter with the interstellar medium.

In order to describe the properties of the corona/wind, we start by considering the equation of hydrostatic equilibrium. In previous chapters, we have found that the concept of HSE *does* apply to certain locations in the Sun, but *it does not apply* in other locations in the Sun. For example, HSE applies in the radiative interior and in the photosphere/chromosphere, but not in the convective zone. Now we raise the question: does HSE apply in the corona?

In order to answer this question, we first note that the word "hydrostatic" includes the prefix "hydro", which refers to a fluid, such as the water in the Earth's oceans or the air existing near the Earth's surface. Both of these fluids are free to respond in bulk to forces arising from gas pressure, gravity, wind forces, and viscosity, and the responses give rise to ocean currents, winds, and waves. However, we have seen (Figures 17.1 and 17.2) that the sun's corona also contains magnetic fields that impose different kinds of forces on the coronal material: these forces have no analog in the Earth's oceans or lower atmosphere. In addressing the concept of HSE, we at first choose to exclude magnetic effects: this simplification allows us to predict certain global properties (density, speed) of the outflow of gas from the Sun, and spacecraft measurements indicate that these predictions are "not too bad". Results to be derived in Sections 18.1–18.5 are based on the assumption that only (nonmagnetic) fluid flow is at work. Once we have dealt with those results, we shall turn in Sections 18.6 and 18.9–18.10 to an examination of how magnetic fields affect the properties of the corona. Of special interest will be to determine (in Section 18.10.1) answers to the following questions: (i) How far out from the Sun do the coronal magnetic fields remain in control of the flow? (ii) Where does the wind eventually break free of those fields and become an independent entity?

## 18.1 GLOBAL BREAKDOWN OF HYDROSTATIC EQUILIBRIUM IN THE CORONA

In order to apply the equation of HSE, $dp/dr = -\rho g$, to the corona and solar wind, we need to allow for the fact that the acceleration due to gravity in *spherical geometry* is given by the formula $g = GM_\odot/r^2$: clearly, $g$ is no longer a constant at all locations in the corona. (In the photosphere, where we consider a *one*-dimensional problem with the vertical height $h$ as the independent variable, it is safe to assume that $g$ remains constant as the height varies within the limited confines of the photosphere: see Equation 5.2). Now, when we deal with the corona, with its *spherical aspects including three dimensions*, the value of $r$ can have values that are significantly larger than $R_\odot$: in such a situation, we must make allowance for the fact that $g$ decreases as we move away from the Sun. This decrease in $g$ with increasing $r$ has a dramatic effect on the solution of the HSE.

To see this, let us assume at first that the coronal material is a perfect gas at a constant temperature: i.e., $p = R_g \rho T/\mu$. This simplifying assumption allows us to write the HSE as

$$\frac{1}{p}\frac{dp}{dr} = -\frac{A}{r^2} \tag{18.1}$$

where $A = GM_\odot/a^2$, and $a = \sqrt{(R_g T/\mu)}$ is the isothermal sound speed. Notice that, when we examine the dimensions of both sides in Equation 18.1, we see that the quantity $A$ must have the dimensions of a *length*.

The integration of Equation 18.1 is straightforward: we find that the pressure at radial location $r$ varies as follows: $\ln(p) = (A/r) + \text{const}$. To evaluate the constant of integration, we consider the conditions at the base of the corona, where $r = r_o$. From our discussion in Chapter 17, we know that $r_o$ is essentially the radial location of the top of the chromosphere, which lies only about 2 Mm above the photosphere. In fact, we can set $r_o = R_\odot$ with a percentage error no larger than 0.5%. At the location $r_o$, the pressure $p_o$ is known to be 0.3–2 dyn cm$^{-2}$ (see Section 17.10).

Using this information, the coronal pressure as a function of radial distance can be written as

$$p(r) = p_o \exp\left[A\left(\frac{1}{r} - \frac{1}{r_o}\right)\right] \tag{18.2}$$

Inspection of Equation 18.2 shows that, as we move farther away from the Sun, i.e., as $r \to \infty$, the pressure $p(r)$ does *not* tend to zero. Instead, the pressure $p(\infty)$ approaches a nonzero asymptotic value of $p_o \exp(-A/r_o)$.

The striking aspect of this solution is that it is a very different result from the one we would get if we were to extend the photospheric (1-D) solution to infinity (Section 5.1). Using the photospheric solution $p(h) = p_o \exp(-h/H)$ where $H$ is the scale height (with $g$=const), we see that as $h \to \infty$, $p(h)$ should go exponentially rapidly to zero. The key difference between photosphere and corona is the fact that in the corona, $g$ is *not* a constant. Instead, $g$ *decreases* as $r$ *increases*: this effectively makes the scale height $H$ become larger and larger as $r$ increases, thereby leading to a reduction in the magnitude of the exponential decline. This aspect of the corona makes the HSE 3-D solution in the corona qualitatively different from the 1-D solution in the photosphere. In the outer corona, the most interesting aspect of the solution of HSE is that there is *no longer* any exponentially rapid decrease of pressure toward zero.

In order to evaluate the asymptotic pressure $p(\infty)$, we need to know the value of the constant $A$. To evaluate $A$, we need to choose a value for $\mu$: what value should we use? In the corona, hydrogen and helium are completely ionized, just as they are in the deep interior; this suggests that we could use the value $\mu = 0.58$ that we used in the hot interior of the Sun (Section 7.8). Inserting $T = 1$ MK and $\mu = 0.58$, we find $a^2 = 1.43 \times 10^{14}$ cm$^2$ sec$^{-2}$. Combining this with the value of $GM_\odot$ (Equation 1.9), we find $A \approx 9 \times 10^{11}$ cm, i.e., $A \approx 13 R_\odot$. For a coronal temperature $T = 2$ MK, we find $A \approx 6.5 R_\odot$.

As a result, *if* the entire corona were to be in HSE, the pressure of the coronal gas at infinity would be smaller than $p_o$ by a factor $e^{-13} \approx 2 \times 10^{-6}$. Inserting $p_o = 0.3$–2 dyn cm$^{-2}$ this would lead to $p(\infty) = (1–4) \times 10^{-6}$ dyn cm$^{-2}$.

The relevant physics question at this point is the following: what are we to compare this value of $p(\infty)$ to? The answer is: the Sun does not exist in a vacuum, but is surrounded by the "interstellar medium" (ISM), which exists in the space between the stars. The ISM contains gas, dust, magnetic fields, and energetic particles ("cosmic rays"). The ISM near the Sun contains hydrogen with number densities $n_a \approx 0.14$ cm$^{-3}$, electrons with $n_e \approx 0.07$ cm$^{-3}$, and temperatures $T$ of order $10^4$ K (e.g., Gayley et al. 1997). The gas pressure in the ISM, $p(\text{ISM}) \approx (n_a + n_e)kT$, is therefore roughly $\approx 3 \times 10^{-13}$ dyn cm$^{-2}$. The mass of dust contributes only about 1% of the mass of gas and contributes negligibly to the pressure. The ISM magnetic fields were estimated in 1997 to have strengths in the range $1.6$–$3 \times 10^{-6}$ G (Gayley et al. 1997): such fields would contribute pressures $B^2/8\pi$ of order

0.1–0.4 × $10^{-12}$ dyn cm$^{-2}$. (More recent direct *measurements* of the ISM field indicate $B = 4.8 \times 10^{-6}$ G [Burlaga and Ness 2016], with a pressure of $\approx 10^{-12}$ dyn cm$^{-2}$.) And the cosmic rays contribute pressures of $\approx 10^{-12}$ dyn cm$^{-2}$ (Ip and Axford 1985). The combined effects of all of the above ISM constituents suggest $p(\text{ISM}) \approx 1.4–1.7 \times 10^{-12}$ dyn cm$^{-2}$, or possibly as large as $2 \times 10^{-12}$ dyn cm$^{-2}$.

Even if the numerical value of $p(\text{ISM})$ were to be uncertain by an order of magnitude or more, one conclusion can be drawn reliably: there exists a large *excess* of $p(\infty)$ over $p(\text{ISM})$: if the solar corona were to be in HSE, the value of $p(\infty)$ would exceed $p(\text{ISM})$ by several orders of magnitude. As a result, it is physically impossible for the ISM to contain the pressure of the solar corona if the latter is in HSE. The conclusion is inevitable: *the solar corona cannot be in HSE*. This important conclusion was announced by Eugene N. Parker (1958).

If the corona does not have the property of being (hydro)*static*, what other option is available? The answer is: the corona must be *dynamic*. This means that the coronal material must undergo expansion of a fluid nature, i.e., the material of the corona *as a whole* must participate in the expansion by flowing outwards. This hydrodynamic process causes material to move from high pressure (at the base of the corona) to low pressure (in the ISM). It is important to note that this expansion of the solar corona *has nothing to do* with the physical process known as "evaporation": in the latter process, the fastest moving particles emerging from a liquid or solid have the ability to escape, but these are typically present as only a small fraction of all available particles, and most of the material "stays behind" and does not escape. In contrast, coronal outflow (assuming no magnetic effects) involves the *bulk flow* of coronal material as a fluid outward from the Sun. This outflow of bulk coronal material from high pressure to low pressure is described by the term *solar wind*, by analogy with winds on Earth, which (in the absence of Coriolis forces) cause the air to move ("blow") from a region of high pressure to another region where the pressure is lower.

## 18.2 LOCALIZED APPLICABILITY OF HSE

Are there any conditions in which the entire corona of the Sun (or of any star) *could* be in HSE? In principle, yes: this could happen if $p(\infty)$ were to have a value no larger than $p(\text{ISM})$. If the Sun (or a star) could achieve that goal, then the coronal pressure *could* in principle be contained by the ISM. One way to achieve that goal would be to reduce $p(\infty)$ by reducing the coronal temperature, thereby increasing the numerical value of $A$. According to Equation 18.2, the value of $p(\infty)$ could be made as small as $p(\text{ISM}) \approx 10^{-12}$ dyn cm$^{-2}$ if $A$ were to have a value as large as $\approx 26–28 R_\odot$. Under what conditions would such values of $A$ be possible? The answer is: the corona would need to have a temperature of $T(HSE) \approx 0.46–0.5$ MK.

Now we see how important it was for us to determine in some detail (Chapter 17) how hot the coronal gas actually is. As Chapter 17 demonstrates, the solar corona is definitely *not* as cool as $T(HSE)$: in fact, there are sound physical reasons (see Section 17.14.3) for the quiet solar corona to have $T = 1–2$ MK. (The active corona is even hotter.) Given the actual values of the physical constants that enter into the electron thermal conductivity and the radiative losses, we simply are not free to make the coronal temperature as low as 0.46–0.5 MK. Therefore, HSE is *not* applicable to the entire corona: given the observed properties of the Sun ($M_\odot$, $R_\odot$) and the empirical coronal temperatures, we are led to the same conclusion as Parker (1958) announced: hydrodynamic expansion is an *intrinsic global property* of the solar corona.

Although HSE is certainly not applicable *in a global sense* to the solar corona, this should not be construed to mean that HSE is *absolutely excluded in each and every locality* of the corona. On the contrary, in certain regions, circumstances *may* allow HSE to be applied locally.

Two examples can be considered. First, in a closed magnetic loop (Section 17.12), magnetic forces prevent ionized gas from escaping across the field lines into the wind: hydrodynamic outflow is not allowed to occur. Within the confines of such a loop, HSE *may* be a good approximation to the profile of density as a function of height.

Second, it can be shown from fluid dynamics that, in the limit where outflow speeds are much less than $a$ (the speed of sound), the hydrodynamic solution approaches the hydrostatic solution. We shall see (Section 18.3) that the flow speed of the solar wind does (eventually) indeed become as large as $a$ (and larger), but this happens only at radial distances that are several $R_\odot$ from the Sun. In view of this, there *does exist* a finite range of radial distances, within a few solar radii of the solar surface, where the wind speed is actually *much less* than $a$. Within this region, HSE *can be used* as a reasonable approximation for the density. We shall take advantage of this result in Section 18.7.

## 18.3 SOLAR WIND EXPANSION: PARKER'S MODEL OF A "THERMAL WIND"

The breakdown of HSE in the corona means that $dp/dr$ *cannot* be equal to $-\rho g$. The imbalance of forces between the pressure gradient and gravity causes the coronal gas to accelerate. E. N. Parker (1958) was the first to report on the quantitative consequences of this force imbalance: since the wind owes its existence mainly to the high *temperature* in the solar corona (i.e., the actual coronal temperature exceeds $T(HSE) = 0.46$–$0.5$ MK: see Section 18.2), the Parker solution is referred to as a "thermal wind".

The conservation of momentum, when applied to unit volume of the corona (in which the mass equals $\rho$, the local density), leads (see Equation 7.1, replacing $z$ with $r$) to the equation

$$\rho \frac{dV}{dt} = -\frac{dp}{dr} - \rho g \tag{18.3}$$

The total time-derivative $d/dt$ in Equation 18.3 can be written as the sum of two terms: $\partial/\partial t + V\partial/\partial r$. In a steady-state situation, where the flow does not depend explicitly on time, only the radial gradient term is present. In a situation where only radial gradients are important, we can write $\partial/\partial r$ as the ordinary derivative $d/dr$. Then inserting $g = GM_\odot/r^2$, Equation 18.3 becomes

$$V \frac{dV}{dr} = -\frac{GM_\odot}{r^2} - \frac{1}{\rho}\frac{dp}{dr} \tag{18.4}$$

It is worthwhile to consider the simplest case, in which the corona is assumed to be isothermal, i.e., $T = $ const at all radial locations. (We will examine in Section 18.4 whether there are physical reasons why this assumption might be "not too bad".) With this assumption, we are in effect greatly simplifying the equation for the conservation of energy. In an isothermal corona, the pressure and density are related at all locations by the formula $p(r) = a^2\rho(r)$. This allows us to rewrite Equation 18.4 as follows:

$$V \frac{dV}{dr} = -\frac{GM_\odot}{r^2} - \frac{a^2}{\rho}\frac{d\rho}{dr} \tag{18.5}$$

Turning now to conservation of mass, we note that, at a radial distance $r$, the rate of mass outflow from the Sun in a spherically symmetric wind is given by $dM/dt = 4\pi r^2 \rho(r) V(r)$. Once the solar wind leaves the corona and flows out into interplanetary space, no further significant mass can be added to the outflow. Therefore, $dM/dt$ is independent of $r$, i.e., $r^2\rho(r)V(r) = $ const. Thus the radial derivative of $r^2\rho(r)V(r)$ is zero. Taking logarithms, this means that

$$\frac{1}{\rho}\frac{d\rho}{dr} + \frac{1}{V}\frac{dV}{dr} + \frac{2}{r} = 0 \tag{18.6}$$

Using Equation 18.6, we can replace the final term in Equation 18.5. Then collecting terms in the radial gradient $dV/dr$, we obtain an equation for $dV/dr$:

$$\left(V - \frac{a^2}{V}\right)\frac{dV}{dr} = \frac{2a^2}{r} - \frac{GM_\odot}{r^2} \tag{18.7}$$

# The Solar Wind

The structure of this equation indicates that $dV/dr$ can be written as the ratio of two terms, $N(r)/D(r)$. The numerator $N(r)$ is the radial function on the right-hand side of Equation 18.7, while the denominator $D(r)$ is the radial function $V(r) - a^2/V(r)$.

At a certain radial location, the function $D(r)$ passes through the value zero. This occurs when the wind speed $V(r)$ becomes equal to the sound speed, $a \approx 120\sqrt{T_6}$ km sec$^{-1}$ (where $T_6$ is the temperature in MK). In coronae with $T_6 = 1$ and 2, the sound speeds are $a \approx 120$ and $a \approx 170$ km sec$^{-1}$, respectively. The radial position where the outflow speed has the particular value $V(r) = a$ is critically important in the wind outflow: this position is referred to as the "sonic point". In order to prevent $dV/dr$ from becoming infinitely large at the sonic point, $N(r)$ must also pass through the value zero at the sonic point.

Setting $N(r) = 0$ at radial location $r = r_s$, we find that the sonic point lies at the radial location

$$r_s = \frac{GM_\odot}{2a^2} \tag{18.8}$$

Inserting the numerical value $GM_\odot = 1.327124 \times 10^{26}$ cm$^3$ sec$^{-2}$ (Equation 1.9), we find that in a corona with $T = 1$ MK (i.e., $a^2 = 1.44 \times 10^{14}$ cm$^2$ sec$^{-2}$), the value of $r_s$ is $4.6 \times 10^{11}$ cm. In a corona with $T = 2$ MK, $r_s \approx 2.3 \times 10^{11}$ cm. Compared with the solar radius, we see that the sonic point lies at radial locations of $r_s \approx 6.6R_\odot$ and $3.3R_\odot$ for coronas with $T = 1$ and 2 MK, respectively.

Thus, in response to the breakdown of HSE, the material in an isothermal corona is accelerated outward, increasing from essentially zero velocity at the base of the corona to a velocity as large as the sound speed at radial locations of a few solar radii. If $T = 1$ MK, the wind reaches a velocity of $\approx 120$ km sec$^{-1}$ at $r \approx 6.6R_\odot$. If $T = 2$ MK, the wind reaches a velocity of $\approx 170$ km sec$^{-1}$ at $r \approx 3.3R_\odot$. The wind acceleration does not stop at the sonic point: at radial locations outside the sonic point, the thermal wind continues to accelerate to even higher speeds. However, outside the sonic point, the acceleration is not as strong as close to the Sun: the wind speed at large radial distances increases only slowly as the radial distance increases (see Section 18.5 for further details).

Now that we have an estimate of the location of the sonic point at $r \approx r_s \approx 6.6R_\odot$, it is worthwhile to recall briefly the results of Dolei et al. (2015) (see Chapter 17) who found that, *inside* one particular helmet streamer (where magnetic field lines remain closed), there was *no evidence* for outflow out to $r \approx 3.5R_\odot$. In contrast, when they observed *outside* the streamer (where field lines are open), the solar wind was found to be expanding freely. Remarkably, in a helmet streamer model, Pneuman (1968) showed that the last closed field line inside the helmet should extend no farther out than $r = 0.5r_s$. Substituting $6.6R_\odot$ for $r_s$, Pneuman's result indicates that at $r \leq 3.3R_\odot$, the field lines inside the helmet are *closed*, thereby preventing the wind from expanding.

The larger the coronal temperature, the faster is the acceleration of the thermal wind. To quantify this, we note that if an increase in velocity by $\Delta V = 120$ (or 170) km sec$^{-1}$ were to occur with constant acceleration over a spatial interval of $\Delta x = 4.6$ (or 2.3) $\times 10^{11}$ cm, the corresponding acceleration ($\approx (\Delta V)^2/(2\Delta x)$) would be roughly 160 (or 630) cm sec$^{-2}$. Thus, the inner solar wind experiences an outward acceleration that, as regards the magnitude, coincidentally is not far from the (downward) acceleration (981 cm sec$^{-2}$) experienced by objects near the Earth's surface.

The concept of a sonic "point" in the Parker wind is highly idealized: it arises because several simplifications have been made along the way, not the least of which is the 1-D assumption of spherical symmetry. In the "real world" of the solar wind, with its highly turbulent 3-D motions, there will certainly be regions of the wind where the radial outflow is supersonic, and other regions where the radial outflow will be subsonic. But the transonic transition is likely to occur at different radial locations above different points on the solar surface. As a result, the sonic "point" of Parker's model will be a sonic "surface" in the real solar wind, and the surface will have a more or less complicated 3-D shape, depending on how hot the gas is at any location. Outside the sonic surface, sound waves cannot propagate towards the Sun: only outward waves are present. But inside the sonic surface, sound waves can propagate both inward and outward.

## 18.4 CONSERVATION OF ENERGY

So far, we have obtained information about the solar wind by explicitly referring to only two conservation laws: one for momentum and one for mass. But there is also a law of conservation of energy. How does that contribute to the thermal wind solution? In fact, our assumption $T =$ const involves a particular solution of the energy equation. The difficulty is that we have not specified, in physical terms, how a constant temperature of 1–2 MK might be *maintained* out to distances of $(3.3$–$6.6)R_\odot$ and beyond. How can the coronal gas remain hot all the way out to these distances?

In order to determine how, in physical terms, the temperature actually varies as a function of radial distance, we need to solve the equation of energy conservation. In order to do that, we would have to include processes that deposit energy in the gas, remove energy from the gas, or distribute energy through the gas.

It is worthwhile to consider one particular physical process that is (as we have already seen, see Section 17.15.1) relevant in the corona: thermal conduction. In spherical geometry, the equation of heat conduction in steady state is described by

$$\frac{d}{dr}\left(r^2 k_{th} \frac{dT}{dr}\right) = 0 \qquad (18.9)$$

We have already seen (Chapter 17, Section 17.14.1) that in coronal plasma, the thermal conductivity $k_{th} = k_o T^{2.5}$. Inserting this in Equation 18.9, we find, after a first integration, that

$$T^{2.5} \frac{dT}{dr} = \frac{const}{r^2} \qquad (18.10)$$

The solution of this equation is $T(r) \sim r^{-2/7}$. This is a rather slow function of radial distance. For example, if $T = 2$ MK at $r = R_\odot$, then at the sonic point distance $(3.3 R_\odot)$, a thermally conducting wind would have $T \approx 1.4$ MK. Thus, contrary to our earlier assumption of constant $T$, the temperature would *not* in fact have remained strictly constant all the way out to the sonic point. On the other hand, it can be admitted that the value $T = 1.4$ MK is not "drastically" cooler than $T = 2$ MK: in fact, according to one perspective, a gas with $T = 1.4$ MK might be regarded as being "almost" as hot as a gas with $T = 2$ MK. Electrons are (as it turns out) very effective at distributing heat through the gas, thereby helping to keep the temperature from falling off too rapidly. From this perspective, a model of the corona that assumes constant temperature (at least out to a few solar radii) is not totally unrealistic.

Another (theoretical) approach to including energy conservation is to *assume* that the solar wind material has the property that the pressure at any radial location is related to the density at that location by a simple relation, $p(r) \sim \rho(\rho)^{(n+1)/n}$. This is nothing less than the "polytrope law" (see Equation 10.1) that we found helpful in describing certain properties of the solar interior. It turns out that a rich variety of solar wind solutions can be derived by considering various values of the polytropic index $n$. The isothermal case corresponds to $n = \infty$. The adiabatic case corresponds to $n = 3/2$. In the latter case, $p \sim \rho^{5/3}$, which for a perfect gas ($p \sim \rho T$) corresponds to $T \sim \rho^{2/3}$. Once the solar wind speed approaches a nearly constant value (see next section), Equation 18.6 implies that $\rho \sim r^{-2}$. In an adiabatic wind, this leads to rapidly declining $T$ as $r$ increases: $T \sim r^{-4/3}$.

There are other possibilities for keeping the corona hot. These include the deposition of energy from wave modes of various kinds (Alfven waves, shocks). Moreover, we shall see (Section 18.9) that the solar wind flow is highly turbulent: dissipation of magnetic fluctuations in the turbulence can add thermal energy to the solar wind, thereby helping to keep the local gas hotter than the $T \sim r^{-4/3}$ prediction in an adiabatic wind (e.g., Smith et al. 2001).

# The Solar Wind

## 18.5 ASYMPTOTIC SPEED OF THE SOLAR WIND: THE MAGNETIC SPIRAL

We have seen that according to Equation 18.7, the wind is already expanding away from the Sun at speeds of 120 (or 170) km sec$^{-1}$ at radial locations of 6.6 (or 3.3) $R_\odot$ for a coronal temperature of 1 (or 2) MK. Let us now consider how the speed behaves as we examine the flow at very large distances from the Sun.

In $N(r)$ (see Equation 18.7), the term in $1/r$ dominates over the term in $1/r^2$ as $r \to \infty$. That is, $N(r) \to 2a^2/r$. Moreover, at large $r$, the wind is supersonic, i.e., $V$ exceeds $a$. As a result, the dominant term in $D(r)$ as $r \to \infty$ is $V$. Therefore, at large $r$, we can approximate Equation 18.7 by $dV/dr = N(r)/D(r) \to 2a^2/rV$. In an isothermal wind, this approximate equation can be integrated to give the solution

$$\frac{1}{2}V^2 = 2a^2 \ln(r) + const. \tag{18.11}$$

The constant of integration can be evaluated by noting that $V = a$ at $r = r_s$. This leads to

$$\left(\frac{V}{a}\right)^2 = 4\ln\left(\frac{r}{r_s}\right) + 1 \tag{18.12}$$

This approximate solution indicates that, at large distances (i.e., $r \gg r_s$), the solar wind speed asymptotically approaches the functional form $V(r) \to 2a\sqrt{(\ln(r/r_s))}$. A logarithmic variation is already a slow function of distance, and so the square root of a logarithmic variation is a very slow function of radial distance. For example, if $T = 1$ MK, by the time the solar wind reaches Earth orbit, i.e., $r = 215.04\ R_\odot$ (Section 1.5), the ratio $r/r_s$ has the value $\approx 33$. According to Equation 18.12, this approximation gives $V(1\ \text{AU}) \approx 3.9a \approx 470$ km sec$^{-1}$. A more accurate solution of Equation 18.7, retaining all terms, indicates that, for a $T = 1$ MK corona, $V(1\ \text{AU}) = 427$ km sec$^{-1}$, while for a $T = 2$ MK corona, $V(1\ \text{AU}) = 674$ km sec$^{-1}$ (Mann et al. 1999).

An important property of the radial profile of the *density* of the solar wind can be identified in the asymptotic limit at great distances: the wind speed in those regions of the wind $V(r)$ varies so slowly with increasing $r$ that one can adopt (without significant error) the limit that $V(r) =$ constant. In this limit, Equation 18.6 indicates that the density of the wind should decrease as $1/r^2$ at large $r$. As a result, with a wind density order 10 cm$^{-3}$ at 1 AU (see Figure 18.1), the density is predicted to fall to a value of order 0.001 cm$^{-3}$ at distances of order 100 AU: we shall return to a consideration of physical processes which occur near 100 AU in Section 18.10.

Turning briefly to energy considerations, we note that if thermal conduction (mainly by electrons) dominates the energy equation in the solar wind, then compared to the coronal temperature at $r = R_\odot$, $T$ at 1 AU would be reduced by factors of $215^{2/7} = 4.6$. Thus, with $T = 1$–2 MK at $r = R_\odot$, $T$ at 1 AU should be 2–4 × 10$^5$ K. On the other hand, if the solar wind were described by an adiabatic polytrope ($n = 3/2$), with $T(r) \sim r^{-4/3}$ (see Section 18.5), the ion temperatures at $r = 1$ AU would be very low, of order 10$^3$ K.

Finally, we note that the material of the solar wind (originating as it does in the corona where the temperature is at least 10$^6$ K) is highly ionized and therefore has a large electrical conductivity. As a result, magnetic field and solar wind are "frozen" together, and the expanding wind drags the magnetic field from the solar surface outward into interplanetary space. Parker (1958) showed that, in the presence of the rotation of the surface of the Sun (see Section 1.11), the radial outflow of the wind will drag the field out in the form of a spiral. By the time the wind reaches the Earth's orbit, the field lines in the spiral are not radial but instead depart from radial by an angle of about 45 degrees. As the distance from the Sun increases, Parker (1958) showed that the departure angle grows increasingly large, until in the outermost reaches of the solar system, the field lines depart from the radial direction by a maximum allowed value close to 90 degrees.

## 18.6 MAGNETIC FIELD EFFECTS: "HIGH-SPEED" WIND AND "SLOW" WIND

Up to this point, we have been considering that the coronal material is a fluid that is *not* subject to any magnetic effects: in such a situation, there would be little reason for the wind to be anything other than spherically symmetric. However, as we have seen (Chapter 17), magnetic fields are definitely observed in multiple regions of the solar surface. It is now time to turn to a discussion of some of the effects that these fields have on the solar wind.

Despite the large amount of variability observed in the solar wind properties (speed, density, temperature, magnetic field) in the vicinity of Earth orbit, two major categories of solar wind have been identified (Zirker 1981): "slow wind", with mean speeds (near Earth) of 330 km sec$^{-1}$, and "high-speed streams (HSS)" (or "fast wind"), with mean speeds of 700 km sec$^{-1}$. The two types of wind differ from each other in more than simply the speed: in fast wind, the proton number density flux is smaller, the electron temperature is smaller, the helium/hydrogen ratio is larger, and the drift speed between helium and hydrogen is larger (Stansby et al. 2020). These differences suggest that different physical mechanisms may be at work in driving the fast wind and the slow wind.

Defining an HSS as one in which the wind speed at first increases by >100 km s$^{-1}$ relative to the preceding wind, and subsequently (after 1–2 days) decreases by > 100 km s$^{-1}$, the years 1996–2008 (i.e., all of cycle 23) were found (Xystouris et al. 2014) to have a total of 710 HSS, with an average rate of 2.5 HSS per month in 1996 and 6.5 per month in 2003. The speeds of the HSS ranged from 400 km s$^{-1}$ to 1199 km s$^{-1}$. The dominant (61%) source of HSS was a coronal hole, while flares also contributed significantly (36%), especially at solar maximum. During a part of cycle 24, i.e., in the years 2009–2016, a total of 303 HSS were reported (Gerontidou et al. 2018). The maximum rate (almost four per month) occurred in 2015, and the maximum speeds were found to be 800–899 km s$^{-1}$. In this cycle, coronal mass ejections (CME's) were the dominant (63%) source of HSS, while coronal holes contributed 37%.

At 1 AU, in the plane of the Earth's orbit, despite the presence of two major components at low and high speeds, it is possible to define an "average" set of plasma characteristics for solar wind flows (e.g., Zirker 1981). The mean values of velocity, density, and ion temperature in the "average" solar wind at 1 AU are found to be 470± 120 km sec$^{-1}$, 8.7±6.6 cm$^{-3}$, and 1.2±0.9 ×10$^5$ K. In view of these "average" properties, Parker's thermal wind model does a good job at predicting at least the *average speed* of the wind near the Earth. And the proton temperatures in the wind as it flows near Earth's orbit are not all that far from the predicted values due to thermal conduction (see Section 18.4).

Where does the fast wind originate? It has been found to emerge preferentially from large coronal holes with widths of >60 heliographic degrees (Krieger et al. 1973). Coronal holes can be most readily detected at the polar caps of the Sun during solar minimum. As we have seen (e.g., Figure 17.1), the magnetic fields that exist in the coronal holes near the north and south poles of the Sun appear to be similar to the patterns that are revealed by iron filings when they are scattered on a sheet of paper sitting above a bar magnet. Since flows of ionized gas are permitted to occur freely *along* magnetic field lines (see discussion following Equation 16.2), the field lines at N and S poles of the Sun provide open channels for ionized gas to flow freely away from the Sun. With no obstacles to slow down the flow, the wind emerging from coronal holes is free to respond to whatever forces are present: the result is a wind with higher speeds than the wind that emerges from most other regions of the Sun. What forces are acting on material in a coronal hole? Parker's idea of the pressure gradient due to locally hot coronal gas is still operative, helping to drive a thermal wind. However, the fact that large-scale open magnetic fields are clearly present in coronal holes suggests that magnetic (Lorentz) forces may also contribute to the wind (Weber and Davis 1967). In the presence of Lorentz forces, a new critical point appears in the outflow, namely, the radial location $r_A$ where the wind speed $V$ grows large enough to be equal *not* to the local sound speed (which occurs at the radial location $r_s$, see eq. 18.8) but instead, $V$ is equal to the local *Alfven speed* $V_A$. The location where $V = V_A$ is referred to in a spherically symmetric wind as the Alfvenic point: with typical values of the parameters in the solar wind, $r_A$ has values that may range from as little as 10–20 $R_\odot$ to perhaps as large as 50 $R_\odot$. By analogy with the sonic point (see Section 18.3), in the "real solar wind" the idealized Alfvenic "point" will be replaced by an Alfvenic surface with a more or less complicated 3-D shape. (For example, $V_A$ may be as large as 1000 km s$^{-1}$ out to $r \approx 10 R_\odot$ at low latitudes, but $V_A$ may be as small as 250 km s$^{-1}$ out to $r \approx 3$–$4 R_\odot$ at high latitudes: see Susino et al. 2015.) Outside the Alfvenic surface, Alfven waves cannot propagate towards the Sun: only outward waves are present. But inside the Alfvenic surface, Alfven waves can propagate both inward and outward.

Since coronal holes are always present to some extent at north and south solar poles, the wind flowing out from high latitudes on the Sun is always "high speed". This was confirmed by the Ulysses spacecraft, which traversed the solar wind emerging from both solar poles (e.g., Phillips et al. 1995). But there are also times when high-speed streams are observed in the plane of the Earth's orbit: these may arise from coronal holes that can sometimes (especially near solar minimum) form at low latitudes. No matter where a coronal hole is formed, the presence of open field lines permits unimpeded escape of the solar wind. This is a primary example of how the Sun's magnetic field imposes its effects on the solar wind.

Where does the slow solar wind (SSW) originate? Following the launch of the Hinode spacecraft, data have suggested that a significant portion (up to 25%) of the SSW *emerges from the edges of active regions* (Abbo et al. 2016). In such active regions (ARs), the upflowing plasma was observed to have speeds of up to 100 km sec$^{-1}$, temperatures of order 1.3 MK, and densities of order $7 \times 10^8$ cm$^{-3}$, and there are rapid changes on time-scales of 5 minutes: apparently the upflow process is very dynamic. ARs are typically sites of closed loops of magnetic field: therefore, if wind is truly escaping from an AR, there must be some physical process that allows closed AR field lines to make (re)connections with open fields in the surrounding regions of the Sun. Abbo et al. (2016) describe several processes that could make such connections.

Although the *speed* of the solar wind shows large (factors 2–3) variations between slow wind and fast wind, the total *flux of energy* in the wind is observed to remain surprisingly constant. The total flux, i.e., the sum of KE and PE, has been evaluated in three distinct data sets: Helios (inner solar wind at ~0.3 AU), Ulysses (out-of-ecliptic solar wind), and Wind (in-ecliptic wind at 1 AU) (Le Chat et al. 2012). Normalizing the fluxes to a common radial distance of 1 AU, the three spacecraft obtained fluxes (in units of mW m$^{-2}$) of $1.4 \pm 0.2$ (Helios), $1.5 \pm 0.4$ (Ulysses), and $1.4 \pm 0.3$ (Wind). These three measurements, obtained over a time span of 24 years, overlap with one another. Le Chat et al. (2012) conclude, "the fast and slow solar winds have the same mean energy flux, either in solar-activity maximum (in 2001) or minimum (in 1996 or 2008)". Thus, the solar wind "energy flux appears as a global solar constant". In years past, solar observers used to refer to a "solar constant" in the context of the radiative output of the Sun. Once radiometers became precise enough to determine that the total solar irradiance actually varied during the 11-year cycle (see Section 1.4), the term "solar constant" fell out of favor, at least in the context of the radiative power. But now, the term appears once more in a very different context: the solar wind. Compared to the radiative output power of the Sun ($1.36 \times$ kW m$^{-2}$ at a radial distance of 1 AU), the solar wind output power is found to be smaller by a factor of about $10^6$. This factor is presumably related to the effectiveness with which the Sun (with its huge radiative power) generates mechanical power to heat the corona, thereby driving off the solar wind.

The amount of coronal heating can also be estimated from the amount of radiation emitted by coronal lines in the EUV: over a time interval of 4–5 decades, when measurements of these spectral regions have become available, the solar EUV power output is typically of order 0.1–1 mW m$^{-2}$, with only very rare excursions to almost 10 mW m$^{-2}$ (Woods et al. 2012). Although the solar radiative output in the EUV can change temporarily by extreme factors of up to 100–1000 during a solar cycle, the typical power output is no more than 1 mW m$^{-2}$. This typical EUV power output, of order $10^{-6}$ times the total solar power output, agrees well with the energy fluxes measured in the solar wind.

It is a major goal of modern solar research to identify precisely the physical process(es) that lead to an effectiveness of no more than $10^{-6}$ for mechanical energy generation (presumably due to [magneto]hydrodynamic processes) compared to radiative energy generation (due to nuclear and electromagnetic interactions).

## 18.7 OBSERVATIONS OF SOLAR WIND PROPERTIES

The simplest model of an isothermal solar wind predicts that wind speeds of several hundred km sec$^{-1}$ and temperatures of up to a few times $10^5$ K should occur at radial locations that lie *near Earth's orbit*. In Parker's original paper (1958), there is a warning to the reader not to take "too literally any of the smooth idealized models which we have constructed in this paper" because instabilities of various kinds are likely to occur in the wind. Nevertheless, it is worthwhile to examine if the observations reveal *any* overlap with the simplest thermal wind model predictions.

### 18.7.1 *In situ* Measurements: ≈ 1 AU and Beyond

Thermal wind predictions have been tested by many spacecraft since the 1960s and found to be not too bad *on average*. (For an informative overview of the early experiments to measure solar wind properties, and the scientists involved, see Hufbauer 1991.) However, the data also indicate that (as Parker mentioned) the solar wind has highly variable properties. Examples of solar wind velocity, density, and proton temperature close to the Earth's orbit are shown for a 27-day interval in Figure 18.1. (The data were obtained by the ACE satellite, which is in an orbit that keeps it always at a distance of about 1.5 million km closer to the Sun than the Earth's orbit.) During the interval of the measurements (in the year 2008), the Sun was at a low level of activity. As can be seen, the speed of the solar wind at 1 AU varies from roughly 300 to roughly 800 km sec$^{-1}$: this range contains the predicted thermal wind speeds at 1 AU from a $T$ = 1–2 MK corona. It is remarkable that such a simple "thermal wind" model does in fact predict wind speeds that actually are observed to occur in the vicinity of 1 AU. The proton temperature ranges from a few times $10^4$ K to a few times $10^5$ K. And the number density of protons in the solar wind near 1 AU is for most of time less than 10 cm$^{-3}$, although there are occasional "spikes" where the number density can rise to values of order 50 cm$^{-3}$.

The properties of the solar wind can be measured *in situ* by spacecraft not merely in the vicinity of the Earth's orbit, but also at radial distances that greatly exceed 1 AU. In terms of the maximum distance at which the wind properties have been transmitted back to Earth, at the time of writing (October 2021), the farthest distance from which any spacecraft has reported data is 154.4 AU away from the Sun (Voyager 1).This distance increases at a rate of 3.6 AU/year, so that Voyager 1 will be 200 AU from the Sun by the year 2034. (However, by then, radio transmissions from Voyager 1 will probably be too weak to be detectable on Earth: the satellite's power supply is a radioactive isotope that has been constantly decaying since Voyager was launched in 1977. Nominally, transmissions after the year 2025 may no longer be strong enough to be reliably detected at Earth.) Measurements that have been made with Voyager 2 (where instruments can still measure plasma speeds) indicate that, *outside* the Earth's orbit, the solar wind does not change much as the radial distance increases from 1 to 50 AU (see Figure 18.2). Applying Equation 18.12, we note that at 1 AU, $r/r_s \approx 33$, while at 50 AU, $r/r_s \approx 1650$. Therefore, $V$(50 AU) is predicted to be larger than $V$(1 AU) by a factor of $\sqrt{(\ln 50)} \approx 2$, *if* the temperature were to remain constant all the way from 1 AU to 50 AU. Measurements, however, show that the temperature is lower at 50 AU than at 1 AU: therefore a predicted increase in $V$ by a factor of two between 1 and 50 AU is certainly an upper limit on the speed of a thermal wind. Inspection of Figure 18.2 shows that the average radial velocity of the wind at 1 AU is of order 350 km sec$^{-1}$, while at 50 AU, the average speed is of order 500 km sec$^{-1}$, an increase by a factor of 1.4: this is consistent with the upper limit of two mentioned earlier. Note also that during the course of the 19 years between the time Voyager 2 was launched until it reached 50 AU, the Sun itself had undergone almost two complete solar cycles: during those cycles, the boundary conditions of the

The Solar Wind

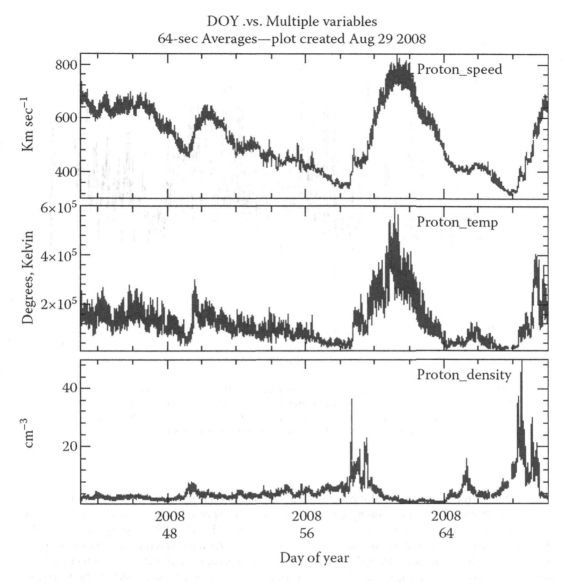

**FIGURE 18.1** Solar wind properties (density, speed, temperature) for a 27-day interval early in the year 2008. DOY=day of year. These (and other) data are publicly available for any time interval between 1998 and 2008 from the ACE spacecraft data archive at www.srl.caltech.edu/ACE/ASC/level2/lvl2DATA_SWEPAM.html.

solar wind in the corona would have undergone significant alterations, and those could also have caused some of the variations in speed in Figure 18.2. But the overall conclusion from Figure 18.2 is that the measured solar wind speed increases only slowly with increasing radial distance. This is consistent with the $\sqrt{(\ln r)}$ dependence predicted by Parker.

Do the properties of the solar wind vary during the solar cycle? Annual averages of solar wind in the ecliptic plane have been plotted by Sokol et al. (2013): although there are minima in the speed during (or near) the solar minima of 1996 and 2008, the largest maxima (in 2004 and 1994) do not correspond to solar maxima. If there exists a solar cycle variation in wind *speed*, it does not appear to be very pronounced. Support for this statement can be obtained by inspection of the top panel in

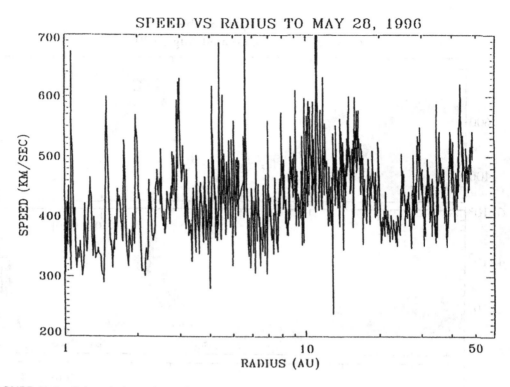

**FIGURE 18.2** Solar wind speed as a function of distance from the Sun, as measured by Voyager 2 in the course of the first 19 years of its operation. Although there are clearly large fluctuations in the speed locally, the ups and downs largely cancel each other out: there is only a slow overall trend towards higher speeds at greater distances from the Sun (Burlaga et al. 1996; used with permission of Springer).

Figure 18.1: there, we see the wind speed during a month interval in early 2008, a year when the Sun was going through a deep and prolonged minimum. The wind speed during this very "quiet" period can be seen to vary from as large as 800 km s$^{-1}$ to as little as 300 km s$^{-1}$: this range spans essentially the entire range of wind speeds that have ever been detected. As regards temporal variations in the densities and magnetic field strengths of the solar wind, Farrugia et al. (2012) used STEREO-A data to determine that, during the deep minimum of 2007–2009, the proton density in the wind was found to be significantly smaller than in the previous solar minimum: ~3–4 cm$^{-3}$ in 2007–2009, compared to ~8 cm$^{-3}$ in 1995–1996. Moreover, the mean magnetic field strength in 2007–2009 was in the range 2.5–4.5 nanotesla (nT), i.e., smaller than the value 5.4 ± 2.5 nT in 1995–1996.

### 18.7.2 IN SITU MEASUREMENTS IN THE INNER WIND: $R < 1$ AU

Let us now consider the properties of the solar wind that have been measured as we move inwards *towards* the Sun. Spacecraft have made extensive *in situ* measurements of wind properties since the early 1960s. In 1962, the Mariner II became the first spacecraft to travel from Earth to Venus, and it sampled the solar wind at radial distances ranging from 1 AU to 0.7 AU. In the course of the flight, during an interval of ~60 days, it was found, "there was always a measurable flow of plasma from the direction of the Sun" (Neugebauer and Conway 1962). The speed of the flow was generally in the range 400–700 km sec$^{-1}$, agreeing "fairly well" with Parker's predictions. Plasma densities were estimated to be in the range of 2.5–4.5 cm$^{-3}$, while temperatures were in the range

(2–7) × $10^5$ K and magnetic field strengths were of order $5 \times 10^{-5}$ G. Combining plasma densities and temperatures, Mariner II obtained an important result: the plasma kinetic energy density exceeded the magnetic energy density by factors of 10 or more. In these conditions, the plasma "wins out", and the magnetic field has to go where the plasma decides (see Section 16.7.7). The magnetic field originates in the Sun, but the solar wind drags the field lines away from the Sun and out into interplanetary space. As it turns out, this action of the wind provides an essential assist to the solar dynamo: in order for the magnetic fields of a new cycle to emerge on the Sun's surface (with reversed polarity), the fields from the "old" cycle have to be removed in some way. The solar wind helps to do this by dragging the field lines far out into space where they eventually disconnect.

For several decades, the closest any spacecraft had approached the Sun was 0.29 AU, i.e., in as close as radial locations of $\approx 70 R_\odot$: this was achieved by Helios in the 1970s.

Starting in 2018, a new era of inner solar wind studies was initiated: the Parker Solar Probe (PSP) was launched in order to make unprecedented *in situ* explorations wind at radial locations that lie closer to the Sun than ever before. PSP is in a highly elliptical orbit with an initial aphelion close to Earth's orbit and a series of perihelia that lie well inside Mercury's orbit (Fox et al. 2016). The orbital changes are caused by a series of seven gravity assists from the planet Venus. At the first three perihelia, PSP approached to a perihelion distance of 35.66 $R_\odot$ from the Sun. When the last three planned perihelia will occur (#22–24 in the year 2025), PSP will approach to 9.86$R_\odot$ (about 22 times closer than Earth's orbit). At closest approach to the Sun, the flux of solar radiant energy $F(PSP)$ will be almost 500 times larger than at Earth, leading to a nominal equilibrium temperature $T_{eq} \sim 1/r^{0.5} \approx$ 4–5 times larger than at Earth's orbit, i.e., $T_{eq} \approx$ 1400 C. To shield the instruments from the Sun's heat, PSP carries a special carbon-foam shield, 2.4 meters wide and 115 mm thick: in the shadow of this shield, the instruments on PSP remain no warmer than 30° C.

The first results from PSP, dealing with data obtained during the first and second perihelion passages, were publicly reported in November 2019. A surprising discovery was the presence of hundreds of "switchbacks" (intermittent reversals) in the magnetic field (Bale et al. 2019): these are kinks or folds in the magnetic field, associated with increases in the wind speed. These features last as short as <1 second and as long as >1 hour, and their origin is (at the time of writing, October 2021) a hotly debated topic. A coronal imaging instrument on PSP reported the presence of small plasma structures which are ejected frequently from the Sun (Howard et al. 2019): these might be evidence for "magnetic islands" created by tearing-mode instability in multiple reconnection sites in the solar wind (see Section 17.19.10). Moreover, Howard et al. reported on hints that the $F$-corona may be *decreasing* in intensity close to the Sun, perhaps due to the expectation that dust (which causes the $F$-corona) cannot survive solar heating at the innermost radial locations where the dust would be heated to the point of sublimation. The subsequent publication (in February 2020) of a large group (52!) of papers based on the first and second perihelion passages of PSP is an indication of the widespread interest in the data. In one of those papers, Moncuquet et al. (2020) reported electron densities of $n_e$ = 100–300 $cm^{-3}$ when PSP was at perihelion. As expected, these values are considerably larger than typical values at 1 AU. If the densities were to scale as $1/r^2$ (appropriate for a wind with constant speed), the density at $r$ = 36$R_\odot$ would exceed those at Earth orbit ($r$ = 215$R_\odot$) by a factor of about 36: using the results of Zirker (1981) for the "average" solar wind at 1 AU, where $n$ = 8.7±6.6 $cm^{-3}$, we would expect $n$ = 310±240 $cm^{-3}$ at the first perihelion of PSP. The fact that this prediction overlaps with the data is an indication that the solar wind speed is indeed almost constant between $r$ = 36$R_\odot$ and Earth's orbit. Moncuquet et al. also reported on the radial profile of electron temperature: they found $T \sim r^{-0.74}$: this is certainly much less steep than adiabatic behavior would predict ($T \sim r^{-1.33}$; see Section 18.4), indicating that heat is being deposited (somehow) into the electrons as they flow out from the Sun. Several of the 52 papers published in 2020 were devoted to theoretical modeling of heating of protons and electrons in the solar wind, especially by turbulent processes.

When PSP eventually makes its closest approach to the Sun in 2025, the probe will still lie formally *outside* the sonic surface for a spherically symmetric $T = 1$ MK corona. However, PSP will eventually be close enough to the Sun to lie *inside* the Alfvenic surface, where the wind speed has accelerated up to the local Alfven speed. In this regard, it is noteworthy that NASA released a claim (in December 2021) that PSP actually passed inside the Alfvenic surface for the first time during its eighth perihelion passage, which occurred in April 2021. (See https://svs.gsfc.nasa.gov/cgi-bin/details.cgi?aid=14045&button=recent)

Therefore, at subsequent perihelion passages, PSP should be in a good location to measure the solar wind properties in a physically significant region, i.e., in sub-Alfvenic wind.

### 18.7.3 Remote Sensing of the Solar Wind

*In situ* measurements are not the only method of obtaining information about solar wind speeds in the inner wind. For several decades, radio astronomers have been providing relevant information using a very different technique. These astronomers performed remote sensing on distant radio sources as the radio waves from those distant sources propagate through the inner solar wind. As the Sun moves through the sky during the year, it passes close to certain natural radio sources at definite times. For example, the Crab Nebula (a bright supernova remnant that radiates strongly at radio wavelengths via synchrotron emission) passes behind the Sun in mid-June each year. Also, certain spacecraft are in orbit around the Sun, and from time to time, they pass behind the Sun (as viewed from Earth).

When a distant source is observed with a radio telescope, the line of sight to that source approaches closest to the Sun at a certain point in space. The radial distance from the Sun's center to that point of closest approach along the line of sight is called the *impact parameter p*. Because the solar wind density falls off with increasing radial distance, a radio observation of that source is most heavily influenced by the properties of the solar wind at the particular radial location $r = p$.

What do the radio data reveal? The most prominent feature is that many radio sources exhibit rapid fluctuations in *intensity* as the source comes closer and closer to the Sun. This phenomenon is known as *interplanetary scintillation (IPS)*. The reason for IPS is that the solar wind is not a smooth flow but a turbulent medium, which does, on average, expand outward. In the wind, turbulence leads to the existence of clumps of matter akin to eddies that form in fast-flowing water or in the jet streams in the atmosphere and ionosphere of the Earth. In any turbulent medium, the clumps may have a range of linear diameters. IPS is caused when a clump of more or less dense material (or more specifically, a region of *electron density* $n_e$ that differs from the mean ambient value by an amount $\delta n_e$) in the solar wind moves across the line of sight between the observer and a distant source. The local change in electron density causes a change in the refractive index of the clump as radio waves propagate through the clump: this change causes the clump to act as a lens to focus the radio waves as seen by an observer downstream of the clump. For clumps located preferentially at $r = p$, the distance $z$ between Earth and clump is about 1 AU: at such a distance, the clumps that are most effective in causing the intensity of the radio source to "twinkle" (scintillate) when observed at wavelength $\lambda$ are those with a diameter $L_c$ of order $\sqrt{(\lambda z)}$: this is referred to as the Fresnel scale. For observations at frequency 327 MHz (i.e., $\lambda \approx 10^2$ cm) (Sasikumar Raja et al. 2019) and inserting $z \approx 1$ AU ($= 1.5 \times 10^{13}$ cm), the Fresnel scale formula indicates that the most effective clump sizes are $L_c \approx 400$ km. At a distance of 1 AU, such clumps are observed to have angular diameters of order 0.5 arc second. Clumps of this size can act as effective lenses, thereby significantly changing the *intensity* of the radio signal recorded at Earth, *provided* that the background source of radio waves has an angular diameter that is *smaller* than 0.5 arcsec: sources with such sizes can be effectively covered entirely by the clump "lens", thereby leading to significant changes in intensity as measured on Earth. (For example, one of the sources observed by Sasikumar Raja et al. 2019 has an angular size of ≤0.01 arcsec and can certainly be effectively "lensed" by the Fresnel clumps.). As the image created by an individual clump moves across a radio telescope, the intensity recorded changes with

# The Solar Wind

an amplitude that is determined by the shape of the clump and by how large is the fractional fluctuation $\delta n_e/n_e$ in the local electron density. If two telescopes, separated on the Earth's surface by a distance $d$, record the same clump (identified by its shape) but are separated in time by an interval $t$, then the speed of the shadow across the Earth's surface can be determined: $d/t$. This is essentially the speed of the clumps transverse to the line of sight, and therefore gives an estimate of the speed of the solar wind perpendicular to the line of sight at the radial location $r = p$.

An example of how the measured clump speed varies as a function of the impact parameter out to roughly $30 R_\odot$ ($\approx 0.15$ AU) is shown in Figure 18.3. The data were obtained using signals from the Venera 15 and Venera 16 spacecraft at wavelengths of 8 cm when those spacecraft (in orbit around Venus) were carried by Venus' orbital motion behind the Sun (as viewed from Earth). There is obviously a lot of scatter. Why? Because the data in Figure 18.3 were obtained over a time interval of several days: as the Sun rotates from one day to the next, the line of sight may shift from passing through a helmet streamer one day, then passing through a coronal hole on the next day, and then passing through a region of quiet corona. Therefore, we cannot expect that the line of sight to a given source will continue to pass through exactly the same coronal material throughout an observing run that lasts several days. It would be surprising if there were *not* considerable scatter in the data. But overall, one can see that the outflow speed is systematically smaller along lines of sight that approach *closest* to the Sun and that the outflow speed *increases* with increasing radial distance. The outflow speed first reaches roughly 120 km sec$^{-1}$ (i.e., the sound speed) at radial distances of about $10 R_\odot$, somewhat larger than the $6.6 R_\odot$ that Parker's thermal wind predicts for a (uniform)

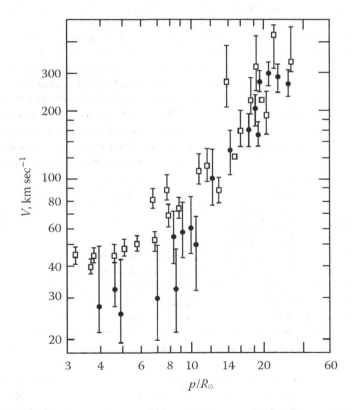

**FIGURE 18.3** Estimates derived from interplanetary scintillations of the outflow speed of solar wind clumps as a function of radial distance from the Sun. The abscissa contains the impact parameter $p$, which is the radial distance from the Sun to the point where the line of sight to the Venera 15 and 16 spacecraft (in orbit around Venus) passes closest to the Sun. (From Yakovlev and Mullan 1996.)

1 MK corona. And by the time $p$ has increased to values of $20R_\odot$ or so, the speed is up to 300–400 km sec$^{-1}$. Thus, at a radial distance of about 0.1 AU from the Sun, the wind is well on its way to reaching a typical speed of 400–500 km sec$^{-1}$ at Earth's orbit (i.e., at $215R_\odot$) (see Section 18.5).

The data in Figure 18.3 were obtained in 1984 when the Sun was near solar minimum. In 2011, a Japanese satellite Akatsuki (also in orbit around Venus) passed behind the Sun and obtained data on the radial profile of solar wind speed. In this case (closer to solar maximum), the sonic point of the wind was found to lie at a radial distance between 4 and 5 $R_\odot$ (Wexler et al. 2020). In the Parker model, this sonic point location corresponds to a coronal temperature of 1.5 MK (see Section 18.3), which is consistent with our conclusion (Section 17.4.1) that gas with a temperature of 1.5 MK is widespread throughout most of the solar corona.

It is not only the *intensity* of the radio signal from a spacecraft that fluctuates when the spacecraft passes behind the Sun: there is also observed to be a fluctuation $\delta f$ in the *frequency* of the signal. (This test does not typically work for natural radio sources, which typically emit only radiation that is spread out over a wide range of frequencies: an exception could occur if the natural source happens to be a maser source.) The values of $\delta f$ behave in a complicated way as a function of the impact parameter $p$ (Pätzold et al. 2012). On the one hand, at large $p$, i.e., $p \geq 10R_\odot$, $\delta f$ is observed to fall off spatially in a more or less smooth manner: as $p$ increases from 10 to $40R_\odot$, $\delta f$ decreases from about 0.2 to 0.02 Hz. Moreover, at large $p$, the value of $\delta f$ at any particular radial location also varies temporally: $\delta f$ is observed to be *smallest at solar minimum*. On the other hand, at small $p$, i.e., $p < 10R_\odot$, the values of $\delta f$ are observed to rise steeply, reaching almost 10 Hz in the case of the closest impacts (2–3 $R_\odot$): moreover, these large values of $\delta f$ in the inner solar corona do *not* change significantly with the solar cycle. These results suggest that a permanent turbulent layer (independent of the phase of the solar cycle) exists in the inner solar corona.

Radio studies of the wind using IPS were the first to provide useful information about the region where the solar wind undergoes its greatest acceleration. As well as information about the wind *speed*, IPS can also provide information as to the *amplitude* of the density fluctuations in the clumps: this amplitude is correlated with an empirical "index" related to the amplitude of scintillations of the IPS signal. Examples of the index obtained at a frequency of 327 MHz are shown in Figure 18.4 (Sasikumar Raja et al. 2019) in addition to sunspots numbers for cycles 21–24. The data show that, during cycle 24, the scintillation index was significantly weaker than in cycles 23 or 22: this suggests that the clumps in the solar wind were less dense (or less abundant, or both) during cycle 24. This reduction in density (or abundance) is probably related to the reduced amount of solar activity in cycle 24, as indicated by the reduced sunspot numbers reported in cycle 24. It is not merely the *number* of sunspots that was reduced in cycle 24; the areas also decreased: if the amplitude of sunspot areas is defined to be 1.0 in cycle 22, Chapman et al. (2014) used data from San Fernando observatory to report that the amplitudes in cycles 23 and 24 were 0.74 and 0.37. Moreover, the mean sunspot areas of individual spots in cycles 22, 23, and 24 were found to be 1643, 1212, and 615 MSH respectively. Furthermore, in the same three cycles, the mean facular areas were 59,000, 45,000, and 27,000 MSH respectively. These results quantify the statement that magnetic activity was definitely at a lower level in cycle 24 than in the two preceding cycles. Further quantitative evidence for reduced solar activity in cycle 24 is that the numbers of M-class and X-class flares in cycle 24 were 32% *lower* than in cycle 23, although the number of C-class flares was 16% *larger* in cycle 24 than in cycle 23 (Kilcik et al. 2020). It is possible that the weakening of the solar fields between cycles 23 and 24 may continue into cycle 25 (Bisoi et al. 2020).

Although the IPS data represent a global integration along the line of sight between Earth and a distant radio source, IPS data can be roughly interpreted in terms of physical conditions in the general vicinity of one particular radial location in space, i.e., at $r \approx p$. On the other hand, *in situ* measurements by spacecraft refer precisely to a very particular point in space, i.e., the point where the spacecraft happened to be when the measurement was made. Although IPS data and *in situ* data involve very different approaches to studying the solar wind, both types of measurements provide

# The Solar Wind

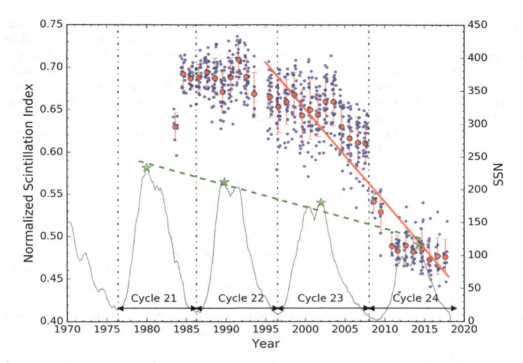

**FIGURE 18.4** Blue/red dots (and left-hand scale): normalized scintillation index (NSI) caused by the solar wind in background radio sources as the Sun passes close to each source at certain times of the year. Blue: annual averages of NSI in individual sources. Red: average NSI for all sources observed in a given year. Observations were made at a frequency of 327 MHz. The index is a measure of the relative amplitude of clumps of electron density in the solar wind. Gray curve (and right-hand scale): sunspot numbers. Green line: decline in peak sunspot number since cycle 21. (From Sasikumar Raja et al. 2019; used with permission of Springer.)

valuable information about some of the conditions of the material in the solar wind. The information provided by IPS has been used for a global survey of multiple radio sources (with angular sizes <0.5 arcsec) distributed all over the sky (Gotwols et al. 1978; Tappin and Howard 2010): these data give a "big picture" snapshot of the multiple clumps of plasma (including CMEs: see Section 18.9) that happen to be emerging in all directions from the Sun at any one instant. A computer code called "CMEchaser" allows astronomers to determine which radio sources in the sky will be occulted by any particular CME (Shaifullah et al. 2020).

Nevertheless, should we really take seriously the wind speed measurements by IPS when they involve integrations along very long lines of sight? Perhaps surprisingly, the answer turns out to be "Yes": by taking annual averages over different data sets that extend across two complete sunspot cycles (1990–2012), Sokol et al. (2013) have demonstrated that, for solar wind in the ecliptic plane, there is "very good agreement" between wind speeds derived from IPS and wind speeds measured by *in situ* spacecraft.

## 18.8 RATE OF MASS OUTFLOW FROM THE SUN

In order to determine how much mass the Sun loses per unit time, we need to know not only the *speed* but also the *density* of the solar wind. We already know that the speed (in an isothermal corona) is equal to the sound speed at the radial location $r = r_s$. Can we also estimate the density at $r = r_s$? This is more difficult, but we can do it, roughly. To do this, we note that between the surface of the Sun and $r = r_s$, the flow speed has not yet reached values as large as the sound speed. In fact, close to the surface, the speed is much smaller than the sound speed. As a result, the *hydrodynamic* terms in the equation of motion are small compared to the *hydrostatic* terms. In other words, close to the surface, the corona, although in principle expanding, is flowing so slowly that *the material is not far from HSE*. To be sure, the closer we get to $r = r_s$, the farther the conditions depart from HSE. And by the time we arrive at $r = r_s$ and beyond, HSE has broken down altogether.

But as a rough approximation, we can use Equation 18.2 (rewritten in terms of particle number density $n$, assuming constant $T$) to evaluate the density at the sonic point:

$$n(r_s) = n_o \exp\left[A\left(\frac{1}{r_s} - \frac{1}{r_o}\right)\right] \tag{18.13}$$

Inserting the empirical values $n_o = 10^{8-9}$ cm$^{-3}$ (see Chapter 17, Section 17.1), we find that in a corona with $T = 1$ MK, where $A = 13$ and $r_s \approx 6.6$ (both in units of $r_0 = R_\odot$), the density at the sonic point $n(r_s) \approx 2 \times 10^{3-4}$ cm$^{-3}$. Repeating the calculation for a corona with $T = 2$ MK, we find a much larger density at the (closer) sonic point $n(r_s) \approx 1 \times 10^{6-7}$ cm$^{-3}$. Outside the sonic point, in the limit where the velocity is varying only slowly with distance, we expect to have $n(r) \sim r^{-2}$. Thus, between $r = r_s$ and the Earth's orbit ($r = 215.04 R_\odot$), the density should decrease by a factor of $(215/6.6)^2 \approx 1100$ (for $T = 1$ MK) and by a factor of $(215/3.3)^2 \approx 4200$ (for $T = 2$ MK). Thus, near the Earth's orbit, we expect to find $n(1\ AU) \approx 2$–$20$ cm$^{-3}$ (if $T = 1$ MK), and $n(1AU) \approx 240$–$2400$ cm$^{-3}$ (if $T = 2$ MK). The empirical densities in the vicinity of 1 AU plotted in Figure 18.1 are consistent with the estimates for an isothermal corona with $T = 1$ MK, but not for $T = 2$ MK.

In the context of the approximate discussion given here (i.e., HSE remains roughly valid out to a radial distance of $r_s$ and constant speed outside $r_s$), we conclude that the Sun may well be able to maintain a coronal temperature of 1 MK out to $r \approx 6.6 R_\odot$, but the Sun is probably *not* able to maintain a temperature as large as 2 MK out to $r \approx 3.3 R_\odot$. Ultimately, the inability of the Sun to maintain coronal gas at 2 MK out to several $R_\odot$ is an indication that the Sun supplies only a finite flux of mechanical energy to the corona. We have already seen in Chapter 17, Section 17.15.3, that the coronal temperature is controlled by the chromospheric pressure, which is in turn controlled by the amount of mechanical energy flux emerging from the Sun.

How large would the mechanical flux have to be in order to maintain a corona at $T = 1$ MK and at $T = 2$ MK? We can estimate a *lower limit* on the necessary energy flux by considering one component only, namely, the kinetic energy (KE). The flux of KE in the wind equals the KE density $0.5\rho V^2$ times the flow speed $V$. At $r = r_s$, this KE flux $F_K(r_s)$ equals $0.5\rho(r_s)a^3$. Transforming back to the base of the corona, this would correspond to an energy flux crossing the surface $r = 1$ solar radius of $F_K(1) = r_s^2 F_K(r_s)$. Inserting numerical values for the case $T = 1$, we find $F_K(1) = 130$–$1300$ ergs cm$^{-2}$ sec$^{-1}$. For the case $T = 2$ MK, the surface flux is found to be much larger, $F_K(1) = 0.5 \times 10^{5-6}$ ergs cm$^{-2}$ sec$^{-1}$. Thus, the energy flux required to heat the corona to 2 MK is at least 100 times the flux required to heat the corona to 1 MK. Now, active regions on the Sun occupy at most 10% of the Sun's area and often much less than 10% (Section 16.2): thus, the quiet Sun typically occupies 10–100 times more surface area than the active regions. In view of this excess of quiet Sun area, it seems plausible to look to the quiet Sun as the source of mechanical energy to power the "typical" solar wind. According to Equation 17.9, the flux of mechanical energy entering the base of the quiet corona is limited to a value of order $F(QS) \approx 5 \times 10^5$ ergs cm$^{-2}$ sec$^{-1}$. Comparing with $F_K(1)$ for the case $T = 1$ MK, we see that the quiet Sun would have no problem in supplying the demands of the KE flux at the base of the corona, with plenty of energy to spare for the radiated flux, plus the internal energy flux ($\sim nTa$), plus the conductive flux ($\sim T^{3.5}$). On the other hand, for the case $T = 2$ MK, even though the coronal temperature has increased by a factor of only two, nevertheless, the flux of KE at the base of the corona, $F_K(1)$, is 1000 times larger than for the case $T = 1$ MK. If the corona were to try to heat its corona to 2 MK, the flux of KE alone would already "soak up" the entire available supply $F(QS)$: there would be nothing available for the radiated flux, for the increased demands on internal energy flux (increased by $> 10^3$ compared to the $T = 1$ MK case), or for the increased demands on conductive flux (increased by 10). For the case $T = 2$ MK, the numerical values suggest that the Sun simply does not generate enough mechanical energy to "go around".

Let us use the empirical data at radial distances $D = 1$ AU to estimate the solar mass loss rate in the "average" solar wind (Zirker 1981), where $V(1\ AU) = 470$ km sec$^{-1}$ and $n(1\ AU) \approx 9$ protons cm$^{-3}$. Allowing for the presence of a few percent helium nuclei as well, the mean gas density at 1 AU is $\rho(1\ AU) \approx 2 \times 10^{-23}$ gm cm$^{-3}$.

Expressing 1 AU in cm (Chapter 1, Section 1.2), we find that the rate of mass outflow of a spherically symmetric wind, $4\pi D^2 \rho(1\ AU)V(1AU)$, is some $3 \times 10^{12}$ gm sec$^{-1}$, i.e., a few million metric tons per second. To be sure, the wind is not altogether spherically symmetric: the polar wind is certainly faster on average than the equatorial wind, while the mean density in the polar wind is smaller. So we do not expect the assumption of spherical symmetry to be reliable to better than a factor of (maybe) two. More commonly, astronomers prefer to express rates of mass loss from stars in units of solar masses per year: in these units, the earlier estimate of a solar mass loss rate of $3 \times 10^{12}$ gm sec$^{-1}$ corresponds to $5 \times 10^{-14}\ M_\odot$ yr$^{-1}$.

As well as mass loss in the solar wind, the Sun is also losing mass as a result of nuclear reactions in the core. The solar power output $L$ of $3.828 \times 10^{33}$ ergs sec$^{-1}$ (Equation 1.11) corresponds to a nuclear mass loss rate $(dM/dt)_{nuc} = L/c^2 \approx 4 \times 10^{12}$ gm sec$^{-1}$. Curiously, this is close to the mass loss rate in the solar wind. Is this a coincidence? It is hard to say: the mass loss rate in the wind is determined by coronal heating processes that we cannot yet fully identify. Perhaps when all mechanisms are better understood, there may be a deep-rooted reason why the Sun has the property of comparable mass loss rates from core and corona.

Given the lifetime of the Sun ($\approx 4.5 \times 10^9$ years, from measurements of meteorite ages), and assuming that the mass loss rates have remained constant over time, we find that in the course of its lifetime, the Sun's mass has decreased by only a few parts in 10,000.

## 18.9 CORONAL MASS EJECTIONS (CMEs)

No matter when spacecraft observations of the solar wind are made in the vicinity of Earth's orbit, they essentially always report the presence of an outflow of matter at speeds of a few hundred km sec$^{-1}$, and with densities which lie typically in the range 1–10 cm$^{-3}$. In the sense that (as Neugebauer and Conway [1962] stated) "there is always something there", the solar wind can be considered to be (in a first approximation) a steady-state phenomenon.

However, from time to time, a major disruption is observed to propagate out through the wind. These events are called "coronal mass ejections", or "CMEs" for short. They originate in the solar corona. An example is illustrated in Figure 18.5, where what appears (in 2-D projection) as a "bulb" of material expands and breaks open, dumping its contents into the solar wind. The contents propagate outward through the solar wind, maintaining identity for a finite distance, and eventually mix in with the ambient wind. In certain cases, CMEs survive as far as Earth's orbit, where they can disturb the Earth's geomagnetic shield, reducing the horizontal component of Earth's field at the equator by a maximum of 0.025 Gauss (Vasyliunas 2011): these disturbances in the Earth's field may give rise to large induced voltages in terrestrial power lines or may trigger the explosion of magnetically sensitive naval mines in a war zone (Knipp et al. 2018). The CME image in Figure 18.5 was obtained by an instrument called LASCO on the SOHO spacecraft: this instrument has been operating since 1996, i.e., for more than 20 years, and is the primary source of much of our information about the properties of CMEs.

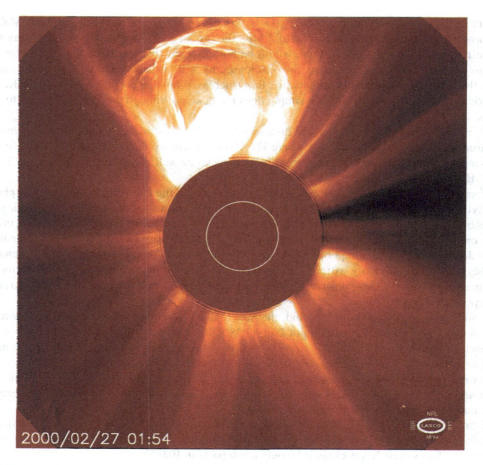

**FIGURE 18.5** A coronal mass ejection lifts off from the Sun. Image obtained by the LASCO instrument on board the SOHO spacecraft. LASCO obscures the brilliant photosphere of the Sun behind the dark mask in the center of the image: the white circle inside the mask represents the location of the Sun's photosphere. (Courtesy of the SOHO/LASCO consortium. SOHO is an international collaboration between ESA and NASA.)

### 18.9.1 RATES OF CME OCCURRENCE

Various research groups have analyzed the records of over 20 years (1996–2018) from LASCO in order to determine statistics of CME properties. The 20+-year interval covers solar cycles 23 and 24. One statistic of interest, which at first sight would appear to be an easy one to measure, is the following: how many CMEs were detected in the years 1996–2018? It turns out that this question is not so easy to answer because CMEs are complex 3-D objects and we observe them commonly only on a 2-D image, i.e., a projection on the plane of the sky. Various algorithms have been developed to measure images automatically and identify a CME according to a variety of imaging criteria: although these algorithms are in principle "objective", the resulting numbers of CMEs differ surprisingly. Lamy et al. (2019) report that four different types of image analyses, when applied to over 20 years of LASCO images, result in the following total numbers of CMEs: 21,452, 27,357, 52,905, and 39,188.

These analyses indicate that the average rate of CMEs may be as small as (roughly) 1000 per year or as large as (roughly) 2500 per year. However, it is more interesting to consider how the CME rate

# The Solar Wind

varies in the course of the solar cycle. Lamy et al. (2019) report that the *monthly* average of CME occurrences reached maximum values of 350–400 in the years 2000–2001 (at the peak of cycle 23) and maximum values of 300–350 in 2014 (at the peak of cycle 24). Thus, when solar activity is at a peak, the average rate of CMEs can be as large as 10–13 *per day*. On the other hand, at solar minimum (in 1996 and 2008), the monthly CME counts fell to 10–30 per month, corresponding to average rates of as low as 0.3 CME per day. Thus, the daily rate of CMEs varies by a factor of order 30–40 between solar maximum and solar minimum.

These data indicate clearly that CMEs owe their existence to the magnetic fields present in greatest abundance in the Sun at solar maximum. We have already seen that the rates of flares in the Sun also change by large factors (~100) between solar minimum and solar maximum (see Section 17.19.2). The fact that both flares and CMEs owe their existence to magnetic fields raises the question: what is the relationship between flares and CMEs? Observations indicate that not every flare is accompanied by a CME (except perhaps in the very largest flares), and not every CME is accompanied by a flare. We shall address the flare–CME relationship in Section 18.9.7.

### 18.9.2 Masses of CMEs

Is there a typical mass associated with CMEs? The data show that small CMEs occur more frequently than large CMEs. According to data obtained for almost 1000 CMEs during the years 1979–1981 at the peak of cycle 21, close to solar maximum (Jackson and Howard 1993), the number of CMEs with mass $M$ (gm) in the sample was found to follow the expression:

$$N(M) = 370\exp(-9.43 \times 10^{-17} M) \qquad (18.14)$$

That is, the number of CMEs decreases exponentially as the mass increases. The existence of an exponential term in the distribution indicates that there is in effect an upper cut-off in the mass distribution: for CMEs with masses in excess of roughly $10^{16}$ gm, the number of CMEs falls off exponentially rapidly compared to those with masses less than $10^{16}$ gm. It appears as if the Sun (at least in the years 1979–1981) was capable of producing CMEs with masses up to (essentially) $10^{16}$ gm, but not much more than that. However, the instrument that was available to Jackson and Howard (1993) was not sensitive enough to allow detection of small CMEs, i.e., those with masses some 10–100 times less massive than $10^{16}$ gm.

More recent data (Lamy et al. 2019), using samples including 20–50 times as many CMEs as those reported by Jackson and Howard (1993), were obtained by instruments (e.g., SOHO/LASCO) that allow researchers to detect CMEs with smaller masses: these show that the most frequent CME masses are $1.5 \times 10^{15}$ gm at both solar maxima (in cycles 23 and 24), and $2.5 \times 10^{14}$ gm at the solar minimum between cycle 23 and cycle 24). Very few CMEs have masses of less than $10^{13}$ gm or greater than $(2-3) \times 10^{16}$ gm. Inspection of CME masses plotted by Lamy et al. (2019: their Figure 44) suggests that the maximum CME mass in their sample is $(5-6) \times 10^{16}$ gm, while the minimum CME mass which has been identified is $10^{12}$ gm. Although the maximum CME mass is not inconsistent with the exponential cut-off reported by Jackson and Howard (1993), Lamy et al. report that the mass distribution does *not* exhibit exponential behavior but instead follows a power law: this suggests that, when a larger CME sample is examined, there is no particular mass that is preferred by CMEs.

### 18.9.3 Speeds of CMEs

Using the four techniques mentioned in Section 18.9.1, Lamy et al. (2019) report on the distributions of speeds of CMEs. In all four techniques, they find that there exists a peak value of speed $v(peak)$ in the distribution, accompanied by a decrease in the number of CMEs with slower and faster speeds. During cycle 24, the peak CME speed found by all four methods of identifying CMEs turned out to be confined within a narrow range: $v(peak)$ = 300–400 km sec$^{-1}$. On the other hand, during cycle 23, the four different methods led to significantly different values of $v(peak)$, ranging from ≈250 km sec$^{-1}$ to ≈600 km sec$^{-1}$. In the combined CME sample of cycles 23 and 24, the distribution of CME speeds is found to have a long "tail" extending up to speeds as large as 2000 km sec$^{-1}$.

Does the CME speed vary during the solar cycle? Obridko et al. (2012) found that the maximum CME speed in cycle 23 occurred around the solar maximum in 2002, while the CME speed was at a minimum during the solar minimum years of 1986, 1996, and 2008.

In a sample of 38 CMEs that could with confidence be associated with flares, the speed of the CME was found (Vasantharaju et al. 2018) to be correlated with the energy flux $F(X)$ from the flare as measured by the GOES X-ray spacecraft: over a range of $F(X)$ values (in units of W m$^{-2}$) extending from $\log F(X) = -5.5$ to $-3$, and over a range of $v(CME)$ extending from 200 km sec$^{-1}$ to 3000 km sec$^{-1}$, the correlation coefficient was found to be 0.49. The existence of such a correlation may be related to the electric field that drives reconnection in the corona (see Section 17.19.1): Hinterreiter et al. (2018) have demonstrated that the stronger the electric field is in an eruptive flare, the faster is the CME that is ejected.

### 18.9.4 Kinetic and Potential Energies of CMEs

What energies are associated with CMEs? The kinetic energy is relatively easy to evaluate. With speeds which are typically in the range 300–1000 km sec$^{-1}$, the KE of a CME with mass $M$ is typically $0.5M \times 10^{15-16}$ ergs. If (as Jackson and Howard [1993] suggested for a small sample of CMEs in the cycle 22 maximum) there exists an upper cut-off in $M = 10^{16}$ gm, then the upper limit on the KE of CMEs is readily estimated: $0.5 \times 10^{31-32}$ ergs.

Taking into account the results reported by Lamy et al. (2019) for a much larger sample of CMEs, each of which has its own value of speed and $M$, the distribution of KEs is found to span a range of four orders of magnitude. At the maximum of cycle 23, the KE of CMEs was found to have an upper limit of $10^{33}$ ergs; the corresponding upper limit on KE at the maximum of cycle 24 was found to be $5 \times 10^{32}$ ergs.

By examining the magnetic field energy in active regions where CMEs originate, Pal et al. (2018) found that the KE of a CME is correlated with the magnetic energy density of the active region. This supports the idea (Section 18.9.1) that CMEs rely on magnetic fields for their existence.

In terms of the KE measured in a sample of 778 CME-related flares, Youssef (2012) reported a correlation coefficient of 0.65 between $KE(CME)$ and $F(X)$. This is a significantly better correlation than that reported between $v(CME)$ and $F(X)$ by Vasantharaju et al. (2018), especially in view of the 20-times larger sample analyzed by Youssef (2012).

As regards the *total* energies of CMEs, it turns out that KE is not the only component that contributes. There is also the potential energy (PE = $GM_\odot M(CME)/R_\odot$) required to raise a CME with mass $M(CME)$ to infinity from the solar surface. Aschwanden (2016) has reported on a sample of 399 M-class and X-class flares observed by SDO during its first 3.5 years of operation: he lists values of KE and PE for events in which a CME was present. Inspection of his table of events indicates that PE and KE do not differ greatly in magnitude, although the PE dominates the KE of the CME in 75% of the cases. The total energy (PE+KE) of a CME is on average about twice the KE: as a result, the upper limit on total CME energy is roughly of order $10^{31-32}$ ergs.

### 18.9.5 Comparison and Contrast between Flares and CMEs

We have already seen (Section 17.19.7) that solar flares have radiated energies that extend up to no more than a few times $10^{32}$ ergs. It is striking that the maximum energy that the Sun releases in a flare is close to the maximum kinetic energy that the Sun releases in a CME. That is, the two most prominent classes of transient energy release from the Sun apparently produce events that, in their largest manifestations, have comparable maximum energies. Possible statistical connections between flares, CME's, and other solar phenomena have been explored extensively (e.g. Munro et al. 1979).

The physical feature that provides a common physical connection between flares and CMEs is the magnetic field. In the presence of solar gravity, and given the amount of plasma in coronal gas, it appears that the magnetic fields that the Sun produces are limited in the maximum amount of energy they can store. If, in a given active region, that limit is exceeded, the field apparently "must" respond by releasing the stored energy. The form that the released energy takes can be either a flare or a CME or a combination of the two, depending on local conditions.

Youssef (2012), in an analysis of 778 CME-related flares, showed that the probability of occurrence of a CME in spatial and temporal relationship with a flare increases strongly as the flare becomes larger: 90% of X-class flares have associated CMEs, whereas only 24% of C-class flares have associated CMEs. Conversely, as regards CMEs with and without flares, Aarnio et al. (2011) used SOHO/LASCO data (for CMEs) and GOES data (for flares) to determine that only 11% of CMEs were associated both temporally and spatially with flares: however, Aarnio et al. pointed out that because of data gaps in LASCO images, this percentage is certainly an underestimate. Despite the apparently low percentage of CMEs with flares, Aarnio et al. found that there is a statistically significant positive correlation between CME mass and the X-ray flux from the flare: this suggests that *some physically significant connection does exist* between flares and CMEs, presumably related to the magnetic field.

Returning to Aschwanden (2016), we note that magnetic data were obtained for each flare in order to allow an estimate of how much magnetic energy $E(diss)$ was actually dissipated in the flare and how this dissipated energy was distributed among the various flare channels. The fraction of $E(diss)$ associated with the CME was found to be in the range 0.01–0.4: this provides a consistency check that the flare has dissipated *more than enough* energy to account for the observed CME energy. Moreover, the CME energy was found to be comparable to the energy that the flare converted into thermal form.

There is undoubtedly a *general* correlation between *some* flares and *some* CMEs (Harrison 1991): they *both* result from the relaxation of complex magnetic topologies and, as such, they *can* occur together, but not necessarily. Nindos and Andrews (2004) studied a sample of 133 M-class and X-class flares and evaluated the magnetic helicity of the active region prior to each flare: they found that when the AR helicity was small (large), the flare was less (more) likely to be accompanied by a CME. That is, the presence of *large helicity* in an AR apparently *favors* CME occurrence. A special corollary of this conclusion was subsequently described by Liu et al. (2016), who reported on an AR (#12192) that had been observed on the Sun in October 2014, close to the maximum of solar cycle 24: this AR produced more than 100 flares during the 15-day interval required to cross the solar disk, including 32 M-class and 6 X-class flares. By any definition, this AR (with six to seven flares per day on average) deserves the classification of "flare-active". Remarkably, during the same time interval, the AR produced only *one* (small) CME. In contrast to AR #12192, with its dearth of CMEs compared to flares, we may cite the counterexample of AR #11158, which appeared in 2011, in the early years of cycle 24 (Kay et al. 2017). In the course of 4 days, this AR had 21 flares (including the first X-class flare of cycle 24), but it also produced 11 CMEs. With a daily flare rate of five to six per day on average, AR #11158 is almost as "flare-active" as AR #12192, and yet the two ARs differed by an order of magnitude in the number of CMEs.

The two ARs just discussed point to an intriguing result: although flares and CMEs both rely on magnetic fields for energy, the conditions for both types of phenomena to occur are *not necessarily the same*. In an attempt to explain this conclusion, Liu et al. pointed to two physical factors that might account for the striking difference in behavior between flares and CMEs in AR #12192: (i) the AR, although possessing large amounts of free magnetic energy, did *not* possess any highly twisted field lines (i.e., the magnetic *helicity* was small); (ii) the ambient magnetic fields overlying the AR were observed to be strong, thereby possibly inhibiting any upward-moving CME material that might have been launched from the AR.

Are flares with and without CMEs statistically different from one another? As regards statistical properties, Yashiro et al. (2006) reported that the distributions of X-ray flares as regards peak flux, fluence, and duration are systematically steeper for flares *without* CMEs than is the case for flares *with* CMEs.

### 18.9.6 CME Contributions to Solar Mass Loss Rates

Do CMEs contribute significantly to the rate of mass loss from the Sun? The answer seems to be "No". An early estimate of CME mass loss rates (Jackson and Howard 1993) found that, even at the maximum of solar cycle 21 (in 1979–1981), the Sun loses mass in the form of CMEs at a rate that is only about 16% of the total mass loss rate. At solar minimum, when the rate of CME occurrence is more than 10 times smaller than at solar maximum, it would be expected that CMEs would contribute even less than 16% of the total mass loss rate.

A sample of more than 6000 CME images from cycle 23 and part of cycle 24 (i.e., between 1996 and 2013) have been analyzed by Cranmer (2017), who finds that, even at solar maximum, "CMEs contribute only about 3% of the background solar wind mass flux". This result is already significantly smaller than the 16% reported by Jackson and Howard (1993) around the time of the solar maximum of cycle 21. Several possible causes for this reduction are listed by Cranmer. Moving on to CMEs at solar minimum, Cranmer (2017) finds that the CME mass flux is found to fall to values as low as 0.1% of the background solar wind mass flux. Based on a somewhat earlier analysis of CME images, Vourlidas et al. (2011) suggested that the CME mass flux from the Sun underwent a real decline in the years between cycle 21 and cycles 23–24. It is interesting that this decline occurs over the same time period when the empirical values of the interplanetary scintillation (IPS) index (see Figure 18.4) has also undergone a more or less steady decline. This might suggest that the small-scale "eddies" in the solar wind with lengths comparable to the Fresnel scale (a few hundred km, i.e., the eddies that give rise to IPS) might possibly be generated when large structures (CMEs) with masses of $10^{13-16}$ gm "plow" through the ambient solar wind, stirring up turbulence.

### 18.9.7 CMEs and Magnetic Helicity

Now that we know that CMEs do not contribute significantly to the solar mass loss rate, we may wonder: do CMEs provide any useful service to the Sun? To address this, we note that the Sun, in its cyclical magnetic behavior, is faced every 11 years (or so) with an important task: the Sun must get rid of the magnetic flux from the old cycle in order to make way for the magnetic flux (with opposite polarity) of the new cycle. And the old cycle fields do not merely have a polarity: they also contain helicity (twistedness, $H_m$) that is organized in such a way that in the northern (southern) hemisphere, the sign of $H_m$ is negative (positive): this organization persists from one 11-year cycle to the next (e.g., Low 1996). Therefore, if the helicity is not removed before the old cycle can give way to the new cycle, then helicity would continue to accumulate to arbitrarily large values in the Sun. We have already seen that the Babcock (1961) model (see Section 16.9) describes how the *polarity* might be reversed by active regions from the old cycle diffusing towards the poles. But how can the old *helicity* be removed?

Rust (1994) noted that filaments on the Sun that are observed to be twisted (i.e., contain helicity) can undergo eruption into the solar wind, i.e., they can become CMEs. Rust found that the sign of $H_m$ of certain eruptive filaments back at the Sun was consistent with the sign of $H_m$ of interplanetary structures that were eventually (after several days) detected at 1 AU. As regards active regions, Cho et al. (2013) showed that in a sample of 34 interplanetary CMEs, the sign of $H_m$ agreed with that of the parent active regions in 30 cases (i.e., 88%): in the remaining four cases, there were extenuating circumstances to explain the apparent lack of consistency. The observations of Rust (1994) and of Cho et al. (2013) suggest that CMEs may be responsible for transporting helicity out of the Sun (see Section 16.11 for quantitative support for this suggestion). In a subsequent theoretical exposition,

Low (1996) also concluded, "CMEs are the means by which accumulated magnetic flux and twist are taken out of the corona". In this regard, it is significant that Liu et al. (2016) found that in an active region that was highly *ineffective* in generating CMEs (but was nevertheless flare-prolific), magnetic helicity was essentially *absent* from the AR, i.e., a basic ingredient of CMEs (helicity) was missing from this particular AR.

Low (1996) has also discussed the interesting question as to the physical relationship between flares and CMEs. Both phenomena are undoubtedly magnetic in nature, since both phenomena are observed to be 10–100 times more abundant at solar maximum than at solar minimum. If both processes relied on the same magnetic properties, then one might expect there to be a close (maybe even one-to-one) correlation between the two. And yet, it is observationally established that there is definitely *not* a one-to-one correspondence between every flare and every CME, even among the largest flares (Youssef 2012). Low (1996) made the suggestion that a key difference between flares and CMEs may have to do with the electrical resistivity of the plasma. Flares rely on *finite* resistivity (so as to quickly discharge magnetic free energy in explosive reconnection), whereas CMEs may be initiated (at least in some cases) in what is known as the "ideal MHD limit", i.e., in a situation where resistivity does *not* play an essential role. In a numerical study of this possibility in a 3-D magnetic field, Rachmeler et al. (2009) show that if a magnetic flux rope (which is surrounded by an overlying magnetic arcade) is twisted enough, the rope may develop a kink that expands up through the arcade and out into space, thereby launching a CME. Rachmeler et al. use the graphic medical term "hernia" to describe how the twisted flux rope can make its way up through the overlying arcade and eventually experience "break-out". Most significantly, Rachmeler et al. demonstrate that this process of "herniation launch" of a CME occurs in the *complete absence* of resistivity: that is, a CME may be initiated as a result of an *ideal* MHD instability. The difference between the physical conditions required for the occurrence of resistive instabilities and the conditions required for the occurrence of nonresistive instabilities may help explain the observational fact that, even though flares and CMEs are both driven by magnetic fields, nevertheless, not every flare (where resistive instability is at work) is accompanied by a CME, nor is every CME (where an ideal instability may be at work) accompanied by a flare.

Rachmeler et al. stress that it is essential for a critical amount of twist, i.e., magnetic helicity $H_m$, to be present if their proposed "herniation" process is to occur. The magnitude of the critical twist depends on (among other factors) the strength and the topology of the magnetic fields in the overlying arcade. If the latter (sometimes referred to as "strapping fields": e.g., Ha and Bellan 2016) are strong enough at great heights, they can prevent a CME from breaking out. A particular example of such a "failed eruption", involving a rising filament that did *not* erupt into a CME, has been captured in detail by Chintzoglou et al. (2017): these authors relied on a carefully planned combination of several satellites, a rocket launch, and several ground-based observatories to demonstrate "for the first time, how magnetic topology can suppress (solar) ejections already in progress".

## 18.10 HOW FAR DOES THE SUN'S INFLUENCE EXTEND IN SPACE?

The solar wind originates in the Sun's corona and eventually flows out past the Earth's orbit. The wind carries with it some of the magnetic field of the Sun: near the Sun, this field has effective control over the wind outflow. But at greater distances from the Sun, the influence of the field becomes weaker and the wind "breaks free": in Section 18.10.1, we discuss where this happens. However, just because the field no longer dominates the wind at a certain point, that does not mean that the Sun has no further influence beyond that point: there is still the ram pressure of the wind that must be dealt with. Far out in the wind, the decreasing density ($\sim 1/r^2$) leads to a reduction of the ram pressure, and eventually this pressure becomes so low that it can be halted by the interstellar medium gas pressure. When that happens, the realm of space that is controlled by the Sun comes effectively to an end. In Section 18.10.2, we discuss the observations that show the radial location where the transition into the ISM takes place.

Disturbances in the wind (e.g., CMEs) can have an effect on our lives on Earth (e.g., radio blackouts, voltage surges). In that sense, the solar wind allows the Sun to "reach out" far beyond the confines of one solar radius and cause certain events on Earth that have nothing to do with the Sun as a source of heat and light.

The solar wind also influences the surroundings of other planets, especially those with magnetic fields. All four giant planets (Jupiter, Saturn, Uranus, and Neptune) were discovered to have strong magnetic fields when the Voyager 2 spacecraft visited them in the decade from 1979 to 1989. The planetary magnetic fields trap charged particles into confined orbits, giving rise to an environment of interesting interactions between fields and plasma. Neptune is on average 30 times farther from the Sun than the Earth is, but even so, the solar wind has an influence on the shape of the magnetosphere even out there.

### 18.10.1 Where Does the "True" Corona End and the "True" Wind Begin?

Is it possible to determine where the corona "ends" and the wind "begins"? This question has been discussed by DeForest et al. (2016) on the basis of extensive processing of images of the solar corona that were obtained by a heliospheric imager on board the STEREO-A spacecraft in 2008, when the Sun was very quiet. Images extending over an elongation range from 4 to 24 degrees from the Sun were obtained of the corona during a 15-day interval: in terms of the effective line-of-sight impact parameter $p$, the aforementioned elongations correspond to $p = 16$–$96 R_\odot$. Using various techniques (unsharp masking, time-shifted image co-addition, suppression of stars in the background sky), DeForest et al. discovered that radial "striae" (including streamers and plumes) are dominant features in the *inner* portions of their images (i.e., close to the Sun). These highly anisotropic striae have properties suggesting that they are controlled by the coronal magnetic fields. A striking feature of the data is that, beyond a certain elongation in the range between 10 and 20 degrees ($p = 44$–$88 R_\odot$), the striae are observed to "fade out", as if the controlling effects of the solar magnetic fields are becoming less effective.

Remarkably, in the same range of elongations where the striae "fade out", DeForest et al. report that more compact brighter features are seen to "fade in" and then persist in the *outer* portions of their images (i.e., far from the Sun): these are locally dense (nearly isotropic) "puffs", which DeForest et al. refer to as "flocculae". The distinction in texture between radial striae and isotropic flocculae, which sets in at elongations of 10–20 degrees ($\approx 44$–$88) R_\odot$, had not been previously reported in the literature (despite the availability of over 10 years of images from STEREO), perhaps because previous analyses did not use such computer-intensive image processing. DeForest et al. make the intriguing suggestion that the flocculae might be density enhancements "associated with the onset of turbulence in the inner heliosphere". They conclude that their results may "mark the profound shift between (on the one hand) the primarily *magnetically structured* corona and (on the other hand) the primarily *hydrodynamic* solar wind". If this conclusion is correct, then one might be permitted to hypothesize that the "true" corona may extend out "only" as far as $(44$–$88)R_\odot$, and then beyond that limit, the "true" solar wind "breaks free" of the coronal fields and becomes an independent turbulent entity. Future observations from the Parker Solar Probe may help to confirm or deny this hypothesis.

### 18.10.2 The Outer Edge of the Heliosphere

How far out does the influence of the solar wind extend? To answer that, we recall (Section 18.1) that in the ISM, the total pressure $p$(ISM) (including magnetic fields and cosmic rays) is $\approx 1.4$–$1.7 \times 10^{-12}$ dyn cm$^{-2}$ (or possibly as large as $2 \times 10^{-12}$ dyn cm$^{-2}$). Now the solar wind outflow, with its energy density $E_w(r) = 0.5\rho(r)V(r)^2$, exerts an outward ram pressure equal to $E_w(r)$. This is the pressure that allows the Sun to "push back" the ISM. But this "pushing back" can work only as long as $E_w(r)$ exceeds $p$(ISM). Therefore, we can estimate the maximum radial extent of the Sun's influence roughly by seeking the radial distance $r_m$ at which $E_w(r_m) \approx p$(ISM).

To solve this, we recall that in an isothermal wind, the speed $V$ in the outer wind varies only very slowly as the distance $r$ from the Sun increases (see discussion following Equation 18.12). Since $V$ is essentially constant, the density $\rho$ must fall off as $1/r^2$. As a result, $E_w(r)$ varies essentially as $\sim 1/r^2$. At $r = 1$ AU, using a mean density of 9 protons cm$^{-3}$ and a mean wind speed of 470 km sec$^{-1}$, we find $E_w(r = 1\text{AU}) \approx 2 \times 10^{-8}$ dyn cm$^{-2}$. At a distance of $r$ AU, this leads to $E_w(r) \approx 2 \times 10^{-8} r^{-2}$ dyn cm$^{-2}$. This pressure becomes equal to $p(\text{ISM})$ at $r_m \approx 100\text{--}120$ AU. Therefore, at distances of roughly 100 AU, the pressure of the ISM should bring the solar wind to a halt. Depending on the local conditions, the halt may be so abrupt that a shock wave is set up: this is referred to as the "termination shock" (TS) of the solar wind. The density of the solar wind material at 100 AU is expected to be smaller than the density at 1 AU by a factor of about $10^4$: with densities at 1 AU of order 10 cm$^{-3}$, we expect densities of solar wind material at the TS to be of order 0.001 cm$^{-3}$.

Thus, the influence of the Sun extends, via the solar wind, to a radial distance that lies well beyond Pluto's orbit, to a distance of order 100 AU. The Sun's "sphere of influence", also known as the "heliosphere", comes to an end at a distance of about 100 AU.

So far, two spacecraft have traveled far enough to make the transition through the TS and communicate their data back to Earth. The Voyager 1 transited the TS in December 2004 at a radial distance $r = 94$ AU. The Voyager 2 transited the TS (in a different direction) in August 2007, at $r = 84$ AU. The different radial distances of the two transitions suggests that the termination shock is not spherical: the Sun moves through the local interstellar medium (LISM) at a finite speed (26 km sec$^{-1}$: Gayley et al. 1997), and models suggest that this leads to a heliosphere with a blunt "head" and a long-drawn out "tail". The two Voyager transitions of the TS occurred at distances not far from the estimate given earlier ($\approx$100 AU).

Beyond the termination shock, there exists a region of space called the "heliosheath" where the lower-density (0.001–0.002 cm$^{-3}$) hotter solar wind adjusts its properties so as to interface (eventually) with the properties of the denser (0.01–0.1 cm$^{-3}$) cooler LISM. At the outer edge of the heliosheath, a second boundary called the "heliopause" (HP) separates "true" solar material (i.e., gas that originated in the solar corona) from "true" LISM material. At the time of writing (October 2021), two spacecraft with working instruments have successfully crossed the HP and entered the LISM: Voyager 1 on August 25, 2012, and Voyager 2 on November 5, 2018. Data subsequently obtained by a plasma wave instrument on Voyager 1 (Gurnett and Kurth 2019) showed that the local density had increased from solar-wind values to an LISM value (0.055 cm$^{-3}$). This increase in density was measured at a distance of 122.6 AU from the Sun. At a later time, after Voyager 2 had crossed the HP, the local density was observed to have increased to an LISM value (0.039 cm$^{-3}$) at a distance of 119.7 AU from the Sun (Gurnett and Kurth 2019). These data suggest that the distance from TS to HP is of order 30 AU. A further indication that Voyager 1 indeed reached the LISM no later than 2013 was provided by magnetic field data (Burlaga and Ness 2016): over the interval 2013–2016, the field strength $B$ measured by Voyager 1 was found to remain essentially constant at a value of $4.8 \times 10^{-6}$ Gauss (i.e., 4.8 µG). Such a value is consistent with long-standing radio data suggesting that the interstellar magnetic field has a random component with a magnitude of $\approx$5 µG (Rand and Kulkarni 1989).

The heliosphere presents a barrier to the galactic cosmic rays (GCR) that are present in the ISM. A fraction of those GCR can reach the Earth's orbit, but only at the expense of "swimming upstream" against the outflowing solar wind, which is carrying along magnetic fluctuations. In the course of penetrating in as far as the Earth's orbit, the GCR are subject to several processes, including diffusion, scattering off magnetic fluctuations in the wind, and adiabatic energy loss. Moreover, because of large-scale spatial variations in the magnetic field intensity, the GCR particles are also subject to drifts that change sign from one 11-year cycle to the next.

As the sunspot numbers go up and down (see the top panel of Figure 18.6), GCR have a harder time *at solar maximum* making their way from the interstellar medium all the way into the Earth's orbit. As a result, the measured fluxes of GCR at Earth are reduced at solar maximum (see the lower panel in Figure 18.6). When sunspots are less abundant, the fluxes of GCR at Earth are larger. As

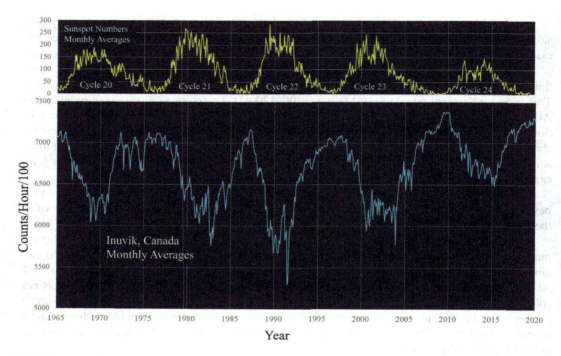

**FIGURE 18.6** Count rates of GCRs over a 55-year interval. Neutron monitors are instruments that respond to GCR with energies of a few GeV. (Plot downloaded from public website http://neutronm.bartol.udel.edu; used with permission of J. Clem.)

a result, the GCR flux is *anti-correlated* with the sunspot number. The amplitude of the relative reduction in GCR flux at solar maximum can be as large as tens of percent (see Figure 18.6). Given that the "typical" solar wind travels at a speed of 470 km sec$^{-1}$ at Earth orbit, and not much more than that at greater distances, we can estimate roughly the time $t_p$ required for solar wind to travel from the Sun to the edge of the heliosphere. With a heliospheric radius $r_m \approx 100$ AU $\approx 1.5 \times 10^{15}$ cm and an assumed constant speed $V = 4.7 \times 10^7$ cm sec$^{-1}$, we find $t_p$ is about one year. As a result, when magnetic conditions change at the Sun, if drifts could be neglected, then it should take a full year (or so) before the heliosphere as a whole "knows" that the magnetic properties of the Sun/wind have changed. This would lead to a phase shift of about 1 year in the GCR count rates relative to the sunspot counts. However, when drifts (which drive particles in opposite directions in even and odd cycles) are taken into account, it turns out that the phase shifts are less than 1 year in even cycles (20, 22, and 24) but are longer than one year in odd cycles (19, 23) (Iskra et al. 2019).

The size of the heliosphere $r_m$ depends on two independent quantities: the solar wind flux (determined by the Sun), and the ISM pressure (determined by conditions in the ISM). The Sun travels at a speed of about 26 km sec$^{-1}$ relative to the LISM (Gayley et al. 1997): with such a motion, the Sun will shift its position in the ISM by almost 100 light-years in a time interval of 1 million years. In the course of the Sun's lifetime (i.e., several billion years), the Sun can travel relative to its starting point more than 100,000 light-years, i.e., essentially all the way across the galaxy. As a result, the properties of the LISM encountered by the Sun will surely have a variety of properties as the Sun ages. Thus, the edge of the heliosphere will certainly (from time to time) approach the Earth more closely than 100 AU. In extreme cases, it is even possible that Earth's orbit might lie *outside* the edge of the heliosphere. (See Exercises 18.1–18.3.)

## EXERCISES

18.1 The Sun is in orbit around the center of the galaxy, taking some 250 My to orbit once. In the course of an orbit, the Sun encounters various types of ISM. In some of these, the local density may rise to values of $n = 10^3$ cm$^{-3}$, $10^5$ cm$^{-3}$, or even $10^7$ cm$^{-3}$. Assuming that the gas temperature remains at the value $T = 100$ K in all clouds, and that all other components of ISM pressure are unchanged, calculate the extent of the heliosphere in each of the aforementioned clouds.

18.2 Also in the various components of ISM, the field strength increases with increasing density $n$ roughly in proportion to $n^{2/3}$. For the values of $n$ given in Exercise 18.1, evaluate the local field strengths (assuming that with $n = 0.1$ cm$^{-3}$, the local $B = 5 \times 10^{-6}$ G). Changes in the density $n$ and in the field strength $B$ combine to change the value of $p$(ISM), thereby altering the location of the TS. Using the new $B$ values associated with increased $n$ values, recalculate the extent of the heliosphere.

18.3 Assuming that only $n$ or $B$ varies, what values of $n$ and $B$ are necessary in order that the extent of the heliosphere shrinks to become as small as the size of the Earth's orbit?

## REFERENCES

Aarnio, A. N., Stassun, K. G., Hughes, W. J., et al., 2011. "Solar flares and CME's", *Solar Phys.* 268, 195.
Abbo, L., Ofman, L., Antiochos, S. K., et al., 2016. "Slow solar wind: Observations and modeling", *Space Sci. Rev.* 201, 55.
Aschwanden, M. J., 2016. "Global energetics of solar flares: IV. CME energetics", *Astrophys. J.* 831, 105.
Babcock, H. W., 1961. "The topology of the Sun's magnetic field and the 22-year cycle", *Astrophys. J.* 133, 572.
Bale, S. D., Badman, S. T., Bonnell, J. W., et al., 2019. "Highly structured slow solar wind emerging from an equatorial coronal hole", *Nature* 576, 237.
Bisoi, S. K., Janardhan, P., & Ananthakrishnan, S., 2020. "Another mini solar maximum in the offing", *J. Geophys. Res. Space Phys.* 125, A027508.
Burlaga, L. F., & Ness, N. F., 2016. "Observations of the interstellar magnetic field in the outer heliosheath: Voyager 1", *Astrophys. J.* 829, 134.
Burlaga, L. F., Ness, N. F., Belcher, J. W., et al., 1996. "Voyager observations of the solar wind in the distant heliosphere", *Space Sci. Rev.* 78, 33.
Chapman, G. A., de Toma, G., & Cookson, A. M., 2014. "An observed decline in the amplitude of recent solar cycle peaks", *Solar Phys.* 289, 3961.
Chintzoglou, G., Vourlidas, A., Savcheva, A., et al., 2017. "The detrimental effects of magnetic topology on solar eruptions", *Astrophys. J.* 843, 93.
Cho, K.-S., Park, S.-H., Marubashi, K., et al., 2013. "Comparison of helicity signs in interplanetary CME's and their solar source regions", *Solar Phys.* 284, 105.
Cranmer, S. R., 2017. "Mass loss rates from CME's", *Astrophys. J.* 840, 114.
DeForest, C. E., Matthaeus, W. H., Viall, N. M., et al., 2016. "Fading coronal structure and the onset of turbulence in the young solar wind", *Astrophys. J.* 828, 66.
Dolei, S., Spadaro, D., & Ventura, R., 2015. "Visible light and UV observations of coronal structures", *Astron. Astrophys.* 577, A34.
Farrugia, C. J., Harris, B., Leitner, M., et al., 2012. "Deep solar activity minimum 2007–2009", *Solar Phys.* 281, 461.
Fox, N. J., Velli, M. C., Bale, S. D., et al., 2016. "The Solar Probe Plus mission: humanity's first visit to our star", *Space Sci. Rev.* 204, 7.
Gayley, K., G., Zank, G. P., Pauls, H. L., et al., 1997. "One-versus two-shock heliosphere: Constraining models with GHRS Ly$\alpha$ spectra towards $\alpha$ Centauri", *Astrophys. J.* 487, 259.
Gerontidou, M., Mavromichalaki, H., & Daglis, T., 2018. "HSS and geomagnetic storms during solar cycle 24", *Solar Phys.* 293, 131.
Gotwols, B. L., Mitchell, D. G., Roelof, E. C., et al., 1978. "Synoptic analysis of IPS radio spectra observed at 34 MHz", *J. Geophys. Res.* 83, 4200.
Gurnett, D. A., & Kurth, W. S., 2019. "Plasma densities near and beyond the heliopause from the Voyager 1 and 2 plasma wave instruments", *Nature Astron.* 3, 1024.

Ha, B. N., & Bellan, P. M., 2016. "Laboratory demonstration of slow rise to fast acceleration of arched magnetic flux ropes", *Geophys. Res. Lett.* 43, 9390.

Harrison, R. A., 1991. "Coronal mass ejections", *Phil. Trans. Roy. Soc. A.* 336, 401.

Hinterreiter, J., Veronig, A. M., Thalmann, J. K., et al., 2018. "Statistical properties of ribbon evolution and reconnection electric fields", *Solar Phys.* 293, 38.

Howard, R. A., Vourlidas, A., Bothmer, B., et al., 2019. "Near-Sun observations of an F-corona decrease and K-corona fine structure", *Nature* 576, 232.

Hufbauer, K., 1991. "The solar wind, 1957–1970", *Exploring the Sun: Solar Science Since Galileo*, Johns Hopkins University Press, Baltimore, MD, pp. 213–258.

Ip, W. H., & Axford, W. I., 1985. "Estimates of galactic cosmic ray spectra at low energies", *Astron. Astrophys.* 149, 7.

Iskra, K., Siluszyk, M., Alania, M., et al., 2019. "Delay time in GCR flux in solar magnetic cycles 1959–2014", *Solar Phys.* 294, 115.

Jackson, B. V., & Howard, R. A., 1993. "A CME mass distribution derived from SOLWIND coronagraph observations", *Solar Phys.* 148, 359.

Kay, C., Gopalswamy, N., Xie, H., et al., 2017. "Deflection and rotation of CME's from AR 11158", *Solar Phys.* 292, 78.

Kilcik, A., Chowdhury, P., Sarp, V., et al., 2020. "Temporal variation in the number of different class X-ray solar flares", *Solar Phys.* 295, 159.

Knipp, D. J., Fraser, B. J., Shea, M. A., et al., 2018. "On the little-known consequences of the 4 August 1972 ultra-fast coronal mass ejecta", *Space Weather* 16, 1635.

Krieger, A. S., Timothy, A. F., & Roelof, E. C., 1973. "A coronal hole and its identification as the source of a high-velocity solar wind stream", *Solar Phys.* 29, 505.

Lamy, P. L., Floyd, O., Boclet, B., et al., 2019. "Coronal mass ejections over solar cycles 23 and 24", *Space Sci. Rev.* 215, 39.

Le Chat, G., Issautier, K., & Meyer-Vernet, N., 2012. "The solar wind energy flux", *Solar Phys.* 279, 197.

Liu, L., Wang, Y., Wang, J., et al., 2016. "Why is a flare-rich active region CME-poor?" *Astrophys. J.* 826, 119.

Low, B. C., 1996. "Solar activity and the corona", *Solar Phys.* 167, 217.

Mann, G., Jansen, F., MacDowall, R. J., et al., 1999. "A heliospheric density model and type III radio bursts", *Astron. Astrophys.* 348, 614.

Moncuquet, M., Meyer-Vernet, N., Issautier, K., et al., 2020. "First *in situ* measurements of electron density and temperature", *Astrophys. J. Suppl.* 246, 44.

Munro, R. H., Gosling, J. T., Hildner, E., et al., 1979. "Association of CME's with other forms of solar activity", *Solar Phys.* 61, 201.

Neugebauer, M., & Conway, C. W., 1962. "The mission of Mariner II: preliminary observations", *Sci.* 138, 1095.

Nindos, A., & Andrews, M. D., 2004. "The associations of big flares and CME's: what is the role of magnetic helicity?" *Astrophys. J. Lett.* 616, L175.

Obridko, V. N., Ivanov, E. V., Ozguc, A., et al., 2012. "CME's and the index of effective solar multipole", *Solar Phys.* 281, 779.

Pal, S., Nandy, D., Srivastava, N., et al., 2018. "Dependence of CME properties on their solar source active region", *Astrophys. J.* 865, 4.

Parker, E. N., 1958. "Dynamics of the interplanetary gas and magnetic fields", *Astrophys. J.* 128, 664.

Pätzold, M., Hahn, M., Tellmann, S., et al., 2012. "Coronal density structures and CME's", *Solar Phys.* 279, 127.

Phillips, J. L., Bame, S. J., Barbes, A., et al., 1995. "Ulysses solar wind plasma observations from pole to pole", *Geophys. Res. Lett.* 22, 3301.

Pneuman, G. W., 1968. "Some general properties of helmeted coronal structures", *Solar Phys.* 3, 578.

Rachmeler, L. A., DeForest, C. E., & Kankelborg, C. C., 2009. "Reconnectionless CME eruption", *Astrophys. J.* 693, 1431.

Rand, R. J., & Kulkarni, S. R., 1989. "The local galactic magnetic field", *Astrophys. J.* 343, 760.

Rust, D. M., 1994. "Spawning and shedding helical magnetic fields in the solar atmosphere", *Geophys. Res. Lett.* 21, 241.

Sasikumar Raja, K., Janardhan, P., Bisoi, S. K., et al., 2019. "Global solar magnetic field and IPS during the past 4 solar cycles", *Solar Phys.* 294, 123.

Shaifullah, G., Tiburzi, C., & Zuca, P., 2020. "CMEchaser: Detecting line-of-sight occultations due to CME's", *Solar Phys.* 295, 136.

Smith, C. W., Matthaeus, W. H., Zank, G. P., et al., 2001. "Heating of the low-latitude solar wind by dissipation of turbulent magnetic fluctuations", *J Geophys. Res.* 106, 825.

Sokol, J. M., Bzowski, M., Tokumaru, M., et al., 2013. "Time variations of solar wind structure from *in situ* measurements and IPS", *Solar Phys.* 285, 167.

Stansby, D., Matteini, L., Horbury, T. S., et al., 2020. "The origin of slow Alfvenic solar wind at solar minimum", *Monthly Not. Roy. Astron. Soc.* 492, 39.

Susino, R., Bemporad, A., & Mancuso, S., 2015. "Physical conditions of coronal plasma at the transit of a shock driven by a CME", *Astrophys. J.* 812, 119.

Tappin, S. J., & Howard, T. A., 2010. "Reconstructing CME structures from IPS observations", *Solar Phys.* 265, 159.

Vasantharaju, N., Vemareddy, P., Ravindra, B., et al., 2018. "Statistical study of magnetic nonpotential measures in confined and eruptive flares", *Astrophys. J.* 860, 58.

Vasyliunas, V. M., 2011. "The largest imaginable magnetic storm", *J. Atmos. Solar Terr. Phys.* 73, 1444.

Vourlidas, A., Howard, R. A., Esfandiari, E., et al., 2011. "Comprehensive analysis of CME mass and energy properties over a full solar cycle", *Astrophys. J.* 730, 59.

Weber, E. J., & Davis, L., 1967. "The angular momentum of the solar wind", *Astrophys. J.* 148, 217.

Wexler, D., Imamura, T., Efimov, A., et al., 2020. "Coronal electron density fluctuations inferred from Akatsuki radio observations", *Solar Phys.* 295, 111.

Woods, T. N., Eparvier, F. G., Hock, R., et al., 2012. "EVE on SDO", *Solar Phys.* 275, 115.

Xystouris, G., Sigala, E., & Mavromichalaki, H., 2014. "A complete catalog of high speed streams during solar cycle 23", *Solar Phys.* 289, 995.

Yakovlev, O. I., & Mullan, D. J., 1996. "Remote sensing of the solar wind using radio waves: Part 2", *Irish Astron. J.* 23, 7.

Yashiro, S., Akiyama, S., Gopalswamy, N., et al., 2006. "Different power-laws for flares with and without CME's", *Astrophys. J. Lett.* 650, L143.

Youssef, M., 2012. "On the relation between CME's and solar flares", *NRIAG J. Astron. Geophys.* 1, 172.

Zirker, J. B., 1981. "The solar corona and the solar wind", *The Sun as a Star*, ed. S. Jordan, NASA, Washington DC, NASA SP-450, p. 135.

# Appendix A: Symbols Used in the Text

Symbols used in the text, their units, and where they are introduced

| | |
|---|---|
| $a$ | isothermal speed of sound (cm sec$^{-1}$) (Section 18.1) |
| $a_{hc}$ | relative acceleration of hot and cold gas (Section 7.1) |
| $a_R$ | radiation density constant (ergs cm$^{-3}$ K$^{-4}$) (Section 2.1) |
| $a_\lambda$ | limb-darkening coefficient (Section 2.2) |
| $A_{ss}$ | area of the visible hemisphere of the Sun (Section 17.5) |
| Å | angstrom (unit of length = $10^{-8}$ cm) (Section 2.1) |
| $b_\lambda$ | limb-darkening coefficient (Section 2.2) |
| $\mathbf{B}$ | magnetic field vector (Section 16.6.1) |
| $B$ | magnitude of $\mathbf{B}$ (Section 16.6.1) |
| $c$ | speed of light (= $3 \times 10^{10}$ cm sec$^{-1}$) (Section 2.1) |
| $c_s$ | adiabatic speed of sound (cm sec$^{-1}$) (Section 7.1) |
| $C_p$ | specific heat at constant pressure (ergs gm$^{-1}$ K$^{-1}$) (Section 6.7.1) |
| $C_v$ | specific heat at constant volume (ergs gm$^{-1}$ K$^{-1}$) (Section 6.7.1) |
| $d_c$ | mass column density (gm cm$^{-2}$) (Section 3.3, 5.1) |
| $d_g$ | angular diameter of convection cells (arc seconds) (Section 6.4) |
| $D$ | distance of Earth from Sun (= $1.5 \times 10^{13}$ cm) (Section 1.2) |
| $D_g$ | horizontal linear diameter of convection cells (cm) (Section 6.4) |
| $e$ | unit of electric charge (= $4.8 \times 10^{-10}$ e.s.u.) (Section 16.6.1) |
| $\mathbf{E}$ | electric field vector (Section 16.6.1) |
| $F(\tau)$ | flux of radiation at optical depth $\tau$ (ergs cm$^{-2}$ sec$^{-1}$) (Equation 2.22) |
| $F_s$ | flux of sound waves emitted by convection (Equation 14.27) |
| $F_\odot$ | flux of radiation at the solar surface (= $6.29 \times 10^{10}$ ergs cm$^{-2}$ sec$^{-1}$) (Section 1.9) |
| $g$ | acceleration due to gravity (cm sec$^{-2}$) (Section 5.1) |
| $g_{ad}$ | adiabatic temperature gradient (K cm$^{-1}$) (Section 6.8, Equation 6.8) |
| $g_i$ | statistical weight of atomic level $i$ (Section 3.3.2) |
| $g_s$ | acceleration due to gravity at the Sun's surface (= 27,420 cm sec$^{-2}$) (Section 1.6) |
| $g_T$ | temperature gradient (K cm$^{-1}$) (Section 6.8) |
| $G$ | Newton's gravitational constant (cm$^3$ gm$^{-1}$ sec$^{-2}$) (Sections 1.1, 1.3) |
| $h$ | height in the atmosphere, increasing *outward* from the Sun (cm) (Section 5.1) |
| $H$ | vertical depth of a granule (Section 6.5) |
| $H_m$ | magnetic helicity (in units of Mx$^2$) (Sections 16.11, 18.9.7) |
| $H_p$ | pressure scale height (cm) (Section 5.1) |
| $i$ | $\sqrt{(-1)}$ (Section 13.5.2) |
| $I$ | ionization potential (electron volts) (Section 4.1) |
| $I_\lambda$ | intensity of radiation per unit wavelength (Section 2.1, Equation 2.4) |
| $I_\nu$ | intensity of radiation per unit frequency (Section 2.1, Equation 2.3) |
| $J$ | mean intensity at arbitrary depth (Equation 2.23) |
| $k$ | Boltzmann's constant (= $1.38 \times 10^{-16}$ ergs K$^{-1}$) (Section 1.8) |
| $k_h$ | horizontal wave number of oscillation mode in the Sun (Section 14.7) |

| | |
|---|---|
| $k_r$ | radial wave number of oscillation mode in the Sun (Section 14.7) |
| $k_\lambda$ | (linear) absorption coefficient at wavelength $\lambda$ (cm$^{-1}$) (Section 2.3) |
| $k_{th}$ | thermal conductivity (ergs cm$^{-1}$ sec$^{-1}$ K$^{-1}$) (Section 8.1) |
| $K$ | degrees Kelvin: unit of temperature (Section 2.1) |
| $K(\tau)$ | related to radiation pressure at optical depth $\tau$ (Equation 2.26) |
| $l$ | angular degree of oscillation mode in the Sun (Section 13.2) |
| $L$ | half-length of coronal loop (cm) (Section 17.15.1) |
| $L_\odot$ | luminosity (=power output) of the Sun (= $3.83 \times 10^{33}$ ergs sec$^{-1}$) (Section 1.4) |
| $m_H$ | mass of the hydrogen atom (= $1.66 \times 10^{-24}$ gm) (Section 1.8) |
| $M$ | Mach number of granule flows (Section 14.8.2) |
| $M_\odot$ | mass of the Sun ($1.99 \times 10^{33}$ gm) (Section 1.3) |
| $n$ | polytropic index (Equation 10.1) |
| $n_a$ | number density of photon absorbers (cm$^{-3}$) (Section 3.3) |
| $n_e$ | number density of electrons (cm$^{-3}$) (Section 4.1) |
| $n_i$ | number density of ions (cm$^{-3}$) (Chapter 17.5) |
| $N$ | number column density (cm$^{-2}$) (Section 3.3) |
| $N$ | number related to adiabatic exponent (Section 14.1) The reader should be careful not to confuse the fundamentally different definitions of $N$ which are used in Sections 3.3 and 14.1 |
| $N(s\text{-}d)$ | number of days with no detectable sunspots (Section 16.1.4) |
| $p$ | gas pressure (dyn cm$^{-2}$) (Section 5.1) |
| $p_c$ | pressure at the center of a polytrope (Section 10.2) |
| $p_e$ | electron pressure (dyn cm$^{-2}$) (Section 4.1) |
| $p_{mag}$ | magnetic pressure (dyn cm$^{-2}$) (Section 16.6.2.1) |
| $P_{ac}$ | acoustic cut-off period (sec) for vertically propagating waves (Equation 13.14) |
| $P_g$ | critical period of gravity modes in the Sun (Section 1.12) |
| $P_p$ | power density in pressure pulse due to granules (Equation 14.26) |
| $q(\tau)$ | slowly varying function of optical depth (Section 2.8) |
| $r$ | radial coordinate (cm) (Section 1.1) |
| $r_g$ | gyroradius (cm) (Equation 16.4) |
| $r_o$ | Emden unit of length (cm) (Equation 10.9) |
| $r_s$ | radial distance to the sonic point in the solar wind (Equation 18.8) |
| $R_g$ | the gas constant (= $8.3 \times 10^7$ ergs mole$^{-1}$ K$^{-1}$) (Section 1.8) |
| $R_\odot$ | radius of the Sun (= $6.96 \times 10^{10}$ cm) (Section 1.5) |
| $S_\lambda$ | the source function at wavelength $\lambda$ (Section 2.4) |
| $t_s$ | sound propagation time between center of a star and its photosphere (Equation 9.6) |
| $T$ | temperature (degrees K) (Section 2.1) |
| $T_b$ | temperature at the base of the convection zone (Section 8.3) |
| $T_{eff}$ | effective temperature (K) (Section 1.9) |
| $u_\nu$ | energy density of radiation per unit frequency (ergs cm$^{-3}$ Hz$^{-1}$) (Section 2.9) |
| $U$ | internal energy (ergs gm$^{-1}$) (Section 6.7.1) |
| $V$ | velocity of vertical gas motion (cm sec$^{-1}$) (Section 7.1) |
| $V_{esc}$ | escape speed from the surface (Section 1.8) |
| $V_{rms}$ | root-mean-square thermal speed of particles (Sections 1.8, 9.2) |
| $w$ | auxiliary pressure variable in oscillation calculations (Equation 14.16) |
| $W_{mag}$ | magnetic energy density (ergs cm$^{-3}$) (Section 16.6.2.1) |

# Appendix A  Symbols Used in the Text

| | |
|---|---|
| $x$ | dimensionless unit of length in the Lane–Emden equation (Section 10.4) |
| $X$ | dimensionless unit of radial displacement in oscillation calculations (Section 14.2) |
| $y$ | Lane–Emden function, which describes polytrope structure (Equation 10.6) |
| $z$ | depth in atmosphere, increasing *inwards* towards the solar center (Section 7.1) |
| $z$ | auxiliary radial displacement variable in oscillation calculations (Equation 14.16) The reader should be careful not to confuse the fundamentally different definitions of $z$ which are used in Section 7.1 and in Equation 14.16 |
| $z_b$ | depth of the base of the convection zone (Section 7.9) |
| $z_r$ | depth below the surface where a $p$-mode is refracted upwards (Equation 14.24) |
| $\alpha$ | mixing length parameter = ratio of $H$ to $H_p$ (Section 6.5) |
| $\alpha$ | dimensionless frequency in oscillation calculation (Equation 14.10). The reader should be careful not to confuse the fundamentally different definitions of Greek alpha which are used in Sections 6.5 and in Equation 14.10 |
| $\beta$ | power-law index in flare distribution (Equation 17.13) |
| $\gamma$ | ratio of specific heats (Section 6.7.1) |
| $\Gamma$ | generalized exponent for adiabatic conditions in ionization zone (Section 6.7.3) |
| $\delta$ | power-law index for nonthermal electron energy distribution (Equation 17.11) |
| $\varepsilon$ | oblateness of the Sun (Section 1.10) |
| $\varepsilon_\lambda$ | radiant emissivity at wavelength $\lambda$ (Section 2.3) |
| $\eta_e$ | magnetic diffusivity (cm$^2$ sec$^{-1}$) (Equation 16.9) |
| $\theta$ | variable in Saha equation inversely related to temperature (Section 4.2) |
| $\theta$ | dimensionless unit of pressure in oscillation calculation (Section 14.2) The reader should be careful not to confuse the fundamentally different definitions of $\theta$ which are used in Sections 4.2 and 14.2 |
| $\xi$ | vertical displacement of a parcel of gas (Section 13.5.1) |
| $\kappa_\lambda$ | opacity (in units of cm$^2$ gm$^{-1}$) at wavelength $\lambda$ (Section 3.1) |
| $\kappa_R$ | Rosseland mean opacity (in units of cm gm) (Equation 3.3) |
| $\lambda$ | wavelength (units of cm, or Å, or μm) (Section 2.1) |
| $\lambda_e$ | de Broglie wavelength of an electron (Section 9.5) |
| $\lambda_p$ | de Broglie wavelength of a proton (Section 11.4.3, Equation 11.5) |
| $\Lambda$ | Coulomb logarithm (Section 11.3) |
| $\Lambda$ | radiative loss function (Section 17.15.2) The reader should be careful not to confuse the fundamentally different definitions of $\Lambda$ which are used in Sections 11.3 and 17.15.2 |
| $\mu$ | $\cos(\psi)$, where $\psi$ = angle between line of sight and Sun's normal (Section 2.2, Figure 2.4) |
| $\mu$ | mean molecular weight (Section 1.8) The reader should be careful not to confuse the fundamentally different definitions of $\mu$ which are used in Sections 1.8 and 2.2 |
| $\nu$ | temporal frequency (sec$^{-1}$) (Section 2.1) |
| $\nu_g$ | critical frequency in the Sun (Section 1.12) |
| $\pi$ | 3.14159 |
| $\rho$ | mass density (gm cm$^{-3}$) (Section 3.1) |
| $\rho_b$ | mass density at the base of the convection zone (Section 8.3) |
| $\rho_c$ | mass density at the center of a polytrope (Section 10.2) |
| $\bar{\rho}$ | mean mass density of the Sun (= 1.4 gm cm$^{-3}$) (Equation 1.14) |
| $\sigma$ | collision cross-section for a particle (cm$^2$) (Section 11.3) |
| $\sigma_B$ | Stefan–Boltzmann constant ($5.67 \times 10^{-5}$ ergs cm$^{-2}$ sec$^{-1}$ K$^{-4}$) (Section 2.1) |
| $\sigma_c$ | Coulomb collision cross-section (Section 11.3) |

| | |
|---|---|
| $\sigma_e$ | electrical conductivity (sec$^{-1}$ [in e.s.u.]) (Section 16.6.2.2) |
| $\sigma_\lambda$ | atomic absorption cross-section (cm$^2$) for a photon with wavelength $\lambda$ (Section 3.3) |
| $\sigma_T$ | Thomson cross-section for photon scattering by a free electron (Equation 3.1) |
| $\tau$ | optical depth (Section 2.3) |
| $\varphi$ | fractional abundance of negative hydrogen atoms (Section 3.4) |
| $\Phi$ | total radiative loss function (ergs cm$^3$ sec$^{-1}$) including lines + continua (Section 17.15.2) |
| $\psi$ | angle between line of sight and Sun's normal (Section 2.2, Figure 2.4) |
| $\omega$ | solid angle (Equation 2.22) |
| $\omega$ | angular frequency of solar oscillations (radians sec$^{-1}$) (Section 14.1) The reader should be careful not to confuse the fundamentally different definitions of $\omega$ which are used in Section 14.1 and in Equation 2.22 |
| $\omega_{ac}$ | acoustic cut-off angular frequency for vertically propagating waves (Equation 13.13) |
| $\Omega(\lambda)$ | angular frequency of solar rotation (radians sec$^{-1}$) at latitude $\lambda$ (Section 1.11) |

# Appendix B: Instruments Used to Observe the Sun

In Figure P. 1 in the Preface of this book, we used acronyms to refer to 15 spacecraft and/or ground-based observatories. These are the sources of many of the measurements of the Sun and/or solar wind described in the text of this book. The following list of the 15 spacecraft and/or ground-based observatories is arranged in alphabetical order.

**DKIST**: Daniel K. Inouye Solar Telescope

Location: Haleakala Mountain, on the island of Maui, in the state of Hawaii, USA. Primary mirror diameter = 4.24 meters. Instruments: (1) broad band imager in certain wavelengths, ranging from 3933 Å to 4861 Å in the blue filter, and from 6563 Å to 7892 Å in the red; (2) a slit-based spectro-polarimeter sensitive to visible wavelengths, to measure all four Stokes parameters $I$, $Q$, $U$, and $V$ in various spectral lines at a location on the Sun's surface that is determined by the location of the slit; (3) a tunable filter to create images of an area of the Sun in one particular spectral line; (4) a spectro-polarimeter sensitive to infrared wavelengths, using optical fibers to collect spectral and spatial information simultaneously at every point in a two-dimensional image. The first image of the Sun to be released publicly by DKIST, showing solar granules at high resolution (see front cover of this book), occurred in January 2020.

**GONG and GONG+**: Global Oscillations Network Group

A ground-based observing program to study the internal structure and dynamics of the Sun using helioseismology. GONG relies on a six-station network of extremely sensitive, and stable, velocity imagers located around the Earth to obtain nearly continuous observations of the Sun's "5-minute" oscillations, or pulsations. The instruments are based on a polarizing Michelson interferometer that observes a Ni I line at 6768 Å, allowing precise measurement of Doppler shift in each 5-arcsec pixel of the entire solar disk. The six identical stations are located at Big Bear Solar Observatory (California, USA), High Altitude Observatory (Mauna Loa, Hawaii, USA), Learmonth Solar Observatory (Western Australia), Udaipur Solar Observatory (Rajasthan, India), El Teide Observatory (Tenerife, Spain), and the Cerro Tololo Interamerican Observatory (Coquimbo region, Chile). The six stations began operations in the course of calendar year 1995, starting with El Teide on March 2 and ending with Udaipur on October 3. For the first 6 years of observations, the solar images were obtained with low-resolution CCDs with $256 \times 256$ pixels. During 2001, all six observatories were converted to higher resolution CCDs ($1024 \times 1024$ pixels): this update is referred to as GONG+. The GONG archives currently contain 124 terabytes of data.

**Hinode**

A joint satellite mission of Japan, US, UK, and Europe, launched September 23, 2006, into a Sun-synchronous orbit at a height of 650 km above Earth's surface. The orbit allows Hinode to observe the Sun continuously for 9 months of the year. In the remaining months, the Sun undergoes eclipses as seen from Hinode, with each eclipse lasting about 10 minutes out of a 98-minute orbit. Instruments include Solar Optical Telescope (SOT), EUV Imaging Spectrometer (EIS), and X-ray/EUV Telescope (XRT). In the wavelength range from 170 to 280 Å, some 30 spectral lines are available to determine differential emission measure in an active region with corresponding temperatures ranging from $\log T = 5.6$ to $\log T = 7.3$.

## IBIS

Location: attached to the Dunn Solar Telescope at Sacramento Peak Observatory, New Mexico, USA. The Interferometric Bidimensional Spectropolarimeter was built in Florence, Italy. First observations were obtained in June 2003. Two Fabry-Perot etalons are used to observe a small region of the Sun with a spectral resolving power of 200,000 over the wavelength range 5500–8600 Å. In spectroscopic mode, it has a circular field of view of 95″ diameter, with a spatial sampling of 0″.098. In dual-beam spectro-polarimetric mode, the field of view is split into two smaller rectangles through a mask, each covering about half of the unobstructed field. The instrument has an additional parallel and synchronized broadband channel with the same image scale as the narrow-band channel that serves as light-level and image distortion reference. Data can be obtained at up to 15 frames per second (fps) in spectroscopic mode at single wavelengths and 10 fps when tuned. In tuned spectro-polarimetric mode, the frame rate is 7 fps.

## IRIS: Interface Region Imaging Spectrograph

IRIS makes use of high-resolution images, data, and advanced computer models to determine how matter, light, and energy move from the sun's 6000 K surface to its million K outer atmosphere or corona. IRIS observes the Sun's chromosphere and the transition ("interface") region, where the chromosphere interfaces with the even hotter corona above. Gas motions, upward and downward, occur in the interface region, causing the interface to occupy a range of altitudes (spanning several thousand km) above the photosphere. IRIS was launched on June 27, 2013, and it travels in a polar, sun-synchronous orbit in such a way that it crosses the equator at the same local time each day. The spacecraft's orbit places it at heights of 390–420 km above Earth's surface. IRIS carries a single instrument: an ultraviolet telescope combined with a spectrograph that images a small portion (~1%) of the Sun's disk in the light of certain selected spectral lines: the lines are formed at temperatures between 5000 and 65,000 K. A new image is obtained every 5–10 seconds. IRIS can resolve features with linear sizes as small as 240 km on the Sun.

## Parker Solar Probe

Launched on August 12, 2018: the goal is to study the Sun's environment at distances that are closer than any spacecraft has ever gone, eventually reaching a distance of 9.86 solar radii (=6.85 million km) from the center of the Sun. To achieve this, the satellite is designed to use seven gravity assists with Venus over an interval of 7 years, gradually changing the orbit so as to reduce the perihelion distance to a minimum in 2025. Instruments on board include (1) FIELDS, designed to measure the scale and shape of electric and magnetic fields in the Sun's atmosphere; FIELDS measures waves and turbulence in the inner heliosphere with high time resolution to understand the fields associated with waves, shocks, and magnetic reconnection; (2) WISPR is the only imaging instrument on board, to detect CMEs, jets, and other ejecta from the Sun; (3) SWEAP is a particle detector to measure velocity, density, and temperature of electrons, protons, and alpha particles (helium nuclei) in the solar corona and wind; (4) ISOIS (pronounced ee-sis) measures solar energetic particle properties for several different elements from hydrogen to iron.

## ROSA: Rapid Oscillations in the Solar Atmosphere

Location: This instrument is attached to the Dunn Solar Telescope at Sacramento Peak Observatory, New Mexico, USA. It allows investigators to image the Sun simultaneously in six wavebands using six cameras: three wavebands are in the blue (Ca K core, G_band, and 4170 Å continuum), and three are in the red (Hα core, and two circular polarized readings at 6302 Å). Images are obtained by cooled CCD cameras built by Andor Technology Ltd in Belfast, Northern Ireland, with low-noise read-out capabilities of up to 200 frames per second. The goal is to investigate oscillatory phenomena (due to e.g., propagating MHD waves) at different altitudes in the solar atmosphere simultaneously at an unprecedented level of detail. Note that Andor Technology also supplied the CCD cameras which are used to gather data by DKIST.

**SDO**: Solar Dynamics Observatory

Launched on February 11, 2010, into an inclined geosynchronous orbit allowing the Sun to be observed continuously and allowing a single dedicated ground station to receive data at an exceptionally high rate (1.5 terabytes of data per day). The goal is to study solar activity and how space weather is affected by that activity. Three scientific experiments are on board, and all three observe the Sun simultaneously: (1) Atmospheric Imaging Assembly (AIA), imaging the Sun in 10 discrete bandpasses centered on wavelengths between 94 Å and 4500 Å, sensitive to a range of temperatures from 4,500 K to as much as 16 million K; images in each band are obtained every 10 seconds; (2) EUV Variability Experiment (EVE), measuring the extreme ultraviolet (from 65 Å to 1050 Å) irradiance of the entire Sun with unprecedented (1 Å) spectral resolution, temporal cadence, and precision (plus a special detector to record the emission in Lyman-$\alpha$; (3) Helioseismic and Magnetic Imager (HMI) provides full-disk coverage at high spatial resolution so as to perform studies of the Sun's seismic and magnetic fields using a line of Fe I at 6173 Å; the field measurements enable the extraction of the full magnetic vector in each pixel at an altitude of about 100 km above the photosphere. Full-Sun images from SDO have angular resolutions of better than 0.5 arcsec, i.e., four times better than SOHO images; moreover, SDO obtains one image every second, whereas SOHO obtains an image once every 12 minutes.

**SOHO**: Solar and Heliospheric Observatory

A project of international collaboration between ESA and NASA to study the Sun from its deep core to the outer corona and solar wind. Launched on December 2, 1995, into an orbit that is centered around a point about 1 million km sunward of the Earth. Instruments on board SOHO were built by teams of researchers from Europe and USA: these include a coronagraph (LASCO), Extreme Ultraviolet Imaging Telescope (EIT: with separate images in four spectral lines), two spectrometers operating in the ultraviolet (SUMER and CDS), two instruments for low-frequency helioseismology (GOLF and VIRGO), SWAN for mapping the global solar wind, and CELIAS for *in situ* measurements of density, speed, and temperature of solar wind plasma. The "very latest SOHO images" obtained by EIT and LASCO can be seen online (as of the time of writing: November 13, 2021) at the website https://sohowww.nascom.nasa.gov/data/realtime-images.html. One of the instruments on board SOHO, the Michelson Doppler Imager, with its line-of-sight magnetograph, has not been used since 2011 when its functions were taken over by the higher resolution HMI instrument (with its vector magnetograph) on board SDO: when the SOHO/MDI was operational, it used a line of Ni I at 6768 Å to measure line-of-sight fields at an altitude of about 125 km above the photosphere.

**Solar Orbiter**

A European satellite launched in February 2020 into an orbit that will bring the spacecraft to a closest distance of 42 million km (=0.28 AU) from the Sun. Goals include obtaining images of the Sun at closer distances than have ever previously been achieved; taking the first ever close-up images of the Sun's polar regions; measuring the composition of the solar wind; and linking the solar wind to its area of origin on the Sun's surface. Instruments include an energetic particle detector, a magnetometer, a solar wind analyzer, radio and plasma wave detectors, a coronagraph, and imagers for EUV, polarimetry, helioseismic, and heliospheric data.

**SOLIS:** Synoptic Optical Long-term Investigations of the Sun

Location: Built by the US National Solar Observatory in Tucson, Arizona, USA, to follow the evolution of solar activity using three instruments: (1) a vector spectromagnetograph creating a full-disk image of vector magnetic fields on the Sun on a daily basis; (2) full-disk patrol that images the entire Sun using tunable filters with 1 arcsec pixels with high temporal cadence (about 10 seconds) in selected spectral lines including H-alpha, Ca II *K*, He I 10830 Å, continuum (white light), and photospheric lines; (3) integrated sunlight spectrometer to obtain high resolution (R = 300,000) spectra of the Sun-as-a-star at selected wavelengths between 3500 Å and 11000 Å. SOLIS was situated at Kitt Peak National Observatory (Arizona, USA) from 2010 until 2014; eventually it is to be located at Big Bear Solar Observatory (California USA), where construction for SOLIS started in summer 2021.

**STEREO:** Solar TErrestrial RElations Observatory

A pair of spacecraft launched on October 26, 2006, into a highly elliptical orbit with apogee near the Moon. A gravity assist from the Moon on December 15, 2006, caused STEREO-A(head) to enter an orbit with a period of 347 days, slightly inside Earth's orbit, such that STEREO A gradually moved ahead of Earth in its orbit. Another gravity assist from the Moon on January 21, 2007, caused STEREO-B(ehind) to enter an orbit with a period of 387 days, slightly outside Earth's orbit, such that STEREO-B gradually lagged behind Earth in its orbit. The two spacecraft moved slowly away from each other at a rate of about 44 degrees per year. On February 6, 2011, STEREO-A and -B were separated by 180 degrees, thereby allowing the full surface of the Sun to be detected simultaneously. The angular separation continues to increase, and the two spacecraft will reach 360-degree separation, and once again come closest to Earth, in the year 2023. Unfortunately, STEREO-B lost contact with Earth on October 1, 2014. Instruments on board both STEREO-A and -B include SECCHI (imaging the inner corona in EUV, imaging the outer corona and the space between Sun and Earth in white light), IMPACT (a solar-wind instrument to make *in situ* measurements of energetic particles and determining the 3-D distribution of electrons and magnetic fields), PLASTIC (measuring plasma properties, including heavy ions), and SWAVES (studying radio bursts propagating from Sun to Earth orbit). Although the STEREO spacecraft were obviously designed to study *the Sun*, nevertheless, the heliospheric imagers (HI1, HI2) are sensitive enough to be capable of detecting background stars also. This allows for calibration of the sensitivity as the spacecraft ages. HI1, observing in a field of view that extends out to about 0.3 AU, could measure the brightnesses of some 1400 stars, while HI2 (extending out to 1 AU) observed some 600 stars. Both cameras were found to be degrading in sensitivity at a rate of about 0.1% per year, about 10 times better than for white-light instruments on other spacecraft (Tappan et al. 2017 *Solar Phys.* 292, 28).

**TRACE:** Transition Region and Coronal Explorer

This NASA Small Explorer Mission (SMEX) was launched in April 1998 in order to explore the 3-D magnetic field in the solar atmosphere, to determine how the field evolves in response to photospheric flows, and to quantify the time-dependent coronal fine structure. The Soft X-ray telescope (SXT) obtained more than 2 million high-resolution images of coronal loops in 12 years of operations. TRACE was the first space-based mission to obtain imaging data spanning an entire cycle of solar activity, studying the Sun at both maximum and minimum. The satellite observed some finescale structures for the first time, including coronal or solar moss, a sponge-like structure found at the base of some coronal loops. On June 21, 2010, TRACE took its last image.

## YOHKOH (also known as Solar-A)

A Japanese satellite mission launched on August 30, 1991, into a near-circular orbit some 675 km above Earth's surface. Instruments included (1) a soft-X-ray telescope (SXT), mainly sensitive to temperatures in excess of 2.5 MK (Yoshida et al. 1995), creating images of the Sun with angular resolution of about 3 arcsec; (2) a hard X-ray telescope (HXT) to create images of the Sun in photons with energies >30 keV; (3) Bragg crystal spectrometer to diagnose very hot plasma; and (4) a wide-band spectrometer to detect photons from soft X-rays to gamma rays. YOHKOH effectively stopped observations in December 2001.

# Index

## A

absorption coefficient, 27–29
  defined, 27
  numerical values, 27
absorption lines
  C-shaped bisector, 63–65
  solar spectrum, 56–66
acceleration due to gravity, 9, 10, 12, 79, 111, 117, 121, 136, 149, 357
  solar wind, 357–382
  vertical, 115–117, 196–197
acoustic dissipation
  balanced by radiation, 247
  rate of, 243–247
acoustic energy flux, 243–244
acoustic waves, *see also* sound waves in sun
  chromospheric heating, 242–251
  in the corona, 335
  cut-off frequency, 199
  cut-off period, 199
  empirical limit in corona, 335
  flux of energy, 243–244
  generated by convection, 223–224, 243, 244, 253
  in polytrope, 140–141, 225
  propagating, 195–199
  propagation time from center to surface, 140, 225
  trapped, 199
active regions, xiv
  coronal density, 324
  coronal magnetic fields, 279
  coronal temperature, 311–316
  defined, 264
  diffusive decay, 267
  latitudes, 266
  localized heating, 293
activity cycle, *see* eleven-year cycle
adiabatic index
  generalized, 103, 108, 123
  ionization, effects of, 102–103
adiabatic oscillations, 207–208
adiabatic processes, 104, 106–108, 120–122
adiabatic temperature gradient, 104, 106–107, 120–122
AIA (Atmospheric Imaging Assembly) instrument on SDO, 317, 319, 342–343, 395
Alfven speed defined, 254, 294
  in chromosphere, 294
  in corona, 294, 307
  in photosphere, 294
  reconnection outflow, 348
  in sunspot umbra, 294
Alfven waves
  chromospheric heating, 293
  into the corona, 294
  coronal heating, 335
  difficult to dissipate, 294
  in the photosphere, 294
  transverse, 294

ambipolar diffusion, 286
amplification of fields
  solar cycle, 296
  time-scale for, 298–299
amplitude
  Alfven waves, 294
  antinode, 214–216
  $g$-modes, 220
  largest $p$-modes, 189
  $p$-modes, 200
  radiant modes, 19
  related to energy flux, 243, 293
  "seeing", 81
  solar irradiance, 6
  solar oscillations, 187–191
  sound waves in photosphere, 243–245
  trapped *vs.* untrapped, 195–199
  turbulence in corona, 323
  variation with height, 245
  velocity differences in convection, 95
angstrom unit, 22
angular degree $l$
  defined, 140, 192
  empirical determination, 192
  horizontal wavelength, 192
angular frequency of small oscillations, 206
angular momentum
  of electron, 271
  of orbit, 272
  of polarized photon, 274
  sublevels in atom, 69
angular radius, 8, 13
angular resolution
  $p$-mode observations, 192
  required for granules, 97
  required for high l modes
angular velocity, 12, 14–16, 226–228
anisotropy of Lorentz force, 287
antenna, 224
antinodes of eigenfunctions, 214–216, 223–224
arches, in X-rays, 327
arcs, *see* arches
artificial satellite, 5
astronomical unit, 4
asymptotic functional forms of oscillation equations, 204–207, 218–220
asymptotic spacing
  in frequency, 194, 213
  in period, 218
  of $p$-modes, 140–141
asymptotic values
  boundary temperature, 37
  coronal pressure, 358
  $L/M$ ratio, 132
  luminosity, 132
  mass, 132
  solar wind speed, 363

atomic energy levels
  bound electrons, 42–44, 46–49, 51, 235, 248, 271–273, 313–315, 330
  in magnetic field, 271–273
atomic mass units, 158
Avogadro's number, 11
azimuthal symmetry, 25

## B

Babcock magnetograph, 275, 279, 282
Balmer lines in Sun, 57, 60–61, 235, 251, 308
barrier penetration, *see* quantum tunneling
base of convection zone, 103, 123–124, 127, 130–132, 135, 139–140, 222–223, 226, 228, 297
base of corona defined, 309–310
  density at, 310
  energy flux at, 374
  pressure at, 310, 325, 332
  velocity at, 307
base of sunspot, 292
Benard, Henri
  convection cells, 94
  study of laboratory convection, 94
beta-decay, 170–171
black-body radiation (= Planck function), 19–23
Bohr model of hydrogenic atom, 42–44, 47, 315
bolometric flux
  umbra *vs.* photosphere, 260
bolometric luminosity, 267
Boltzmann's constant, 11, 20, 101, 332
Boltzmann distribution, 20
Boltzmann factor, 17, 48, 69–71, 73
boundary conditions, xv, 34, 115, 135, 142
  in corona, 282
  Lane-Emden equation, 149
  oscillation equations, 210, 211
boundary of polytrope, 150
boundary temperature, 37, 82
bound-bound transition, 47, 48, 82, 127, 250
bound electron, 43, 47, 72
bound-free transition, 47, 48, 50, 54–56, 82, 127
bound state
  negative hydrogen ion, 50
breaking waves
  defined, 245
  local heating, 246
bridging the Coulomb gap, 165–166
Brunt-Vaisala frequency, 220
buoyancy
  in granules, 116
  in magnetic field, 297

## C

calcium lines (H and K) ionized calcium, 56, 66, 77, 235, 237
"captive audience" loops, 327
  not in CH, 327
cavity
  radiation, 19–22
  sound waves, 190, 200
CDS (= Coronal Diagnostic Spectrometer), instrument on SOHO, 322, 395

cell, center of supergranule, 109, 224, 237–240, 242–244, 246–247, 249, 269, 276, 293, 295
cell *vs.* network differences, 240–242
  energy supplies, 245–247
central condensation, 154–155
central density
  in a polytrope, 146, 148, 151, 154–155, 208
  in the Sun, 10, 138, 142, 151, 206
central pressure
  in a polytrope, 152
  in the Sun, 142
central temperature in a polytrope, 147
  in hydrostatic equilibrium, xvi, 8, 117, 147, 166
  in the Sun, xvi, 137, 166
channels for flare energy, 345
characteristic gravitational period in the Sun, 16, 148, 208
charged particle in a magnetic field, 283–286
checking solar theories, xvi, xvii, 11, 80, 118, 140–141, 175, 187, 225, 245, 295, 379
chemical composition
  in corona, 319, 331
  in photosphere, 79, 87, 243, 331
  solar interior, 119, 123–125, 137
Cherenkov radiation
  chlorine (cleaning fluid), 181
  neutrinos, 181–185
chromosphere, 233–255
  acoustic flux, 243–245
  CaK observations, 235, 237
  calculating the temperature increase, 247–249
  color, 234
  definition, 234
  deposition of energy, 245–247
  dissipation length, 246, 334
  eclipse image, 234
  H$\alpha$ observations, 238–240
  heating by untrapped modes, 196
  helium ionization, 75
  hotter than photosphere, 236, 241
  input energy flux, 243–245
  low chromosphere, 248–249, 286, 289
  mechanical work, 242–243
  middle chromosphere, 249–251
  network, 95–96, 237–244, 246–247, 249, 252, 254, 266, 269–270, 293–296
  not heated by p-modes, 243
  observations on the disk, 236–240
  opacity power-law, 56, 248
  plage, 98, 252, 262, 264, 269–270, 324
  radiative cooling, 247–249
  supergranules, 108–111
  temperature profiles (empirical), 241
  thermostatic effect of hydrogen, 251
  thickness, 236
  two components, 240
  upper chromosphere, 73–74, 242, 245, 249, 251, 253–255, 289, 294, 315, 323, 328, 332–333
  volumetric rate of energy deposition, 245–247
chromospheric heating
  excess due to Alfven waves, 294
  network, 293–295
  plage, 293
CH, *see* coronal holes
circular path particle in magnetic field, 283

# Index

circular polarization, 273–275
circulation time granules, 98–99
closed loops, 327, 329, 330, 365
    densities in, 327
    trapped gas in, 327
clumps in solar wind, 370–373
CME's, *see* coronal mass ejections (CME's)
CNO cycle, 158, 160, 177, 180–181
collisions between particles
    defined, 160
    distant, 161
    frequency in solar core, 161
column density, 44, 45, 49, 51, 80, 89, 310
computational procedure ("step-by step")
    convection zone, 122–123
    interior, 135–137
    oscillations, 211–213
    photosphere, 82–86
    polytrope, 152–154
conducting fluid magnetic interactions, 286
conduction of heat
    kinetic theory, 128
    molecular process, 94
conductivity
    electrical, 106, 287, 318
        in photosphere, 289
        in solar atmosphere, 289, 363
        in sunspot umbra, 289
    Spitzer value, 288, 289
    thermal, 106, 329
        electrons, 329, 359
        photons, 127–129
conservation
    energy, 19, 145, 175, 197, 360, 362
    magnetic helicity, 280
    mass, 145, 146, 197, 207, 360, 362
    momentum, 116, 142, 145, 146, 175, 196–197, 206, 360, 362
continuous neutrino energy spectrum, 177–178
contrast heating of chromosphere *vs.* corona, 334
convection in laboratory, 94
convection in Sun
    empirical properties (*see* granules)
    inhibited in umbra, 106, 163, 290
convection modeling
    3-D modeling, 98–99, 103, 110, 118, 122
    critical temperature gradient, 104–105
    energy flux, 83, 98
        above the photosphere, 119
        below the photosphere, 119
        in the photosphere, 119
    generalized exponent, 103, 108, 123
    mixing length theory, 117–119
    model computation, 122–123
    onset, 107
    power laws of temperature, 121, 122
    speeds, 95
    step-by-step, 122–123
    vertical acceleration, 115–116
    vertical length scales, 98, 116–117
convection speeds
    depth dependence, 224
    determining factors, 115
convection zone, 115–125
    acceleration due to gravity, 123
    base location, 103, 123–124, 127, 130–132, 135–136, 139–140, 222, 226, 228, 297
    deep regions, 122
    depth, 124–125
    differential rotation, 227–228
    $p$-mode excitation, 223–224
    power law behavior, 121–122
    spherical shell, 123–124, 151
    superadiabatic region, 118, 224
    uppermost layers, 107, 121
convective envelope, 124–125, 151–152, 185, 201, 218, 279
convective inhibition in umbra, 263, 290
convective instability/stability, 105
    ionization effects, 103, 107, 122
convective turbulence and sunspot erosion, 267
convergence of rotation curves, 228
cooling rate in corona
    conductive, 331
    radiative, 331
cooling time-scale
    continuum radiation, 247–248
coordinate space, 70
core of the Sun, xvi, 7, 11, 24, 139–140, 159–162, 165–166, 169–170, 172–173, 175–176, 179, 185, 187, 347
corona, 307–352
    abrupt transition from chromosphere, 325, 333
    active region parameters, 324
    Alfven speed, 294
    asymptotic hydrostatic pressure, 358
    densities, 309–310
    eclipse images, Figs. 15.1, 17.1–17.4
    electrical conductivity, 289
    electron temperatures, 311–315, 324–325
    emission measure, 319–321
    energy fluxes, 333–334
    expansion (see solar wind)
    fields or gas dominate, 295–296
    gas pressures, 325
    ion temperatures, 323–324
    magnetic loops, 271, 276–277, 326–328, 337
    maximum brightness, 307
    polar fields, Figs. 17.2, 17.4, 271, 281–282, 296, 298
    polarized light, 310
    quiet Sun parameters, 324
    radiative losses, 331–332
    radio polarization, 270, 279
    scale height, 310, 320, 325, 334, 358
    spatial structure, 307, 325, 326
    temperature of line formation, 315–316
    thermal conduction, 329–330
    trapped gas in loops, 328
    volumetric energy deposition rate, 333–334
    white-light corona, 283, 309–312, 326–327
    X-ray image, 327
    X-ray line strength, 319–320
coronal analysis to determine $Ne$, 324–325
coronal density
    Edlen's limit, 312–313
    radial profile, 310
coronal emission lines
    Edlen, 311
    green line, 311, 316

red line, 311
  in X-rays, 313–315
coronal heating
  magnetic carpet, 336–337
  volumetric rate of energy deposition, 333–334
  wave fluxes required, 334
  waves, acoustic, 335
  waves, Alfven, 335
coronal holes, xiii, 245, 266, 317, 319, 324, 326–329, 331, 364–365, 371
  high-speed wind, 364
  low density, 327–328
  open fields, 327–328
coronal mass ejections (CME's), xv, 270, 300, 346, 351, 364, 375–378
  characteristic mass, 377
  vs. flares, 378, 381
  ideal MHD instability, 381
  mass distribution, 377
  mass loss rate, 380
  maximum mass, 377
  solar cycle variations, 376–377
coronal radiation, output power, 334
coronal shape and 11-year cycle, 307
coronal streamers, 280, 307, 309, 324, 326, 382
coronal temperatures
  empirical, 311–315
  theoretical estimate, 328–332
Coulomb cross-section, 161–162, 329
Coulomb effects in conductivity, 289
Coulomb gap
  bridging the gap, 165–166
  defined, 164
Coulomb logarithm, 161, 329–330
Coulomb repulsion in negative H ion, 50
Cowling approximation, 202, 205–208, 213–214, 218, 225, 230
Crab Nebula, remote sensing of solar wind, 370
critical frequency
  acoustic, 199
  gravitational, 16
critical radius, location of peak gravity, 136
critical temperature gradient, onset of convection, 104–107
cross-product of vectors, 283
cross-section
  Coulomb, 161, 329
  HeI edge, 51
  Lyman edge, 46–48
  negative H ion, 50–51
  neutrino, 176
  Thomson, 46–48, 310
C-shaped bisector, 63–65
  in umbrae, 263
current density, 286
curvature of field lines, 287, 291
cut-off
  acoustic frequency, 199
  acoustic period, 199
  neutrino energy, 177–178

## D

damped solutions
  $g$-modes, 220

nuclear wave function, 168
sound waves, 198–199
De Broglie wavelength, 23–24, 47, 141–142, 163–166, 168
decay of magnetic field, 288
  in a pore, 289
  in a spot, 289
deep interior of the sun, 120, 127
degenerate electrons, 141, 151–152
degree of oscillation mode $l$, 192
degree of polarization, 279, 310
DEM (= Differential Emission Measure), 318–320, 342, 393
density, exponential profile, 80
density differential, 116
departures from spherical symmetry
  cell vs. network, 240
deposition of energy
  in chromosphere, 245, 248
  in corona, 333–334
  heating caused by, 248
  by sound waves, 245
depth scale
  linear, 29
  optical, 29
depths for p-modes excitation, 223–224
  penetration, 222
detectors of solar neutrinos, 181–184
deuteron, 158–159, 167, 170
differential rotation
  effects on fields, 298–299
  latitudinal, 13, 228
  magnitude, 15
  radial, 16, 228
diffusion
  of magnetic field, 288
  of photons from core, 139
  of radiant energy, 128
diffusive decay
  active regions, 299
  time-scale for, 299
di-proton, instability, 170
discrete frequencies, solar oscillations, 193
dispersion relation, 221
displacement of gas, vertical sound wave, 196
dissipation length (waves)
  chromosphere, 246
  corona, 334
DKIST (= Daniel K Inouye Solar Telescope), 91
$D$ lines (sodium), 56, 313
dominance, field or gas, 289, 295
Doppler effect, 12, 14–15, 58–59, 63–65, 96, 108–109, 190, 227–228, 244, 253, 255, 324, 393
downflows
  convection, 95–96
  in intergranular lanes, 95, 290
Dreicer field, 351
dynamics of convection, 115–117

## E

earth
  atmosphere, 81, 91, 93, 286
  earthquakes, xv, 187
  magnetic field, 282, 375
  mass, 5

# Index

orbit, 1–3
oxygen, 56
radius, 5
eclipse of Sun, 308–309
$E$ corona, 307
eddies in turbulent flow, 95
Eddington approximation, 32, 35
Eddington-Barbier relationship, 32–33
Eddington relation, 37
Eddington solution
    applicability, 35, 82, 87, 89
    non-applicability, 242, 246
"edge" (in spectrum)
    Balmer, 49
    bound-free transition, 47
    HeI, HeII, 51
    Lyman, 47–49, 51–52, 54–55
    negative H ion, 50
    Paschen, 49
edge of Sun's disk, sharp, 81
effective polytropic index, 146, 152
effective temperature
    photosphere, 11, 36, 89, 99, 167
    umbra, 260
efficiency of sound emission, 224
eigenfrequency, 140, 198, 211–214, 220, 225–226
eigenfunction, 9, 140, 194–195, 206, 214–216, 222–224, 229
eigenmodes
    $g$-modes, 216–217
    $p$-modes, 140, 211–212
EIS (Extreme-ultraviolet Imaging Spectrograph), instrument on Hinode, 318, 322, 324–325, 393
EIT (= Extreme-ultraviolet Imaging Telescope), instrument on SOHO, 14, 316–318, 321–322, 324, 395
electrical conductivity, 106, 287, 289, 363
electrical resistivity, 381
electric field, motional, 283, 288, 351
electron
    charge, 283
    magnetic moment, 271
    spin, 271
electron degeneracy, 141–142, 151
electron density
    in corona, 309–310, 324
    radial profile, 310
electron pressure, 44, 71, 74–75, 77, 90, 315
electrons
    bound, 43, 47, 55, 69, 72, 127, 176, 236, 311, 316, 319, 328
    free, 40, 50–51, 54, 69, 75, 236, 310, 316, 319, 328
    nonrelativistic, 151, 171
    relativistic, 43–44, 66–67, 151–152, 163
electron scattering coronal light, 307
    Thomson cross-section, 40, 46–47, 53, 176, 310
electron volt, 42
eleven-year cycle, xiv, 8, 228, 264, 266–268, 297, 300, 322, 365, 380, 383
Emden unit of length, 148, 152, 208
emission lines
    in chromospheric spectrum, 235
    in coronal spectrum, 307–308, 312
    Edlen, 311
emission measure
    defined, 319–321
emissivity, 27–28

end-point (cut-off) energy (neutrino), 177
energy build-up (pre-flare), 343
energy change in displacement, 104–105
    algebraic sign, 105
energy deficit (in spots), 260
energy density
    black-body radiation, 20–22, 34, 36
    flares, 347
    kinetic energy, 9, 295, 369
    magnetic field, 287, 290, 369, 378
    radiation, 22, 34, 36, 129
    solar wind, 369, 382
    thermal, 248, 251, 287
    wave, 243
energy deposition rates (volumetric)
    in chromosphere, 246
    in corona, 333–334
energy distribution (flares), 343
energy equation, 142, 145–146, 152, 207–208, 362–363
energy-generating core, 131, 137–140, 152, 157, 159–162, 165–167, 169–173, 175–179, 185, 187, 218, 222, 226, 228
energy input
    to chromosphere, 243–245
    to corona, 333–334
entropy and ionization, 121
equation of state
    perfect gas, 115, 118, 131, 141–142, 146–147
    polytrope, 145–147, 150, 210
equilibrium
    gas spheres, 146
    hydrostatic, xvi, 79–80, 82, 89, 91, 97, 107–108, 115–116, 127, 131, 135, 137, 139, 141, 146, 149, 151–152, 166, 196–197, 206, 249, 310, 325, 357–361, 374
    radiative, 34, 91, 104, 242
    thermodynamic, 21, 89, 250
equivalent width, 58
escape speed, 11, 137, 141, 166
escape time from Sun's center
    neutrinos, 176
    photons, 139–140
evacuation of gas (spots), 291
EVE (= Extreme-ultraviolet Variability Experiment), instrument on SDO, 51–52, 320–321, 351, 395
exact solution of RTE, 35
excitation depth of p-modes, 223
excited states
    atoms/ions, 42, 46, 48, 52, 54, 56
    nuclei, 166, 179–180
exclusion principle, 70, 141
exothermic reactions, 159
extent of Sun's influence, 381–384

# F

faculae
    defined, 268, 293
    excess brightness, 7, 268, 295
    near the limb, 268, 271
    and pores, 293
    Wilson depression, 293
fast particles from flares
    electrons, 343–345
    protons, 345

fast solar wind
    from coronal holes, 364
    high ion temperatures, 324
F corona, 307
Fermi (unit of length), 163
Fermi, Enrico, and neutrinos, 175
Fick's law of diffusion, 128
"fields" and "hedgerows"
    in Ca K, 238
    in H$\alpha$, 239
    on white light, 93
five-minute oscillations, 189
flares, xiv, xv, xvii, 337–352
    amount of energy, 345–347
    areas, 343
    vs. CME's, 378, 381
    electron densities, 341
    energy densities, 347
    in H$\alpha$, 338, 340
    light curves, 339–340
    linear scales, 343
    locations, 343
    magnetic energy, 347
    magnetic field strengths, 347
    magnetic reconnection, 347–351
    maximum energy release, 346–347
    numbers detected, 339–341
    resistive instability, 381
    size distribution, 341
    SOC (= self-organized criticality), 341, 349
    spatial location, 343
    temperatures, 341–342
    white light, 337–339
flash spectrum, 235–237, 240, 251, 308, 313
flux of
    mechanical energy, 243–244, 334
    neutrinos, 177–178
    radiant energy, 6–7, 11, 22, 34, 36
follower sunspot, 259
    magnetic polarity, 279
foot-points of loops, 300
    active regions, 276–277, 326
    coronal fields, 300, 327
    separation, 307
forbidden lines, 41, 180, 311–312, 325
forces of nature, strong and weak, 157
Fraunhofer lines, 57
free-free opacity, 48, 50, 55, 82
frequency of solar oscillations
    asymptotic spacing, 217–218
    determinant of p-mode spacing, 140–141
    empirical results, 190, 194
"frozen" field, 281, 285–286, 288–290, 295–297, 343
    gives mass to field line, 294
    in reconnection, 347–348
functional forms, asymptotic, 211
fusion, nuclear, xvi, 137, 157–159, 162, 164, 167
"fuzzy glow" corona, 317, 322, 324, 326, 328, 335–337

## G

galactic cosmic rays (GCR), 267, 383–384
    11-year cycle, 384
    phase shift relative to sunspots, 384
    solar wind protects Earth, 383
Galileo, xiii, 12, 96, 200–201, 259, 265
GALLEX (neutrino detector), 183
gallium neutrino reaction, 183
gamma-rays, polarized, 274
Gamow factor
    defined, 168
    value in the Sun, 169
    velocity-sensitive, 169
gas constant, 11, 79, 101
GCR, see galactic cosmic rays
Gegenbauer polynomials, 15
generalized adiabatic exponent, 103, 108, 123
GLE (= Ground level event), 345
global magnetic field of Sun, 280–282, 301
global sound propagation, 119, 212
g-modes, 154, 187, 201–202, 205, 210, 216–220, 229
    asymptotic spacing in period, 217–218, 220
    non-existence if n< N, 210
    in radiative interior, 201
GOES X-ray satellite, 266, 339–343, 345, 351, 378–379
GONG (= Global Oscillations Network Group), 9, 189, 195, 202, 231, 393
gradient
    critical, 105–108
    of temperature, 87, 89–90, 94, 102, 104–105, 106
granules, 63, 93–100
    acoustic effects in, 99
    circulation pattern, 98–99
    circulation time, 98
    empirical properties, 93–100
    energy flux, 100–101
    intensity differentials, 99–100
    lifetime, 94, 223, 244
    and magnetic carpet, 337
    shape, 93–95
    sizes, 96–98
    temperature differentials, 99–100
    velocities, 95–96
    vertical depth, 98, 116–117
gravitational constant, Newton's, 2, 6
gravitational potential, 11–12, 149
gray atmosphere, 33–36, 50, 82, 104
ground state-
    atomic, 42, 44, 46–48, 56, 69, 77, 315
    nuclear, 179–181
gyrofrequency, 279, 285
gyroradius, 285, 289, 348
    fluid analog, 287
    numerical values, 285

## H

Hale's polarity law, 279, 298
half-width of line, 324
H-alpha (H$\alpha$)
    filaments, 269
    in flares, 338, 340
    H and K lines (ionized calcium), 59, 61–62, 74, 77, 235, 237, 244, 264, 269, 338
    observing the chromosphere, 238–240, 269–270
hard X-rays, 313, 338–339, 341, 343–344, 346, 397
heavy water (neutrino detector), 183–185

# Index

"hedgerows" and "fields"
    in Ca K, 238
    in H$\alpha$, 239
    in white light, 93
helioseismology, xv, 15, 125, 147, 185, 187, 393, 395
    age of the Sun, 214
    global sound propagation, 140–141
    probing the "far side" of the Sun, 295
    radial profile of the sound speed, 226
    solar rotation, 15–16, 226, 228
    testing a solar model, 124, 225
    testing spot models, 293
heliosphere, 382–384
helium
    in chromosphere, 51, 75, 221
    ionization temperature, 75–76
helix, motion of charged particle, 285
helmet streamers, 280, 307
    extent, 307
    motions inside, 361
    motions outside, 361
    X-ray loops, 326
high speed wind, 364–365
    coronal holes, 364
HINODE satellite, 97, 99, 109, 235, 262, 276, 278, 291, 318, 322, 324–325, 338, 365, 393
HMI (= Helioseismic Magnetic Imager, on SDO), xiv, 8, 109–110, 192, 230, 276, 277, 295, 351, 395
horizontal wavelength, 192, 202, 221
HSE, see hydrostatic equilibrium
hydrodynamic expansion, 8
    global coronal property, 359
hydrogen
    atom, 11, 42–44
    dominant emitter, 234–235, 239, 251
    edges, 48–49, 54
    ionization fraction in photosphere, 75
    ionization strips, 72
    negative ion, 50, 56, 76, 82, 89
    spectral lines, 57, 60–61, 235, 251, 308
    upper level populations, 49, 73, 250
hydrostatic equilibrium
    center of Sun, xvi, 8, 137, 166
    chromosphere/corona transition, 325
    degenerate electrons, 151–152
    global breakdown in corona, 357–359
    inner solar wind, 357, 359
    photosphere, 79–81, 115, 127, 139
    in a polytrope, 146
    polytrope, 149, 151–152
    radiative interior, 127
    spherical solution, 357–359
    stratified atmosphere, 196
    in the umbra, 292

## I

ideal MHD instability in CME's, 381
imbalanced forces in convection, 116
IMF (spiral magnetic field), 301, 363
impact parameter, 370–372, 382
inhibiting convection in umbra, 263, 290
*in situ* measurements of magnetic field, 366, 368–370, 372–373, 396

instability, convective, 105, 107, 223
intensity of radiation, 12, 19–21
    black-body (Planck) radiation, 11, 18–19, 21, 23, 54–56, 250
    incoming, 35, 127
    outgoing, 35, 127
    per unit frequency, 20, 21
    per unit wavelength, 21
interface convection zone/radiative interior, 226
intergranular lanes, 93, 95, 97–100, 102, 121, 253, 261, 290
internal energy chromosphere, 247
    convection zone, 101–105, 118
    corona, 374–375
    ionizing gas, 102–103, 124
    neutral gas, 101
internal structure of the Sun, how to check on, xvi, xvii, 11, 80, 118, 140–141, 175, 187, 225, 245, 295, 379
International Astronomical Union (IAU), 4
International Union of Geodesy and Geophysics, 5
interplanetary magnetic field (= IMF), 301, 363
interstellar medium pressure, 382–383
ion inertial length, 256
ionization, 69–77
    in chromosphere, 73–75
    degree of, 72–73, 75–77, 79, 103, 108, 115, 121, 289–290
    effects of, 121–122
    equilibrium, 51, 69
    facilitates convection, 107
    in photosphere, 75, 79, 289
    Saha equation, 71–73
    umbra, 289
ionization potential, 51, 69, 71, 77, 102, 315, 331
ionization strips, 72, 77, 102
ionized gas, "tied" to field lines, 281, 285–286, 288–290, 294–297, 343, 347–348, 363
ion temperatures, 323–324, 363
IRIDIUM 175 (artificial satellite), 5
IRIS (= Interface Region Imaging Spectrograph), 251–253, 394
iron, in corona, 311, 314–315
iron filings in magnetic field, 271, 281, 307–308, 364
ISM, see interstellar medium pressure
isospin doublet, 171
isothermal atmosphere, 80, 197
isothermal perfect gas, 80

## J

jet, from reconnection, 351

## K

Kamiokande neutrino detector, 183–184
$K$ corona, 307
Kepler, Johannes, 2, 3
Kepler satellite, 218
kinetic theory, 73, 128, 141
K line (ionized calcium), 15, 56, 59, 61–62, 74, 237, 264, 269
knock-on electrons, in Cherenkov detector, 182–184
Kramers "law" of opacity, 55, 76, 130, 132, 139, 145, 155

## L

laminar flow, 81, 94, 99
lande $g$-factor, 273
Lane-Emden equation, 147–151, 152–154, 210, 220
Lane-Emden function, 147, 149, 220
Laplacian operator, 207
large separation, 213–214, 218
latitudes of active regions, 237, 266, 270, 349
latitudinal structure index $l$, 191, 207, 226, 276
leader sunspot, 259, 279, 283, 298
    magnetic field, 279, 283
Legendre functions, 192
leptons, 185
lifetime of Sun, 161, 162, 214, 375
light curve, 339
limb brightening, 26, 30, 33
limb darkening, 8, 25–26, 30–33, 35, 39, 50, 240–241
linear absorption coefficient, *see* absorption coefficient
linear polarization, 65, 276
line broadening, 59, 244
line profile, 57–58, 60, 62, 63–66, 95, 275
line width
    $p$-modes, 190
    spectral lines, 59, 66, 271, 324
longitudinal (line-of-sight) field, 273, 275–277, 395
longitudinal structure index $m$, 191–192, 207, 227
loops, in X-rays, xiv, 327, 396
    active regions, 271, 276–277, 279, 289, 301, 324, 326–327, 337, 345, 349, 365
    spatial scales, 326, 327, 333
Lorentz force
    anisotropic, 283, 286–287
    direction, 283–284, 286
    magnitude, 261, 283, 287
low chromosphere
    defined, 241
    onset of temperature increase, 246, 249
    and temperature minimum, 62, 74, 241–242, 246–249
lower boundary, $p$-mode cavity, 190, 200, 221–222
lower hemisphere, radiative transfer, 30, 35
lower photosphere, 39, 49, 79
    hydrogen ionization, 75
    negative H ion, 77
luminosity of the Sun, 6–9, 11, 79, 115, 120, 131, 135, 157, 160, 167, 267–268, 343
Lyman edge, 47–49, 51–52, 54–55
Lyman lines, 46, 315

## M

Mach number, 224
magnetic activity, 292
    and coronal shape, 307
    defined, 270
magnetic carpet and coronal heating, 336–337
magnetic diffusivity, 288
magnetic energy and reconnection, 347–349, 352
magnetic field direction and polarization of Zeeman lines, 273–275
magnetic fields, 259–301
    active regions, 254, 271, 294, 301, 326, 343, 347, 378
    Alfven waves, 233, 255, 294, 326, 362, 365
    amplification, 296

Babcock magnetograph, 275
chromospheric network, 270
coronal heating, 326, 334, 337, 341
cycle, 298–300
diffusion, 289
dominant over gas, 295
dominated by gas, 295
effects in spectral line, 65, 271–276
energy density, 287, 290, 347, 369, 378
enhanced energy supply, 243, 254, 294–295, 327–328, 347
free energy, 280, 300, 343, 351–352, 381
global field, 271, 280–282, 298, 301, 307
maximum strength in Sun, xiv, 277–278
measurements, 270–281
MHD (*see* magnetohydrodynamics)
polarization of light, 273–277
pressure, 254, 286–287, 291–292
quiet Sun, 282
radio polarization, 279
reconnection (*see* flares)
release of magnetic energy (*see* flares)
strength
    in active regions, 282, 294
    in compact flux ropes, 282
    in coronal plasma, 279, 294
    in flares, 347
    in IMF, 266, 281
    in ISM, 359
    maximum permitted in Sun, 297
    in photosphere, 285, 294
    at solar poles, 271, 280–282, 298, 301
    in sunspot umbrae, 277–278
    in toroidal structures, 296–297
stronger in network/plage, 249, 293–294
sunspot flux blocking, 260
swept into network, 269
tension, 286
umbra, 187, 259–262
vertical, 260
wave energy, 294–295
wave modes, 233, 245, 247, 253–256, 294, 296, 326, 328, 362
Zeeman effect, 271–276
magnetic flux, 240, 282, 288, 291, 296, 301–302, 327, 338, 343, 347–348, 380–381
magnetic interactions
    with charged particles, 283
    with conducting fluid, 286
magnetic islands, 348–349, 369
magnetic moment, 271
    electron, 272
    orbital, 272
magnetic reconnection
    coronal heating, 337
    flares, 347–352
    motional electric field, 351
magnetohydrodynamics
    defined, xiii, 287
    interactions with charged particles, 283
    interactions with conducting fluids, 286–289
    reconnection, 347–351
main sequence, 167
mass ejections, *see* coronal mass ejections (CME's)

# Index

mass loss rate from the Sun
    in nuclear reactions, 7, 375
    in solar wind, 8, 375
mass of Sun's core, 161
mass profile of Sun, 132
mass-radius relationship, 151
maximum acceleration due to gravity in Sun, 137
maximum effectiveness of tunneling, 172
Maxwellian velocity distribution, 169
Maxwell's equations, 286–287
MDI (= Michelson Doppler Imager on SOHO), 9–10, 13, 109–110, 192–194, 202, 226–227, 264, 277–278, 395
mean free path, 21, 128–129, 139, 175–177, 329
mean free time, 129, 139, 173, 312–313
mean intensity (radiation), 34
mean thermal energy, 20, 157, 172
mean thermal speed, 70, 128–129, 142, 172, 285, 316, 329, 351
    protons in Sun's core, 170
mechanical energy deposition rate (volumetric)
    chromosphere, 246
    corona, 333–334
mechanical energy fluxes
    chromosphere, 244
    corona, 334
    finite supply, 374
mechanical properties of star, 146, 152
megameter, 10, 121
megatons of TNT, 346
metals, 51, 53, 55, 79, 127, 331
    ionized in photosphere, 75
MHD (= magnetohydrodynamics), *see* magnetic fields
microflares, 346
microturbulence
    in corona, 324
    in photosphere, 59, 243–244, 324
    sound wave amplitudes, 244
middle chromosphere defined, 242
    hydrogen thermostat, 251
    lack of fit by model, 249
millionths, units of sunspot area, 263
"missing depth" of convection model, 124
mixing length theory, 98, 117–119, 131, 224
    convective energy flux, 100–103
    mixing length parameter, 98, 117
    temperature differential, 99–100
MLT, *see* mixing length theory
model of the Sun, 19, 91, 135, 137, 140, 142, 152
    chromosphere, 243–251
    mechanical, 137, 141
    photosphere, 88
molecular weight, mean, 53, 79, 87, 101, 115, 119, 123, 310
    in convection zone, 123
    in deep interior, 123, 137
moments of radiation intensity, 34–35
momentum-changing collisions, 161–162
momentum space, 70–71
monatomic gas, 101–103, 108, 120–121, 207
moon diameter, 8, 234
    motion during eclipse, 236
motional electric field, 283, 288, 351
motion parallel to magnetic field, 284
multipole expansion, 224

## N

nanoflares, 341, 346
National Aeronautics and Space Administration (NASA), xiv, xvii, 1, 6–7, 109, 189, 226, 317, 318, 322, 327, 336, 370, 376, 395, 396
National Institute of Standards and Technology (NIST), 2, 158
negative hydrogen ion, 50–51, 56, 76–77, 82, 89
    network, chromospheric, 237
    bright in Ca K, 238
    dark in H$\alpha$, 239
    enhanced heating, 293–295
    magnetic fields, 269
network *vs* cell, 240
    energy supplies, 243–245
    temperature, 110, 241
neutral gas in magnetic field, 285–286
neutrinos, 139–140, 158–160, 175–185
    coming from Sun, 183
    continua, 177–178, 180–181
    cross-section, 176
    detection, 181–185
    flavor mixing, 185
    fluxes at Earth, 177–178
    lines, 177–178
    rest mass, 176, 185
neutron decay, 170
neutron monitor detectors, 345, 384
Newton's second law, 116, 196–197, 283–289
nodes of eigenfunctions
    in latitude, 140, 192, 210, 220, 221
    in radius, 140, 194–195, 214–216
nonadiabatic processes, 120–121
nonanalytic solutions, 151
nonthermal electrons, 343–346, 349, 351
nonthermal motions, 324
nonuniform brightness, 91–93
northern lights, xiv
nuclear force, 164–166, 169
    strength of, 164
nuclear reactions, xvi, 8, 24, 137, 142, 157–173, 175, 180, 185, 187, 290, 347, 375
    bridging the Coulomb gap, 165–168
    CNO cycle, 158, 160, 177, 180–181
    conditions required, 162–166
    energy generation, 137, 145–146, 157–173
    pp-I chain, 158–160
    pp-II chain, 178–179
    pp-III chain, 178–180
    probability of occurrence, 167, 171
    quantum tunneling, 168–173
    rates in Sun, 160–172
    temperature sensitivity, 171–173
    weak interaction effects, 171, 175, 274
numerical modeling, *see* computational procedure ("step-by step")

## O

oblateness, 12–14, 259
Ohm's law, 287–288
onset of convection, 104–107, 260
opacity, xvi, 33, 35, 39–66

effectiveness of bound electrons, 47, 55, 72
in gray atmosphere, 33, 35, 50
Kramers "law", 55, 76, 130, 132–133, 139, 143, 145, 155
limiting behavior, 53–55
Lyman edge, 47–49, 51–52, 54–55
maximum values, 55
in photosphere, 35, 45, 49, 77, 82–86, 88, 89
power-law fits, 55–56, 76, 130, 132, 139, 145, 248
radiative interior, 127–132, 139
related to radiative loss function, 319, 323, 330–331
Rosseland mean, 42, 51–56
sources of, 44–51
    absorption lines, 56–60
    bound electrons, 46–47, 55, 72, 127, 176
    bound-free transitions, 47–48, 50–51, 54–56, 82, 127
    electron scattering, 40, 46, 53–54, 307
    free-free transitions, 48, 51, 55, 82, 279, 319, 323, 344
    helium, 47, 49, 51, 53, 72, 75
    hydrogen, excited states, 48–49
    hydrogen, ground state, 42, 44, 46–48
    negative hydrogen ion, 50–51
in strong lines, 33, 43, 56, 59–60, 62, 251, 315
units, 45–46, 52
open magnetic fields, coronal holes, 364
optical depth, 8, 27, 29–34, 36–37, 39–40, 44–45, 49, 51, 60–61, 81–82, 87, 89–90, 97, 99, 104, 107, 128, 235, 240, 246
in chromosphere, 248
and cooling time, 248
in photosphere, 88
zero point, 29
Orbiting Solar Observatories (OSO), 313
oscillations in the Sun, observations, 187–202
frequency spacings, 140–141, 190, 194, 218
long period, 201
short period ("5-minute oscillations"), 189
spatial structure, 191–194
temporal variability, 188–191
trapped and untrapped waves, 195–200
oscillations in the Sun, theory, 205–229
asymptotic behavior, 218–220
derivation of equations, 205–209
eigenfrequencies, 213–214
eigenfunctions, 214–216
g-mode period spacing, 220
penetration of modes below surface, 221–222
p-mode frequency spacing, 218–220
preferred excitation of certain modes, 223–224
overshooting, convective, 226

# P

pairs of sunspots
east-west alignment, 259, 279, 296, 298
toroidal field origin, 296
parcel of gas, 49, 100, 104–105, 116–118, 196–198, 247, 288, 297
Parker, E. N., xv, 282, 292, 301, 341, 347–348, 359–361, 363, 366–367, 372
Parker Solar Probe, 265, 369, 382, 394

"patchy" chromosphere, 234–236
"patchy" corona, in hottest gas, 317, 322, 327
Pauli, Wolfgang, 70, 141, 175
penumbra, 106, 187, 200–201, 259–261, 263, 282, 291
absent from pores, 261, 263, 291
inclined fields, 187, 200, 244, 259, 282
perfect gas
behavior, 90, 101, 111, 141–142, 145, 149, 151, 198, 220
equation of state, 79–80, 108, 115, 121, 132, 135, 139, 146–147, 149, 358, 362
internal energy, 101, 103–105, 118, 247, 374–375
material at center of Sun, 141–142
period of g-mode oscillations
asymptotic spacing, 218, 220
phase space, 70, 141
photons
conductivity, 127–129
escape time, 139–140
heat transporters, 128
ionization by, 49–50
photosphere, xiii
acoustic energy flux, 243–244, 254
Alfven waves, 233, 255, 294, 326, 335
antinodes near, 214–215, 222–224
base, 121, 125
calculating a model, 82–87
column density, 49, 51, 80
convective flux, 101–103, 119, 262
convective turbulence, 250, 293, 297, 301, 328, 335
definition, xiii, 29
density in, 10–11, 89, 206, 310
electrical conductivity, 289
energy densities of flows, 290
images (showing granules), 92, 261
linear extent in depth (radius), 89, 233
lower, 39, 49, 75, 77, 79
magnetic decay time, 289
magnetic fields, 271, 285
mass column density in, 44–45, 49, 80, 89
microturbulence in, 59, 243–244
mixing length parameter, 98, 117–118
model of, 82–90
nearly gray opacity, 35, 50
negative H ion, 50–51, 76–77
opacity, 83–86, 88
pressure in, 75, 89, 196, 291
principal source of opacity, 50–51, 76–77
scale height, 79–81, 97–98, 117, 140, 196, 198
sound speed, 101, 224, 245, 294
sound travel time from core, 140–141
sound wave amplitudes, 245
temperature in, 89
upper, 39, 49, 54, 74–75, 77, 79, 81, 102, 199–200, 245–246
plages, 98, 252, 262, 264, 269–270, 324
excess heating, 293–295
Planck, Max, 20–23
constant, 20, 70
function, 21–22, 36–37, 42, 51–55, 81, 99, 247, 250
plasma, 48, 51, 63, 162, 167, 185, 254, 270–271, 279–281, 286–288, 300–301, 311, 313, 315–317, 319, 321, 324, 326–328, 331, 342–343, 345, 347–348, 351, 362, 364–366, 368–369, 373, 379, 381–383
plasmoids, 348–349

# Index

plateau in chromosphere, 74, 241–242, 250–251
$p$-modes
    asymptotic spacing in frequency, 194, 218
    degree $l$, 110, 191–192, 207, 227
    depth of penetration below surface, 221–222
    excitation, 223
    largest amplitudes, 189, 195
    preferred spacing, 190
    pressure dominates, 210, 216
    radial order, 191, 194–195, 214–215, 219, 222
    trapped, 190, 195–196, 199–201, 243, 282
    wavenumber, horizontal, 221
    wavenumber, radial, 221
Poisson's equation, 147
polarized light
    circular, 65, 273, 275–276, 279, 284
    from corona, 310
    in Hanle effect, 271
    linear, 65, 274–276
    in Zeeman lines, 273–275
polarized radio emission, 279
polar regions of the Sun, 14, 281–282, 299, 310, 395
    reversals of field, 281, 298–299
poloidal magnetic field, 296–298, 307–309
polynomial fit, 26, 31, 33, 44
polytropes, 145–154
    adiabatic, 145–146, 151, 328
    analytical solutions, 149–151
    central condensation, 138, 146, 151, 154–155, 212
    defined, 146
    Emden unit of length, 148, 152, 208
    equation of state, 145
    Lane-Emden equation, 147–151
    numerical computation, 152–154
    oscillations in, 205–206, 208–218, 220, 225
    relevance to "real" stars, 151–152, 154, 155
    series expansion, 149, 211
    surface, 147, 149–155
polytropic index, 146, 148–149, 152–153, 195, 207, 210, 220, 225, 362
polytropic "star" and oscillations, 208
pores, 261–263, 277, 291, 295
    decay time, 289
    and granules, 263, 289, 291
position vector, 1–2
post-tunneling processes, 169, 171, 173, 179
potential well
    gravitational, 11
    nuclear, 164, 170
power-law
    conductivity, 329
    electron distribution, 343
    flare energy distribution, 341
    opacity fits, 55–56, 248
    polytrope, 145–146
    pressure, 108
power output
    radiation ("luminosity"), 6–8
    sound, 223–224
power spectrum
    of single $l$ value, 193
    temporal, 188–191
    two-dimensional, 191–194
    velocities in Sun, 188–189

pp-cycle, *see* nuclear reactions
pressure
    comparison chromosphere/corona, 325
    electron, 71, 74–75, 77, 90, 315
    fluctuations, 209, 216, 223
    gas, xiii, 11
    waves, 101, 189–190
pressure pulse from granule, 223
pressure scale height, *see* scale height
pressure-temperature relation
    adiabatic, 108
    polytrope, 146
probability of nuclear reaction, 167, 171
probability of quantum tunneling, 171
    peak value, 172
prominences, 269–270
    support by horizontal field, 293
    on the disk, 270
propagation of sound waves, 196

## Q

quadrupole emission from convection, 224
quality factor, 190
quantum effects, 141, 166
quantum mechanics, 41, 43, 49, 58, 141, 165, 167–168, 320
quantum tunneling, 168–169, 171–172, 179
quarks, 185
quiet Sun
    coronal density, 309–311, 324–325
    coronal temperature, 311–313, 324–325
    ubiquitous 1–2MK gas, 317, 328

## R

race against time, solar cycle, 298
radar, 3–4, 323
radial component of oscillation displacement, 205–207
radial eigenfunction, 9, 140, 194–195, 214, 223
radial order of eigenmode, 191, 194–195, 214–215, 219, 222
radial profiles of parameters in corona, 310, 363, 369, 372
    in polytropes, 146, 148–149, 205, 214, 216
    in the Sun, 19, 27, 39, 135–137, 220, 225–227, 259
radiation
    density constant, 22, 36, 129
    flow through the solar atmosphere, 19, 21
    intensity, 12, 19–21, 24, 26–28, 30–32, 34–35, 38, 40, 49, 52, 56–59, 61, 65, 81, 99–100, 110, 127, 187, 190, 240, 246–247, 260, 307, 310, 313, 319, 325, 328, 339, 342–343, 369–370, 372
    transfer equation, 27–32
radiation pressure
Eddington approximation, 34
    ratio to energy density, 34
    ratio to gas pressure, 138–139
radiative cooling
    chromosphere, 247
    corona, 330
    hydrogen lines, 250
radiative equilibrium, 34, 91, 104, 242
    inapplicable to chromosphere, 242
radiative-hydrodynamic code, 100

radiative interior, 7, 37, 55, 127–132, 135–139, 143, 145–146, 155, 201, 217, 222, 226, 228, 296–297, 357
    pressure gradient, 131–132
    temperature gradient, 131
radiative leakage, 102, 104
radiative loss function (optically thin), 319, 323, 330–331
radiative probability of forbidden lines, 312
radiative transfer equation (= RTE), 27–35, 65, 91
    special solutions, 30–32
radioactive decay, and the neutrino, 175
radio astronomy, xiii, 279, 322–323, 370
radio emission mechanisms, 279, 323
radiometer, 6, 365
radio observations, magnetic fields, 279
radius of polytropic "star", 148
ramp (in penumbra), 187, 200–201, 255
ram pressure, 223, 381–382
random walk
    field lines, 299
    photons, 139–140
ranges of parameters, surface to center, 142
Rayleigh-Jeans law, 20
Rayleigh scattering, 46
realistic solar model, eigenfunctions, 215–216
"real" stars and polytropes, 145, 148, 151, 152
reconnection (magnetic), *see* flares
remote sensing
    magnetic fields, 270, 279
    solar wind, 370
resistive dissipation, 288–289
resistive instability in flares, 381
resolving power, 65, 273, 394
resonant cavity, 190, 200
rest-mass energy, 1 a.m.u., 140
reversal of polar fields, 279
RHESSI (= Ramaty High Energy Solar Spectroscopic Imager), 80, 321, 338, 341–342, 344, 351
ribbons in flare, 338, 343, 347
ridges in power spectrum, 194, 222
    loci of constant $n_r$, 195
ringing of the Sun, multiple tones, 190
rise-time for flux tube, 297
r.m.s. (= root mean square) thermal speed, 11, 137, 141, 165–167
ROSA (= Rapid Oscillations in Solar Atmosphere) instrument, 394
rose color of chromosphere, 234
    due to Hα line emission, 251
Rosseland mean opacity definition, 42, 51–56
    table of, 83–86
Rotation of the Sun
    interactions with magnetic fields, 298–300
    interior, 226–228
    surface, 12–15, 214
    velocity, 12
roughness of solar surface, 333
RTE, *see* radiative transfer equation (= RTE)
runaway temperature, 251, 328, 351
Runge-Kutta numerical scheme, 153, 210–211

# S

SAGE neutrino detector, 183
Saha equation, 69–77
    helium in the chromosphere, 75
    helium in the interior, 76
    highly ionized elements, 315
    hydrogen in the chromosphere, 73, 315
    hydrogen in the interior, 75
    ionization strips, 72
    negative hydrogen ion, 76–77
Sargent rule in particle decay, 171
scale height, 79–81, 97–98, 117, 140, 196, 198, 246, 285
    in corona, 310, 320, 325, 334, 358
Schrodinger equation, 43–44, 168
scintillation of radio sources, 370–373, 380
    probe solar wind acceleration, 372
SDO (= Solar Dynamics Observatory), xiv, 8, 10, 51–52, 109–110, 192, 202, 230, 252, 276–277, 279, 295, 317, 319–321, 342–343, 351, 378, 395
"seeing", 81, 91, 93
self-gravitating sphere, 147
shape of spectrum and the mean opacity, 41–42
sharp edge of disk, xiii, 81
shock formation, 245–246
shock heating, 246
simple harmonic motion, 198
sinusoid, 220
slab, finite, 30–31
slow solar wind, 364–365
    emerges from active region edges, 365
small separation of p-modes, 213–214, 216
SNO (= Sudbury Neutrino Observatory), 183–186
SNU (= solar neutrino unit), 181–183
sodium D lines, fine structure, 313
soft X-rays, 313, 339–340, 396–397
SOHO (= Solar and Heliospheric Observatory), 6, 9–10, 13–14, 109–110, 189, 192–194, 202, 226–227, 264, 277–279, 316–318, 321–322, 375–377, 379, 395
    images, 317–318, 376
    solar oscillations, 189, 193
solar activity, xv, 96, 110, 229, 267, 270, 276, 289, 300, 307, 321, 365, 372, 377, 395–396
solar chromosphere, *see* chromosphere
solar core, xvi, 7, 11, 24, 139–140, 157, 159–162, 165–166, 169–173, 175–179, 185, 187, 347
solar cycle, *see* eleven-year cycle
solar disk sextant, 13
Solar Dynamics Observatory, *see* SDO (= Solar Dynamics Observatory)
solar flares, *see* flares
solar interior, probe with sound waves, xv, 187–202, 205–228
solar irradiance, largest at solar maximum, 7, 293
    and faculae, 293
solar maximum and minimum defined, 264
solar model, 10, 39, 88, 103, 123, 132, 135–137, 140, 142, 145, 148, 152, 181, 183–185, 213, 215, 224–226, 242–243, 249
solar neutrino problem, 182, 185–186
Solar Orbiter, 395
solar oscillations, *see* oscillations in the Sun
solar polar fields, 271, 281–282, 296, 298
solar spectrum
    acoustic power, 244
    photosphere, 12, 19, 23, 49–50, 57
    X-rays, 313–316, 319

# Index

solar wind
  acceleration, evidence for, 361, 371–372
  asymptotic speed, 363
  average wind at 1AU, 364
  defined, 359
  density at 1AU, 364
  density at sonic point, 374
  energy equation, 362
  fast (high-speed) wind, 324, 364
  hot ions in fast wind, 324
  hydrodynamic outflow, 359–360
  hydrostatic equilibrium, global, 357–359
  hydrostatic equilibrium, local, 359–360
  kinetic energy (KE) flux, 374
  magnetic fields, 281, 363
  rate of mass outflow, 373–375
  slow wind, 364, 365
  sonic point, 361–362, 365, 372, 374
  spatial extent of Sun's influence, 381–384
  speed at 1AU, 363–364, 375, 383
  steady-state flow, 360
  temperature at 1AU, 363–364, 367
solid body rotation, departures from, see differential rotation
SOLIS (= Synoptic Optical Long-term Investigations of the Sun), 276, 396
sound speed, 101, 107–108, 140, 147, 196–198, 220, 224–226, 233, 245, 254–255, 292, 294, 310, 358, 361, 364, 371, 373–374
  adiabatic, 197, 220
  isothermal, 358
  in photosphere, 101
  radial profile in Sun, 225
sound travel time in Sun, 140
sound waves in Sun, see also acoustic waves
  ability to do work, 243, 245
  amplitudes in chromosphere, 245
  amplitudes in photosphere, 245
  flux generated by convection, 223–224, 243–245
  flux reaching the chromosphere, 244–245
  important for chromospheric heating, 233
  propagation in stratified atmosphere, 195–199
  reflection (trapped), 200
  refraction, 221
source function, 29, 31–34, 36–37, 65, 81, 99, 240, 249
  exponential form, 32–33
  Planck function, 81, 99
  polynomial form, 31, 33
South Pole, observing solar oscillations, 187–188
spatial structure in corona, 308–309, 312
specific entropy, 121
  and ionization, 121
specific heat
  constant pressure, 100–101
  constant volume, 101, 128, 329
  in convection zone, 123
  energy transport, 100
  ionizing medium, 102–103
  ratio, 103
spectral lines
  absorption, 56–60
  emission, 62, 235, 307
speed of light, 12
speed of sound, see sound speed

spherical harmonics, 191
SphinX, 321, 339, 342
spicule properties, 61, 235–240, 242, 249, 333
  and the network, 242, 249
Spitzer formula for conductivity, 289
spots, see sunspots
"squeezed" eigenfunctions, high-$l$, 192, 222, 224
standard solar model, 137, 181, 183–185
statistical weight
  bound levels, 48, 69
  free electrons, 69–71
steep temperature gradient, upper chromosphere, 251
Stefan-Boltzmann constant, 9, 22, 36, 129, 248
step-by-step modeling, see computational procedure ("step-by step")
STEREO spacecraft, xvi, 222, 280, 295, 321, 324, 368, 382, 396
storage of pre-flare energy, 343
stratified atmosphere
  cut-off frequency, 199
  cut-off period, 199
  wave propagation, 196–199
stretching field lines, 296–297, 341
strong force, see nuclear force
sub-modes with index $m$, 227
Sun "as a star", 188, 191, 396
Sun, global parameters
  acceleration due to gravity, 10
  angular radius, 8, 13
  central density, 10, 138, 142, 146, 148, 151
  critical (gravity) frequency, 16
  distance from Earth, 3–4
  effective temperature, 9, 11–12, 36, 89, 99, 167
  effects on life, xiv
  energy flux at surface, 11, 101, 120, 130, 244
  escape speed, 11, 137, 141, 166
  irradiance (TSI = "solar constant"), 6–8, 23, 293, 365, 395
  lifetime, 7, 214
  limb, xiii, 8–9, 12–14, 25–26, 30–33, 35, 39, 50, 65, 81, 92, 234–236, 238–241, 293
  linear scale corresponding to 1 arc sec, 4
  luminosity, 6–8, 9, 11, 79, 115, 120, 135, 157, 160, 267–268
  mass, 6–7, 10–11, 16
  mean density, 10–11, 16, 138, 142, 154
  power output (see luminosity)
  radius, 8–13
  shape, 12–13
  sharpness of the disk edge, 81
  solar constant (= total solar irradiance, TSI), 6–8
  sunspot cycle, xiv, 6–10, 15, 110, 202, 228, 259–260, 264–267, 277–282, 297–302, 307, 320–322, 340, 346, 364–365, 367, 369, 372–373, 377–380, 383, 396
  variable power output, 7
  visible surface, xiv–xv, 19, 22, 27, 29, 142, 233
sunspots, xiii, 6–7, 10, 12, 14, 16, 96, 109, 202, 237, 259–270, 277–279, 281, 291–293, 295–298, 317–318, 328, 343, 372, 383
  Alfven speed, 294
  angular diameters, 263
  areas, 263–264
  changes in luminosity, 6–7, 267

chromosphere, 237
cycle (see sunspot cycle)
direction of field, 275
effective temperature, 260
electrical conductivity in umbra, 289
energy deficits, 260, 268, 293
evidence for umbral magnetic field, 272
faculae associated with, 7, 268, 270, 293, 295
follower (spot), 259, 279, 298–299
Hale's polarity law, 279, 298
horizontal (nearly) field in penumbra, 260, 282
images, 19, 224, 244, 245, 248, 249, 252
inclined fields in penumbra, 187, 200, 244, 259, 282
inhibition of convection, 106, 290–291, 293
internal gas pressure, 291–292
leader (spot), 259, 279, 283, 298
lifetimes, 267
magnetic activity, 202, 268, 270
magnetic field strengths, 262–263, 273, 277–278, 282, 287
magnetic pressure, 254, 286–287, 291
numbers, 264–267
pairs, 259, 276, 296, 298
penumbra, 187, 200–201, 259–261, 263, 282, 291
pores, 263, 289, 291, 295
radiant intensity relative to photosphere, 260, 293
reduced density in umbra, 292
shallowness of, 292–293
surrounded by plage, 264
temperatures in, 260
umbra, 187, 259–263, 268, 272–273, 275, 277, 282, 287, 289–292, 294
vertical field in umbra, 106, 187, 260, 277, 282, 290–291
wavelength dependence of intensity, 260
Wilson depression, 291–292
superadiabatic gradient, 118, 120
superadiabaticity, 120
supergranules
active region decay, 278
circulation time, 110
defined, 109, 238
flows in, 109
lifetime, 110
linear size, 109
number of granules contained, 110
observed in Ca K, 237, 239
observed in Hα, 238–240
rotational effects, 110
weak convective flux, 110
super-hot component in flares, 342
supernova, 152, 370

T

tearing instability, 348, 369
temperature
boundary (Eddington model), 37, 241
chromospheric, 240–242
Edlen's work, 311
effective, 9, 11, 36, 89, 99, 167, 218, 260
electron, in corona, 311–314
extremely high, in fast wind, 324
ions, 323–324
of line formation, 315

photospheric, 88–89
quiet corona, 328–331
solar wind, 367
temperature differences
empirical, in granules, 99
theory, in MLT convection, 118–119
temperature gradient
adiabatic, 104–105
critical value, 105
related to luminosity, 131
related to pressure, 131–132
temperature minimum, 62, 74, 241–242, 246–249
associated with shock heating, 245–246
density at, 249
temperature sensitivity
of thermonuclear reactions, 171–173
temperature vs height
chromosphere, 240–242
differences cell vs network, 240–242
temporal variability, 188–191
termination shock, 383
location of, 383
thermal conduction
in corona, 329
by electrons, 329
in solar interior, 128–129
in solar wind, 362
thermal convection
in granules, 100–102
not in supergranules, 110
thermal energy
in flare, 345
thermal energyrelease, 104
thermal motions, 59, 63, 102, 287, 324
thermal pool, 27, 160
thermal population, 157, 159, 166, 170
thermal properties
of stellar interior, 146, 287
thermal protons bridging the Coulomb gap, 166
thermal speed, 70, 128–129, 141–142, 165, 170, 172, 285, 329, 351
thermal velocities iron in corona, 323–324
thermal velocity distribution, 169
thermodynamics, 147
thermonuclear reactions, 157, 162, 164, 166–167, 169, 171–172
Coulomb barrier penetration, 165–167
Gamow factor, 168
maximum effectiveness, 172
rates of, 169
regulated by weak force, 171
sensitivity to temperature, 171–173
thermostat
in middle chromosphere, 251
thin slab, 27, 31, 44
Thomson cross-section, 40, 46, 53, 176, 310
Thomson scattering in
corona, 310
three-D modeling of granules, 98–99
including 3-D radiative transfer, 100, 103
"tied" to the field, see "frozen" field
time-scale
for flare, 339, 347
time-scale of magnetic decay, 288

Index

time series solar velocity, 188
topology
    of CaK chromosphere, 237–238
    of granules, 93
toroidal field, 296–298
TRACE (= Transition Region and Coronal Explorer), 396
transfer of energy
    into p-mode, 223–225
transition region
    chromosphere/corona, 332–333
    conduction, 333
    "discontinuity", 325
    thickness, 325
transmission of Alfven waves
    density jump, 335–336
trapped modes, 195–200
"true" corona ends, 382
"true" wind begins, 382
turbulence
    in Earth's atmosphere, 81
    effects on acoustic power, 243–244
turbulent flow, 94–95
turbulent stresses, 228
turnover time of convection cell, 98–99
twenty-two year cycle
    magnetic polarity, 282, 299
two-stream approximation, 35

## U

ultraviolet catastrophe, 20
Ulysses spacecraft, 365
umbra, 106, 187, 244, 259–263, 272–273
    darkness of, 269
    field strength, 277–278, 282, 287, 290
    inhibiting convection, 106, 290–291
    vertical field, 106, 260, 291
unit vector of magnetic field, 287
untrapped waves, 195–200
upflows
    associated with bright granules, 95–96
    convective, 59, 102
upper boundary convection zone, 107
    p-mode cavity, 200
upper chromosphere
    definition, 242
    density in, 249
    hydrogen ionized, 251
    temperature runaway, 251
upper hemisphere, 30
upper photosphere, 39, 49, 54, 74–75, 77, 79, 81, 102, 199, 245
    cut-off period, 199
    hydrogen ionization, 75
    negative hydrogen ion, 77
    and temperature minimum, 246

## V

variability
    luminosity, 7
    polar fields, 281
    of single p-mode, 191
    sunspot numbers, 264–267

variability of Sun
    timescale of days/years, 187
    timescale of minutes, 187
vector displacement in oscillation, 205
vector magnetograph, 276–277
velocities in granules
    difference downflows/upflows, 95
    horizontal, 96
    vertical, 95
velocity broadening, 59
venus, 3, 6, 8, 368, 369, 371, 372, 394
vertical acceleration, 115, 116, 117, 196
vertical displacement, 104, 196, 202
vertical field in umbra, 282, 290, 291
vertical length scale, 116
vertical sound waves, 196
viscosity, 94, 106, 228, 300
volumetric rates
    conductive losses, 330
    energy deposition, 246, 331
    radiative losses, 247, 331

## W

wave equation in stratified atmosphere, 196
wave heating
    chromosphere, 245–247
    corona, 305–307
wavelength of sound waves, 224
wavelength shift due to Zeeman effect, 271–272
wave modes, magnetic, 294
wavenumber, 221
    horizontal, 221
    radial, 221
waves
    acoustic, 195
    in a stratified atmosphere, 196
    longitudinal (sound), 245
    magnetic, in corona, 324
    in solar atmosphere, 59
    inside the sun, 187
    transverse (Alfven), 252, 294
weak force (= weak interactions), 157
    controls pp-rate, 161–162
    and neutrinos, 159, 160
    strength of, 170
    in the pp reaction, 170
white dwarf stars, 151–152
white-light corona, 309
why 11 years for the solar cycle, 298–300
Wien maximum, 22, 52, 54, 56
Wien's law, 54
Wilcox Solar Observatory, 282
Wilson depression
    in faculae, 293
    in spots, 291–292
wind, *see* solar wind
wings of absorption line, 58
work against gravity, 104

## X

X-ray astronomy, xiii, 44, 80, 309
X-rays coronal, 313

hard, 313
soft, 313
spectrum, 314
temperature of line formation, 315–316

## Y

year, sidereal, 1
YOHKOH, 317, 321, 326, 327, 344, 349, 397

## Z

Zeeman effect anomalous, 273
   longitudinal, 273
   normal, 271
   transverse, 275
Zeeman splitting, 251
   circular polarization, 273–275
   linear polarization, 275
Zurich sunspot number, 264